Protein and Nucleic Acid Structure and Dynamics

Protein and Nucleic Acid Structure and Dynamics

Jonathan King, *Editor*

Department of Biology
Massachusetts Institute of Technology

A volume in the Annual Reviews Special Collections Program

The Benjamin/Cummings Publishing Company, Inc.
Menlo Park, California • Reading, Massachusetts • Wokingham, U.K.
Don Mills, Ontario • Amsterdam • Sydney • Singapore
Tokyo • Mexico City • Bogota • Santiago • San Juan

Front Cover: Computer graphic representation of a bovine trypsin inhibitor-trypsin complex. Surfaces and skeletons shown. (Courtesy of Robert Langridge, Computer Graphics Laboratory, University of California, San Francisco. Copyright © Regents of the University of California.)

Copyright © 1985 by The Benjamin/Cummings Publishing Company, Inc.

ISBN 0-8053-5403-4

ABCDEFGHIJ-MA-8987654

The Benjamin/Cummings Publishing Company, Inc.
2727 Sand Hill Road
Menlo Park, California 94025

Annual Reviews Sources

Rossmann, M.G. and P. Argos, from *Ann. Rev. Biochem.* 1981. 50: 497-532. E.E. Snell, P.D. Boyer, A. Meister, and C.C. Richardson (Eds.). Copyright © 1981 by Annual Reviews Inc. All rights reserved.

Chothia, C., from *Ann. Rev. Biochem.* 1984. 53: 537-572. C.C. Richardson, P.D. Boyer and A. Meister (Eds.). Copyright © 1984 by Annual Reviews Inc. All rights reserved.

Blundell, T. and S. Wood, from *Ann. Rev. Biochem.* 1982. 51: 123-154. E.E. Snell, P.D. Boyer, A. Meister and C.C. Richardson (Eds.). Copyright © 1982 by Annual Reviews Inc. All rights reserved.

Markley, J.L. and E.L. Ulrich, from *Ann. Rev. Biophys. Bioeng.* 1984. 13: 493-521. D.M. Engelman, C.R. Cantor and T.D. Pollard (Eds.). Copyright © 1984 by Annual Reviews Inc. All rights reserved.

Kim, P.S. and R.L. Baldwin, from *Ann. Rev. Biochem.* 1982. 51: 459-489. E.E. Snell, P.D. Boyer, A. Meister and C.C. Richardson (Eds.). Copyright © 1982 by Annual Reviews Inc. All rights reserved.

Karplus, M. and J.A. McCammon, from *Ann. Rev. Biochem.* 1983. 53: 263-300. E.E. Snell, P.D. Boyer, A. Meister, and C.C. Richardson (Eds.). Copyright © 1983 by Annual Reviews Inc. All rights reserved.

Eisenberg, D., from *Ann. Rev. Biochem.* 1984. 53: 595-623. C.C. Richardson, P.D. Boyer and A. Meister (Eds.). Copyright © 1984 by Annual Reviews Inc. All rights reserved.

Harrington, W.F. and M.E. Rodgers, from *Ann. Rev. Biochem.* 1984. 53: 35-73. C.C. Richardson, P.D. Boyer and A. Meister (Eds.). Copyright © 1984 by Annual Reviews Inc. All rights reserved.

Publisher's Foreword

The proliferation of scientific information in recent years has been so rapid that carefully written and well referenced reviews are of greater importance to scientists than ever before. Such reviews are critical resources for students just entering a research field and for researchers whose interests are broadening, as well as essential references for many specialists. Recognizing the broad utility of review coverage at this level, Benjamin/Cummings has joined with Annual Reviews Inc. to provide access to its outstanding scientific reviews in new formats: The Benjamin/Cummings—Annual Reviews Special Collections Program. Each volume in this program is dedicated to a single topic of current scientific interest and consists of articles taken from one or more of the Annual Review series. Compiled and introduced by an eminent scientist, the articles in each volume provide review coverage and exhaustive referencing of the original literature in the area discussed. By bringing together the rigorous scholarly standards of Annual Reviews articles and Benjamin/Cummings' worldwide resources and commitment to educational publishing in science, we believe this unique program will be of real utility to those active in science today as well as those who will be active tomorrow.

James W. Behnke
Editor-in-Chief
The Benjamin/Cummings Publishing Company, Inc.
Menlo Park, California
November 1984

Preface

Molecular biology is progressing from the static description of protein and nucleic acid conformation at the atomic level to an understanding of the forces that determine these structures and their dynamics. These form and function relationships underlie catalysis, the regulation of gene expression, the formation of mature structures from their precursors, and, in short, govern much of the normal activity of the cell.

In recent years *Annual Reviews* has published a number of excellent and authoritative reviews covering this emerging area. This volume collects them in one place both for ease of access, and for the ways in which the articles illuminate each other when considered together. Advances in protein structure, folding, and dynamics are covered first, followed by nucleic acid structure and dynamics, and then by a final section examining aspects of protein-nucleic acid interactions.

Protein Structure

The structures of a large number of water soluble proteins have by now been determined by X-ray crystallography. From this body of information some general features of the organization of units of secondary structure into structures of higher order have emerged. Rossmann and Argos catalogue classes of helices, sheets, and turns, and then consider their organization into super-secondary structures. Through comparison of specific domains among families of proteins these authors consider the convergence and divergence of structural features during evolution. The detailed patterns of packing of alpha helices and beta sheets are analyzed by Chothia, with considerable attention to the interacting surfaces between secondary structure units. Despite the apparent variety of interactions, a number of rules governing the packings of these units and the topology of the links between them can be derived.

The conformational flexibility of polypeptide chains is particularly clear from the study of polypeptide hormones. Despite their variation between carrier and receptor environments, a number have been crystallized and these are reviewed by Blundell and Wood. These systems offer the opportunity to identify which features of a peptide interact with the external environment to determine chain conformation. Nuclear magnetic resonance has emerged as a complementary technique to X-ray diffraction, providing details of atomic interactions in polypeptides in the solution state. New insights through NMR concerning the conformation and motions of the atoms in polypeptide chains in solution are covered by Markley and Ulrich.

Though proteins have been traditionally represented as static structures, their various atoms, helices, and domains are in motion. Karplus and McCammon review the motions and modes of subunits at various levels including atoms, side chains, loop displacement, hinge motions, as well as collective motions of the entire molecules. The biological significance of these movements is made particularly clear by the authors' discussion of the entry of oxygen into the myoglobin and hemoglobin heme cavities and of the effects of ligands on this accessibility.

Despite detailed knowledge of both the structures and amino acid sequences of numerous proteins, the mechanisms through which amino acid sequence determines protein structure remain obscure. Solving this problem requires knowledge of the actual folding pathways for polypetide chains. Kim and Baldwin summarize our knowledge of defined refolding pathways for polypeptides and the character of those intermediate stages which have been identified.

Membrane proteins have been much less accessible to X-ray analysis than water soluble enzymes due to the lack of suitable crystals. Eisenberg's review describes the structure of bacterio-rhodopsin and of a group of small peptide toxins which bind to or insert into membranes. Though few crystal structures of membrane proteins have been determined, a considerable amount of sequence data has accumulated and these sequences are analyzed in terms of the location and orientation of hydrophobic residues.

Myosin is representative of a very different kind of molecule from either the membrane proteins or the carrier proteins. Harrington and Rodgers succinctly summarize an extensive body of non-crystallographic data on the structure and conformation of the myosin molecule and relate that to the assembly of myosin molecules into thick filaments, and to the mechanism of the force-generating event.

Nucleic Acid Structure

The last few years have seen a change of views concerning DNA. The realization that DNA structure is varied under physiological conditions and that this variation is a feature of its biological activity have come as recent surprises. Associated with this has been a more detailed understanding of the interactions both along the chain and between chains in B-form DNA, as summarized in the Zimmerman article. Record et al provide a critical analysis of the manner in which solution conditions and interaction with ligands cause the collapse of DNA into a folded or condensed state.

Considerable excitement has come from the identification and characterization of left-handed Z-DNA. Rich, Nordheim and Wang summarize the characteristics of Z-DNA with special emphasis on its physiological role within cells. Recognition that the flexibility and conformation of the DNA double helix are properties of its sequence and its context opens up new views on modes of gene expression and regulation.

The role of nucleic acids in providing structural, rather than coding, information is manifest in the two large RNA molecules that serve as the backbones of all ribosomes. Noller summarizes the evidence for a very complex secondary and tertiary structure of the ribosomal RNA. These structures probably not only define sites of interaction with ribosomal proteins, but are involved in the motions of different parts of the ribosome during the steps in the protein synthetic cycle.

The simplest of the nucleic acid molecules under consideration is transfer RNA. Studies of its molecular biology reveal that very subtle modifications of bonding and structure alter its biological activity both as a receptor for the RNA synthetases, and in the translation process itself. Reid provides a clear account of the use of NMR spectroscopy to identify the base-pairing interactions within tRNA molecules and the use of these as reporters for tRNA conformation.

Protein-Nucleic Acid Interactions

Although most of our images of protein and nucleic acid structure come from analysis of molecules not undergoing interactions with other macromolecules, such interactions are the crux of biological activity. Using the X-ray structure of isolated phage lambda repressors, Sauer and Pabo examine models of repressor-DNA interactions. These models suggest a wrapping of the protein arms around the DNA as part of the binding interaction, and identify regions of the protein likely to be involved in sequence recognition, confirmed by studies of mutant repressors. The models assume little alteration in DNA conformation, but data from crystals of protein-DNA complexes should soon be able to assess this.

The small RNA-protein complexes involved in RNA processing appear to play a pivotal role in gene expression in higher organisms. Though the structures of such complexes have not been solved at the atomic level, their sequences have been determined. Busch et al summarize these results and discuss the role of these species in RNA processing and other transcriptional and post-transcriptional processes.

The best studied and most general of DNA-protein complexes is the nucleosome, the basic unit of chromosome structure in higher organisms. McGhee and Felsenfeld synthesize the considerable body of work on histone-histone and histone-DNA interactions, and explore how there interactions influence the higher order structure of chromatin and the processes of gene expression and replication.

Jonathan King
Cambridge, Massachusetts
November 1984

CONTENTS

I

Protein Structure

Ann. Rev. Biochem. 1981. 50:497–532

1

PROTEIN FOLDING

Michael G. Rossmann and Patrick Argos

Department of Biological Sciences, Purdue University, West Lafayette,
Indiana 47907

CONTENTS

PERSPECTIVE

Linderstrøm-Lang and his co-workers (1) were the first to recognize structural levels of organization within a protein. They introduced the terms primary, secondary, and tertiary structure. Although a variety of helical secondary structures had been proposed (cf 2, 3), it was Pauling (4, 5) who recognized the α-helix and β-pleated sheet, which provide an acceptable interpretation of Astbury's α- and β-diffraction patterns for fibrous proteins. Nevertheless, details of the α-helix were not seen at high resolution until the advent of the myoglobin structure (6), while the first atomic resolution observation of a β-sheet as a small antiparallel segment in lysozyme was not published until 1965 (7). Since that time well over 100 distinct structures have been determined. This wealth of information has led to a detailed examination of structural hierarchy as displayed by folded polypeptide chains.

The notion was entertained, even in the 1930s, that a protein would spontaneously refold after in vitro denaturation (8, 9). While three-dimensional structures demonstrate the uniqueness of a general fold with respect to a given protein, they do not directly discern the folding pathways. It was not until the 1960s, when the properties of proteins were better understood,

3

that the concept of spontaneous renaturation enjoyed wide acceptance. The pivotal work was that of Anfinsen and his co-workers, who "scrambled" ribonuclease, with its eight sulfhydryl groups, by allowing the reduced protein to reoxidize under denaturing conditions of 8 M urea (10). Removal of the denaturant and addition of mercaptoethanol resulted in a stable, functionally active conformation, though "unscrambling" frequently took hours to complete, an obvious discrepancy with in vivo rates. Denaturation-renaturation investigations have since been performed on a variety of other proteins including myoglobin (11), staphylococcal nuclease (12), lysozyme (13), and pancreatic trypsin inhibitor (14–16). The most detailed work has centered on disulfide proteins, where the covalent formation of S–S bonds can be used to characterize intermediates. In this way Creighton (17–19) was able to draw a folding pathway for bovine pancreatic trypsin inhibitor (Figure 1). It is noteworthy that essential intermediates exhibit some incorrect S–S pairing (17).

With the prompting of crystallographic results that revealed the form of folded proteins and renaturation experiments that demonstrated the spontaneity of refolding, a significant understanding emerged of the physical principles underlying the folding operation. Elementary principles of thermodynamics state that folding in a constant physiological environment

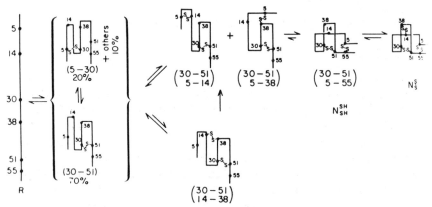

Figure 1   Schematic diagram of the pathway of folding and unfolding of normal bovine pancreatic trypsin inhibitor. The solid line represents the polypeptide backbone, with the positions of the cysteine residues indicated. The configurations of species N $_{SH}^{SH}$ and N $_S^S$ approximate the conformation of the native inhibitor; those of the others are relatively arbitrary except for the relative positions of the cysteine residues involved in disulfide bonds.

The brackets around the single-disulfide intermediates indicate that they are in rapid equilibrium; only the two most predominant species are depicted. The + between intermediates (30–51,5–14) and (30–51,5–38) signifies that both are formed directly from the single-disulfide intermediates, that both are converted directly to N $_{SH}^{SH}$, and that either or both are intermediates in the arrangement of (30–51,14–38) to N $_{SH}^{SH}$. [Reprinted with permission from Creighton (17). Copyright by Academic Press Inc. (London) Ltd.]

must be synonymous with the reduction of Gibbs free energy, though the folded protein may not have attained a "global" minimum (20, 21). Physically this implies burying of hydrophobic groups within the folded molecular core, creation of ion pairs and hydrogen bonds, and reduction of molecular surface. However, such considerations do not provide any information on pathway, since they are concerned with energy. Accordingly, the complex interactions of the polypeptide chain with itself and the environment must be considered. This requires a great simplification of the appearance of the polypeptide (20, 22–24). Alternative simplifications, suggested by light scattering and hydrodynamic measurements, assume nucleation centers, such as helices, around which the polypeptide can condense.

Methods to predict secondary structure from the primary amino acid sequence have been developed to avoid the difficult thermodynamic and statistical calculations. The predictive algorithms are statistical and rely on known protein structures. Perhaps the best known and easiest to apply is the technique of Chou & Fasman (25–27) who rank the amino acids as helix, sheet, and turn formers. They then elaborate on the number and kind of residues required to nucleate and terminate a given structural element. Two international competitions have been held to determine the accuracy of various techniques in the prediction of the adenylate kinase (28) and phage lysozyme (29) secondary structures. It is clear that these methods work better on some proteins than others, but they generally predict with a moderate degree of accuracy (21). Nevertheless, secondary structural predictions have found wide applicability in the analysis of amino acid sequences where the structure is unknown, and are particularly valuable when other functional properties of the structures are known (30). Recently, attempts have been made to extend these methods to predict tertiary structure by analyzing such variables as the packing of α-helices (31, 32) or the frequency of topological arrangements within β-sheets (cf 33).

The relationship between sequences and fold is not rigorous. The code that relates sequence to structure is highly degenerate, and yet remains responsive to the protein solvent. Furthermore there appear to be only a limited number of amino acid sequences that can provide a unique structure in a given environment; all others are nonsense. This may in part be the basis for the apparently small number of folds or architectural classes that have so far been observed. Albeit, with only a few exceptions (34), these observations have been confined to aqueous-soluble proteins.

Enzymes often utilize to their advantage a change of environment to alter conformation, as is implied by the term "induced fit" (35, 36). This is typified by the movement of the loop in lactate dehydrogenase (LDH), which is controlled by NAD binding (37, 38). Huber (39) has drawn attention to the order-disorder phenomenon that can occur in the formation of the trypsin specificity pocket or in the Fc fragment of the immunoglobulins.

Similarly, certain sections of viral coat proteins may fold as α-helices only in the presence of RNA (40, 41). The disordered segments of a polypeptide chain frequently start and end with glycines and contain few, if any, aromatic residues.

The degeneracy of the relationship between sequence and structure is essential in the process of evolution, as it permits an alteration of specific amino acids without destruction of the fold and, hence, without loss of function [see Lesk & Chothia (42) who analyze this degeneracy for the globin structure]. Indeed, the great conservation of residues in the active center of enzymes and the associated conservation of fold clearly demonstrate that function is a controlling aspect in protein evolution [see Doolittle (43) for a recent discussion of protein evolution]. Quaternary interactions can have a regulating effect on the function of each subunit. The presence of functional globin monomers such as lamprey hemoglobin or sperm whale myoglobin is one of many examples that show that the evolution of monomers usually precedes the subsequent evolution of allosteric oligomeric proteins as hemoglobin.

Excellent reviews on protein structure and fold include those of Jane Richardson (44), Schulz & Schirmer (21) and Cantor & Schimmel (45). The valuable book by Dickerson & Geis (46), an updated version of which is soon to be published by Benjamin, has been a standard text for most students and scholars in this area. In addition, there are a long line of reviews on protein structure in the *Annual Review of Biochemistry* (e.g. 47–52) and in other journals (e.g. 53, 54). Mention should also be made of reviews relevant to the dynamics of folding (e.g. 55–64). The present discourse confines itself to the analysis of folds of known protein structures.

SUMMARY

After some general remarks on protein structure, there follows a discussion on primary, secondary, and tertiary organization. The account of primary structure includes a discussion of the conformation of disulfide bonds. Types of helices, sheets, and turns are described in the section on secondary structure, followed by a discussion of super-secondary structure and the effects of metals and prosthetic groups on protein fold.

The crux of the review lies in an examination of tertiary structure, or specifically of domains that are defined, in part, as functional units within a polypeptide chain. An assembly of domains can in turn result in a protein whose function is quite sophisticated. Some consideration of domain recognition is given in the section on taxonomy and in the appendix. The key part of the tertiary structure section concentrates on a taxonomic protein classification dependent not only on structure but also on function. A discussion of the requirements imposed by quaternary structure on a fold are omitted in this review. Finally, no review of this kind can escape a discussion of evolutionary convergence and divergence.

THE ORGANIZATION OF A POLYPEPTIDE

This section concerns the static organization of proteins as found by crystal structure analysis, and is, therefore, limited to those protein domains with unique structures, as opposed to "random coil" conformations. The latter may often be observed for proteins, particularly under solvent denaturing conditions, or for segments of polypeptide chain within a crystal.

The gross structures of proteins are not affected by crystallization. For instance, the structure of trypsin when crystallized on its own (65, 66) or when complexed with inhibitor (67) changes very little. Nor is the detailed structure greatly perturbed, as shown by numerous NMR studies of proteins in solution (e.g. 68) and by the potential for enzyme activity in the crystal (e.g. 69). Furthermore, the fold of many functionally different proteins is often similar, e.g. the structures of triose phosphate isomerase, of the central domain of pyruvate kinase (70), and of bacterial aldolase (71, 72). Sizable forces are usually necessary to upset the unique native structure of a protein. Nevertheless, minor changes do occur between a specific crystalline conformation and the large number of related conformations likely to exist in solution. At times these small structural changes can lead to appreciable differences in the properties of the protein (73, 74). This review focuses primarily on the overall pattern of folding and not on details related to specific side chains.

Primary Structure

The polypeptide backbone consists of a series of planar *trans* linkages with mean dimensions shown in Figure 2 (75–78). The precise bond lengths and angles are derived from single crystal studies of small peptides, where data can be gathered to at least 0.7 Å resolution. It is usually necessary to assume these dimensions in analyzing the conformation of a complete protein. It is only the rare protein that is sufficiently ordered to provide data beyond 2.5 Å resolution. Departure of the peptide unit from planarity has been examined by Ramachandran and co-workers (79, 80). The occurrence of *cis* peptides, discussed by Ramachandran & Mitra (81), has been observed only when proline is one of the residues (82). The nomenclature of the dihedral angles in a peptide linkage is shown in Figure 3. Their definition can be found, for instance, in Dickerson & Geis (46) who give due caution on conventions. The angles ψ and ϕ represent rotations about single bonds and are constrained by steric hindrance. The angle ω is usually assumed to

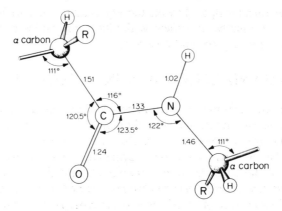

Figure 2 Dimension of planar *trans* peptide linkage.

be zero for a *trans* planar peptide. The backbone angle τ is affected by strain and is somewhat dependent on the type of side chain. Possible combinations of ϕ and ψ are nevertheless limited, and these constraints are best visualized as an orthogonal plot of these variables, known as a Ramachandran diagram (83).

The interrelationship of backbone structure and the sequence of amino acid side chain is clearly vital in the determination of fold. Kendrew et al (84), for instance, pointed out that serine and threonine have a preference

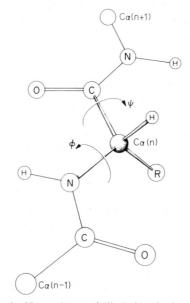

Figure 3 Nomenclature of dihedral angles in peptides.

for hydrogen bonding to backbone NH groups. Preferred orientation of side chain groups in known protein structures has been analyzed by Janin et al (85), Warme & Morgan (86) and Crippen & Kuntz (87). They show a reasonable correlation with conformations predicted from energy calculations (e.g. 62).

An in-depth study of about 60 disulfide bonds observed in proteins has been made by Richardson (44). The most important characteristic is the dihedral angle about the S–S bond, which determines whether the Cys–Cys conformation will be a right-or left-handed spiral. However the dihedral angles on each cysteine side chain also have a limited repertoire because of steric hindrance. Richardson was able to characterize four distinct types of conformations. The common conformations have distances of 6 and 5 Å between the C_α atoms for left- and right-handed bonding systems, respectively.

Secondary Structure

The formation of secondary structure largely relies on hydrogen bonding networks. The existence of a hydrogen bond in a protein structure is usually inferred from the positions of oxygen or nitrogen atoms. In general a hydrogen bond is roughly linear and 2.8–3.0 Å long (88, 89). Thus when the positional error of protein atoms is more than about 0.4 Å, the identification cannot always be made with certainty. In the absence of hydrogen bonds, electrostatic attraction may have some stabilizing effect on secondary structures if charges are closer than about 4 Å.

HELICAL STRUCTURES Two types of ring systems can be defined within a polypeptide chain:

$$\underset{|}{H}\!\!-\!\!-\!\!-\!\!-\!\!-\!\!-\!\!-\!\!-\!\!-\!\!-\!\!-\!\!-\!\!-\!\!-\!\!-\!\!-\!\!-\!\!\underset{|}{O}$$

$$-\,N\,-\,(CO\,-\,CHR\,-\,NH)_n\,-\,C\,- \qquad \text{with } m = 3n + 4, \text{ or}$$

$$\underset{|}{O}\!\!-\!\!-\!\!-\!\!-\!\!-\!\!-\!\!-\!\!-\!\!-\!\!-\!\!-\!\!-\!\!-\!\!-\!\!-\!\!-\!\!\underset{|}{H}$$

$$-\,C\,-\,(CHR\,-\,NH\,-CO)_n\,-CHR\,-\,N\,- \qquad \text{with } m = 3n + 5.$$

Any particular structure can then be characterized by three symbols: S, m, and r or l. S denotes the number of residues per turn of the helix; m, the number of atoms in the hydrogen bonded ring; and r or l according to whether the helix is right- or left-handed. A helix is thus designated by the symbol $S_m r$ or $S_m l$. The most stable helix type is the right-handed α-helix, which is denoted $3.6_{13} r$. The only other type of helix that is found with any significant frequency in globular proteins is the $3_{10} r$ helix, which usually appears as single turns ending an α-helix. Variations and intermediates of these conformations have been discussed by Némethy et al (90) who introduced the terms α_I and α_{II} for helices whose carbonyl groups point toward or away from the helix axis. A collagen-like, left-handed helix with proline in every third position has been found in cytochrome c_{551} (91). A left-handed δ-helix $(4.3_{14}l)$ has been proposed for the 22 amino-terminal residues of the *lac* repressor (92).

Aggregates of α-helices form left-handed, three-stranded "coiled-coils" in keratin (93). One such short segment has been found in the structure of southern bean mosaic virus (SBMV; 40) where parallel α-helices from three polypeptides twist about threefold axes. A similar left-handed twist can be seen in the packing arrangement of α-helices within the hemoglobin molecule (Figure 4), hemerythrin (94, 95), and cytochrome b_{562} (96).

Figure 4   The hemoglobin molecule, showing the left-handed twist of α-helices. (Reprinted with permission from Max F. Perutz.)

The packing of α-helices was first considered by Crick (97). He found that the side chains can intercollate between each other only when the helix axes cross at an angle of about −70° or +20°. His ideas were modified and extended by Chothia et al (98) whose packing analysis resulted in angles of −82°, −60° and +19° (classes I, II, and III, respectively). These results are reasonably well born out in the analysis of known structures (98). Richards and co-workers (31, 32) have investigated the way α-helices may aggregate and use this to "predict" the structure of myoglobin. Such predictions can be substantially improved when some chemically derived inter-residue distances are known (99). Attempts have been made to show how the four helices in tobacco mosaic virus (TMV) structure (100) and the eight-helical myoglobin structure (101) may self-assemble to produce the known structure. Argos et al (102) have pointed out that four-helical clusters are fairly common and may represent a "super-secondary structure," while Blow et al (103) discuss all possible connectivities between four α-helices formed from a single polypeptide. Whether such structures form due to the selectivity of helix aggregation [convergent evolution to a stable structure (see 104)] or as a result of divergence from a primordial structure is unclear.

SHEET STRUCTURES   Both parallel and antiparallel β-pleated sheet structures, with different characteristic hydrogen bond formations, were proposed by Pauling (105). It has recently been demonstrated that they are also differentiated by their amino acid compositions (106, 107). Chothia (108) was the first to discuss in print the characteristic twist of sheets, whether parallel or antiparallel (Figure 5). Successive strands within a sheet are twisted in a left-handed manner while the succeeding planes of peptide bonds in a given strand are twisted right-handedly. Neighboring strands may rotate from 0° to 30° with respect to each other. Weatherford & Salemme (109) propose that the twist is induced by the deformation of the peptide nitrogen toward a tetrahedral conformation, which causes slight nonplanarity of the peptide and optimizes the hydrogen bond geometry.

Richardson et al (110) investigated anomalies within β-pleated sheets. They found, mostly in antiparallel sheet, one important distortion, a "β-bulge," in which an extra residue is introduced between the closely spaced hydrogen bonds (Figure 6). This disrupts the sheet and can, therefore, only occur on the edge of a β-sheet. Two main types of bulges were explicitly identified: the classic and the G1 type. The values of the ψ and ϕ angles for residues in the bulge represent distinct conformations (110). The G1 bulge is nearly always associated with a glycine in position 1 of the bulge (Figure 6) and is very often associated with a type II tight turn (see section on turns). The latter requires a glycine in its third position, which corresponds to the first position of the G1 β-bulge.

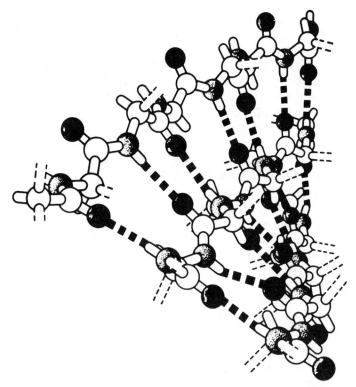

Figure 5 A β-pleated sheet formed from chains with a right-handed twist. The first chain is antiparallel to the second which is parallel to the third. View along the chain illustrates the right-hand twist. [Reprinted with permission from Chothia (108). Copyright by Academic Press Inc. (London) Ltd.]

Both Richardson (111) and Sternberg & Thornton (112, 113) have searched for systematic preferences in the sequential laying-out of strands within parallel, antiparallel, or mixed sheets (Table 1). The principal trend is for strands near each other in the sheet to be sequential along the polypeptide chain. Attempts have been made to predict (115) the sequence of strands within a sheet from these observed natural preferences. However, it would be necessary to predict the occurrence of β-strands within a polypeptide from amino acid sequence data, itself a risky procedure, to further predict tertiary structure (116). Ptitsyn and co-workers (117) have suggested folding pathways, using the principle that all strands must be laid down next to an immediate neighbor. They discuss the formation of both antiparallel (118) and parallel sheets (119).

It has been necessary to establish a nomenclature that permits an easy description of the connectivity within a β-pleated sheet. The procedure of

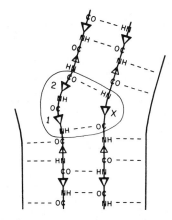

Figure 6   A β-bulge (outlined region) at the edge of an antiparallel β-sheet. Smaller triangles represent side chains that are below the sheet, larger triangles those that are above it. [Reprinted with permission from Richardson et al (110).]

Richardson (114) seems particularly elegant (see Figure 7). Successive segments of the polypeptide chain that form strands of a β-sheet are designated ±nx. The numeral n (Figure 7) designates the number of strands between consecutive portions of the polypeptide chain; the x, if present, designates parallel β-pleated sheet with the required cross-over between successive strands; and the sign can be used to designate whether the next strand is laid down in a positive or negative direction from the current strand.

Table 1   Frequency of antiparallel and parallel strands within a β-pleated sheet

Antiparallel[a]		Parallel[b]			
Turn type	Frequency	Turn type (right)	Frequency	Turn type (left)	Frequency
1	105	1x	44	1x	1
2	9	2x	24	2x	1
3	12	3x	11		
4	2	4x	3		
5	3	5x	—		
6	—	6x	—		
7	1	7x	1		
8	—	8x	1		
		12x	1		

[a] Nomenclature and data are due to Richardson (111). The actual results are strongly affected by the available structural sample, but the trend for strands near each other within the sheet to be sequential along the polypeptide chain is clear.
[b] Nomenclature and data are due to Richardson (114).

(a)

(b)

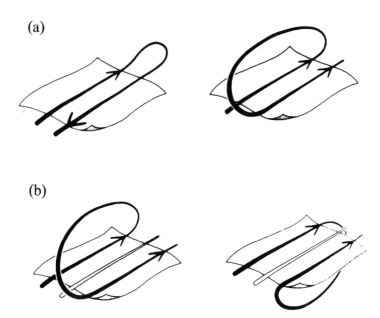

Figure 7 (*a*) Illustration of the two main classes of topological connection in β-sheets: (*left*) "hairpin", "plain", or "same-end" connection; this specific example is type ±1; (*right*) "cross-over", "cross", or "opposite-end" connection; this specific example is type ±1x. (*b*) (*left*) A right-handed ±2x cross-over connection; (*right*) left-handed ±2x cross-over connection. Direction is not indicated for the skipped strand, since it may be either parallel or antiparallel to the others. [Reprinted with permission from Richardson (114).]

TURNS   Different secondary structural elements are frequently connected by sharp turns, particularly at the surface of the molecule (120). These turns are called reverse turns, β-turns, β-bends, etc. They generally have distinctly recognizable conformations and are often restricted in their amino acid composition (120a). Crawford et al (121) point out that about one third of all amino acids in globular proteins are in turns. Venkatachalam (122), Crawford et al (121), Lewis et al (123), and Chou & Fasman (124) have all examined the conformation of turns. Their consensus on the definition of a turn is in terms of four residues, connected by three peptide linkages, stabilized by a hydrogen bond between the carbonyl of the first and amide of the fourth group. However, the actual formation of the hydrogen bond is frequently absent, although the distance between the appropriate O and N atoms is reasonably small (less than 4 Å). Some authors (125, 126) have defined turns by using C_a carbons only. This leads to a more general definition that is not as intimately involved with the chemical requirements of hydrogen bonding or steric hindrance.

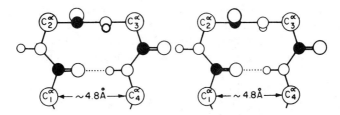

Figure 8   The two types of chain reversal that have hydrogen bonding between the carbonyl of the first and amide of the fourth residue. The four alpha carbons are not meant to appear planar. Type I is on the left, and Type II is on the right. [Reprinted with permission from Crawford et al (121).]

Different types of β-turn conformations were designated I, II, and III and I', II', and III' by Lewis et al (123), a nomenclature that is now often adopted. The primed conformations represent the converse dihedral angles to the unprimed. Conformations I and II are related by a 180° flip of the central peptide (Figure 8). Conformation III cannot readily be distinguished from I, as the dihedral angles differ by less than 30°. Other conformations, IV to VII, were also defined, but they represent special cases (e.g. VI has a *cis* proline at position 3) or encompass unusual situations. Table 2 shows the properties of the four major conformations and the preference for glycine in certain positions because of steric hindrance (Figure 8). This preference had been predicted by Venkatachalam (122), particularly for position 3 in the type II bend. In practice the requirement for glycine does not appear to be as absolute as his predictions would suggest, because of departures from the mean dihedral angles. The possible peptide conformations in antibiotics, where D amino acids are common, have been reviewed by Chandrasekaran & Prasad (127) but lie outside the scope of this review.

SUPER-SECONDARY STRUCTURES   The term super-secondary structure was coined by Rao & Rossmann (128) to describe a recurring fold consisting of a number of secondary structural elements and yet not comprising the complete tertiary structure of a molecule or domain (see section on domain

Table 2   Mean dihedral angles for the two types of chain reversal (123)[a]

	ϕ_2	ψ_2	ϕ_3	ψ_3	Comment
I	−60	−30	−90	0	
II	−60	120	80	0	Glycine likely in position 3
I'	60	30	90	0	Glycine preferred in position 2
II'	60	−120	−80	0	Glycine preferred in position 2

[a] See Figure 8 for definition of subscripts to dihedral angles and turn identifications.

definition). The occurrence of such folds is likely to be the result of pressure to find energetic stability through packing of these elements. Rao & Rossmann considered, in particular, the unique hand and fold of the mononucleotide binding structure present in dehydrogenases and many other proteins.

Two β-strands with their accompanying cross-over connections have been analyzed (114, 129). Both analyses observe that the connection between strands within a parallel β-pleated sheet almost always (Table 1) retains the same hand, which was defined as right-handed (see Figure 7). It is therefore to be concluded that right-handed cross-over structures are stable for reasons about which Richardson (114) and Sternberg & Thornton (129) have speculated. The cross-over connection is often a helix resulting in a β-α-β super-secondary structure.

A series of super-secondary structures can often form a functional domain. For instance, the NAD binding domain in dehydrogenases may be described as +1x, +1x, -3x, -1x, -1x. Another example is the triose phosphate isomerase structure (130; Figure 9), which occurs also in pyruvate

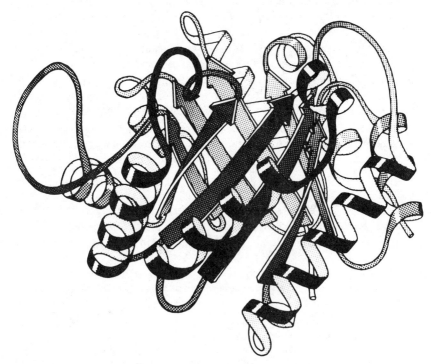

Figure 9   Schematic drawing of a triose phosphate isomerase subunit. Copyright by Jane S. Richardson. [Reprinted with permission from Richardson (71). Copyright by Academic Press, Inc.]

kinase (70, 131) and 2-keto-3-deoxy-6-phosphogluconate aldolase (71, 72), and can be described as $(+1x)_8$. Indeed β-α-β structures form the largest group of observed protein structures. Many of these have been "catalogued" (111, 132–134). The dipoles created by the parallel α-helices on either side of the central parallel sheet produce a significant electrostatic field, which provides suitable orientation and binding forces for the negatively charged phosphates of nucleotides (135).

Schulz (136) introduced the term "β-meander," which represents a +1, +1 three-stranded antiparallel sheet. This is a common fold (Table 1) and must therefore be classified as a super-secondary structure. A mononucleotide fold plus a β-meander comprises a domain that is twice repeated in the tertiary structure of glutathione reductase (136).

METALS AND PROSTHETIC GROUPS IN PROTEINS   It is not clear whether the presence of metals or prosthetic groups affects the unique fold of a polypeptide. Only minor conformational changes occur in metal-free concanavalin A, relative to the native structure (137). It is possible, in many instances, to remove the prosthetic group without damage (e.g. the heme in globins), whereas in other cases the protein becomes denatured [e.g. the structural Zn in liver alcohol dehydrogenase (LADH) or the heme in catalase]. In any event the fold must adapt itself to the requirement of binding a specific metal. Carp myogen (Table 3) displays three "E–F" hands each of which consists of two sequential helices joined by a right angular turn. Six ligands, situated close to each other within the turn, bind a calcium ion. Kretsinger suggests (244) that the various examples of this fold may all have originated from a common ancestor. In many metal-containing proteins the polypeptide chain makes a tight turn about a metal ion so that most of the liganding residues are situated close to each other along the chain. Examples are carbonic anhydrase, thermolysin, and the Fe–S cage in ferredoxin (48).

Tertiary Structure of Domains

DOMAIN—DEFINITION   The word domain was probably first used to describe the repetition of homologous sequences within the light and heavy chains of an immunoglobulin molecule (245). It was subsequently recognized that similar sequences within a complete polypeptide chain determine similar folds. Such structural domains were often found to be spatially well separated, giving the molecule an appearance of being bi- or multilobal. Later it was observed that differing domains within a single molecule might also have completely different architecture, such as a parallel β-pleated sheet acting as one domain and a "β-barrel" (44) acting as the other. Furthermore, a domain in one molecule might also occur in a quite different

Table 3 Taxonomy of known domains[a] in globular proteins

I. All α structures

  A. Heme-binding proteins—type A: oxygen and electron carriers (Figure 10)[b]

   1. Globins: α chain of hemoglobin (141, 142); β chain of hemoglobin (141, 142); single-chain hemoglobins in worm (143); single-chain hemoglobins in lamprey (144); single-chain hemoglobins in root nodules of legumes (145); myoglobin (146)
   2. Cytochrome b_5 and related structures such as domains in cytochrome b_2 and sulfite oxidases (147, 148)
   3. Cytochromes c: cytochrome c (149, 150); cytochrome c_2 (151); cytochrome c_{550} (152); cytochrome c_{551} (91); cytochrome c_{555} (153)

  B. Four-helical protein domains[c]

   1. Oxygen carriers without a heme group, but using Fe: hemerythrin (95); myohemerythrin (154)
   2. Heme-binding proteins—type B, electron carriers: cytochrome b_{562} (96); cytochrome c' (155)
   3. TMV protein (156)
   4. Ferritin (157)
   5. Second domain in Tyr-tRNA synthetase (158)
   6. FB fragment of protein A (159)
   7. First domain of papain (160)

  C. Calcium-binding proteins[d]: carp myogen (161); troponin C (162)

  D. Miscellaneous α structures: uteroglobin (163); purple membrane protein[e] (34); pig heart citrate synthetase[f] (165); second domain of thermolysin (166); glucagon[g] (167); 6-phosphogluconate dehydrogenase (168); first and third domains of hemagglutinin (169[h])

II. All β structures

  A. Single sheets of antiparallel strands: carbonic anhydrase[i] (170); *Streptomyces* subtilisin inhibitor[j] (171); second domain of p-hydroxybenzoate hydroxylase (172); second domain of papain[k] (173); bacteriochlorophyll a-protein[l] (174)
  B. Greek key β-barrels with 6–8 strands (Figure 11)

   1. Serine proteases[m]: chymotrypsin and chymotrypsinogen (175, 176); trypsin and trypsinogen (177–179); elastase (180); *Myxobacter 495* α-lytic protease (181); *Streptomyces griseus* protease A and protease B (182)

   2. Acid proteases[m] (183): penicillopepsin (184); *Rhizopus* pepsin (185); *Endothia* pepsin (185, 186); pepsin (187)

   3. "Immunoglobulin domains"

     a. In immunoglobulins[n] (52)
     b. Superoxide dismutase[o] (188)

     c. Plastocyanin[p] (189)
     d. Azurin[p] (190)

Table 3 *(Continued)*

II. All β structures *(continued)*

    4. Miscellaneous barrel structures: soybean trypsin inhibitor (191, 192); prealbumin (193); staphylococcal nuclease (194); second domain of pyruvate kinase (131); DNA-unwinding protein (195); first domain of thermolysin (166)

  C. "Jelly roll" structures (196)

    1. Coat proteins of some spherical viruses

      a. Both domains of tomato bushy stunt virus (197)
      b. SBMV (40)

      c. Middle domain of hemagglutinin (169[h])
    2. Concanavalin A (198, 199)
    3. cAMP receptor protein[q]

III. α/β structures

  A. Domains that bind nucleotides toward the carboxyl end of a predominantly parallel β-pleated sheet

    1. NAD binding domains in dehydrogenases: first large domain of LDH (200); first domain of malate dehydrogenase (201); second domain of LADH (202); first domain of glyceraldehyde-3-phosphate dehydrogenase (203)

    2. Kinases[r]: adenylate kinase (205); two similar domains of phosphoglycerate kinase (206, 207); hexokinase (208); both domains of phosphofructokinase (209)

    3. NADP-dependent dehydrogenases: dihydrofolate reductase (210); first two domains of glutathione reductase (136); NADP binding domain of 6-phosphogluconate dehydrogenase (211[s]); first domain of p-hydroxybenzoate hydroxylase (172)

    4. Pyridoxal phosphate-binding enzymes: first domain of aspartate aminotransferase (212, 213); second domain of phosphorylase (214)

    5. Other nucleotide-binding enzymes: flavodoxin[t] (215); second domain of phosphorylase[u] (214); Tyr-tRNA synthetase[v] (158); catalytic domain of aspartate transcarbamylase[w] (216); Tu elongation factor[v] (217)

    6. Enzymes that bind phosphates: phosphoglycerate mutase (218); glucose 6-phosphate isomerase (219)

  B. Structures based on a single, primarily parallel, β-pleated sheet, with active sites toward the carboxyl end of the sheet, that do not, however, bind nucleotides: carboxypeptidase (220); subtilisin (221, 222); thioredoxin (223); glutathione peroxidase (224)

  C. Structures that contain two domains of primarily parallel β-pleated sheet with the carboxyl ends of the sheet directed toward each other: phosphoglycerate kinase (206); hexokinase (208); phosphofructokinase (209); rhodanese (225); arabinose binding protein (226); phosphorylase (214); catalytic domain of aspartate transcarbamylase (216); glucose 6-phosphate isomerase (219)

  D. Structures with eight successive 1x cross-over turns that form a barrel (Figure 9): triose phosphate isomerase (130); aldolase (71, 72); first domain of pyruvate kinase (131); glycolate oxidase (227); first domain of Taka-amylase A (227a)

Table 3 *(Continued)*

IV. Small α + β proteins

    A.   Structures based primarily on β-sheets that frequently contain disulfide bonds. Most of these proteins are extracellular (228)

        1.   Neurotoxins and similar structures

                a. Sea snake neurotoxin (229)

                b. Erabutoxin (230)

                c. Wheat germ agglutinin (231)

        2.   Structures that can bind polysaccharides (232, 233): hen egg white lysozyme (234); phage lysozyme (235); α-lactalbumin (236)

        3.   Miscellaneous structures: ribonuclease (237); papain (173); insulin (238); pancreatic trypsin inhibitor (67); phospholipase A_2 (239)

    B.   Fe-S proteins: rubredoxin (240); both domains of ferredoxin[x] (232, 241); high potential iron protein[y] (242)

    C.   Multiheme cytochrome c_3 (243)

V. Large α + β proteins: catalytic domain of LDH (200); catalytic domain of glyceraldehyde 3-phosphate dehydrogenase (203); catalytic domain[z] of LADH (202)

---

[a] Inside bracket indicates strong functional and structural similarity. Outside bracket indicates good structural equivalence. Outside dashed lines indicate some structural similarity.

[b] Structural and functional relationships among these proteins have been discussed by Argos & Rossmann (138). The structurally common part corresponds to a central exon in the globin gene (138–140).

[c] The four α-helices run mostly antiparallel and are connected sequentially (102, 103). The topology of the connectivity is the same in categories 1–4.

[d] These consist of three repeated domains of α-helices, AB, CD, and EF, related by roughly 90°. Calcium atoms are bound into the elbows of the last two domains. It has been suggested that this domain is a fairly universal calcium-binding domain in muscle proteins (161) and troponin C (162).

[e] This protein is primarily in a lipid membrane environment and thus may have different folding rules to water-soluble proteins. There are seven helices in this structure. Their individual polarities have not been unambiguously established, but the helix axes are essentially parallel and lie perpendicular to the membrane surface (164).

[f] There are at least 12 α-helices in one subunit of this molecule with no β-structure.

[g] A single helix.

[h] D. C. Wiley, personal communication.

[i] Ten antiparallel strands.

[j] 5 Antiparallel strands.

[k] 6 Antiparallel strands.

[l] 15 Antiparallel strands with the ends curled around to make a cavity for seven chlorophyll moieties.

[m] Two similar domains in tandem.

[n] Variable-heavy, variable-light, constant-heavy, constant-light domains in tandem in immunoglobulins, in Bence-Jones protein, and in Fc and Fab fragments of the immunoglobulins. For general review see Amzel & Poljak (52).

[o] Cu-Zn–containing.

[p] Cu–containing.

[q] T. A. Steitz, personal communication.

[r] Most of these are bilobal and slightly change the lobal separation during catalysis (204).

[s] M. J. Adams, personal communication.

[t] Binds FMN.

[u] Binds AMP.

[v] Binds ADP and ATP.

[w] Binds CTP.

[x] Each binds one Fe_4S_4 cluster.

[y] With one Fe_4S_4 cluster.

[z] The catalytic domain consists of two segments; the major portion is at the beginning of the polypeptide and the minor portion at the end, with the NAD binding domain in between.

molecule. For instance, the NAD binding domain has been found to occur in all NAD-linked dehydrogenases whose structures have been determined (246, 247), and, as mentioned earlier, the triose phosphate isomerase structure (Figure 9) occurs in pyruvate kinase (70, 131) and probably in bacterial aldolase (71, 72). This has quite naturally led to the proposal of divergent evolution from a common ancestral structure whose corresponding gene has been copied and fused with a variety of different genes. Domain structures would then be maintained fairly faithfully by the demands of particular functions and structural stability.

The following properties can generally be ascribed to a domain:

1. Similar domain structures or their amino acid sequences can be found either within the same polypeptide or in a different molecule.
2. Domains within a polypeptide are spatially separated from each other, or at least form a compact "glob" or cluster of residues.
3. Domains have a specific function, such as binding a nucleotide or polysaccharides.
4. The active center of a molecule is at the interface between domains, which permits the simple function of each to be brought together to form a more sophisticated molecule (248).

A corollary of these definitions, particularly the spatial separation of compact domains, is that proteolytic cleavage can frequently be used to identify and separate domains. Examples are the separation of the heme and FMN binding domains of cytochrome b_2 (249), the separation of the two elastase domains (250), and the recognition of domains in the β subunit of E. coli tryptophan synthetase (251). It is noteworthy that, in the elastase case, the carboxyl-terminal domain retains its ability to bind substrate, but there is no catalytic activity, since the catalytic residues are in part on the amino-terminal domain. Whether excised domains in general retain their fold and function is subject to dispute. The excision may expose large hydrophobic patches, which might destabilize the domain structure in an aqueous environment.

Not all the above listed properties of a domain need be applicable simultaneously. There have been attempts to define domains entirely on the basis of structure using only the second criterion above (252, 253). An exemplary systematic procedure to recognize domains by their spatial separation has been proposed by Janin & Wodak (254). They first devise an algorithm to compute the surface area of any portion of the protein (255). Then they compute the surface area generated by cleaving the polypeptide chain at all possible positions. When no significantly large area is so generated, the globular nature of the two parts would be indicated. Others (132, 253, 256, 257) have assigned domains on a rather more intuitive basis.

Blake (258) and Gilbert (259) have suggested that domain integrity is preserved in eucaryotes by their coding within exons. This is supported by the common helical folding pattern observed in globin, cytochrome c_{551}, and cytochrome b_5 (Figure 10), which has a heme binding function and corresponds to the middle exon of the β globin chain (138). Similarly, conserved specific functional properties can be ascribed to this portion of the globin chain (260). Indeed, Craik et al (261) have been able to excise

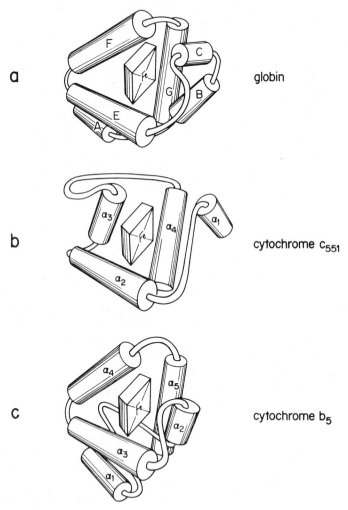

Figure 10   Diagrammatic representation of the similar polypeptide chain topology in (a) hemoglobin β chain, (b) cytochrome c_{551}, and (c) cytochrome b_5. [Reprinted with permission from Argos & Rossmann (138). Copyright by the American Chemical Society.]

this portion of the globin chain and demonstrate its ability to bind heme. Since domains may thus be the result of independent genetic development, they may also represent the basis for protein folding. It is therefore reasonable to assume that the domain will fold independently of other domains in the complete polypeptide and act as a nucleation center. The final tertiary structure will be attained as an assembly of the folded domains. Exons could provide a mode for rapid evolutionary development within the context of energetically favorable folds.

TAXONOMY   It had been generally assumed prior to 1959 that every protein structure would be radically different to every other protein structure. The three-dimensional arrangements of secondary structural elements were thought to be almost infinite. Although myoglobin and the α- and β-hemoglobin chains were observed to have similar folds, this could be readily understood in terms of a common precursor for heme-dependent oxygen carriers. In 1971, some 12 years later, it was discovered that malate dehydrogenase and LDH have similar tertiary, if not quaternary, structures (262). This appeared reasonable in that both these enzymes have similar substrates, although their functions are related to different metabolic pathways. Then followed the realization that LDH and flavodoxin share some structural similarity (128) including the approximate orientation and position for coenzyme binding (246). This structural equivalence was a more radical observation, as both molecules show only weak functional relationships and the structural similarity is confined to a portion of the molecule (Table 4). Other surprising similarities, such as lysozyme with α lactalbumin (236) or superoxide dismutase with the immunoglobulin domain (188), provided further enigmas. A debate ensued between proponents of divergence and those supporting convergence to stable super-secondary structures. On the other hand, the concept of convergence in the active center region of molecules had indeed been well illustrated in chymotrypsin and subtilisin (263), which have totally different folds. Nevertheless, it has remained difficult to differentiate between convergence and divergence in the evolution of similar protein folds (see next section).

Once the initial shock of structural similarity in protein structures had passed, it became possible to construct taxonomic classifications. The first comprehensive attempt by Levitt & Chothia (132) used four major classifications: 1. all α proteins with only α-helix secondary structure (e.g. myoglobin); 2. all β proteins with mainly β-sheet secondary structure (e.g. superoxide dismutase or chymotrypsin); 3. $\alpha + \beta$ proteins with α-helices and β-strands that tend to segregate into all α and all β segments along the peptide chain (e.g. papain or thermolysin); and 4. α/β proteins with mixed or approximately alternating segments of α-helical and β-strand secondary

Table 4 Various criteria for assessing structural equivalence between functional domains in a variety of protein comparisons

Comparison		Number of residues		Number of equivalenced residues	Percentage of equivalenced residues		rms deviation (Å)	MBC/C[a]
Molecule 1	Molecule 2	Molecule 1	Molecule 2		Molecule 1	Molecule 2		
Globins:								
Hemoglobin β	Hemoglobin α	146	141	139	82.7	91.0	1.6	0.70
Heme binding proteins:								
Hemoglobin β	Cytochrome b_5	146	93	48	32.9	51.6	4.1	1.29
Hemoglobin β	Cytochrome c_{551}	146	82	49	33.6	59.8	3.5	1.33
Cytochrome b_5	Cytochrome c_{551}	93	82	41	44.1	50.0	4.9	1.60
Nucleotide binding domains:								
LDH[b]	GAPDH[c]	144	148	83	57.5	56.1	2.9	1.12
LDH	Flavodoxin	144	138	32	22.2	23.2	2.4	1.23
Immunoglobulin folds:								
Superoxide dismutase[d]	Immunoglobulin domain (C_L)[d]	151	110	42	27.8	38.2	1.9	—
"Jelly roll" structure:								
TBSV(S)[e]	TBSV(P)[f]	167	110	69	41.3	62.8	3.8	—
Con A[g]	TBSV(P)	237	110	68	28.7	61.8	3.4	—
Con A	TBSV(S)	237	167	82	34.7	49.2	3.2	—
SBMV-C[h]	TBSV(S)-C[h]	219	200	182	82.7	91.0	2.7	—
Lysozymes:								
T4 phage	Hen egg white	164	129	64	39.1	49.7	4.1	1.53

[a] Minimum base change per codon.
[b] Lactate dehydrogenase.
[c] Glyceraldehyde-3-phosphate dehydrogenase.
[d] Taken from Richardson et al (188).
[e] Tomato bushy stunt virus shell domain.
[f] Tomato bushy stunt virus protruding domain.
[g] Concanavalin A.
[h] "C" refers to the C subunit within SBMV and TBSV.

structure (e.g. flavodoxin or kinases). Partial classifications were also concurrently attempted by Sternberg & Thornton (133), and Richardson (111). Three especially useful categories were introduced by Richardson (111) for "β-barrel" structures (Figure 11). These were the "Indian basket structures" as in papain or soybean trypsin inhibitor, the "Greek key structures" as in chymotrypsin or superoxide dismutase, and the "lightning structures" as in triose phosphate isomerase. The unique hand of all these common

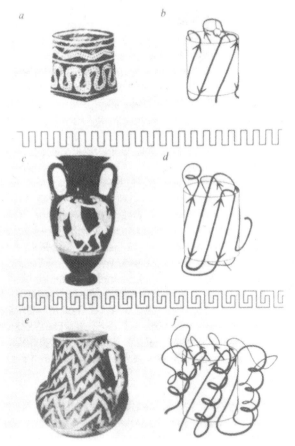

Figure 11   Comparison of geometric motifs common on Greek and American Indian weaving and pottery with the backbone folding patterns found for cylindrical β-sheet structures in globular proteins. (*a*) Indian polychrome cane basket from Louisiana: (*b*) Polypeptide backbone of rubredoxin; (*c*) Red-figured Greek amphora showing Cassandra and Ajax (about 450 BC); (*d*) Polypeptide backbone of prealbumin; (*e*) Early Anasazi Indian redware pitcher from New Mexico. (*f*) Polypeptide backbone of triose phosphate isomerase. [Reprinted with permission from Richardson (111). Copyright by Macmillan Journals Limited.]

structures was noted. Richardson's analogy with art objects is particularly appropriate. The differences in architecture of the various β-barrels is as diverse and recognizable as the different cultures pertaining to the pots and baskets in Figure 11. Just as an art historian can discern the history and function of a particular vase, so the molecular taxonomist should be able to determine the history and function of a polypeptide fold.

A more quantitative approach to the classification of β-barrels has been suggested by McLachlan (264). He defines the shear, S, as the number of residues on a given strand when the hydrogen bonding network is followed around the barrel starting at residue i and finishing at residue $i \pm S$. If the distance along a strand between successive residues is a (usually 3.5 Å), the distance from one hydrogen-bonded strand to the next is b (usually 4.7 Å), and the slope of the strands to the cylindrical barrel axis is α, then $S = [(nb/a)\tan\alpha]$, and $R = \{b/[2\sin(\pi/n)\cos\alpha]\}$ where there are n strands in the barrel. It follows that, if the shear stays constant but the number of strands increase, then the strands become progressively more parallel to the barrel axis.

The most recent and most complete taxonomic classification of protein structures is that of Richardson (44). This is in part dependent on the number of "layers" within a structure. Regrettably, none of the current classification schemes takes any significant account of function. That is, the pendulum has swung far from the early preoccupation with functional relationships at a time when no structural comparisons between diverse molecules was conceivable. An attempt has been made to fuse these two approaches in Table 3. The major classifications follow roughly those suggested by Levitt & Chothia (132). Subdivisions are then created with special sensitivity to the function of domains within the polypeptide chain. Taxonomic classifications can be ambiguous whether they are designed to distinguish botanical specimens or molecules. Hence on some occasions molcules are listed under more than one heading. References generally refer to the most up-to-date information on the molecular structure or to a pertinent review.

CONVERGENCE, DIVERGENCE, GENE DUPLICATION, AND GENE FUSION   Evolutionary studies have had considerable difficulty differentiating between convergence toward a functionally useful structure or divergence from an ancestral structure whose functions have been refined and specialized. A sufficient fossil record usually alleviates this problem. However, in molecular structure there are no fossils; the molecules are those of today. The molecular taxonomist thus faces the perplexing task of reconstructing history from the similarities and variations found in today's specimens (265).

The generally accepted technique for differentiation between convergence and divergence involves a count of the characters that two molecules share or by which they differ. If the majority are the same, it would seem probable that the two molecules (or organisms) have diverged from a common ancestor with a few characters altered. However, such statements are only probabilistic; as the number of similar and dissimilar characters approach each other, it becomes impossible to make a useful differentiation.

The molecular evolutionists have primarily utilized the amino acids within protein sequences for taxonomic characters; but, when divergence has extensively progressed, amino acid sequence comparisons become insensitive. Fortunately, the protein fold is conserved to a far greater extent than the amino acid sequence and thus provides taxonomic properties for examination. However, both amino acid sequence and structural comparisons neglect function as an important characteristic that can delineate evolutionary history. On the other hand, function can be emphasized at the expense of structural information, as exemplified by the work of Dickerson on the evolution of bacterial respiratory processes involving cytochrome c–like molecules (266). For instance, function is not conserved in the divergence of haptoglobin from serine proteases in spite of good homology in sequence (267) and perhaps fold (268).

The first example of the use of structure as a quantitative tool in measuring evolutionary distance was given by Schulz & Schirmer (269) who used the sequence of strands in nucleotide binding folds. On this basis they computed the probability that a particular structure could occur more than once by chance. These values were later corrected (129) by allowing for the apparent preference of right-handed over left-handed cross-over structures. Given such measurements it is possible to construct networks (that can usually be related to phylogenetic trees) by standard techniques (cf 270–272) that permit quantitative testing of all postulated trees. Qualitative assessment of evolutionary relationships (192, 273) must be regarded with suspicion.

Many morphological features (e.g. 232, 270, 274, 275) of molecular structure have been utilized to measure evolutionary distances (see Table 4). A list of such features would include: the minimum base change per codon of residues associated with equivalenced C_α atoms; the rms separation of C_α atoms for spatially superimposed molecules; the similarity of protein folding topology; the number of structural insertions and deletions and their fraction of the structures superimposed; the coincidence of substrate positioning and orientation relative to the compared protein folds; commonality of function; shared active center geometries; axes of symmetry relating two domains; and the like. In projecting convergent or divergent probabilities, it is difficult to assign weights to the various criteria. For example, there could be unequal rates of convergence of different characters toward a functional protein; alternatively, divergence from a common ancestor could have occurred sufficiently early in biological time so that the remaining similarity is obscured. Nonetheless, there are many rich examples of structural comparison (e.g. 276).

Emphasis has been given in the construction of Table 3 to the study of the divergence of polypeptide folds with specific functions and the convergence of fold to energetically favorable structures. A likely example of divergence is found in the structural comparison of the functionally related globins (42). Another example is the relationship among eucaryotic and procaryotic cytochrome c–like molecules (266), where 80% or more of the C_α atoms can be spatially equivalenced with a mean rms separation of approximately 2.0 Å and a minimum base change per codon of around 1.0. The good topological and spatial agreement between the folds of the functionally diverse superoxide dismutase and the immunoglobulin domain (188) provides a likely example of convergence to a stable fold. Here the separation of equivalenced atoms was 1.9 Å with no obvious correlation of amino acid sequences.

Salient enigmas are the β-α-β folds found in various proteins. In dehydrogenases, the six parallel strands and associated helices are well preserved; NAD binding provides the obvious functional link, and the minimum base change per codon value is far from random. However, when manifested in other structures, the number of parallel strands may be less than six, insertions and deletions often occur, antiparallel strands can intrude, and the cofactor is not accurately positioned at the carboxyl end of the sheet (275). The best known example of active center convergence is that of chymotrypsin and subtilisin (263), where 23 catalytic site atoms superimpose to within an rms deviation of 1 Å. A further illustration of particular and perhaps unexpected similarity is the environment of the essential Zn atom in carbonic anhydrase, carboxypeptidase, thermolysin, and LADH (277, 278).

A reasonably common phenomenon is the repeat of a domain along the same polypeptide chain, as for example in the light and heavy chains of immunoglobulins. In some cases (Table 5) these domains are related by a twofold axis (273) as if they were identical subunits in an oligomeric protein with a proper point group (283). The active center may be associated with one of the domains (rhodanese) or both (acid protease). A conceivable mechanism for this phenomenon has been discussed by Tang et al (183) in terms of gene fusion. Obvious examples of fusion of unlike genes is implied by the occurrence of greatly different domains within subunits of NAD-dependent dehydrogenases (284).

Table 5 Gene duplication within the same polypeptide chain with domains related by diad[a]

Protein	Rotation	Translation	Reference
1. Rhodanese	180° ± 1°	< 1 Å	225
2. Calcium-binding protein	180° ± 7.5°	none	279
3. Ferredoxin	180° ± 2°24'	none	232
4. Hexokinase	180° ± 17°	2.1 Å	275
5. Phosphoglycerate kinase	180° ± 12°	16 Å	275
6. Chymotrypsin and other serine proteases	an approximate two-fold screw axis		264
7. Acid proteases	180° ± 12°	0.0 Å	185, 280
8. Myohemerythrin	180° ± 1.6°	3.6 Å	281, 282
9. TMV	180° ± 8.8°	2.9 Å	282
10. Glutathione reductase	180° ± 43°	3.9 Å	136

[a] There are other examples (e.g. arabinose binding protein) where the nature of the relation between the two similar domains has not yet been established quantitatively.

CONCLUSION

With the advent of a relatively large catalogue of structures for water-soluble proteins, order is emerging from diversity. A variety of molecular structural characteristics have been recognized and these have been incorporated in various taxonomic classifications. The "domain," within polypeptides, has generally been recognized as a stable folding unit. Larger and more sophisticated enzymes are built of domains whose evolutionary history precedes that of the enzyme itself. The study of protein folding is thus likely to concentrate on viable domain structures, their cataloguing, and folding pathways.

APPENDIX—DETERMINING STRUCTURAL EQUIVALENCE

In the discussion and classification of folds it is necessary to distinguish between like and unlike structures. It is easy to recognize structures that are very similar or totally different. However in the presence of sizable insertions or deletions, the difficulty of visualizing a three-dimensional object may obscure the underlying similarities. Once equivalent atoms have been selected in two molecules, then quantitative methods are readily available for finding their best relative superpositions and hence the rms value between the superimposed structures (128, 285–288). The problem arises in recognizing significant similarity in structure and determining which atoms are to be assigned as structurally equivalent.

A number of authors have used "diagonal" or "distance" plots for rapid visual recognition of structural domains (see 248, 289, 290). The distance between pairs of atoms, usually only the C_α atoms, are plotted as a symmetrical square matrix. Atoms close together along the polypeptide chain (as in an α-helix) will plot along the diagonal, while parallel and antiparallel strands within a sheet appear as streaks perpendicular and parallel to the diagonal. A complete domain will have a particular pattern and can be readily recognized in such two-dimensional plots.

Further quantitation was not truly attempted until Rossmann & Argos (232) produced a three-dimensional search procedure. Any rigid body can be superimposed onto any other rigid body by a suitable combination of three rotation angles and three translation vectors. Rossmann and Argos searched all orientations of a rigid body with another. (Figure 12 shows the search function whose largest peak corresponds to the superposition seen in Figure 13.) When the proteins have similar folds and orientation, then the vectors between equivalenced atoms will be roughly parallel and of equal length. A search for the orientation with the maximum number of

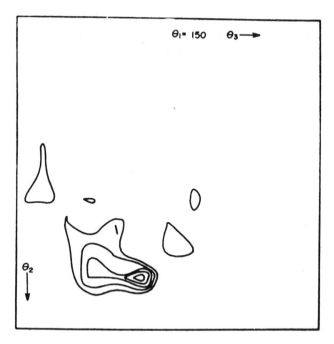

Figure 12 Comparison of phage lysozyme with hen egg white lysozyme showing section $\theta_1=150°$. Contours are drawn at levels 29, 34, 39, ... equivalent amino acid residues. [Reprinted with permission from Rossmann & Argos (232). Copyright by Academic Press Inc. (London) Ltd.]

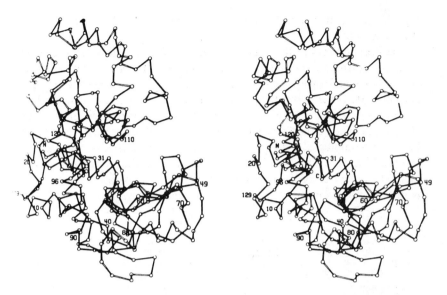

Figure 13   Stereo view of the α-carbon backbone of hen egg white lysozyme (dark lines) superimposed on the phage lysozyme structure (thin double lines). The numbering refers to the hen egg white lysozyme sequence. This superposition corresponds to the large peak in the search function shown in Figure 12. [Reprinted with permission from Rossmann & Argos (232). Copyright by Academic Press Inc. (London) Ltd.]

parallel and equal vectors between atoms selected progressively along the polypeptides provides not only a test for topological equivalence but also identification of insertions and deletions (232, 285). It remained, however, to be seen whether the size and complexity of the equivalenced protein sections were sufficient to permit the description "similar structures" or even "divergence from a common ancestor." A variety of criteria have been proposed by Rossmann & Argos (275) to differentiate these possibilities.

An alternative procedure has been described by Remington & Matthews (232). This method can be rapid if certain computational techniques are adopted (264). It is the three-dimensional equivalent of the Jukes & Cantor (291) method for finding one-dimensional sequence analogies. A section of polypeptide chain of a given number of residues (say 60) is selected in both molecules, and the C_α atoms are then equivalenced sequentially. A least-squares procedure is used to minimize the distance between equivalenced atoms. If the variance is small, then the fit must be good. The procedure is systematically repeated after moving a given number of residues along one chain. All possible starting combinations are tried in turn. The variance can be plotted against the starting residue number for one of the chains. If a good likeness of fold has been found, the variance will be small all along the polypeptide. The Remington-Matthews method does not readily allow for insertions or deletions within the trial structural segment (292). Thus the test segment should be small. However, if it is too small, the answers become trivial, since fits will confine themselves to individual secondary structural elements. The best segment length must be found empirically (292) and is usually around 40 residues. The method's utility lies in its straightforwardness, speed, and lack of a complex algorithm. The significance of a comparison can be analyzed in both the Remington-Matthews and the Rossmann-Argos method by the size of the respective minimum or maximum value of the search function relative to the background variation.

Literature Cited

1. Linderstrøm-Lang, K. U., Schellman, J. A. 1959. *Enzymes* 1:443–510. 2nd ed.
2. Bragg, L., Kendrew, J. C., Perutz, M. F. 1950. *Proc. R. Soc. London Ser. A* 203:321–57
3. Huggins, M. L. 1943. *Chem. Rev.* 32:195–218
4. Pauling, L., Corey, R. B., Branson, H. R. 1951. *Proc. Natl. Acad. Sci. USA* 37:205–11
5. Pauling, L., Corey, R. B. 1951. *Proc. Natl. Acad. Sci. USA* 37:241–51
6. Kendrew, J. C., Dickerson, R. E., Strandberg, B. E., Hart, R. G., Davies, D. R., Phillips, D. C., Shore, V. C. 1960. *Nature* 185:422–27
7. Blake, C. C. F., Koenig, D. F., Mair, G. A., North, A. C. T., Phillips, D. C., Sarma, V. R. 1965. *Nature* 206:757–61
8. Anson, M. L., Mirsky, A. E. 1934. *J. Gen. Physiol.* 17:399–408
9. Northrop, J. H. 1932. *J. Gen. Physiol.* 16:323–37
10. Anfinsen, C. B., Haber, E., Sela, M., White, F. H. Jr. 1961. *Proc. Natl. Acad. Sci. USA* 47:1309–14
11. Schechter, A. N., Epstein, C. J. 1968. *J. Mol. Biol.* 35:567–89
12. Epstein, H. F., Schechter, A. N., Chen, R. F., Anfinsen, C. B. 1971. *J. Mol. Biol.* 60:499–508
13. Ristow, S. S., Wetlaufer, D. B. 1973. *Biochem. Biophys. Res. Commun.* 50:544–50
14. Creighton, T. E. 1974. *J. Mol. Biol.* 87:563–77
15. Creighton, T. E. 1974. *J. Mol. Biol.* 87:579–602
16. Creighton, T. E. 1974. *J. Mol. Biol.* 87:603–24
17. Creighton, T. E. 1977. *J. Mol. Biol.* 113:275–93
18. Creighton, T. E. 1977. *J. Mol. Biol.* 113:295–312
19. Creighton, T. E. 1977. *J. Mol. Biol.* 113:313–28
20. Némethy, G., Scheraga, H. A. 1977. *Q. Rev. Biophys.* 10:239–352
21. Schulz, G. E., Schirmer, R. H. 1979. *Principles of Protein Structure.* New York: Springer. 314 pp.
22. Levitt, M., Warshel, A. 1975. *Nature* 253:694–98
23. Levitt, M. 1976. *J. Mol. Biol.* 104:59–107
24. Hagler, A. T., Honig, B. 1978. *Proc. Natl. Acad. Sci. USA* 75:554–58
25. Chou, P. Y., Fasman, G. D. 1974. *Biochemistry* 13:211–22
26. Chou, P. Y., Fasman, G. D. 1974. *Biochemistry* 13:222–45
27. Fasman, G. D., Chou, P. Y., Adler, A. J. 1976. *Biophys. J.* 16:1201–38
28. Schulz, G. E., Barry, C. D., Friedman, J., Chou, P. Y., Fasman, G. D., Finkelstein, A. V., Lim, V. I., Ptitsyn, O. B., Kabat, E. A., Wu, T. T., Levitt, M., Robson, B., Nagano, K. 1974. *Nature* 250:140–42
29. Matthews, B. W. 1975. *Biochim. Biophys. Acta* 405:442–51
30. Wootton, J. C. 1974. *Nature* 252:542–46
31. Richmond, T. J., Richards, F. M. 1978. *J. Mol. Biol.* 119:537–55
32. Cohen, F. E., Richmond, T. J., Richards, F. M. 1979. *J. Mol. Biol.* 132:275–88
33. Cohen, F. E., Sternberg, M. J. E., Taylor, W. R. 1980. *Nature* 285:378–82

34. Henderson, R., Unwin, P. N. T. 1975. *Nature* 257:28–32
35. Koshland, D. E. Jr. 1958. *Proc. Natl. Acad. Sci. USA* 44:98–104
36. Koshland, D. E. Jr. 1973. *Sci. Am.* 229(4):52–64
37. Rossmann, M. G., Adams, M. J., Buehner, M., Ford, G. C., Hackert, M. L., Lentz, P. J. Jr., McPherson, A. Jr., Schevitz, R. W., Smiley, I. E. 1972. *Cold Spring Harbor Symp. Quant. Biol.* 36:179–91
38. Musick, W. D. L., Rossmann, M. G. 1979. *J. Biol. Chem.* 254:7611–20
39. Huber, R. 1979. *Trends Biochem. Sci.* 4:271–76
40. Abad-Zapatero, C., Abdel-Meguid, S. S., Johnson, J. E., Leslie, A. G. W., Rayment, I., Rossmann, M. G., Suck, D., Tsukihara, T. 1980. *Nature* 286: 33–39
41. Argos, P. 1981. *Proc. Aharon Katzir-Katchalsky Conf. Structural Aspects of Recognition and Assembly in Biological Macromolecules, 7th, Nof Ginossar, Israel, 1979.* In press
42. Lesk, A. M., Chothia, C. 1980. *J. Mol. Biol.* 136:225–70
43. Doolittle, R. F. 1979. In *The Proteins,* ed. H. Neurath, R. L. Hill, 4:1–118. New York: Academic. 679 pp. 3rd ed.
44. Richardson, J. S. 1981. *Adv. Protein Chem.* In press
45. Cantor, C. R., Schimmel, P. R. 1980. *Biophysical Chemistry. Part I: The Conformation of Biological Macromolecules.* San Francisco: Freeman. 365 pp.
46. Dickerson, R. E., Geis, I. 1969. *The Structure and Action of Proteins.* New York: Harper & Row. 120 pp.
47. Jensen, L. H. 1974. *Ann. Rev. Biochem.* 43:461–74
48. Liljas, A., Rossmann, M. G. 1974. *Ann. Rev. Biochem.* 43:475–507
49. Davies, D. R., Padlan, E. A., Segal, D. M. 1975. *Ann. Rev. Biochem.* 44:639–67
50. Kretsinger, R. H. 1976. *Ann. Rev. Biochem.* 45:239–66
51. Salemme, F. R. 1977. *Ann. Rev. Biochem.* 46:299–329
52. Amzel, L. M., Poljak, R. J. 1979. *Ann. Rev. Biochem.* 48:961–97
53. Matthews, B. W. 1977. See Ref. 43, 3:403–590
54. Blake, C. C. F. 1972. *Prog. Biophys. Mol. Biol.* 25:85–130
55. Tanford, C. 1968. *Adv. Protein Chem.* 23:121–282
56. Tanford, C. 1970. *Adv. Protein Chem.* 24:1–95
57. Brandts, J. F. 1969. In *Structure and Stability of Biological Macromolecules,* ed. S. N. Timasheff, G. D. Fasman, pp. 213–90. New York: Dekker. 694 pp.
58. Hermans, J. Jr., Lohr, D., Ferro, D. 1972. *Adv. Polymer Sci.* 9:229–83
59. Wetlaufer, D. B., Ristow, S. 1973. *Ann. Rev. Biochem.* 42:135–58
60. Ptitsyn, O. B., Lim, V. I., Finkelstein, A. V. 1972. *FEBS Fed. Eur. Biochem. Soc. Meet.* 25:421–31
61. Baldwin, R. L. 1975. *Ann. Rev. Biochem.* 44:453–75
62. Anfinsen, C. B., Scheraga, H. A. 1975. *Adv. Protein Chem.* 29:205–300
63. Creighton, T. E. 1978. *Prog. Biophys. Mol. Biol.* 33:231–97
64. Jaenicke, R., ed. 1980. *Protein Folding.* Amsterdam: Elsevier
65. Fehlhammer, H., Bode, W. 1975. *J. Mol. Biol.* 98:683–92
66. Chambers, J. L., Stroud, R. M. 1977. *Acta Crystallogr. Sect. B* 33:1824–37
67. Deisenhofer, J., Steigemann, W. 1975. *Acta Crystallogr. Sect. B* 31:238–50
68. Sloan, D. L., Young, J. M., Mildvan, A. S. 1975. *Biochemistry* 14:1998–2008
69. Fletterick, R. J., Madsen, N. B. 1980. *Ann. Rev. Biochem.* 49:31–61
70. Levine, M., Muirhead, H., Stammers, D. K., Stuart, D. I. 1978. *Nature* 271:626–30
71. Richardson, J. S. 1979. *Biochem. Biophys. Res. Commun.* 90:285–90
72. Mavridis, I. M., Tulinsky, A. 1976. *Biochemistry* 15:4410–17
73. Quiocho, F. A., McMurray, C. H., Lipscomb, W. N. 1972. *Proc. Natl. Acad. Sci. USA* 69:2850–54
74. Johansen, J. T., Vallee, B. L. 1973. *Proc. Natl. Acad. Sci. USA* 70:2006–10
75. Pauling, L., Corey, R. B. 1951. *Proc. Natl. Acad. Sci. USA* 37:235–41
76. Corey, R. B., Pauling, L. 1953. *Proc. R. Soc. London Ser. B* 141:10–20
77. Marsh, R. E., Donohue, J. 1967. *Adv. Protein Chem.* 22:235–56
78. Ramachandran, G. N., Kolaskar, A. S., Ramakrishnan, C., Sasisekharan, V. 1974. *Biochim. Biophys. Acta* 359:298–302
79. Ramachandran, G. N., Lakshminarayanan, A. V., Kolaskar, A. S. 1973. *Biochim. Biophys. Acta* 303:8–13
80. Ramachandran, G. N., Kolaskar, A. S. 1973. *Biochim. Biophys. Acta* 303: 385–88
81. Ramachandran, G. N., Mitra, A. K. 1976. *J. Mol. Biol.* 107:85–92
82. Huber, R., Kukla, D., Bode, W., Schwager, P., Bartels, K., Deisenhofer,

J., Steigemann, W. 1974. *J. Mol. Biol.* 89:73–101
83. Ramachandran, G. N., Sasisekharan, V. 1968. *Adv. Protein Chem.* 23:283–438
84. Kendrew, J. C., Watson, H. C., Strandberg, B. E., Dickerson, R. E., Phillips, D. C., Shore, V. C. 1961. *Nature* 190:666–70
85. Janin, J., Wodak, S., Levitt, M., Maigret, B. 1978. *J. Mol. Biol.* 125:357–86
86. Warme, P. K., Morgan, R. S. 1978. *J. Mol. Biol.* 118:289–304
87. Crippen, G. M., Kuntz, I. D. 1978. *Int. J. Peptide Protein Res.* 12:47–56
88. Ramakrishnan, C., Prasad, N. 1971. *Int. J. Protein Res.* 111:209–31
89. Hamilton, W. C., Ibers, J. A. 1968. *Hydrogen Bonding in Solids.* New York: W. A. Benjamin. 284 pp.
90. Némethy, G., Phillips, D. C., Leach, S. J., Scheraga, H. A. 1967. *Nature* 214:363–65
91. Almassy, R. J., Dickerson, R. E. 1978. *Proc. Natl. Acad. Sci. USA* 75:2674–78
92. Chandrasekaran, R., Jardetzky, T. S., Jardetzky, O. 1979. *FEBS Lett.* 101:11–14
93. Crick, F. H. C. 1952. *Nature* 170:882–83
94. Ward, K. B., Hendrickson, W. A., Klippenstein, G. L. 1975. *Nature* 257:818–21
95. Stenkamp, R. E., Sieker, L. C., Jensen, L. H., McQueen, J. E. Jr. 1978. *Biochemistry* 17:2499–504
96. Mathews, F. S., Bethge, P. H., Czerwinski, E. W. 1979. *J. Biol. Chem.* 254:1699–706
97. Crick, F. H. C. 1953. *Acta Crystallogr.* 6:689–97
98. Chothia, C., Levitt, M., Richardson, D. 1977. *Proc. Natl. Acad. Sci. USA* 74:4130–34
99. Cohen, F. E., Sternberg, M. J. E. 1980. *J. Mol. Biol.* 137:9–22
100. Lim, V. I., Efimov, A. V. 1976. *FEBS Lett.* 69:41–44
101. Ptitsyn, O. B., Rashin, A. A. 1975. *Biophys. Chem.* 3:1–20
102. Argos, P., Rossmann, M. G., Johnson, J. E. 1977. *Biochem. Biophys. Res. Commun.* 75:83–86
103. Blow, D. M., Irwin, M. J., Nyborg, J. 1977. *Biochem. Biophys. Res. Commun.* 76:728–34
104. Efimov, A. V. 1979. *J. Mol. Biol.* 134:23–40
105. Pauling, L., Corey, R. B. 1951. *Proc. Natl. Acad. Sci. USA* 37:729–40
106. Lifson, S., Sander, C. 1980. *J. Mol. Biol.* 139:627–39

107. Lifson, S., Sander, C. 1979. *Nature* 282:109–11
108. Chothia, C. 1973. *J. Mol. Biol.* 75:295–302
109. Weatherford, D. W., Salemme, F. R. 1979. *Proc. Natl. Acad. Sci. USA* 76:19–23
110. Richardson, J. S., Getzoff, E. D., Richardson, D. C. 1978. *Proc. Natl. Acad. Sci. USA* 75:2574–78
111. Richardson, J. S. 1977. *Nature* 268:495–500
112. Sternberg, M. J. E., Thornton, J. M. 1977. *J. Mol. Biol.* 110:285–96
113. Sternberg, M. J. E., Thornton, J. M. 1977. *J. Mol. Biol.* 115:1–17
114. Richardson, J. S. 1976. *Proc. Natl. Acad. Sci. USA* 73:2619–23
115. Sternberg, M. J. E., Thornton, J. M. 1977. *J. Mol. Biol.* 113:401–18
116. Sternberg, M. J. E., Thornton, J. M. 1978. *Nature* 271:15–20
117. Ptitsyn, O. B., Finkelstein, A. V., Falk (Bendzko), P. 1979. *FEBS Lett.* 101:1–5
118. Finkelstein, A. V., Ptitsyn, O. B., Bendzko, P. 1979. *Biofizika* 24:21–26
119. Ptitsyn, O. B., Finkelstein, A. V. 1979. *Biofizika* 24:27–31
120. Kuntz, I. D. 1972. *J. Am. Chem. Soc.* 94:4009–12
120a. Ashida, T., Tanaka, I., Yamane, T., Kakudo, M. 1980. In *Biomolecular Structure, Conformation, Function, and Evolution,* ed. R. Srinivasan, 1:607–20. Oxford/New York: Pergamon
121. Crawford, J. L., Lipscomb, W. N., Schellman, C. G. 1973. *Proc. Natl. Acad. Sci. USA* 70:538–42
122. Venkatachalam, C. M. 1968. *Biopolymers* 6:1425–36
123. Lewis, P. N., Momany, F. A., Scheraga, H. A. 1973. *Biochim. Biophys. Acta* 303:211–29
124. Chou, P. Y., Fasman, G. D. 1977. *J. Mol. Biol.* 115:135–75
125. Rose, G. D. 1978. *Nature* 272:586–90
126. Levitt, M., Greer, J. 1977. *J. Mol. Biol.* 114:181–239
127. Chandrasekaran, R., Prasad, B. V. V. 1978. In *CRC Crit. Rev. Biochem.* 5:125–61
128. Rao, S. T., Rossmann, M. G. 1973. *J. Mol. Biol.* 76:241–56
129. Sternberg, M. J. E., Thornton, J. M. 1976. *J. Mol. Biol.* 105:367–82
130. Phillips, D. C., Sternberg, M. J. E., Thornton, J. M., Wilson, I. A. 1978. *J. Mol. Biol.* 119:329–51
131. Stuart, D. I., Levine, M., Muirhead, H., Stammers, D. K. 1979. *J. Mol. Biol.* 134:109–42

132. Levitt, M., Chothia, C. 1976. *Nature* 261:552–58
133. Sternberg, M. J. E., Thornton, J. M. 1977. *J. Mol. Biol.* 110:269–83
134. Janin, J. 1979. *Bull. Inst. Pasteur Paris* 77:337–73
135. Hol, W. G. J., van Duijnen, P. T., Berendsen, H. J. C. 1978. *Nature* 273: 443–46
136. Schulz, G. E. 1980. *J. Mol. Biol.* 138:335–47
137. Shoham, M., Yonath, A., Sussman, J. L., Moult, J., Traub, W., Kalb (Gilboa), A. J. 1979. *J. Mol. Biol.* 131:137–55
138. Argos, P., Rossmann, M. G. 1979. *Biochemistry* 18:4951–60
139. Nishioka, Y., Leder, P. 1979. *Cell* 18:875–82
140. Blake, C. C. F. 1978. *Nature* 273:267
141. Perutz, M. F. 1978. *Sci. Am.* 239(6): 92–125
142. Perutz, M. F. 1979. *Ann. Rev. Biochem.* 48:327–86
143. Padlan, E. A., Love, W. E. 1974. *J. Biol. Chem.* 249:4067–78
144. Hendrickson, W. A., Love, W. E., Karle, J. 1973. *J. Mol. Biol.* 74:331–61
145. Vainshtein, B. K., Harutyunyan, E. H., Kuranova, I. P., Borisov, V. V., Sosfenov, N. I., Pavlovsky, A. G., Grebenko, A. I., Konareva, N. V. 1975. *Nature* 254:163–64
146. Watson, H. C. 1969. In *Prog. Stereochem.* 4:299–333
147. Mathews, F. S., Levine, M., Argos, P. 1972. *J. Mol. Biol.* 64:449–64
148. Guiard, B., Lederer, F. 1979. *J. Mol. Biol.* 135:639–50
149. Swanson, R., Trus, B. L., Mandel, N., Mandel, G., Kallai, O. B., Dickerson, R. E. 1977. *J. Biol. Chem.* 252:759–75
150. Ashida, T., Tanaka, N., Yamane, T., Tsukihara, T., Kakudo, M. 1973. *J. Biochem. Tokyo* 73:463–65
151. Salemme, F. R., Freer, S. T., Xuong, N. H., Alden, R. A., Kraut, J. 1973. *J. Biol. Chem.* 248:3910–21
152. Timkovich, R., Dickerson, R. E. 1976. *J. Biol. Chem.* 251:4033–46
153. Korszun, Z. R., Salemme, F. R. 1977. *Proc. Natl. Acad. Sci. USA* 74:5244–47
154. Klotz, I. M., Klippenstein, G. L., Hendrickson, W. A. 1976. *Science* 192:335–44
155. Weber, P. C., Bartsch, R. G., Cusanovich, M. A., Hamlin, R. C., Howard, A., Jordan, S. R., Kamen, M. D., Meyer, T. E., Weatherford, D. W., Xuong, N. H., Salemme, F. R. 1980. *Nature* 286:302–4
156. Bloomer, A. C., Champness, J. N., Bricogne, G., Staden, R., Klug, A. 1978. *Nature* 276:362–68
157. Banyard, S. H., Stammers, D. K., Harrison, P. M. 1978. *Nature* 271:282–84
158. Irwin, M. J., Nyborg, J., Reid, B. R., Blow, D. M. 1976. *J. Mol. Biol.* 105:577–86
159. Deisenhofer, J., Jones, T. A., Huber, R., Sjödahl, J., Sjöquist, J. 1978. *Hoppe-Seylers Z. Physiol. Chem.* 359:975–85
160. Drenth, J., Jansonius, J. N., Koekoek, R., Sluyterman, L. A. A., Wolthers, B. G. 1970. *Phil. Trans. R. Soc. London Ser. B* 257:231–36
161. Kretsinger, R. H. 1972. *Nature New Biol.* 240:85–88
162. Kretsinger, R. H. 1981. *CRC Crit. Rev. Biochem.* In press
163. Mornon, J. P., Fridlansky, F., Bally, R., Milgrom, E. 1980. *J. Mol. Biol.* 137:415–29
164. Engelman, D. M., Henderson, R., McLachlan, A. D., Wallace, B. A. 1980. *Proc. Natl. Acad. Sci. USA* 77:2023–27
165. Wiegand, G., Kukla, D., Scholze, H., Jones, T. A., Huber, R. 1979. *Eur. J. Biochem.* 93:41–50
166. Colman, P. M., Jansonius, J. N., Matthews, B. W. 1972. *J. Mol. Biol.* 70:701–24
167. Sasaki, K., Dockerill, S., Adamiak, D. A., Tickle, I. J., Blundell, T. L. 1975. *Nature* 257:751–57
168. Adams, M. J., Helliwell, J. R., Bugg, C. E. 1977. *J. Mol. Biol.* 112:183–97
169. Wiley, D. C., Skehel, J. J. 1977. *J. Mol. Biol.* 112:343–47
170. Kannan, K. K., Notstrand, B., Fridborg, K., Lövgren, S., Ohlsson, A., Petef, M. 1975. *Proc. Natl. Acad. Sci. USA* 72:51–55
171. Mitsui, Y., Satow, Y., Watanabe, Y., Iitaka, Y. 1979. *J. Mol. Biol.* 131:697–724
172. Wierenga, R. K., de Jong, R. J., Kalk, K. H., Hol, W. G. J., Drenth, J. 1979. *J. Mol. Biol.* 131:55–73
173. Drenth, J., Jansonius, J. N., Koekoek, R., Wolthers, B. G. 1971. *Enzymes* 3:485–99 3rd ed.
174. Matthews, B. W., Fenna, R. E., Bolognesi, M. C., Schmid, M. F., Olson, J. M. 1979. *J. Mol. Biol.* 131:259–85
175. Blow, D. M. 1971. *Enzymes* 3:185–212. 3rd ed.
176. Kraut, J. 1971. *Enzymes* 3:165–83. 3rd ed.
177. Bode, W., Schwager, P. 1975. *J. Mol. Biol.* 98:693–717
178. Bode, W., Huber, R. 1978. *FEBS Lett.* 90:265–69

179. Kossiakoff, A. A., Chambers, J. L., Kay, L. M., Stroud, R. M. 1977. *Biochemistry* 16:654–64
180. Hartley, B. S., Shotton, D. M. 1971. *Enzymes* 3:323–73. 3rd ed.
181. Brayer, G. D., Delbaere, L. T. J., James, M. N. G. 1979. *J. Mol. Biol.* 131:743–75
182. James, M. N. G., Delbaere, L. T. J., Brayer, G. D. 1978. *Can. J. Biochem.* 56:396–402
183. Tang, J., James, M. N. G., Hsu, I. N., Jenkins, J. A., Blundell, T. L. 1978. *Nature* 271:618–21
184. James, M. N. G., Hsu, I. N., Delbaere, L. T. J. 1977. *Nature* 267:808–13
185. Subramanian, E., Swan, I. D. A., Liu, M., Davies, D. R., Jenkins, J. A., Tickle, I. J., Blundell, T. L. 1977. *Proc. Natl. Acad. Sci. USA* 74:556–59
186. Wong, C. H., Lee, T. J., Lee, T. Y., Lu, T. H., Yang, I. H. 1979. *Biochemistry* 18:1638–40
187. Andreeva, N. S., Gustchina, A. E. 1979. *Biochem. Biophys. Res. Commun.* 87:32–42
188. Richardson, J. S., Richardson, D. C., Thomas, K. A., Silverton, E. W., Davies, D. R. 1976. *J. Mol. Biol.* 102:221–35
189. Colman, P. M., Freeman, H. C., Guss, J. M., Murata, M., Norris, V. A., Ramshaw, J. A. M., Venkatappa, M. P. 1978. *Nature* 272:319–24
190. Adman, E. T., Stenkamp, R. E., Sieker, L. C., Jensen, L. H. 1978. *J. Mol. Biol.* 123:35–47
191. Blow, D. M., Janin, J., Sweet, R. M. 1974. *Nature* 249:54–57
192. McLachlan, A. D. 1979. *J. Mol. Biol.* 133:557–63
193. Blake, C. C. F., Geisow, M. J., Oatley, S. J., Rérat, B., Rérat, C. 1978. *J. Mol. Biol.* 121:339–56
194. Arnone, A., Bier, C. J., Cotton, F. A., Day, V. W., Hazen, E. E. Jr., Richardson, D. C., Richardson, J. S., Yonath, A. 1971. *J. Biol. Chem.* 246:2302–16
195. McPherson, A., Jurnak, F. A., Wang, A. H. J., Molineux, I., Rich, A. 1979. *J. Mol. Biol.* 134:379–400
196. Argos, P., Tsukihara, T., Rossmann, M. G. 1980. *J. Mol. Evol.* 15:169–79
197. Harrison, S. C., Olson, A. J., Schutt, C. E., Winkler, F. K., Bricogne, G. 1978. *Nature* 276:368–73
198. Reeke, G. N. Jr., Becker, J. W., Edelman, G. M. 1975. *J. Biol. Chem.* 250:1525–47
199. Hardman, K. D., Ainsworth, C. F. 1972. *Biochemistry* 11:4910–19
200. Holbrook, J. J., Liljas, A., Steindel, S.

J., Rossmann, M. G. 1975. *Enzymes* 11:191–292. 3rd ed.
201. Banaszak, L. J., Bradshaw, R. A. 1975. *Enzymes* 11:269–96 3rd ed.
202. Brändén, C. I., Jörnvall, H., Eklund, H., Furugren, B. 1975. *Enzymes* 11:103–90 3rd ed.
203. Moras, D., Olsen, K. W., Sabesan, M. N., Buehner, M., Ford, G. C., Rossmann, M. G. 1975. *J. Biol. Chem.* 250:9137–62
204. Pickover, C. A., McKay, D. B., Engelman, D. M., Steitz, T. A. 1979. *J. Biol. Chem.* 254:11323–29
205. Pai, E. F., Sachsenheimer, W., Schirmer, R. H., Schulz, G. E. 1977. *J. Mol. Biol.* 114:37–45
206. Banks, R. D., Blake, C. C. F., Evans, P. R., Haser, R., Rice, D. W., Hardy, G. W., Merrett, M., Phillips, A. W. 1979. *Nature* 279:773–77
207. Bryant, T. N., Watson, H. C., Wendell, P. L. 1974. *Nature* 247:14–17
208. Steitz, T. A., Anderson, W. F., Fletterick, R. J., Anderson, C. M. 1977. *J. Biol. Chem.* 252:4494–500
209. Evans, P. R., Hudson, P. J. 1979. *Nature* 279:500–4
210. Matthews, D. A., Alden, R. A., Freer, S. T., Xuong, N., Kraut, J. 1979. *J. Biol. Chem.* 254:4144–51
211. Abdallah, M. A., Adams, M. J., Archibald, I. G., Biellmann, J. F., Helliwell, J. R., Jenkins, S. E. 1979. *Eur. J. Biochem.* 98:121–30
212. Eichele, G., Ford, G. C., Glor, M., Jansonius, J. N., Mavrides, C., Christen, P. 1979. *J. Mol. Biol.* 133:161–80
213. Borisov, V. V., Borisova, S. N., Sosfenov, N. I., Vainshtein, B. K. 1980. *Nature* 284:189–90
214. Sprang, S., Fletterick, R. J. 1979. *J. Mol. Biol.* 131:523–51
215. Smith, W. W., Burnett, R. M., Darling, G. D., Ludwig, M. L. 1977. *J. Mol. Biol.* 117:195–225
216. Monaco, H. L., Crawford, J. L., Lipscomb, W. N. 1978. *Proc. Natl. Acad. Sci. USA* 75:5276–80
217. Morikawa, K., la Cour, T. F. M., Nyborg, J., Rasmussen, K. M., Miller, D. L., Clark, B. F. C. 1978. *J. Mol. Biol.* 125:325–38
218. Campbell, J. W., Watson, H. C., Hodgson, G. I. 1974. *Nature* 250:301–3
219. Shaw, P. J., Muirhead, H. 1977. *J. Mol. Biol.* 109:475–85
220. Hartsuck, J. A., Lipscomb, W. N. 1971. *Enzymes* 3:1–56. 3rd ed.
221. Drenth, J., Hol, W. G. J., Jansonius, J. N., Koekoek, R. 1972. *Eur. J. Biochem.* 26:177–81

222. Wright, C. S., Alden, R. A., Kraut, J. 1969. *Nature* 221:235–42
223. Söderberg, B. O., Sjöberg, B. M., Sonnerstam, U., Brändén, C. I. 1978. *Proc. Natl. Acad. Sci. USA* 75:5827–30
224. Ladenstein, R., Epp, O, Bartels, K., Jones, A., Huber, R., Wendel, A. 1979. *J. Mol. Biol.* 134:199–218
225. Ploegman, J. H., Drent, G., Kalk, K. H., Hol, W. G. J. 1978. *J. Mol. Biol.* 123:557–94
226. Newcomer, M. E., Miller, D. M. III, Quiocho, F. A. 1979. *J. Biol. Chem.* 254:7529–33
227. Lindqvist, Y., Brändén, C. I. 1979. *J. Biol. Chem.* 254:7403–4
227a. Matsuura, Y., Kusunoki, M., Harada, W., Tanaka, N., Iga, Y., Yasuoka, N., Toda, H., Narita, K., Kakudo, M. 1980. *J. Biochem. Tokyo* 87:1555–58
228. Drenth, J., Low, B. W., Richardson, J. S., Wright, C. S. 1980. *J. Biol. Chem* 255:2652–55
229. Tsernoglou, D., Petsko, G. A. 1977. *Proc. Natl. Acad. Sci. USA* 74:971–74
230. Kimball, M. R., Sato, A., Richardson, J. S., Rosen, L. S., Low, B. W. 1979. *Biochem. Biophys. Res. Commun.* 88:950–59
231. Wright, C. S. 1977. *J. Mol. Biol.* 111:439–57
232. Rossmann, M. G., Argos, P. 1976. *J. Mol. Biol.* 105:75–95
233. Remington, S. J., Matthews, B. W. 1978. *Proc. Natl. Acad. Sci. USA* 75:2180–84
234. Ford, L. O., Johnson, L. N., Machin, P. A., Phillips, D. C., Tjian, R. 1974. *J. Mol. Biol.* 88:349–71
235. Remington, S. J., Anderson, W. F., Owen, J., Ten Eyck, L. F., Grainger, C. T., Matthews, B. W. 1978. *J. Mol. Biol.* 118:81–98
236. Brew, K., Vanaman, T. C., Hill, R. L. 1967. *J. Biol. Chem.* 242:3747–49
237. Wyckoff, H. W., Tsernoglou, D., Hanson, A. W., Knox, J. R., Lee, B., Richards, F. M. 1970. *J. Biol. Chem.* 245:305–28
238. Cutfield, J. F., Cutfield, S. M., Dodson, E. J., Dodson, G. G., Emdin, S. F., Reynolds, C. D. 1979. *J. Mol. Biol.* 132:85–100
239. Dijkstra, B. W., Drenth, J., Kalk, K. H., Vandermaelen, P. J. 1978. *J. Mol. Biol.* 124:53–60
240. Watenpaugh, K. D., Sieker, L. C., Jensen, L. H. 1979. *J. Mol. Biol.* 131:509–22
241. Adman, E. T., Sieker, L. C., Jensen, L. H. 1976. *J. Biol. Chem.* 251:3801–6
242. Freer, S. T., Alden, R. A., Carter, C. W. Jr., Kraut, J. 1975. *J. Biol. Chem.* 250:46–54
243. Haser, R., Pierrot, M., Frey, M., Payan, F., Astier, J. P., Bruschi, M., Le Gall, J. 1979. *Nature* 282:806–10
244. Kretsinger, R. H. 1975. In *Calcium Transport in Secretion and Contraction,* ed. E. Carafoli, F. Clementi, W. Drabikowski, A. Margreth, pp. 469–78. Amsterdam: North Holland
245. Hill, R. L., Delaney, R., Fellows, R. E. Jr., Lebovitz, H. E. 1966. *Proc. Natl. Acad. Sci. USA* 56:1762–69
246. Rossmann, M. G., Moras, D., Olsen, K. W. 1974. *Nature* 250:194–99
247. Ohlsson, I., Nordström, B., Brändén, C. I. 1974. *J. Mol. Biol.* 89:339–54
248. Rossmann, M. G., Liljas, A. 1974. *J. Mol. Biol.* 85:177–81
249. Pompon, D., Lederer, F. 1976. *Eur. J. Biochem.* 68:415–23
250. Ghelis, C., Tempete-Gaillourdet, M., Yon, J. M. 1978. *Biochem. Biophys. Res. Commun.* 84:31–36
251. Goldberg, M. E., Zetina, C. R. 1980. See Ref. 64, pp. 469–84
252. Rose, G. D. 1979. *J. Mol. Biol.* 134:447–70
253. Crippen, G. M. 1978. *J. Mol. Biol.* 126:315–32
254. Rashin, A. A., Janin, J., Wodak, S. J. 1981. *J. Mol. Biol.* In press
255. Wodak, S. J., Janin, J. 1980. *Proc. Natl. Acad. Sci. USA* 77:1736–40
256. Wetlaufer, D. B. 1973. *Proc. Natl. Acad. Sci. USA* 70:697–701
257. Wetlaufer, D. B., Rose, G. D., Taaffe, L. 1976. Biochemistry 15:5154–57
258. Blake, C. C. F. 1979. *Nature* 277:598
259. Gilbert, W. 1981. In *Introns and Exons: Playground of Evolution, ICN-UCLA Symp., Los Angeles, 1979.* In press
260. Eaton, W. A. 1980. *Nature* 284:183–85
261. Craik, C. S., Buchman, S. R., Beychok, S. 1980. *Proc. Natl. Acad. Sci. USA* 77:1384–88
262. Hill, E., Tsernoglou, D., Webb, L., Banaszak, L. J. 1972. *J. Mol. Biol.* 72:577–91
263. Kraut, J., Robertus, J. D., Birktoft, J. J., Alden, R. A., Wilcox, P. E., Powers, J. C. 1972. *Cold Spring Harbor Symp. Quant. Biol.* 36:117–23
264. McLachlan, A. D. 1979. *J. Mol. Biol.* 128:49–79
265. Sokal, R. R., Sneath, P. H. A. 1963. *Principles of Numerical Taxonomy.* San Francisco: Freeman. 359 pp.
266. Dickerson, R. E. 1980. *Sci. Am.* 242(3):136–53

267. Kurosky, A., Barnett, D. R., Lee, T. H., Touchstone, B., Hay, R. E., Arnott, M. S., Bowman, B. H., Fitch, W. M. 1980. *Proc. Natl. Acad. Sci. USA* 77:3388–92

268. Greer, J. 1980. *Proc. Natl. Acad. Sci. USA* 77:3393–97

269. Schulz, G. E., Schirmer, R. H. 1974. *Nature* 250:142–44

270. Eventoff, W., Rossmann, M. G. 1975. *CRC Crit. Rev. Biochem.* 3:111–40

271. Fitch, W. M., Margoliash, E. 1967. *Science* 155:279–84

272. Farris, J. S., Kluge, A. G., Eckardt, M. J. 1970. *Syst. Zool.* 19:172–89

273. McLachlan, A. D. 1976. In *Taniguchi Symp. Biophys., Mishima, Japan, Oct., pp. 208–39*

274. Buehner, M., Ford, G. C., Moras, D., Olsen, K. W., Rossmann, M. G. 1973. *Proc. Natl. Acad. Sci. USA* 70:3052–54

275. Rossmann, M. G., Argos, P. 1977. *J. Mol. Biol.* 109:99–129

276. Rossmann, M. G., Argos, P. 1978. *Mol. Cell. Biochem.* 21:161–82

277. Argos, P., Garavito, R. M., Eventoff, W., Rossmann, M. G., Brändén, C. I. 1978. *J. Mol. Biol.* 126:141–58

278. Kester, W. R., Matthews, B. W. 1977. *J. Biol. Chem.* 252:7704–10

279. McLachlan, A. D. 1972. *Nature New Biol.* 240:83–85

280. Blundell, T. L., Sewell, B. T., McLachlan, A. D. 1979. *Biochim. Biophys. Acta* 580:24–31

281. Hendrickson, W. A., Ward, K. B. 1977. *J. Biol. Chem.* 252:3012–18

282. McLachlan, A. D., Bloomer, A. C., Butler, P. J. G. 1980. *J. Mol. Biol.* 136:203–24

283. Monod, J., Wyman, J., Changeux, J. P. 1965. *J. Mol. Biol.* 12:88–118

284. Rossmann, M. G., Liljas, A., Brändén, C. I., Banaszak, L. J. 1975. *The Enzymes* 11:61–102. 3rd ed.

285. Rossmann, M. G., Argos, P. 1975. *J. Biol. Chem.* 250:7525–32

286. Diamond, R. 1976. *Acta Crystallogr. Sect. A* 32:1–10

287. Kabsch, W. 1976. *Acta Crystallogr. Sect. A* 32:922–23

288. Hendrickson, W. A. 1979. *Acta Crystallogr. Sect. A* 35:158–63

289. Nishikawa, K., Ooi, T. 1974. *J. Theor. Biol.* 43:351–74

290. Némethy, G., Scheraga, H. A. 1979. *Proc. Natl. Acad. Sci. USA* 76:6050–54

291. Jukes, T. H., Cantor, C. R. 1969. In *Mammalian Protein Metabolism,* ed. H. N. Munro, 3:21–132. New York: Academic. 571 pp.

292. Remington, S. J., Matthews, B. W. 1980. *J. Mol. Biol.* 140:77–99

Ann. Rev. Biochem. 1984. 53:537–72

2

PRINCIPLES THAT DETERMINE THE STRUCTURE OF PROTEINS

Cyrus Chothia

Christopher Ingold Laboratory, University College London,
20 Gordon Street, London WC1H 0AJ, England; and
Medical Research Council, Laboratory of Molecular Biology,
Hills Road, Cambridge CB2 2QH, England

CONTENTS

PERSPECTIVES AND SUMMARY

The principles governing the structure of globular proteins were not revealed, as it was hoped they would be, by the first atomic descriptions of protein molecules. The early protein structures did confirm certain central ideas already present in the 1950s (1), but significant progress beyond that has occurred only in recent years. This review outlines these recent advances. Certain aspects of this work have been reviewed previously (2–6).

Here I deal with the static structure of soluble proteins built from α helices and β-pleated sheets. The intrinsic properties of polypeptides mean that many of the principles reviewed also apply to structural and membrane proteins, and to the very small number of proteins that do not contain α helices or β sheets.

The principles described in this review are concerned with six different levels or aspects of protein structure:

1. The local folding of the polypeptide chain—it was established in the 1950s and 1960s that the principle underlying the structure of helices, sheets, and turns is the simultaneous formation of hydrogen bonds by buried peptide groups and the retention of conformations close to those of minimum energy. Recent work has given precise information on the preferred conformations of secondary structures and side chains. It has also shown how variations within allowed conformational regions permit β sheets to twist, coil, bend, and bulge, and α helices to kink.

2. The association of secondary structures—α helices and β sheets in proteins close pack. The shape of their surfaces is determined, to a first approximation, by the conformation of the main chain. As a consequence, α helices and β sheets usually pack in one of a small number of relative orientations.

3. The links between secondary structures tend to be right-handed and short, and do not form knots. These features probably arise from the kinetics of the folding process.

4. A consequence of the principles outlined in 1–3 is that protein chains usually fold to give secondary structures arranged in one of a few common patterns. Thus, there are families or classes of proteins that have a similar tertiary structure but no evolutionary or functional relationship.

5. The stability of protein structures arises from the reduction in the surface accessible to solvent that occurs on folding and the formation of intramolecular hydrogen bonds. The chemical nature of the surfaces buried by residues and by secondary structures shows clear regularities. The total surface buried within proteins is a function of their molecular

weight. The principles outlined so far arise from the intrinsic chemical and physical properties of polypeptides, the thermodynamics of their structure, and the kinetics of the folding process. Quite different principles are involved in the following.

6. Functional determinants of protein structure—of the protein structures that are possible on thermodynamic and kinetic grounds, only some contain crevices appropriate for the formation of active sites.

CONFORMATION OF HELICES, PLEATED SHEETS, AND SIDE CHAINS

The refinement of protein structures using X-ray data of very high resolution has resulted in accurate descriptions of the conformations of helices, β sheets, and side chains. These confirm the dominating influence on conformation of the steric limits on torsion angles and of the need for buried polar groups to be hydrogen bonded (7–9). There are only rare instances of torsion angles outside the allowed regions or of buried polar groups without hydrogen bonds (10–15).

Hydrogen Bonds

Surveys of hydrogen bonds in parallel and antiparallel β sheets and in α helices show little or no significant difference in the principal features of their geometry. For all three secondary structures the mean bond lengths and angles are close to N...O, 2.9 Å; NH...O, 2.0 Å; NH...Ô=C, 155°; and N-Ĥ...O, 160°. Hydrogen bonds in 3_{10} helices and reverse turns are ~0.2 Å longer and a little more bent (12, 16, 17). The standard deviation of the NH...Ô=C angle is lower in α helices, ~6°, than it is in β sheets, ~12°. This is because the peptides in α helices tilt so that the carbonyl points ~15° away from the helix axis (18, 19), while in β sheets the peptides tilt both above and below the plane of the β sheets (J. Janin, unpublished information).

The constraints that hydrogen bonds place on protein structure principally arise from the need to keep bond lengths close to the optimum: Their standard deviations are ~0.15 Å. The constraints due to angular geometry are much less stringent.

α Helices

In α helices the mean values of main-chain torsion angles are very close to $\langle\phi\rangle = -65°$, $\langle\psi\rangle = -41°$ and $\langle\omega\rangle = 178°$ (12, 13, 16, 17, 19–22). The standard deviations of $\langle\phi\rangle$ and $\langle\psi\rangle$ are usually ~6°.

In two structures, determined at 1.2 Å and 0.98 Å respectively, it was noted that hydrogen bonds on the buried side of an α helix were about 0.2 Å shorter than those on the side accessible to solvent (22, 23). The shortness of the interior hydrogen bonds is related to their peptide groups being less tilted away from the helix axis than those on the surface, whose carbonyl oxygens form additional hydrogen bonds to the solvent (18, 22, 24).

The residues at the ends of α helices are usually irregular in conformation. This commonly involves one or two residues at the N termini and two to four at the C termini. At the C termini a turn of 3_{10} or α_{II} helix is often found and occasionally a turn of π helix (12, 16, 17, 26).

The bending of α helices has been analyzed. Of the helices with 15 residues or more, two fifths had bends of 10°–30°. These bends are produced by small cumulative changes in a few $\phi\psi$ values in one region of the helix: Helices kink rather than bend smoothly. Bends of more than 20° involve the loss of one or two hydrogen bonds (A. M. Lesk, C. Chothia, unpublished).

Left-Handed α Helices, Polyproline Helices, and Turns

A single turn of a left-handed α helix is formed by residues 226–229 in thermolysin (27). This turn has the sequence –Asp–Asn–Gly–Gly– and its mean $\phi\psi$ values are 64°(15), 42°(9). This conformation is also found for single residues at the end of some α helices (28) and for a significant fraction of Asn residues (28a).

Helices with a conformation close to that of polyproline are observed in four proteins (13, 20, 22, 28b). The regions involved include residues 7–9 in pancreatic trypsin inhibitor, Gln–Pro–Pro; 59–63 in cytochrome c_{551}, –Ile–Pro–Met–Pro–Pro–; and 2–8 in avian pancreatic polypeptide, –Pro–Ser–Gln–Pro–Thr–Tyr–Pro– (13, 20, 22). Their mean $\phi\psi$ values are $-69°(6)$, $150°(6)$; $-78°(18)$, $140°(15)$; and $-72°(8)$, $140°(14)$ respectively; these $\phi\psi$ values are close to those of the polyproline II (9) helix, $-78°$, $149°$.

The steric limits on main-chain torsion angles restricts the number of conformations possible for the turns in the protein chain that are formed by three or four residues (29–33). As noted in two detailed reviews (34, 35) the observed conformations are close to those expected. Turns nearly always occur on the protein surface as they contain several polar groups that do not form intramolecular hydrogen bonds (36, 37). The rare turns that are buried within proteins have trapped water molecules hydrogen bonded to these polar atoms (38).

β Sheets: Twisting and Coiling

The $\phi\psi$ values observed for β sheets are not localized as they are for helices but spread over most of the region normally allowed for extended polypeptide chains. This spread of $\phi\psi$ values gives the β sheets a twist that is right handed when they are viewed along their strands (39).

The twisted conformation of β sheets arises from a combination of factors

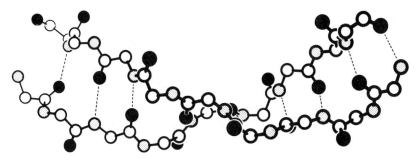

Figure 1   A two-stranded strongly twisted β sheet in which the strands are coiled in right-handed direction. Only main-chain atoms are shown. The structure illustrated here is formed by residues 4–12 and 16–24 in thermolysis (27). [Reprinted with permission from (51).]

acting at different levels. Model energy calculations, of differing degrees of sophistication, have shown that a conformation with a right-hand twist is favored for individual peptides (9, 39), by intrastrand forces (for polyalanine and polyvaline but not for polyisoleucine) and by interstrand forces (41–43). The twist is also favored at the tertiary structure level. β Sheets with a right-hand twist have surfaces complementary to those of α helices (see below). For β sheets packed together, a mean $\phi\psi$ value near the center of the normally allowed region permits the greatest degree of local accommodation. These factors act to differing extents in different protein structures and the degree of the twist in β sheets varies. The dihedral angle between adjacent strands varies between $-50°$ and $+10°$ and has a mean of $\sim -20°$ (44–46).

In strongly twisted β sheets the strands must be coiled as well as twisted otherwise they would splay apart (47). Such strands give coiled-twisted β sheets (Figure 1). The properties of model β sheets built from straight and coiled strands have been investigated (48–50). These models suggest the extent of twist is limited by hydrogen-bond geometries and that antiparallel β sheets can have greater conformational diversity than parallel. The fitting of models to observed structures showed that the strands of strongly twisted β sheets are indeed coiled. For strands to coil in the right-handed direction appropriate for the formation of strongly twisted β sheets, their $\phi\psi$ values must have the following relations (51):

$$\psi_i \simeq -\phi_{i+1}, \psi_{i+1} > -\phi_{i+2}, \psi_{i+2} \simeq -\phi_{i+3}, \psi_{i+3} > -\phi_{i+4}, \ldots$$

β Sheets: β Bends and β Bulges

β Sheets not only twist and coil; they can also fold upon themselves so the strand direction changes by $\sim 90°$. This occurs, for example, in orthogonal

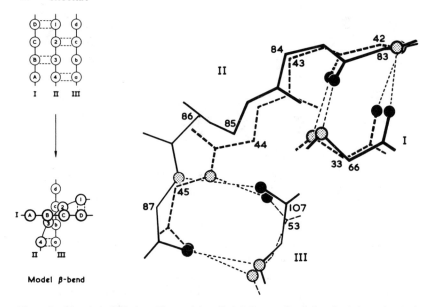

Model β-bend

Figure 2 β bends and β bulges. The model on the left shows a flat β sheet (*top*) that twists and folds on itself so strands I and II are at right angles to each other and strand II makes a right-handed bend of ∼90°. Hydrogen bonds are maintained. The bend in the central strand can be made by a residue in a polyproline conformation producing a local coiling of the strand [a β-bend (46)] or by the insertion of a residue in an α-conformation [a β-bulge (6, 51)]. On the right a β bend and a β bulge from elastase (11) are shown superposed. Main-chain atoms equivalent to model strand II and model residues ID and IIIa are shown. The superposition demonstrates the similar effect of β bends and β bulges on strands of β sheet. [Reprinted with permission from (46).]

β-sheet packings (see below and Figure 10). This local bending is usually produced in one of two ways: by β bends (46) or β bulges (6, 52). In β bends a residue with a conformation near that of polyproline produces a local coiling of the strand. In β bulges the insertion of a residue in an α conformation makes the change in strand direction. The net effect of both of these is to produce in the β sheets a right-handed bend of ∼90° (Figure 2).

Side Chains

Energy calculations have been used to calculate the preferred conformations of side chains and the results have been compared with the conformations observed in protein structures (53–62). There is very good agreement between the expected and observed conformations (17, 59, 60).

To a large extent, difference in side-chain conformation involves rotations around single bonds. Steric hindrance between substituents on the first and fourth atoms and in the case of Ser and Thr hydrogen bonding to main-chain polar groups mean that there are a small number of preferred

Table 1   Principal side-chain configurations observed in proteins[a]

Side Chain	Configurations[b] (%)				
Met, Glu, Gln, Lys, Arg	g^+t33,	tt28,	g^+g^+14,	tg^-8,	g^-t7
Leu	g^+t38,	tg^-19			
Ile	g^+t46,	g^+g^+16,	g^-t14,	tt10	
Val	g^+66,	g^-21,	t13		
Thr	e^+48,	g^-39,	t13		
Cys	g^+57,	t27,	g^-16		
Ser	g^-38,	g^+34,	t28		
Asn, Asp	g^+51,	t34,	g^-15		
Trp, Tyr, Phe	g^+57,	t31,	g^-13		
His	g^+45,	t45,	g^-10		

[a] Data taken from (59).
[b] The g^+, g^-, and t conformations are defined in Figure 3.

conformations for each side chain. These are listed in Table 1 (59). Inspection of five very accurate protein structures shows that the range of torsion angles within these different conformations is small (17).

The roles of the peptide backbone and the protein environment in side-chain conformation have been determined (59, 60). For a dipeptide the different preferred conformations (Table 1) have only small differences in energy. There is a weak relationship between secondary structure and side-chain conformation. The χ_1 values are on average 54% g^+, 35% t, and 11% g^- (59). (See Figure 3 for definition of t, g^+, and g^- conformations.) In

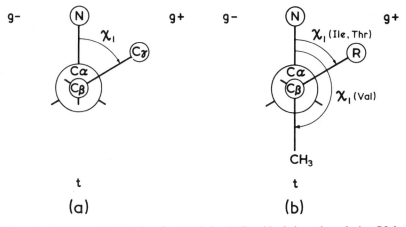

Figure 3   Torsion angle definitions for side chains. (*a*) For side chains unbranched at $C\beta$ the ideal g^+, t, and g^- conformations are equivalent to χ_1 values of $-60°$, $180°$, and $+60°$. (*b*) For Ile and Thr the ideal g^+, t and g^- conformations are also equivalent to χ_1 values of about $-60°$, $180°$, and $+60°$. For Val, however, the ideal g^+, t, and g^- conformations analogous to those of Ile and Thr are given by χ_1 values of $180°$, $+60°$, and $-60°$.

helices the t conformation restricts the $\phi\psi$ helical region and three-quarters of side chains have χ_1 in the g^+ conformation (59). Interaction of the side chain with the rest of the protein selects one preferred conformation and sharpens the potential around it (60).

CLOSE PACKING IN PROTEINS

The efficiency of residue packing in proteins—their packing density— has important implications for their structural, thermodynamic, and mechanical properties. The packing density of a molecule or residue is the ratio of the volume enclosed by its van der Waals envelope to the volume it actually occupies in a crystal, liquid, or protein interior. For organic solids, except those formed from molecules of unusual shape, this ratio has values in the range 0.68–0.80. For liquids the packing density is about 15% lower.

Protein Interiors

The controversy over whether the protein interior is more accurately described as an oil drop (63) or a crystalline molecule (64) has been decided in favor of the latter.

The comparison, for six proteins, of the sum of van der Waals volumes of their constituent amino acids with their partial specific volume gives packing densities of 0.72–0.77 (65). These values are typical of organic solids. The use of partial specific volumes in this calculation is not quite correct (see below), but the general conclusion has been confirmed by all latter work.

The introduction of the Voronoi polyhedra procedure to calculate, from atomic coordinates, the volume occupied by protein atoms gave the first quantitative description of packing in proteins (66). The application of this method and how detailed results vary with different parameters has been discussed by several authors (3, 66–69). For residues buried in the protein interior the results are unambiguous and accurate within the limits of the atomic coordinates used.

The initial use of the Voronoi procedure showed that the packing density in the interior of ribonuclease and lysozyme is the same as that of organic solids (66, 67). The mean volume occupied by residues in the interior of nine proteins was calculated (70). The results are given in Table 2, and show that the mean volume of a residue in the interior proteins is the same as that in crystals of its amino acid (Figure 4). The standard deviation of the mean residue volumes is $\sim 6\%$. A significant but unknown part of this is due to the inaccuracies of the atomic coordinates used in the calculations.

Within the protein interior the main-chain regions that form hydrogen bonds have a higher density than the side-chain regions that are packed by

Table 2   The mean volumes of residues buried in proteins[a]

Residue	Volume (Å³)	Residue	Volume (Å³)	Residue	Volume (Å³)
Gly	66	Ser	99	His	167
Ala	92	Thr	122	Asn	135
Pro	129	Cys	106	Gln	161
Val	142	Cyh	118	Asp	125
Ile	169	Met	171	Glu	155
Leu	168	Tyr	204	Lys	171
Phe	203	Trp	238	Arg	202

[a] Data from (70) except for the Arg value that comes from (156).

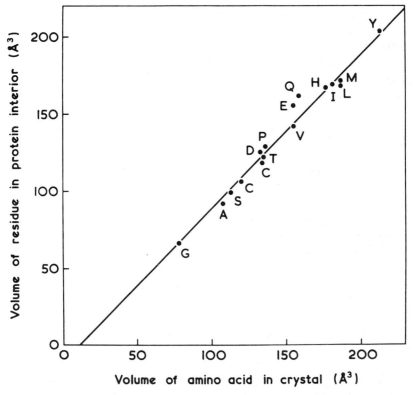

Figure 4   The mean volume of residues buried in protein interiors plotted against the volume of their amino acid in crystals. The line has a slope of 1 and an intercept of 11 Å² — the volume lost by an amino acid on forming a residue. Drawn from data in (70).

van der Waals forces (71). Most of the contacts formed by residues are to others close in the sequence or, if it is in a secondary structure, to those in the same α helix or β sheet (72).

Even with perfect close packing, $\sim 25\%$ of the total volume is not occupied by protein atoms. Nearly all this space is in the form of very small cavities. Cavities the size of water molecules or larger do occasionally occur (73, 74) but form only a small percentage of the total protein volume. The size of the cavities varies with thermal fluctuations of their surrounding atoms. Such fluctuations, or mobile defects, facilitate in proteins dynamic behavior such as the rotation of side chains (74–76).

Protein Surfaces

The Voronoi procedure can be used to estimate contact areas, surface roughness, and solvent accessibility (68). For residues on the surface of proteins the calculation of atomic volumes depends upon the structure of the solvent shell. This has not been determined for any protein used in these calculations and the use of artificial solvent shells gives results of uncertain accuracy.

If the total volume of a protein is calculated from its residue composition using the volumes in Table 2, a value is obtained that is about 10% greater than the experimentally determined partial specific volume. The reason for this discrepancy is not known. A plausible explanation is that the rough protein surface occupies interstices in the open water structure. Semiquantitative calculations support this idea (77).

PACKING OF α HELICES AND β SHEETS

Two factors have dominating influence on the packing of α helices and β sheets in proteins. First, residues in the protein interior close pack; second, packed secondary structures have a conformation close to the minimum free-energy conformation of the isolated secondary structures (see previous sections). These factors imply that the manner in which α helices and β sheets pack depends upon the shape of their surfaces (78).

The geometrical principles governing the packing of secondary structures have been embodied in a set of simple models. These models assume that, to a first approximation, surface shape arises from the main-chain conformation of α helices and β sheets. Size, shape, and conformation of side chains determine which particular class of packing occurs and modulates its exact geometry. Comparison of the geometrical predictions of the models with the features observed in real packing shows that this assumption is usually correct.

Here I outline only the different packing models. Details of the comparison between these models and packings observed in known protein structures can be found in the references cited. I do not discuss the attempts to predict the three-dimensional structure using packing and chain topology rules [this work has been recently reviewed (79)].

Packing of α Helices

The problem of how α helices pack together was first discussed in 1953 when a model colloquially known as "knobs into holes" was developed (80). With the availability of atomic coordinates of a large number of globular proteins it became apparent that helices do not generally pack with their axes inclined at +20° and −70° as expected from the knobs into holes model. The observed distribution of angles covers the whole range of possible values though there is a sharp peak at −50° with a shoulder at +20° (81; see Figure 5). Several analyses of helix packing have been published (78, 80–85) and a comparison of their results is given in (81).

The "ridges into grooves" model was developed to explain the observed distribution of interaxial angles and the patterns formed by residues that make interhelix contacts (78, 81; Figure 6). In this model, residues on the surfaces of helices form ridges separated by grooves. Helices pack together by the ridges of one helix packing into the grooves of the other and vice versa. The ridges on the helix surface are usually formed by the side chains of residues four separate in the sequence: i, $i+4$, $i+8$,... and $i+1$, $i+5$, $i+9$, ... (the $\pm 4n$ ridges). Occasionally they are formed by residues whose separation is three ($\pm 3n$ ridges), and, rarely, by adjacent residues (the $\pm 1n$

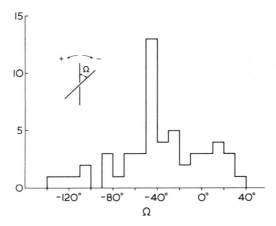

Figure 5   The relative orientations observed for packed α helices in proteins. The angle Ω describes the relative orientation of packed secondary structures. It is defined as the angle between the strands of β sheet and/or helix axes when projected onto their plane of contact. Data for this figure is taken from (81) and supplemented by my unpublished calculations.

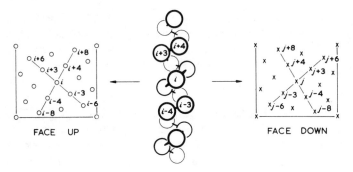

FACE UP

FACE DOWN

CLASS 1-4
$i \pm 4n$ $j \pm 1n$

CLASS 4-4
$i \pm 4n$ $j \pm 4n$

CLASS 3-4
$i \pm 4n$ $j \pm 3n$

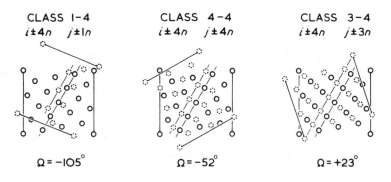

$\Omega = -105°$

$\Omega = -52°$

$\Omega = +23°$

ridges). Intercalculation of different ridges and grooves by packed helices inclines the axis at different characteristic angles: $\sim -50°$ when both helices use $\pm 4n$ ridges (class 4-4); $\sim +20°$ when one helix uses $\pm 4n$ ridges and the other $\pm 3n$ ridges (class 3-4), and so on (78, 81; Figure 6). These different packing classes give different patterns for the contacts between residues at the interface.

The use of particular ridges ($\pm 4n$, $\pm 3n$ or $\pm 1n$) by a helix depends upon the size and conformation of side chains, and upon the surface of the other helix. The $\pm 4n$ ridges are more commonly involved in packing because side chains with this separation splay apart less than do side chains in $\pm 3n$ ridges or $\pm 1n$ ridges (Figure 6).

The general validity of this packing model can be demonstrated by direct inspection of helix interfaces observed in protein structures (81). The actual Ω values depart somewhat from ideal values because of heterogeneity in the size of side chains and because of variations in helix geometry. Larger departures occasionally occur when, in a ridge, a very small residue is next to a large one, as the gap produced can allow ridges to cross each other. Thus, the wide distribution of observed Ω values arises from the different packing classes and the spread of values within particular classes (Figure 7). The dominance of the $\pm 4n$ ridges explains the peak at $\Omega \simeq -50°$ as this is the orientation when helices pack with both using the $\pm 4n$ ridges.

In helix packings the distance between the helix axes varies between 6.8 Å and 12.0 Å. This variation is principally a function of the size of the side chains at the center of the interface (81). The mean interaxial distance for packed helices is 9.4 Å (81). The mean interpretation of atoms at the interface is 2.3 Å (84). Thus, the contacts between packed helices mainly involve the ends of side chains. This observation is important in considering how helices can move relative to each other (86).

α-Helix–β-Sheet Packing

Analysis of observed protein structures shows that there is a strong tendency for the helices packed onto sheets to have their axes nearly parallel

---

Figure 6   The ridges into grooves model for the packing of α helices. Helical nets can be used to describe the relative position and the packing of residues. The nets are made by marking the position of residues on a piece of paper wrapped around a helix. Placing a face-down net over a face-up net gives a representation of the residue arrangements that can occur at helix interfaces. The central part of the figure shows how packing the $j \pm 1n$, $j \pm 4n$, and $j \pm 3n$ rows of one helix between the $i \pm 4n$ rows of a second helix inclines their axis at ideal angles of $-105°$, $-52°$, and $+23°$. The $i \pm 4n$ rows are more common in helix packings because residues in these rows splay apart less than those in other rows (see lower part of figure). On going from $i - 4n$ to $i + 4n$ we rotate 80° around the helix axis; on going from $i - 3n$ to $i + 3n$ we rotate 120°. [This figure is a modified version of two figures in (81).]

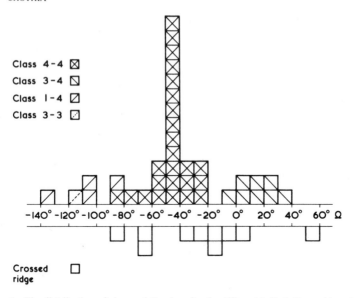

Class 4 - 4 ⊠
Class 3 - 4 ◨
Class 1 - 4 ◪
Class 3 - 3 ⊡

−140° −120° −100° −80° −60° −40° −20° 0° 20° 40° 60° Ω

Crossed ridge   ☐

Figure 7   The distribution of observed Ω values for the different helix-helix packing classes. The letter Ω is defined in the caption of Figure 5 and the packing classes in Figure 6 and text. [Reprinted with permission from (81).]

to the sheet strands (44, 78, 87; Figure 8). A model for the packing of α helices onto β sheets (44, 78) is illustrated in Figure 9. I call this the complementary twist model for helix-sheet packing.

The β sheets observed in globular proteins have a right-handed twist

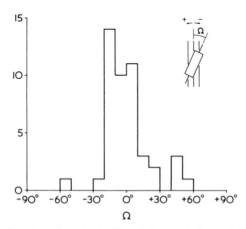

Figure 8   The relative orientations observed for α helices packed onto β sheets. The letter Ω is the angle between the helix axes and β strands (see caption to Figure 5). Data taken from (44, 87).

when viewed in a direction parallel to the polypeptide chain. Now in an α helix the adjacent two rows of residues $i, i+4, i+8, \ldots$ and $i+1, i+5, i+9, \ldots$ form a surface that also has a right-handed twist (Figure 9). The essential feature of the packing model is that α helices pack onto β sheets with their axes parallel to the β-sheet strand because in this orientation these two rows

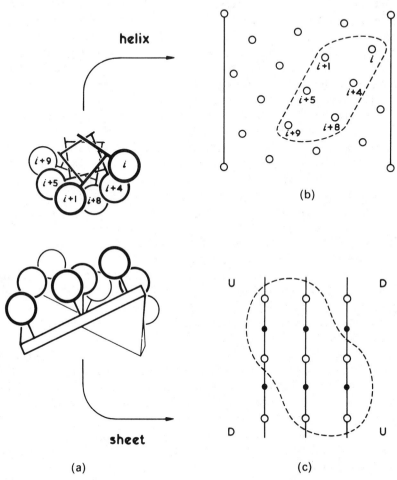

Figure 9   The complementary twist model for α-helix to β-sheet packing. Side chains on one side of an α helix and a twisted β sheet are shown as open circles (*a*). When the helix axis is parallel to the strand direction, the helix residues $i, i+1, i+4, i+5, i+8, i+9, \ldots$ form a twisted surface that is complementary to that of the twisted β sheet. The patterns formed by the residues in contact at the interface are shown on flattened projections of the α helix (*b*) and β sheet (*c*). U and D mark the corners of the β sheet that move up and down from the plane when the β sheet is twisted. Note that the β-sheet contact residues cluster about a line joining U to U, the concave diagonal of the β sheet. [Reprinted with permission from (44).]

of residues have a surface complementary to that of the twisted β sheet (Figure 9).

Two detailed analyses of helix-sheet packings (44, 87) have demonstrated the validity of the complementary twist model. The model predicts that the helix residues forming the interface with the β sheet will have the sequential relationship of the type $i, i+1, i+4, i+5, i+8, \ldots$ (Figure 9). For observed packing $\sim 90\%$ of the helix interface residues do have relationships of this type (44, 87). A subset of these, $i+1, i+4, i+5$, and $i+8$, forms the larger part of the contact (87). On the β-sheet side of the interface residues are expected to cluster about a line parallel to the concave diagonal of the β sheet (Figure 9), and this is indeed observed. Note that these oblique patterns for the interface residues occur even though the α-helix axis and β-sheet strands are parallel.

If two α helices pack side by side on the face of a β sheet its twist will rotate one helix relative to the other. From the mean values of β-sheet twist ($-19°$ between strands ~ 4.5 Å apart) and of the distance between helix axis (9.4 Å), we would expect such helices to be inclined at $\sim -40°$ to each other. This implies that the interface between helices must be formed by the $\pm 4n$ ridges (class 4-4), as use of other ridges would produce a quite different relative orientation (see previous section). Inspection of the helices that pack onto β sheets and against each other shows that four fifths pack as expected (44). The other fifth are in other packing classes; they have one end of the helix in contact with the β sheet and then splay away.

There is one issue over which the two analyses of helix-sheet packing have led to different conclusions. This concerns the details of the side-chain packing. As the periodicity of residues in α helices is different to that in β sheets, there cannot be the regular overall pattern of intercalation found in helix-helix packing. One group (44) argues that the helix and sheet surfaces can be regarded as essentially smooth with only occasional irregular intercalation where particularly large or small residues occur. The other group (87) argues that regular local intercalation occurs at the center of the contact with 4 helix residues, $i+1, i+4, i+5$, and $i+8$, surrounding one β-sheet residue. Further investigation is required.

Helices are sometimes associated with orthogonal β-sheet packings (see below). No analysis or discussion of this minor class of helix-sheet packing has been published.

Packing of β Sheets

β Sheets, in those proteins formed by their association, are of two types. In one, the β sheets are just twisted. In the other, they are both twisted and folded upon themselves (78; Figure 10). This particular folding of β sheets is produced by a local coiling, or a β bulge, bending the strands by $\sim 90°$. The

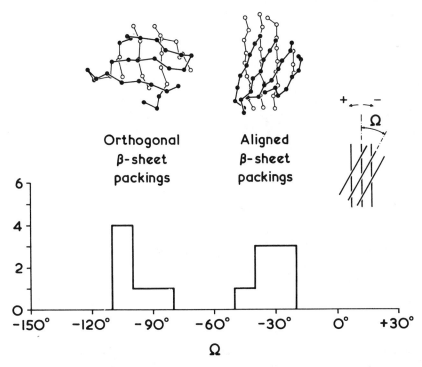

Figure 10 The relative orientation of packed β sheets in proteins in the aligned and orthogonal classes. The letter Ω is defined in the caption to Figure 5. The upper part of the figure shows an example of each class. Circles, representing the Cα atoms of residues in β sheets, are closed for the β sheet above the plane of the page and are open for the β sheet below the plane. Note how in the orthogonal packing strands move from one β sheet to the next without interruption. Data in this figure were taken from (45, 46).

two types of β sheet pack differently giving *aligned β-sheet packings* and *orthogonal β-sheet packings* respectively (Figure 10; 46).

ALIGNED β-SHEET PACKINGS   Structures in this class are formed by the face-to-face packing of β sheets that can be regarded as essentially independent: In nearly all cases the links between the β sheets involve several residues in a non β-sheet conformation (45). In this class the main-chain direction in one β sheet is at an angle of $\sim -30°$ to the main-chain direction in the other β sheet (78; Figure 10).

Models for, and analyses of, aligned β-sheet packings have been published (45, 88). The models relate the relative orientation of the packed β sheets to their right-handed twist. Figure 11 shows this relation. On the left of this figure I show two hypothetical untwisted β sheets packed with the rows of side chains at their interface aligned. On the right I show what

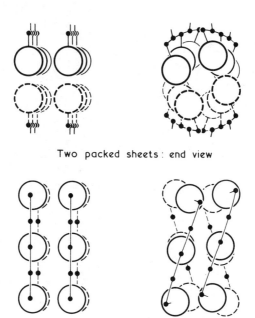

Two packed sheets: end view

Two packed sheets: top view

Figure 11   A model for the aligned packing of β sheets. The main chain is represented by small closed circles and the side chain at the interface by large open circles. The figure shows views of packed β sheets with two strands each. On the left the β sheets are flat and residues at the interface aligned. On the right we show what happens when the β sheets twist about an axis in the plane of their interface. The ends of the side chains maintain their alignment but the main chains now point in different directions.

happens when the β sheets are given a right-handed twist about an axis parallel to the strands and in the plane of their interface. Though the ends of the side chains that form the contact across the interface maintain their alignment, the main chains of the two β sheets do not. They point in different directions. If one looks from on top, the main chains make an angle (Ω) that is negative. If the main chains are 10 Å apart and the twist along the chain 4°/Å (typical values), Ω is −35°. The observed values of Ω are in the range −20° to −50°, departure from the model value arising from variation in the extent of the twist and from the imprecise alignment of side chains at the interface.

The residues that form the β-sheet–β-sheet interface have an anti-complementary pattern and this has also been related to their twist (88).

ORTHOGONAL β-SHEET PACKINGS   The packings in this class are formed by β sheets folded on themselves. As in the aligned class, the β sheets are twisted and pack face to face. They differ in that the direction of the main chains in the different layers are at ~90° and in that the strands at one corner, or two diagonally opposite corners, pass from one layer to the next without interruption (46; Figure 10).

A model for orthogonal β-sheet packings (46) is shown in Figure 12. The principal features arise from the right-handed twist of the β sheets and from the manner in which they fold upon themselves. The folding is produced by the strands at the corners that go directly from one layer to the next by right-hand 90° bends. These bends are usually produced by a local coiling of the strands, a β bend, or by a β bulge (46, 52). The right-handed geometry of these conformations arises from the values normally allowed for $\phi\psi$ (see a previous section).

The effect of the β sheets in the two layers being twisted and inclined at 90° is that they are only in contact along one diagonal. Along the other diagonal they splay apart (Figure 12). The space between the splayed corners is usually filled by helices or large side chains in loops (46).

The β-sheet hydrogen bonding in these structures is often shown on

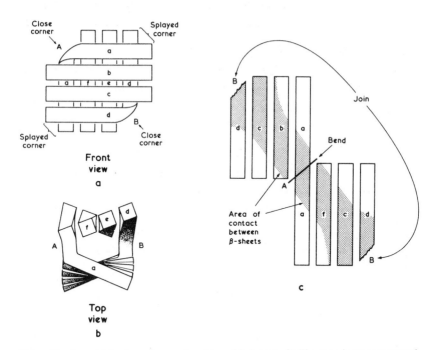

Figure 12 A model for the orthogonal packing of β sheets (a, b). The strands are represented by rods with square cross-section. The β sheets have the normal right-handed twist so that two corners come close together (A and B) and two splay apart. At the close corners, strands pass from one β sheet to the next through β bends or β bulges (see Figure 2 and text). If the packing is opened out by straightening bend A and breaking the chain at B we get a single β sheet with staggered strands (c). [Reprinted with permission from (46).]

two-dimensional plans obtained by the equivalent of straightening and breaking the chains that go between layers (Figure 12). These plans show the structure as a single β sheet with staggered strands and hydrogen bonds going right round. This latter feature, and the appearance of these structures in certain projections, has led to their being referred to as barrels or cylinders. Such descriptions are inaccurate and possibly misleading (46).

Residue Composition and Packing in β Sheets

The β-sheet surfaces that form interfaces to other secondary structures are predominantly formed by five residues—Val, Leu, Ile, Phe, and Ala. In Table 3 we give the proportion of these residues in β sheets, in the interface regions of β sheets, and for buried residues in proteins. For β sheets, the five residues form about two thirds of the interface regions with just three—Val, Leu, and Ile—forming half (44–46).

The hydrophobicity of the inner strands of β sheets and the high proportion of Val and Ile in β sheets, particularly in parallel β sheets, has been noted by several authors (89–93). It has been suggested that this arises from the intrinsic nature of interaction between strands in the β sheets. However, the high Val, Leu, and Ile content is only a characteristic of the interface regions. Parallel β sheets usually have both faces covered by α helices with two thirds of the residues being part of an interface (44, 87). Antiparallel β sheets are usually packed against just one other β sheet and only one third of their residues are part of an interface. The more prominent

Table 3   Common nonpolar residues in β sheets, α helices, and their contact regions

Residue	Parallel β sheets[a]		Antiparallel β sheets[b]		α Helices[c]	
	Total β sheet (%)	Contact region[d] (%)	Total β sheet (%)	Contact regions[d] (%)	Total α helices (%)	Contact regions[d] (%)
Val	18	24	14	22	7	12
Leu	9	10	8	16	7	12
Ile	11	14	8	15	6	9
Phe	4	6	4	11	3	6
Ala	8	7	8	8	9	12
Total	50	61	42	72	32	51

[a] Data from (44). Similar values for total composition are given in (92, 93).
[b] Data from (45, 46). Similar values for total composition are given in (92, 93).
[c] Data from (81).
[d] Contact regions are those parts of the β sheets or α helices that make contacts with other β sheets or α helices.

part played by interface residues in parallel β sheets accounts for their high overall Val and Ile content (45, 46).

The common occurrence of Val, Ile, and Leu in interface regions probably arises from their packing requirements. The spacing of residues along the strands (~ 7.0 Å) is different to that between strands (~ 4.5 Å), so in orthogonal packings and in any but the most ideal aligned packing there could not be a regular pattern of intercalation. This also applies to helix-sheet interfaces (see above). Well-packed surfaces favor Val, Ile, and Leu residues that make good interstrand contacts. Intercalation for particularly large or small residues is occasional and irregular (44–46).

CHAIN TOPOLOGY IN PROTEINS

The examination of the topology of protein chains (37, 94–107) has shown clear regularities. These regularities have been described by different authors in different ways. A comparison of their results suggests that the topological regularities so far observed can be expressed as three rules.

1. *Pieces of secondary structure that are adjacent in the sequence are also often in contact in three dimensions* (37, 94, 97, 102–104, 107). Three possible structures for sequentially linked pieces of secondary structure are shown in Figure 13: $\alpha\alpha$, two antiparallel packed α helices; $\beta\beta$, two antiparallel β-sheet strands; and $\beta\alpha\beta$, a helix packed against two adjacent parallel β-sheet strands. An examination of 31 proteins showed that two thirds of the pieces of secondary structure were part of an $\alpha\alpha$, $\beta\beta$, or $\beta\alpha\beta$ unit (37). These substructures were named folding units. The number of folding units in the 31 proteins is *close* to the maximum possible. An examination of more recent protein structure shows that $\alpha\beta\beta$ and $\beta\beta\alpha$ folding units (a helix packed against two adjacent antiparallel strands) occur to a lesser but still significant extent. This close connectivity of strands and helices is very significantly different to that which would be expected if connections were made randomly (37).

   The common occurrence of $\beta\alpha\beta$ units had previously been noted and called super secondary structures (96). An extensive analysis of the connections within β sheets has been made (102–104, 107).

2. *The connections in β-X-β units* (where the βs are parallel strands in the same β sheet, though not necessarily adjacent, and X is an α helix, a strand in a different β sheet, or an extended piece of polypeptide) *are right-handed* (Figure 13). This observation, made for certain $\beta\alpha\beta$ units (96), was later shown to be generally true for proteins (98–101). There are over a hundred cases that satisfy this rule and only three exceptions (6).

3. *The connections between secondary structures neither cross each other nor make knots in chain.* This rule is partly or totally stated in somewhat

different ways in a number of papers (102–105a). The only clear exception at present known to this rule is the knot found in the chain of carbonic anhydrase. It is a rather trivial exception in that the knot is formed by the N terminus of the chain looping round the C terminus (102). However, loop penetration is occasionally observed in protein structures (105a; D. F. Dykes, quoted in 105b).

Topological rules in addition to these three have been put forward. In all cases so far they can be shown to be the same as, or arise from a combination of, the three given here. This is the case with the left-handed

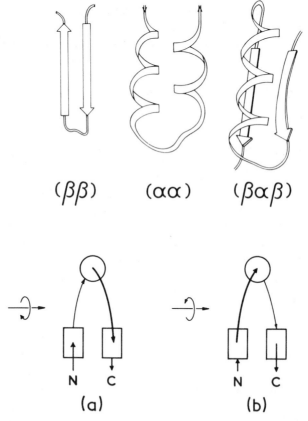

(ββ)   (αα)   (βαβ)

(a)   (b)

Figure 13   Topological feature of proteins: αα, ββ and βαβ are folding units, or super-secondary structures, frequently found in protein (32). βαβ Units are represented in (a) and (b) in an end-on view with circles representing an α helix and rectangles the β-sheet strands (see 37). This unit can be right-handed (a) or left-handed (b). In nearly all cases they are observed to be right-handed (6, 96, 98–101). This is also true when the α helix is replaced by a β-sheet strand or an extended chain.

"Greek Key" and "Jelly Roll" topologies (6), which can be shown to arise from a combination of the rules given above (C. Chothia, unpublished).

The origin of the topological regularities in proteins is uncertain. There are no apparent thermodynamic origins and explanations have centered around the kinetics of protein folding. It has been suggested that the limits on the rates of diffusion of chain segments favor the association of those close in the sequence (rule 1). Estimates of the actual rate of diffusion (108), however, make this explanation unlikely. Another explanation is that, during folding, intermediates made up from segments close in the sequence are used because they have a lower entropic cost than those formed by segments distant in the sequence.

CLASSIFICATION OF PROTEIN STRUCTURES

The packing models and topology rules embody the basic principle determining the arrangement of secondary structures in nearly all proteins. Actual structures are built by a particular set of variations on, and combinations of, these models and rules. (They also have sufficient stability and a function—see later sections.) The small number of common packing classes and the limitations on chain topologies mean that certain arrangements of secondary structures occur frequently. Thus, proteins quite unrelated by evolution or function may have similar structures (37, 98–102). This also means that a system for the structural classification of proteins or protein domains can be made on the basis of the arrangement of their secondary structures (4–6, 37).

The original system (37) divided proteins into four classes:

1. All-α proteins are composed mainly of α helices. One common motif is the "four α-helical bundle" (109)—four are nearly parallel but with a small left-hand twist because class 3-4 helix packings put their axes at $\sim +20°$ to each other (81). Proteins built by other classes of helix packing tend to be larger and more heterogeneous, presumably because it is difficult to make stable proteins with a few α helices inclined at large angles.

2. All-β proteins are composed mainly of β sheets. The large majority of proteins in this class contain aligned or orthogonal β-sheet packings with connection between the strands limited by the topology rules.

3. α/β Proteins have α helices and β-sheet strands that approximately alternate along the chain. These structures usually contain a central β sheet with α helices packed on both sides. The right-handed nature of the $\beta\alpha\beta$ units (see above) is important for the function as well as the structure of this class of protein (see a later section).

4. $\alpha + \beta$ Proteins contain α helices and β-sheet strands that tend to

segregate rather than alternate along the chain. The amount of α helix or β sheet in these proteins varies, making their structure heterogeneous. If the amount of either is appreciable the two secondary structures pack as in the all-α or all-β classes. The contact between α helices and β sheets may then involve packings of the kind found in α/β or in the splayed corners of orthogonal β-sheet packings (46).

The original four classes have been divided by some authors (5, 6), who give lists of the proteins they consider as belonging to each subclass. These subclasses are based on function, possible evolutionary relationships (5), or the details of chain topology (6). This more elaborate approach can produce inconsistencies. For example, the system that has a strong emphasis on chain topology puts in the same subclass proteins that would be in separate subclasses if their packing was emphasized.

The great variation possible within normal packing and topology rules, together with the rare exceptions to these rules, means that any classification systems that cover the large majority of protein structures can only be approximate. It can be a reasonable but not exact guide to the secondary structures present in the protein, the nature of their packings, and the topology of the chain.

PROTEIN STABILITY : HYDROGEN BONDS, HYDROPHOBICITY, AND SURFACE AREAS

The previous sections have discussed the principles governing the conformation and packing of secondary structures and the topology of the links between them. To be viable, however, a protein must also have overall stability. The net energy gained from hydrophobicity, hydrogen bonds, van der Waals forces, and electrostatic interactions must compensate for the loss of conformational entropy and any conformational strain.

Although the earlier emphasis on hydrogen bonding as the major source of protein stability (7) later shifted to hydrophobicity (1, 63), it is clear that both forces have very important effects on the structure of proteins. Recently the contribution of electrostatics has been discussed. Though electrostatics is very important for protein function, the magnitude of its contribution to protein stability and its influence on actual structure is not clear at the moment. As we have already seen, van der Waals forces (packing) determine the manner in which secondary structures associate. They may also contribute to stability (110, 111).

Hydrogen Bonds

Estimates of the energy of hydrogen bonds between protein atoms relative to those between protein atoms and water were made by measuring the association in water of small molecules containing groups similar to those found in proteins, e.g. urea and N-methylacetamide (112–116). These experiments gave very small energy differences, suggesting that hydrogen bonds made only a marginal contribution to protein stability. Recently it has been pointed out that, for entropic reasons, intramolecular interactions can have substantially greater free energies than intermolecular, so the contribution of hydrogen bonds can be substantial (117). This also accounts for the lower enthalpy of the folded state (110, 111, 117).

An early survey of protein structures showed that the proportion of polar groups that form intramolecular hydrogen bonds is essentially constant and close to 0.5. As proteins have similar amino acid compositions, this implies that the contribution made by hydrogen bonds to protein stability is proportional to molecular weight (70).

Hydrophobicity and Surface Areas

The hydrophobic contribution to protein stability arises of course from the folded protein making fewer contacts with water than the unfolded. With the introduction of algorithms to calculate the surface areas of protein atoms (73, 118–120), it became possible to calculate changes in the areas of protein-solvent and protein-protein contact that occur in folding or association. Three different measures of the protein surface have been used : accessible surface area, accessibility, and contact surface area. These are defined and the relations between them discussed in (3). For the same structure, the value for the accessible surface area is *approximately* 3.5 times as large as that for the contact surface area.

For amino acids there is a good correlation between their surface area and the energy of transfer from water to organic solvents (121, 122). [A very similar correlation is found for other molecules (123–128).] The correlation suggests that the changes in surface area that occur in any process can be used to calculate the contribution of hydrophobic free energy.

Four important features of the relation between surface area and hydrophobic free energy include the following:

1. For any particular set of transfer experiments the correlation is very high, > 0.99 (121, 122).
2. The magnitude of the unitary free energy derived from the correlation is dependent upon the nature of the organic solvent involved in the transfer process (129). The value obtained from amino acid data varies from 86 (Ethanol) to 131 (Carbon Tetrachloride) cal/Å^2 of contact surface with a mean 100 cal/Å^2 (122). For accessible surface areas a mean value of 25 cal/Å^2 has been used (70).
3. The relationship is also found for the transfer of amino acids from water to 8 M urea and 6 M guanidine hydrochlorite (130) and shows that the

hydrophobic interactions are reduced relative to water by 7.1 and 8.3 cal/$Å^2$ of accessible surface. This can explain how urea and guanidine hydrochloride unfold protein and the greater effectiveness of the latter (130).

4. The different correlations show small threshold values that vary between ~ 10 and ~ 40 $Å^2$ of accessible surface area [(128, 130) and inspection of data in (121, 122)].

Changes in surface area in proteins have been widely used to calculate hydrophobic free energies. If these results are used without some estimate of the other nonhydrophobic energy changes, they can be misleading.

Objections have been made to the application to proteins of the values obtained from amino acid transfer experiments. The values are only a half or a third of that derived from macroscopic concepts of surface chemistry (131). Surface area calculations do not account for solvent effects on the dimerization of methane nor on the gauche/trans equilibrium of butane (132).

Residue Surfaces and Environments

Accessible surface areas and accessibilities have been published for residues in proteins, model tripeptides, (Gly–X–Gly and Ala–X–Ala) and dipeptides (73, 118, 133, 134). Surface areas calculated for residues and side chains in different conformations show small variations (118, 134). The extent to which residue types are buried in proteins has been determined (133, 135) as has the composition of the inside and surface of proteins (135). Some of these results are given in Table 4. They show that about half of the nonpolar residues are buried within the protein, about a quarter of the residues with one polar atom, and about one tenth of the residues with two polar atoms or a charged atom. The residues in this last group are, when buried, usually involved in the coordination of metals, part of domain interfaces, or part of the active site.

The partition, between the inside and outside of proteins, of these *groups* of residues correlates well with the hydration energy of their side chains (136), i.e. their affinity for water. Within the nonpolar group hydrophobicity differentiates the lower extent to which Gly and Ala are buried but it does not differentiate between Val, Ile, Leu, and Phe.

The distribution of residues observed in protein structures has been used to develop "empirical hydrophobicity scales." In one of these, a residue's hydrophobicity is related to its average distance from the protein center of mass and the average orientation of its side chain (137–140). In another, the average surroundings of residue types are used to characterize their hydrophobicity (141–143).

A "hydropathy scale" has been developed (143a) in which the value for

Table 4 Residue surface areas and hydrophobicities and the composition of buried and accessible surfaces in proteins

| Residue | Accessible surface area (Å²)[a] | Proportion buried in proteins[b] (%) | Hydrophobicity[c] (kcal·mol⁻¹) relative to Gly | | Molar fraction (%) | | |
			Transfer free energy[d]	Hydration free energy[e]	Average protein[f]	Accessible residues[g]	Buried residues[g]
Gly	75	36	0.00	0.00	8.4	6.7	11.8
Ala	115	38	0.5	0.45	8.6	6.2	11.2
Val	155	54	1.5	0.40	6.6	4.5	12.9
Ile	175	60	—	0.24	4.5	2.8	8.6
Leu	170	45	1.8	0.11	7.4	4.8	11.7
Phe	210	50	2.5	3.15	3.6	2.4	5.1
Pro	145	18	—	—	5.2	4.8	2.7
Ser	115	22	-0.3	7.45	7.0	9.4	8.0
Thr	140	23	0.4	7.27	6.1	7.0	4.9
Cys	135	48	—	3.63	2.9	0.9	4.1
Met	185	40	1.3	3.87	1.7	1.0	1.9
Tyr	230	15	2.3	8.50	3.4	5.1	2.6
Trp	255	27	3.4	8.27	1.3	1.4	2.2
His	195	17	0.5	12.66	2.0	2.5	2.0
Asn	160	12	-0.2	12.08	4.3	6.7	2.9
Gln	180	7	-0.3	12.08	3.9	5.2	1.6
Asp	150	15	—	13.34	5.5	7.7	2.9
Glu	190	18	—	12.59	6.0	5.7	1.8
Lys	200	3	—	11.91	6.6	10.3	0.5
Arg	225	1	—	22.31	4.9	4.5	0.5

[a] Value for the residue R in the extended tripeptide Gly–R–Gly (133).
[b] Residues were taken to be buried if they had 5% or less of their potential surface accessible to solvent (133).
[c] Hydrophobic is a misnomer: There is no hydrophobia between water and hydrocarbons; there is just not enough hydrophilia to pry the water molecules apart (168).
[d] Measured for transfer from water to ethanol or dioxane (169).
[e] Measured for transfer from water to vapor (136).
[f] From (170).
[g] From (135). Here buried residues were those with an accessible surface area of less than 20 Å².

each residue is based upon both the average extent to which it is buried and on its energy of transfer from water to vapor. The mean hydropathy of proteins does not change with molecular weight; large proteins are not more or less hydrophobic than small.

The distribution of residues in protein sequences has been surveyed (144, 144a), as have the interactions between residues within proteins (145, 146). Certain distributions and interactions deviate from random expectations. The reason for this is not understood.

Secondary-Structure Surfaces and Residue Composition

One calculation has suggested that the observed probabilities of residues being found in α helices and β sheets is proportional to the surface that becomes buried when they are incorporated into these secondary structures (147). The calculation carried out is the following: Different residues, X, in an extended chain, $Ala_3-X-Ala_3$, were incorporated into an α helix, $Ala_4-X-Ala_4$, and into a β sheet, $Ala_7/Ala_3-X-Ala_3/Ala_7$. For each residue, X, the loss of contact surface that occurs in these two processes was determined. Residues were then ordered on the size of the area losses to give two lists, one for the chain to α-helix process and one for the chain to β-sheet process. The two lists were then compared to lists of the probability of the residues being found in α helices or β sheets (148). For β sheets the rank correlation coefficient between the two lists is 0.74 (Pro excluded); for α helices it is 0.65 (Pro, Tyr, and Glu excluded) (147).

All residues occur to some extent in α helices, β sheets, and turns (148, 148a). This means that prediction of amino acids based just on the residue composition of regions of polypeptide can only have limited success (144a). The best of such schemes predict correctly the secondary structure of not more than 56% of residues (148b). More accurate results are obtained with those schemes that include consideration of both local and long-range interactions (148c, 148d).

When an average residue goes from a chain in an extended conformation to an α helix or two stranded β sheets it buries about half its potential accessible surface (~ 80 Å^2). If the β sheet is extended by additional strands the inner residues bury another quarter (133). The effect on individual residue surfaces of secondary-structure formation and packing has been described in some detail (84, 87, 88).

Chemical Character of Buried and Accessible Surfaces

Of the total surface buried within proteins, about three fifths is buried within secondary structures and two fifths between them (88, 133, 149). The formation of hydrogen bonds between peptides makes half the surface buried *within* secondary structures polar (133). The surfaces buried *between* secondary structures are very hydrophobic (1, 73, 133). Quantitative expressions of this are given by the calculation of their hydropathy (143a) or hydrophobic moment (149a). Two thirds of the surface buried between secondary are formed by nonpolar atoms and more than half of the polar part is formed by groups that hydrogen bond within their own secondary structure (133). Thus, the net general effect is that, on folding, both polar and nonpolar surfaces are reduced by similar amounts, approximately three quarters (73, 118, 133). On a more detailed level, this general statement has one important qualification: As proteins increase in size they bury an increasing proportion of their surface (see next section). This relative increase in buried surface occurs at the expense of just the protein's nonpolar surface (133).

The polar surface buried in proteins is very largely formed by peptide groups while the nonpolar part is mainly formed by side chains (73, 118, 133). However, a large proportion of nonpolar side chains remains on the surface (Table 4). The role of a particular nonpolar side chain in protein folding depends upon which other side chains are in its vicinity on the formation of the secondary structures. If the neighbors are also nonpolar, the residue is usually part of the surfaces buried within the protein; if they are polar it is usually on the surface, whatever its intrinsic hydrophobicity.

Extent of the Accessible and Buried Surfaces

The accessible surface area, A_s, of small and medium size monomeric protein (50–320 residues) is a function of their molecular weight, M_r (70). The relationship can be expressed by Equation 1 (150, 151):

$$A_s = 11.1 \, M_r^{2/3}. \qquad\qquad 1.$$

For all but a few cases this equation gives A_s values to within a few per cent of those observed. No data for monomeric proteins with more than 320 residues have been published.

The implications of Equation 1 for the shape of those proteins to which it applies have been discussed (152). The ratio of the protein's actual surface area, A_s, to the surface area of a smooth ellipsoid of the same volume increases with molecular weight. For small proteins ($M_r \simeq 6,000$) the ratio is 1.4:1; for larger proteins ($M_r \simeq 35,000$) it is 1.65:1. The greater surface area of real proteins is, of course, due to the textured nature of their surface. The actual ratios have two interesting implications (152): First, they imply that larger proteins must be more highly textured or more aspherical than smaller proteins, as had been predicted much earlier (153). Second, the ratios are large enough to allow a great variety of protein shapes. Spheres, spheres with large clefts, and ellipsoids or rods or discs with axial ratios of 1 to 2, when they have the same volume, differ in surface area by less than

10%. This analysis (152) disproves previous suggestions that proteins have surface area/volume ratios that approach minimum values characteristic of a sphere.

For oligomeric proteins the surface area of isolated monomers with 50–250 residues is given quite accurately by Equation 1 (70, 150, 151). For oligomers whose monomers have 330–840 residues, this is not the case. They have surface areas that are approximately proportional to molecular weight, and are 20%–50% greater than the values given by equation 1 (70, 135, 154).

The accessible surface area of protein chains in an extended conformation, A_t, is quite accurately given by Equation 2 (70, 151):

$$A_t = 1.45 \, M_r. \qquad\qquad 2.$$

Thus, for those proteins whose accessible surface area is given by Equation 1, the potential surface buried by the protein's folding, A_B, will be approximately given by Equation 3:

$$A_B = 1.45 \, M_r - 11.1 \, M_r^{2/3}. \qquad\qquad 3.$$

This means that small proteins, ~ 50 residues, will bury $\sim 90 \, \text{Å}^2$ for each residue and that large proteins, ~ 320 residues, will bury $\sim 120 \, \text{Å}^2$. As just noted, Equation 1 does not apply to large oligomeric proteins but the values so far published for their A_s values (70, 120, 154) show that they bury a constant proportion of their surface that is also equivalent to $\sim 120 \, \text{Å}^2$ per residue.

The extent of the buried (and therefore accessible) surfaces in proteins has been rationalized on the basis of the following argument: The area buried on the folding of a protein is directly related to the hydrophobic energy that helps compensate for the loss of conformational entropy (70, 155). The loss of entropy by the main chain will be proportional to its length, i.e. its molecular weight. The loss by the side chains will be proportional to the number buried. For monomeric and small oligomeric proteins, this increases with molecular weight. The number of buried residues, n_B, is related to the number in a protein, n, by the equation $n_B = (n^{1/3} - k)^3$ (135). For large oligomeric proteins, the proteins, the proportion is constant (154) and the same as that found in monomers of 300 residues.

Surfaces Buried Between Protein Subunits

For oligomeric proteins the accessible surface area that becomes buried, when two monomers are brought together, varies between 1000 Å^2 and 5500 Å^2 (120, 154–158). The lower values, 1000 Å^2 to 2000 Å^2, are found in oligomers where the structure of the isolated and associated monomers are very similar, and it has been argued that the size of the buried surfaces can account for the stability of the associations (156–158). The larger values are found where the monomers have their structure stabilized or changed by the association. In some cases the interface is formed by loops or chain termini that would be flexible in the isolated monomer (158). In other cases the active structure of the monomers appears only on association (159).

FUNCTIONAL DETERMINANTS OF STRUCTURE: DOMAINS IN PROTEINS

The proteins found in biological systems are, of course, present because they perform some function. Many specific examples of the relation between structure and function have been described (see any recent text book). Here I briefly mention just two general features of protein structure that are selected by function: They relate to the topology of α/β proteins and the domain structure found in many proteins.

In α/β proteins a clear relationship has been demonstrated between the general topology of the structure and the position and geometry of the active sites (160). α/β Proteins have a β sheet with its strands connected by α helices that pack onto the faces of the β sheet (37). In all these proteins, substrates and cofactors bind in crevices formed by the connections two adjacent β-sheet strands make to α helices on *opposite* sides of the β sheet (Figure 14). The occurrence and position of such crevices are determined by

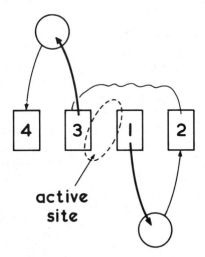

Figure 14   The active site in α/β proteins is formed at a position where the direction of the strand order changes (160): between strands 1 and 3 in the example shown here. The twist of the β sheet, not shown here, increases the crevice nature of the space between the loops that come from strands 1 and 3.

the right-hand nature of the $\beta\alpha\beta$ units and the strand order of the β sheet. They can occur only where the direction of the strand order changes: between strands 3 and 1 in the β sheet with strand order 4312 in Figure 14. Thus, the requirement to form an active site discriminates between the arangements of $\beta\alpha\beta$ units that are possible on thermodynamic and kinetic grounds (160).

Many proteins have a domain structure—globular units separated by a cleft. [The definitions of domains and the methods for their location have recently been reviewed (161).] It has been suggested that domains have a structural role in proteins in facilitating folding (94). In large proteins, at least, experimental evidence had previously shown the independent refolding of fragments [(162); more recent work is reviewed in (163)].

Domains are clearly of central importance to protein function: Many active sites occur at domain interfaces; in enzymes with more than one substrate the different binding sites usually occur on different domains; in proteins with more than one function different domains usually perform the different functions (161). In certain proteins the domain structure seems to have been created by gene duplication or gene fusion. Gene duplication is indicated by a protein's domains being similar in structure. This gives the molecule pseudo-symmetry (164, 165). Gene fusion is indicated by domains with the same function and similar structure being found in different proteins in association with other domains of quite different structure (166).

CONCLUSION

The seeming complexity of the first protein structures created shock of the kind expressed in the contemporary description of myoglobin (171) as "almost nothing but a complicated set of rods [of polypeptide] sometimes going straight for a distance then turning a corner and going off in a new direction . . . much more complicated and irregular than most of the early theories of the structure of proteins had suggested." The work described here has shown that the apparently complex structure of proteins is in fact governed by a set of relatively simple principles. Individual proteins arise from particular combinations of and variations on these principles. An analogous situation is found in linguistics, where a set of simple grammatical rules govern the generation of different, and sometimes complex, sentences.

Although the principles described here do not, in most cases, predict exactly what will occur for a particular protein, they do strongly predict the sorts of things that are to be expected and suggest how exceptions may arise. As more protein structures become known, the principles will be refined and extended.

Much remains to be done. The detailed energetics of protein structure are not clearly understood. The relation between the folding process and the final structure is, at this moment, mostly speculative. The recent achievements, however, together with those of the 1950s and the 1960s, have laid the foundations of a comprehensive theory for protein structure.

ACKNOWLEDGMENTS

I thank John Cresswell for drawing the figures and Arthur Lesk and Joël Janin for their comments on the manuscript. The author's work described in this review has been supported by the Royal Society, National Institute of General Medical Sciences, Science Research Council, and Medical Research Council.

Literature Cited

1. Perutz, M. F., Kendrew, J. C., Watson, H. C. 1965. *J. Mol. Biol.* 13:669–78
2. Schulz, G. E., Schirmer, R. H. 1979. *Principles of Protein Structure.* New York: Springer-Verlag. 314 pp.
3. Richards, F. M. 1977. *Ann. Rev. Biophys. Bioeng.* 6:151–76
4. Janin, J. 1979. *Bull. Inst. Pasteur* 77:337–73
5. Rossmann, M. G., Argos, P. 1981. *Ann. Rev. Biochem.* 50:497–532
6. Richardson, J. S. 1981. *Adv. Prot. Chem.* 34:167–339
7. Pauling, L., Corey, R. B., Branson, H. R. 1951. *Proc. Natl. Acad. Sci. USA* 37:205–11
8. Pauling, L., Corey, R. B. 1951. *Proc. Natl. Acad. Sci. USA* 37:729–40
9. Ramachandran, G. N., Sasisekharan, V. 1968. *Adv. Prot. Chem.* 23:283–437
10. Moews, P. C., Kretsinger, R. H. 1975. *J. Mol. Biol.* 91:201–28
11. Sawyer, L., Shotton, D. M., Campbell, J. W., Wendell, P. L., Muirhead, H., et al. 1978. *J. Mol. Biol.* 118:137–208
12. Baker, E. N. 1980. *J. Mol. Biol.* 141:441–84
13. Matsuura, Y., Takano, T., Dickerson, R. E. 1982. *J. Mol. Biol.* 156:389–409
14. Remington, S. J., Wiegand, G., Huber, R. 1982. *J. Mol. Biol.* 158:111–52
15. Holmes, M. A., Matthews, B. W. 1982. *J. Mol. Biol.* 160:623–39
16. Steigemann, W., Weber, E. 1979. *J. Mol. Biol.* 122:309–38
17. James, M. N. G., Sielecki, A. R. 1983. *J. Mol. Biol.* 163:299–361
18. Watson, H. C. 1969. *Prog. Sterochem.* 4:229–333
19. Artymiuk, P. J., Blake, C. C. F. 1981. *J. Mol. Biol.* 152:737–62
20. Deisenhofer, J., Steigemann, W. 1975. *Acta Crystallogr. Sect. B* 31:238–50
21. Takano, T., Dickerson, R. E. 1981. *J. Mol. Biol.* 153:79–94
22. Glover, I., Haneef, I., Pitts, J., Wood, S., Moss, D., et al. 1983. *Biopolymers* 22:293–304
23. Sakabe, N., Sakabe, K., Sasaki, K. 1981. *Structural Studies on Molecules of Biological Interest,* ed. G. Dodson, J. P. Glusker, D. Sayre, pp. 509–26. Oxford: Clarendon. 610 pp.
24. Bolin, J. T., Filman, D. J., Matthews, D. A., Hamlin, R. C., Kraut, J. 1982. *J. Biol. Chem.* 257:13650–72
25. Deleted in press
26. Fermi, G., Perutz, M. F. 1981. *Atlas of Molecular Structures in Biology. Vol. 2. Haemoglobin and Myoglobin,* ed. D. C. Phillips, F. R. Richards. Oxford: Clarendon. 104 pp.
27. Matthews, B. W., Weaver, L. H., Kester, W. R. 1974. *J. Biol. Chem.* 249:8030–44
28. Schellman, C. 1980. See Ref. 167, pp. 53–61
28a. Ravichandran, V., Subramanian, E. 1981. *Int. J. Peptide Prot. Res.* 18:121–26
28b. Marquart, M., Deisenhofer, J., Huber, R. 1980. *J. Mol. Biol.* 141:369–91
29. Venkatachalam, C. M. 1968. *Biopolymers* 6:1425–36
30. Lewis, P. N., Momany, F. A., Scheraga, H. A. 1973. *Biochim. Biophys. Acta* 303:211–39
31. Nemethy, G., Printz, M. P. 1972. *Macromolecules* 5:755–58
32. Crawford, J. L., Lipscomb, W. N., Schellman, C. G. 1973. *Proc. Natl. Acad. Sci. USA* 70:538–42
33. Huber, R., Steigemann, W. 1974. *FEBS Lett.* 48:235–37
34. Chou, P. Y., Fasman, G. D. 1977. *J. Mol. Biol.* 115:135–75

35. Smith, J. A., Pease, L. G. 1980. *CRC Crit. Rev. Biochem.* 8:315–99
36. Kuntz, I. D. 1972. *J. Am. Chem. Soc.* 94:4009–12
37. Levitt, M., Chothia, C. 1976. *Nature* 261:552–58
38. Rose, G. D., Young, W. B., Gierasch, L. M. 1983. *Nature* 304:655–57
39. Chothia, C. 1973. *J. Mol. Biol.* 75:295–302
40. Scott, R. A., Scheraga, H. A. 1966. *J. Chem. Phys.* 45:2091–2101
41. Chou, K. C., Pottle, M., Nemethy, G., Ueda, Y., Scheraga, H. A. 1982. *J. Mol. Biol.* 162:89–112
42. Chou, K. C., Scheraga, H. A. 1982. *Proc. Natl. Acad. Sci. USA* 79:7047–51
43. Chou, K. C., Nemethy, G., Scheraga, H. A. 1983. *J. Mol. Biol.* 168:389–407
44. Janin, J., Chothia, C. 1980. *J. Mol. Biol.* 143:95–128
45. Chothia, C., Janin, J. 1981. *Proc. Natl. Acad. Sci. USA* 78:4146–50
46. Chothia, C., Janin, J. 1982. *Biochemistry* 21:3955–65
47. Nishikawa, K., Scheraga, H. A. 1976. *Macromolecules* 9:365–407
48. Salemme, F. R., Weatherford, D. W. 1981. *J. Mol. Biol.* 146:101–17
49. Salemme, F. R., Weatherford, D. W. 1981. *J. Mol. Biol.* 146:119–41
50. Salemme, F. R. 1981. *J. Mol. Biol.* 146:143–56
51. Chothia, C. 1983. *J. Mol Biol.* 163:107–17
52. Richardson, J. S., Getzoff, E. D., Richardson, D. C. 1978. *Proc. Natl. Acad. Sci. USA* 75:2574–78
53. Chandrasekaran, R., Ramachandran, G. N. 1970. *Int. J. Prot. Res.* 2:223–33
54. Ponnuswamy, P. K., Sasisekharan, V. 1971. *Biopolymers* 10:565–82
55. Sasisekharan, V., Ponnuswamy, P. K. 1971. *Biopolymers* 10:583–92
56. Nemethy, G., Scheraga, H. A. 1977. *Quart. Rev. Biophys.* 10:239–352
57. Finkelstein, A. V. 1976. *Mol. Biol.* 10:879–86
58. Finkelstein, A. V., Ptitsyn, O. B. 1977. *Biopolymers* 16:469–95
59. Janin, J., Wodak, S., Levitt, M., Maigret, B. 1978. *J. Mol. Biol.* 125:357–86
60. Gelin, B. R., Karplus, M. 1979. *Biochemistry* 18:1256–68
61. Bhat, T. N., Sasisekharan, V., Vijayan, M. 1979. *Int. J. Peptide Prot. Res.* 13:170–84
62. Thornton, J. M. 1981. *J. Mol. Biol.* 151:261–87
63. Kauzmann, W. 1959. *Adv. Prot. Chem.* 14:1–64
64. Liquori, A. M. 1966. *Principles of Biomolecular Organization,* ed. G. E. W. Wolstenholme, M. O'Connor, pp. 40–68. London: Churchill. 491 pp.
65. Klapper, M. H. 1971. *Biochem. Biophys. Acta* 229:557–66
66. Richards, F. M. 1974. *J. Mol. Biol.* 82:1–14
67. Finney, J. L. 1975. *J. Mol Biol.* 96:721–32
68. Finney, J. L. 1978. *J. Mol. Biol.* 119:415–41
69. Gellatly, B. J., Finney, J. L. 1982. *J. Mol. Biol.* 161:305–22
70. Chothia, C. 1975. *Nature* 254:304–8
71. Kuntz, I. D., Crippen, G. M. 1979. *Int. J. Peptide Prot. Res.* 13:223–28
72. Crippen, G. M., Kuntz, I. D. 1978. *Int. J. Peptide Prot. Res.* 12:47–56
73. Lee, B., Richards, F. M. 1971. *J. Mol. Biol.* 55:379–400
74. Richards, F. M. 1979. *Carlsberg Res. Commun.* 44:47–63
75. Lumry, R., Rosenberg, A. 1975. *Collect. Int. CNRS L'eau Syst. Biol.* 246:55–63
76. McCammon, J. A., Lee, C. Y., Northrup, S. H. 1983. *J. Am. Chem. Soc.* 105:2232–37
77. Liquori, A. M., Sadun, C. 1981. *Int. J. Biol. Macromol.* 3:56–59
78. Chothia, C., Levitt, M., Richardson, D. 1977. *Proc. Natl. Acad. Sci. USA* 74:4130–34
79. Sternberg, M. J. E. 1983. *Computing in Biological Science,* ed. M. Gerson, A. Barrett, pp. 143–77. Amsterdam: Elsevier/North-Holland
80. Crick, F. H. C. 1953. *Acta Crystallogr.* 6:689–97
81. Chothia, C., Levitt, M., Richardson, D. 1981. *J. Mol. Biol.* 145:215–50
82. Efimov, A. V. 1977. *Dokl. Acad. Nauk SSSR* 235:699–702
83. Dunker, A. K., Zaleske, D. J. 1977. *Biochem. J.* 163:45–57
84. Richmond, T., Richards, F. M. 1978. *J. Mol. Biol.* 119:537–55
85. Efimov, A. V. 1979. *J. Mol. Biol.* 134:23–40
86. Chothia, C., Lesk, A. M., Dodson, G. G., Hodgkin, D. C. 1983. *Nature* 302:500–5
87. Cohen, F. E., Sternberg, M. J. E., Taylor, W. R. 1982. *J. Mol. Biol.* 156:821–62
88. Cohen, F. E., Sternberg, M. J. E., Taylor, W. R. 1981. *J. Mol. Biol.* 148:253–72
89. Sternberg, M. J. E., Thornton, J. M. 1977. *J. Mol. Biol.* 115:1–17
90. von Heijne, G., Blomberg, C. 1978. *Biopolymers* 17:2033–37
91. Toniolo, C. 1978. *Macromolecules* 11:437–38

92. Lifson, S., Sander, C. 1979. *Nature* 282:109–11
93. Lifson, S., Sander, C. 1980. *J. Mol. Biol.* 139:627–39
94. Wetlaufer, D. E. 1974. *Proc. Natl. Acad. Sci. USA* 70:697–701
95. Schulz, G. E., Schirmer, R. H. 1974. *Nature* 250:142–44
96. Rao, S. T., Rossmann, M. G. 1973. *J. Mol. Biol.* 76:241–56
97. Richardson, J. S., Richardson, D. C. Thomas, K. A., Silverton, E. W., Davies, D. R. 1976. *J. Mol. Biol.* 102:221–35
98. Richardson, J. S. 1976. *Proc. Natl. Acad. Sci. USA* 73:2619–23
99. Sternberg, M. J. E., Thornton, J. M. 1976. *J. Mol. Biol.* 105:367–82
100. Sternberg, M. J. E., Thornton, J. M. 1977. *J. Mol. Biol.* 110:269–83
101. Nagano, N. 1977. *J. Mol. Biol.* 109:235–50
102. Richardson, J. S. 1977. *Nature* 268:495–500
103. Finkelstein, A. V., Ptitsyn, O. B., Bendzko, P. 1979. *Biofizika* 24:21–26
104. Ptitsyn, O. B., Finkelstein, A. V. 1979. *Biofizika* 24:27–31
105. Connolly, M. L., Kuntz, I. D., Crippen, G. M. 1980. *Biopolymers* 19:1167–82
105a. Klapper, M. H., Klapper, I. Z. 1980. *Biochim. Biophys. Acta* 626:97–105
105b. Creighton, T. E. 1974. *J. Mol. Biol.* 87:603–24
106. Ptitsyn, O. B., Finkelstein, A. V. 1980. *Quart. Rev. Biophys.* 13:339–86
107. Efimov, A. V. 1982. *Mol. Biol.* 16:799–806
108. Baldwin, R. L. 1980. See Ref. 167, pp. 369–85
109. Weber, P. C., Salemme, F. R. 1980. *Nature* 287:82–84
110. Bello, J. 1977. *J. Theor. Biol.* 68:139–42
111. Bello, J. 1978. *Int. J. Peptide Prot. Res.* 12:38–41
112. Schellman, J. A. 1955. *C.R. Trav. Lab. Carlsberg Ser. Chim.* 29:223–29
113. Klotz, I. M., Franzen, J. S. 1962. *J. Am. Chem. Soc.* 84:3461–66
114. Susi, H., Timasheff, S. N., Ard, J. S. 1964. *J. Biol Chem.* 239:3051–54
115. Klotz, I. M., Farnham, S. B. 1968. *Biochemistry* 7:3879–82
116. Kresheck, G. C., Klotz, I. M. 1969. *Biochemistry* 8:8–12
117. Creighton, T. E. 1983. *Biopolymers* 22:49–58
118. Shrake, A., Rupley, J. A. 1973. *J. Mol. Biol.* 79:351–71
119. Greer, J., Bush, B. L. 1978. *Proc. Natl. Acad. Sci. USA* 75:303–7
120. Wodak, S. J., Janin, J. 1980. *Proc. Natl. Acad. Sci. USA* 77:1736–40
121. Chothia, C. 1974. *Nature* 248:338–39
122. Gelles, J., Klapper, M. H. 1978. *Biochim. Biophys. Acta* 533:465–77
123. Miller, K. W., Hildebrand, J. H. 1968. *J. Am. Chem. Soc.* 90:3001–4
124. Sinanoglu, O. 1968. *Molecular Associations in Biology*, ed. B. Pullman, pp. 427–45. New York: Academic
125. Hermann, R. B. 1972. *J. Phys. Chem.* 76:2754–59
126. Harris, M. J., Higuchi, T., Rutting, J. H. 1973. *J. Phys. Chem.* 77:2694–2703
127. Reynolds, J. A., Gilbert, D. B., Tanford, C. 1974. *Proc. Natl. Acad. Sci. USA* 71:2925–27
128. Hermann, R. B. 1977. *Proc. Natl. Acad. Sci. USA* 74:4144–45
129. Nandi, P. 1976. *Int. J. Peptide Prot. Res.* 8:253–64
130. Creighton, T. E. 1979. *J. Mol. Biol.* 129:235–64
131. Tanford, C. 1979. *Proc. Natl. Acad. Sci. USA* 76:4175–76
132. Karplus, M. 1980. *Biophys. J.* 32:45–46
133. Chothia, C. 1976. *J. Mol. Biol.* 105:1–14
134. Manavalan, P., Ponnuswamy, P. K. 1977. *Biochem. J.* 167:171–82
135. Janin, J. 1979. *Nature* 277:491–92
136. Wolfenden, R., Anderson, L., Cullis, P. M., Southgate, C. C. B. 1981. *Biochemistry* 20:849–55
137. Rackovsky, S., Scheraga, H. A. 1977. *Proc. Natl. Acad. Sci. USA* 74:5248–51
138. Wertz, D. H., Scheraga, H. A. 1978. *Macromolecules* 11:9–15
139. Meirovitch, H., Rackovsky, S., Scheraga, H. A. 1980. *Macromolecules* 13:1398–1405
140. Meirovitch, H., Scheraga, H. A. 1980. *Macromolecules* 13:1406–14
141. Manavalan, P., Ponnuswamy, P. K. 1977. *Arch. Biochem. Biophys.* 184:476–87
142. Manavalan, P., Ponnuswamy, P. K. 1978. *Nature* 275:673–74
143. Ponnuswamy, P. K., Prabhakaran, M., Manavalan, P. 1980. *Biochim. Biophys. Acta* 623:301–16
143a. Kyte, J., Doolittle, R. F. 1982. *J. Mol. Biol.* 157:105–32
144. Black, J. A., Harkins, R. N., Stenzel, P. 1976. *Int. J. Peptide Prot. Res.* 8:125–30
144a. Klapper, M. H. 1977. *Biochem. Biophys. Res. Commun.* 78:1018–24
145. Warme, P. K., Morgan, R. S. 1978. *J. Mol. Biol.* 118:273–87
146. Warme, P. K., Morgan, R. S. 1978. *J. Mol. Biol.* 118:289–304
147. Richards, F. M., Richmond, T. 1978. *Molecular Interactions and Activity in Proteins*, pp. 23–45. Amsterdam: Excerpta Medica. 279 pp.

148. Chou, P. Y., Fasman, G. D. 1978. *Adv. Enzymol.* 47:45–148
148a. Levitt, M. 1978. *Biochemistry* 17: 4277–85
148b. Kabsch, W., Sander, C. 1983. *FEBS Lett.* 151:179–82
148c. Ptitsyn, O. B., Finkelstein, A. V. 1983. *Biopolymers* 22:15–25
148d. Taylor, W. R., Thornton, J. M. 1983. *Nature* 301:540–42
149. Lesk, A. M., Chothia, C. 1980. *Biophys. J.* 32:35–47
149a. Eisenberg, D., Weiss, R. M., Terwilliger, T. C. 1982. *Nature* 299:371–74
150. Janin, J. 1976. *J. Mol. Biol.* 105:13–14
151. Teller, D. C. 1976. *Nature* 260:729–31
152. Gates, R. E. 1979. *J. Mol. Biol.* 127:345–51
153. Fisher, H. F. 1964. *Proc. Natl. Acad. Sci. USA* 51:1285–91
154. Sprang, S., Yang, D., Fletterick, R. J. 1979. *Nature* 280:333–35
155. Janin, J., Chothia, C. 1979. *FEBS: 12th Meet. Dresden 1978*, ed. E. Hofmann, W. Pfeil, H. Aurich, 52:227–37. Oxford: Pergamon. 522 pp.
156. Chothia, C., Janin, J. 1975. *Nature* 256:705–8
157. Janin, J., Chothia, C. 1976. *J. Mol. Biol.* 100:197–221
158. Chothia, C., Wodak, S., Janin, J. 1976.

Proc. Natl. Acad. Sci. USA 73:3793–97
159. Jaenicke, R., Rudolph, R. 1980. See Ref. 167, pp. 525–46
160. Bränden, C. I. 1980. *Quart. Rev. Biophys.* 13:317–38
161. Janin, J. Wodak, S. J. 1983. *Prog. Biophys. Mol. Biol.* 42:21–78
162. Goldberg, M. E. 1969. *J. Mol. Biol.* 46:441–46
163. Wetlaufer, D. B. 1981. *Adv. Prot. Chem.* 34:61–92
164. McLachlan, A. D. 1979. *J. Mol. Biol.* 128:49–79
165. McLachlan, A. D. 1980. See Ref. 167, pp. 79–99
166. Rossmann, M. G., Liljas, A., Bränden, C. I., Banaszak, L. J. 1975. *Enzymes* 11:61–102
167. Jaenicke, R., ed. 1980. *Protein Folding.* Amsterdam: Elsevier/North-Holland. 587 pp.
168. Hildebrand, J. H. 1979. *Proc. Natl. Acad. Sci. USA* 76:1
169. Nozaki, Y., Tanford, C. 1971. *J. Biol. Chem.* 246:2211–17
170. Dayhoff, M. D. 1976. *Atlas of Protein Sequence and Structure*, Vol. 5, Suppl. 2. Washington DC: Natl. Biomed. Res. Found.
171. Kendrew, J. C. 1961. *Sci. Am.* 205(6): 96–110

Ann. Rev. Biochem. 1982. 51:123–54

3

THE CONFORMATION, FLEXIBILITY, AND DYNAMICS OF POLYPEPTIDE HORMONES

Tom Blundell and Stephen Wood

Laboratory of Molecular Biology, Department of Crystallography, Birkbeck College, University of London

CONTENTS

PERSPECTIVES AND SUMMARY

Polypeptide hormones are manufactured and often stored in specialized endocrine cells and are circulated to specific target tissues. In many cases the hormone message is known to be conveyed to the target cell via a cell surface receptor for the hormone, and rapid degradation, often mediated by

75

receptor binding and internalization, seems to be a common feature. We clearly need to define the conformation of polypeptide hormones during this complex life cycle if we are to understand the various processes at the molecular level.

Analysis of conformation, flexibility, and dynamics presents a range of challenging problems both experimentally and conceptually, as Schwyzer was the first to realise (1). Small polypeptide hormones of less than 30 amino acids will generally be flexible, and their conformations will depend on their concentration, on the solvent, and on other molecules in solution (1, 2). For instance, glucagon has little defined structure in dilute aqueous solutions (3), but secondary structure is stabilized by self-association at high concentrations (2, 4) and by the presence of nonaqueous solvents and lipid micelles (5). The conformations of some small hormones such as oxytocin may be limited by disulfide bridges and consequently they may have better defined mainchain conformations, but even these are flexible and comprise several conformers in equilibrium in aqueous solutions (6). Many larger polypeptide hormones such as insulin, and probably glycoprotein hormones and growth hormones, have "globular" structures with hydrophobic cores (7). Their preferred conformations are probably retained in aqueous solutions, in oligomeric forms, and in the crystalline state, although even insulin shows large changes under extreme conditions such as high salt concentrations, and the conformation and flexibility in dilute solutions has been the subject of some debate (8, 9). The smallest polypeptide hormone with a hydrophobic core appears to be pancreatic polypeptide, with 36 amino acids (10), but many larger polypeptides, such as β lipotropin, have more flexible structures. Clearly, stability of the tertiary structure of the globular polypeptide hormones makes experimental data in vitro more easily relevant to events in vivo.

The conformational analysis of a polypeptide hormone cannot be considered complete until a detailed structure has been defined in crystals and in solutions of different properties corresponding to the environment in storage granules, in circulation, and in the receptor complex. It would be most advantageous to study at high resolution the granules themselves and the isolated receptor complex, but these both present great experimental difficulties. Perhaps the best that has been done so far is with glucagon: a low resolution electron microscopal and diffraction study has been made of granules (11); X-ray analysis has defined the conformation in the crystalline state (2, 12); NMR has been used to detail the conformation and dynamics of monomers in dilute aqueous solutions (3) and in hormone micelle complexes (5), and of trimers in more concentrated aqueous solutions (4); and a range of other spectroscopic probes such as circular dichroism (13) and tryptophan fluorescence (14) have given supplementary information on the conformation in aqueous and nonaqueous solutions and in the presence of lipid micelles.

X-ray crystallography can provide the most detailed time-averaged conformational description of a polypeptide hormone in the crystalline state, given that crystals are obtained. For the larger hormones, like insulin, where the crystals contain considerable amounts of solvent, the method of multiple isomorphous replacement has been quite successful. Some of the smaller hormones, like oxytocin, produce crystals containing little solvent and provide problems at the limit of the more traditional methods of structure analysis of small molecules. Recently, the refinement of protein structures at high resolution and the definition of local disorder and molecular flexibility have been receiving increased attention from protein crystallographers. Pancreatic polypeptide (PP) has now been refined at 0.98 Å resolution, which allows a detailed analysis of the thermal vibrations. These indicate a flexible C terminus and a more ordered globular region (10). Glucagon on the other hand is relatively disordered even in the crystals with a limiting resolution of $\sim$ 3.0 Å (12).

Spectroscopic methods provide important complementary information on the solution conformation of polypeptides, particularly with respect to the dynamics. Development of high resolution NMR techniques have allowed an extensive assignment of resonances even in molecules as complex as glucagon (3, 4). Chemical shifts and nuclear Overhauser effects have been analyzed to define proximity of side groups, and distance geometry algorithms have been used to generate conformations consistent with these data (3). The recent analysis of glucagon bound to deuterated lipid micelles is very encouraging (5) and may represent a method that can be extended to hormone receptor complexes.

For the larger hormones, circular dichroism provides an overall impression of secondary structure composition and flexibility. Calculation of optical activity from crystal atomic coordinates may permit the extraction of more detailed information from circular dichroism measurements (8). Spectroscopic methods provide particularly important data to supplement prediction of structures from amino acid sequence. Predictions based on sequence information only give indications of secondary structure, but more detailed predictions have been achieved by model building members of protein families where detailed structural information is available for at least one member. This type of approach has proved particularly useful for insulin-related growth factors where scarcity of materials has limited other experimental investigations (16).

Much of our knowledge of the tertiary structures of polypeptide hormones, especially of the larger molecules such as the growth hormones and glycoproteins, derives from studies of accessibility to chemical probes and cross-linking agents. Unfortunately, space limitation permits these approaches to be discussed only in relation to the results of spectroscopic and diffraction techniques, which provide the central theme of this review.

Although hormones are defined in terms of their endocrine origins and their action via the circulation, it is now evident that there are no distinct frontiers between them and locally acting regulatory factors, the so-called paracrine agents, neurotransmitters, and growth factors (17). Indeed, insulin may act as a growth factor by binding to somatomedin C (insulin-like growth factor) receptors; vasoactive intestinal polypeptide (VIP) is primarily a neurotransmitter in the central nervous system rather than a gut hormone; and somatostatin, while acting as a local inhibitory agent both in the pancreas and in the pituitary, also has action on distant tissues through the circulation. For these reasons we consider some polypeptide paracrine agents, growth factors, and neurotransmitters in our discussion. We avoid too much emphasis on classification due to origin—pancreatic hormones, pituitary hormones, etc—but try to discuss the molecules in terms of families related by sequence homologies and conformational preferences (17).

INSULIN-LIKE POLYPEPTIDES

Insulin Structures in Crystals

The conformation of insulin has been studied using a wide selection of techniques, the most penetrating of which have been X-ray analyses of the crystal structures of different oligomeric forms. The first successful analyses were carried out on rhombohedral porcine 2Zn insulin crystals in which insulin hexamers are assembled from three equivalent dimers (18, 19). Each dimer is coordinated to two zinc ions lying 16 Å apart on a threefold axis, and each zinc is coordinated by three equivalent B10 histidines. There are two crystallographically independent molecules (I and II) within each dimer. In the medium resolution analysis several parts of the mainchain and many sidechains were poorly defined, but these difficulties have been largely overcome by high resolution refinement at 1.5 Å resolution by Dodson et al (20, 21), at 1.1 Å resolution by Sakabe et al (23–26), and at 1.8 Å resolution by Liang et al (19). Errors in atomic positions in these structures range between 0.03 Å for many well-defined, usually buried mainchain atoms and 1.0Å for some surface sidechains and solvent molecules. The structures of bovine (S. P. Wood and S. Bedarkar, unpublished results) and human (semisynthesized) 2Zn insulin have also been studied by medium resolution X-ray analysis, and crystals of recombinant DNA synthesized human insulin are under study (A. Cleasby, J. E. Pitts, S. P. Wood, and T. L. Blundell, unpublished results). In the presence of high chloride ion concentrations insulin crystallizes in a rhombohedral form containing four zinc ions bound to the insulin hexamer in which the differences between molecules I and II are greater than in 2Zn insulin (21, 27, 28). The structure of this form was determined initially at 2.8 Å resolution using isomorphous replacement (27) and refined at 1.5 Å resolution (21, 28). A cubic crystal form containing symmetrical insulin dimers and no zinc ions can be prepared from slightly alkaline solutions, and this structure has been determined using the methods of molecular replacement (29). Insulin from the hagfish does not bind zinc ions and produces tetragonal crystals of dimers containing a twofold axis. This structure was determined initially by isomorphous replacement to 3.1 Å resolution, and the refinement to 1.9 Å is in progress (21, 30).

A comparison of the conformations of the protomers (see Figure 1) in various crystallographic forms reveals much about the flexibility of the insulin molecule that may be relevant to its structure in solution and its biological properties (21). Each protomeric structure is a fairly compact globular structure with two essentially nonpolar surfaces that are involved in intermolecular contacts leading to the formation of dimers and hexamers. With the exception of molecule I of the 4Zn hexamers, each protomer involves right-handed helical segments A2–A8, A13–A19, and B9–B19, extended regions B1–B7 and B24–B30, and turns hinging on glycines at B8, B20, and B23. The structures of molecule II of both the 2Zn and the 4Zn hexamers and the hagfish insulin resemble each other closely (21), and this structure may therefore represent a stable conformer.

The conformational differences between molecules I and II of the 2Zn form appear to be concerted, and presumably derive from intermolecular interactions in the crystals, for the hagfish dimeric structure shows that they could be identical with perfect twofold symmetry. The different crystallographic environments of A8, A10, and B5 lead to: differences in the helices A2–A7; a rotation around the A5–A6 bond of 40°; a change of the A7–B7 conformation; and movement of the B-chain C-terminal residues, of which B25 Phe adopts a different conformation in molecule I (21, 28).

In 4Zn insulin the asymmetry between the two protomers of the dimer is greater, particularly toward the B-chain N terminus of molecule I, where the B-chain α helix starts at residue 2 rather than at residue 9 as in molecule II (21, 28). This allows B5 and B10 of adjacent dimers each to bind one zinc ion, thus generating three sites around the threefold axis in addition to the one zinc site on the threefold axis coordinated by three histidines at B10 of molecule II. In fact there is disorder in this part of the molecule, with a percentage of the histidines at B10 coordinating a partially occupied zinc ion on the threefold axis. This is consistent with the observation of transitions between the 2Zn and 4Zn insulin forms in the crystalline state (31).

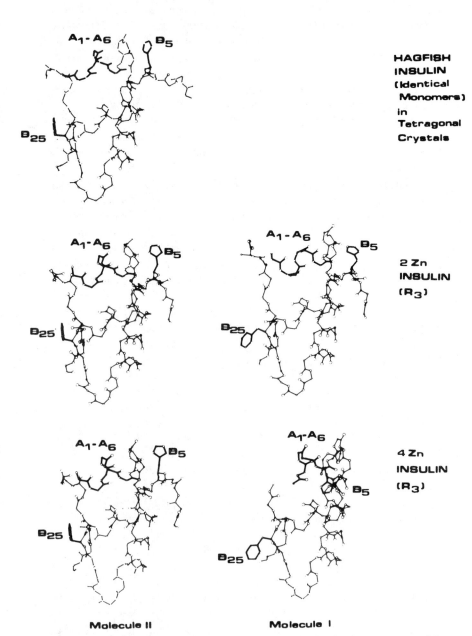

Figure 1   Comparison of five independent insulin molecules showing the arrangement of the N terminus of the A chain (A1–A6), B5 histidine, and B25 phenylalanine. Reproduced by permission of Dr. G. Dodson from (21).

The hydrophobic core of the insulin molecule in all five conformers (Figure 1) receives a rather similar contribution from B-chain sidechains; but A-chain contributions are more varied. With the exception of B25 phenylalanine, the dimer-forming residues in all the insulins are rather similar in spite of the conformational changes occurring elsewhere, and this highlights the strength and specificity of this interaction. Certain regions of the insulin structure are disordered even in the crystals. For instance, B22 Arg, a residue apparently important to structure and activity, is now seen to occupy two positions, forming a salt bridge with either A21 carboxylate as proposed earlier, or with A17 glutamate sidechain carboxylate.

The comparison of the five conformers (2Zn I and II, 4Zn I and II, and hagfish) refined structures has highlighted areas of flexibility in the insulin molecule. Although it is possible that molecule II of 2Zn insulin has a unique stability, the capacity of the insulin molecule to change its conformation may be relevant to the generation of biological activity at the receptor.

In the very high resolution (1.1 Å) refinement of the 2Zn insulin structure at 4°C by Sakabe et al (23–26), some hydrogen atoms bound to atoms with high thermal parameters could not be seen in the electron density difference maps, but many were defined by positive regions, as shown in Figure 2 for B10 His. Detailed analysis of the hydrogen bonding network of insulin has shown that in the B-chain helix the bond lengths vary between 2.8 and 3.2

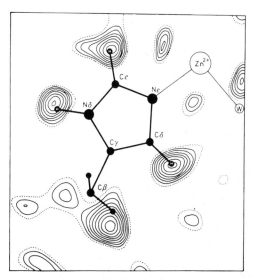

Figure 2   A difference electron density map of the imidazole ring plane on B10 histidine in molecule 1 of 2Zn insulin hexamers. The phases were calculated without contributions of the hydrogen atoms on the basis of the structure refined at 1.1 Å resolution. Reproduced by permission of Drs. N. and K. Sakabe from (25).

Å, and the rms difference between molecules of the dimer is 0.04 Å in bond length and less than 2° in the defining angles. Consistently shorter bond lengths for buried as opposed to solvent exposed hydrogen bonds indicate that the B-chain helix tends to curve against the interior of the molecule. In the β sheet of the dimer, the two interior of the four hydrogen bonds are shorter and straighter, which indicates some cooperativity in the hydrogen bonding. About 150 water molecules lie within 3.5 Å of protein atoms and make up the first hydration shell; 73% of these are hydrogen bonded to polar insulin atoms.

Insulins with reduced receptor affinities due to addition of thiazolidine (32, 33), t-butoxycarbonyl (32, 33), and glutamyl [(34); S. Bedarkar, unpublished] groups at A1 Gly have been crystallized in the 2Zn crystal form and studied by difference Fourier and refinement techniques. The addition of these groups appears to cause rotation and some disordering of the residues A2–A5, and very small movements in the adjacent B-chain residues. Larger changes appear in insulin crosslinked between A1 and B29 with diaminosuberimidate (36). Several insulins with shortened B-chain C termini have been crystallized (37–39), and crystal forms of despentapeptide insulin have been solved by molecular replacement (39), which indicates a similar general tertiary fold for this molecule, together with some movement of the B-chain N-terminal residues. This is the first monomeric insulin to be studied by X-ray techniques.

Dimeric and 2Zn hexameric forms of proinsulin have been crystallized (40). Preliminary molecular replacement studies on large crystals of the dimeric form suitable for high resolution X-ray analysis indicate a twofold axis (T. L. Blundell, G. Godley, J. E. Pitts, S. P. Wood, I. J. Tickle and G. L. Taylor, unpublished results).

Insulin Structure in Solution

Circular dichroism spectroscopy has been used widely in the study of insulin conformation in solution. However, many early interpretations have been confused by the effects of self-association in both the near UV (due to restriction of rotation of aromatic sidechains) and the far UV (where changes in β sheet and perhaps helix affect the spectra). Clearly insulins at equivalent states of association must be compared (42). Helix estimations (43) are probably made with most confidence, and for insulin these agree fairly well with the crystal structure. Strickland & Mercola (44) and more recently, Wollmer and co-workers (8, 45) have used the crystal atomic coordinates to calculate the tyrosyl circular dichroism of insulin resulting from electric dipole-electric dipole coupling employing either a distributed monopole or a dipole approximation to obtain the interaction energies. The calculations generated the correct sign for the 275 nm band observed experimentally and provided a fairly good estimate of the magnitude in different states of association. Experimentally this band is enhanced about fourfold in the hexamer compared to the monomer.

However, there are still considerable technical barriers in the study of the circular dichroism of insulin at the high dilutions characteristic of monomer formation. Recently Pocker & Biswas (9) have reported circular dichroism spectra for insulin at concentrations of 1 μM in the near UV and down to 60 nM in the far UV using an instrument in which a high frequency elasto-optic modulator is used in place of the pockel cells used in older instruments and in which the dynode voltage is artificially suppressed during recordings at extreme dilutions. On dilution to an insulin concentration of 60 nM the far UV spectrum corresponds to 28% helix content compared with 45% estimated from the crystal structure. Assuming that the concentrations are not confused by absorption of insulin on to surfaces and that the measurement techniques are valid, this may represent further evidence for flexibility in the insulin structure.

A similar change in the circular dichroism is observed in guinea pig and porcupine insulins, which form only monomers (42, 46, 47). However, this is not unexpected, as B22 Arg is replaced by a glutamic acid group in these insulins (see below). In casiragua insulin (48), which is also monomeric, the far UV circular dichroism is consistent with a structure containing a percentage of helix closer to that of the crystal structure.

NMR spectroscopy has not made major contributions to our understanding of insulin conformation in solution, probably as a result of the complexity of the dimeric and hexameric species and their low solubility at physiological pH. However, Bradbury & Brown have described NMR studies of the amino groups (49), and Williamson & Williams (50) have reported a 270 MHz proton NMR study of insulin oligomers and zinc binding. Having assigned key reference resonances of methyl groups of isoleucines, valines, and leucines and aromatic protons they were able to follow conformational changes on the assembly of 2Zn insulin hexamers and 4Zn insulin hexamers and even predict a 6Zn hexamer that has yet to be identified crystallographically. ^{13}C NMR measurement of insulin carbamylated with ^{13}C enriched potassium cyanate (51) showed two closely spaced carbamyl glycine lines in conditions where dimers were the dominant species in solution. This finding has been held to reflect the imperfect twofold axis seen in crystals, but may be due to different conformations of monomeric and dimeric forms coexisting under the conditions of the experiment.

Hystricomorph Insulins

Many insulin and insulin-like polypeptides have not been available in sufficient quantities for detailed crystal or solution studies of their conformations. However, the amino acid sequences show a conservation of features necessary to achieve an insulin-like fold in spite of often very low total homology (17), and recent developments in interactive computer graphics allow detailed prediction of their conformations.

The sequences of hystricomorph insulins show unusual substitutions that are of interest, as they lead to reduction of potency (42, 46–48). Models of guinea pig and porcupine (M. Bajaj, R. Horuk, S. P. Wood and T. L. Blundell, unpublished) and casiragua (53) and coypu (53) insulins show that porcine insulin-like dimers and zinc hexamers cannot form. Furthermore, substitution of B22 Arg by glutamate in porcupine and guinea pig insulins would introduce a negative charge in close proximity to the A21 carboxylate, the repulsion between which may lead to the difference of conformation indicated by circular dichroism. Consistent with this idea, the circular dichroism spectra of guinea pig and porcupine insulins become more porcine-insulin-like at low pH when the charge is neutralized (46).

Insulin-Like Growth Factors and Relaxins

Insulin-like growth factors (I and II) have also been modeled (see Figure 3) on the basis of their homology with insulin, which includes identical disulfide dispositions, hydrophobic cores, and glycines equivalent to B8, B20, and B23 (17, 54; S. Bedarkar, T. L. Blundell and R. E. Humbel, unpublished results). The shorter C peptides (8 and 12 residues) have high β-turn potential and are easily accommodated, connecting residues equivalent to A1 and B30 of insulin without affecting the general features of the fold. The A-chain extensions of 8 (I) and 6 (II) residues must also lie on the surface, and here again, there is probably a β turn. Surface residues give rise to extensive charge networks. The lack of B10 His, the existence of B17 Phe, and the charge at B4 Glu make it unlikely that insulin-like growth factors form zinc insulin hexamers. However, the unusual hydrophobic surface patch, especially that involving B14 Ala, B17 Phe, B18 Val, and A13 Leu of insulin-like growth factor I, would be available for interaction with a binding or carrier protein. The models show that the antigenic regions of insulin, which usually include B1–B4 and A8–A10, are quite different in the insulin-like growth factors, which explains their nonsuppressibility with anti-insulin antibodies.

Although the primary structure of pig ovary relaxin indicates a two-chain, disulfide bonded protein with little homology to insulin, the main-chain A1–20 and B6–21 can adopt a conformation identical to that of insulin, so that intrachain and interchain disulfides occupy equivalent spatial positions to those of insulin (56, 57). Residues B6 Leu and B14 Ala, which are close together in insulin, have their respective positions reversed to B6 Ala and B14 Leu in porcine relaxin. A similar effect occurs with A2

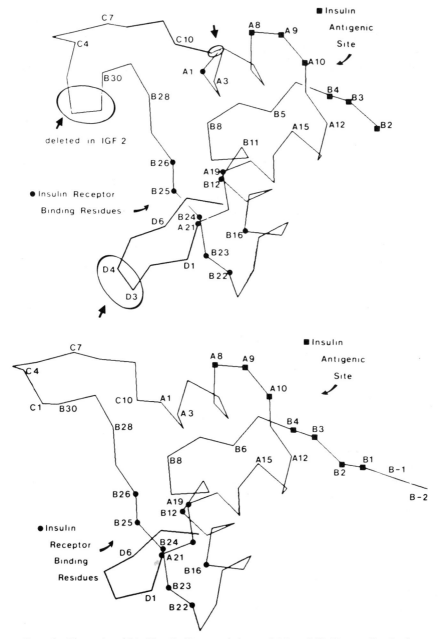

Figure 3 The tertiary fold of insulin-like growth factors (*a*) I, and (*b*) II as predicted using interactive computer graphics model building and the atomic coordinates of porcine insulin (17, 54). The α carbons only are shown. The numbering is as porcine insulin for the A- and B-chain regions. The connecting peptide residues are numbered with a prefix C and the residues at the extension of the A chain with D.

Leu and A16 Ile of relaxin. These pairs of sidechains point toward the core from opposite sides, thus maintaining the core volume. B15 Trp is almost completely buried in the core, and the tryptophan at B24 probably compensates for the lack of other aromatic residues equivalent to B25 Phe and B26 of insulin. Circular dichroism has confirmed the possibility of an insulin-like conformation for relaxin (58). The sequences of rat and sand tiger shark relaxins show numerous differences from porcine relaxin but features crucial to the insulin fold are maintained (59, 60).

Insulin Structure in Storage and at the Receptor

In the B-cell storage granule, the high concentration of insulin and the presence of zinc and calcium ions favor the presence of zinc insulin hexamers. Electron micrographs of storage granules from several species indicate crystals of many morphologies with sharp edges and regular repeats, which reflects the diversity of crystal forms grown in vitro. For instance, the rat and mouse produce two insulins of different sequence; in vitro, one crystallizes in a rhombohedral form (Rat I), while the other (Rat II), produces octahedral crystals with space group $P4_232$ with a cubic cell dimension of ~ 67 Å (61). In the electron microscope, sections of granules consistent with both forms are observed, and in some cases a 50 Å repeat (hexamer diameter) is observed. Granules of high quality from the grass snake, after fixing and sectioning, are clearly rhombic dodecahedra, and probably contain zinc insulin hexamers packed with cubic symmetry (62, 63). Such a packing could give rise to the 90 Å repeat seen in micrographs.

The exact metal ion concentrations and pH within the storage granule are not certain, and variation of these factors could have important consequences for the state of the stored protein. For instance, if the pH were around 7, then zinc insulin hexamers would bind extra zinc ions at a number of sites on the hexamer surface and possibly also at the B13 glutamate residues in the central channel, sites that have been defined by X-ray analysis of crystals. It has also been demonstrated that calcium ions can crosslink adjacent hexamers (through A4 and B30) and stabilize the crystal lattice, and this may represent a further means of rapidly removing freshly prepared insulin into the solid storage form (63).

The formation of amorphous or crystalline aggregates would be an effective form of concentration of the hormone and would ensure that hydrophobic surface regions were unavailable for interaction with the membranes of the vesicles (2, 7, 17). Aggregated forms of insulin are also less susceptible to proteolytic attack during prolonged storage. Some species, however, do not bind zinc ions or form hexamers, and in some of these animals (for instance, guinea pigs and porcupines) B22 arginine, which is a susceptible site for proteolytic degradation in the insulin monomer, has been substituted. Although most of the amino acid substitutions in guinea pig, coypu, and casiragua insulin would stabilize the monomeric form by making it more hydrophilic, they lead to a great loss of affinity for insulin receptors. When released into the circulation, the insulin granules experience a rapid dilution to $\sim 10^{-10}$ M, where the insulin monomer is the prevalent form. It therefore seems that the conformation of the insulin monomer is important for receptor binding, perhaps because some region of its surface has a capacity to recognize complementary regions on the cell surface receptor (7, 17, 18). Analysis of the hormone receptor affinity constant, the kinetics of the association and dissociation processes, and comparative studies of various hormone analogues have been employed to probe the nature of the interaction. The first proposal for the nature of the receptor binding region involved an insulin conformer similar to that defined by X-ray analysis, with receptor interactions mediated by a largely invariant surface region, part of which is involved in dimer formation (18). The model has been more clearly defined by extending the analogy of receptor binding and insulin dimerization; the latter process involves a complementarity in shape, charge, hydrophobicity, and hydrogen bonding capacity, as one might expect of the receptor interaction (32). The insulin surface region is larger than the region involved in dimer formation, which is consistent with thermodynamic analysis of the interaction (64). It includes a central hydrophobic region (B24 Phe, B25 Phe, B26 Tyr, B12 Val, B16 Tyr, and A19 Tyr) and polar side groups (A1 Gly, A4 Glu, A5 Gln, A21 Asn, B9 Ser, B10 His, B13 Glu, B21 Glu, B22 Arg, and B27 Thr) on the periphery. Even if the insulin molecule is somewhat less-ordered in solution than in the crystal, the ease of attainment of the crystal structure that occurs on dimerization could well find a parallel in the receptor docking process, and this would explain the observed importance of the tertiary structure and dimer-forming residues for full receptor affinity. The model that assumes a receptor binding region larger, but inclusive of the region involved in dimer formation, requires a decrease in receptor binding if the dimerization is reduced, but the converse is not necessarily true.

For insulin-like growth factors monomeric species are likely to bind the receptor. Part of the receptor binding region of insulin, especially that involved in dimer formation, is retained in insulin-like growth factors, but much is inaccessible due to the A-chain extension and the connecting peptide (17, 33). As this effect is less marked in insulin-like growth factor II, its stronger binding to the insulin receptor is predicted by the model. For relaxin, extensive sequence changes throughout the proposed binding region explain the absence of insulin-like biological activity. The receptor binding region of insulin-like growth factors may involve some residues in common with insulin, as insulins show a small but measurable ability to bind growth factor receptors. Certain hystricomorph rodent insulins are more potent growth factors than expected from their poor ability to bind insulin receptors (65). Work on definition of residues and conformers important for growth promoting effects will clearly be an area of intense activity in the near future.

GROWTH HORMONE FAMILY

Comparison of the amino acid sequences of growth hormone (somatotropin), prolactin, and placental lactogen (chorionic somatomammotropin) shows extensive homologies, which indicates that they can be usefully considered as members of a protein family (for review see 66). They are single chain proteins with molecular weights in the range 20,000–22,000. Growth hormone and placental lactogen have two equivalent intrachain disulfide bridges, while prolactin has a third. Study of the structure of these hormones has been limited so far to circular dichroism, fluorescence (67), sequence based predictive methods, and chemical accessibility measurements. Several crystal forms of placental lactogen have been reported (33), and recently the quality of one of these forms has been significantly improved, which indicates that a crystal structure should be available soon (68). Circular dichroism measurements suggest a high proportion of α helix (45–60%) (67) that persists in a range of conditions, which indicates stable globular structures. Prediction methods indicate rather less helix ($\sim$37%) and likely regions of β sheet and β turns (69, 70).

Enzymatic digestion experiments have been performed in efforts to identify active fragments of growth hormone. Plasmin digestion removes a hexapeptide (comprising residues 135–140) without affecting activity. Subsequent breakage of the disulfide bonds yields two peptides (comprising residues 1–134 and 141–191) that have little or no activity. However, full activity has been restored by noncovalent recombination of these fragments (71). Recently this approach has been extended to the preparation of recombinants of growth hormone and placental lactogen fragments (72) containing at least one reformed disulfide bond. The amino-terminal fragment (residues 1–134) was found to dominate the circular dichroism and biological characteristics of either recombinant that were close to the native values. However, cleavage of placental lactogen into two large fragments at the single tryptophan residue 86, while maintaining the disulfide bond of the above recombinants, led to a complete loss of activity and major structural changes (73).

Extracts of human pituitary glands contain a lower molecular weight variant of growth hormone where residues 32–46 are absent. Comparative

NMR spectroscopy, circular dichroism, spectrophotometric titration, and tyrosine nitration experiments (74) indicate that in whole growth hormone, residues 32–46 are mainly helical. If the secondary structures of the residues in common to the two molecules are identical then the difference spectrum is consistent with the contribution expected for residues 32–46 in human growth hormone from a Chou & Fasman analysis (70).

GLYCOPROTEIN HORMONES

This family of hormones comprises the gonadotrophins: lutropin, follitropin, and thyrotropin from the anterior pituitary and chorionic gonadotrophin (75). They are relatively large hormones (molecular weight 28,000–38,000) and comprise two subunits (α and β). The α chain is slightly smaller (96 residues) and is common to all the members of the group from one animal. Hormone specificity resides in the β subunit, but even here, considerable homology is observed. There are several carbohydrate attachment sites distributed over both subunits, and the sugar accounts for 16–45% of the weight of the hormones. The α subunits contain five disulfide bonds and the β subunit six, but the exact pairing of the cysteine residues is not yet clear. The noncovalent interaction between subunits is very strong, and therefore the associated form is probably present in circulation and involved in receptor binding.

Much information has been gathered on the properties of the isolated subunits that contributes to the picture we have of the conformation of these hormones (76, 77). The structural and functional integrity of the α subunit of lutropin can be regained following complete reduction and reoxidation of the disulfide bonds, but this is not the case for the β subunit. Sequential enzymatic digestion of the α subunit of lutropin yields a "core" containing all the disulfide bonds, somewhat reminiscent of snake venom α neurotoxins. Circular dichroism measurements have been reported by many workers, and there is agreement that β sheet ($\sim$30%) is the predominant feature of secondary structure. Sequence based predictions indicate a much higher β structure content ($\sim$60%) and a considerable β-turn contribution. This abundance of β sheet and turns, coupled with hydrodynamic indications of a compact structure and the high proline and disulfide content, all contribute to a general view of a globular protein. Circular dichroism studies of tyrosine exposure on subunit dissociation, IR spectroscopy on shielding of exchangeable peptide hydrogens on association, and NMR estimates of the pKa's of imidazole protons have been reported (reviewed in 76, 77), but a detailed crystal structure analysis is needed to add cohesion to these data.

PANCREATIC POLYPEPTIDE

Crystal Structure

The structure of the 36 amino acid avian (turkey) pancreatic polypeptide was determined by X-ray analysis, initially at 2.1 Å resolution using isomorphous replacement and anomalous scattering measurements. Subsequently the phase information was extended to 1.4 Å resolution, and the structure has been refined at 0.98 Å resolution, providing a detailed description of the molecule and its association mechanism (10, 78–80). The molecule comprises two well-defined secondary structures lying approximately antiparallel, their connection including two β turns (Figure 4). Residues 1–8, including prolines at 2, 5, and 8, form a polyproline-like helix, the proline rings of which all contribute toward hydrophobic interactions, interdigitating with nonpolar sidechains of an α helix formed by residues 14–32. The polyproline-like helix bends gently around the α helix in a way that optimizes contacts of Pro 2 with Val 30 and Tyr 27; of Pro 5 with Tyr 27, Leu 24, and Phe 20; and of Pro 8 with Phe 20 and Leu 17. The aromatic

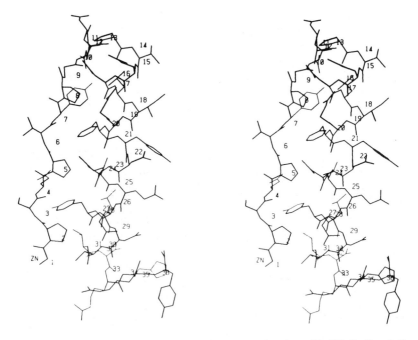

Figure 4   A stereo view of the structure of avian pancreatic polypeptide (80). Prolines 2, 5, and 8 are part of a polyproline-helix type conformation, and they interdigitate with the hydrophobic sidechains of the α helix in residues 18–32. The C terminus is flexible.

sidechains of Tyr 7 and Phe 20 produce a further hydrophobic contact. Residues 33 and 34 form a turn, orienting the less-well-defined and highly flexible C-terminal residues away from the helix axis.

An extensive hydrophobic surface region of the molecule, which involves both the helices, is in contact with a twofold symmetry-related molecule in the crystal, which defines a dimer. The axes of the two α helices are separated by about 11 Å and make an angle of about 150°, which allows sidechains of one helix to fit the grooves between sidechains of the second. The remainder of the surface of the molecule is polar.

Dimers of avian pancreatic polypeptide coordinate zinc ions in the crystal. Each zinc ion has five ligands including the N-terminal nitrogen and carbonyl oxygen of Gly 1 from one dimer, the amide of Asn 23 from a second, and Nϵ of the imidazole ring of His 34 from a third dimer. A water molecule is the fifth ligand in a distorted trigonal bipyramidal geometry of coordination.

Only sixteen residues of avian pancreatic polypeptide are conserved in the bovine molecule, but computer graphics model building and predictive methods show that a rather similar structure can be achieved (79). The major difficulty is experienced in the replacement of His 34 by proline. This change may be expected to alter the flexibility and conformation of the C-terminal residues. The bovine molecule would be expected to form dimers, but the substitutions at His 34 and Asn 23 of avian pancreatic polypeptide indicate that zinc-linked oligomers are unlikely. Similar conclusions apply to the human pancreatic polypeptide molecule.

Structure in Solution

Centrifuge and gel filtration, and UV spectra experiments have been reported for the avian and bovine polypeptides in solution (81–83) that indicate that pancreatic polypeptides form dimers. At neutral pH, avian pancreatic polypeptide shows a $K_D \simeq 5 \times 10^{-8}$ M; the strength of this association and its temperature dependence is consistent with the extensive nonpolar interactions indicated in the crystal dimer. The association is weaker for the bovine polypeptide; this permits a study of the near UV circular dichroism, which shows a concentration-dependent diminution between 10^{-4} and 10^{-6} M, indicative of the release of aromatic residues from restricted environments in the dimer. Tyr 7, Tyr 20, and Tyr 27 would all be expected to experience such effects on disruption of the dimer. Difference UV absorption measurements over this concentration range also show perturbation of one or more tyrosine residues. In the far UV the spectra for both polypeptides indicate a considerable amount of helix. Of particular importance is the finding that for bovine polypeptide the helix signal is maintained at high dilution where there is an appreciable monomer population.

In conclusion, it appears that in spite of the small size of this molecule, it can adopt an essentially globular structure, major elements of which are retained in the monomeric state in solution. This presumably is the species involved in binding to target tissue receptors, but aggregated states, in particular zinc linked oligomers, may be more important at sites of storage in the endocrine cell.

GLUCAGON FAMILY

Pancreatic glucagon (29 amino acids) is a member of a large family of pancreatic polypeptides found, amongst other places, in the alimentary tract and the central nervous system. This family also includes secretin (the first hormone to be discovered), vasoactive intestinal polypeptide, and glucose-dependent insulinotropic peptide (formerly known as gastric inhibitory peptide). Evidence for the existence of further homologous intestinal peptides indicates that the family may be very extensive (17).

Crystal Structure of Glucagon

Although porcine glucagon crystals are most easily obtained at high pH, where glucagon is most soluble, Sasaki et al (12) showed that these crystals undergo a phase change on lowering the pH to between 6 and 7 that involves a retention of cubic symmetry, but a shortening of the cubic cell parameter from 47.9 Å to 47.1 Å. Similar crystals are obtained by careful crystallization at pH 6.5; less-ordered crystals with a cell parameter of 48.7 Å form at lower pH ($\sim$3) (84, 85). These observations reflect the flexibility of glucagon (85).

In the cubic crystals, residues 6–28 are approximately α helical, although Gly 4 endows the N terminus with flexibility (12, 85). The helical conformer brings together the hydrophobic groups in two patches comprising Phe 6, Tyr 10, Tyr 13, and Leu 14 toward the N terminus, and Ala 19, Phe 22, Val 23, Trp 25, Leu 26, and Met 27 toward the C terminus. These hydrophobic patches are responsible for the intermolecular interactions that stabilize the cubic crystals involving a series of trimers that are not mutually exclusive, each of which contains a perfect threefold axis. Trimer 1 involves heterologous hydrophobic interactions between N- and C-terminal hydrophobic patches, while trimer 2 (Figure 5) has contacts between the C-terminal hydrophobic patches alone. Although trimer 1 involves a greater decrease of accessibility of the helical conformer (18.4%) compared with trimer 2 (14%), trimer 1 involves a greater necessary formation of helix (between residues 6 and 26) compared with trimer 2, which requires helix formation only between residues 16 and 27. Trimer 1, which involves charged interactions between Asp 9 and Asp 21 and positively charged guanidinium groups, as well as a buried tyrosyl hydroxyl group (Tyr 10), would not be expected to be stable at very low or very high pH, an observation that is consistent with the greater solubility of glucagon in an alkaline solution (85).

Solution Structure of Glucagon

In dilute aqueous solution, glucagon exists as a monomer (3). A largely unstructured and flexible chain is indicated by recent proton NMR studies on dilute aqueous solutions, which show narrow line widths and a single set of chemical shifts that, with the exception of the aromatic tyrosines, are independent of temperature and phosphate concentration (3, 4). A mainly

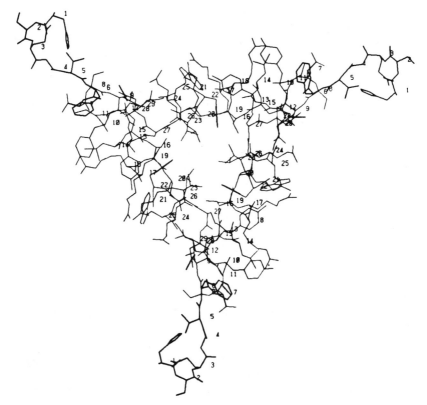

Figure 5   A glucagon trimer defined by X-ray analysis has hydrophobic contacts between C-terminal residues (85). In solution, NMR indicates that the helical region exists only in the C-terminal residues involved in the contacts.

flexible and nonspherical structure is also indicated by viscosity measurements (86), although not by the earlier results (87); the intrinsic viscosity is found to be insensitive to denaturing agents such as urea (86). Also, most NMR chemical shifts are not affected by 8 M urea and are similar to those expected for random coil, except those of Val 23 methyl resonances, Val 23 Hβ, and Leu 26 Hγ. These have been attributed (3) to a proximity between the sidechain for Val 23, with the aromatic ring of Trp 25 in about 20% of the population of conformers; a similar interaction occurs in a peptide fragment corresponding to residues 22–26; these residues cannot have a helical conformation. There may also be some structure in a percentage of the conformers involving Tyr 10 and Tyr 13, which unfold with increasing temperature (86).

The concentration dependence of most spectroscopic probes indicates that glucagon self-associates, with an accompanying change in secondary structure. Circular dichroism and optical rotary dispersion (13) indicate that conformers of 35% α-helical content are induced in concentrated solutions. Fluorescence studies of rhodamine 6G dye bound to glucagon at pH 10.6 (88), and optical detection of magnetic resonance (14) favor preferential formation of secondary structure in the C-terminal region close to the tryptophan (Trp 25). Proton NMR studies in concentrated solutions at high pH also suggest that aggregation involves C-terminal residues 22–29, and the concentration dependence of the chemical shifts can be fully accounted for by a two-state model involving a monomer-trimer equilibrium (4).

The constants for trimerization (Kα) at pH 10.6 for glucagon in D$_2$O with 0.2 M sodium phosphate, are 1.08 ($\pm$.42) $\times$ 10^6 M^{-2} at 30°C, and 5.86 ($\pm$1.07) $\times$ 10^4 M^{-2} at 50°C (4). Similar values are obtained from the concentration dependence of the circular dichroism at different temperatures (89). Analysis of the temperature dependence of the association constants (4, 89) and calorimetry (90) show that glucagon association is characterized by large negative values of both $\Delta H°$ and $\Delta S°$, whose temperature dependence decreases with increasing temperature. $\Delta Cp°$ has a large negative value, which indicates hydrophobic interactions (89). These data are consistent with a dominant negative entropy of a coil-helix transition on association that is greater than the positive entropy change resulting from hydrophobic interactions. The negative enthalpy change makes the overall free energy change favorable. The data indicate a trimeric structure similar to trimer 2 that involves hydrophobic interactions, with a well-defined structure only in the C-terminal region. This is supported by evidence from nuclear Overhauser enhancement indicating no interactions between residues widely separated in the glucagon sequence of the kind found in trimer 1 (86).

Although the evidence is clear that trimer 1 does not exist in acidic or alkaline solutions, a small percentage of such conformers may exist in neutral conditions. The increase in oligomer formation on decreasing the pH corresponds to the pK of the tyrosine, and the same tyrosine involvement in self-association is observed at lower pH. However, Wagman et al (4) find no evidence for an abnormal pK for the tyrosines, which would be expected of trimer 1.

On standing or warming acid solutions, long fibrils form; IR spectra indicate that these are antiparallel β sheet. The ability of glucagon to form both α-helical and β-sheet conformers is reflected in the prediction of secondary structure (91) that the sequence favors both α-helix and β-sheet formation, and suggests that the conformation is delicately balanced between these two conformations.

Micelle-Bound Glucagon

As glucagon is amphipathic in a helical conformation, hydrophobic surfaces favor helix formation. Indeed fluorescence and circular dichroism studies (92–94) show that chloroethanol, detergents, and surfactant micelles bind glucagon, with a concomitant formation of ordered secondary structure that is possibly α helix and probably mainly in the C-terminal region.

More recently, Wüthrich and co-workers (5) have undertaken a complete conformational analysis of glucagon bound to micelles, using high resolution NMR techniques. They show that 1 glucagon molecule binds to 40 detergent molecules (perdeuterated dodecylphosphocholine) with a well-defined and extended conformation. Electron paramagnetic resonance and NMR studies indicate that the molecule is parallel to the micelle surface (95). The combined use of proton-proton Overhauser enhancements and distance geometry algorithms using upper limits for selected proton-proton distances define possible conformations of the glucagon molecules bound to the micelle (5). Several structures meet the distance criteria for the segment Ala 19 to Met 27. Although certain distances are inconsistent with the existence of the helical conformer found in the crystals, it is possible to build a helical conformer with these interatomic distances by moving the sidechain of Met 27, which is restricted by intermolecular interactions in trimer 2, and by rotating some sidechains through small angles that are not precluded at 3.0 Å resolution by the X-ray analysis. The details of the conformer bound to the micelles will be further defined as the NMR analysis proceeds.

Secretin

The difficulties in obtaining quantities of natural secretin suitable for structural studies have meant that synthetic secretin and analogues have been used (96–101). The circular dichroism of secretin between pH 5 and pH 8 varies little and indicates perhaps a 30% α-helical conformation. In more acid solutions the ellipticity at 225 nm decreases, which indicates some loss of structure, but the helix is stabilized by lowering the temperature. Circular dichroism of synthetic analogues indicates that four residues at the N terminus do not contribute toward stabilization of the conformation. Ion pair interactions between Glu 9 and Asp 15 and the arginyl groups appear to be critical as the helical character decreases on modifying the carboxylates to carboxamides (98, 99). However, it is apparent that the helical stretch is most likely in the C-terminal half, as it is in glucagon (98). This would bring the hydrophobic residues Leu 19, Leu 22, Leu 23, Leu 26, and Val 27 into a hydrophobic patch that may interact with hydrophobic groups such as Phe 6, Leu 10, and Leu 13 in the same molecule. This structure may be stabilized by ion pair interactions. However, intermolecular interactions as in glucagon may also occur.

Glucagon in Storage Granules and at the Receptor

Glucagon is stored in granules that are usually amorphous in character. However, the glucagon granules of teleosts are crystalline rhombic dodecahedra (11, 102, 103) with cell dimensions that vary between 41 and 48 Å. Values smaller than that observed for porcine glucagon crystals bathed in their solvent of crystallization at pH 6.5 are expected as a result of the dehydration necessary for electron microscopy. These crystalline granules almost certainly have a structure similar to that of the crystals studied by X-ray analysis, and contain helical glucagon trimers. Amorphous granules may also contain trimers (2). At neutral pH, glucagon is very insoluble, and the high rate of precipitation makes crystals very difficult to obtain. Thus it is not surprising that the granules are often amorphous in mammalian A cells, even though conditions (and possibly the chemical sequence) may be conducive to crystallization in the teleost.

Amorphous or crystalline granules containing trimers are an effective way of concentrating the glucagon molecules for storage. The existence of trimers increases the thermodynamic stability and makes the hormone less available to degradation by proteolytic enzymes. Thus there is a strong parallel in the roles of zinc insulin hexamers of insulin and trimers and higher oligomers of glucagon in the storage of these hormones (17).

Glucagon storage granules, like those of insulin, become unstable at high dilutions in circulation, and circulating glucagon must exist largely as monomers with little defined secondary structure. Study of the data on receptor binding and activity of glucagon suggests that almost the entire molecule is required for full biological potency. Des-His[1]-glucagon and other glucagons modified at the N terminus are partial agonists with reduced affinity for the glucagon receptor, which suggests that this region might be involved in hormone action, whereas the major part of the molecule, including the C-terminal hydrophobic region, might be responsible for enhancing receptor affinity (2). Evidence that interaction with the receptor is entropy-driven suggests a model in which a helical conformer, stabilized by hydrophobic interactions, is induced or selected at the receptor from a population of conformers.

NEUROHYPOPHYSEAL HORMONES

Many of the recent conformational studies of neurohypophyseal hormones concern developments of the model for oxytocin, originally derived by Urry & Walter (104) on the basis of proton NMR studies in dimethyl sulfoxide. A β turn leads to a hydrogen bond between the mainchain carbonyl of Tyr 2 and the mainchain NH of Asn 5, and the 20-membered ring is closed by a right-handed disulfide bridge. The C-terminal residues form a further β turn with a hydrogen bond between the carbonyl of Cys 2 and NH of Gly 9. It was proposed that the structure is made compact by folding the C-terminal residues so that the terminal amide is close to the protonated amino terminus and the sidechain of Asn 5 hydrogen bonds to NH of Leu 8. This model has been held to be consistent with hydrogen deuterium exchange studies of oxytocin in dimethylsulfoxide, with the temperature dependence of the peptide proton chemical shifts, with the coupling constants for the NH–CH dihedral angles, and with solvent perturbation studies (105). Calculations show that this conformation is one of several low energy conformers, most of which show no preference for hydrogen bonds (106). In fact, NMR studies carried out in aqueous solutions (107) and at different pHs show that the molecule is flexible and sensitive to the environment. Alternative structures that may coexist in solution have been predicted using interactive computer graphics model building techniques (108).

NMR, circular dichroism, and laser Raman spectroscopic studies (109, 110) show that the gross conformations of the mainchains of the tocin and pressin 20-membered rings, including the disulfide bridges, appear to be similar, although in vasopressin there is the possibility of interaction between Tyr 2 and Phe 3 aromatic rings, which become more closely associated in water than in dimethylsulfoxide. The vasopressin molecule, however, appears to be more flexible than oxytocin, and the C-terminal residues are less tightly associated with the 20-membered ring structure. This is reflected in the easier solvent perturbation in 8-lysine-vasopressin than in oxytocin. Part of the flexibility appears to derive from the presence of two positive charges, the α-amino group of Cys 1 and the ϵ-amino group of Lys 8; if one of these is deleted, the conformation tightens up. Conformational changes also occur in neurohypophyseal hormones when the pH is lowered to deprotonate the α-amino group.

Earlier discussion of the relation of conformation to the biological action of neurohypophyseal hormones considered a relatively rigid structure binding to the receptor (111). Modifications were considered to be of three types. Group A included those modifications that disturbed the conformation, such as replacement of Asn 5, the sidechain of which is postulated to be involved in an intramolecular interaction with Leu 8 NH, replacement of Ile 3 with proline (112), replacement of the sulfur by selenium, or even replacement of the two half cystines by alanines. These modifications lead to a change in all bioactivities. Group B includes modifications such as at positions 3 or 8, which maintain the mainchain conformation but tend to have differential effects on either "tocin" or "pressin" receptor binding. Group C includes modifications that maintain conformation and affinity but decrease intrinsic activity leading to antagonist action, for example, in 2–O–ethyltyrosine oxytocin. More recently, Hruby and collaborators (6) have emphasized the importance of flexibility and the dynamics of receptor interactions. A ^{15}N NMR study of oxytocin in water has shown that although the tocin ring is rigid on the time scale of 10^{-10} sec, there is dynamic interconversion between conformers, with only Gln 4 and Asn 5 peptide nitrogens in relatively restricted backbone conformations (113). ^{13}C NMR (114), circular dichroism (115), and laser Raman spectroscopy (115) in aqueous solutions show that the antagonist [1-penicillamine]-oxytocin, which has two methyl groups substituted on the Cys 1 sulfur, is relatively more rigid than oxytocin itself. Thus the antagonist action may arise from the inability of the modified oxytocin molecule, once bound to the receptor, to explore those conformations that are important in eliciting a biological response.

Although flexibility of the neurohypophyseal hormone structures makes solution studies and therefore NMR studies of critical importance, some ambiguities in conformational parameters would certainly be clarified by a crystal structure analysis. Although oxytocin in various complexes, deamino oxytocin, deamino-1-seleno-oxytocin, and deamino-6-seleno-oxytocin, have all been crystallized, the X-ray analyses have proved complicated. Recently, new three-dimensional X-ray data sets have been measured in our laboratory with a view to using the sulfurs as heavy atoms, and anomalous scatterers to solve the phase problem. X-ray analysis of the crystalline C-terminal tetrapeptide Cys -Pro-Leu-Gly-NH$_2$ revealed a Gly-9-NH to Cys-6-CO hydrogen bond, as predicted in the model of Walter & Urry (104), although solutions contain a percentage of *cis* Cys-Pro peptide not found in oxytocin.

SOMATOSTATIN

The conformation of the hypothalamic tetradecapeptide, somatostatin, has been investigated using various spectroscopic techniques, but no crystallographic results have been published. The first tentative model presented by Holladay & Puett (116) was based on circular dichroism spectra and their dependence on guanidinium hydrochloride concentrations, which indicate the presence of ordered secondary structure in the monomer, and secondary structure prediction methods, which show residues 6–12 to have a high β-structure potential. The model consists of an elongated hairpin loop with several residues in antiparallel β structure. One end of the molecule is hydrophobic, and the aromatic rings partially shield the peptide-peptide hydrogen bonds. Subsequently, similar techniques were used to investigate perturbations of the proposed β turn by changes in amino acid sequence (117). Long-lived (D-Trp[8])-somatostatin showed a distinct conformational difference from somatostatin. Laser Raman spectroscopy allows estimation of disulfide torsion angles and indicates β-sheet structure, but does not distinguish between inter-and intramolecular hydrogen bonding, as good spectra were only recorded at very high peptide concentrations (~9%) (118).

NMR studies on fragments (119, 120), analogues (121), and whole somatostatin (122) have been hindered by severe line broadening, aggregation, and signal assignment problems and it is clear that a complete assignment of observed resonances to a 14 amino acid molecule is still not a trivial process. Hallenga et al (119, 120) have reported NMR spectra for peptide fragments in the region 10–13 and 9–14 and noted conformational effects that are also present in the whole hormone. The inequivalence of Thr 10 and Thr 12 resonances was attributed to a shielding of Thr 10 sidechain by the Phe 11 aromatic ring, a conformer favored in dimethylsulfoxide. Semiempirical energy calculations indicate an averaged conformation in which semiextended and folded structures are important, but only a portion of the folded forms are capable of explaining some chemical shift differences. Subsequently, two almost complete NMR assignments of somatostatin have been presented that employ a wide selection of assignment techniques (122, 123). However, there are some significant discrepancies between the assignments. Both studies suggest that somatostatin could exist in aqueous solution in a conformational equilibrium between several low energy conformations. The amides of residues 8, 9, and 10 must be hydrogen bonded or solvent shielded, and while types I and II β turns are excluded, participation of β-II turns in the conformational equilibrium is possible. In particular, a β-II turn from Phe-7-Thr-10 would permit a close approach of Lys 9 and Trp 8 sidechains, thus explaining the upfield shift of the Lys 9 γ-CH$_2$ resonances. Substitution of D-tryptophan at residue 8 should stabilize this conformation, and for this analogue a much larger upfield shift of the Lys 9 resonance is observed (121). On the basis of studies with this analogue it has been suggested that the biologically active conformation of somatostatin at its receptor is not the predominant conformer in solution, but rather a conformer favored by D-tryptophan substitution. It is known that this substitution greatly prolongs the life of this analogue in vivo, but its high potency persists in in vitro assays where degradation is minimized.

Veber and co-workers (124) have extended the study of analogues of this type by introducing covalent conformational constraints and eliminating apparently unimportant residues. The analogue cyclo(Aha-Cys-Phe-DTrp-Lys-Thr-Cys) where Aha is 7-aminoheptanoic acid, has equivalent or higher activity than somatostatin in inhibiting growth hormone, insulin, and glucagon release. The diminished molecular flexibility of this analogue has been defined by NMR measurements. Using a computer modeling system, the feasibility of replacing the Cys-Aha-Cys bridge by various dipeptides in the conformations defined for the various types of β turn was tested. Suitable cyclic hexapeptides were synthesized and bioassayed, and NMR spectra were recorded. The characteristic upfield γ-CH$_2$ protons of Lys 9 were apparent in most of the analogues, but the activity was augmented substantially by closing the ring with Phe-Pro, which provided hydrophobicity and a residue with restricted conformational preferences. The NMR spectrum of this hexapeptide indicated a high degree of molecular rigidity, suiting conformational analysis. The analogue was almost twice as potent as somatostatin in inhibiting growth hormone release and was degraded slowly. Thus nine of the amino acids of native somatostatin were replaced with a single proline residue, which favors a distinct conformer with an enhanced receptor affinity.

ENDORPHIN AND ADRENOCORTICOTROPIN FAMILIES

The sequences of ACTH, melanocyte stimulating hormone, β lipotropin and β endorphin (the C-terminal 31 residues of β lipotropin) are contained within their pituitary precursor, proopiomelanocortin. [Met^5]-enkephalin, a natural opiate of the brain, is contained in residues 61–65 of β lipotropin, although this peptide is synthesized in its own precursor. The polypeptides of this group all show great conformational flexibility in aqueous solutions; but secondary structure can be induced in the hormones and in fragments by nonaqueous solvents, lipid micelles, and salts.

The studies of Schwyzer and his colleagues (1) have some time ago established the flexibility of melanocyte stimulating hormone (MSH) and ACTH in aqueous solutions. Although α helix is not easily induced in the whole molecule, theoretical predictions and circular dichroism indicate that fragments such as [Pro^1, Ala^2, Ala^3, Val^4]-ACTH (10) can be highly helical, especially in trifluoroethanol (125).

The peptides [Leu^5]-enkephalin (Tyr Gly Gly Phe Leu) and [Met^5]-enkephalin (Tyr Gly Gly Phe Met) bind to the same receptor site as the rigid opiate compounds of the morphine and oripavine family. Their small size has allowed synthesis of many analogues and the production of large quantities for NMR, circular dichroism, Raman spectroscopy, and X-ray crystallographic analysis. The conformation is very flexible, and the existence of a preferred conformer depends critically on whether the molecule is uncharged, cationic, or zwitterionic; on the concentration; and on the solvent. ^{1}H and ^{13}C NMR and Raman spectroscopy (126–129) show that the zwitterionic form favors a type I β turn at residues Gly-3-Phe-4 of [Leu^5]-enkephalin, one of several low-energy forms indicated by conformational analysis (130–133). This interpretation has, however, been disputed. The cationic form is in an unfolded state in dimethyl sulfoxide (134). In aqueous solutions the molecule exists as an ensemble of conformers with Tyr 1 relatively rigidly fixed, although there is little evidence for a folded structure (135–138).

Attempts at crystal structure analysis have confirmed the existence of several different conformers. When crystallized from alcohol [Leu^5]-enkephalin (S. Bedarkar, T. L. Blundell, L. Hearn, B. Morgan and I. J. Tickle, unpublished results) is arranged in hydrogen bonded sheets probably comprised of head-to-tail β-sheet dimers with an approximate twofold axis (space group P2; a = 11.46, b = 15.59, c = 16.73 Å, β = 92.2°; Z = 4). The exact nature of the hydrogen bonding has yet to be defined. The [Met^5]-enkephalin forms similar sheets in a related crystal cell (space group P2; a = 11.61, b = 17.99, c = 16.52Å, β = 91.2°; Z = 4). These may correspond to the associated forms observed in solution at high concentrations (134–140). However, in aqueous ethanol, quite different crystals form, with four molecules in the asymmetric unit, each containing a distorted β turn at residues Gly-2-Gly-3 (141, 142). The four molecules are arranged in two pairs, with similar tyrosine sidechain positions. Very small conformational differences distinguish the molecules with similar tyrosine sidechain orientations (142).

The variety of conformers observed confirms the existence of molecular flexibility. Identification of the receptor bound conformer awaits study of the enkephalin-receptor complex, but some understanding has been obtained by comparing conformations with morphine and oripavine structures (143) and by conformational study of analogues with restricted flexibility, such as a cyclic enkephalin analogue in which the COOH-terminal carboxyl group of Leu 5 is cyclized to the γ-amino moeity of α,γ -diaminobutyric acid substituted in position 2 (144). This analogue prevents the realization of many of the conformational features ascribed to native enkephalin in solution or in the crystalline state, and its high in vitro activity suggests a model for the enkephalin bound at the receptor.

Circular dichroism spectra of β endorphin and β lipotropin (145–147) in water show little evidence of secondary structure, and intrinsic viscosities and sedimentation coefficients of the two polypeptides indicate that neither is compact or globular. However, methanol, trifluorethanol, dioxane, or sodium dedecylsulfate promote the formation of up to 50% α-helical structure. Similar α-helical structures are induced by aqueous solutions of various lipids including cerebroside sulfate, ganglioside GM1, phosphatidyl serine, and phosphatidic acid, but not cerebroside or phosphatidyl choline (148). The helices are broken up by Ca^{2+} ions. It is suggested that similar interactions with lipids may be of importance in allowing orientation of the NH_2-terminal pentapeptide (Met-enkephalin) for binding and activation of the receptor.

CONCLUSIONS

The interaction of a polypeptide hormone with its receptor must involve a complementarity in terms of hydrophobicity, charge distribution, and hydrogen bonding capacity of the two molecular surfaces. In this respect, compared to a flexible molecule, a more rigid hormone molecule might appear to be advantageous in that there is a smaller loss of entropy on formation of the receptor hormone complex. This must be the main advantage of the relatively stable conformers of the globular polypeptide hormones such as insulin, glycoprotein hormones, and growth hormones, and of cyclic peptides such as oxytocin. Furthermore, the high biological potency of the cyclic analogues of somatostatin and enkephalin may be due to their being held in a rigid conformation complementary to the receptor. However, studies of polypeptide hormones indicate that many smaller molecules exist as an ensemble of conformers, especially in aqueous solutions, and even the larger molecules such as insulin can undergo conformational changes on modification of the environment. Flexibility may have several advantages. First it may accelerate the hormone receptor recognition process by allowing more interactions to be explored; second, conformational changes in the receptor and the hormone may be required once the hormone is bound in order to activate the biological response. The greater flexibility of the N terminus of glucagon, which is important for full potency, and the discovery that oxytocin antagonists are often more rigid than the native hormone may be reflections of this process. Third, flexibility of the hormone may be important for its clearance from the circulation or for its deactivation by proteolysis.

However, the molecular biology of polypeptide hormones is complex; it involves biosynthesis, transport within the endoplasmic reticulum, storage, circulation, receptor binding, and clearance. The roles of conformation in these processes may be varied, reflecting different ways of optimizng through evolution their overall usefulness.

Literature Cited

1. Schwyzer, R. J. 1973. *J. Mondial Pharm.* 3:254–60
2. Blundell, T. L. 1979. *Trends Biochem. Sci.* 4:80–83
3. Boesch, C., Bundi, A., Oppliger, M., Wüthrich, K. 1978. *Eur. J. Biochem.* 91:209–14
4. Wagman, M. E., Dobson, C. M., Karplus, M. 1980. *FEBS Lett.* 119:265–70
5. Braun, W., Boesch, C., Brown, L. R., Gō, N., Wüthrich, K. 1981. *Biophys. Biochim. Acta.* In press
6. Meraldi, J. P., Hruby, V. J., Brewster, A. I. 1977. *Proc. Natl. Acad. Sci. USA* 74:1373–77
7. Blundell, T. L. 1979. *Trends Biochem. Sci.* 4:51–54
8. Wollmer, A., Strassburg, W., Hoenjet, E., Glatter, U., Fleischhauer, J., Mercola, D. A., De Graaf, R. A. G., Dodson, E. J., Dodson, G. G., Smith, D. G., Brandenburg, D., Danho, W. 1980. *Insulin: Chemistry, Structure and Function of Insulin and Related Hormones,* pp. 27–38. Berlin: Walter De Gruyter. 752 pp.
9. Pocker, Y., Biswas, S. B. 1980. *Biochemistry* 19:5043–49
10. Tickle, I. J., Glover, I. D., Pitts, J. E., Wood, S. P., Blundell, T. L. 1981. *Acta Crystallogr.* Sect 1, Subsect. 4–5 (Suppl.)
11. Lange, R. H., Kobayashi, K. 1980. *J. Ultrastruct. Res.* 72:20–26
12. Sasaki, K., Dockerill, S., Adamiak, D. A., Tickle, I. J., Blundell, T. L. 1975. *Nature* 257:751–57
13. Srere, P. A., Brooks, G. C. 1969. *Arch. Biochem. Biophys.* 129:708–10
14. Ross, J. B. A., Rousslang, K. W., Deranleau, D. A., Kwiram, A. L. 1977. *Biochemistry* 16:5398–402
15. Deleted in proof
16. Blundell, T. L., Bedarkar, S., Rinderknecht, E., Humbel, R. E. 1978. *Proc. Natl. Acad. Sci. USA* 75:180–84

17. Blundell, T. L., Humbel, R. E. 1980. *Nature* 287:781–87
18. Blundell, T. L., Dodson, G. G., Hodgkin, D. C., Mercola, D. A. 1972. *Adv. Protein Chem.* 26:279–402
19. Peking Insulin Structure Research Group. 1981. *Structural Studies on Molecules of Biological Interest,* pp. 501–8. Oxford: Clarendon Press. 610 pp.
20. Dodson, E. J., Dodson, G. G., Hodgkin, D. C., Reynolds, C. D. 1979. *Can. J. Biochem.* 57:469–79
21. Cutfield, J. F., Cutfield, S. M., Dodson, E. J., Dodson, G. G., Reynolds, C. D., Vallely, D. 1981. See Ref. 19, pp. 501–8
22. Deleted in proof
23. Sakabe, K., Sakabe, N., Sasaki, K. 1980. *Water and Metal Cations in Biological Systems,* pp. 117–27. Tokyo: Japan Scientific Societies Press
24. Sakabe, N., Sakabe, K., Sasaki, K. 1978. *Proinsulin, Insulin and C-Peptide,* pp. 73–83. Amsterdam: Excerpta Medica
25. Sakabe, N., Sakabe, K., Sasaki, K. 1981. See Ref. 19, pp. 509–26
26. Sakabe, N., Sasaki, K., Sakabe, K. 1981. *Acta Crystallogr.* 02.7–01 (Suppl.)
27. Bentley, G. A., Dodson, E. J., Dodson, G. G., Hodgkin, D. C., Mercola, D. A. 1976. *Nature* 261:166–69
28. Dodson, E. J., Dodson, G. G., Reynolds, C. D., Vallely, D. G. 1980. See Ref. 8, pp. 9–16
29. Dodson, E. J., Dodson, G. G., Lewitova, A., Sabesan, M. 1978. *J. Mol. Biol.* 125:387–96
30. Cutfield, J. F., Cutfield, S. M., Dodson, E. J., Dodson, G. G., Emdin, S. F., Reynolds, C. D. 1979. *J. Mol. Biol.* 132: 85–100
31. Bentley, G., Dodson, G. G., Lewitova, A. 1978. *J. Mol. Biol.* 126:871–75
32. Pullen, R. A., Lindsay, D. G., Wood, S. P., Tickle, I. J., Blundell, T. L., Wollmer, A., Krail, G., Brandenburg,

D., Zahn, H., Gliemann, J., Gammeltoft, S. 1976. *Nature* 259:369–73

33. Pullen, R. A., Jenkins, J. A., Tickle, I. J., Wood, S. P., Blundell, T. L. 1975. *Mol. Cell. Biochem.* 8:5–20

34. Friesen, H-J., Brandenburg, D., Diaconescu, C., Gattner, H-G., Naithani, V. K., Nowak, J., Zahn, H., Dockerill, S., Wood, S. P., Blundell, T. L. 1977. *Proc. Am. Peptide Symp., 5th,* pp. 136–46

35. Deleted in proof

36. Dodson, G. G., Cutfield, S., Hoenjet, E., Wollmer, A., Brandenburg, D. 1980. See Ref. 8, pp. 17–26

37. Peking Insulin Structure Research Group. 1976. *Sci. Sin.* 19:358–63

38. Lu, Z., Yu, R. 1980. *Sci. Sin.* 13:1592–99

39. Bi, R. C., Cutfield, S., Dodson, E. J., Dodson, G. G. 1981. *Acta Crystallogr.* 01.3–01 (Suppl.)

40. Rosen, L. S., Fullerton, W. W., Low, B. W. 1972. *Arch. Biochem. Biophys.* 152:569–733

41. Deleted in proof

42. Wood, S. P., Blundell, T. L., Wollmer, A., Lazarus, N. R., Neville, R. W. J. 1975. *Eur. J. Biochem.* 55:531–42

43. Hennessey, J. P., Johnson, W. C. 1981. *Biochemistry* 20:1085–94

44. Strickland, E. H., Mercola, D. 1976. *Biochemistry* 15:3875–83

45. Wollmer, A., Fleischhauer, J., Strassburger, W., Thiele, H., Brandenburg, D., Dodson, G., Mercola, D. 1977. *Biophys. J.* 20:233–43

46. Horuk, R., Blundell, T. L., Lazarus, N. R., Neville, R. W. J., Stone, D., Wollmer, A. 1980. *Nature* 286:822–24

47. Horuk, R., Wood, S. P., Blundell, T. L., Lazarus, N. R., Neville, R. W. J., Raper, J. H., Wollmer, A. 1980. *Hormones and Cell Regulation,* 4:123–39. Amsterdam: Elsevier/North Holland Biomed. Press

48. Horuk, R., Wood, S. P., Blundell, T. L., Lazarus, N., Neville, R. 1980. *Actual. Chim. Therap.* 7:15–25

49. Bradbury, J. H., Brown, L. R. 1977. *Eur. J. Biochem.* 76:573–80

50. Williamson, K. L., Williams, R. J. P. 1979. *Biochemistry* 18:5966–72

51. Led, J. J., Grant, D. M., Horton, W. J., Sundby, F., Vilhelmsen, K. 1975. *J. Am. Chem. Soc.* 97:5997–6002

52. Deleted in proof

53. Blundell, T. L., Horuk, R. 1981. *Hoppe-Seylers Z. Physiol. Chem.* 362:727–37

54. Blundell, T. L., Bedarkar, S., Rinderknecht, E., Humbel, R. E. 1978. *Proc. Natl. Acad. Sci. USA* 75:180–84

55. Deleted in proof

56. Bedarkar, S., Turnell, W. G., Blundell, T. L., Schwabe, C. 1977. *Nature* 270:449–51

57. Isaacs, N., James, R., Niall, H. 1978. *Nature* 271:278–81

58. Schwabe, C., Harmon, S. J. 1978. *Biochem. Biophys. Res. Commun.* 84: 374–80

59. Gowan, L. K., Reinig, J. W., Schwabe, C., Bedarkar, S., Blundell, T. L. 1981. *FEBS Lett.* 129:80–82

60. Bedarkar, S., Blundell, T. L., Gowan, L. K., Schwabe, C. 1981. *New York Acad. Sci.* 637: In press

61. Wood, S. P., Tickle, I. J., Blundell, T. L., Wollmer, A., Steiner, D. F. 1978. *Arch. Biochem. Biophys.* 186:175–83

62. Cao, Q., Li, T., Peng, X., Zhang, Y. 1980. *Sci. Sin.* 23:1309–13

63. Pitts, J. E., Wood, S. P., Horuk, R., Bedarkar, S., Blundell, T. L. 1980. See Ref. 8, pp. 673–82

64. Waelbroeck, M., van Obberghen, E., De Meyts, P. 1979. *J. Biol. Chem.* 254: 7736–45

65. King, G. L., Kahn, R. 1981. *Nature* 292:644–46

66. Wallis, M. 1975. *Biol. Rev.* 50:35–98

67. Aloj, S., Edelhoch, H. 1971. *J. Biol. Chem.* 246:5047–52

68. Moffat, K. 1980. *Int. J. Peptide Protein Res.* 15:149–53

69. Chen, C. H., Sonenberg, M. 1977. *Biochemistry* 16:2110–18

70. Hartman, P. G., Chapman, G. E., Moss, T., Bradbury, E. M. 1977. *Eur. J. Biochem.* 88:363–71

71. Li, C. H., Bewley, T. A. 1976. *Proc. Natl. Acad. Sci. USA* 73:1476–79

72. Russell, J., Sherwood, L. M., Kowalski, K., Schneider, A. B. 1981. *J. Biol. Chem.* 256:296–300

73. Russell, J., Katzhendler, J., Kowalski, K., Schneider, A. B., Sherwood, L. M. 1981. *J. Biol. Chem.* 256:304–7

74. Chapman, G. E., Rogers, K. M., Brittain, T., Bradshaw, R. A., Bates, O. J., Turner, C., Cary, P. D., Crane-Robinson, C. 1981. *J. Biol. Chem.* 256:2395–2401

75. Pierce, J. G., Parsons, T. F. 1981. *Ann. Rev. Biochem.* 50:465–95

76. Guidice, L. C., Pierce, J. G. 1978. *Structure and Function of Gonadotrophins,* Ch. 4. New York: Plenum

77. Garnier, J. 1978. See Ref. 76, Ch. 17

78. Wood, S. P., Pitts, J. E., Blundell, T. L., Tickle, I. J., Jenkins, J. A. 1977. *Eur. J. Biochem.* 78:119–26

79. Pitts, J. E., Blundell, T. L., Tickle, I. J., Wood, S. P. 1979. *Proc. Am. Peptide Symp., 6th,* pp. 1011–16

80. Blundell, T. L., Pitts, J. E., Tickle, I. J., Wood, S. P., Wu, C-W. 1981. *Proc. Natl. Acad. Sci. USA* 78:4175–79
81. Noelken, M. E., Chang, P. J., Kimmel, J. R. 1980. *Biochemistry* 19:1838–43
82. Chang, P. J., Noelken, M. E., Kimmel, J. R. 1980. *Biochemistry* 19:1844–49
83. Pitts, J. E. 1980. *Structures and function of pancreatic polypeptide hormones.* PhD thesis. Univ. Sussex, UK. 210 pp.
84. Blundell, T. L., Dockerill, S., Pitts, J. E., Wood, S. P., Tickle, I. J. 1978. *Membrane Proteins,* pp. 249–57. Oxford: Pergamon
85. Blundell, T. L. 1981. *Glucagon.* Berlin: Springer. In press
86. Wagman, M. 1981. *Proton NMR studies of glucagon association in solution.* PhD thesis. Harvard Univ. Cambridge, Mass. 403 pp.
87. Epand, R. M. 1971. *Can. J. Biochem.* 49:166–69
88. Formisano, S., Johnson, M. L., Edelhoch, H. 1978. *Biochemistry* 17:1468–73
89. Formisano, S., Johnson, M. L., Edelhoch, H. 1978. *Proc. Natl. Acad. Sci. USA* 74:3340–44
90. Johnson, R. E., Hruby, V. J., Rupley, J. A. 1979. *Biochemistry* 18:1176–79
91. Chou, P. Y., Fasman, G. D. 1975. *Biochemistry* 14:2536–41
92. Gratzer, W. B., Beaven, G. H., Rattle, H. W. E., Bradbury, E. M. 1968. *Eur. J. Biochem.* 3:276–83
93. Epand, R. M., Jones, A. J. S., Sayer, B. 1977. *Biochemistry* 16:4360–68
94. Wu, C. S. C., Yang, J. T. 1980. *Biochemistry* 19:2117–22
95. Brown, L. R., Boesch, C., Wüthrich, K. 1981. *Biochim. Biophys. Acta.* In press
96. Van Zon, A., Beyerman, H. C. 1976. *Helv. Chim. Acta* 59:1112–19
97. Jäger, E., Filippi, B., Knof, S., Lehnert, P., Moroder, L., Wünsh, E. 1979. *Hormone Receptors in Digestion and Nutrition,* pp. 25–35 Amsterdam: North-Holland
98. Bodanszky, M., Fink, M. L. 1976. *Bioorg. Chem.* 5(3):275–82
99. Bodanszky, M., Fink, M. L., Funk, K. W., Said, S. I. 1976. *Clin. Endocrinol.* 195S–205S (Suppl.)
100. Bodanszky, M. 1976. *J. Am. Chem. Soc.* 98:974–77
101. Yanahara, N., Yanahara, C., Kubota, M., Sakagami, M., Otsuki, M., Baba, S., Shiga, M. 1979. *Peptides,* pp. 539–42. Rockford, Ill: Pierce Chem. Co.
102. Lange, R. H. 1976. *Endocrine Gut and Pancreas,* pp. 167–78. Amsterdam: Elsevier/North-Holland Biomed. Press
103. Lange, R. H. 1979. *Eur. J. Cell. Biol.* 20:71–75
104. Urry, D. W., Walter, R. 1971. *Proc. Natl. Acad. Sci. USA* 68:956–58
105. Walter, R., Glickson, J. D. 1973. *Proc. Natl. Acad. Sci. USA* 70:1199–1203
106. Kotelchuck, D., Scheraga, H. A., Walter, R. 1972. *Proc. Natl. Acad. Sci. USA* 69:3629–33
107. Brewster, A. I. R., Hruby, V. J. 1973. *Proc. Natl. Acad. Sci. USA* 70:3806–9
108. Honig, B., Kabat, E. A., Katz, L., Levinthal, C., Wu, T. T. 1973. *J. Mol. Biol.* 80:277–95
109. Wyssbrod, H. R., Fischman, A. J., Live, D. H., Hruby, V. J., Argawal, N. W., Upson, D. A. 1979. *J. Am. Chem. Soc.* 101:4037–43
110. Tu, A. T., Lee, J., Deb, K. K., Hruby, V. J. 1979. *J. Biol. Chem.* 254:3272–78
111. Walter, R., Schwartz, I. L., Darnell, J. H., Urry, D. W. 1971. *Proc. Natl. Acad. Sci. USA* 68:1355–59
112. Deslauriers, R., Smith, I. C. P., Levy, G. C., Orlowski, R., Walter, R. 1978. *J. Am. Chem. Soc.* 109:3912–17
113. Live, D. H., Wyssbrod, H. R., Fischman, A. J., Agosta, W. C., Bradley, C. H., Cowburn, D. 1979. *J. Am. Chem. Soc.* 101:474–79
114. Hruby, V. J., Deb, K. K., Spatola, A. F., Upson, D. A., Yamamoto, D. 1979. *J. Am. Chem. Soc.* 101:202–12
115. Hruby, V. J., Deb, K. K., Fox, J., Bjornason, J., Tu, A. T. 1978. *J. Biol. Chem.* 253:6060–67
116. Holladay, L. A., Puett, D. 1976. *Proc. Natl. Acad. Sci. USA* 73:1199–1202
117. Holladay, L. A., Rivier, J., Puett, D. 1977. *Biochemistry* 16:4895–900
118. Han, S. L., Rivier, J. E., Scheraga, H. A. 1980. *Int. J. Peptide Protein Res.* 15:355–64
119. Knappenberg, M., Brison, J., Dirkx, J., Hallenga, K., Deschrijver, P., van Binst, G. 1979. *Biochim. Biophys. Acta* 580:266–76
120. Hallenga, K., van Binst, G., Knappenberg, M., Brison, J., Michel, A., Dirkx, J. 1979. *Biochim. Biophys. Acta* 577:82–101
121. Arison, B. H., Hirschmann, R., Veber, D. F. 1978. *Bioorg. Chem.* 7(5):447–51
122. Buffington, L., Garsky, V., Massiot, G., Rivier, J., Gibbons, W. A. 1980. *Biophys. Res. Commun.* 93(2):376–84
123. Hallenga, K., van Binst, G., Scarso, A., Michel, A., Knappenberg, M., Dremier, C., Brison, J., Dirkx, J. 1980. *FEBS Lett.* 119:47–52
124. Veber, D. F., Freidinger, R. M., Perlow, D. S., Paleveda, W. J. Jr., Holly, F.

W., Strachan, R. G., Nutt, R. F., Arison, B. H., Homnick, c., Randall, W. C., Glitzer, M. S., Saperstein, R., Hirschmann, R. 1981. *Nature* 292: 55–56

125. Mutter, H., Mutter, M., Bayer, E. 1979. *Z. Naturforsch. Teil. B* 34:874–85

126. Stimpson, E. R., Meinwald, Y. C., Scheraga, H. A. 1979. *Biochemistry* 18:1661–71

127. Khaled, M. A., Urry, D. W., Bradley, R. J. 1979. *J. Chem. Soc. London Perkin Trans. II* 12:1693–99

128. Niccolai, N., Garsky, V., Gibbons, W. A. 1980. *J. Am. Chem. Soc.* 102: 1517–20

129. Jones, C. R., Garsky, V., Gibbons, W. A. 1977. *Biochem. Biophys. Res. Commun.* 76:619–25

130. Isogai, Y., Nemethy, G., Scheraga, H. A. 1977. *Proc. Natl. Acad. Sci. USA* 74:414–18

131. Momany, F. A. 1977. *Biochem. Biophys. Res. Commun.* 75:1098–1103

132. DeCoen, J. L., Humblet, C., Koch, M. H. J. 1977. *FEBS Lett.* 73:38–42

133. Balodis, Y. Y., Nikifarovich, G. V., Grinsteine, I. V., Vegner, R. E., Chipens, G. I. 1978. *FEBS Lett.* 86:239–42

134. Higashijima, T., Kobayashi, J., Nagai, U., Miyazawa, T. 1979. *Eur. J. Biochem.* 97:43–57

135. Bleich, H. E., Day, A. R., Freer, R. J., Glasel, J. A. 1979. *Biochem. Biophys. Res. Commun.* 87:1146–53

136. Fischman, A. J., Riemen, M. W., Cowburn, D. 1978. *FEBS Lett.* 94:236–40

137. Kobayashi, J., Higashijima, T., Nagai, U., Miyazawa, T. 1980. *Biochim. Biophys. Acta* 621:190–203

138. Kobayashi, J., Nagai, U., Higashijima, T., Miyazawa, T. 1979. *Biochim. Biophys. Acta* 577:195–206

139. Deleted in proof

140. Khaled, M. A., Long, M. M., Thompson, W. D., Bradley, R. J., Brown, G. B., Urry, D. W. 1977. *Biochem. Biophys. Res. Commun.* 76:224–29

141. Smith, G. D., Griffin, J. F. 1978. *Science* 199:1214–16

142. Blundell, T. L., Hearn, L., Tickle, I. J., Palmer, R. A., Morgan, B. A., Smith, G. D., Griffin, J. F. 1979. *Science* 205:220

143. Gorin, F. A., Balasubramanian, T. M., Barry, C. D., Marshall, G. R. 1978. *J. Supramol. Struct.* 9:27–39

144. Di Maio, J., Schiller, P. W. 1980. *Proc. Natl. Acad. Sci. USA* 77:7162–66

145. St.-Pierre, S., Gilardeau, C., Chretien, M. 1976. *Can J. Biochem.* 54:992–98

146. Yang, J. T., Brewley, T. A., Chen, G. C., Li, C. H. 1977. *Proc. Natl. Acad. Sci. USA* 74:3235–38

147. Hollosi, M., Kajtar, M., Grat, L. 1977. *FEBS Lett.* 74:185–88

148. Wu, C. S. C., Lee, N. M., Loh, H. H., Yang, J. T., Li, C. H. 1979. *Proc. Natl. Acad. Sci. USA* 76:3656–59

Ann. Rev. Biophys. Bioeng. 1984. 13 : 493–521

4

DETAILED ANALYSIS OF PROTEIN STRUCTURE AND FUNCTION BY NMR SPECTROSCOPY : Survey of Resonance Assignments

John L. Markley and Eldon L. Ulrich

Department of Chemistry, Purdue University, West Lafayette, Indiana 47907

PERSPECTIVES AND SUMMARY

The first reported nuclear magnetic resonance (NMR) study of a protein was the publication in 1957 of a 40 MHz ^{1}H NMR spectrum of bovine pancreatic ribonuclease A (497), which revealed four distinct regions of overlapping peaks. These features were interpreted later the same year (267) in terms of the spectra of the constituent amino acids of the protein molecule. It was soon recognized that NMR spectroscopy offered the potential for providing information about the structural environment, chemical properties, and motional characteristics of defined atoms and functional groups in protein molecules. The logical prerequisites for obtaining this information would be the resolution of spectral signals from individual groups and their assignment to particular sites in a defined amino acid residue or prosthetic group of the protein. That this could be accomplished, at least in a limited sense, was demonstrated by the detection of hyperfine-shifted ^{1}H resonances from individual groups in oxidized cytochrome c (309), ring-current shifted ^{1}H resonances in spectra of lysozyme and cytochrome c (391), and individual histidine C_ε–H groups in ribonuclease A (69) along with other proteins (395). The pH titration shifts of the histidine peaks provided pK'_a values for known individual side chains in the molecule (review 377). Detection of site-specific chemical shifts demonstrated that NMR could be used to follow protein conformational changes including denaturation-renaturation processes (527); and the sharpening of resonances on denaturation (178) indicated that NMR would be useful for quantitating mobilities of protein constituents.

The early NMR experiments considered only chemical shifts, peak areas, and peak widths. Measurable parameters have been expanded to include relaxation times, nuclear Overhauser enhancements (NOEs), NOE connectivities, coupling constants, and coupling connectivities. Nuclei in proteins that are accessible to NMR analysis have been extended from ^{1}H to ^{31}P (227) (phosphoproteins), ^{13}C (9, 340), ^{15}N (211), ^{2}H (441, 599) and ^{3}H (596) (both with enrichment), plus several metal nuclei in metalloproteins. The development of techniques for the assignment of individual resonances to specific protein groups has progressed in parallel with the ability to detect and resolve them. Milestones along this path were the development of superconducting solenoids for high-field NMR (178), which now go as high as 600 MHz for ^{1}H NMR (63), selective isotopic substitution strategies allowing the resolution of individual resonances used first for ^{2}H-labeling (385) and later for ^{13}C-, ^{15}N-, or ^{3}H-enrichment, application of pulse-Fourier transform NMR in one (9) and two dimensions (421), and spectral-simplification and resolution-enhancement methods discussed below. Incorporation of ^{19}F-labeled amino acids into a protein (538) and insertion of ^{113}Cd^{3+} into a protein metal binding site (28) have led to numerous studies involving substitutions of these atoms that are not normally found in biomolecules. New approaches for the insertion of non-perturbing NMR probes appear regularly. For example, the rare ^{57}Fe isotope has been substituted recently into the heme of myoglobin, and a ^{13}C-^{57}Fe coupling constant has been observed on binding ^{13}CO (338).

The current literature reveals a number of trends. (a) Two-dimensional NMR methods for extending assignments are now yielding nearly complete assignments in ^{1}H NMR spectra of small proteins (molecular weight under 7,000). (b) NMR studies of families of closely related proteins are allowing assignments to be made in larger proteins and provide the potential for comprehensive investigations of the structural and dynamical consequences of sequence variations. (c) Multinuclear NMR approaches are generating information pertinent to mechanisms of enzyme catalysis and inhibition. (d) NMR results have given valuable insights into the origin of cooperative (positive and negative) effects in proteins. (e) Recent detection of resonances from "exotic" metal nuclei such as ^{25}Mg (179), ^{43}Ca (15, 180), and ^{6}Li or ^{7}Li (468) bound to proteins (along with ^{113}Cd mentioned above) is opening up new possibilities in the study of the role of metal ions in proteins. (f) NMR is being used to investigate very large protein-containing assemblies such as protein aggregates, multienzyme complexes, ribosomes, and virus particles. Highly mobile groups in these complexes give rise to sharp high-resolution NMR signals; and with suitable enrichment, solid state NMR techniques can be applied to the study of their more-rigid regions. (*g*) Considerable success has been achieved in NMR investigations of interactions between proteins and other classes of biomolecules: for example, nucleic acids, lipids, and carbohydrates.

The repertoire of techniques and number of applications has expanded to the point where a review of this size can touch on only a few highlights. Our literature search identified over 200 kinds of proteins, including 75 enzymes, whose structures have been investigated by NMR spectroscopy (over 450 different proteins have been studied if species variants are counted). We include here a catalog of recent reviews, a brief survey of current methodology of protein NMR spectroscopy in solution, and a discussion of proteins for which extensive NMR assignments have been made. The major classes of proteins that have been investigated in detail are presented in Table 1. We have excluded from this review studies of peptides (although the distinction we make between proteins and peptides is somewhat arbitrary), nonfunctional fragments or constituents of proteins, and protein model compounds. The field has reached the stage where it would be valuable to institute a data bank for the deposition of protein NMR parameters for individual, assigned resonances obtained under defined conditions: included should be chemical shifts, coupling constants, NOE and relaxation values along with the dependence of these parameters on variables such as pH, temperature, and, for redox proteins, oxidation state.

OTHER REVIEWS

In focusing this review on structural studies of the proteins themselves in solution, we only cite recent reviews on the related topics of the use of NMR to determine the kinetics of chemically exchanging systems involving proteins (465) or to follow enzyme-catalyzed isotopic exchange reactions (123). Solid-state NMR investigations of proteins have been reviewed recently (192, 439, 473).

The general topic of protein NMR spectroscopy has been covered in several older reviews (92, 114, 349, 354, 462, 463, 463a, 486, 602, 607). Recent monographs that treat NMR in molecular biology (270) or NMR studies of the conformations of biomolecules (204) contain sections on proteins. Wüthrich's book on peptide and protein NMR (601), although dated, still presents a useful compendium of chemical shifts, pH titration shifts, and coupling constants of ^1H, ^{13}C, and ^{15}N nuclei in peptides. Two series of volumes (44, 121) regularly have chapters on NMR applications to proteins. Recent edited volumes (64, 154, 440, 518) also contain relevant chapters. NMR spectroscopy of specific kinds of nuclei in proteins is

Table 1   Survey of proteins studied by NMR spectroscopy[a]

Proteins[b]	Nuclei studied	References[c]
A. Proteinase inhibitors and related proteins		
1. Bovine pancreatic trypsin inhibitor (PTI) (Kunitz) family		
PTI	^{1}H, ^{13}C	567, 608, *605*, 571, 572, 568, 420, 475, 611, 471, 152, 613, 474, 617, 415, 79, 476, 454, 213, 78, 459, 79, 522
colostrum inhibitor	^{1}H	574
toxin E	^{1}H	35
toxin K	^{1}H	288
isoinhibitor K	^{1}H	575
2. Pancreatic secretory trypsin inhibitor (PSTI) (Kazal) family		
PSTI [3]	^{1}H	139, 138, 137
seminal plasma inhibitor	^{1}H	375, 532, 531, 530
ovomucoid [7]	^{1}H, ^{13}C	387, 132, 584, 431
3. *Streptomyces* subtilisin inhibitor (SSI) family		
SSI	^{1}H, ^{13}C, ^{15}N	277, 5, 3, 4, 187, 186
4. Soybean trypsin inhibitor (STI) (Kunitz) family		
STI	^{1}H, ^{13}C	39, 241, 382
B. Toxins		*88, 89, 252*
1. Bee venom toxins		
apamin	^{1}H, ^{13}C	582, *89*, 90, 433, 87
melittin	^{1}H	77, 81, 80, 342, 341, 75, 529
2. *Laticauda semifasciata* toxins		
erabutoxins [3]	^{1}H	402, 251, 246, 254, 542, 253, 250
laticauda semifasciata III	^{1}H	248, 247
3. *Cobra* toxins		
Naja toxins [11]	^{1}H	237, 402, 218, 170, *89*, 33, 524, 169, 168, 190, 34, 32
4. *Bungarus* toxins		
α-bungarotoxin	^{1}H	402, 169
5. Lytic factor		
Haemachatus	^{1}H	524
C. Enzymes		
1. Oxidoreductases		
[1.1.1.1] alcohol dehydrogenase [2]	^{113}Cd	53
[1.1.3.4] glucose oxidase	^{31}P	260
[1.5.1.3] tetrahydrofolate dehydrogenase (dihydrofolate reductase) [6]	^{1}H, ^{13}C, ^{19}F	*478*, 364, 207, 175, 206, 177, 176, 388, 457, 458, 301, 48, 119

Table 1—*continued*

Proteins[b]	Nuclei studied	References[c]
[1.11.1.5] cytochrome *c* peroxidase	^{1}H	496, 495, 494, 328, 493, 325
[1.11.1.7] peroxidase [4]	^{1}H	329, 327, 328, 332, 325, 330, 331, 326
[1.15.1.1] superoxide dismutase [5]	^{1}H	41, 219, 528, 102, 359, 101
2. Transferases		
[2.4.1.1] phosphorylase, glycogen [3]	^{31}P	232
[2.7.3.2] creatine kinase	^{1}H	298, 487
[2.7.3.9] PEP-phosphotransferase [2]	^{1}H	500, 484
[2.7.4.3] adenylate kinase [3]	^{1}H	280, 393
[2.7.5.1] phosphoglucomutase	^{1}H, ^{6}Li, ^{7}Li, ^{31}P, ^{113}Cd	470, 469
3. Hydrolases		
[3.1.1.4] phospholipase a$_2$ [5]	^{1}H, ^{13}C, ^{43}Ca	163, 16, 164, 264, 263
[3.1.3.1] alkaline phosphatase	^{13}C, ^{31}P, ^{113}Cd	127, 200, 445, 444, 442, 54, 116, 240, 28, 50
[3.1.27.3] ribonuclease T$_1$ [2]	^{1}H, ^{13}C	249, 22, 21, 302, 189
[3.1.27.5] ribonuclease (A, B, S, S') [7]	^{1}H, ^{13}C, ^{19}F	316, 317, 318, 17, 46, 578, 122, 492, 255, 352, 259, 579, 428, 51, 515, 162, 18, 67, 516, 76, 514, 70, 449, 450, 451, 378, 377, 104, 398, 394
[3.1.31.1] micrococcal nuclease (staphylococcal nuclease) [2]	^{1}H	269, 172, 268
[3.2.1.17] lysozyme [3]	^{1}H, ^{2}H, ^{13}C	147, 503, 235, 581, 468, 312, 142, 461, 599, 351, 515, 144, 11, 453, 93, 438, 435, 8
[3.4.21.1] chymotrypsin	^{1}H, ^{13}C, ^{19}F	415, 460, 381, 374, 481, 482, 480
[3.4.21.4] pancreatic trypsin [2]	^{1}H, ^{13}C	415, 455, 475, 460, 459, 384, 481
[3.4.21.5] thrombin (kringle)	^{1}H	546, 231, 135
[4.21.12] α-lytic proteinase	^{1}H, ^{13}C, ^{15}N	585, 37, 460, 38, 481, 242
[3.4.21.14] subtilisin	^{1}H	276, 481
[3.4.22.2] papain	^{1}H	357, 272
[3.5.2.6] β-lactamase	^{1}H	40
4. Lyases		
[4.2.1.1] carbonic anhydrase [6]	^{1}H, ^{113}Cd	286, 96, 534, 28, 95
[4.2.1.14] D-serine dehydratase	^{31}P	501
5. Isomerases		
[5.3.1.1] triosephosphate isomerase [2]	^{1}H, ^{13}C, ^{31}P	274, 82
6. Ligases		
[6.2.1.5] succinyl-CoA synthetase	^{31}P	564, 563

Table 1—*continued*

Proteins[b]	Nuclei studied	References[c]
D. Cofactor proteins		
pancreatic colipase	^{1}H	99, 587, 97, 98, 588
α-lactalbumin [5]	^{1}H	43, 541
E. Heme proteins		*339, 294*
1. Cytochromes		*619, 404*
a. *c*-type		
cytochrome *c* [19]	^{1}H, ^{2}H, ^{13}C	506, 58, 591, 600, 85, 291, 59, 408, 410, 409, 411, 499, 129, 292, 203, 293, 413, 438, 148, 155, 143, 289, 435, 392, 434, 526, 466
cytochrome c_{551} [4]	^{1}H	507, 291, 293, 129, 406, 295
cytochrome c_{552} [2]	^{1}H	233, 291, 129, 296
cytochrome c_{553} [5]	^{1}H	551
cytochrome c_{556} [2]	^{1}H	405
cytochrome c_{557}	^{1}H	290, 289
cytochrome cd_1	^{1}H	543
cytochrome *c'* [3]	^{1}H	405, 167, 166
cytochrome c_2	^{1}H	521
b. other cytochromes		
cytochrome P-450	^{3}H	596
cytochrome b_5	^{1}H	324
cytochrome b_{562}	^{1}H	620
2. Myoglobins and hemoglobins		
myoglobin [9]	^{1}H, ^{2}H, ^{13}C, ^{19}F	*210*, 197, 297, 329, 328, 66, 303, 325, 488, 133, 333, 65, 321, 322, 432, 326, 243, 60, 61, 323, 590, 595, 589, 438, 389, 519
leghemoglobin [2]	^{1}H	368, 307, 19, 66, 333, 545, 273
Chironomus hemoglobin [3]	^{1}H	*198*, 320, 337, 335
human hemoglobin [19]	^{1}H, ^{13}C	*229*, 256, 401, 489, 540, 491, 416, 502, 539, 490, 369, 334, 555, 36, 436, 191, 299
F. Copper proteins		*552*
plastocyanin [10]	^{1}H, ^{13}C, ^{113}Cd	306, 171, 303, 128, 184, 386
azurin [2]	^{1}H, ^{13}C, ^{113}Cd	171, 400, 1, 49, 221, 550, 549, 220
stellacyanin	^{1}H, ^{113}Cd	171

Table 1—*continued*

Proteins[b]	Nuclei studied	References[c]
G. Non-heme Iron Proteins		*91*
1. Ferredoxins		
2Fe·2S ferredoxin [13]	^{1}H, ^{13}C	419a, 109, 107, 110, 112, 111, 113
4Fe·4S ferredoxin [9]	^{1}H, ^{13}C	419a, 447, 412, 448, 413
2. Respiratory pigments		
hemerythrin	^{1}H	624
H. Nucleic acid binding proteins		
gene 5 protein [2]	^{1}H, ^{19}F	430, 13, 12, 194, 193, 126, 125
lac repressor	^{1}H, ^{19}F	*285*, 498, 31, 271, 57, 472, 30, 84
cro protein	^{1}H, ^{19}F	29, 315, 304, 257
elongation factor Tu [2]	^{1}H	425, 424
adenosine cyclic 3′,5′-phosphate receptor protein (CRP or CAP)	^{1}H	118
ubiquitin	^{1}H	100
high mobility group chromosomal protein	^{1}H	100
I. Metal binding proteins		
calmodulin [9]	^{1}H, ^{113}Cd	13a, 13b, 245, 244, 311, 14a, 15, 224, 223, 222, 181, 504
parvalbumin [7]	^{1}H, ^{13}C, ^{113}Cd	347a, 15, 347, 346, 427, 426
metallothionein [12]	^{113}Cd	*27*, *446*, 73, 443
transferrin [2]	^{1}H	598
S-100 protein [2]	^{1}H	372, 305
intestinal calcium binding protein	^{1}H	511
J. Hydrophobic and membrane-associated proteins		
crambin	^{1}H	343, 136
thionin	^{1}H	344
myelin basic protein [5]	^{1}H	397a, 397, 396, 239, 361, 360
lipoprotein from *Escherichia coli*	^{13}C, ^{19}F	348
acyl carrier protein	^{1}H	389a, 191a
phosphocarrier protein, HPR [5]	^{1}H, ^{31}P	279, 278, 483, 500, 195
K. Rhodopsins		
bacteriorhodopsin	^{2}H, ^{13}C	212, 623
rhodopsin	^{13}C	517

Table 1—*continued*

Proteins[b]	Nuclei studied	References[c]
L. Hormone binding proteins and protein hormones		
neurophysin [2]	^{1}H	560, 366, 52, 365
lutropin [2]	^{1}H	370, 74
insulin [3]	^{1}H, ^{113}Cd	533, 68, 594
choriogonadotropin	^{1}H	183
serum gonadotropin	^{1}H	74
epidermal growth factor	^{1}H	139a
M. Flavoproteins		*414*
flavodoxin [8]	^{1}H, ^{13}C	156, 403, 554, 553, 261
N. Plasma proteins		
immunoglobulins [16]	^{1}H, ^{31}P	*150*, 512, 199, 201, 258, 20, 513, 24, 25, 577, 202
plasma albumin	^{13}C	115
O. Glycoproteins		
antifreeze glycoproteins [4]	^{13}C	45
ovalbumin	^{31}P	562
P. Muscle proteins		
myosin	^{1}H, ^{31}P	216, 308
tropomyosin [3]	^{1}H, ^{31}P	160, 159, 157, 158, 371
troponin-C [3]	^{1}H, ^{19}F, ^{31}P, ^{25}Mg, ^{43}Ca	541a, 15, 224, 223, 14, 222, 523, 182, 157, 47, 505

[a] Texts or abstracts of over 1700 publications on NMR of proteins were screened in constructing this table. Unfortunately, each paper could not be read thoroughly; and unintentional omissions are bound to have occurred. References were included if they met two criteria: (*i*) A second stage resonance assignment was made (or corrected) or detailed studies of a protein were carried out using previous assigments. (*ii*) The assigned resonance in question arises from a natural part of the protein; prosthetic groups were included, but inhibitors or other small ligands were not.

[b] Enzymes are identified by numbers (in parentheses) and names recommended by the Nomenclature Committee of the International Union of Biochemistry (1978). (Nomenclature Committee of the International Union of Biochemistry, 1979. *Enzyme Nomenclature 1978*. New York: Academic. 606 pp.) Commonly used names are given in parentheses. Numbers in square brackets indicate the number of species variants that have been studied by NMR; references to all of these may not be included.

[c] References are listed chronologically beginning with the most recent; references to review articles are in italics.

covered in a number of reviews: nuclei other than ^{1}H (120), ^{13}C (6, 7, 10, 161, 238, 363, 486), ^{31}P (561), ^{15}N (282), ^{19}F (196, 535, 537), ^{113}Cd (26, 151), ^{17}O and ^{18}O labeling (547, 548), ^{43}Ca in proteins that bind calcium ions (180), the study of various metal nuclei in metalloproteins (345, 556, 558, 559), and ^{3}H (106). General reviews of protein dynamics (390), hydrogen exchange (597), and protein folding (300) contain discussions of contributions of NMR spectroscopy. A plethora of articles have reviewed the important contributions of NMR to detecting and quantitating internal mobility in proteins by NMR relaxation (265, 362) or by a combination of techniques (592, 593) with the primary subjects being PTI, the bovine pancreatic trypsin inhibitor (Kunitz) (566, 573, 606, 612, 614–616), and lysozyme (141, 147).

Recent NMR texts (e.g. 214, 510) discuss modern pulse-Fourier transform techniques or instrumentation (188). Several methods of relevance to protein NMR have been treated in depth in separate monographs or reviews: two-dimensional Fourier transform NMR (42, 185), multiple quantum NMR (165), and the nuclear Overhauser effect (62, 429, 485). NMR methods as applied specifically to proteins have been reviewed in a number of cases: two-dimensional Fourier transform NMR (417, 422, 610) ring current shifts and their calculation (452), NMR in aqueous solutions (464), photochemically induced nuclear polarization (234, 283, 284), uses of chromium and cobalt nucleotide complexes (557), carbamate reactions with proteins (209), and the sequential assignment of individual ^{1}H resonances in proteins (609).

Detailed reviews have appeared on NMR studies of particular classes of proteins: collagen and elastin (544), membrane proteins (437), organization and dynamics of plasma lipoproteins (287)— including low-density lipoproteins (106, 217) and high-density lipoproteins (71), flavins and flavoproteins (414), active centers of iron-sulfur proteins (91), electron carrier proteins from sulfate reducing bacteria (622), blue-copper proteins (552), neurotoxins (88, 89, 252), histones (130), iron-containing proteins (173), and NMR studies of metal nuclei in metalloenzymes (345, 558). NMR studies of heme proteins have been reviewed extensively (225, 228, 294, 336, 339, 603) along with several subtopics: model compounds as aids in interpreting NMR spectra of heme proteins (319), ^{1}H NMR of hemoglobins (198, 226, 229), NMR studies of myoglobins (210), and cytochromes (404, 407, 619, 621). NMR investigations of calcium-binding proteins (310, 355, 565) and in particular parvalbumin (103) have been summarized. Reviews on individual proteins include PTI (608), its internal motions (567), and a comparison of its solution and crystal structures (604, 605), the *lac* repressor (367) and its interaction with *lac* operator DNA (285), gene 5 protein and its interaction with oligonucleotides (125), metallothionein (27, 446, 583), and an antibody binding site (150, 153).

General reviews on NMR of enzymes have treated the geometry of enzyme-bound substrates and analogues (208, 399) and the active sites of enzyme complexes (124) or phosphoryl transfer enzymes (558). The serine proteinases have been the subject of three reviews (379, 383, 525). Reviews that focus on individual enzymes and describe NMR studies include those on alkaline phosphatase (127), dihydrofolate reductase (174, 477–479), thymidylate synthase (86), creatine kinase (298), succinyl-coenzyme A synthetase (564), and lysozyme (146) and its hydrogen exchange properties (147).

CURRENT NMR METHODOLOGY

One-Dimensional NMR

The methods used for one-dimensional NMR spectroscopy of proteins are now fairly well standardized. When ^2H_2O is used as the solvent, pulse-Fourier transform spectroscopy is generally the method of choice. Selective enhancement of sharp lines over broader lines (resolution enhancement) can be achieved by use of a Carr-Purcell sequence (94) or more commonly by apodization (point-by-point multiplication of the free induction decay) with a sine bell (sinusoidal window function) (140) or similar function. Complex spectra may be simplified by selecting subspectra on the basis of their coupling multiplicity, by use of a J-modulated spin echo pulse sequence (94) or by the related APT (attached proton test) pulse sequence for heteronuclear NMR (113), by use of a summed spin-echo spectrum (55), or by multiple quantum coherence (236). Interproton distance information can be obtained from nuclear Overhauser experiments (528). Truncated-driven NOE measurements (152) are usually superior to steady-state NOE measurements because the method provides a way of limiting spin-diffusion effects (2, 281, 536). Factors influencing the time development of NOEs in proteins have been investigated (149). Solvent suppression for ^1H NMR in solutions containing ^1H_2O is commonly achieved by single-frequency irradiation to saturate the water signal, which, however, leads to transfer of saturation to signals from rapidly exchanging –OH and –NH groups. In cases where the rapidly exchangeable protons are the objects of study, their resonances can be detected through use of the pulse sequences developed by Redfield and co-workers (467) or Plateau & Gueron (456). Rapid-scan correlation spectroscopy (134) also may be used conveniently in situations where there are large solvent or buffer peaks; improvements have been suggested (23) for the processing of correlation NMR data sets. The standard relaxation methods used to determine correlation times (see the reviews above) have been augmented recently by a technique involving the measurement of the rotating frame spin-lattice relaxation in the presence of an off-resonance radio-frequency field (262), which extends the measurements to longer correlation times. It has been pointed out that interatomic distances used for dipolar relaxation calculations should include vibrational corrections and that such corrections can influence calculated correlation times by as much as a factor of 2 (145). The importance of spin diffusion (cross relaxation) on NMR relaxation in paramagnetic proteins has been delineated (520).

Newer multipulse strategies for broadband decoupling incorporating cyclic sequences of a train of composite spin inversion pulses (580), such as the "MLEV" (508) or "WALTZ" (509) supercycles, require less power than older methods. This is important for heteronuclear NMR studies of proteins at high field strengths where sample heating by the proton decoupler can present serious problems.

A novel NMR variant of solvent-perturbation spectroscopy is afforded by photochemical dynamic polarization (CIDNP) experiments using photo-excited external dyes to polarize transitions from the aromatic side chains of histidine, tyrosine, and tryptophan (for a review see 284) or from other groups by means of covalently attached dye molecules (353). The technique has been applied to a large number of proteins and has been used to correct and extend assignments as well as to probe the solvent accessibility of protein groups under various conditions.

Two-Dimensional NMR

Many of the two-dimensional NMR techniques originally developed with organic molecules have been applied to small proteins. The first experiments with proteins utilized two-dimensional J-resolved ^1H spectroscopy (421) along with cross-sections and projections (418). Heteronuclear (^1H, ^{13}C) two-dimensional J-spectroscopy has been demonstrated with a protein (113), but since the $^1H-^{13}C$ coupling constants are similar for nearly all residues, its utility is limited to detection of abnormal coupling constants such as for the histidine ring C–H groups. Heteronuclear (^1H, ^{113}Cd) 2D J-spectroscopy has been carried out with ^{113}Cd, Zn metallothionein (26) with the advantage that a projection could be used to eliminate $^{113}Cd-^{113}Cd$ coupling.

The various kinds of two-dimensional spectroscopy based on chemical shift connectivity can be broken down into two classes: J-connectivity and NOE- or exchange-connectivity experiments. Homonuclear (^1H, ^1H) J-connectivity methods include two-dimensional spin-echo correlated spectroscopy (SECSY) (423), foldover-corrected correlated spectroscopy (FOCSY) (419), and two-dimensional correlated spectroscopy (COSY)

(571). COSY is the most versatile method where data accumulation is not limited by available digitization. Three-bond couplings normally are observed in COSY, but longer-range couplings can be studied. The four-bond connectivity between the C_δ–H and C_ε–H of histidine rings in proteins has been reported (303). Phase-sensitive COSY, although less sensitive, appears to afford superior resolution of peaks near the diagonal and may be advantageous for measurement of ^1H–^1H spin-spin coupling constants (375). Two-dimensional double-quantum ^1H NMR spectroscopy has been applied to a protein (576). This method does not give rise to peaks on the diagonal and therefore is ideally suited for resolving connectivities between resonances with very similar chemical shifts. It is also useful for analyzing remote connectivities. Another approach to determining remote connectivities is two-dimensional relayed coherence transfer spectroscopy, which has been used to assign spin systems in proteins (303a, 566a).

Of the various heteronuclear J-connectivity experiments, only (^1H, ^{13}C) J-connectivity has been demonstrated thus far with a protein. The earlier applications used ^{13}C-enrichment (108, 306), but the experiment can be performed at natural abundance provided that sufficient protein is available (387). Two-dimensional (^{13}C, ^{13}C) J-connectivity (387) and (^{15}N, ^1H) and (^{15}N, ^{13}C) (205) J-connectivity experiments have been carried out with amino acids or peptides. Experimental evidence showing a large dispersion in (^1H, ^{15}N) (350) and (^{13}C, ^{15}N) (277) connectivities of peptide units in proteins (obtained by one-dimensional NMR experiments) indicates that these heteronuclear two-dimensional NMR approaches should be of great value if they become feasible for proteins.

Two-dimensional NOE spectroscopy has been shown only for proton-proton interactions (NOESY) in proteins (313), where it is of utility for elucidating cross-relaxation pathways or extending NMR assignments, particularly in ^1H_2O solution (314); the solvent resonance can be suppressed by saturation methods (586). A comparison has been made of selective NOE measurements made by one-dimensional and two-dimensional methods (56). Spin-diffusion problems appear to limit present methods for two-dimensional NOE spectroscopy to proteins with molecular weights below 20,000.

RESONANCE ASSIGNMENTS IN PROTEINS

Assignment Strategies

A useful methodological distinction (92, 146) has been made between first stage assignments, i.e. determination of the kind of amino acid giving rise to the resonance, and second stage assignments, i.e. identification of the particular residue in the peptide sequence giving rise to the resonance. At present, an independently determined and accurate amino acid sequence is required before NMR assignments can be contemplated. First stage assignments usually can be made on the basis of characteristic chemical shifts (and occasionally their pH dependence) and coupling pattern (spin system). Second stage assignments generally require additional information if more than one residue of a given type is present in the protein. Numerous strategies can be followed in making assignments. Isotopic substitution offers a reliable, but somewhat tedious method. Specific labeling can be achieved by isotopic exchange (380, 394), chemical modification and reversal (259, 275, 499), peptide synthesis (105), enzymatic semisynthesis (39, 475), or biosynthesis (83, 376). A $[^{13}C, ^{15}N]$ double-labeling procedure for peptide groups where assignments are made on the basis of dipeptide patterns appears very attractive (277). Selective ^{17}O labeling of an enzymic phosphate has been used to assign a ^{31}P resonance in the NMR spectrum of a phosphoprotein (470). With the advent of NMR methods for extending assignments, it is convenient to distinguish primary resonance assignments, that is assignments made directly to particular atoms or groups in the protein (the second stage assignment above), from secondary assignments, assignments made by extensions of these primary assignments by NMR methods (generally NOE or J-connectivity experiments) (618).

At present nearly all of the ^1H resonances in the spectrum of a small protein that gives sharp NMR peaks can be resolved and assigned. The paradigm for complete ^1H NMR spectral analysis of a protein is the bovine pancreatic trypsin inhibitor (6,500 mol wt), which has been studied for several years and served as the model for the development of homonuclear (^1H) two-dimensional NMR techniques for the identification of spin systems of amino acid side chains (420) and peptide sequential extensions of NMR assignments (568).

Homologous or Variant Proteins

Protein variants provide a reliable means of making second stage (primary) assignments in larger proteins that still yield resolvable peaks, as shown first with staphylococcal nuclease (16,800 mol wt) (268). Modern genetic methods offer the means of extending this assignment approach to virtually any residue (271) as well as providing ways of producing large quantities of interesting proteins. From the standpoint of elucidating the function of single residues in proteins, a comparison of proteins with single residue replacements is ideal (229, 387). Several amino acid replacements can be acceptable, or even desirable in certain cases, if the object is to use the spectral differences to assign resonances. More distantly related proteins have been compared in order to determine common features such as the pK'_a value of the active site residues in serine proteinases (383) or the pattern of electron delocalization or ligand stereochemistry in cytochromes (292, 405, 506, 507, 551). Several interesting families of proteins or protein domains are beginning to be studied in detail: proteins homologous to the Kazal secretory proteinase inhibitor (138, 387, 431, 530–532); "kringle" fragments of plasminogen (135, 231, 546); thionins and crambin (344); myoglobins from various species (60, 61); myelin basic proteins (397a); ferredoxins (107, 109–111, 419a); and PTI and related proteinase inhibitors and toxins (35, 288, 571, 575).

INTERPRETATION OF NMR PARAMETERS

The availability of large numbers of assigned peaks in proteins whose structures have been determined by X-ray crystallography should shortly lead to advances in our understanding of the origins of chemical shifts and coupling constants in proteins. Preliminary attempts have been made to use protein NMR results to determine secondary and tertiary structural features of proteins in solution. One approach has been to use distance information generated by two-dimensional NOE experiments as input data (72) for the distance geometry algorithm of Crippen and co-workers (131, 215). Another method has been to rely on short-range distances determined by NOE data along with amide proton exchange rates and ϕ-angle-dependent $^3J_{^1HN-C_a^1H}$ coupling constants (582). Complete determination of the solution structure of a protein by NMR means still is an unrealized goal.

A sound theoretical basis for NMR chemical shifts in proteins is still lacking. In spite of the fact that progress has been made in evaluating ring-current contributions to chemical shifts (452), a general means of predicting diamagnetic anisotropy or electric field shifts of peptide groups is unavailable (117). A severe complication in calculating NMR parameters arises from the mobility of protein groups (266). Picosecond molecular dynamics simulations will not prove helpful here because NMR chemical shifts and coupling constants are averaged over a much longer time scale (230). Environmental shifts play a proportionately smaller role in ^{13}C and ^{15}N chemical shifts than in ^1H shifts. Large ^{13}C chemical shift differences can be interpreted in terms of differences in bond order (39); and ^{13}C chemical shifts can be more reliable than ^1H NMR chemical shifts for assignment purposes (306). It has been postulated that the chemical shift of protein-bound ^{113}Cd can be used to distinguish types of metal ligands, but recent results with blue-copper proteins casts some doubt on this (171). Several pK_a' values of surface groups determined experimentally by NMR methods have been evaluated rather successfully by a modified form of the Tanford-Kirkwood electrostatic theory (373).

Progress has been made in incorporating the results of molecular dynamics simulations of protein motions into an analysis of NMR relaxation rates. The molecular dynamics trajectories carried out to date have been limited by the speed and cost of computations to about 100 ps. Nevertheless, the rapid motions modeled by these trajectories are expected to have a measurable effect on relaxation of protonated (358) and nonprotonated (356) carbons.

The sensitivity of NMR spectral parameters to minor changes in protein sequence or structure or the presence of contaminants presents special demands on the purity of protein samples. Apart from this difficulty, multiple (interconvertible) forms of proteins have been reported on several occasions on the basis of NMR evidence. Three of the better-understood examples are (a) a pH-dependent conformational equilibrium in azurin (1, 550), which is dependent on the protonation state of His-35 (49) and appears to explain anomalies in the kinetics of electron transfer between azurin and cytochrome c_{551}; (b) the detection of doubling of ^{13}C resonances from two tryptophan residues of *Streptococcus faccium* dihydrofolate reductase (364), which indicates the presence of three interconvertible forms of the enzyme and may explain the complex kinetics of coenzyme and inhibitor binding observed for the homologous enzymes from other species (478); and (c) the observation of "heme disorder" in hemoproteins that stems from the existence of two molecular forms, which differ by a 180° rotation of the heme in its binding pocket (323).

SUMMARY

Techniques are now available for making extensive assignments in NMR spectra of proteins of moderate size (molecular weight 20,000 or less). Such assignments provide the first step for experiments designed to extract the full complement of NMR parameters for each group in a protein. The stage is set for exciting research scenarios in protein chemistry involving, for example, the determination of hydrogen exchange kinetics at all exchangeable positions whose half times are on the order of 100 ms (277) or longer than a few minutes (316, 569, 570); the characterization of intermediates in protein folding pathways (318); measurement of the distribution of internal motions within a protein molecule (573); a detailed description of the biophysical consequences of single amino acid replacements in small proteins (387); elucidation of the mechanisms of conformational transitions in proteins; and multiparametric characterization of the parts of an enzyme that participate in catalytic mechanisms. Small proteins for which extensive ^{1}H NMR assignments have been made include lysozyme, several cytochromes, ferredoxins, myelin basic proteins, PTI and related proteinase inhibitors, proteinase inhibitors from seminal plasma and avian eggs, apamin, and several snake venom neurotoxins. (References are given in Table 1.)

ACKNOWLEDGMENTS

We thank numerous colleagues for supplying reprints and preprints of their work. Preparation of this review has been supported by the National Institutes of Health (GM 09077, RR 01077) and the United States Department of Agriculture Competitive Research Grants Office, Cooperative State Research Service, Science and Education (82-CRCR-1-1045).

Literature Cited

1. Adman, E. T., Canters, G. W., Hill, H. A. O., Kitchen, N. A. 1982. *FEBS Lett.* 143:287–92
2. Akasaka, K. 1983. *J. Magn. Reson.* 51:14–25
3. Akasaka, K., Fujii, S., Hatano, H. 1982. *J. Biochem. Tokyo* 92:591–98
4. Akasaka, K., Fujii, S., Kaptein, R. 1981. *J. Biochem. Tokyo* 89:1945–49
5. Akasaka, K., Hatano, H., Tsuji, T., Kainosho, M. 1982. *Biochim. Biophys. Acta* 704:503–8
6. Allerhand, A. 1978. *Acc. Chem. Res.* 11:469–74
7. Allerhand, A. 1979. *Methods Enzymol.* 61:458–549
8. Allerhand, A., Childers, R. F., Oldfield, E. 1973. *Biochemistry* 12:1335–41
9. Allerhand, A., Cochran, D. W., Doddrell, D. 1970. *Proc. Natl. Acad. Sci. USA* 67:1093–96
10. Allerhand, A., Dill, K., Goux, W. J. 1979. See Ref. 440, pp. 31–50
11. Allerhand, A., Norton, R. S., Childers, R. F. 1977. *J. Biol. Chem.* 252:1786–94
12. Alma, N. C. M., Harmsen, B. J. M., Hilbers, C. W., Van der Marel, G., Van Boom, J. H. 1981. *FEBS Lett.* 135:15–20
13. Alma, N. C. M., Harmsen, B. J. M., Van Boom, J. H., Van der Marel, G., Hilbers, C. W. 1982. *Eur. J. Biochem.* 122:319–26
13a. Andersson, A., Drakenberg, T., Thulin, E., Forsén, S. 1983. *Eur. J. Biochem.* 134:459–65
13b. Andersson, A., Forsén, S., Thulin, E., Vogel, H. J. 1983. *Biochemistry* 22:2309–13
14. Andersson, T., Drakenberg, T., Forsén, S., Thulin, E. 1981. *FEBS Lett.* 125:39–43
14a. Andersson, T., Drakenberg, T., Forsén, S., Thulin, E. 1982. *Eur. J. Biochem.* 126:501–5
15. Andersson, T., Drakenberg, T., Forsén, S., Thulin, E., Swärd, M. 1982. *J. Am. Chem. Soc.* 104:576–80
16. Andersson, T., Drakenberg, T., Forsén, S., Wieloch, T., Lindstrom, M. 1981. *FEBS Lett.* 123:115–17
17. Andini, S., D'Alessio, G., Di Donato, A., Paolillo, L., Piccoli, R., Trivellone, E. 1983. *Biochim. Biophys. Acta* 742:530–38
18. Antonov, I. V., Gurevich, A. Z., Dudkin, S. M., Karpeiskii, M. Ya., Sakharovskii, V. G., Yakovlev, G. I. 1978. *Eur. J. Biochem.* 87:45–54
19. Appleby, C. A., Blumberg, W. E., Bradbury, J. H., Fuchsman, W. H., Peisach, J., Wittenberg, B. A., et al. 1982. See Ref. 228, pp. 435–41
20. Arata, Y., Honzawa, M., Shimizu, A. 1980. *Biochemistry* 19:5130–35
21. Arata, Y., Kimura, S., Matsuo, H., Narita, K. 1979. *Biochemistry* 18:18–24
22. Arata, Y., Kimura, S., Matsuo, H., Narita, K. 1980. *Biochem. Biophys. Res. Commun.* 73:133–40
23. Arata, Y., Ozawa, H., Ogino, T., Fujiwara, S. 1978. *Pure Appl. Chem.* 50:1273–80
24. Arata, Y., Shimizu, A. 1979. *Biochemistry* 18:2513–20
25. Arata, Y., Shimizu, A., Matsuo, H. 1978. *J. Am. Chem. Soc.* 100:3230–32
26. Armitage, I. M., Otvos, J. D. 1982. See Ref. 44, 4:79–144
27. Armitage, I. M., Otvos, J. D., Briggs, R. W., Boulanger, Y. 1982. *Fed. Proc.* 41:2974–80
28. Armitage, I. M., Pajer, R. T., Schoot Uiterkamp, A. J. M., Chlebowski, J. F., Coleman, J. E. 1976. *J. Am. Chem. Soc.* 98:5710–12
29. Arndt, K., Boschelli, F., Cook, J., Takeda, Y., Tecza, E., Lu, P. 1983. *J. Biol. Chem.* 258:4177–83

30. Arndt, K. T., Boschelli, F., Lu, P., Miller, J. H. 1981. *Biochemistry* 20:6109–18

31. Arndt, K., Nick, H., Boschelli, F., Lu, P., Sadler, J. 1982. *J. Mol. Biol.* 161:439–57

32. Arseniev, A. S., Balashova, T. A., Utkin, Y. N., Tsetlin, V. I., Bystrov, V. F., Ivanov, V. T., Ovchinnikov, Yu. A. 1976. *Eur. J. Biochem.* 71:595–606

33. Arseniev, A. S., Pashkov, V. S., Pluzhnikov, K. A., Rochat, H., Bystrov, V. F. 1981. *Eur. J. Biochem.* 118:453–62

34. Arseniev, A. S., Surin, A. M., Utkin, Y. N., Tsetlin, V. I., Bystrov, V. F., Ivanov, V. T., Ovchinnikov, Yu. A. 1978. *Bioorg. Khim.* 4:197–207

35. Arseniev, A. S., Wider, G., Joubert, F. J., Wüthrich, K. 1982. *J. Mol. Biol.* 159:323–51

36. Asakura, T., Adachi, K., Wiley, J. S., Fung, L. W.-M., Ho, C., Kilmartin, J. V., Perutz, M. F. 1976. *J. Mol. Biol.* 104:185–95

37. Bachovchin, W. W., Kaiser, R., Richards, J. H., Roberts, J. D. 1981. *Proc. Natl. Acad. Sci. USA* 78:7323–26

38. Bachovchin, W. W., Roberts, J. D. 1978. *J. Am. Chem. Soc.* 100:8041–47

39. Baillargeon, M. W., Laskowski, M. Jr., Neves, D. E., Porubcan, M. A., Santini, R. E., Markley, J. L. 1980. *Biochemistry* 19:5703–10

40. Baldwin, G. S., Galdes, A., Hill, H. A. O., Smith, B. E., Waley, S. G., Abraham, E. P. 1978. *Biochem. J.* 175:441–47

41. Bannister, J. V., Bannister, W., Cass, A. E. G., Hill, H. A. O., Johansen, J. T. 1980. *Dev. Biochem.* 11A:284–89

42. Bax, A. D. 1982. *Two-Dimensional Nuclear Magnetic Resonance in Liquids.* Amsterdam: Reidel. 200 pp.

43. Berliner, L. J., Kaptein, R. 1981. *Biochemistry* 20:799–807

44. Berliner, L. J., Reuben, J., eds. 1978, 1980, 1981, 1982. *Biological Magnetic Resonance*, Vols. 1, 2, 3, 4. New York: Plenum. 345 pp., 351 pp., 268 pp., 340 pp.

45. Berman, E., Allerhand, A., DeVries, A. L. 1980. *J. Biol. Chem.* 255:4407–10

46. Biringer, R. G., Fink, A. L. 1982. *J. Mol. Biol.* 160:87–116

47. Birnbaum, E. R., Sykes, B. D. 1978. *Biochemistry* 17:4965–71

48. Blakley, R. L., Cocco, L., London, R. E., Walker, T. E., Matwiyoff, N. A. 1978. *Biochemistry* 17:2284–93

49. Blaszak, J. A., Ulrich, E. L., Markley, J. L., McMillin, D. R. 1982. *Biochemistry* 21:6253–58

50. Block, J. L., Sheard, B. 1975. *Biochem. Biophys. Res. Commun.* 66:24–30

51. Blum, A. D., Smallcombe, S. H., Baldwin, R. L. 1978. *J. Mol. Biol.* 118:305–16

52. Blumenstein, M., Hruby, V. J., Viswanatha, V. 1979. *Biochemistry* 18:3552–57

53. Bobsein, B. R., Myers, R. J. 1980. *J. Am. Chem. Soc.* 102:2454–55

54. Bock, J. L., Kowalsky, A. 1978. *Biochim. Biophys. Acta* 526:135–46

55. Bolton, P. H. 1981. *J. Magn. Reson.* 45:418–21

56. Bösch, C., Kumar, A., Baumann, R., Ernst, R. R., Wüthrich, K. 1981. *J. Magn. Reson.* 42:159–63

57. Boschelli, F., Jarema, M. A. C., Lu, P. 1981. *J. Biol. Chem.* 256:11595–99

58. Boswell, A. P., Eley, C. G. S., Moore, G. R., Robinson, M. N., Williams, G., Williams, R. J. P., et al. 1982. *Eur. J. Biochem.* 124:289–94

59. Boswell, A. P., Moore, G. R., Williams, R. J. P., Chien, J. C. W., Dickinson, L. C. 1980. *J. Inorg. Biochem.* 13:347–52

60. Botelho, L. H., Friend, S. H., Matthew, J. B., Lehman, L. D., Hanania, G. I. H., Gurd, F. R. N. 1978. *Biochemistry* 17:5197–5205

61. Botelho, L. H., Gurd, F. R. N. 1978. *Biochemistry* 17:5188–96

62. Bothner-By, A. A. 1979. See Ref. 518, pp. 177–219

63. Bothner-By, A. A., Dadok, J. 1979. See Ref. 440, pp. 169–202

64. Bothner-By, A. A., Glickson, J. D., Sykes, B. D., eds. 1982. *Biochemical Structure Determination by NMR.* New York: Dekker. 232 pp.

65. Bradbury, J. H., Carver, J. A., Parker, M. W. 1981. *J. Chem. Soc. Chem. Commun.* 1981:208–9

66. Bradbury, J. H., Carver, J. A., Parker, M. W. 1982. *FEBS Lett.* 146:297–301

67. Bradbury, J. H., Crompton, M. W., Teh, J. S. 1977. *Eur. J. Biochem.* 81:411–22

68. Bradbury, J. H., Ramesh, V., Dodson, G. 1981. *J. Mol. Biol.* 150:609–13

69. Bradbury, J. H., Scheraga, H. A. 1966. *J. Am. Chem. Soc.* 88:4240–46

70. Bradbury, J. H., Teh, J. S. 1975. *J. Chem. Soc. Chem. Commun.* 1975:936–37

71. Brasure, E. B., Henderson, T. O. 1981. In *High-Density Lipoproteins*, ed. C. F. Day, pp. 73–93. New York: Dekker

72. Braun, W., Bösch, C., Brown, L., Go, N., Wüthrich, K. 1981. *Biochim. Biophys. Acta* 667:377–96

73. Briggs, R. W., Armitage, I. M. 1982. *J. Biol. Chem.* 257:1259–62

74. Brown, F. F., Parsons, T. F., Sigman, D. S., Pierce, J. G. 1979. *J. Biol. Chem.* 254:4335–38

75. Brown, L. R. 1979. *Biochim. Biophys. Acta* 557:135–48

76. Brown, L. R., Bradbury, J. H. 1976. *Eur. J. Biochem.* 68 : 227–35
77. Brown, L. R., Braun, W., Kumar, A., Wüthrich, K. 1982. *Biophys. J.* 37 : 319–28
78. Brown, L. R., De Marco, A., Richarz, R., Wagner, G., Wüthrich, K. 1978. *Eur. J. Biochem.* 88 : 87–95
79. Brown, L. R., De Marco, A., Wagner, G., Wüthrich, K. 1976. *Eur. J. Biochem.* 62 : 103–7
80. Brown, L. R., Lauterwein, J., Wüthrich, K. 1980. *Biochim. Biophys. Acta* 622 : 231–44
81. Brown, L. R., Wüthrich, K. 1981. *Biochim. Biophys. Acta* 647 : 95–111
82. Browne, C. A., Campbell, I. D., Kiener, P. A., Phillips, D. C., Waley, S. G., Wilson, I. A. 1976. *J. Mol. Biol.* 100 : 319–43
83. Browne, D. T., Kenyon, G. L., Packer, E. L., Sternlicht, H., Wilson, D. M. 1973. *J. Am. Chem. Soc.* 95 : 1316–23
84. Buck, F., Rüterjans, H., Kaptein, R., Beyreuther, K. 1980. *Proc. Natl. Acad. Sci. USA* 77 : 5145–48
85. Burns, P. D., La Mar, G. N. 1981. *J. Biol. Chem.* 256 : 4934–39
86. Byrd, R. A., Dawson, W. H., Ellis, P. D., Dunlap, R. B. 1978. *Dev. Biochem.* 4 : 367–70
87. Bystrov, V. F., Arseniev, A. S., Gavrilov, Yu. D. 1978. *J. Magn. Reson.* 30 : 151–84
88. Bystrov, V. F., Arseniev, A. S., Kondakov, V. I., Maiorov, N., Okhanov, V., Ovchinnikov, Yu. A. 1983. In *Neurotoxins as Tools in Neurochemistry, Proc. USSR-Berlin (West) Symp., March 22–24*, ed. Yu. Ovchinnikov, F. Hucho. Berlin : de Gruyter
89. Bystrov, V. F., Ivanov, V. T., Okhanov, V. V., Miroshnikov, A. I., Arseniev, A. S., Tsetlin, V. I., Pashkov, V. S. 1981. *Advances in Solution Chemistry*, ed. I. Bertini, L. Lunazzi, A. Dei, pp. 231–51. New York : Plenum
90. Bystrov, V. F., Okhanov, V. V., Miroshnikov, A. I., Ovchinnikov, Yu. A. 1980. *FEBS Lett.* 119 : 113–17
91. Cammack, R., Dickson, D. P. E., Johnson, C. E. 1977. In *Iron-Sulfur Proteins*, ed. W. Lovenberg, 3 : 283–330. New York : Academic
92. Campbell, I. D. 1977. See Ref. 154, pp. 33–49
93. Campbell, I. D., Dobson, C. M., Williams, R. J. P. 1975. *Proc. R. Soc. London Ser. A* 345 : 23–40
94. Campbell, I. D., Dobson, C. M., Williams, R. J. P., Wright, P. E. 1975. *FEBS Lett.* 57 : 96–99
95. Campbell, I. D., Lindskog, S., White, A. I. 1975. *J. Mol. Biol.* 98 : 597–614
96. Campbell, I. D., Lindskog, S., White, A. I. 1977. *Biochim. Biophys. Acta* 484 : 443–52
97. Canioni, P., Cozzone, P. J. 1979. *Biochimie* 61 : 343–54
98. Canioni, P., Cozzone, P. J. 1979. *FEBS Lett.* 97 : 353–57
99. Canioni, P., Cozzone, P. J., Sarda, L. 1980. *Biochim. Biophys. Acta* 621 : 29–42
100. Cary, P. D., King, D. S., Crane-Robinson, C., Bradbury, E. M., Rabbani, A., Goodwin, G. H., Johns, E. W. 1980. *Eur. J. Biochem.* 112 : 577–88
101. Cass, A. E. G., Hill, H. A. O., Smith, B. E., Bannister, J. V., Bannister, W. H. 1977. *Biochemistry* 16 : 3061–66
102. Cass, A. E. G., Hill, H. A. O., Smith, B. E., Bannister, J. V., Bannister, W. H. 1977. *Biochem. J.* 165 : 587–89
103. Cavé, A., Parello, J. 1981. In *Immun. Intercell., Les Houches, Ec. Ete Phys. Theor., 3rd,* ed. R. Balian, M. Chabre, P. F. Devaux, pp. 197–227. Amsterdam : North-Holland
104. Chaiken, I. M., Cohen, J. S., Sokoloski, E. A. 1974. *J. Am. Chem. Soc.* 96 : 4703–5
105. Chaiken, I. M., Freedman, M. H., Lyerla, J. R. Jr., Cohen, J. S. 1973. *J. Biol. Chem.* 248 : 884–91
106. Chambers, V. M. A., Evans, E. A., Elvidge, J. A., Jones, J. R. 1978. *Rev. Radiochem. Cent.* 19 : 1–68
107. Chan, T.-M., Hermodson, M. A., Ulrich, E. L., Markley, J. L. 1983. *Biochemistry* 22 : 5988–95
108. Chan, T.-M., Markley, J. L. 1982. *J. Am. Chem. Soc.* 104 : 4010–11
109. Chan, T.-M., Markley, J. L. 1983. *Biochemistry* 22 : 5982–87
110. Chan, T.-M., Markley, J. L. 1983. *Biochemistry* 22 : 5996–6002
111. Chan, T.-M., Markley, J. L. 1983. *Biochemistry* 22 : 6008–10
112. Chan, T.-M., Ulrich, E. L., Markley, J. L. 1983. *Biochemistry* 22 : 6002–7
113. Chan, T.-M., Westler, W. M., Santini, R. E., Markley, J. L. 1982. *J. Am. Chem. Soc.* 104 : 4008–10
114. Chapman, G. E. 1977. *Nucl. Magn. Reson.* 6 : 154–73
115. Chen, T.-C., Knapp, R. D., Rohde, M. F., Brainard, J. R., Gotto, J. M. Jr., Sparrow, J. T., Morrisett, J. D. 1980. *Biochemistry* 19 : 5140–46
116. Chlebowski, J. M., Armitage, I. M., Coleman, J. E. 1977. *J. Biol. Chem.* 252 : 7053–61
117. Clayden, N. J., Williams, R. J. P. 1982. *J. Magn. Reson.* 49 : 383–96
118. Clore, G. M., Gronenborn, A. M. 1982. *Biochemistry* 21 : 4048–53

119. Cocco, L., Blakley, R. L., Walker, T. E., London, R. E., Matwiyoff, N. A. 1978. *Biochemistry* 17:4285–90
120. Cohen, J. S. 1978. *CRC Crit. Rev. Biochem.* 5:25–43
121. Cohen, J. S., ed. 1980. *Magnetic Resonance in Biology*, Vol. 1. New York: Wiley. 309 pp.
122. Cohen, J. S., Niu, C.-H., Matsuura, S., Shindo, H. 1980. *Dev. Biochem.* 10:3–16
123. Cohn, M. 1982. *Ann. Rev. Biophys. Bioeng.* 11:23–42
124. Cohn, M., Reed, G. H. 1982. *Ann. Rev. Biochem.* 51:365–94
125. Coleman, J. E., Armitage, I. M. 1977. See Ref. 154, pp. 171–200
126. Coleman, J. E., Armitage, I. M. 1978. *Biochemistry* 17:5038–45
127. Coleman, J. E., Armitage, I. M., Chlebowski, J. F., Otvos, J. D., Schoot Uiterkamp, A. J. M. 1979. See Ref. 518, pp. 345–95
128. Cookson, D. J., Hayes, M. T., Wright, P. E. 1980. *Nature* 283:682–83
129. Cookson, D. J., Moore, G. R., Pitt, R. C., Williams, R. J. P., Campbell, I. D., Ambler, R. P., et al. 1978. *Eur. J. Biochem.* 83:261–75
130. Crane-Robinson, C. 1978. See Ref. 44, 1:33–90
131. Crippen, G. M. 1979. *Int. J. Pept. Protein Res.* 13:320–26
132. Croll, D. H. 1982. PhD thesis. Purdue Univ., West Lafayette, Ind. 218 pp.
133. Cutnell, J. D., La Mar, G. N., Kong, S. B. 1981. *J. Am. Chem. Soc.* 103:3567–72
134. Dadok, J., Sprecher, R. F. 1974. *J. Magn. Reson.* 13:243–48
135. De Marco, A., Hochschwender, S. M., Laursen, R. A., Llinás, M. 1982. *J. Biol. Chem.* 257:12716–21
136. De Marco, A., Lecomte, J. T. J., Llinás, M. 1981. *Eur. J. Biochem.* 119:483–90
137. De Marco, A., Menegatti, E., Guarneri, M. 1979. *Eur. J. Biochem.* 102:185–94
138. De Marco, A., Menegatti, E., Guarneri, M. 1982. *Biochemistry* 21:222–29
139. De Marco, A., Menegatti, E., Guarneri, M. 1982. *J. Biol. Chem.* 257:8337–42
139a. De Marco, A., Menegatti, E., Guarneri, M. 1983. *FEBS Lett.* 159:201–6
140. De Marco, A., Wüthrich, K. 1976. *J. Magn. Reson.* 24:201–4
141. Delepierre, M., Dobson, C. M., Hoch, J. C., Olejniczak, E. T., Poulsen, F. M., Ratcliffe, R. G., Redfield, C. 1981. In *Biomol. Stereodyn., Proc. Symp.*, ed. R. H. Sarma, 2:237–53. Guilderland, NY: Adenine Press
142. Delepierre, M., Dobson, C. M., Poulsen, F. M. 1982. *Biochemistry* 21:4756–61

143. Dickenson, L. C., Chien, J. C. W. 1975. *Biochemistry* 14:3534–42
144. Dill, K., Allerhand, A. 1977. *J. Am. Chem. Soc.* 99:4508–11
145. Dill, K., Allerhand, A. 1979. *J. Am. Chem. Soc.* 101:4376–78
146. Dobson, C. M. 1977. See Ref. 154, pp. 77–94
147. Dobson, C. M. 1982. *Jerusalem Symp. Quantum Chem. Biochem.* 15:481–95
148. Dobson, C. M., Moore, G. R., Williams, R. J. P. 1975. *FEBS Lett.* 51:60–65
149. Dobson, C. M., Olejniczak, E. T., Poulsen, F. M., Ratcliffe, R. G. 1982. *J. Magn. Reson.* 48:97–110
150. Dower, S., Dwek, R. A. 1979. See Ref. 518, pp. 271–303
151. Drakenberg, T., Lindman, B., Cavé, A., Parello, J. 1978. *FEBS Lett.* 92:346–50
152. Dubs, A., Wagner, G., Wüthrich, K. 1979. *Biochim. Biophys. Acta* 577:177–94
153. Dwek, R. A. 1977. See Ref. 154, pp. 125–56
154. Dwek, R. A., Campbell, I. D., Richards, R. E., eds. 1977. *NMR in Biology, Proc. Br. Biophys. Soc., March.* London: Academic. 381 pp.
155. Eakin, R. T., Morgan, L. O., Matwiyoff, N. A. 1975. *Biochem. J.* 152:529–35
156. Edmondson, D. E., James, T. L. 1982. *Dev. Biochem.* 21:111–18
157. Edwards, B. F. P., Lee, L., Sykes, B. D. 1978. In *Biomol. Struct. Funct., Symp.*, ed. P. F. Agris, pp. 275–93. New York: Academic
158. Edwards, B. F. P., Sykes, B. D. 1978. *Biochemistry* 17:684–89
159. Edwards, B. F. P., Sykes, B. D. 1980. *Biochemistry* 19:2577–83
160. Edwards, B. F. P., Sykes, B. D. 1981. *Biochemistry* 20:4193–98
161. Egan, W., Shindo, H., Cohen, J. S. 1977. *Ann. Rev. Biophys. Bioeng.* 6:383–417
162. Egan, W., Shindo, H., Cohen, J. S. 1978. *J. Biol. Chem.* 253:16–17
163. Egmond, M. R., Hore, P. J., Kaptein, R. 1983. *Biochim. Biophys. Acta* 744:23–27
164. Egmond, M. R., Slotboom, A. J., De Haas, G. H., Dijkstra, K., Kaptein, R. 1980. *Biochim. Biophys. Acta* 623:461–66
165. Emid, S. 1983. *Bull. Magn. Reson.* 4:99–104
166. Emptage, M. H., Xavier, A. V., Wood, J. M. 1980. *Cienc. Biol. Coimbra* 5:133–35
167. Emptage, M. H., Xavier, A. V., Wood, J. M., Alsaadi, B. M., Moore, G. R., Pitt, R. C., et al. 1981. *Biochemistry* 20:58–64
168. Endo, T., Inagaki, F., Hayashi, K.,

Miyazawa, T. 1979. *Eur. J. Biochem.* 102:417–30

169. Endo, T., Inagaki, F., Hayashi, K., Miyazawa, T. 1981. *Eur. J. Biochem.* 120:117–24

170. Endo, T., Inagaki, F., Hayashi, K., Miyazawa, T. 1982. *Eur. J. Biochem.* 122:541–47

171. Engeseth, H. R., Otvos, J. D. 1983. *Am. Chem. Soc. Abstr. Pap., 186th Meet., Aug. 28–Sep. 21, Inorg. Div., Abstr. No. 53*

172. Epstein, H. F., Schechter, A. N., Cohen, J. S. 1971. *Proc. Natl. Acad. Sci. USA* 68:2042–46

173. Fairhurst, S. A., Sutcliffe, L. H. 1978. *Prog. Biophys. Mol. Biol.* 34:1–79

174. Feeney, J. 1978. *Jerusalem Symp. Quantum Chem. Biochem.* 11:297–310

175. Feeney, J., Birdsall, B., Albrand, J. P., Roberts, G. C. K., Burgen, A. S. V., Charlton, P. A., Young, D. W. 1981. *Biochemistry* 20:1837–42

176. Feeney, J., Roberts, G. C. K., Kaptein, R., Birdsall, B., Gronenborn, A., Burgen, A. S. V. 1980. *Biochemistry* 19:2466–72

177. Feeney, J., Roberts, G. C. K., Thomson, J. W., King, R. W., Griffiths, D. V., Burgen, A. S. V. 1980. *Biochemistry* 19:2316–21

178. Ferguson, R. C., Phillips, W. D. 1967. *Science* 157:257–67

179. Forsén, S., Andersson, A., Drakenberg, T., Teleman, E., Thulin, E., Vogel, H. J. 1983. *4th Int. Symp., Calcium Binding Proteins in Health and Disease, Trieste, Italy, May 16–19,* ed. B. de Bernard. Amsterdam: Elsevier

180. Forsén, S., Andersson, T., Drakenberg, T., Thulin, E., Swärd, M. 1982. *Fed. Proc.* 41:2981–86

181. Forsén, S., Thulin, E., Drakenberg, T., Krebs, J., Seamon, K. 1980. *FEBS Lett.* 117:189–94

182. Forsén, S., Thulin, E., Lilja, H. 1979. *FEBS Lett.* 104:123–26

183. Frankenne, F., Maghuin-Rogister, G., Birdsall, B., Roberts, G. C. K. 1983. *FEBS Lett.* 151:197–200

184. Freeman, H. C., Norris, V. A., Ramshaw, J. A. M., Wright, P. E. 1978. *FEBS Lett.* 86:131–35

185. Freeman, R., Morris, G. A. 1979. *Bull. Magn. Reson.* 1:5–26

186. Fujii, S., Akasaka, K., Hatano, H. 1980. *J. Biochem.* 88:789–96

187. Fujii, S., Akasaka, K., Hatano, H. 1981. *Biochemistry* 20:518–23

188. Fukushima, E., Roeder, S. B. W. 1981. *Experimental Pulse NMR. A Nuts and Bolts Approach.* Reading, Mass: Addison-Wesley. 539 pp.

189. Fülling, R., Rüterjans, H. 1978. *FEBS Lett.* 88:279–82

190. Fung, C. H., Chang, C. C., Gupta, R. K. 1979. *Biochemistry* 18:457–60

191. Fung, L. W.-M., Ho, C. 1975. *Biochemistry* 14:2526–35

191a. Galley, H. U., Spencer, A. K., Armitage, I. M., Prestegard, J. H., Cronan, J. E. Jr. 1978. *Biochemistry* 17:5377–82

192. Ganesh, K. N. 1982. *Curr. Sci.* 51:866–74

193. Garssen, G. J., Kaptein, R., Schoenmakers, J. G. G., Hilbers, C. W. 1978. *Proc. Natl. Acad. Sci. USA* 75:5281–85

194. Garssen, G. J., Tesser, G. I., Schoenmakers, J. G. G., Hilbers, C. W. 1980. *Biochim. Biophys. Acta* 607:361–71

195. Gassner, M., Stehlik, D., Schrecker, O., Hengstenberg, W., Maurer, W., Rüterjans, H. 1977. *Eur. J. Biochem.* 75:287–96

196. Gerig, J. T. 1978. See Ref. 44, 1:139–203

197. Gerig, J. T., Klinkenborg, J. C., Nieman, R. A. 1983. *Biochemistry* 22:2076–87

198. Gersonde, K. 1978. In *Symp. Pap.-IUPAC Int. Symp. Chem. Nat. Prod., 11th,* ed. N. Marekov, I. Ognyanov, A. Orahovats, 4:90–104. Sofia, Bulgaria: Izd. BAN

199. Gettins, P., Boyd, J., Glaudemans, C. P. J., Potter, M., Dwek, R. A. 1981. *Biochemistry* 20:7463–69

200. Gettins, P., Coleman, J. E. 1983. *J. Biol. Chem.* 258:396–407

201. Gettins, P., Dwek, R. A. 1981. *FEBS Lett.* 124:248–52

202. Gettins, P., Potter, M., Rudikoff, S., Dwek, R. A. 1977. *FEBS Lett.* 84:87–91

203. Gordon, S. L., Wüthrich, K. 1978. *J. Am. Chem. Soc.* 100:7094–96

204. Govil, G., Hosur, R. V. 1982. *Conformation of Biological Molecules, NMR Basic Principles and Progress,* Vol. 20. Berlin: Springer-Verlag. 216 pp.

205. Gray, G. A. 1983. *Org. Magn. Reson.* 21:111–18

206. Gronenborn, A., Birdsall, B., Hyde, E. I., Roberts, G. C. K., Feeney, J., Burgen, A. S. V. 1981. *Biochemistry* 20:1717–22

207. Gronenborn, A., Birdsall, B., Hyde, E. I., Roberts, G. C. K., Feeney, J., Burgen, A. S. V. 1981. *Nature* 290:273–74

208. Gupta, R. K., Mildvan, A. S. 1978. *Methods Enzymol.* 54:151–92

209. Gurd, F. R. N., Matthew, J. B., Wittebort, R. J., Morrow, J. S., Friend, S. H. 1980. In *Biophys. Physiol. Carbon Dioxide Symp.,* ed. C. Bauer, G. Gros, H. Bartels, pp. 89–101. Berlin: Springer-Verlag

210. Gurd, F. R. N., Wittebort, R. J., Rothgeb, T. M., Neireiter, G. Jr. 1982. See Ref. 64, pp. 1–29
211. Gust, D., Moon, R. M., Roberts, J. D. 1975. *Proc. Natl. Acad. Sci. USA* 72:4696–700
212. Harbison, G. S., Herzfeld, J., Griffin, R. G. 1983. *Biochemistry* 22:1–5
213. Harina, B. M., Dyckes, D. F., Willcott, M. R. III, Jones, W. C. Jr. 1978. *J. Am. Chem. Soc.* 100:4897–99
214. Harris, R. K. 1983. *Nuclear Magnetic Resonance. A Physiochemical View.* London: Pitman. 250 pp.
215. Havel, T. F., Crippen, G. M., Kuntz, I. D. 1979. *Biopolymers* 18:73–81
216. Henry, G. D., Dalgarno, D. C., Marcus, G., Scott, M., Levine, B. A., Trayer, I. P. 1982. *FEBS Lett.* 144:11–15
217. Herak, J. N., Pifat, G., Brnjas-Kralijevic, J., Jurgens, G., Holasek, A. 1980. *Period. Biol.* 82:351–55
218. Hider, R. C., Drake, A. F., Inagaki, F., Williams, R. J. P., Endo, T., Miyazawa, T. 1982. *J. Mol. Biol.* 158:275–91
219. Hill, H. A. O., Lee, W. K., Bannister, J. V., Bannister, W. H. 1980. *Biochem. J.* 185:245–52
220. Hill, H. A. O., Leer, J. C., Smith, B. E., Storm, C. B., Ambler, R. P. 1976. *Biochem. Biophys. Res. Commun.* 70:331–38
221. Hill, H. A. O., Smith, B. E. 1979. *J. Inorg. Biochem.* 11:79–93
222. Hincke, M. T., Hagen, S., Sykes, B. D., Kay, C. M. 1980. *Dev. Biochem.* 14:315–17
223. Hincke, M. T., Sykes, B. D., Kay, C. M. 1981. *Biochemistry* 20:3286–94
224. Hincke, M. T., Sykes, B. D., Kay, C. M. 1981. *Biochemistry* 20:4185–93
225. Ho, C., Fung, L. W.-M., Wiechelman, K. J. 1978. *Methods Enzymol.* 54:192–223
226. Ho, C., Fung, L. W.-M., Wiechelman, K. J., Pifat, G., Johnson, M. F. 1975. In *Erythrocyte Structure and Function*, pp. 43–64. New York: Liss
227. Ho, C., Kurland, R. J. 1966. *J. Biol. Chem.* 241:3002–7
228. Ho, C., Lam, C. H. J., Takahashi, S., Viggiano, G. 1982. In *Hemoglobin and Oxygen Binding, Int. Symp. Interact. Iron Proteins Oxygen Electron Transp.*, ed. C. Ho, pp. 141–49. New York: Elsevier
229. Ho, C., Russu, I. M. 1981. *Methods Enzymol.* 76:275–312
230. Hoch, J. C., Dobson, C. M., Karplus, M. 1982. *Biochemistry* 21:1118–25
231. Hochschwender, S. M., Laursen, R. A., De Marco, A., Llinás, M. 1983. *Arch. Biochem. Biophys.* 223:58–67
232. Hoerl, M., Feldmann, K., Schnackerz,

K. D., Helmreich, E. J. M. 1979. *Biochemistry* 18:2457–64
233. Hon-Nami, K., Kihara, H., Kitagawa, T., Miyazawa, T., Oshima, T. 1980. *Eur. J. Biochem.* 110:217–23
234. Hore, P. J., Kaptein, R. 1982. In *ACS Symp. Ser. 191, NMR Spectrosc.: New Methods Appl.*, pp. 285–318
235. Hore, P. J., Kaptein, R. 1983. *Biochemistry* 22:1906–11
236. Hore, P. J., Scheek, R. M., Kaptein, R. 1983. *J. Magn. Reson.* 1983:339–42
237. Hosur, R. V., Wider, G., Wüthrich, K. 1983. *Eur. J. Biochem.* 130:497–508
238. Howarth, O. W., Lilley, D. M. J. 1978. *Prog. Nucl. Magn. Reson. Spectrosc.* 12:1–40
239. Hughes, D. W., Stollery, J. G., Moscarello, M. A., Deber, C. M. 1982. *J. Biol. Chem.* 257:4698–700
240. Hull, W. E., Halford, S. E., Gutfreund, H., Sykes, B. D. 1976. *Biochemistry* 15:1547–61
241. Hunkapiller, M. W., Forgac, M. D., Yu, E. H., Richards, J. H. 1979. *Biochem. Biophys. Res. Commun.* 87:25–31
242. Hunkapiller, M. W., Smallcombe, S. H., Whitaker, D. R., Richards, J. H. 1973. *J. Biol. Chem.* 248:8306–8
243. Ikeda-Saito, M., Inubushi, T., McDonald, G. G., Yonetani, T. 1978. *J. Biol. Chem.* 253:7134–37
244. Ikura, M., Hiraoki, T., Hikichi, K., Mikuni, T., Yazawa, M., Yagi, K. 1983. *Biochemistry* 22:2573–79
245. Ikura, M., Toshifumi, H., Hikichi, K., Mikuni, T., Yazawa, M., Yagi, K. 1983. *Biochemistry* 22:2568–72
246. Inagaki, F., Boyd, J., Campbell, I. D., Clayden, N. J., Hull, W. E., Tamiya, N., Williams, R. J. P. 1982. *Eur. J. Biochem.* 121:609–16
247. Inagaki, F., Clayden, N. J., Tamiya, N., Williams, R. J. P. 1981. *Eur. J. Biochem.* 120:313–22
248. Inagaki, F., Clayden, N. J., Tamiya, N., Williams, R. J. P. 1982. *Eur. J. Biochem.* 123:99–104
249. Inagaki, F., Kawano, Y., Shimada, I., Takahashi, K., Miyazawa, T. 1981. *J. Biochem.* 89:1185–95
250. Inagaki, F., Miyazawa, T., Hori, H., Tamiya, N. 1978. *Eur. J. Biochem.* 89:433–42
251. Inagaki, F., Miyazawa, T., Tamiya, N., Williams, R. J. P. 1982. *Eur. J. Biochem.* 123:275–82
252. Inagaki, F., Miyazawa, T., Williams, R. J. P. 1981. *Biosci. Rep.* 1:743–55
253. Inagaki, F., Tamiya, N., Miyazawa, T. 1980. *Eur. J. Biochem.* 109:129–38
254. Inagaki, F., Tamiya, N., Miyazawa, T., Williams, R. J. P. 1981. *Eur. J. Biochem.* 118:621–25

255. Inagaki, F., Watanabe, K., Miyazawa, T. 1979. *J. Biochem.* 86:591–94
256. Inubushi, T., Ikeda-Saito, M., Yonetani, T. 1983. *Biochemistry* 22:2904–7
257. Iwahashi, H., Akutsu, H., Kobayashi, Y., Kyogoku, Y., Ono, T., Koga, H., Horiuchi, T. 1982. *J. Biochem.* 91:1213–21
258. Jackson, W. R. C., Leatherbarrow, R. J. 1981. *Biochemistry* 20:2339–45
259. Jaeck, G., Benz, F. W. 1979. *Biochem. Biophys. Res. Commun.* 86:885–92
260. James, T. L., Edmondson, D. E. 1981. *Biochemistry* 20:617–21
261. James, T. L., Ludwig, M. L., Cohn, M. 1973. *Proc. Natl. Acad. Sci. USA* 70:3292–95
262. James, T. L., Matson, G. B., Kuntz, I. D. 1978. *J. Am. Chem. Soc.* 100:3590–94
263. Jansen, E. H. J. M., Meyer, H., De Haas, G. H., Kaptein, R. 1978. *J. Biol. Chem.* 253:6346–47
264. Jansen, E. H. J. M., Van Scharrenburg, G. J. M., Slotboom, A. J., De Haas, G. H., Kaptein, R. 1979. *J. Am. Chem. Soc.* 101:7397–99
265. Jardetzky, O. 1979. See Ref. 440, pp. 141–67
266. Jardetzky, O. 1980. *Biochim. Biophys. Acta* 621:227–32
267. Jardetzky, O., Jardetzky, C. D. 1957. *J. Am. Chem. Soc.* 79:5322–23
268. Jardetzky, O., Markley, J. L. 1970. *Il Farmaco, Ed. Sci.* 25:894–97
269. Jardetzky, O., Markley, J. L., Thielmann, J., Arata, Y., Williams, M. W. 1972. *Cold Spring Harbor Symp. Quant. Biol.* 36:257–61
270. Jardetzky, O., Roberts, G. C. K. 1981. *NMR in Molecular Biology.* New York: Academic. 681 pp.
271. Jarema, M. A. C., Lu, P., Miller, J. H. 1981. *Proc. Natl. Acad. Sci. USA* 78:2707–11
272. Johnson, F. A., Lewis, S. D., Shafer, J. A. 1981. *Biochemistry* 20:44–48
273. Johnson, R. N., Bradbury, J. H., Appleby, C. A. 1978. *J. Biol. Chem.* 253:2148–54
274. Jones, R. B., Waley, S. G. 1979. *Biochem. J.* 179:623–30
275. Jones, W. C. Jr., Rothgeb, T. M., Gurd, F. R. N. 1976. *J. Biol. Chem.* 251:7452–60
276. Jordan, F., Polgar, L. 1981. *Biochemistry* 20:6366–70
277. Kainosho, M., Tsuji, T. 1982. *Biochemistry* 21:6273–79
278. Kalbitzer, H. R., Deutscher, J., Hengstenberg, W., Roesch, P. 1981. *Biochemistry* 20:6178–85
279. Kalbitzer, H. R., Hengstenberg, W., Roesch, P., Muss, P., Bernsmann, P.,

Engelmann, R., et al. 1982. *Biochemistry* 21:2879–85
280. Kalbitzer, H. R., Marquetant, R., Roesch, P., Schirmer, R. H. 1982. *Eur. J. Biochem.* 126:531–36
281. Kalk, A., Berendsen, H. J. C. 1976. *J. Magn. Reson.* 24:343–66
282. Kanamori, K., Roberts, J. D. 1983. *Acc. Chem. Res.* 16:35–41
283. Kaptein, R. 1980. *Curso Reson. Magn. Nucl.: Reson. Magn. Nucl. Pulsos, Alta Resoluc., 1st, Madrid,* pp. 385–407
284. Kaptein, R. 1982. See Ref. 44, 4:145–91
285. Kaptein, R., Scheek, R. M., Zuiderweg, E. R. P., Boelens, R., Klappe, K. J. M., van Boom, J. H., et al. 1983. In *Structure and Dynamics: Nucleic Acids and Proteins,* ed. E. Clementi, R. H. Sarma, pp. 209–25. New York: Adenine Press
286. Kaptein, R., Wyeth, P. 1980. *Cienc. Biol.* 5:125–27
287. Keim, P. 1979. *Biochem. Discuss.* 7:9–50
288. Keller, R. M., Baumann, R., Hunziker-Kwik, E. H., Joubert, F. J., Wüthrich, K. 1983. *J. Mol. Biol.* 163:623–46
289. Keller, R. M., Pettigrew, G. W., Wüthrich, K. 1973. *FEBS Lett.* 36:151–56
290. Keller, R. M., Picot, D., Wüthrich, K. 1979. *Biochim. Biophys. Acta* 580:259–65
291. Keller, R. M., Schejter, A., Wüthrich, K. 1980. *Biochim. Biophys. Acta* 626:15–22
292. Keller, R. M., Wüthrich, K. 1978. *Biochim. Biophys. Acta* 533:195–208
293. Keller, R. M., Wüthrich, K. 1978. *Biochem. Biophys. Res. Commun.* 83:1132–39
294. Keller, R. M., Wüthrich, K. 1981. See Ref. 44, 3:1–52
295. Keller, R. M., Wüthrich, K., Pecht, I. 1976. *FEBS Lett.* 70:180–84
296. Keller, R. M., Wüthrich, K., Schejter, A. 1977. *Biochim. Biophys. Acta* 491:409–15
297. Keniry, M. A., Rothgeb, T. M., Smith, R. L., Gutowsky, H. S., Oldfield, E. 1983. *Biochemistry* 22:1917–26
298. Kenyon, G. L., Reed, G. H. 1982. *Adv. Enzymol.* 54:367–426
299. Kilmartin, J. V., Breen, J. J., Roberts, G. C. K., Ho, C. 1973. *Proc. Natl. Acad. Sci. USA* 70:1246–49
300. Kim, P. S., Baldwin, R. L. 1982. *Ann. Rev. Biochem.* 51:459–89
301. Kimber, B. J., Feeney, J., Roberts, G. C. K., Birdsall, B., Griffiths, D. V., Burgen, A. S. V., Sykes, B. D. 1978. *Nature* 271:184–85
302. Kimura, S., Matsuo, H., Narita, K. 1979. *J. Biochem.* 86:301–10

303. King, G., Wright, P. E. 1982. *Biochem. Biophys. Res. Commun.* 106:559–65
303a. King, G., Wright, P. E. 1983. *J. Magn. Reson.* 54:328–32
304. Kirpichnikov, M. P., Kurochkin, A. V., Skryabin, K. G. 1982. *FEBS Lett.* 150:407–10
305. Klevit, R. E., Girard, P., Esnouf, M. P., Williams, R. J. P. 1981. In *Calcium Phosphate Transp. Biomembrane, Int. Workshop,* ed. F. Bronner, M. Peterlik, pp. 25–29. New York: Academic
306. Kojiro, C. L., Markley, J. L. 1983. *FEBS Lett.* 162:52–56
307. Kong, S. B., Cutnell, J. D., La Mar, G. N. 1983. *J. Biol. Chem.* 258:3843–49
308. Koppitz, B., Feldmann, K., Heilmeyer, L. M. G. Jr. 1980. *FEBS Lett.* 117:199–202
309. Kowalsky, A. 1965. *Biochemistry* 4:2382–88
310. Krebs, J. 1981. *Cell Calcium* 2:295–311
311. Krebs, J., Carafoli, E. 1982. *Eur. J. Biochem.* 124:619–27
312. Krishnamoorthy, G., Prabhananda, B. S. 1982. *Biochim. Biophys. Acta* 709:53–57
313. Kumar, A., Ernst, R. R., Wüthrich, K. 1980. *Biochem. Biophys. Res. Commun.* 95:1–6
314. Kumar, A., Wagner, G., Ernst, R. R., Wüthrich, K. 1980. *Biochem. Biophys. Res. Commun.* 96:1156–63
315. Kurochkin, A. V., Kirpichnikov, M. P. 1982. *FEBS Lett.* 150:411–15
316. Kuwajima, K., Baldwin, R. L. 1983. *J. Mol. Biol.* 169:281–97
317. Kuwajima, K., Baldwin, R. L. 1983. *J. Mol. Biol.* 169:299–323
318. Kuwajima, K., Kim, P. S., Baldwin, R. L. 1983. *Biopolymers* 22:59–67
319. La Mar, G. N. 1979. See Ref. 518, pp. 305–43
320. La Mar, G. N., Anderson, R. R., Budd, D. L., Smith, K. M., Langry, K. C., Gersonde, K., Sick, H. 1981. *Biochemistry* 20:4429–36
321. La Mar, G. N., Budd, D. L., Smith, K. M. 1980. *Biochim. Biophys. Acta* 622:210–18
322. La Mar, G. N., Budd, D. L., Smith, K. M., Langry, K. C. 1980. *J. Am. Chem. Soc.* 102:1822–27
323. La Mar, G. N., Budd, D. L., Viscio, D. B., Smith, K. M., Langry, K. C. 1978. *Proc. Natl. Acad. Sci. USA* 75:5755–59
324. La Mar, G. N., Burns, P. D., Jackson, J. T., Smith, K. M., Langry, K. C., Strittmatter, P. 1981. *J. Biol. Chem.* 256:6075–79
325. La Mar, G. N., Cutnell, J. D., Kong, S. B. 1981. *Biophys. J.* 34:217–25
326. La Mar, G. N., De Ropp, J. S. 1979.

Biochem. Biophys. Res. Commun. 90:36–41
327. La Mar, G. N., De Ropp, J. S. 1982. *J. Am. Chem. Soc.* 104:5203–6
328. La Mar, G. N., De Ropp, J. S., Chacko, V. P., Satterlee, J. D., Erman, J. E. 1982. *Biochim. Biophys. Acta* 708:317–25
329. La Mar, G. N., De Ropp, J. S., Latos-Grazynski, L., Balch, A. L., Johnson, R. B., Smith, K. M., et al. 1983. *J. Am. Chem. Soc.* 105:782–87
330. La Mar, G. N., De Ropp, J. S., Smith, K. M., Langry, K. C. 1980. *J. Am. Chem. Soc.* 102:4833–35
331. La Mar, G. N., De Ropp, J. S., Smith, K. M., Langry, K. C. 1980. *J. Biol. Chem.* 255:6646–52
332. La Mar, G. N., De Ropp, J. S., Smith, K. M., Langry, K. C. 1981. *J. Biol. Chem.* 256:237–43
333. La Mar, G. N., Kong, S. B., Smith, K. M., Langry, K. C. 1981. *Biochem. Biophys. Res. Commun.* 102:142–48
334. La Mar, G. N., Nagai, K., Jue, T., Budd, D. L., Gersonde, K., Sick, H., et al. 1980. *Biochem. Biophys. Res. Commun.* 96:1172–77
335. La Mar, G. N., Overkamp, M., Sick, H., Gersonde, K. 1978. *Biochemistry* 17:352–61
336. La Mar, G. N., Smith, K. M., De Ropp, J. S., Burns, P. D., Langry, K. C., Gersonde, K., et al. 1982. See Ref. 228, pp. 419–24
337. La Mar, G. N., Smith, K. M., Gersonde, K., Sick, H., Overkamp, M. 1980. *J. Biol. Chem.* 255:66–70
338. La Mar, G. N., Viscio, D. B., Budd, D. L. 1978. *Biochem. Biophys. Res. Commun.* 82:19–23
339. Latos-Grazynski, L., Balch, A. L., La Mar, G. N. 1982. *Adv. Chem. Ser.* 201:661–74
340. Lauterbur, P. C. 1970. *Appl. Spectrosc.* 24:450–52
341. Lauterwein, J., Bösch, C., Brown, L. R., Wüthrich, K. 1979. *Biochim. Biophys. Acta* 556:244–64
342. Lauterwein, J., Brown, L. R., Wüthrich, K. 1980. *Biochim. Biophys. Acta* 622:219–20
343. Lecomte, J. T. J., De Marco, A., Llinás, M. 1982. *Biochim. Biophys. Acta* 703:223–30
344. Lecomte, J. T. J., Jones, B. L., Llinás, M. 1982. *Biochemistry* 21:4843–49
345. Lee, L., Sykes, B. D. 1980. *Adv. Inorg. Biochem., Methods for Determining Metal Ion Environments in Proteins: Structure and Functions of Metalloproteins,* ed. D. W. Darnall, R. G. Wilkins, 2:183–210. New York: Elsevier. 324 pp.

346. Lee, L., Sykes, B. D. 1980. *Biophys. J.* 32:193–210
347. Lee, L., Sykes, B. D. 1980. *J. Magn. Reson.* 41:512–14
347a. Lee, L., Sykes, B. D. 1983. *Biochemistry* 22:4366–73
348. Lee, N., Inouye, M., Lauterbur, P. C. 1977. *Biochem. Biophys. Res. Commun.* 78:1211–18
349. Leigh, J. S. 1976. In *Introduction to the Spectroscopy of Biological Polymers*, ed. D. W. Jones, pp. 189–219. London: Academic
350. LeMaster, D., Richards, F. M. 1983. *Proc. 8th Meet. Int. Soc. Magn. Reson., August 22–26*, p. 104 (Abstr.)
351. Lenkinski, R. E., Dallas, J. L., Glickson, J. D. 1979. *J. Am. Chem. Soc.* 101:3071–77
352. Lenstra, J. A., Bolscher, B. G. J. M., Stob, S., Beintema, J. J., Kaptein, R. 1979. *Eur. J. Biochem.* 98:385–97
353. Lerman, C. L., Cohn, M. 1980. *Biochem. Biophys. Res. Commun.* 97:121–25
354. Levine, B. A., Moore, G. R., Ratcliffe, R. G., Williams, R. J. P. 1979. *Int. Rev. Biochem.* 24:77–141
355. Levine, B. A., Williams, R. J. P., Fullmer, C. S., Wasserman, R. H. 1977. *Calcium-Binding Proteins Calcium Funct., Proc. Int. Symp., 2nd*, ed. R. H. Wasserman, pp. 29–37. New York: Elsevier
356. Levy, R. M., Dobson, C. M., Karplus, M. 1982. *Biophys. J.* 39:107–13
357. Lewis, S. D., Johnson, F. A., Shafer, J. A. 1981. *Biochemistry* 20:48–51
358. Lipari, G., Szabo, A., Levy, R. M. 1982. *Nature* 300:197–98
359. Lippard, S. J., Burger, A. R., Ugurbil, K., Pantoliano, M. W., Valentine, J. S. 1977. *Biochemistry* 16:1136–41
360. Littlemore, L. A. T. 1978. *Aust. J. Chem.* 31:2387–98
361. Littlemore, L. A. T., Ledeen, R. W. 1979. *Aust. J. Chem.* 32:2631–36
362. London, R. E. 1980. See Ref. 121, pp. 1–69
363. London, R. E., Avitabile, J. 1978. *J. Am. Chem. Soc.* 100:7159–65
364. London, R. E., Groff, J. P., Cocco, L., Blakley, R. L. 1982. *Biochemistry* 21:4450–58
365. Lord, S. T., Breslow, E. 1978. *Biochem. Biophys. Res. Commun.* 80:63–70
366. Lord, S. T., Breslow, E. 1979. *Int. J. Pept. Protein Res.* 13:71–77
367. Lu, P., Jarema, M. A., Rackwitz, H. R., Friedman, R. L. 1979. See Ref. 440, pp. 59–66
368. Mabbutt, B. C., Wright, P. E. 1983. *Biochim. Biophys. Acta* 744:281–90
369. Maciel, G. E., Shatlock, M. P., Houtchens, R. A., Caughey, W. S. 1980.

J. Am. Chem. Soc. 102:6884–85
370. Maghuin-Rogister, G., Degelaen, J., Roberts, G. C. K. 1979. *Eur. J. Biochem.* 96:59–68
371. Mak, A., Smillie, L. B., Bárány, M. 1978. *Proc. Natl. Acad. Sci. USA* 75:3588–92
372. Mani, R. S., Shelling, J. G., Sykes, B. D., Kay, C. M. 1983. *Biochemistry* 22:1734–40
373. March, K. L., Maskalick, D. G., England, R. D., Friend, S. H., Gurd, F. R. N. 1982. *Biochemistry* 21:5241–51
374. Mariano, P. S., Glover, G. I., Petersen, J. R. 1978. *Biochem. J.* 171:115–22
375. Marion, D., Wüthrich, K. 1983. *Biochem. Biophys. Res. Commun.* 113:967–74
376. Markley, J. L. 1972. *Methods Enzymol.* 26:605–27
377. Markley, J. L. 1975. *Acc. Chem. Res.* 8:70–80
378. Markley, J. L. 1975. *Biochemistry* 14:3546–54
379. Markley, J. L. 1979. See Ref. 518, pp. 397–461
380. Markley, J. L., Cheung, S.-M. 1973. *Proc. Int. Conf. on Stable Isotopes in Chem., Biol., Med., Argonne, Ill.*, pp. 103–18
381. Markley, J. L., Ibañez, I. B. 1978. *Biochemistry* 17:4627–40
382. Markley, J. L., Kato, I. 1975. *Biochemistry* 14:3234–37
383. Markley, J. L., Neves, D. E., Westler, W. M., Ibañez, I. B., Porubcan, M. A., Baillargeon, M. W. 1980. *Dev. Biochem.* 10:31–61
384. Markley, J. L., Porubcan, M. A. 1976. *J. Mol. Biol.* 102:487–509
385. Markley, J. L., Putter, I., Jardetzky, O. 1968. *Science* 161:1249–51
386. Markley, J. L., Ulrich, E. L., Krogmann, D. W. 1977. *Biochem. Biophys. Res. Commun.* 78:106–14
387. Markley, J. L., Westler, W. M., Chan, T.-M., Kojiro, C. L., Ulrich, E. L. 1984. *Fed. Proc.* In press
388. Matthews, D. A. 1979. *Biochemistry* 18:1602–10
389. Mayer, A., Ogawa, S., Shulman, R. G., Yamane, T., Calvero, J. A., Gonsalves, A. M. d'A. R., et al. 1974. *J. Mol. Biol.* 86:749–56
389a. Mayo, K. H., Tyrell, P. M., Prestegard, J. H. 1983. *Biochemistry* 22:4485–93
390. McCammon, J. A., Karplus, M. 1983. *Acc. Chem. Res.* 16:187–93
391. McDonald, C. C., Phillips, W. D. 1967. *J. Am. Chem. Soc.* 89:6332–41
392. McDonald, C. C., Phillips, W. D. 1973. *Biochemistry* 12:3170–86
393. McDonald, G. C., Cohn, M., Noda, L. 1975. *J. Biol. Chem.* 250:6947–54
394. Meadows, D. H., Jardetzky, O., Epand,

R. M., Rüterjans, H. H., Scheraga, H. A. 1968. *Proc. Natl. Acad. Sci. USA* 60:766–72

395. Meadows, D. H., Markley, J. L., Cohen, J. S., Jardetzky, O. 1967. *Proc. Natl. Acad. Sci. USA* 58:1307–13

396. Mendz, G. L., Moore, W. J., Carnegie, P. R. 1982. *Aust. J. Chem.* 35:1979–2006

397. Mendz, G. L., Moore, W. J., Martenson, R. E. 1983. *Biochim. Biophys. Acta* 742:215–23

397a. Mendz, G. L., Moore, W. J., Martenson, R. E. 1983. *Biochim. Biophys. Acta* 748:168–75

398. Migchelsen, C., Beintema, J. J. 1973. *J. Mol. Biol.* 79:25–38

399. Mildvan, A. S., Gupta, R. K. 1978. *Methods Enzymol.* 49:322–59

400. Mitra, S., Bersohn, R. 1982. *Proc. Natl. Acad. Sci. USA* 79:6807–11

401. Miura, S., Ho, C. 1982. *Biochemistry* 21:6280–87

402. Miyazawa, T., Endo, T., Inagaki, F., Hayashi, K., Tamiya, N. 1982. *Biopolymers* 22:139–45

403. Moonen, C. T. W., Hore, P. J., Mueller, F., Kaptein, R., Mayhew, S. G. 1982. *FEBS Lett.* 149:141–46

404. Moore, G. R., Huang, Z.-X., Eley, C. G. S., Barker, H. A., Williams, G., Robinson, M. N., Williams, R. J. P. 1982. *Faraday Discuss. Chem. Soc.* 74:311–29

405. Moore, G. R., McClune, G. J., Clayden, N. J., Williams, R. J. P., Alsaadi, B. M., Angstrom, J., et al. 1982. *Eur. J. Biochem.* 123:73–80

406. Moore, G. R., Pitt, R. C., Williams, R. J. P. 1977. *Eur. J. Biochem.* 77:53–60

407. Moore, G. R., Williams, R. J. P. 1977. *FEBS Lett.* 79:229–32

408. Moore, G. R., Williams, R. J. P. 1980. *Eur. J. Biochem.* 103:493–502

409. Moore, G. R., Williams, R. J. P. 1980. *Eur. J. Biochem.* 103:503–12

410. Moore, G. R., Williams, R. J. P. 1980. *Eur. J. Biochem.* 103:533–41

411. Moore, G. R., Williams, R. J. P., Chien, J. C. W., Dickson, L. C. 1980. *J. Inorg. Biochem.* 12:1–15

412. Moura, J. J. G., Xavier, A. V., Bruschi, M., Le Gall, J. 1977. *Biochim. Biophys. Acta* 459:278–89

413. Moura, J. J. G., Xavier, A. V., Cookson, D. J., Moore, G. R., Williams, R. J. P., Bruschi, M., Le Gall, J. 1977. *FEBS Lett.* 81:275–80

414. Mueller, F., Van Schagen, C. G., Van Berkel, W. J. H. 1980. In *Flavins Flavoproteins, Proc. Int. Symp., 6th*, ed. K. Yagi, T. Yamano, pp. 359–71. Tokyo: Jpn. Sci. Soc.

415. Muszkat, K. A., Weinstein, S., Khait, I.,

Vered, M. 1982. *Biochemistry* 21:3775–79

416. Nagai, K., La Mar, G. N., Jue, T., Bunn, H. F. 1982. *Biochemistry* 21:842–47

417. Nagayama, K. 1981. *Adv. Biophys.* 14:139–204

418. Nagayama, K., Bachmann, P., Wüthrich, K., Ernst, R. R. 1978. *J. Magn. Reson.* 31:133–48

419. Nagayama, K., Kumar, A., Wüthrich, K., Ernst, R. R. 1980. *J. Magn. Reson.* 40:321–34

419a. Nagayama, K., Ozaki, Y., Kyogoku, Y., Hase, T., Matsubara, H. 1983. *J. Biochem. Tokyo* 94:893–902

420. Nagayama, K., Wüthrich, K. 1981. *Eur. J. Biochem.* 114:365–74

421. Nagayama, K., Wüthrich, K., Bachmann, P., Ernst, R. R. 1977. *Biochem. Biophys. Res. Commun.* 78:99–105

422. Nagayama, K., Wüthrich, K., Bachmann, P., Ernst, R. R. 1977. *Naturwissenschaften* 64:581–83

423. Nagayama, K., Wüthrich, K., Ernst, R. R. 1979. *Biochem. Biophys. Res. Commun.* 90:305–11

424. Nakano, A., Miyazawa, T., Nakamura, S., Kaziro, Y. 1979. *Arch. Biochem. Biophys.* 196:233–38

425. Nakano, A., Miyazawa, T., Nakamura, S., Kaziro, Y. 1980. *Biochemistry* 19:2209–15

426. Nelson, D. J., Opella, S. J., Jardetzky, O. 1976. *Biochemistry* 15:5552–60

427. Nelson, D. J., Theoharides, A. D., Nieburgs, A. C., Murray, R. K., Gonzalez-Fernandez, F., Brenner, D. S. 1979. *Int. J. Quantum. Chem.* 16:159–74

428. Niu, C.-H., Matsuura, S., Shindo, H., Cohen, J. S. 1979. *J. Biol. Chem.* 254:3788–96

429. Noggle, J. H., Schirmer, R. E. 1971. *The Nuclear Overhauser Effect*. New York: Academic. 259 pp.

430. O'Conner, T. P., Coleman, J. E. 1983. *Biochemistry* 22:3375–81

431. Ogino, T., Croll, D. H., Kato, I., Markley, J. L. 1982. *Biochemistry* 21:3452–60

432. Ohms, J. P., Hagenmaier, H., Hayes, M. B., Cohen, J. S. 1979. *Biochemistry* 18:1599–1602

433. Okhanov, V. V., Afanasev, V. A., Gurevich, A. Z., Elyakova, E. G., Miroshinikov, A. I., Bystrov, V. F., Ovchinnikov, Yu. A. 1980. *Sov. J. Bioorg. Chem.* (English transl.) 6:840–60

434. Oldfield, E., Allerhand, A. 1973. *Proc. Natl. Acad. Sci. USA* 70:3531–35

435. Oldfield, E., Allerhand, A. 1975. *J. Am. Chem. Soc.* 97:221–24

436. Oldfield, E., Allerhand, A. 1975. *J. Biol. Chem.* 250:6403–7

437. Oldfield, E., Janes, N., Kinsey, R., Kintanar, A., Lee, R. W. K., Rothgeb, T. M., et al. 1981. *Biochem. Soc. Symp.* 46:155–81
438. Oldfield, E., Norton, R. S., Allerhand, A. 1975. *J. Biol. Chem.* 250:6381–402
439. Opella, S. J. 1982. *Ann. Rev. Phys. Chem.* 33:533–62
440. Opella, S. J., Lu, P., eds. 1979. *NMR and Biochemistry: A Symposium Honoring Mildred Cohn.* New York: Dekker. 434 pp.
441. Oster, O., Neireiter, G. W., Clouse, A. O., Gurd, F. R. N. 1975. *J. Biol. Chem.* 250:7990–96
442. Otvos, J. D., Alger, J. R., Coleman, J. F., Armitage, I. M. 1979. *J. Biol. Chem.* 254:1778–80
443. Otvos, J. D., Armitage, I. M. 1979. *Experientia Suppl.* 34:249–57
444. Otvos, J. D., Armitage, I. M. 1980. *Biochemistry* 19:4021–30
445. Otvos, J. D., Armitage, I. M. 1980. *Biochemistry* 19:4031–43
446. Otvos, J. D., Armitage, I. M. 1982. See Ref. 64, pp. 65–96
447. Packer, E. L., Rabinowitz, J. C., Sternlicht, H. 1978. *J. Biol. Chem.* 253:7722–30
448. Packer, E. L., Sweeney, W. V., Rabinowitz, J. C., Sternlicht, H., Shaw, E. N. 1977. *J. Biol. Chem.* 252:2245–53
449. Patel, D. J., Canuel, L. L., Bovey, F. A. 1975. *Biopolymers* 14:987–97
450. Patel, D. J., Canuel, L. L., Woodward, C., Bovey, F. A. 1975. *Biopolymers* 14:959–74
451. Patel, D. J., Woodward, C., Canuel, L. L., Bovey, F. A. 1975. *Biopolymers* 14:975–86
452. Perkins, S. J. 1982. *Biol. Magn. Reson.* See Ref. 44, 4:193–336
453. Perkins, S. J., Johnson, L. N., Phillips, D. C., Dwek, R. A. 1977. *FEBS Lett.* 82:17–22
454. Perkins, S. J., Wüthrich, K. 1978. *Biochim. Biophys. Acta* 536:406–20
455. Perkins, S. J., Wüthrich, K. 1980. *J. Mol. Biol.* 138:43–64
456. Plateau, P., Gueron, M. 1982. *J. Am. Chem. Soc.* 104:7311–12
457. Poe, M., Hoogsteen, K., Matthews, D. A. 1979. *J. Biol. Chem.* 254:8143–52
458. Poe, M., Wu, J. K., Short, C. Jr., Florance, J., Hoogsteen, K. 1979. *Dev. Biochem.* 4:483–88
459. Porubcan, M. A., Neves, D. E., Rausch, S. K., Markley, J. L. 1978. *Biochemistry* 17:4640–47
460. Porubcan, M. A., Westler, W. M., Ibañez, I. B., Markley, J. L. 1979. *Biochemistry* 18:4108–16
461. Poulson, F. M., Hoch, J. C., Dobson, C. M. 1980. *Biochemistry* 19:2597–2607
462. Rattle, H. W. E. 1976. *Amino-Acids Pept. Proteins* 8:199–208
463. Rattle, H. W. E. 1979. *Amino-Acids Pept. Proteins* 10:221–37
463a. Rattle, H. W. E. 1983. *Amino-Acids Pept. Proteins* 14:242–61
464. Redfield, A. G. 1978. *Methods Enzymol.* 49:253–70
465. Redfield, A. G. 1978. *Methods Enzymol.* 49:359–69
466. Redfield, A. G., Gupta, R. K. 1972. *Cold Spring Harbor Symp. Quant. Biol.* 36:405–11
467. Redfield, A. G., Kunz, S. D., Ralph, E. K. 1975. *J. Magn. Reson.* 19:114–17
468. Redfield, C., Poulsen, F. M., Dobson, C. M. 1982. *Eur. J. Biochem.* 128:527–31
469. Rhyu, G. I., Markley, J. L., Ray, W. J. Jr. 1983. See Réf. 350, p. 110 (Abstr.)
470. Rhyu, G. I., Ray, W. J. Jr., Markley, J. L. 1984. *Biochemistry.* 23:In press
471. Ribeiro, A. A., King, R., Restivo, C., Jardetzky, O. 1980. *J. Am. Chem. Soc.* 102:4040–51
472. Ribeiro, A. A., Wemmer, D., Bray, R. P., Wade-Jardetzky, N. G., Jardetzky, O. 1981. *Biochemistry* 20:823–29
473. Rice, D. M., Blume, A., Herzfeld, J., Wittebort, R. J., Huang, T. H., DasGupta, S. K., Griffin, R. G. 1981. See Ref. 141, pp. 255–70
474. Richarz, R., Tschesche, H., Wüthrich, K. 1979. *Eur. J. Biochem.* 102:563–71
475. Richarz, R., Tschesche, H., Wüthrich, K. 1980. *Biochemistry* 19:5711–15
476. Richarz, R., Wüthrich, K. 1978. *Biochemistry* 17:2263–69
477. Roberts, G. C. K. 1977. In *Drug Action Mol. Level, Rep. Symp.*, ed. G. C. K. Roberts, pp. 127–50. Baltimore: Univ. Park Press
478. Roberts, G. C. K. 1983. In *Pteridines and Folic Acid Derivatives, Chemical, Biological and Chemical Aspects*, ed. J. A. Blair. New York: de Gruyter
479. Roberts, G. C. K., Feeney, J., Birdsall, B., Kimber, B., Griffiths, D. V., King, R. W., Burgen, A. S. V. 1977. See Ref. 154, pp. 95–109
480. Robillard, G., Shulman, R. G. 1972. *J. Mol. Biol.* 71:507–11
481. Robillard, G., Shulman, R. G. 1974. *J. Mol. Biol.* 86:519–40
482. Robillard, G., Shulman, R. G. 1974. *J. Mol. Biol.* 86:541–58
483. Roesch, P., Kalbitzer, H. R., Schmidt-Aderjan, U., Hengstenberg, W. 1981. *Biochemistry* 20:1599–1605
484. Roossien, F. F., Dooyewaard, G., Robillard, G. T. 1979. *Biochemistry* 18:5793–97

485. Roques, B. P., Rao, R., Marion, D. 1980. *Biochimie* 62:753–73
486. Rosenthal, S. N., Fendler, J. H. 1976. *Adv. Phys. Org. Chem.* 13:279–424
487. Rosevear, P. R., Desmeules, P., Kenyon, G. L., Mildvan, A. S. 1981. *Biochemistry* 20:6155–64
488. Rothgeb, T. M., Oldfield, E. 1981. *J. Biol. Chem.* 256:1432–46
489. Russu, I. M., Ho, C. 1982. *Biochemistry* 21:5044–51
490. Russu, I. M., Ho, N. T., Ho, C. 1980. *Biochemistry* 19:1043–52
491. Russu, I. M., Ho, N. T., Ho, C. 1982. *Biochemistry* 21:5031–43
492. Santoro, J., Juretschke, H. P., Rüterjans, H. 1979. *Biochim. Biophys. Acta* 578:346–56
493. Satterlee, J. D., Erman, J. E. 1981. *J. Am. Chem. Soc.* 103:199–200
494. Satterlee, J. D., Erman, J. E. 1983. *Biochim. Biophys. Acta* 743:149–54
495. Satterlee, J. D., Erman, J. E., La Mar, G. N., Smith, K. M., Langry, K. C. 1983. *Biochim. Biophys. Acta* 743:246–55
496. Satterlee, J. D., Erman, J. E., La Mar, G. N., Smith, K. M., Langry, K. C. 1983. *J. Am. Chem. Soc.* 105:2099–2104
497. Saunders, M., Wishnia, A., Kirkwood, J. G. 1957. *J. Am. Chem. Soc.* 79:3289–90
498. Scheek, R. M., Zuiderweg, E. R. P., Klappe, K. J. M., Van Boom, J. H., Kaptein, R., Rüterjans, H., Beyreuther, K. 1983. *Biochemistry* 22:228–35
499. Schejter, A., Lanir, A., Vig, I., Cohen, J. S. 1978. *J. Biol. Chem.* 253:3768–70
500. Schmidt-Aderjan, U., Roesch, P., Frank, R., Hengstenberg, W. 1979. *Eur. J. Biochem.* 96:43–48
501. Schnackerz, K. D., Feldmann, K., Hull, W. E. 1979. *Biochemistry* 18:1536–39
502. Scholberg, H. P. F., Fronticelli, C., Bucci, E. 1980. *J. Biol. Chem.* 255:8592–98
503. Schramm, S., Oldfield, E. 1983. *Biochemistry* 22:2908–13
504. Seamon, K. B. 1980. *Biochemistry* 19:207–15
505. Seamon, K. B., Hartshorne, D. J., Bothner-By, A. A. 1977. *Biochemistry* 16:4039–46
506. Senn, H., Eugster, A., Wüthrich, K. 1983. *Biochim. Biophys. Acta* 743:58–68
507. Senn, H., Wüthrich, K. 1983. *Biochim. Biophys. Acta* 743:69–81
508. Shaka, A. J., Frenkiel, T., Freeman, R. 1983. *J. Magn. Reson.* 52:159–63
509. Shaka, A. J., Keeler, J., Frenkiel, T., Freeman, R. 1983. *J. Magn. Reson.* 52:335–38
510. Shaw, D. 1976. *Fourier Transform NMR Spectroscopy.* Amsterdam: Elsevier. 357 pp.
511. Shelling, J. G., Sykes, B. D., O'Neil, J. D. J., Hofmann, T. 1983. *Biochemistry* 22:2649–54
512. Shimizu, A., Honzawa, M., Ito, S., Miyazaki, T., Matsumoto, H., Nakamura, H., et al. 1983. *Mol. Immunol.* 20:141–48
513. Shimizu, A., Honzawa, M., Yamamura, Y., Arata, Y. 1980. *Biochemistry* 19:2784–90
514. Shindo, H., Cohen, J. S. 1976. *J. Biol. Chem.* 251:2648–52
515. Shindo, H., Egan, W., Cohen, J. S. 1978. *J. Biol. Chem.* 253:6751–55
516. Shindo, H., Hayes, M. B., Cohen, J. S. 1976. *J. Biol. Chem.* 251:2644–47
517. Shriver, J., Mateescu, G., Fager, R., Torchia, D., Abrahamson, E. W. 1977. *Nature* 270:271–74
518. Shulman, R. G., ed. 1979. *Biological Applications of Magnetic Resonance.* New York: Academic. 595 pp.
519. Shulman, R. G., Wüthrich, K., Yamane, T., Antonini, E., Brunori, M. 1969. *Proc. Natl. Acad. Sci. USA* 63:623–28
520. Sletten, E., Jackson, J. T., Burns, P. D., La Mar, G. N. 1983. *J. Magn. Reson.* 52:492–96
521. Smith, G. M. 1979. *Biochemistry* 18:1628–34
522. Snyder, G. H., Rowan, R. III, Karplus, S., Sykes, B. D. 1975. *Biochemistry* 14:3765–77
523. Sperling, J. E., Feldman, K., Meyer, H., Jahnke, U., Heilmeyer, L. M. G. Jr. 1979. *Eur. J. Biochem.* 101:581–92
524. Steinmetz, W. E., Moonen, C., Kumar, A., Lazdunski, M., Visser, L., Carlsson, F. H. H., Wüthrich, K. 1981. *Eur. J. Biochem.* 120:467–75
525. Steitz, T. A., Shulman, R. G. 1982. *Ann. Rev. Biophys. Bioeng.* 11:419–44
526. Stellwagen, E., Shulman, R. G. 1973. *J. Mol. Biol.* 75:683–95
527. Sternlicht, H., Wilson, D. 1967. *Biochemistry* 6:2881–92
528. Stoesz, J. D., Malinowski, D. P., Redfield, A. G. 1979. *Biochemistry* 18:4669–75
529. Strom, R., Crigo, C., Viti, V., Guidoni, L., Podo, F. 1978. *FEBS Lett.* 96:45–50
530. Štrop, P., Čechova, D., Wüthrich, K. 1983. *J. Mol. Biol.* 166:669–76
531. Štrop, P., Wider, G., Wüthrich, K. 1983. *J. Mol. Biol.* 166:641–67
532. Štrop, P., Wüthrich, K. 1983. *J. Mol. Biol.* 166:631–40
533. Sudmeier, J. L., Bell, S. J., Storm, M. C., Dunn, M. F. 1981. *Science* 212:560–62
534. Sudmeier, J. L., Perkins, T. G. 1977. *J. Am. Chem. Soc.* 99:7732–33

535. Sykes, B. D., Hull, W. E. 1978. *Methods Enzymol.* 49:270–95
536. Sykes, B. D., Hull, W. E., Snyder, G. H. 1978. *Biophys. J.* 21:137–46
537. Sykes, B. D., Weiner, J. H. 1980. In *Magnetic Resonance in Biology*, ed. J. S. Cohen, pp. 171–96. New York: Wiley
538. Sykes, B. D., Weingarten, H. I., Schlesinger, M. J. 1974. *Proc. Natl. Acad. Sci. USA* 71:469–73
539. Takahashi, S., Lin, A. K. L. C., Ho, C. 1980. *Biochemistry* 19:5196–5202
540. Takahashi, S., Lin, A. K. L. C., Ho, C. 1982. *Biophys. J.* 39:33–40
541. Takesada, H., Nakanishi, M., Tsuboi, M., Ajisaka, K. 1976. *J. Biochem.* 80:969–74
541a. Teleman, O., Drakenberg, T., Forsén, S., Thulin, E. 1983. *Eur. J. Biochem.* 134:453–57
542. Thiery, C., Nabedryk-Viala, E., Menez, A., Fromageot, P., Thiery, J. M. 1980. *Biochem. Biophys. Res. Commun.* 93:889–97
543. Timkovich, R., Cork, M. S. 1982. *Biochemistry* 21:5119–23
544. Torchia, D. A., Batchelder, L. S., Fleming, W. W., Jelinski, L. W., Sarkar, S. K., Sullivan, C. E. 1983. *Ciba Found. Symp.* 1983(93):98–111
545. Trewhella, J., Wright, P. E. 1980. *Biochim. Biophys. Acta* 625:202–20
546. Trexler, M., Banyai, L., Patthy, L., Pluck, N. D., Williams, R. J. P. 1983. *FEBS Lett.* 154:311–18
547. Tsai, M.-D. 1982. *Methods Enzymol.* 87:235–79
548. Tsai, M.-D., Bruzik, K. 1984. In *Biological Magnetic Resonance*, Vol. 5, ed. L. J. Berliner, J. Reuben. New York: Plenum. In press
549. Ugurbil, K., Bersohn, R. 1977. *Biochemistry* 16:3016–23
550. Ugurbil, K., Norton, R. S., Allerhand, A., Bersohn, R. 1977. *Biochemistry* 16:886–94
551. Ulrich, E. L., Krogmann, D. W., Markley, J. L. 1982. *J. Biol. Chem.* 257:9356–64
552. Ulrich, E. L., Markley, J. L. 1978. *Coord. Chem. Rev.* 27:109–40
553. Van Schagen, C. G., Mueller, F. 1981. *Eur. J. Biochem.* 120:33–39
554. Van Schagen, C. G., Mueller, F. 1981. *FEBS Lett.* 136:75–79
555. Viggiano, G., Wiechelman, K. J., Chervenick, P. A., Ho, C. 1978. *Biochemistry* 17:795–99
556. Villafranca, J. J. 1982. *Fed. Proc.* 41:2959–60
557. Villafranca, J. J. 1982. *Methods Enzymol.* 87:180–97
558. Villafranca, J. J., Raushel, F. M. 1980. *Ann. Rev. Biophys. Bioeng.* 9:363–92
559. Villafranca, J. J., Raushel, F. M. 1982. *Fed. Proc.* 41:2961–73
560. Virmani-Sardana, V., Breslow, E. 1983. *Int. J. Pept. Protein. Res.* 21:182–89
561. Vogel, H. J. 1984. In *Phosphorus-31 NMR, Principles and Applications*, ed. D. Gorenstein. New York: Academic
562. Vogel, H. J., Bridger, W. A. 1982. *Biochemistry* 21:5825–31
563. Vogel, H. J., Bridger, W. A. 1982. *J. Biol. Chem.* 257:4834–42
564. Vogel, H. J., Bridger, W. A. 1983. *Biochem. Soc. Trans.* 11:315–23
565. Vogel, H. J., Drakenberg, T., Forsén, S. 1983. In *NMR of Newly Accessible Nuclei*, ed. P. Laszlo. New York: Academic. In press
566. Wagner, G. 1982. *Comments Mol. Cell. Biophys.* 1:261–80
566a. Wagner, G. 1983. *J. Magn. Reson.* 55:151–56
567. Wagner, G. 1983. *Q. Rev. Biophys.* 16:1–57
568. Wagner, G., Kumar, A., Wüthrich, K. 1981. *Eur. J. Biochem.* 114:375–84
569. Wagner, G., Wüthrich, K. 1979. *J. Mol. Biol.* 130:31–37
570. Wagner, G., Wüthrich, K. 1979. *J. Mol. Biol.* 134:75–94
571. Wagner, G., Wüthrich, K. 1982. *J. Mol. Biol.* 155:347–66
572. Wagner, G., Wüthrich, K. 1982. *J. Mol. Biol.* 160:343–61
573. Wagner, G., Wüthrich, K. 1983. *Naturwissenschaften* 70:105–14
574. Wagner, G., Wüthrich, K., Tschesche, H. 1978. *Eur. J. Biochem.* 86:67–76
575. Wagner, G., Wüthrich, K., Tschesche, H. 1978. *Eur. J. Biochem.* 89:367–77
576. Wagner, G., Zuiderweg, E. R. P. 1983. *Biochim. Biophys. Acta* 113:854–60
577. Wain-Hobson, S., Dower, S. K., Gettins, P., Givol, D., McLaughlin, A. C., Pecht, I., et al. 1977. *Biochem. J.* 165:227–35
578. Walters, D. E., Allerhand, A. 1980. *J. Biol. Chem.* 255:6200–4
579. Wang, F.-F. C., Hirs, C. H. W. 1979. *J. Biol. Chem.* 254:1090–93
580. Waugh, J. S. 1982. *J. Magn. Reson.* 50:30–49
581. Wedin, R. E., Delepierre, M., Dobson, C. M., Poulsen, F. M. 1982. *Biochemistry* 21:1098–1103
582. Wemmer, D., Kallenbach, N. R. 1983. *Biochemistry* 22:1901–6
583. Weser, U., Rupp, H. 1979. *Top. Environ. Health* 2:267–83
584. Westler, W. M., Bogard, W. C. Jr., Laskowski, M. Jr., Markley, J. L. 1982. *Fed. Proc.* 41:1188 (Abstr.)
585. Westler, W. M., Markley, J. L.,

Bachovchin, W. W. 1982. *FEBS Lett.* 138:233–35

586. Wider, G., Hosur, R. V., Wüthrich, K. 1983. *J. Magn. Reson.* 52:130–35
587. Wieloch, T., Borgstrom, B., Falk, K.-E., Forsén, S. 1979. *Biochemistry* 18:1622–28
588. Wieloch, T., Falk, K.-E. 1978. *FEBS Lett.* 85:271–74
589. Wilbur, D. J., Allerhand, A. 1976. *J. Biol. Chem.* 251:5187–94
590. Wilbur, D. J., Allerhand, A. 1977. *FEBS Lett.* 79:144–46
591. Williams, G., Eley, C. G. S., Moore, G. R., Robinson, M. N., Williams, R. J. P. 1982. *FEBS Lett.* 150:293–99
592. Williams, R. J. P. 1978. *Biochem. Soc. Trans.* 6:1123–26
593. Williams, R. J. P. 1981. *Biochem. Soc. Symp.* 46:57–72
594. Williamson, K. L., Williams, R. J. P. 1979. *Biochemistry* 18:5966–72
595. Wittebort, R. J., Rothgeb, T. M., Szabo, A., Gurd, F. R. N. 1976. *Proc. Natl. Acad. Sci. USA* 76:1059–63
596. Woods, L. F. J., Wiseman, A., Libor, S., Jones, J. R., Elvidge, J. A. 1980. *Biochem. Soc. Trans.* 8:98–99
597. Woodward, C., Simon, I., Tüchsen, E. 1982. *Mol. Cell. Biochem.* 48:135–60
598. Woodworth, R. C., Williams, R. J. P., Alsaadi, B. M. 1977. In *Proteins Iron Metab., Proc. Int. Meet., 3rd*, ed. E. B. Brown, J. Fielding, pp. 211–18. New York: Grune & Stratton
599. Wooten, J. B., Cohen, J. S. 1979. *Biochemistry* 18:4188–91
600. Wooten, J. B., Cohen, J. S., Vig, I., Schejter, A. 1981. *Biochemistry* 20:5394–5402
601. Wüthrich, K. 1976. *NMR in Biological Research: Peptides and Proteins.* New York: Am. Elsevier. 379 pp.
602. Wüthrich, K. 1977. In *NATO Adv. Study Inst. Ser., Ser. B*, pp. 347–60
603. Wüthrich, K. 1977. In *NATO Adv. Study Inst. Ser., Ser. B*, pp. 361–74
604. Wüthrich, K. 1977. See Ref. 154, pp. 51–62
605. Wüthrich, K. 1980. In *Front. Biorg. Chem. Mol. Biol., Proc. Int. Symp.*, ed. S.

N. Anachenko, pp. 161–68. Oxford: Pergamon
606. Wüthrich, K. 1981. *Biochem. Soc. Symp.* 46:17–37
607. Wüthrich, K. 1981. *Macromol. Chem. Phys. Suppl.* 5:234–52
608. Wüthrich, K. 1982. *NATO Adv. Study Inst. Ser., Ser. A 45*, pp. 215–35
609. Wüthrich, K. 1983. *Biopolymers* 22:131–38
610. Wüthrich, K., Nagayama, K., Ernst, R. R. 1979. *Trends Biochem. Sci.* 4:N178–81
611. Wüthrich, K., Roder, H., Wagner, G. 1980. In *Protein Folding, Proc. Conf. Ger. Biochem. Soc., 28th*, ed. R. Jaenicke, pp. 549–64. Amsterdam: Elsevier
612. Wüthrich, K., Wagner, G. 1978. *Trends Biochem. Sci.* 3:227–30
613. Wüthrich, K., Wagner, G. 1979. *J. Mol. Biol.* 130:1–18
614. Wüthrich, K., Wagner, G. 1981. In *Biomol. Struct. Conform. Funct. Evol., Proc. Int. Symp.*, ed. R. Srinivasan, E. Subramanian, N. Yathindra, 2:23–29. Oxford: Pergamon
615. Wüthrich, K., Wagner, G. 1983. *Ciba Found. Symp.* 93:310–28
616. Wüthrich, K., Wagner, G., Richarz, R., Braun, W. 1980. *Biophys. J.* 32:549–60
617. Wüthrich, K., Wagner, G., Richarz, R., Perkins, S. J. 1978. *Biochemistry* 17:2253–63
618. Wüthrich, K., Wider, G., Wagner, G., Braun, W. 1982. *J. Mol. Biol.* 155:311–19
619. Xavier, A. V. 1983. *NATO Adv. Study Inst. Ser., Ser. C* 100:291–311
620. Xavier, A. V., Czerwinski, E. W., Bethge, P. H., Mathews, F. S. 1978. *Nature* 275:245–47
621. Xavier, A. V., Moura, I., Moura, J. J. G., Santos, M. H., Villalain, J. 1982. *NATO Adv. Study Inst. Ser., Ser. C* 89:127–41
622. Xavier, A. V., Moura, J. J. G. 1978. *Biochimie* 60:327–38
623. Yamaguchi, A., Unemoto, T., Ikegami, A. 1981. *Photochem. Photobiol.* 33:511–16
624. York, J. L., Millett, F. S., Minor, L. B. 1980. *Biochemistry* 19:2583–88

Ann. Rev. Biochem. 1982. 51:459–89

5

SPECIFIC INTERMEDIATES IN THE FOLDING REACTIONS OF SMALL PROTEINS AND THE MECHANISM OF PROTEIN FOLDING[1]

Peter S. Kim and Robert L. Baldwin

Department of Biochemistry, Stanford University School of Medicine, Stanford, California 94305

CONTENTS

[1]Abbreviations used for proteins are: BPTI, bovine pancreatic trypsin inhibitor; cyt *c,* horse heart cytochrome *c;* RNase A, bovine pancreatic ribonuclease A; RNase S, a derivative of RNase A cleaved at the peptide bond between residues 20 and 21; S peptide, residues 1–20 of RNase S; S protein, residues 21–124 of RNase S. Other abbreviations are: N, native; U, unfolded; I, intermediate; I_N, quasinative intermediate; U_F and U_S, fast- and slow-folding species, respectively; k, rate constant; GuHCl, guanidinium chloride; t_m, temperature at the midpoint of an unfolding transition; ΔC_P, change in heat capacity at constant pressure; θ, mean residue ellipticity. Lysozyme refers to hen egg white lysozyme and myoglobin to sperm whale myoglobin.

PERSPECTIVES AND SUMMARY

The stage is set for determining the pathway of folding of representative small proteins by characterizing the structures of well-populated folding intermediates. Structural intermediates accumulate in kinetic experiments under conditions (especially low temperatures) where the intermediates are stable relative to the unfolded form. The major problem appears to be in tracing the folding pathway for a single unfolded form, because the *cis-trans* isomerization of proline residues about X-Pro peptide bonds often gives multiple unfolded forms, with different rates of refolding. The role of proline isomerization in refolding is beginning to be understood. Covalent intermediates have been trapped and characterized in the refolding process, which accompanies reoxidation of the disulfide bonds for two small proteins. Equilibrium intermediates have been found and characterized for some unusual small proteins, and it appears that the unfolding reactions induced by certain salts or methanol yield equilibrium intermediates even for proteins that normally show highly cooperative folding.

The pathway of folding should reveal the mechanism of folding and help in determining the code by which the amino acid sequence of a protein specifies its tertiary structure. This will aid in engineering changes in protein structure using recombinant DNA techniques. Knowledge of the pathway can show at which stage an amino acid substitution causes a change in folding.

Earlier work searching for equilibrium intermediates with small, single-domain proteins such as RNase A and lysozyme gave chiefly negative results and led to the use of the two-state approximation (N $\rightleftharpoons$ U, N = native, U = unfolded), in which folding intermediates are neglected. Populated intermediates were thought to be ruled out by the success of the two-state approximation. However, such highly cooperative folding is found inside the folding transition zone where intermediates are only marginally stable. Folding can be measured kinetically in conditions where intermediates are more stable.

The evidence is now essentially complete that multiple forms of an unfolded protein are produced by the *cis-trans* isomerization of proline residues about peptide bonds after unfolding. Slow-folding (U_S) and fast-folding (U_F) forms of an unfolded protein can be recognized in refolding experiments by the fact that native protein is formed in separate slow ($U_S \rightarrow N$) and fast ($U_F \rightarrow N$) refolding reactions. They also can be recognized in unfolding experiments by refolding assays made during the fast (N $\rightarrow U_F$) and slow ($U_F \rightleftharpoons U_S$) phases of unfolding. The kinetic properties of the $U_F \rightleftharpoons U_S$ reaction, measured during unfolding, match those of proline isomerization in such specific aspects as catalysis by strong acid and cleavage of X–Pro bonds by an enzyme specific for the *trans* proline isomer. Folding in strongly native conditions occurs before proline isomerization and can go almost to completion. An enzymatically active intermediate that still contains a wrong proline isomer is found when RNase A folds at 0°–10°C. Partial folding increases the rate of proline isomerization, possibly because strain is produced in folding intermediates. The unexpectedly low proportion of U_S species in several unfolded proteins suggests that not all proline residues produce slow-folding species.

Information about the folding pathway is preliminary in all cases but, surprisingly, S–S bond formation in BPTI proceeds via obligatory two-disulfide intermediates each having a nonnative S–S bond. The major folding reaction occurs in a single S–S rearrangement. Both kinetic and equilibrium results for folding with S–S bonds intact are consistent with a framework model in which the H-bonded secondary structure is formed early in folding. The tertiary structures of α-lactalbumin, penicillinase, and carbonic anhydrase are disrupted before their secondary structures unfold, in denaturant-induced unfolding. Kinetically, secondary structure is formed early in the folding of RNase A and RNase S, as judged by stopped-flow CD studies and protection of NH protons against exchange with solvent. The kinetic mechanism of folding appears to be sequential folding with defined intermediates. Folding probably proceeds along a pathway determined by the most stable intermediates. Independently folding domains have been demonstrated via fragment isolation for the small proteins ovomucoid and elastase, the α subunit of tryptophan synthase, and for larger proteins. For multidomain proteins, the pathway of folding involves independent folding of individual domains followed by domain interaction. Mutants blocked kinetically in the folding and assembly of a trimeric protein, the tail spike protein of phage P22, have been demonstrated.

INTRODUCTION

Statement of the Problem

Small, monomeric proteins fold to thermodynamically stable structures, as judged by reversibility of folding. Nevertheless, folding is very rapid in most cases (seconds or less). The number of possible conformations is astronomical. If the pathway is under thermodynamic control, how does the protein find the most stable structure so quickly? If the pathway is under kinetic control, how are incorrectly folded structures avoided?

The amino acid sequence codes for the folding of a protein, but amino acid substitutions (changes in the code) are allowed at almost all residue positions without drastic changes in the folding pattern. X-ray structures have been determined for globins that are related only distantly through evolution and have only a few amino acids in common; yet the "globin fold" is strikingly similar in each case. The qualitative features of folding appear to be the same in horse cyt c and yeast cytochrome c, even though they differ by 46% in amino acid sequence. The code for folding is not a simple code like the mRNA triplet code.

The complexity of the code probably reflects the complexity of the folding process. Determination of the pathway of folding may be the chief means of solving the code, because the specific interactions can be measured at different stages in folding. This can be illustrated by a current model for folding, according to which α helices and β sheets form at their correct locations in the otherwise unfolded chain. Amino acid substitutions may be allowed because the choice between helix, sheet, or no folding is averaged over several residues, so that one substitution need not tip the balance, or because only a few specific interactions determine the locations of the α helices and β sheets. At the second stage in folding, the model predicts that α helices and β sheets interact via special pairing sites,[2] which are coded by only a few residues, thus allowing substitutions at other residue positions. A decade ago it was difficult to understand how an amino acid substitution could be tolerated in the interior of a protein because the side chains are closely packed and the protein structure was thought to be rigid. Today it is known, especially from NMR studies of tyrosine and phenylalanine ring flips, that the interior of a protein is flexible.

---

[2]Alternatively, the pairing between α helices and/or β strands may be determined by their relative positions in the chain (1a).

The practical problems in determining the folding pathway are severe. The methods that give structural information about protein folding (e.g. X-ray, NMR, and CD) are intrinsically slow, so that equilibrium intermediates are needed. Moreover, the intermediates must be well-populated, since these methods require reasonably pure materials. The pathway of folding should first be determined for the simplest case, that of small, "single-domain" proteins like BPTI, RNase A, myoglobin, or staph nuclease. But folding of these small proteins is highly cooperative in most cases, and equilibrium intermediates are not populated. On the other hand, although kinetic intermediates may be well populated, steps in folding are often fast (1–100 msec). It is necessary to find methods of trapping intermediates in a stable form, so that they can be studied at leisure, of slowing down folding, or of adapting spectroscopic methods so that they will give structural information rapidly. Some solutions to these problems have been found and are discussed here.

Other Reviews

An excellent summary of work on protein folding, as of September 1979, is contained in a set of symposium papers collected by Jaenicke (1) into a book, *Protein Folding*. It includes a separate review of recent experimental work. A new monograph on protein folding is being prepared by Ghélis & Yon (2). The study of folding intermediates was reviewed in 1978 by Creighton (3) and in 1975 by Baldwin (4). Wetlaufer (5) has just reviewed the folding of protein fragments. Structural studies of folding, and the possible relationships between the final structure and the pathway of folding have been reviewed recently by Richardson (6), Ptitsyn & Finkelstein (7), Thomas & Shechter (8), and Rossmann & Argos (9). Thermodynamic data on the energetics of folding, obtained by calorimetric studies of folding transitions, have been reviewed by Privalov in 1979 (10). Nemethy & Scheraga consider both theoretical and experimental aspects in a 1977 review (11), which stresses the possible stereochemical determinants of folding. The use of packing principles and surface areas in analyzing the folding process was reviewed in 1977 by Richards (12). Ikegami (13) has just reviewed work on interpreting folding transitions by a cluster model. Earlier general reviews of folding have been given by Anfinsen & Scheraga in 1975 (14) and by Wetlaufer & Ristow in 1973 (15). The nature of protein folding transitions and of the unfolded state was discussed by Tanford (16, 17) in a pair of classic reviews.

MODELS FOR THE FOLDING PATHWAY

These models are of two kinds: kinetic and structural. Since structural data on folding intermediates are only now starting to be available, both classes of models still consist of guesses about the folding process. Present structural models are based chiefly on reflection about the X-ray structures of native proteins. The aim of a structural model is to give the actual structures of intermediates as well as their order on the pathway, without being too specific about the factors that control the rate of folding. The aim of a kinetic model is to indicate the dominant intermediates and give the factors that control the rate of folding without being too specific about the structures of the intermediates.

Kinetic Models

1. BIASED RANDOM SEARCH   The possibility that proteins might fold by a purely random search of all possible conformations was considered by Levinthal (18) and then dismissed, because the time required for folding would be impossibly long [10^{50} years for a chain of 100 residues (19)]. However, it is possible that a biased random search could occur in a reasonable time. By means of a computer-simulated folding for a lattice model, it is possible to show that the number of possible chain conformations is drastically reduced if only self-avoiding (sterically possible) conformations are allowed (M. Levitt, personal communication, 1978). Levitt also found that a significant fraction of the self-avoiding conformations are fairly compact, and he suggested that rapid folding might begin whenever the unfolded chain assumes a backbone conformation sufficiently like that of the native protein, because the major free energy barrier to folding (the necessary reduction in entropy of the polypeptide chain) has been overcome [compare (20)]. This result, taken together with the possibility of a rapid, nonspecific collapse when refolding is initiated, [compare (21)] suggests that a biased random search could play an important role in early stages of folding. However, it may not be easy to test the prediction that a nonspecific collapse precedes specific folding, because specific structures can be formed very rapidly: α helix formation in model systems occurs in 10^{-5}–10^{-7} sec (22–24).

2. NUCLEATION-GROWTH   The term nucleation has been used with two quite different meanings to describe models for protein folding. In both cases the nucleus is the structure formed at the beginning of folding, which guides subsequent steps. With the first meaning of nucleation, folding is sequential, and early folding intermediates may be populated: the nucleated molecule may be directly observable in kinetic folding experiments. With the second meaning, the folding reaction proceeds rapidly as soon as a nucleus is provided (as in crystallization of a supercooled liquid after seeding with a crystal, or as in α helix formation) and the nucleated molecule is not observable as a populated species either because folding occurs rap-

idly after nucleation or because the nucleus is unstable by itself and breaks down if it is not stabilized by further folding. The second meaning of nucleation is the classical one in chemistry, and the first meaning is a special usage that has developed gradually in protein folding work. The term was used originally in its correct (second) sense, but when it became apparent that the "nucleated" molecules might nevertheless be observable in kinetic folding experiments, the term was still retained by several workers. We suggest that the term nucleation now be dropped in protein folding studies unless it is used with its classical meaning, and that the other type of folding be referred to as *sequential folding* in which the first structure formed is the *kernel* (see Model 4).

In the nucleation-growth model, folding cannot start until an initial reaction occurs (nucleation), and subsequent folding takes place rapidly compared to the observed folding reaction. Folding intermediates are not populated, because folding is too fast once it starts, and the rate of folding is determined by the nucleation reaction. Therefore, demonstration of a populated intermediate in folding rules out the nucleation-growth model. Populated kinetic intermediates have been demonstrated in the folding of several proteins (see section on kinetic intermediates), so it appears that protein folding is not a nucleation-limited reaction.

3. **DIFFUSION-COLLISION-ADHESION**   In this *microdomain coalescence* model, short segments of the unfolded chain fold independently into microdomains. These are unstable, but they diffuse, collide, coalesce, and become stable (19, 25). Karplus & Weaver (19, 26) calculated the folding rate for a diffusion-collision model in which the diffusional collision is rate limiting, and computer simulations for the folding of apoMb have been made on the assumption that diffusion is rate limiting (27). However, adhesion or coalescence of the two microdomains may be the rate-limiting step, in this model. The two steps can be written as:

$$\text{A}\underline{\quad}\text{B} \quad \underset{k_{21}}{\overset{k_{12}}{\rightleftharpoons}} \quad \text{A·B} \quad \underset{k_{32}}{\overset{k_{23}}{\rightleftharpoons}} \quad \text{C}$$

where A___B are the two microdomains, linked by the polypeptide chain, which diffuse together to form the encounter complex A·B, and C is the product formed by adhesion. There are two limiting cases: (*a*) If k_{21}, the rate of dissociation of A·B, is large compared to k_{23}, the rate of adhesion, then $v = (k_{12}/k_{21})k_{23}$, where v is the overall rate of forming C from A___B. In this case, it is the equilibrium ratio of A·B to A___B and not the diffusion-controlled rate of forming A·B that enters into the overall rate expression. This situation is found commonly for reactions in solution (28) because k_{21} is always large (of the order of 10^{10} s^{-1}). (*b*) In special cases, such as proton transfer reactions, k_{23} can be very large (10^{12} s^{-1} for proton transfer) so that $k_{23} \gg k_{21}$ and $v = k_{12}$. This is the situation envisaged in the original diffusion-collision model.

The diffusion-collision model (19, 26) has been tested (29, 30) by asking if the folding rate of RNase A depends on solvent viscosity in the direct $(U_F \rightarrow N)$ folding reaction. Model compound studies have shown that diffusion of one segment of a chain molecule relative to another is dependent on solvent viscosity (31). The overall folding rate of RNase A was found to be independent of solvent viscosity (29) when either glycerol or sucrose was added, which demonstrates that diffusion is not rate limiting. Recently, a fast reaction (msec) of RNase A has been found (32) that is strongly affected by solvent additives that change the viscosity. Since it can be measured at temperatures far below the transition zone for unfolding, its relation to the folding process is not yet clear.

4. SEQUENTIAL FOLDING   Folding occurs in a unique and definite sequence of steps, analogous to a metabolic pathway. Intermediates may be populated in suitable conditions of folding. To demonstrate sequential folding, it is necessary to show that there are specific, well-populated intermediates. This test has been satisfied for the folding at low temperatures (0°–10°C) of the major slow-folding species of RNase A (33–38) and also of RNase S (30, 39, 40). Proline isomerization is one step in these folding reactions, but folding can proceed to an enzymatically active, native-like form of RNase A before proline isomerization occurs (33, 36–38). Therefore, the folding process is probably similar (i.e. sequential) for both the fast and slow folding species.

Structural Models

The increasing number of X-ray crystal structures has stimulated the proposal of many structural models for protein folding. The possible relationship between the folding pathway and the final structure of a protein has been the subject of several recent reviews (6–9, 12, 25, 41–46) and we do not review these models here.

Some models for the folding of an entire class of proteins postulate that folding begins by forming a "primitive" H-bonded structure that breaks down to generate the observed structure. The primitive postulated by Ptitsyn & Finkelstein (7) for all β proteins is a long two-stranded antiparallel β structure with a central hairpin loop. The Greek key pattern of connections between β strands (47) then results from breaking this hairpin helix into shorter segments by opening unpaired loops. The particular "swirl" of the Greek key (only one is found, and not its isomer) results from the right-handed twist of the β sheet. An α-helical folding primitive has been postulated by Lim (45).

WORKING MODELS   In comparing experimental results with models, experimentalists have two choices: either to take an existing model and to test it against their results, or else to extract a working model from the experimental data. Since present structural information is "low resolution," such working models are necessarily low resolution. Two generalized working models are being tested currently. The first is the *framework model* in which the H-bonded secondary structure is formed early in folding. The second is *modular assembly,* in which essentially complete folding of any part of a protein occurs at one time, although different parts of the protein fold at different times (folding by parts). Note that the folding process may combine features of both models: formation of H-bonded secondary structure may precede tertiary interactions, as in the framework model, while separate subdomains (each capable of forming its own secondary structure) may fold at different times, as in modular assembly.

PROLINE ISOMERIZATION AND SLOW-FOLDING SPECIES

Formation of Slow-Folding Species

Proline isomerization as a slow step in protein folding was suggested by Brandts and co-workers (48) as a possible explanation for the two unfolded forms of RNase A (49). The folding of RNase A shows biphasic kinetics: a fast phase (50 msec at 25°C) precedes a major (80%) slow phase (20 sec at 25°C); both the fast- and the slow-folding reactions produce native enzyme (49). In unfolding experiments, the fast-folding species is formed rapidly, and at least two slow-folding species (U_S^{II}, U_S^I) are formed slowly (38, 48, 50–52). A quantitative study of the unfolding and refolding kinetics of RNase A (53) has demonstrated that the 3-species mechanism

$$N \longleftrightarrow U_F \longleftrightarrow U_S$$

explains the unfolding and refolding kinetics in the folding transition zone, where N is only marginally stable, and folding intermediates are not well populated. The fast-folding species (U_F) of RNase A is not a partly folded or nucleated molecule, since its concentration (20% of the unfolded molecules) is not affected by high temperature or strong denaturants such as 6 M GuHCl or 8.5 M urea (49, 54). Recently, the same tests used for RNase A have demonstrated the existence of both fast- and slow-folding molecules in hen lysozyme (55, 56) and in horse cyt c (57). Urea-gradient electrophoresis experiments at 2°C show that unfolded chymotrypsinogen and α-chymotrypsin also contain both slow- and fast-folding molecules

(58). Unfolding occurs in two stages, consistent with a $N \rightarrow U_F \rightleftharpoons U_S$ unfolding mechanism for a BPTI derivative (59), yeast isocyt c (60), and pepsinogen (61).

The first good evidence that the $U_F \rightleftharpoons U_S$ reaction of RNase A is proline isomerization was based on a comparison of the kinetics of the $U_F \rightleftharpoons U_S$ reaction in the unfolded protein (51) with the $cis \rightleftharpoons trans$ isomerization of prolyl residues in model compounds (48, 50, 62–65), and it made use of the conclusion that both U_F and U_S are completely unfolded (49, 54). The most striking characteristics, which are common to both reactions, are (51): (a) a high activation enthalpy (~ 20 kcal/mol); (b) catalysis by strong acids; and (c) kinetics that are independent of GuHCl concentration, which confirm that the interconversion of U_S and U_F does not involve residual structure (223). RNase A has three nitratable tyrosine groups including Tyr 115, which follows Pro 114; the kinetics of the $U_F \rightleftharpoons U_S$ reaction for nitrotyrosyl RNase A have been correlated with the pK changes during unfolding (66), and it has been suggested that NO_2-Tyr-115 provides an optical probe monitoring isomerization of Pro 114.

In recent work (L. -N. Lin and J. F. Brandts, personal communication, 1981), the appearance of a specific wrong proline isomer during unfolding has been correlated directly with the formation of a slow-folding species. Enzymatic cleavage specific for the $trans$ X-Pro bond has been used to break the peptide bond between Tyr 92 and Pro 93 during the unfolding of RNase A, and the results can be correlated with the formation of U_S^{II}. Pro 93 is cis in native Rnase A and $trans$ in U_S^{II} (the major unfolded species).

Proline Isomerization During Folding

Although the $U_F \rightleftharpoons U_S$ reaction of RNase A is almost certainly proline isomerization, the refolding of U_S^{II} has kinetic properties very different from proline isomerization. The activation enthalpy for the $U_S \rightarrow N$ reaction is small (< 5 kcal/mol at pH 6 and 20–40°C) as compared with 20 kcal/mol for proline isomerization (50). Furthermore, the rate of the $U_S \rightarrow N$ reaction is strongly dependent on the GuHCl concentration, unlike proline isomerization (50). Thus, proline isomerization is not the initial and rate-limiting step in the folding of the major U_S species of RNase A, in contrast to the original proposal (48).

At low temperatures (0°–10°C), a native-like intermediate (I_N) is formed in the folding of U_S^{II}. I_N has nearly the same tyrosine absorbance and enzymatic activity as the native protein (33, 38), but it differs from N in having a wrong proline isomer. I_N unfolds to give U_S, whereas N unfolds to give U_F (33). Proline isomerization appears to be the final or nearly final step in folding ($I_N \rightarrow N$). Proline isomerization can be 20 to 40 times faster in I_N than in the unfolded protein (33), perhaps because the non-native proline residues are under strain in I_N. The rate of proline isomerization in a cyclic pentapeptide is six times faster than in the corresponding blocked linear peptide, probably because of strain (67). In the case of U_S^{II}, an incorrect proline isomer does not block folding in moderate or strongly native folding conditions, but rather slows down the folding process: the probable explanation is that folding intermediates are less stable with a non-native proline isomer.

The kinetics of folding for three different carp parvalbumins provide further evidence for the role of proline isomerization in protein folding (68, 69). They have similar amino acid sequences and spectroscopic properties, but one of the parvalbumins contains a proline residue and the other two do not (69). All three proteins show complex folding kinetics; however, the parvalbumin that contains a proline residue shows an additional slower phase not seen in the other two proline-free proteins, and this phase has some kinetic properties like those of proline isomerization (69). However, a similar comparison of cytochrome c molecules from two different species has not given comparable results (70). The probable existence of nonessential proline residues complicates this kind of comparison. The role of proline isomerization has been studied in the refolding kinetics of a specific fragment of procollagen (71).

Nonessential Proline Residues

Not all prolines in a protein may affect the kinetics of folding: there may be "essential" and "nonessential" proline residues (51). The X-ray crystal structure of RNase S suggests that Pro 114 may be accommodated in either the *cis* or *trans* configuration [H. W. Wyckoff, quoted in (51)]. Levitt (72), used conformational energy calculations to study the effect of non-native proline isomers in BPTI, which has four *trans* proline residues. Proline residues can be classified into three groups, based on the energy difference between the native protein and the minimum energy structure with a wrong proline isomer (72). He suggested that these types of proline residues should produce different types of folding reactions (59, 72). Type I (small energy difference) should not affect the rate of folding, type II (intermediate energy difference) should slow down but not block folding, and type III (large energy difference) should block folding in the manner originally suggested by Brandts and co-workers (48).

So far, only type II prolines have been characterized (59). The "type" of folding reaction will depend on the folding conditions. In the folding of RNase A at 25°C, the activation enthalpy changes from 3 kcal/mol (type

II folding) in 0.1 M GuHCl to 18 kcal/mol (type III folding) in 2 M GuHCl (37, 50).

Wüthrich and co-workers have studied proline-containing linear oligopeptides and shown that the *cis/trans* ratio and the isomerization rate of the X-Pro bond depend on the charge and nature of the amino acid preceding the proline residue (62, 63, 67, 73, 74).

Nonproline peptide bond isomerization may be an important factor in the folding of some proteins. For example, the crystal structure of carboxypeptidase A shows three *cis* peptide bonds that are not N-terminal to prolyl residues (75).

KINETIC INTERMEDIATES

Multiple Unfolded Forms

The existence of multiple unfolded forms of a protein, arising from proline isomerization, is discussed above. It presents a serious problem in working out the kinetic pathway of folding, since the pathway must be studied separately for each unfolded form. In the case of RNase A, it has been possible to study the major slow-folding species U_S^{II} [(80% of the total slow folding species (38)]. In the case of hen lysozyme, it is possible to study the direct folding reaction ($U_F \rightarrow N$) of the species with correct essential proline isomers (55, 56).

The standard test for the presence of a kinetic intermediate is the existence of two kinetic phases: whenever more than one phase is observed, at least three species must be involved. However, the three species could be U_F, U_S, and N, and the two phases could be $U_F \rightarrow N$ and $U_S \rightarrow N$, so that a structural intermediate need not be present.

Tests for Structural Intermediates

The standard test for a structural intermediate is the *kinetic ratio test,* whose application to the problem of protein folding has been discussed (39). If the folding reaction shows different kinetics when measured by two different probes, then a structural intermediate must be present, provided that all unfolded species appear alike when measured by each probe. If two (or more) well-resolved kinetic phases are found, and the two probes change differently in different phases, then there is at least one structural intermediate that can be studied readily. Three probes that are particularly informative are: (*a*) stopped-flow CD (76, 77); (*b*) enzyme activity, measured by combination with specific ligands (38, 39, 56, 78) or measured directly (38, 49); (*c*) protection of NH protons against exchange with solvent (34, 35). The test of *specific combination* between fragments has been applied to the folding of RNase S (39). The principle is that, if combination between S peptide (residues 1–20) and S protein (residues 21–124) occurs early in the folding of S protein, then a structural intermediate must be present to provide the combining site.

RNase A and RNase S

The present minimal mechanism for the folding of the U_S^{II} species of RNase A at 0°–10°C is:

$$U_S \rightleftharpoons I_1 \rightleftharpoons I_N \rightleftharpoons N$$

but it is probable that additional intermediates are populated between I_1 and I_N. I_1 has been observed by protection of NH protons against exchange with solvent (34, 35). The protected protons are trapped early and remain trapped throughout folding. The average degree of protection in I_1 is at least 100 (pH 7.5, 10°C) (35) for the 50 most protected NH protons of native RNase A. In native proteins, some NH protons are protected by as much as 10^8 (79, 80). I_N is highly folded as measured by tyrosine absorbance or binding of the specific inhibitor 2'-CMP (33, 38), and I_N is enzymatically active (38). Nevertheless, I_N still contains a wrong proline isomer (33) and the $I_N \rightleftharpoons$ reaction can also be followed by a fluorescence change (36, 37).

The unfolding pathway of RNase A contains an additional intermediate I_U (53):

$$N \rightleftharpoons I_U \rightleftharpoons U_F \rightleftharpoons U_S$$

The refolding kinetics of I_U have been measured, using a sequential stopped-flow apparatus (52), and it is known that proline isomerization does not occur freely in I_U (81). Thus far I_U has been studied only in unfolding conditions, and it is not known whether I_U is also populated in refolding experiments.

The refolding kinetics of RNase S are more complex (39, 40, 82) than those of RNase A, but they contain additional information about the role of the S-peptide moiety (residues 1–20) in folding. Recent stopped-flow CD measurements on the folding of RNase S (U_S) show that sequential steps in folding can be resolved: β-sheet formation precedes the S-peptide α-helix formation (A. M. Labhardt, personal communication, 1981). Enzymatic activity, as measured by the ability to bind 2'-CMP, is regained together with the α-helix formation. It is not yet known when proline isomerization occurs.

Lysozyme

The refolding kinetics of hen egg white lysozyme (83) have recently been reinvestigated by Utiyama and co-workers (55, 56). Their results demonstrate that unfolded lysozyme also exists in a mixture of fast and slow folding species ($U_F:U_S$ ratio of 90:10) [cf Hagerman (84)]. The refolding kinetics outside the transition zone are biphasic, and both phases produce native enzyme (56). The unfolding kinetics are monophasic, and the formation of U_F precedes U_S (56). Evidence for an early intermediate in folding is based on an absorbance change that occurs in the dead time of the stopped-flow instrument (20 msec), after correction for solvent effects (55). The activation enthalpy for the folding of U_S is only 11 kcal/mol (vs 20 kcal/mol for proline isomerization), which suggests that some folding occurs before proline isomerization in U_S (56).

Cytochrome c

The refolding kinetics of horse Fe(III) cyt c (85) have been reinvestigated recently (57). As in RNase A and lysozyme, unfolded cyt c consists of an equilibrium mixture of U_F and U_S (in a 78:22 ratio), which gives rise to biphasic refolding kinetics, with native enzyme formed in both phases (57). An earlier suggestion that the fast refolding reaction is the formation of an abortive intermediate (85) has been ruled out. Sequential unfolding/refolding ("double jump") experiments demonstrate that the unfolding of cyt c produces U_F, which then isomerizes to U_S (57). Similar results have been obtained with yeast iso-2 cyt c (60). An intermediate has been identified in the $U_F \rightarrow N$ reaction of cyt c; the Soret absorbance change precedes the recovery of the native 695 nm band spectrum. In the $U_S \rightarrow N$ reaction, an ascorbate-reducible intermediate is formed before native enzyme is produced (57). Several kinetic intermediates have been found in unfolding by monitoring heme absorbance (86).

Carbonic Anhydrase

This protein (mol wt 29,000) is about twice as large as the other proteins discussed above. The folding of bovine carbonic anhydrase is much slower than the folding of RNase A, lysozyme, or cyt c. Both refolding and unfolding kinetics show multiple phases (78, 87–89), but their relationship to possible multiple forms of the unfolded protein has not yet been investigated (carbonic anhydrase has 20 prolines). Carbonic anhydrase contains Zn^{2+} and the presence or absence of Zn^{2+} strongly affects the refolding kinetics (90, 91). There is evidence for early formation of the H-bonded framework: changes in θ_{222} (secondary structure) precede changes in θ_{270} (tertiary structure) (88). A spin-label study of carbonic anhydrase shows that an intermediate is formed within 0.1 sec after the start of refolding (89). A late intermediate can bind a specific inhibitor, but does not have enzymatic activity (78). Interestingly, the fastest observed phase in the refolding of carbonic anhydrase has a rate that increases with increasing GuHCl concentration, which suggests that there is an early abortive intermediate in a rapid preequilibrium with the unfolded state (88). However, carbonic anhydrase is known to precipitate easily during attempts at renaturation (92–94), and it is possible that this is aggregation dependent.

Other Proteins

Creighton (58) has recently introduced urea gradient electrophoresis as a method of studying folding intermediates. A linear gradient of urea perpendicular to the direction of migration is used, and the migration pattern is observed as a function of time; the patterns obtained with native and unfolded proteins are compared (58). The temperature of electrophoresis is low (2°C), to decrease the rates of folding and of proline isomerization. Slow-folding (U_S) forms of the unfolded protein have been demonstrated with RNase A, chymotrypsinogen, and α-chymotrypsin (58). Slow-folding species have not been detected in several other small proteins, including lysozyme and cyt c. However, this is not surprising, since reactions with half times less than 8 min are too fast to measure by this method, and the fraction of U_S molecules is small in these proteins (55–57). Several examples of compact kinetic intermediates in folding have been demonstrated with this method (58).

Kinetic and urea gradient electrophoresis experiments also demonstrate the existence of at least two unfolded forms in the α subunit of tryptophan synthase, in addition to two rapidly formed intermediates (95, 96). Moreover, binding of a substrate analogue during refolding displays biphasic kinetics; the rates for the two phases are identical with those observed for folding in the absence of the analogue (96). The α subunit can be cleaved proteolytically into two fragments, each of which can fold by itself, but neither one alone can bind the substrate analogue (97). The relationship between these fragments and the two kinetic intermediates of the intact α subunit is not yet known.

The unfolding of apomyoglobin has been studied by stopped-flow CD and by fluorescence (77) as has the folding of the β chain of hemoglobin (98). Both studies suggest the existence of specific interactions apart from helix formation. The helices of apomyoglobin break down more rapidly in unfolding, as judged by CD, than interactions detected by a fluorescence probe (77). In refolding, the β chain interacts rapidly and specifically with the heme, followed by slower helix formation (98).

EQUILIBRIUM INTERMEDIATES

Tests for Intermediates

The equilibrium unfolding transitions of most small, globular proteins are highly cooperative, and the two-state approximation ($N \rightleftharpoons U$) is usually a good working model for equilibrium studies. However, in the past few years, several examples of proteins with populated equilibrium intermediates have been reported. There are two tests for an equilibrium intermediate based on the use of probes (16) (Figure 1): (a) a biphasic transition as measured by a single probe and (b) noncoincident transitions as measured by different probes. Either one of these observations is sufficient evidence for an equilibrium intermediate, and they are not mutually exclusive. There is also a calorimetric test for intermediates (99, 100): $\Delta H_{vH} < \Delta H_{cal}$, where ΔH_{cal} is the calorimetrically determined enthalpy of unfolding and ΔH_{vH} is the apparent enthalpy computed from the temperature dependence of the $N \rightleftharpoons U$ equilibrium constant, by the van't Hoff relation. Equilibrium measurements cannot demonstrate that an intermediate is actually on the pathway of folding: it may be an abortive intermediate or an alternative native form. Also, aggregation of an unfolded protein is known to occur inside the unfolding transition zone in some cases.

Modular Assembly Versus Framework Formation

A biphasic transition (Figure 1 a) is evidence for modular assembly, or folding by parts. To use this as evidence for the mechanism of folding, it is necessary to know whether or not the molecule contains two or more stable domains. If so, the folding must be judged complex: the first goal is to understand the mechanism of folding for small "single-domain" proteins. Noncoincident transition curves (Figure 1 b), which show that the secondary structure is more stable than the tertiary structure, provide evidence for the framework model. Far-UV CD (210–240 nm) has been used as a probe of secondary structure, and either enzyme activity, specific ligand binding, or spectroscopic bands of aromatic residues (270–300 nm) can be used as probes of tertiary structure.

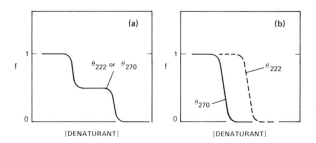

Figure 1   (a) Different probes of secondary and tertiary structure each give similar biphasic transition curves, which is interpreted as evidence for modular assembly or domain folding. (b) The secondary structure is more resistant than the tertiary structure to denaturant-induced unfolding, as measured by CD, which is interpreted as evidence for a framework model of folding.

α-Lactalbumin

The three-state (N $\rightleftharpoons$ I $\rightleftharpoons$ U) equilibrium unfolding transition of α-lactalbumin has been well characterized by Kuwajima, Sugai, and co-workers (101–108) and recently also by Ptitsyn and co-workers (224). The GuHCl unfolding transition shows non-coincident changes in CD at different wavelengths; aromatic signals (θ_{270} and θ_{296}) show an unfolding transition at a lower GuHCl concentration than the unfolding of the secondary structure (θ_{222}) (102, 104). Three important properties of I are: (a) I is in rapid equilibrium with U (a time constant of less than 1 msec), (b) I is present even when the protein's disulfide bonds are reduced, and (c) I has a far-UV CD spectrum close to that of N (102, 104). The fast interconversion (U $\rightleftharpoons$ I) may be compared to the fast helix-coil transition of synthetic polypeptides (22–24). In contrast, the I $\rightleftharpoons$ N reaction can be measured in seconds (101, 103).

Similar intermediates (as judged by CD) are formed in 2 M GuHCl or 3 M $NaClO_4$ (105), and in unfolding by acid (102, 104) or base (108). Unfolding by acid or base involves the titration of ionizable groups that have abnormal pKs in N but normal pKs in I. These groups with abnormal pKs are responsible for a 10^4-fold increase in the unfolding rate (N $\rightarrow$ I) at low and high pH (103, 108).

α-lactalbumin is not very stable (102). This may explain why an equilibrium intermediate is populated. The secondary structure may be intrinsically stable, while the tertiary structure is weak and easily disrupted by extremes of pH or by moderate concentrations of GuHCl or $NaClO_4$. It is interesting to compare α-lactalbumin with lysozyme, since these two proteins are believed to be structurally homologous (109). Lysozyme is a more stable protein, with a t_m approaching 80° at pH 5 (110). The high stability of its tertiary structure may prevent a comparable intermediate from being observed for lysozyme, because more drastic conditions are needed to unfold lysozyme than α-lactalbumin. Recently α-lactalbumin has been found to bind Ca^{2+} (107), and the stability of its tertiary structure is markedly increased by Ca^{2+} binding.

Penicillinase

Penicillinase is a single-chain protein (mol wt 29,000) with no disulfide bonds. Studies of the GuHCl-induced unfolding of penicillinase by Pain, Robson and co-workers (111–116) demonstrate at least one well-populated equilibrium intermediate and were initially interpreted as providing evidence for a framework model (111). The equilibrium transition curves fall into two categories (111, 112): (a) UV absorbance, aromatic CD, enzymatic activity, and viscosity measurements all show the same transition whereas

(b) far-UV CD and ORD changes occur at much higher GuHCl concentrations. The intermediate is a monomer, has significant secondary structure as judged by CD, and is expanded (112, 114). The existence of at least one equilibrium intermediate has been verified by NMR (116) and another intermediate has been found by urea-gradient electrophoresis (115). There are several similarities with results found for α-lactalbumin. (a) The intermediate has little tertiary structure (as measured with spectroscopic probes), but has a far-UV CD spectrum like that of N. (b) N is not very stable. (c) The intermediate is formed too rapidly to be measured with manual methods (i.e. within 30 sec), whereas the formation of N is slow (several minutes) (112, 116). Unlike penicillinase, the intermediate of α-lactalbumin is compact (224).

There is evidence for folding by domains (or by subdomains) in penicillinase. Digestion of the protein with CNBr results in three large fragments (mol wts 10,500, 9,500, and 8,500) that have been isolated. They combine to form a compact globular complex whose far-UV CD spectrum is indistinguishable from that of native penicillinase (113, 116). The reassembled complex has a different aromatic CD, and does not have enzymatic activity. The isolated fragments combine specifically with antibodies directed against native penicillinase, but it is not known whether they constitute stable domains by the test of showing a folding transition. The folding of penicillinase probably involves both framework formation and subdomain assembly.

Ovomucoid

Ovomucoid is a small (186-residue) protease inhibitor that is clearly a three-domain protein. The equilibrium unfolding transition is complex when measured by a single probe (117–121). Sequencing studies first indicated the presence of three homologous domains (121, 122). This explained earlier observations on the inhibitory properties of different avian ovomucoids; some have one site for trypsin inhibition, others have two sites, one each for trypsin and chymotrypsin; and still others have three sites, two for trypsin and one for chymotrypsin (123–125).

Domains can be isolated by digesting ovomucoid at interdomain sites (121, 126, 127). The isolated domains can refold independently into their native conformations. The isolated fragments: (a) have significant secondary and tertiary structure as judged by CD (128), (b) show both acid and thermal unfolding transitions (119, 121), (c) react with antisera to the native protein (126), (d) retain their inhibitory activity (121, 122), and (e) reform the correct disulfide bonds after reduction (120). One of the domains has recently been crystallized and the structure refined to 2.5 Å resolution (129). Recent analyses of the DNA and mRNA for ovomucoid demonstrate that the gene segments coding for each domain are separated from each other by intervening sequences (130). The DNA sequence of the ovomucoid gene suggests that it evolved from a primordial ovomucoid gene by two separate intragenic duplications (130).

The unfolding transition of intact ovomucoid induced by GuHCl or urea is biphasic; both phases of the transition can be monitored by viscosity or tyrosine absorbance (117, 118). Since these probes usually monitor tertiary structure, it is not yet clear whether the two phases of the transition arise from independent unfolding of domains or from a two-step unfolding of the entire molecule (e.g. step 1: disruption of domain/domain contacts; step 2: unfolding of the individual domains).

Other Proteins

The GuHCl unfolding transition of growth hormone (mol wt 22,000) follows a framework model; changes in tertiary structure (A_{290}) occur at significantly lower GuHCl concentrations than changes in secondary structure (θ_{222}) (131). Moreover, as with α-lactalbumin, the intermediate is very similar to acid-denatured growth hormone (131, 132). Equilibrium studies on the GuHCl unfolding of carbonic anhydrase (mol wt of $\sim$ 29,000) in the absence of Zn^{2+} also support a framework model (133), as do kinetic studies on this protein (78, 88). The equilibrium unfolding transition of cyt c shows spectrally measurable intermediates (134–139); loosening of the polypeptide chain around the heme precedes the unfolding of the remainder of the molecule. However, the calorimetric criterion for a two-state transition is satisfied with cyt c (99). The thermal unfolding transition of the lac repressor headpiece has been monitored by NMR; the changes in chemical shifts show that it is clearly not a two-state transition (140).

Proteins that have equilibrium properties suggesting domain assembly include: the α subunit of tryptophan synthase (141, 142), the β subunit of tryptophan synthase (143, 144), phosphorylase b (145), Bence Jones proteins (146), phosphoglycerate kinase (116), paramyosin (147), and myosin (148). In addition, papain, which has two structural domains separated by a deep cleft, gives a ratio of ΔH measured calorimetrically to the van't Hoff value ($\Delta H_{cal}/\Delta H_{vH}$) of 1.8, which suggests that the two domains fold almost independently (149).

Salt- and Methanol-Induced Unfolding

The role of neutral salts in stabilizing or denaturing folded proteins is complex (150–154). The effects of salts have been broken down into effects on the peptide group and effects on the nonpolar side chains (151, 153). Certain salts (e.g. LiCl, LiClO$_4$, and CaCl$_2$) induce unfolding of proteins at high concentrations. These unfolding transitions appear to be incomplete, based on comparisons of the physical properties of the salt-denatured and GuHCl-denatured protein (16, 155, 156). Recently, it has been found that the addition of urea to salt-denatured RNase A produces a second transition (156).

Alcohols can also induce the unfolding of proteins (16). In particular, adding methanol or ethanol lowers the t_m of RNase A, and decreases the cooperativity of the transition (157, 158). Recently, proton NMR measurements at low temperatures in MeOH-H_2O mixtures have shown that equilibrium intermediates are well populated (158); the results have been interpreted by a framework model.

ENERGETICS OF FOLDING

These questions about the energetics of folding are of particular importance for understanding the mechanism of folding. (*a*) Which is the most stable structure of those proposed for the initial stage in folding? (*b*) Can the stabilities of possible folding intermediates be correlated with some property, such as water-accessible surface area, that can be computed directly from the structure (12, 41, 159)? (*c*) Are there specific structural interactions (e.g. H bonds or salt bridges) that are important energetically and that guide the formation of structure? Definitive answers will come from the structures of actual folding intermediates. Meanwhile, some specific questions are being answered from studies of model compounds and protein fragments. Also, accurate data are being obtained from studies of intact proteins on the factors that affect protein stability.

Model Compound Studies

In the last decade Scheraga and co-workers have determined the stability constant (s) and nucleation constant (σ) for α helix formation for most of the amino acid residues. A given residue is incorporated randomly as a "guest" in a water-soluble, helix-forming polypeptide "host" (derivatives of poly-L-glutamine). The results give the helix-forming propensities of the different amino acid residues as a function of temperature. A short, informative review of the method and results has been given (160).

The striking fact that emerges is that short α helices (10–20 residues) are intrinsically unstable in water. The ratio of helix to random coil is given approximately by: (helix)/(coil) $= \sigma s^{n-1}/(s-1)^2$ for short helices (175), where n is the number of amino acid residues. The largest value of s measured for any residue is 1.3 (Met at 0°C) (160). With $\sigma \leqslant 10^{-4}$, it is clear that a short helix of any composition should be unstable in water.

This conclusion had been predicted from studies of the stability of the amide (–NH...O–C–) H bond in aqueous solution (161, 162). Using an infrared technique, H-bonded dimers and higher oligomers could barely be detected in aqueous solutions of N-methylacetamide (161). A lactam (δ-valerolactam), which forms two amide H bonds per dimer, forms a stronger dimer in water (163, 164), with an association constant at 25°C of 0.01 M^{-1} (164). This reaction of dimer formation shows a substantial, favorable enthalpy change: −3 kcal/mol H bond (164).[3]

A stronger H bond, which is easily measured in water by an NMR technique, is the charge stabilized H bond (−COO⁻...HN−), formed when the side chain of a glutamic acid residue bends back to bond with its own peptide NH (165). The bond is disrupted by protonation of the carboxylate anion.

The free energy change for burying an amino acid side chain inside a protein as the protein folds up has been estimated from the free energy of transfer of the side chain from water to organic solvents (166). This transfer free energy has been correlated with the water-accessible surface area of hydrophobic side chains (12, 167, 168). Dispersion forces may make an important contribution to binding of hydrophobic amino acids to tRNA synthetases (169) and, similarly, they may be important in stabilizing protein folding. A theoretical study of the hydrophobic effect (170, 171) indicates that the correlation between reduction in water-accessible surface area and transfer free energy is not general and should not be extended to folding intermediates (171) without further justification. Transfer free energy data for the peptide group (151, 162) indicate that it strongly prefers to remain in water, and this should be considered in estimating the stability of a possible folding intermediate.

Studies with small molecules show that it is possible to demonstrate ion pairs in aqueous solution (e.g. guanidinium⁺··· carboxylate⁻ pairs) in a bimolecular reaction, but the strength of the interaction is not large (172, 173).

Protein Fragments

A decade ago several workers tested the possibility of using protein fragments as models for folding intermediates. If a protein folds first into microdomains, and these then coalesce into subdomains, and so on, then appropriately chosen fragments should fold at least partially, and their structures should give clues about the folding process. It is well documented

---

The free energy change for forming the peptide H bond in water has been estimated from the s and σ values for α helix formation (225). Since three residues are fixed in a helical conformation in the nucleation reaction without forming H bonds, we may take the free energy of the nucleation reaction (+ 5.6 kcal/mole if $σ = 10^{-4}$) and divide by 3 to get the free energy change per residue (+2 kcal/mole) when it adopts the helical conformation without forming an H bond. Since the s values obtained by the host-guest technique are close to 1, the overall free energy change including the H bond is close to 0, and the free energy change per H bond is −2 kcal/mole. This estimate does not distinguish between the entropic and enthalpic contributions to H bond formation.

that separate domains of larger proteins fold independently or nearly so [see below and also (5)]. However, most studies of subdomain fragments have given the disappointing result that the fragment is predominantly unfolded in aqueous solution.

Nevertheless, an intriguing example has been reported of a "microdomain" that shows partial helix formation in water (174, 175). The C peptide of RNase A (residues 1–13) is essentially unfolded in aqueous solution at 25°C but not at 1°C (174). Aggregation occurs above 2 mg/ml but the helix forms intramolecularly, and helix formation is observed by CD at concentrations as low as 40 μg/ml. NMR data indicate that all, or nearly all, residues participate in helix formation (175). Up to $\sim 30\%$ helix content can be observed in water (175), which is 1000 times greater than the helix content predicted by the host-guest studies. pH titration shows that protonation of His 12, and deprotonation of either Glu 2 or Glu 9 or of both are required for significant helix formation (175). The results suggest that specific salt bridge(s) (e.g. His 12^+...Glu 9^-) nucleate the helix by stabilizing the first turn.

Intact Proteins

Salt bridges have been suspected of being important for folding since their discovery in α-chymotrypsin (176) and hemoglobin (177–179). Recently, salt bridges have been demonstrated in several proteins, including phosphorylase (180) and BPTI (181). Estimates of the strength of individual salt bridges range from –1 to –3 kcal/mol (179, 181, 182).

An advance has been made in treating the electrostatic properties of proteins. A simple discrete charge model, which makes use of X-ray structures to give the locations and solvent accessibilities of the ionizing groups, predicts rather well the pKs of individual groups observed by NMR [(183, 184), see however (185)]. In doing this the model also predicts the electrostatic contribution to the free energy of folding.

The overall thermodynamics of folding are now known accurately from calorimetry for several proteins (10). The data confirm that the folded structures of globular proteins are only marginally stable ($\Delta G = -5$ to -15 kcal/mol), and demonstrate that the thermodynamics of folding as a function of temperature are dominated by the large and approximately constant value of ΔC_P: both ΔH and ΔS are strongly temperature dependent (10, 186).

The striking conclusion that emerges from these studies is that there is a large and favorable contribution to the enthalpy of unfolding that cannot come from hydrophobic interactions (10). Privalov argues that this nonhydrophobic contribution to the unfolding enthalpy is nearly temperature independent and that it arises from H bonds and dispersion forces (10). If one assumes that it arises entirely from H bond enthalpy, then its magnitude corresponds to about -1.7 kcal/mol H bond for several proteins (10), which can be compared with the -3 kcal/mol H bond estimated for the dimerization of δ-valerolactam (163, 164). These results suggest that peptide H bonds may make a major contribution to the stabilization of the native structure.

RELATED TOPICS

The following topics are closely related to our review but limitations of space prevent their review here. We give references to reviews and to some recent papers, and comment on the relationships to work reviewed here.

Disulfide Intermediates

The best evidence for specific intermediates in protein folding experiments comes from the studies by Creighton of trapped disulfide intermediates in BPTI. The work has been reviewed recently (3, 187). Some basic properties of the system are as follows. (*a*) There are multiple intermediates and multiple pathways: however, there is a "most-favored" pathway and obligatory intermediates. (*b*) The obligatory two-disulfide intermediates each have one non-native S–S bond. Moreover, an abortive, or dead-end, intermediate has two native S–S bonds. Theorists working with computer simulation of folding have sought to explain these surprising facts. (*c*) The spectrum of intermediates narrows down toward the most-favored pathway in conditions favoring folding (188) (i.e. low temperatures, or the presence of stabilizing Hofmeister anions such as SO_4^{2-}). This increases the overall rate of the refolding/reoxidation process. Thus, the most-favored pathway proceeds via the most stable intermediates, and the rate of the overall process depends on how well these intermediates are populated. (*d*) The one-disulfide intermediates equilibrate with each other before the second S–S bond is formed and so do the two-disulfide intermediates, albeit more slowly [compare the recent study of RNase A by Konishi et al (189)]. To a first approximation, there is "thermodynamic control" of the refolding/reoxidation pathways. (*e*) The major folding process occurs in a single S–S rearrangement, near the end of the pathway. The study of earlier trapped intermediates has shown chiefly that it is difficult to detect and characterize any specific structure (190). Nevertheless, the importance of specific interactions is shown by the close correlation between a narrow or broad spectrum of these intermediates and whether the folding conditions are strongly native or marginally so (188). (*f*) Unfolding/reduction and refolding/reoxidation can be studied in the same conditions by varying the ratio of oxidant to reductant. Since the conditions are the same and the process is reversible, the pathways of unfolding and refolding are the same.

Domain Folding and Exons

Stable domains that show folding transitions have been isolated after limited proteolysis from numerous large proteins, and the roles of these domains in folding have been investigated in a few cases, including the β_2 subunit of tryptophan synthase (143, 144, 191), the "double-headed" enzyme aspartokinase-homoserine dehydrogenase I of *E. coli*, which has both enzyme activities joined in a single chain (192–194), immunoglobulins, including Bence-Jones proteins (146, 195, 196), elastase (197), and the λ repressor (198). A generalization from these studies is that domains often fold independently in kinetic terms, and are thermodynamically stable, but that a subsequent slow rearrangement, which involves interactions between domains, is commonly required for full activity. We have already discussed domain folding in the case of the small proteins ovomucoid, penicillinase, and the α subunit of tryptophan synthase.

Gilbert (199, 200) has proposed that exons code for protein domains and that genetic recombination within introns speeds up the evolution of new proteins. Introns that separate domain coding regions have been demonstrated for immunoglobulins (201, 202) and for ovomucoid (130). A large intron occurs inside the coding region for the C peptide, thus separating the A and B chains of rat insulin (203). Introns occur within the coding regions for α helices in the globin genes (204): however, the polypeptide fragment coded by the central exon does bind heme specifically and tightly (205).

Oligomeric Proteins

A systematic study of the folding and assembly of several oligomeric enzymes has been made by Jaenicke and co-workers; this work has been reviewed recently (206). Some central points are as follows. (*a*) Only in rare cases is thermodynamic equilibrium ever reached between unfolded monomer and folded oligomer. Kinetic studies give information about a folding pathway that is not readily reversible in most cases. Nevertheless, refolding can give native enzyme in almost quantitative yield in special conditions. (*b*) Aggregation of partly folded chains is the major technical problem. It can be minimized by special procedures, including folding conditions that stabilize folded monomers. (*c*) In general, folded monomers are inactive; rabbit muscle aldolase is an exception. (*d*) Specificity of association is high: mixtures of closely related enzymes (e.g. lactate and malate dehydrogenase) do not refold to give "chimeric" species (207).

The pathways of assembly of aspartate transcarbamylase from catalytic and regulatory subunits have been studied by Schachman & co-workers (208–210). Pulse-chase experiments with radioactively labeled subunits, followed by separation of intermediates in electrophoresis, have been used to give a model for the assembly process.

Temperature-sensitive mutants in the assembly or folding of the phage P22 tail spike protein have been studied by King and co-workers (211–213). These mutants appear to be kinetically blocked in folding or assembly at restrictive temperatures. When synthesized at the permissive temperature, the mutants are as stable as the wild-type protein, as measured by the kinetics of irreversible thermal denaturation. When synthesized at the restrictive temperature, many of the mutant proteins can be reactivated in the absence of new protein synthesis by a shift to the permissive temperature, which indicates that the mutant chains that accumulate at the restrictive temperature are capable of folding and assembly (213).

Fragment Exchange and Local Unfolding Reactions

Certain pairs of polypeptide fragments, which individually are unfolded, can combine to form a native-like, enzymatically active, complex. This type of complementation has been studied extensively by Taniuchi and co-workers for two small proteins, staph nuclease (20) and cyt c (214). The probability of successful complementation increases if the fragments are overlapping. Even three fragments can combine to form a complex (215). The dissociation of these complexes is of particular interest as a model system for studying local unfolding reactions. Local unfolding has often been proposed as a chief means of allowing amide proton exchange in native proteins (216), and local unfolding reactions could be part of the overall unfolding pathway.

Current views of protein flexibility indicate that water can penetrate readily into the interior of a globular protein, and it has been suggested that minor perturbations of the folded structure may permit amide proton exchange (217–219). To decide between "deep breathing" and highly local breathing as dominant mechanisms of exchange, it is necessary to measure the kinetics and equilibria of the breathing reactions by independent methods. Fragment exchange studies can yield both kinetic and equilibrium data and, since the contact regions between fragments are extensive, it is fairly certain that deep breathing reactions are required to break the contacts. A major conclusion from the fragment exchange studies is that the rates and equilibrium constants are in a range where they can be expected to contribute significantly to amide proton exchange in native proteins. This has been demonstrated directly for the dissociation of S peptide from RNase S (220, 221) where amide proton exchange of ^{3}H-labeled S peptide combined with S protein is concentration dependent, and therefore occurs partly by dissociation to free S peptide, even at 0°C, pH 7. A local unfolding reaction in staph nuclease has been demonstrated directly by using antibodies directed against the unfolded protein (222).

CONCLUDING REMARKS

The kinetic mechanism of protein folding proves to be *sequential folding* with defined intermediates in two cases: the major folding reaction of RNase A ($U_S^{II} \rightarrow N$) and the refolding/reoxidation of reduced BPTI or of reduced RNase A. It is likely that sequential folding is a general mechanism. The main objection to it has been the high cooperativity of folding measured by equilibrium experiments inside the folding transition zone, which implies that all folding intermediates are unstable. This objection does not apply if the kinetics of folding are studied outside the transition zone, where intermediates can be stable relative to the unfolded protein. Finding small proteins that do show equilibrium intermediates (especially α-lactalbumin, which is probably a single-domain protein) adds to the evidence that the cooperativity of folding is marginal, not absolute.

Understanding the role of *proline isomerization* in the kinetics of folding has been an essential part of the recent progress. In order to study the kinetic pathway of folding, it is necessary to identify the different unfolded species of a protein and to study the folding of each one separately.

Most present results support a *framework model* of folding in which the H-bonded secondary structure is formed at an early stage, and they are not consistent with strict *modular assembly* of small proteins, or folding by parts, in which both the secondary and tertiary structures of any part of a protein are formed at the same time. However, the H-bonded secondary structure even of a small protein may itself be formed in distinct stages. Modular assembly may apply to the domain folding of larger, multidomain proteins; however, present evidence suggests that a structural rearrangement occurs after the initial folding of separate domains and before the full activities of the native protein are regained. The framework model was first suggested by finding that the secondary structures of some unusual small proteins (e.g. penicillinase, carbonic anhydrase, and α-lactalbumin) are more resistant to unfolding by denaturants than are their tertiary structures. Kinetic intermediates consistent with a framework model were found by ^{3}H trapping experiments with RNase A, which showed that many NH protons are protected from exchange with solvent early in folding and then throughout the folding process, as expected if H-bonded secondary structure is formed early in folding. Stopped-flow CD measurements appear capable of resolving stages in the formation of α helices and β sheets, during the $U_S \rightarrow N$ folding reaction of RNase S.

Well-populated intermediates in the folding of small proteins are now a reality. Characterization of these intermediates will provide a benchmark for theorists working on prediction of folding from sequence and the elucidation of the folding pathway.

ACKNOWLEDGMENTS

We thank Drs. K. Kuwajima and O. B. Ptitsyn for discussion of this review, and many colleagues for sending manuscripts not yet in print. P. S. Kim is a predoctoral fellow of the Medical Scientist Training Program supported by the National Institutes of Health (GM 07365). This work has been supported by grants to R. L. Baldwin from the National Science Foundation (PCM 77-16834) and the National Institutes of Health (2 RO1 GM 19988-21). Use of the Stanford Magnetic Resonance Facility (supported by NSF Grant GP 23633 and NIH Grant RR 00711) is gratefully acknowledged.

Literature Cited

1. Jaenicke, R., ed. 1980. *Protein Folding,* Proc. 28th Conf. German Biochem. Soc. Amsterdam: Elsevier/North-Holland Biomed. Press. 587 pp.
1a. Ptitsyn, O. B. 1981. *FEBS Lett.* 131: 197–202
2. Ghélis, C., Yon, J. 1982. *Protein Folding.* New York: Academic. In press
3. Creighton, T. E. 1978. *Prog. Biophys. Mol. Biol.* 33:231–97
4. Baldwin, R. L. 1975. *Ann. Rev. Biochem.* 44:453–75
5. Wetlaufer, D. B. 1981. *Adv. Protein Chem.* 34:61–92
6. Richardson, J. S. 1981. *Adv. Protein Chem.* 34:167–339
7. Ptitsyn, O. B., Finkelstein, A. V. 1980. *Q. Rev. Biophys.* 13:339–86
8. Thomas, K. A., Schechter, A. N. 1980. In *Biological Regulation and Development,* ed. R. F. Goldberger, 2:43–100. New York: Plenum
9. Rossmann, M. G., Argos, P. 1981. *Ann. Rev. Biochem.* 50:497–532
10. Privalov, P. L. 1979. *Adv. Protein Chem.* 33:167–241
11. Némethy, G., Scheraga, H. A. 1977. *Q. Rev. Biophys.* 10:239–352
12. Richards, F. M. 1977. *Ann. Rev. Biophys. Bioeng.* 6:151–76
13. Ikegami, A. 1981. *Adv. Chem. Phys.* 46:363–413
14. Anfinsen, C. B., Scheraga, H. A. 1975. *Adv. Protein Chem.* 29:205–300
15. Wetlaufer, D. B., Ristow, S. 1973. *Ann. Rev. Biochem.* 42:135–58
16. Tanford, C. 1968. *Adv. Protein Chem.* 23:121–282
17. Tanford, C. 1970. *Adv. Protein Chem.* 24:1–95
18. Levinthal, C. 1968. *J. Chim. Phys.* 65: 44–45
19. Karplus, M., Weaver, D. L. 1976. *Nature* 260:404–6
20. Taniuchi, H. 1973. *J. Biol. Chem.* 248: 5164–74
21. Levitt, M., Warshel, A. 1975. *Nature* 253:694–98
22. Hammes, G. G., Roberts, P. B. 1969. *J. Am. Chem. Soc.* 91:1812–16
23. Barksdale, A. D., Stuehr, J. E. 1972. *J. Am. Chem. Soc.* 94:3334–38
24. Gruenewald, B., Nicola, C. U., Lustig, A., Schwarz, G., Klump, H. 1979. *Biophys. Chem.* 9:137–47
25. Ptitsyn, O. B., Rashin, A. A. 1975. *Biophys. Chem.* 3:1–20
26. Karplus, M., Weaver, D. L. 1979. *Biopolymers* 18:1421–37
27. Cohen, F. E., Sternberg, M. J. E., Phillips, D. C., Kuntz, I. D., Kollman, P. A. 1980. *Nature* 286:632–34
28. Hammes, G. G. 1978. *Principles of Chemical Kinetics.* New York: Academic. 268 pp.
29. Tsong, T. Y., Baldwin, R. L. 1978. *Biopolymers* 17:1669–78
30. Baldwin, R. L. 1980. See Ref. 1, pp. 369–85
31. Haas, E., Katchalski-Katzir, E., Steinberg, I. Z. 1978. *Biopolymers* 17:11–31
32. Tsong, T. Y. 1982. *Biochemistry.* In press
33. Cook, K. H., Schmid, F. X., Baldwin, R. L. 1979. *Proc. Natl. Acad. Sci. USA* 76:6157–61
34. Schmid, F. X., Baldwin, R. L. 1979. *J. Mol. Biol.* 135:199–215
35. Kim, P. S., Baldwin, R. L. 1980. *Biochemistry* 19:6124–29
36. Schmid, F. X. 1980. See Ref. 1, pp. 387–400
37. Schmid, F. X. 1981. *Eur. J. Biochem.* 114:105–9

38. Schmid, F. X., Blaschek, H. 1981. *Eur. J. Biochem.* 114:111–17
39. Labhardt, A. M., Baldwin, R. L. 1979. *J. Mol. Biol.* 135:231–44
40. Labhardt, A. M. 1980. See Ref. 1, pp. 401–25
41. Rashin, A. A. 1979. *Stud. Biophys.* 77:177–84
42. Wako, H., Saitô, H. 1978. *J. Phys. Soc. Jpn.* 44:1931–38
43. Abe, H., Gō, N. 1981. *Biopolymers* 20: 1013–31
44. Gō, N., Abe, H. 1981. *Biopolymers* 20: 991–1011
45. Lim, V. 1980. See Ref. 1, pp. 149–66
46. Lesk, A. M., Rose, G. D. 1981. *Proc. Natl. Acad. Sci. USA* 78:4304–8
47. Richardson, J. S. 1977. *Nature* 268: 495–500
48. Brandts, J. F., Halvorson, H. R., Brennan, M. 1975. *Biochemistry* 14:4953–63
49. Garel, J.-R., Baldwin, R. L. 1973. *Proc. Natl. Acad. Sci. USA* 70:3347–51
50. Nall, B. T., Garel, J.-R., Baldwin, R. L. 1978. *J. Mol. Biol.* 118:317–30
51. Schmid, F. X., Baldwin, R. L. 1978. *Proc. Natl. Acad. Sci. USA* 75:4764–68
52. Hagerman, P. J., Schmid, F. X., Baldwin, R. L. 1979. *Biochemistry* 18: 293–97
53. Hagerman, P. J., Baldwin, R. L. 1976. *Biochemistry* 15:1462–73
54. Garel, J.-R., Nall, B. T., Baldwin, R. L. 1976. *Proc. Natl. Acad. Sci. USA* 73: 1853–57
55. Kato, S., Okamura, M., Shimamoto, N., Utiyama, H. 1981. *Biochemistry* 20:1080–85
56. Kato, S., Shimamoto, N., Utiyama, H. 1982. *Biochemistry.* 21:38–43
57. Ridge, J. A., Baldwin, R. L., Labhardt, A. M. 1981. *Biochemistry* 20:1622–30
58. Creighton, T. E. 1980. *J. Mol. Biol.* 137:61–80
59. Jullien, M., Baldwin, R. L. 1981. *J. Mol. Biol.* 145:265–80
60. Nall, B. T., Landers, T. A. 1981. *Biochemistry* 20:5403–11
61. McPhie, P. 1982. *J. Biol. Chem.* 257:689–93
62. Grathwohl, C., Wüthrich, K. 1976. *Biopolymers* 15:2025–41
63. Grathwohl, C., Wüthrich, K. 1976. *Biopolymers* 15:2043–57
64. Cheng, H. N., Bovey, F. A. 1977. *Biopolymers* 16:1465–72
65. Steinberg, I. Z., Harrington, W. F., Berger, A., Sela, M., Katchalski, E. 1960. *J. Am. Chem. Soc.* 82:5263–79
66. Garel, J.-R. 1980. *Proc. Natl. Acad. Sci. USA* 77:795–98
67. Grathwohl, C., Wüthrich, K. 1981. *Biopolymers* 20:2623–33
68. Brandts, J. F., Brennan, M., Lin, L.-N. 1977. *Proc. Natl. Acad. Sci. USA* 74: 4178–81
69. Lin, L.-N., Brandts, J. F. 1978. *Biochemistry* 17:4102–10
70. Babul, J., Nakagawa, A., Stellwagen, E. 1978. *J. Mol. Biol.* 126:117–21
71. Bruckner, P., Bächinger, H. P., Timpl, R., Engel, J. 1978. *Eur. J. Biochem.* 90, 595–604
72. Levitt, M. 1981. *J. Mol. Biol.* 145: 251–63
73. Wüthrich, K., Grathwohl, C. 1974. *FEBS Lett.* 43:337–40
74. Hetzel, R., Wüthrich, K. 1979. *Biopolymers* 18:2589–606
75. Rees, D. C., Lewis, M., Honzatko, R. B., Lipscomb, W. N., Hardman, K. D. 1981. *Proc. Natl. Acad. Sci. USA* 78: 3408–12
76. Luchins, J., Beychok, S. 1978. *Science* 199:425–26
77. Kihara, H., Takahashi, E., Yamamura, K., Tabushi, I. 1980. *Biochem. Biophys. Res. Commun.* 95:1687–94
78. Ko, B. P. N., Yazgan, A., Yeagle, P. L., Lottich, S. C., Henkens, R. W. 1977. *Biochemistry* 16:1720–25
79. Willumsen, L. 1971. *CR Trav. Lab. Carlsberg* 38:223–95
80. Karplus, S., Snyder, G. H., Sykes, B. D. 1973. *Biochemistry* 12:1323–29
81. Rehage, A., Schmid, F. X. 1982. *Biochemistry.* In press
82. Labhardt, A. M., Baldwin, R. L. 1979. *J. Mol. Biol.* 135:245–54
83. Tanford, C., Aune, K. C., Ikai, A. 1973. *J. Mol. Biol.* 73:185–97
84. Hagerman, P. J. 1977. *Biopolymers* 16:731–47
85. Ikai, A., Fish, W. F., Tanford, C. 1973. *J. Mol. Biol.* 73:165–84
86. Tsong, T. Y. 1973. *Biochemistry* 12: 2209–14
87. Stein, P. J., Henkens, R. W. 1978. *J. Biol. Chem.* 253:8016–18
88. McCoy, L. F., Rowe, E. S., Wong, K.-P. 1980. *Biochemistry* 19:4738–43
89. Carlsson, U., Aasa, R., Henderson, L. E., Jonsson, B.-H., Lindskog, S. 1975. *Eur. J. Biochem.* 52:25–36
90. Yazgan, A., Henkens, R. W. 1972. *Biochemistry* 11:1314–18
91. Wong, K.-P., Hamlin, L. M. 1975. *Arch. Biochem. Biophys.* 170:12–22
92. Nilsson, A., Lindskog, S. 1967. *Eur. J. Biochem.* 2:309–17
93. Wong, K.-P., Hamlin, L. M. 1974. *Biochemistry* 13:2678–83

94. Ikai, A., Tanaka, S., Noda, H. 1978. *Arch. Biochem. Biophys.* 190:39–45
95. Matthews, C. R., Crisanti, M. M. 1981. *Biochemistry* 20:784–92
96. Crisanti, M. M., Matthews, C. R. 1981. *Biochemistry* 20:2700–6
97. Higgins, W., Fairwell, T., Miles, E. W. 1979. *Biochemistry* 18:4827–35
98. Leutzinger, Y., Beychok, S. 1981. *Proc. Natl. Acad. Sci. USA* 78:780–84
99. Privalov, P. L., Khechinashvili, N. N. 1974. *J. Mol. Biol.* 86:665–84
100. Privalov, P. L. 1974. *FEBS Lett.* 40: (Suppl) S140–53
101. Kuwajima, K., Nitta, K., Sugai, S. 1975. *J. Biochem. Tokyo* 78:205–11
102. Kuwajima, K., Nitta, K., Yoneyama, M., Sugai, S. 1976. *J. Mol. Biol.* 106: 359–73
103. Kita, N., Kuwajima, K., Nitta, K., Sugai, S. 1976. *Biochim. Biophys. Acta* 427:350–58
104. Kuwajima, K. 1977. *J. Mol. Biol.* 114: 241–58
105. Maruyama, S., Kuwajima, K., Nitta, K., Sugai, S. 1977. *Biochem. Biophys. Acta* 494:343–53
106. Nozaka, M., Kuwajima, K., Nitta, K., Sugai, S. 1978. *Biochemistry* 17: 3753–58
107. Hiraoka, Y., Segawa, T., Kuwajima, K., Sugai, S., Murai, N. 1980. *Biochem. Biophys. Res. Commun.* 95:1098–1104
108. Kuwajima, K., Ogawa, Y., Sugai, S. 1981. *J. Biochem. Tokyo* 89:759–70
109. Browne, W. J., North, A. C. T., Phillips, D. C., Brew, K., Vanaman, T. C., Hill, R. L. 1969. *J. Mol. Biol.* 42:65–86
110. Khechinashvili, N. N., Privalov, P. L., Tiktopulo, E. I. 1973. *FEBS Lett.* 30:57–60
111. Robson, B., Pain, R. H. 1973. In *Conformation of Biological Molecules Polymers, 5th Jerusalem Symp.* ed. E. D. Bergmann, B. Pullman, pp. 161–72. New York: Academic
112. Robson, B., Pain, R. H. 1976. *Biochem. J.* 155:331–44
113. Carrey, E. A., Pain, R. H. 1977. *Biochem. Soc. Trans.* 5:689–92
114. Carrey, E. A., Pain, R. H. 1978. *Biochim. Biophys. Acta* 533:12–22
115. Creighton, T. E., Pain, R. H. 1980. *J. Mol. Biol.* 137:431–36
116. Adams, B., Burgess, R. J., Carrey, E. A., Mackintosh, I. R., Mitchinson, C., Thomas, R. M., Pain, R. H. 1980. See Ref. 1, pp. 447–67
117. Waheed, A., Qasim, M. A., Salahuddin, A. 1977. *Eur. J. Biochem.* 76:383–90
118. Baig, M. A., Salahuddin, A. 1978. *Biochem. J.* 171:89–97
119. Matsuda, T., Watanabe, K., Sato, Y. 1981. *Biochem. Biophys. Acta* 669: 109–12
120. Matsuda, T., Watanabe, K., Sato, Y. 1981. *FEBS Lett.* 124:185–88
121. Kato, I., Kohr, W. J., Laskowski, M. 1978. *FEBS Meet.* 47:197–206
122. Kato, I., Schrode, J., Wilson, K. A., Laskowski, M. 1976. In *Protides of the Biological Fluids, 23rd,* ed. H. Peeters, pp. 235–43. Oxford: Pergamon. 699 pp.
123. Rhodes, M. B., Bennett, N., Feeney, R. E. 1960. *J. Biol. Chem.* 235:1686–93
124. Osuga, D. T., Bigler, J. C., Uy, R. L., Sjøberg, L., Feeney, R. E. 1974. *Comp. Biochem. Physiol. B* 48:519–33
125. Laskowski, M., Kato, I. 1980. *Ann. Rev. Biochem.* 49:593–626
126. Beeley, J. G. 1976. *Biochem. J.* 155:345–51
127. Bogard, W. C., Kato, I., Laskowski, M. 1980. *J. Biol. Chem.* 255:6569–74
128. Watanabe, K., Matsuda, T., Sato, Y. 1981. *Biochem. Biophys. Acta* 667: 242–50
129. Weber, E., Papamokos, E., Bode, W., Huber, R., Kato, I., Laskowski, M. 1981. *J. Mol. Biol.* 149:109–23
130. Stein, J. P., Catterall, J. F., Kristo, P., Means, A. R., O'Malley, B. W. 1980. *Cell* 21:681–87
131. Holladay, L. A., Hammonds, R. G., Puett, D. 1974. *Biochemistry* 13: 1653–61
132. Burger, H. G., Edelhoch, H., Condliffe, P. G. 1966. *J. Biol. Chem.* 241:449–57
133. Wong, K.-P., Tanford, C. 1973. *J. Biol. Chem.* 248:8518–23
134. Schejter, A., George, P. 1964. *Biochemistry* 3:1045–49
135. Myer, Y. P. 1968. *Biochemistry* 7: 765–76
136. Stellwagen, E. 1968. *Biochemistry* 7: 2893–98
137. Kaminsky, L. S., Miller, V. J., Davison, A. J. 1973. *Biochemistry* 12:2215–21
138. Drew, H. R., Dickerson, R. E. 1978. *J. Biol. Chem.* 253:8420–27
139. Myer, Y. P., MacDonald, L. H., Verma, B. C., Pande, A. 1980. *Biochemistry* 19:199–207
140. Wemmer, D., Ribeiro, A. A., Bray, R. P., Wade-Jardetzky, N. G., Jardetzky, O. 1981. *Biochemistry* 20:829–33
141. Yutani, K., Ogasahara, K., Suzuki, M., Sugino, Y. 1979. *J. Biochem. Tokyo* 85:915–21
142. Yutani, K., Ogasahara, K., Sugino, Y. 1980. *J. Mol. Biol.* 144:455–65
143. Goldberg, M. E., Zetina, C. R. 1980. See Ref. 1, pp. 469–84

144. Zetina, C. R., Goldberg, M. E. 1980. *J. Mol. Biol.* 137:401–14
145. Chignell, D. A., Azhir, A., Gratzer, W. B. 1972. *Eur. J. Biochem.* 26:37–42
146. Azuma, T., Hamaguchi, K., Migita, S. 1972. *J. Biochem. Tokyo* 72:1457–67
147. Riddiford, L. M. 1966. *J. Biol. Chem.* 241:2792–802
148. Tsong, T. Y., Karr, T., Harrington, W. F. 1979. *Proc. Natl. Acad. Sci. USA* 76:1109–13
149. Tiktopulo, E. I., Privalov, P. L. 1978. *FEBS Lett.* 91:57–58
150. von Hippel, P. H., Wong, K. Y. 1965. *J. Biol. Chem.* 240:3909–23
151. Nandi, P. K., Robinson, D. R. 1972. *J. Am. Chem. Soc.* 94:1299–1315
152. Schrier, E. E., Schrier, E. B. 1967. *J. Phys. Chem.* 71:1851–60
153. von Hippel, P. H., Peticolas, V., Schack, L., Karlson, L. 1973. *Biochemistry* 12:1256–64
154. Melander, W., Horváth, C. 1977. *Arch. Biochem. Biophys.* 183:200–15
155. Sharma, R. N., Bigelow, C. C. 1974. *J. Mol. Biol.* 88:247–57
156. Ahmad, F., Bigelow, C. C. 1979. *J. Mol. Biol.* 131:607–17
157. Fink, A. L., Grey, B. L. 1978. In *Biomolecular Structure and Function* eds. P. F. Agris, R. N. Loeppky, B. D. Sykes, pp. 471–77. New York: Academic. 614 pp.
158. Biringer, R. G., Fink, A. L. 1982. *J. Mol. Biol.* In press
159. Lesk, A. M., Chothia, C. 1980. In *Biophysical Discussions, Proteins and Nucleoproteins, Structure, Dynamics, and Assembly,* ed. V. A. Parsegian. pp. 35–47. New York: Rockefeller Univ. Press. 676 pp.
160. Scheraga, H. A. 1978. *Pure Appl. Chem.* 50:315–24
161. Klotz, I. M., Franzen, J. S. 1962. *J. Am. Chem. Soc.* 84:3461–66
162. Klotz, I. M., Farnham, S. B. 1968. *Biochemistry* 7:3879–82
163. Susi, H., Timasheff, S. N., Ard, J. S. 1964. *J. Biol. Chem.* 239:3051–54
164. Susi, H. 1969. In *Structure and Stability of Biological Macromolecules,* eds. S. N. Timasheff, G. D. Fasman, pp. 2:575–663. New York: Dekker. 694 pp.
165. Bundi, A., Wüthrich, K. 1979. *Biopolymers* 18:299–311
166. Nozaki, Y., Tanford, C. 1971. *J. Biol. Chem.* 246:2211–17
167. Chothia, C. 1975. *Nature* 254:304–8
168. Reynolds, J. A., Gilbert, D. B., Tanford, C. 1974. *Proc. Natl. Acad. Sci. USA* 71:2925–27
169. Fersht, A. R., Shindler, J. S., Tsui, W.-C. 1980. *Biochemistry* 19:5520–24
170. Pratt, L. R., Chandler, D. 1977. *J. Chem. Phys.* 67:3683–704
171. Karplus, M. 1980. See Ref. 159, pp. 45–46
172. Tanford, C. 1954. *J. Am. Chem. Soc.* 76:945–46
173. Springs, B., Haake, P. 1977. *Bioorg. Chem.* 6:181–90
174. Brown, J. E., Klee, W. A. 1971. *Biochemistry* 10:470–76
175. Bierzynski, A., Kim, P. S., Baldwin, R. L. 1982. *Proc. Natl. Acad. Sci. USA.* In press
176. Birktoft, J. J., Blow, D. M. 1972. *J. Mol. Biol.* 68:187–240
177. Perutz, M. F., TenEyck, L. F. 1972. *Cold Spring Harbor Symp. Quant. Biol.* 36:295–310
178. Kilmartin, J. V., Hewitt, J. A. 1972. *Cold Spring Harbor Symp. Quant. Biol.* 36:311–14
179. Perutz, M. F. 1978. *Science* 201: 1187–91
180. Sprang, S., Fletterick, R. J. 1980. See Ref. 159, pp. 175–92
181. Brown, L. R., DeMarco, A., Richarz, R., Wagner, G., Wüthrich, K. 1978. *Eur. J. Biochem.* 88:87–95
182. Fersht, A. R. 1972. *Cold Spring Harbor Symp. Quant. Biol.* 36:71–73
183. Shire, S. J., Hanania, G. I. H., Gurd, F. R. N. 1975. *Biochemistry* 14:1352–58
184. Matthew, J. B., Hanania, G. I. H., Gurd, F. R. N. 1979. *Biochemistry* 18:1919–36
185. Kilmartin, J. V., Fogg, J. H., Perutz, M. F. 1980. *Biochemistry* 19:3189–93
186. Schellman, J. A., Hawkes, R. B. 1980. See Ref. 1, pp. 331–43
187. Creighton, T. E. 1980. See Ref. 1, pp. 427–46
188. Creighton, T. E. 1980. *J. Mol. Biol.* 144:521–50
189. Konishi, Y., Ooi, T., Scheraga, H. A. 1981. *Biochemistry* 20:3945–55
190. Kosen, P. A., Creighton, T. E., Blout, E. R. 1980. *Biochemistry* 19:4936–44
191. Högberg-Raibaud, A., Goldberg, M. E. 1977. *Proc. Natl. Acad. Sci. USA* 74:442–46
192. Garel, J.-R., Dautry-Varsat, A. 1980. See Ref. 1, pp. 485–99
193. Garel, J.-R., Dautry-Varsat, A. 1980. *Proc. Natl. Acad. Sci. USA* 77:3379–83
194. Dautry-Varsat, A., Garel, J.-R. 1981. *Biochemistry* 20:1396–401
195. Isenman, D. E., Lancet, D., Pecht, I. 1979. *Biochemistry* 18:3327–36
196. Goto, Y., Hamaguchi, K. 1981. *J. Mol. Biol.* 146:321–40

197. Ghélis, C., Tempete-Gaillourdet, M., Yon, J. M. 1978. *Biochem. Biophys. Res. Commun.* 84:31–36
198. Pabo, C. O., Sauer, R. T., Sturtevant, J. M., Ptashne, M. 1979. *Proc. Natl. Acad. Sci. USA* 76:1608–12
199. Gilbert, W. 1978. *Nature* 271:501
200. Gilbert, W. 1980. *J. Supramolec. Structure* Suppl. 4, p. 115. (Abstr. 9th ICN-UCLA Symp.)
201. Brack, C., Tonegawa, S. 1977. *Proc. Natl. Acad. Sci. USA* 74:5652–56
202. Brack, C., Hirama, M., Lenhard-Schuller, R., Tonegawa, S. 1978. *Cell* 15:1–14
203. Lomedico, P., Rosenthal, N., Efstratiadis, A., Gilbert, W., Kolodner, R., Tizard, R. 1979. *Cell* 18:545–58
204. Lawn, R. M., Fritsch, E. F., Parker, R. C., Blake, G., Maniatis, T. 1978. *Cell* 15:1157–74
205. Craik, C. S., Buchman, S. R., Beychok, S. 1980. *Proc. Natl. Acad. Sci. USA* 77:1384–88
206. Jaenicke, R., Rudolph, R. 1980. See Ref. 1, pp. 525–48
207. Jaenicke, R., Rudolph, R., Heider, I. 1981. *Biochem. Int.* 2:23–31
208. Bothwell, M., Schachman, H. K. 1974. *Proc. Natl. Acad. Sci. USA* 71:3221–25
209. Bothwell, M. A., Schachman, H. K. 1980. *J. Biol. Chem.* 255:1962–70
210. Bothwell, M. A., Schachman, H. K. 1980. *J. Biol. Chem.* 255:1971–77
211. Smith, D. H., Berget, P. B., King, J. 1980. *Genetics* 96:331–52
212. Goldenberg, D. P., King, J. 1981. *J. Mol. Biol.* 145:633–51
213. Smith, D. H., King, J. 1981. *J. Mol. Biol.* 145:653–76
214. Hantgan, R. R., Taniuchi, H. 1978. *J. Biol. Chem.* 253:5373–80
215. Andria, G., Taniuchi, H., Cone, J. L. 1971. *J. Biol. Chem.* 246:7421–28
216. Englander, S. W., Downer, N. W., Teitelbaum, H. 1972. *Ann. Rev. Biochem.* 41:903–24
217. Woodward, C. K., Hilton, B. D. 1979. *Ann. Rev. Biophys. Bioeng.* 8:99–127
218. Richards, F. M. 1979. *Carlsberg Res. Commun.* 44:47–63
219. Wagner, G., Wüthrich, K. 1979. *J. Mol. Biol.* 134:75–94
220. Schreier, A. A., Baldwin, R. L. 1976. *J. Mol. Biol.* 105:409–26
221. Schreier, A. A., Baldwin, R. L. 1977. *Biochemistry* 16:4203–9
222. Furie, B., Schechter, A. N., Sachs, D. H., Anfinsen, C. B. 1975. *J. Mol. Biol.* 92:497–506
223. Schmid, F. X., Baldwin, R. L. 1979. *J. Mol. Biol.* 133:285–87
224. Dolgikh, D. A., Gilmanshin, R. I., Brazhnikov, E. V., Bychkova, V. E., Semisotnov, G. V., Venyaminov, S. Y., Ptitsyn, O. B. 1981. *FEBS Lett.* 136:311–15
225. Ptitsyn, O. B. 1972. *Pure Appl. Chem.* 31:227–44

Ann. Rev. Biochem. 1983. 53:263–300
Copyright © 1983 by Annual Reviews Inc. All rights reserved

6

DYNAMICS OF PROTEINS: ELEMENTS AND FUNCTION

M. Karplus and J. A. McCammon

Department of Chemistry, Harvard University, Cambridge, Massachusetts 02138

Department of Chemistry, Fleming Building, University of Houston, Houston, Texas 77004

CONTENTS

INTRODUCTION

The classic view of proteins has been static in character, primarily because of the dominant role of the information provided by high-resolution X-ray crystallography for these very complex systems. The intrinsic beauty and

169

remarkable detail of the drawings of protein structures led to an image in which each protein atom is fixed in place; an article on lysozyme by Phillips (1), the books by Dickerson & Geis (2), and by Perutz & Fermi (3), and the review by Richardson (4) give striking examples. Stating clearly the static viewpoint, Tanford (5) suggested that as a result of packing considerations "the structure of native proteins must be quite rigid." Phillips (6) wrote recently ". . . the period 1965–75 may be described as the decade of the rigid macromolecule. Brass models of double helical DNA and a variety of protein molecules dominated the scene and much of the thinking."

Most attempts to explain enzyme function have been based on the examination of the average structure obtained from crystallography; e.g. the high specificity of enzymes for their substrates has been likened to the complementarity of two pieces of a jigsaw puzzle. Cases in which conformational changes were known from X-ray data to be induced by ligand or substrate binding (e.g. the allosteric transition in hemoglobin) were generally treated as abrupt transitions between otherwise static structures.

The static view of protein structure is being replaced by a dynamic picture. The atoms of which the protein is composed are recognized to be in a state of constant motion at ordinary temperatures. From the X-ray structure of a protein, the average atomic positions are obtained, but the atoms exhibit fluidlike motions of sizable amplitudes around these average positions. Crystallographers have acceded to this viewpoint and have come so far as to sometimes emphasize the parts of a protein molecule they do not see in a crystal structure as evidence of motion or disorder (7).

The new understanding of protein dynamics subsumes the static picture in that use of the average positions still allows discussion of many aspects of protein function in the language of structural chemistry. However, the recognition of the importance of fluctuations opens the way for more sophisticated and accurate interpretations of protein function. The dynamic picture incorporates a variety of phenomena known to be involved in the biological activity of proteins, but whose detailed description was not possible under the static view. Transient packing defects due to atomic motions play an essential role in the penetration of oxygen to the heme-binding site in myoglobin and hemoglobin (8, 9). Functional interactions of flexible ligands with their binding sites often require conformational adjustments in both the ligand and the binding protein; the ligands involved include drugs, hormones, and enzyme substrates (10, 11). The structural changes in the binding proteins regulate the activity of many of these molecules through induced fit and allosteric effects (12–14). The chemical transformations of substrates by enzymes typically involve significant atomic displacements in the enzyme-substrate complexes. The mechanisms and rates of such transformations are sensitive to the dynamic properties of these systems; for example, the differences in the vibrational modes of the initial and transition states affect the free energies of activation and catalytic rates (13–15). Electron transfer processes may depend strongly on vibronic coupling and fluctuations that alter the distance between the donor and acceptor (16–19). The relative motion of distinct structural domains is important in the activities of myosin (20–22), other enzymes (23–25), and antibody molecules (26–28), as well as in the assembly of supramolecular structures such as viruses (29).

Any attempt to understand the function of proteins requires an investigation of the dynamics of the structural fluctuations and their relation to activity and conformational change. The review deals primarily with theoretical approaches to protein dynamics. This rapidly developing field of study is founded on efforts to supplement our understanding of protein structure with concepts and techniques from modern chemical theory, including reaction dynamics and quantum and statistical mechanics. From a knowledge of the potential energy surface, the forces on the component atoms can be calculated and used to determine phase space trajectories for a protein molecule at a given temperature. Such molecular dynamics simulations, which have been successfully applied to gases and liquids containing a large number of atoms, provide information concerning the thermodynamic properties and the time-dependence of processes in the system of interest (30). More generally, statistical mechanical techniques have succeeded very well in characterizing molecular motion and chemical reaction in condensed phases (31–33). The application of these methods to proteins is natural in that proteins contain many atoms, are densely packed, and function typically in liquid environments (34).

In this review we present first a brief overview of the wide range of motions that occur in proteins. We then outline the methods that can be used to study the various motions, and review the results obtained so far. We emphasize the role of the motions in the biological activity and compare with experiments where data exist. We conclude with an outlook for the future of this new and exciting field.

A number of reviews of the theory of protein dynamics has already appeared (35–39a). Specific aspects of protein dynamics, including the rapidly growing body of experimental data, have been reviewed (36, 40–50). The proceedings of a Ciba Foundation meeting (March 2–4, 1982) devoted to *Protein Motion and Its Relation to Function* are to be published (51). Two detailed reviews surveyed protein folding recently from the structural (52) and dynamic viewpoints (53).

OVERVIEW

Globular proteins have a wide variety of internal motions. They can be classified for convenience in terms of their amplitude, energy, and time scale, or by their structural type. Table 1 lists the ranges involved for these quantities; Careri, Fasella & Gratton (54) give a complementary summary. One expects an increase in one quantity (e.g. the amplitude of the fluctuation) to correspond to an increase in the others (e.g. a larger energy and a longer time scale). This is often true, but not always. Some motions are slow because they are complex, involving the correlated displacements of many atoms. An example might be partial-to-total unfolding transitions, in which the correlation of amplitude, energy, and time scale is expected to hold. However, in much more localized events, often involving small displacements of a few atoms, the motion is slow because of a high activation barrier; an example is the aromatic ring flips in certain proteins (55–60). In this case the macroscopic rate constant can be very slow ($k \sim 1$ sec^{-1} at 300°K), not because an individual event is slow (a ring flip occurs in $\sim 10^{-12}$ sec), but because the probability is very small ($\sim 10^{-12}$) that a ring has sufficient energy to get over an activation barrier on the order of 16 Kcal.

At any given time, a typical protein exhibits a wide variety of motions; they range from irregular elastic deformations of the whole protein driven by collisions with solvent molecules to chaotic librations of interior groups driven by random collisions with neighboring atoms in the protein. Considering only typical motions at physiological temperatures, the smallest effective dynamical units in proteins are those that behave nearly as rigid bodies because of their internal covalent bonding. Examples include the phenyl group in the side chain of phenylalanine, the isopropyl group in the side chains of valine or leucine, and the amide groups of the protein backbone. Except for the methyl rotations in the isopropyl group, these units display only relatively small internal motions owing to the high energy cost associated with deformations of bond lengths, bond angles, or dihedral angles about multiple bonds. The important motions in proteins involve relative displacements of such groups associated with torsional motions about the

Table 1 Classification of internal motions of globular proteins

Scales of motions (300° K)	
Amplitude	0.01 to 100 Å
Energy	0.1 to 100 Kcal
Time	10^{-15} to 10^3 sec

Types of motions	
Local	Atom fluctuations, side chain oscillations, loop and "arm" displacements
Rigid body	Helices, domains, subunits
Large-scale	Opening fluctuation, folding and unfolding
Collective	Elastic-body modes, coupled atom fluctuations, soliton and other non-linear motional contributions

rotationally permissive single bonds that link the groups together. High frequency vibrations occur within the local group, but these are not of primary importance in the relative displacements.

Most groups in a protein are tightly encaged by atoms of the protein or of the surrounding solvent. At very short times ($\lesssim 10^{-13}$ s), such a group may display a rattling motion in its cage, but such motions are of relatively small amplitude ($\lesssim 0.2$ Å). More substantial displacements of the group occur over longer time intervals; these displacements involve concomitant displacements of the cage atoms. Broadly speaking, such "collective" motions may have a local or rigid-body character. The former involves changes of the cage structure and relative displacements of neighboring groups, while the latter involves relative displacements of different regions of the protein but only small changes on a local scale.

The presence of such motional freedom implies that a native protein at room temperature samples a range of conformations. Most are in the general neighborhood of the average structure, but at any given moment an individual protein molecule is likely to differ significantly from the average structure. This in no way implies that the X-ray structure, which corresponds to the average in the crystal, is not important. Rather, it suggests that fluctuations about that average are likely to play a role in protein function. In a protein, as in any polymeric system in which rigidity is not supplied by covalent cross-links, significant fluctuations cannot be avoided; they must, therefore, have been taken into account in the evolutionary development.

Although the existence of the fluctuations is now well established, our understanding of their biological role in specific areas is incomplete. Both conformational and energy fluctuations with local to global character are expected to be important. In a protein, as in other nonrigid condensed systems, structural changes arise from correlated fluctuations. Perturbations, such as ligand binding, that produce tertiary or quaternary alterations do so by introducing forces that bias the fluctuations in such a way that the protein makes a transition from one structure to another. Alternatively, the fluctuations can be regarded as searching out the path or paths along which the transition takes place.

In considering the internal motions of proteins, one must separate the dynamic from the thermodynamic aspects; in the latter, the presence of flexibility is important (e.g. entropy of binding), while in the former the directionality and time scale play a role. Another way of categorizing the two is that in the second, equilibrium behavior is the sole concern, while in the first, the dynamics is the essential element. In certain cases, some aspects of the dynamics may be unimportant because they proceed on a time scale that is much faster than the phenomenon of interest. An example might be the fast local relaxation of atoms involved in a much slower hinge bending motion; here only the time scale of the latter would be expected to be involved in determining a rate process, though the nature of the former would be of considerable interest. In other situations, the detailed aspects of the atomic fluctuations may be the essential factor.

DYNAMICS METHODOLOGY

To study theoretically the dynamics of a macromolecular system, one must have a knowledge of the potential energy surface, the energy of the system as a function of the atomic coordinates. The potential energy can be used directly to determine the relative stabilities of the different possible structures of the system (30). The forces acting on the atoms of the systems are obtained from the first derivatives of the potential with respect to the atom positions. These forces can be used to calculate dynamical properties of the system, e.g. by solving Newton's equations of motion to determine how the atomic positions change with time (30, 31, 61). From the second derivatives of the potential surface, the force constants for small displacements can be evaluated and used to find the normal modes (62); this serves as the basis for an alternative approach to the dynamics in the harmonic limit (62, 63).

Although quantum mechanical calculations can provide potential surfaces for small molecules, empirical energy functions of the molecular mechanics type (64–67) are the only possible source of such information for proteins and their solvent surroundings. Since most of the motions that occur at ordinary temperatures leave the bond lengths and bond angles of the polypeptide chains near their equilibrium values, which appear not to vary significantly throughout the protein (e.g. the standard dimensions of the peptide group first proposed by Pauling et al in 1951; 68), the energy-function representation of the bonding can be hoped to have an accuracy on the order of that achieved in the vibrational analysis of small molecules. Where globular proteins differ from small molecules is that the contacts among nonbonded atoms play an essential role in the potential energy of the folded or native structure. From the success of the pioneering conformational studies of Ramachandran et al in 1963 (69), which used hardsphere nonbonded radii, it is likely that relatively simple functions (Lennard-Jones nonbonded potentials supplemented by a special hydrogen-bonding term and electrostatic interactions) can adequately describe the interactions involved.

The energy function used for proteins are generally composed of terms representing bonds, bond angles, torsional angles, van der Waals interactions, electrostatic interactions, and hydrogen bonds. The resulting expression has the form (64–67, 70):

$$E(\mathbf{R}) = \frac{1}{2} \sum_{\text{bonds}} K_b(b - b_0)^2 + \frac{1}{2} \sum_{\substack{\text{bond} \\ \text{angles}}} K_\theta (\theta - \theta_0)^2 \qquad 1.$$

$$+ \frac{1}{2} \sum_{\text{torsional}} K_\phi [1 + \cos(n\phi - \delta)]$$

$$+ \sum_{\substack{nb \text{ pairs} \\ r < 8 \text{ Å}}} \frac{A}{r^{12}} - \frac{C}{r^6} + \frac{q_1 q_2}{Dr} + \sum_{\substack{\text{H} \\ \text{bonds}}} \frac{A'}{r^{12}} - \frac{C'}{r^{10}} .$$

The energy is a function of the Cartesian coordinate set, $\mathbf{R}$, specifying the positions of all the atoms involved, but the calculation is carried out by first evaluating the internal coordinates for bonds (b), bond angles (θ), dihedral angles (ϕ), and interparticle distances (r) for any given geometry, $\mathbf{R}$, and using them to evaluate the contributions to Equation 1, which depends on the bonding energy parameters K_b, K_θ, K_ϕ, Lennard-Jones parameters A and C, atomic charges q_i, dielectric constant D, hydrogen-bond parameters A' and C', and geometrical reference values b_0, θ_0, n, and δ. For most protein atoms an extended atom representation is used; i.e., one extended atom replaces a nonhydrogen atom and any hydrogens bonded to it. However, although the earliest studies employed the extended atom representation for all hydrogens, present calculations treat hydrogen-bonding hydrogens explicitly and generally use a more accurate function to represent hydrogen bonding interactions (e.g. angular terms are included) than that given in Equation 1 (70).

Given a potential-energy function, one may take any of a variety of approaches to study protein dynamics. The most exact and detailed information is provided by molecular-dynamics simulations, in which one uses a computer to solve the Newtonian equations of motion for the atoms of the protein and any surrounding solvent (70–73). With currently available computers, it is possible to simulate the dynamics of small proteins for up to a few hundred ps. Such periods are long enough to characterize completely the librations of small groups in the protein and to determine the dominant contributions to the atomic fluctuations. To study slower and more complex processes in proteins, it is generally necessary to use methods other than straightforward molecular dynamics simulation. A variety of dynamical approaches, such as stochastic dynamics (74–78), harmonic dynamics (63, 79–81), and activated dynamics (59, 82–86), can be introduced to study particular problems.

Molecular Dynamics

To begin a dynamical simulation, one must have an initial set of atomic coordinates and velocities. These are obtained from the X-ray coordinates of the protein by a preliminary calculation that serves to equilibrate the system (72, 73). The X-ray structure is first refined using an energy-minimization algorithm to relieve local stresses caused by nonbonded atomic overlaps, bond length distortions, etc. The protein atoms are then assigned velocities at random from a Maxwellian distribution corresponding to a low temperature, and a dynamical simulation is performed for a period of a few ps. The equilibration is continued by alternating new velocity assignments, chosen from Maxwellian distributions corresponding to successively increased temperatures, with similar intervals of dynamical relaxation. The temperature, T, for this microcanonical ensemble is measured in terms of the mean kinetic energy for the system composed of N atoms as:

$$\frac{1}{2} \sum_{i=1}^{N} m_i < v_i^2 > = \frac{3}{2} N k_B T. \qquad 2.$$

In this equation, m_i and $<v_i^2>$ are the mass and average velocity squared of the ith atom, and k_B is the Boltzmann constant. Any residual overall translational and rotational motion can be removed to simplify analysis of the subsequent conformational fluctuations. The equilibration period is considered finished when no systematic changes in the temperature are evident over a time of about 10 ps (slow fluctuations could be confused with continued relaxation over shorter intervals). It is necessary also to check that the atomic momenta obey a Maxwellian distribution and that different regions of the protein have the same average temperature. The actual dynamical simulation results (coordinates and velocities for all the atoms as a function of time) for determining the equilibrium properties of the protein are then obtained by continuing to integrate the equations of motion for the desired length of time.

Several different algorithms for integrating the equations of motion in Cartesian coordinates are used in protein molecular dynamics calculations. Most common are the Gear predictor-corrector algorithm, familiar from small molecule trajectory calculations (72) and the Verlet algorithm, widely used in statistical mechanical simulations (87).

Stochastic Dynamics

In certain cases it is advantageous to simplify the dynamical treatment by separating the system under study into two parts. One part is that whose dynamics are to be examined and the other serves as a heat bath for the first;

this could be a protein in a solvent or one portion of a protein with the surrounding protein serving as the heat bath. In such an analysis (e.g. of a tyrosine sidechain in a protein) the displacement of the part whose dynamics is to be studied relative to its neighbors is presumed to be analogous to molecular diffusion in a liquid or solid. The allowed range of motion can be characterized by an effective potential-energy function termed the "potential of mean force" (30, 72); this potential corresponds to the free energy of displacement of the elements being studied in the average field due to surrounding bath atoms. The motion of the group under study is determined largely by the time variation of its nonbonded interactions with the neighboring atoms. These interactions produce randomly varying forces that act to speed or slow the motion of the group in a given direction. In favorable cases, these dynamical effects can be represented by a set of Langevin equations of motion (30, 72, 74). For a particle in one dimension, we can write:

$$m\frac{d^2x}{dt^2} = F(x) - f\frac{dx}{dt} + R(t) \qquad\qquad 3.$$

Here, m and x are the mass and position of the particle, respectively, and t is the time; thus, the term on the left is simply the acceleration of the particle. The term $F(x)$ represents the systematic force on the particle derived from the potential of mean force. The terms $-f dx/dt$ and $R(t)$ represent the effects of the varying forces caused by the bath acting on the particle; the first term is the average frictional force caused by the motion of the particle relative to its surroundings (f is the friction coefficient), and $R(t)$ represents the remaining randomly fluctuating force. The Langevin equation and its generalized forms are phenomenological in character but they are consistent with more detailed models for the atomic dynamics.

The Langevin equation also provides a useful focal point in the discussion of large-scale motions (88, 89). For displacements of whole sections of polypeptide chain away from protein surface (local denaturation), the terms corresponding to the one on the left of Equation 3 are typically negligible in comparison to the others (78). The motion then has no inertial character and the chain displacements have the particularly erratic character of Brownian motion. For elastic deformations of the overall protein shape, such as those involved in interdomain or hinge-bending motions, the potential of mean force may have a simple Hooke's law or springlike character (88). Finally, the larger-scale structural changes involved in protein folding (e.g. the coming together of two helices connected by a coil region to form part of the native structure) are also likely to have Brownian character (90, 91).

Harmonic Dynamics

Harmonic dynamics provides an alternative approach to the dynamics of a protein or one of its constituent elements (e.g. an α-helix). Early attempts to examine dynamical properties of proteins or their fragments used the harmonic approximation. They were motivated by vibrational spectroscopic studies (92), in which the calculation of normal mode frequencies from empirical potential functions has long been a standard step in the assignment of infrared spectra (62). One assumes that the vibrational displacements of the atoms from their equilibrium positions are small enough that the potential energy can be approximated as a sum of terms that are quadratic in the displacements. The coefficients of these quadratic terms form a matrix of force constants which, together with the atomic masses, can be used to set up a matrix equation for the vibrational modes of the molecule (62). For a molecule composed of N atoms, $3N$-6 eigenvalues provide the internal vibrational frequencies of the molecule; the associated eigenvectors give the directions and relative amplitudes of the atomic displacements in each normal mode.

Although the harmonic model may be incomplete because of the contribution of anharmonic terms to the potential energy (Equation 1), it is nevertheless of considerable importance because it serves as a first approximation for which the theory is highly developed. Further, the harmonic model is essential for some quantum mechanical treatments of vibrational contributions to the heat capacity and free energy (81, 93) and for certain approaches to unimolecular reactions (94).

Activated Dynamics

Enzyme catalyzed reactions generally involve some processes in which the rate is limited by an energy barrier. In many cases the phenomenological time scale of such activated events is a microsecond or longer. Such processes that are intrinsically fast but occur rarely (i.e. with an average frequency much less than 10^{11} sec^{-1}) are not observed often enough for adequate characterization in an ordinary molecular dynamics simulation. To study such processes, alternative dynamical methods can be employed.

It is often possible to identify the particular character of the structural change involved (e.g. the reaction path) and then to approximate the associated energy changes. In the adiabatic-mapping approach, one calculates the minimized energy of the protein consistent with a given structural change (57, 95, 96). Minimization allows the remainder of the protein to relax in response to the assumed structural change, so that the resulting energy provides a rough approximation to the potential of mean force. Accurate potentials of mean force can be calculated by means of specialized molecular-dynamics calculations (59), but the computational requirements are greater. To analyze the time-dependence of the process, the potential of mean force is incorporated into a model for the dynamics such as the familiar transition state theory (57, 83). A more detailed understanding of the process can be obtained by analyzing trajectories chosen to sample the barrier region (59, 85, 86). The trajectory analysis displays the space and time correlations of the atomic motions involved and provides experimentally accessible quantities such as rate constants and activation energies.

Simplified Model Dynamics

To simulate processes that are intrinsically complicated (i.e. that involve the sampling of many configurations), it is sometimes possible to use simplified models for the structure and energetics of the protein. In one model of this kind, each residue in the protein is represented by a single interaction center and these centers are linked by virtual bonds (97, 98). The energy function for this model is obtained by averaging interresidue interactions over all the local atomic configurations within each residue (78, 98–100). Thus, the model incorporates the assumption of separated time scales for local and overall chain motions. The reduced number of degrees of freedom allows rapid calculation of the energy and forces, so that significantly longer dynamical simulations are possible than with a more detailed model. Such an approach may be particularly useful for studying local unfolding or folding of proteins and their secondary structural elements (78).

ATOMIC FLUCTUATIONS

Figure 1 gives a qualitative picture of the fluctuations observed in the molecular dynamics simulation of the basic pancreatic trypsin inhibitor (PTI), a small protein with 58 amino acids and 454 heavy atoms; only the α-carbon atoms plus the three disulfide bonds are shown. The left-hand drawing represents the X-ray structure and the right-hand drawing an instantaneous picture of the equilibrated structure after 3 ps (71). The two structures are very similar, but there are small differences throughout. The largest displacements appear in the C-terminal end, which interacts with a neighboring molecule in the crystal, and in the loop in the lower left, which has rather weak interactions with the rest of the molecule. Corresponding behavior and deviations from the X-ray structure would be observed in "snap shots" taken at any other time during the simulation.

Mean-Square Fluctuations and Temperature Factors

A more quantitative measure of the motions is obtained from the mean-square fluctuations of the atoms from their average positions. These can be

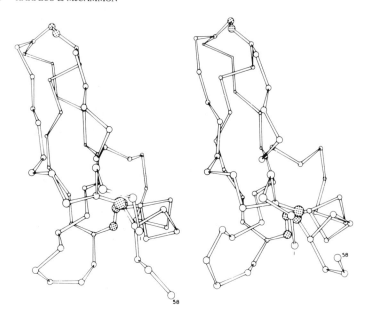

Figure 1   Drawing of α-carbon skeleton plus S–S bonds of PTI; left-hand drawing is the X-ray structure and right-hand drawing is a typical "snapshot" during the simulation.

related to the atomic temperature or Debye-Waller factors, B, determined in an X-ray diffraction study of a protein crystal (101–105). The mean-square positional fluctuation, $<\Delta r^2>_{dyn}$, with the assumption of isotropic and harmonic motion can be written:

$$<\Delta r^2>_{dyn} = \frac{3B}{8\pi^2} - <\Delta r^2>_{dis}. \qquad 4.$$

$<\Delta r^2>_{dis}$ is the contribution to B from lattice disorder and other effects that are difficult to evaluate experimentally. For a number of proteins at ambient temperatures (101–105), the measured value of $(3B/8\pi^2)$ averaged over all of the nonsurface atoms of the protein is in the range 0.48–0.58 Å². Comparison of this result with the mean value of $<\Delta r^2>_{dyn}$ from protein simulations (0.28–0.36 Å²; 101, 106, 107) suggests that the nonmotional contribution to the B-factor $<\Delta r^2>_{dis}$, is in the range 0.20–0.25 Å². The only experimental estimate of $<\Delta r^2>_{dis}$ is from Mössbauer data for the heme iron in myoglobin (102); for that one atom a somewhat smaller value (0.14 Å²) was obtained. Thus, in the cases examined, approximately half of the experimental B-factor is associated with thermal fluctuations in the atomic positions and half with other sources. However, some protein crystals, particularly those with a high percentage of water, appear to have a larger disorder contribution (109).

There is generally an increase in the magnitude of the experimental and theoretical fluctuations with distance from the center of the molecule. The magnitudes of the rms fluctuations range from ~ 0.4 Å for backbone atoms to ~ 1.5 Å for the ends of long sidechains. The hydrogen-bonded secondary structural elements (α-helices, β-sheets) tend to have smaller fluctuations than the random coil parts of the protein (106, 108). The magnitude of the fluctuations vary widely throughout the protein interior, suggesting that the system is inhomogeneous and that some regions are considerably more flexible than others.

To examine the importance of bond length and bond angle fluctuations, simulations were performed on PTI in which the bond lengths or both the bond lengths and the bond angles were fixed at their average values (73). It was found that use of fixed bond lengths (normal fluctuations ± 0.03 Å) does not significantly alter the dynamical properties on a time scale longer than 0.05 ps, but that constraint of the bond angles (normal fluctuations, $\pm 5°$) reduces the mean amplitude of the atomic motions by a factor of two. This result demonstrates that in a closely packed system, such as a protein in its native configuration, the excluded volume effects of repulsive van der Waals interactions introduce a strong coupling between the dihedral angle and bond-angle degrees of freedom.

Figure 2 shows a comparison of the calculated and experimental rms fluctuations on a residue-by-residue basis for reduced cytochrome c (101). The experimental values were corrected for an estimated disorder contribution by subtracting from all of them $<\Delta r^2>_{dis} = 0.25$ Å^2, obtained from the average calculated results for the protein interior. There is generally good agreement between the experimental and theoretical values. This correlation confirms the reliablity of both the simulation results and the temperature factors as detailed measures of the internal mobility of proteins.

Since most of the molecular dynamics simulations have been done for a protein in vacuum, it is expected that, particularly for the exterior residues, the results will be in error owing to the absence of solvent and, with regard to X-ray temperature factors, the absence of the crystal environment. For cytochrome c, the most prominent differences between theoretical and experimental mean displacements (Figure 2) involve the residues calculated to have very large fluctuations; these are all charged sidechains (particularly lysines) that protrude from the protein and so are not correctly treated in the vacuum simulation. This result is confirmed by molecular-dynamics simulations of PTI in a Lennard-Jones solvent and in a crystal environment. The simulations show that the motion of the outside residues is significantly perturbed by the surrounding medium (107, 110); in particular, the interaction between charged sidechains of a given protein and its crystal neighbors can produce a reduction in the rms values (66, 110). Such results for the external residues contrast with those for the protein interior, where the environmental effects on the amplitude of fluctuations are found to be small. The dominant medium effect on the equilibrium properties of the PTI molecule is that the average structure in the solvent or crystal field is significantly closer to the X-ray structure than is the vacuum result (e.g. for the C^α-atoms, the vacuum simulation has an rms deviation from the X-ray structure equal to 2.2 Å, while those for the solvent and crystal simulations are 1.35 Å and 1.52 Å, respectively).

Recently, a crystal simulation of PTI, including the water molecules, was completed (111). Although the simulation is too short (12 ps) for definitive conclusions, the magnitude of the fluctuations corresponded to those found in the earlier simulations of PTI, while the dynamic average structure was somewhat closer to the X-ray result (C^α-atom rms deviation is 1.06 Å). That the integrity of the dynamic average structure is necessary to obtain the

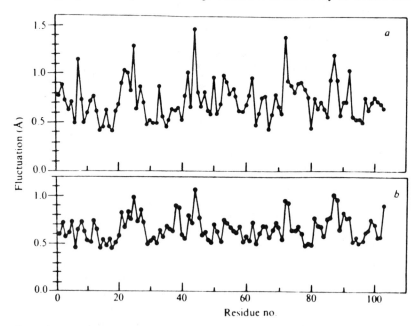

Figure 2   Calculated and experimental rms fluctuations of ferrocytochrome c; residue averages are shown as a function of residue number: (*a*) molecular-dynamics simulation; (*b*) X-ray temperature factor estimation corrected for mean disorder contribution.

correct rms fluctuations was noted in a simulation of rubredoxin (112);the agreement between calculated and experimental (113) temperature factors was poor, apparently because of a significant perturbation in the structure during the simulation.

Of interest also are the results from the dynamic simulation concerning deviations of the atomic motions from the isotropic, harmonic behavior assumed in most X-ray analyses of proteins. The motions of many of the atoms were found in the simulations to be highly anisotropic and somewhat anharmonic. The rms fluctuation of an atom in its direction of largest displacement is typically twice that in its direction of smallest displacement; larger ratios are not uncommon (106, 110, 114, 115). It is sometimes possible to rationalize these directional preferences in terms of local bonding, e.g. torsional oscillation of a small group around a single bond (115). In most cases, however, the directional preferences appear to be determined by larger-scale collective motions involving the atom and its neighbors (115–118). The atom fluctuations are generally also anharmonic; that is, the potentials of mean force for the atom displacements deviate from the simple parabolic forms that would obtain at sufficiently low temperature (106, 110, 119). The most markedly anharmonic atoms are those having multiple minima in their potentials of mean force. The shape of the PTI potential surface in the region of the native structure indicates that the anharmonicity is primarily associated with the softest collective modes of displacement in the protein (120).

Time-Dependence: Local and Collective Effects

Analyses of the time development of the atomic fluctuations were made for PTI (117, 118) and cytochrome c (116). The atomic fluctuations that contribute to the temperature factor (thermal ellipsoid) can be separated into local oscillations superposed on motions with a more collective character. The former have a subpicosecond time scale; the latter, which can involve only a few neighboring atoms, a residue, or groups of many atoms in a given region of the protein, have time scales ranging from 1–10 ps or longer ($v \cong 3$ to 30 cm^{-1}). By following the time development of the atomic fluctuations of PTI from 0.2–25 ps, it was shown that the high-frequency oscillations, which contribute about 40% of the average rms fluctuations of mainchain atoms, tend to be uniform over the structure. It is the longer-time-scale, more collective motions that introduce the variations in the fluctuation magnitudes that characterize different parts of the protein structure (117, 118). The correlations of anisotropy with local bonding are often destroyed by these large amplitude collective motions (115, 116).

The time dependence of atom and group motions in proteins can be characterized more fully by calculating appropriate time correlation function (72, 117, 118). The time correlation function of a fluctuating quantity describes the average manner in which a typical fluctuation decays (30, 31). For the positional fluctuations of individual atoms in the protein interior, the time correlation function has a partial loss of amplitude within the first 0.2 ps, followed by much slower decay on a time scale of several ps; the slow component has significant oscillations in many cases (72, 116–118). The decay times of the correlation functions are increased by including external solvent in the dynamic simulation; this effect is most pronounced for atoms at the protein surface (110, 117, 118). An analysis of the relaxation times for the atoms in PTI plus solvent yields a wide range (0.45–10 ps); in vacuum, the times shift to somewhat shorter values (0.2–6 ps).

Although there is no direct experimental measure of the time scale of the atomic fluctuations, it has been shown that NMR-relaxation parameters (T_1, T_2, and NOE values) are sensitive to picosecond motions. Of particular relevance are ^{13}C-NMR data since for protonated carbons, the C–H bond reorientation provides the dominant relaxation mechanism. Preliminary comparisons of ^{13}C-relaxation data for PTI suggest that the α-carbon mobility has a small effect on the relaxation parameters and that increasing effects are expected as the observed carbon is further out along a sidechain (121, 122). The effect of internal motions on other NMR parameters, such as chemical shifts (123) and vicinal-coupling constants, was also examined by molecular-dynamics simulations.

Biological Function

Although many of the individual atom fluctuations observed in the simulations or obtained from temperature factors may in themselves not be important for protein function, they contain information that is of considerable significance. The calculated fluctuations are such that the conformational space available to a protein at room temperature includes the range of local structural changes observed on substrate or inhibitor binding for many enzymes. There may be a correlated directional character to the active-site fluctuations that play a role in catalysis. Further, the small amplitude fluctuations are essential to all other motions in proteins; they serve as the "lubricant" which makes possible larger-scale displacements, such as domain motions (see Table I), on a physiological time scale. It may be possible to extrapolate from the short time fluctuations to larger-scale protein motions. This is suggested by the approximate correspondence between the rms fluctuations of hydrogen bond lengths in a dynamical simulation of PTI (124) and the relative exchange rate of the hydrogens as measured by NMR (42). Changes in the fluctuations induced by perturbations, e.g. ligand binding, are likely to be important as well, e.g. the entropy differences, for the study of which molecular-dynamics techniques were developed (125), may make a significant contribution to the binding free energy (15).

The collective modes are likely to be of particular significance in the biological function; they may be involved in the displacements of side-chains, loops or other structural units required for the transition from an inactive to the active configuration of a globular protein and in the correlated fluctuations that play a direct role in enzyme catalysis. Further, the extended nature of these motions makes them more sensible to the environment, e.g., differences in the simulation results between vacuum and solution results for PTI (117, 118). Because they involve sizable portions of the protein surface, the collective motions may be involved in transmitting external solvent effects to the protein interior (45). They might also be expected to be quenched at low temperature by freezing of the solvent. Their contribution to the mean-square fluctuations could explain the transition observed near 200°K in the temperature dependence of the fluctuations in proteins like in myoglobin (126).

SIDECHAIN MOTIONS

The motions of aromatic sidechains serve as a convenient probe of protein dynamics. The sidechain motions span a time range from picoseconds, during which local oscillations occur, to milliseconds or longer required for 180° rotations. To cover this range of motions requires use of a variety of approaches that complement each other in the analysis of protein dynamics. Further, the results obtained are typical of a class of motional phenomena that play a significant role.

Tyrosines in PTI

The torsional librations of buried tyrosines in PTI were studied in some detail (72). We focus on a particular aromatic sidechain, Tyr 21, whose ring is surrounded by and has a significant nonbonded interaction with atoms of its own backbone and of surrounding residues that are more distant along the polypeptide chain. Figure 3 shows a potential energy contour map for the sidechain dihedral angles χ_1 and χ_2 of Tyr 21 in the free dipeptide (*top*) and in the protein (*bottom*) (57). The minimum energy conformations are very similar in the two cases; this appears to be true for most interior residues of proteins. Where the plots differ is that the sidechain is much more rigidly fixed in position by its nonbonded neighbors in the protein than it is by interactions with the backbone of the chain in the dipeptide.

Figure 4 (*top*) shows the torsional fluctuations of Tyr 21 observed during a PTI simulation (72); the quantity plotted is $\Delta\phi = \phi - <\phi>$ where $<\phi>$ is the time average of the ring torsional angle. Figure 4 (*bottom*) shows corresponding torsional fluctuation history for the ring in an isolated tyrosine fragment simulation.

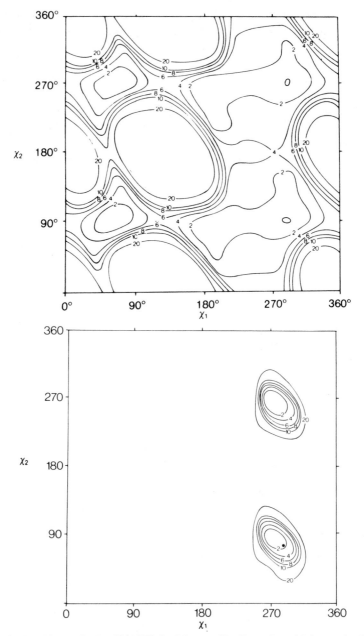

Figure 3 (κ_1, κ_2)) maps for Tyr-21 in **PTI**: (*top*) free peptide; (*bottom*) peptide in protein; the black dot corresponds to the X-ray value for (κ_1, κ_2) in the protein; energy contours in kcal/mol.

The torsional motion of the ring is less regular when it is surrounded by the protein matrix than in the separated fragment. In PTI, the rms fluctuation of the Tyr 21 torsion angle is 12°, while that for the tyrosine fragment is 15°. This relatively small difference in amplitudes as compared with the forms of the rigid rotation potentials (Figure 3) indicates that protein relaxation involving correlated fluctuations must play an important role in

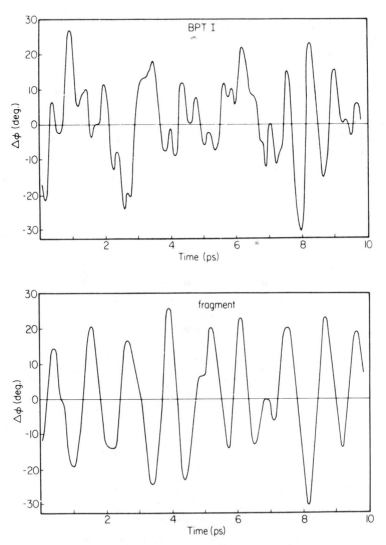

Figure 4 Evolution of the Tyr-21 ring torsional angle during 9.8 ps of dynamical simulation: (*top*) in the protein; (*bottom*) in the isolated tyrosine fragment.

the ring oscillations. The short time, local motion in the protein is consistent with a torsional Langevin equation that contains a harmonic restoring force (see Equation 3). The frictional random force terms are similar to those expected for ring rotation in an organic solvent; this is consistent with the hydrophobic environments of the rings in the protein. The time correlation functions for the torsional fluctuations decay to small values in a short time (~ 0.2 ps). However, the quantities involved in the relaxation times (110, 127) measured in fluorescence depolarization (trigonometric functions of the angles) decay much more slowly. For the tyrosine rings in PTI there is rapid partial decay in less than a picosecond to a plateau value equal to about 75% of the initial value; this behavior was recently confirmed by fluorescent depolarization measurements (128). Corresponding calculations (130) for the fluorescent depolarization of the tryptophan residues in lysozyme based on a molecular dynamics simulation (106) indicate a wide range of variation in the depolarization behavior. Since there are six tryptophans in a variety of environments, their behavior is expected to correspond to that which occurs more generally in proteins (129). Certain interior tryptophans have almost no decay over the time scale of the simulation while one in the active site (Trp 62) has its anisotropy reduced to 0.6 after 5 ps.

Tyrosine and phenylalanine ring rotations by 180° were studied by NMR in proteins (40, 43, 44, 56). Such ring "flips" occur very infrequently because of the large energy barrier due to steric hindrance (57–59). The long time intervals separating flips preclude systematic study by conventional molecular-dynamics methods. A modified molecular-dynamics method was recently developed to handle such local activated processes (59). This method is similar to adiabatic mapping in that one starts with an assumed "reaction coordinate" that defines the fundamental structural changes involved. It differs from the adiabatic method in that it involves consideration of all thermally accessible configurations and not just the minimum energy one for each value of the reaction coordinate. Also it provides a detailed description of the structural and dynamical features of the process. In this method, one calculates separately the factors in the rate constant expression (82, 131):

$$k = \frac{1}{2}\kappa <|\dot{\xi}|> [\rho(\xi^\dagger)/\int_i \rho(\xi)d\xi].$$ 5.

Here, ξ is the reaction coordinate, $\dot{\xi} = d\xi/dt$, and $\xi^\dagger$ is the value of ξ in the transition state region for the process. The factor in square brackets is the probability that the system will be in the transition state region, relative to the probability that it is in the initial stable state. This quantity corresponds roughly to the term $\exp(-\Delta G^\dagger/RT)$ in more familiar expressions for rate constants; it can be calculated by carrying out a sequence of simulations in which the system is constrained to stay near particular values of ξ. The remaining factors can be evaluated by analysis of trajectories initiated in the transition state region (59, 85, 86). The transmission coefficient κ is equal to one in ideal transition state theory (equilibrium populations maintained in the stable states and uninterrupted crossings through the transition state region); for real systems κ is less than one.

Application of this modified molecular dynamics method to the flipping of a tyrosine ring in PTI shows that the rotations themselves required only 0.5–1.0 ps (85, 86). At the microscopic level, the processes responsible for flipping are the same as those responsible for the smaller amplitude librations. The ring goes over the barrier not as the result of a particularly energetic collision with some cage atom, but as the result of a transient decrease in frequency and intensity of collisions that would drive the ring away from the barrier. These alterations of the collision frequency are caused by small, transient packing defects (86). The packing defects help to initiate ring rotation, but they are much too small to allow free rotation of the ring by a simple vacancy or free-volume mechanism (86, 132). The ring tends to be tightly encaged even in the transition state orientation. Collisions with cage atoms in the transition state produce frictional forces similar to those that occur in the stable state librations; these frictional effects reduce the transition rate to about 20% of the ideal transition-state theory value (59). As to the free energy of activation, the calculations suggest that the activation enthalpy contribution is similar to that found by adiabatic mapping techniques (57, 58) and that the activation entropy is small.

Although no enzyme has yet been studied by the techniques applied to the tyrosine ring flips, the methodology is applicable to the activated processes central to most enzymatic reactions. Further, many of the qualitative features found for the tyrosines (e.g. lowering of the potential of mean force by cage relaxation, alteration of the rate by frictional effects) should be present in general.

Ligand-Protein Interaction in Myoglobin

A biological problem where sidechain fluctuations are important concerns the manner in which ligands like carbon monoxide and oxygen are able to get from the solution through the protein matrix to the heme group in myoglobin and hemoglobin and then out again. The high-resolution X-ray structure of myoglobin (8, 9, 133, 134) does not reveal any path by which ligands such as O_2 or CO can move between the heme-binding site and the outside of the protein. Since this holds true both for the unliganded and liganded protein, i.e. myoglobin (133) and oxymyoglobin (134), structural fluctuation must be involved in the entrance and exit of the ligands. Empirical energy function calculations (96) showed that the rigid protein would have barriers on the order of 100 kcal/mol; such high barriers would make the transitions infinitely long on a biological time scale. Figure 5, panel I gives the nonbonded potential contour lines seen by a test particle representing an O_2 molecule in a plane (xy) parallel to the heme and displaced 3.2 Å from it in the direction of the distal histidine; the coordinate system in this and related figures has the iron at the origin and the z-axis normal to the heme plane. The low potential-energy minimum corresponds to the observed position of the distal O atom of an O_2 molecule forming a bent Fe–O–O bond (134). The shortest path for a ligand from the heme pocket to the exterior (the low energy region in the upper left of the figure) is between His E7 and Val E11. However, this path is not open in the X-ray geometry because the energy barriers due to the surrounding residues indicated in the figure are greater than 90 kcal/mol.

To analyse pathways available in the thermally fluctuating protein, ligand trajectories were calculated with a test molecule of reduced effective diameter to compensate for the use of the rigid protein structure (96). A trajectory was determined by releasing the test molecule with substantial kinetic energy (15 kcal/mol) in the heme pocket and following its classical motion for a suitable length of time. A total of 80 such trajectories were computed; a given trajectory was terminated after 3.75 ps if the test molecule had not escaped from the protein. Slightly more than half the test molecules failed to escape from the protein in the allowed time; 25 molecules remained trapped near the heme-binding site, while another 21 were trapped in two cavities accessible from the heme pocket. Most of the molecules that escaped did so between the distal histidine (E7) and the sidechains of Thr E10 and Val E11 (see Figure 5, panel I) A secondary pathway was also found; this involves a more complicated motion along an extension of the heme pocket into a space between Leu B10, Leu E4, and Phe B14, followed by squeezing out between Leu E4 and Phe B14. Figure 6 shows a typical model trajectory following this path. Additional, more complicated pathways also exist, as indicated by the range of motions observed in the trial trajectories.

In the rigid X-ray structure, the two major pathways have very high barriers for a thermalized ligand of normal size. Thus, it was necessary to study the energetics of barrier relaxation to determine whether either of the pathways had acceptable activation enthalpies. Local dihedral rotations of key sidechains, analogous to the tyrosine sidechain oscillations described above, were investigated; it was found that the bottleneck on the primary pathway could be relieved at the expense of modest strain in the protein by rigid rotations of the sidechains of His E7, Val E11, and Thr E10. The reorientation of these three sidechains and the resultant opening of the

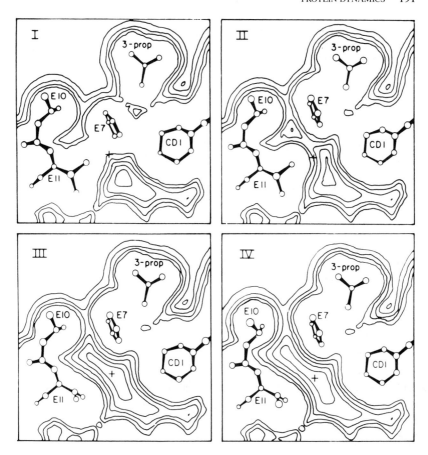

Figure 5 Myoglobin-ligand interaction contour maps in the heme (x,y) plane at $z = 3.2$ Å (the iron is at the origin) showing protein relaxation; a cross marks the iron atom projection onto the plane. Distances are in Å and contours in kcal; the values shown correspond to 90, 45, 10, 0, and -3 kcal/mol relative to the ligand at infinity. The highest contours are closest to the atoms whose projections onto the plane of the figure are denoted by circles. *Panel I:* X-ray structure; *panels II–IV:* sidechain rotations discussed in the text.

pathway to the exterior is illustrated schematically in Figure 5; Panel I shows the X-ray structure; in Panel II the distal histidine (E7) was rotated to $\chi_1 = 220°$ at an energy cost of 3 kcal/mol; in Panel III, Val E11 was also rotated to $\chi_1 = 60°$ (~ 5 kcal/mol); and Panel IV has the additional rotation of Thr E10 to $\chi_1 \cong 305°$ (< 1 kcal/mol). In this manner a direct path to the exterior was created with a barrier of ~ 5 kcal/mol at an energy cost to the protein of ~ 8.5 kcal/mol, as compared with the X-ray structure value of nearly 100 kcal/mol. On the secondary path, however, no simple torsional motions reduced the barrier due to Leu E4 and Phe B14, since the necessary rotations led to larger strain energies. A test sphere was fixed at each of the bottlenecks and the protein was allowed to relax by energy minimization (adiabatic limit), in the presence of the ligand (57, 96).

Approximate values for the relaxed barrier heights were 13 kcal/mol and 6 kcal/mol for the two primary path positions and 18 kcal/mol for the secondary path position. These barriers are on the order of those estimated in the photolysis, rebinding studies for CO myoglobin by Frauenfelder et al (45, 135, 136). Further, a path suggested by the energy calculations was found to correspond to a high mobility region in the protein as determined by X-ray temperature factors (102).

The type of ligand motion expected for such a several-barrier problem can be determined from the trajectory studies mentioned earlier. What happens is that the ligand spends a long time in a given well, moving around in and undergoing collisions with the protein walls of the well (see Figure 6). When there occurs a protein fluctuation sufficient to significantly lower

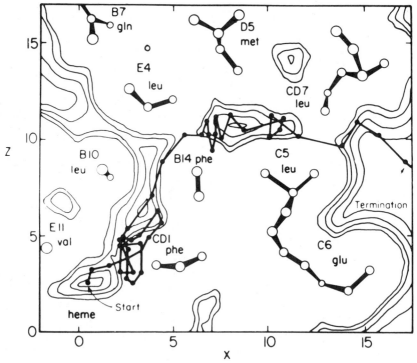

Figure 6 Diabatic-ligand trajectory following a secondary pathway (see text); a projection of the trajectory on the plane of the figure [(x,z) plane at $y = 0.5$ Å)] is shown with the dots at 0.15 ps intervals. The start of the trajectory at the heme iron and the termination point exterior to the protein are indicated by arrows.

the barrier, or the ligand gains sufficient excess energy from collisions with the protein, or more likely both at the same time, the ligand moves rapidly over the barrier and into the next well where the process is repeated. In a completely realistic trajectory involving a fluctuating protein and ligand-protein energy exchange, the time spent in the wells would be much longer than that found in the diabatic model calculations (Figure 6). Further, from the complexity of the range of pathways in the protein interior, it is likely that the motion of the ligand will have a diffusive character.

The analysis of myoglobin suggests that the native structure of a protein is often such that the small molecules that interact with the protein cannot enter or leave if the atoms are constrained to their average positions. Conse-quently, sidechain and other fluctuations may be required for ligand binding by proteins and for the entrance of substrates and exit of products from enzymes. Some analyses of the effects of such "gated" accessibility on the observed kinetics were made (137–139).

Exterior Sidechain and Loop Motions

In several enzymes, a displacement of surface sidechains or entire loops on substrate binding occurs. In carboxypepdidase A (140), for example, when the substrate binds, the structural changes include a large displacement of the sidechain of Tyr 248, which moves through more than 10 Å toward the active site. Another example is provided by an external loop in triophos-phate isomerase, which was shown by X-ray diffraction to fold over the substrate when it is bound (141). If surface residues are involved, as is often the case, the motion is best treated by stochastic dynamics.

To study the motion of aliphatic sidechains in solution (76), the end of the chain attached to the macromolecule is held fixed and the Langevin equations of motion (Equation 3) for the atoms of the chain are solved simultaneously for periods of up to a microsecond. The methyl and methy-lene groups of the chain are treated as single extended atoms with a friction coefficient corresponding to methane in water and a generalized empirical potential energy function is used to represent the intramolecular interac-tions (nonbonded and torsional) in the usual way, except that they are slightly modified to take into account the presence of solvent; that is, a potential of a mean force replaces the isolated molecule potential function (142). It is found that the motion with respect to a given torsion angle separates into two time scales (76). The shorter time motion, on the order of tenths of picoseconds, corresponds to torsional oscillations within a potential well, and the longer, on the order of two hundred picoseconds, corresponds to transitions from one potential well to another; the torsional barrier used in the potential function is ~ 2.8 kcal/mol. Thus, analogous to the above description for an oxygen molecule moving through myoglobin,

the sidechain spends most of the time oscillating about a single conformation (i.e. with each dihedral angle remaining in a given well) and only rarely
makes a transition from one conformation to another. To test the validity
of this type of calculation, comparisons of the stochastic trajectory results
with NMR relaxation measurements (e.g. ^{13}C NMR) were made (122).

RIGID BODY MOTIONS

A type of motion that plays an important role in proteins is referred to as
a rigid-body motion (Table I). It involves the displacement of one part of
a protein relative to another such that each moving portion can be approximated as a rigid body. However, smaller fluctuations must accompany the
rigid-body motions to reduce the required energy and permit them to
proceed at a sufficiently rapid rate.

Hinge Bending

Many enzymes (23–25) and other protein molecules (e.g. immunoglobulins)
consist of two or more distinct domains connected by a few strands of
polypeptide chain that may be viewed as "hinges." In lysozyme, for example, it was noted in the X-ray structure (143) that when an active-site
inhibitor is bound, the cleft closes down somewhat as a result of relative
displacements of the two globular domains that surround the cleft. Other
classes of proteins (kinases, dehydrogenases, citrate synthase) have considerably larger displacements of the two lobes on substrate binding than
does lysozyme (23–25).

    In the theoretical analysis of lysozyme (88), the stiffness of the hinge was
evaluated by the use of an empirical energy function (66, 88). An angle-
bending potential was obtained by rigidly rotating one of the globular
domains relative to a bending axis which passes through the hinge and
calculating the changes in the protein conformational energy. This procedure overestimates the bending potential, since no allowance is made for the
relaxation of the unfavorable contacts between atoms generated by the
rotation. To take account of the relaxation, an adiabatic potential was
calculated by holding the bending angle fixed at various values and permitting the positions of atoms in the hinge and adjacent regions of the two
globular domains to adjust themselves so as to minimize the total potential
energy. As in a previous adiabatic ring rotation calculation (57), only small
(<0.3 Å) atomic displacements occurred in the relaxation process. Localized motions involving bond angle and local dihedral angle deformations
occur. The frequencies associated with them (>100 cm^{-1}) are much greater
than the hinge-bending frequency (≈ 5 cm^{-1}), so that the use of the adiabatic-bending potential is appropriate.

The bending potentials were found to be approximately parabolic, with the restoring force constant for the adiabatic potential about an order of magnitude smaller than that for the rigid potential (see Figure 7). However, even in the adiabatic case, the effective force constant is about 20 times as large as the bond-angle bending force constant of an α carbon (i.e. $N-C_\alpha-C$); the dominant contributions to the force constant come from repulsive nonbonded interactions involving on the order of fifty contacts. If the adiabatic potential is used and the relative motion is treated as an angular harmonic oscillator composed of two rigid spheres, a vibrational frequency of about 5 cm^{-1} is obtained. This is a consequence of the fact that, although the force constant is large, the moments of inertia of the two lobes are also large.

Although fluctuations in the interior of the protein, such as those considered in myoglobin, may be insensitive to the solvent (because the protein matrix acts as its own solvent), the domain motion in lysozyme involves two lobes that are surrounded by the solvent. To take account of the solvent effect in the simplest possible way, the Langevin equation (Equation 3) for a damped harmonic oscillator was used. The friction coefficient for the solvent damping term was evaluated by modeling the two globular domains as spheres (144). From the adiabatic estimate of the hinge potential and the magnitude of the solvent damping, it was found that the relative motion of the two globular domains in lysozyme is overdamped; i.e. in the absence of

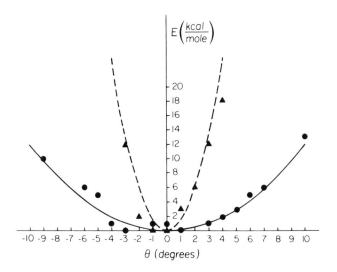

Figure 7 Change of conformational energy produced by opening ($\theta<0$) and closing ($\theta>0$) the lysozyme cleft; calculated values are for the rigid bending potential (triangles) and for the adiabatic-bending potential (circles); the origins for the two calculations are superposed.

driving forces the domains would relax to their equilibrium positions without oscillating. The decay time for this relaxation was estimated to be about 2×10^{-11}s. Actually, the lysozyme molecule experiences a randomly fluctuating driving force owing to collisions with the solvent molecules, so that the distance between the globular domains fluctuates in a Brownian manner over a range limited by the bending potential; a typical fluctuation opens the binding cleft by 1 Å and lasts for 20 ps.

The methodology developed in the lysozyme study is now being applied to a number of other proteins; they include antibody molecules (145), L-arabinose binding protein (146), and liver alcohol dehydrogenase (147). For the L-arabinose binding protein, calculations and experiment both suggest that the binding site is open in the unliganded protein but is induced to close by a hinge-bending motion upon ligation (146, 148). In the case of liver alcohol dehydrogenase, the open structure is stable in the crystal for the apoenzyme (149). Adiabatic energy-minimization calculations (147) suggest that the apoenzyme is highly flexible as far as its hinge-bending mode (rotation of the catalytic relative to the coenzyme binding domain) is concerned and that normal thermal fluctuations would lead to a closed structure (rotation of $\sim 10°$) similar to that found in the holoenzyme (150).

Since the hinge-bending motion in lysozyme and in other enzymes involves the active-site cleft, it is likely to play a role in the enzymatic activity of these systems. In addition to the possible difference in the binding equilibrium and solvent environment in the open and closed state, the motion itself could result in a coupling between the entrance and exit of the substrate and the opening and closing of the cleft (137–139). The interdomain mobility in immunoglobins may be involved in adapting the structure to bind different macromolecular antigens and, more generally, it may play a role in the cross-linking and other interactions required for antibody function. In the coat protein of tomato bushy stunt virus, a two-domain structure with a hinge peptide was identified from the X-ray structure (29, 151) and rotations about the hinge were shown to be involved in establishing different subunit interactions for copies of the same protein involved in the assembly of the complete viral protein shell.

Quaternary Structural Change

A classic case where large-scale motion plays an essential role is the allosteric transition in the hemoglobin tetramer (152–154). It is clear from the X-ray data that the subunits move relative to each other (quaternary change) and that more localized atomic displacements occur within each subunit (tertiary change). The coupling between these two types of structural changes is an essential factor in the cooperative mechanism of hemoglobin. As a first step in unraveling the nature of the motions involved, a reaction path for the tertiary structure change induced by ligand binding within a subunit was worked out by the use of empirical energy calculations (155, 156). The results are in good agreement with limited structural data available for intermediates in the ligation reaction (157). The calculations show how the perturbation introduced in the heme by the binding of ligand leads to displacements in the protein atoms, so that alterations appear in surface regions in contact with other subunits. This provides a basis for the coupling between tertiary and quaternary structural change, although the details of the motions leading from the unliganded to the liganded quaternary structure have yet to be worked out.

As is clear from the above, most of the information available on rigid motions comes from high-resolution crystal structures of proteins. Low angle X-ray scattering analyses of solutions provided evidence for radius of gyration changes that are in accord with the crystal results where available (158) or provide evidence for structural changes in cases where only one structure is known (148, 159). However, there is almost no experimental evidence on the time scale of the rigid-body motions. Fluorescence depolarization studies of labeled antibody molecules show that the time scale for internal motions is consistent with the diffusional displacements of flexibly hinged domains (145, 160). It would be of great interest to have corresponding data on the domain motions of enzymes.

α-HELIX MOTION: HARMONIC AND SIMPLIFIED MODEL DYNAMICS

Early evidence for motion in the interior of proteins or their fragments comes from analyzing vibrational spectroscopic studies. It is generally assumed in interpreting such data that a harmonic potential and the resulting normal-mode description of the motions is adequate. This approximation is most likely correct for the tightly bonded secondary structural elements, like α-helices and β-sheets. The fluctuations of a finite α-helix (hexadecaglycine) were determined from the normal modes of the system (63). At 300°K, the rms fluctuations of mainchain dihedral angles (ϕ and ψ) about their equilibrium values are equal to $\sim 12°$ in the middle of the helix and somewhat larger near the ends. The dihedral angle fluctuations are significantly correlated over two neighboring residues; these correlations tend to localize the fluctuations (63, 71, 79, 80). Fluctuations in the lengths between adjacent residues (defined as the projection onto the helix axis of the vector connecting the centers of mass of adjacent residues) ranged from about 0.15 Å in the middle of the helix to about 0.25 Å at the ends. These length fluctuations are negatively correlated for residue pairs $(i-1, i)$ and $(i, i+1)$ so as to preserve the overall length of the helix; positive correlations are observed for the pairs (4, 5), (8, 9) and (8, 9), (12, 13), suggesting that the motion of residue 8 is coupled to the motions of residues 4 and 12 to retain optimal hydrogen bonding.

Recently a full molecular-dynamics calculation was performed (81) for a decaglycine helix as a function of temperature between 5°–300°K and the results compared with those obtained in the harmonic approximation (63). For the mean-square positional fluctuations, $<\Delta r^2>$, of the atoms, the harmonic approximation is valid in the classical limit below 100°K, but there are significant deviations above that temperature; e.g. at 300°K, the average value of $<\Delta r^2>$ obtained for the α-carbons from the full dynamics is more than twice that found in the harmonic model. Quantum effects on the fluctuations are found to be significant only below 50°K. The temperature dependence of the fluctuations in the simulations is similar to that observed for α-helices in myoglobin between 80–300°K by X-ray diffraction (161).

As an approach to the helix-coil transition in α-helices, a simplified model for the polypeptide chain was introduced to permit a dynamic simulation on the submicrosecond time scale appropriate for this phenomenon (78): each residue is represented by a single interaction center ("atom") located at the centroid of the corresponding sidechain and the residues are linked by virtual bonds (97), as described earlier in the section on simplified model dynamics. The diffusional motion of the chain "atoms" expected in water was simulated by using a stochastic dynamics algorithm based on the Langevin equation with a generalized force term, Equation 3. Starting from an all-helical conformation, the dynamics of several residues at the end of a 15-residue chain were monitored in several independent 12.5-ns simulations at 298°K. The mobility of the terminal residue was quite large, with a rate constant $\approx 10^9\ s^{-1}$ for the transitions between coil and helix states. This mobility decreased for residues further into the chain; unwinding of an interior residue required simultaneous displacements of residues in the coil, so that larger solvent frictional forces were involved. The coil region did not move as a rigid body, however; the torsional motions of the chain were correlated so as to minimize dissipative effects. Such concerted transitions are not consistent with the conventional idea that successive transitions occur independently. Analysis of the chain diffusion tensor showed that the frequent occurrence of the correlated transition results from the relatively small frictional forces associated with these motions (100).

PERSPECTIVE

Theoretical protein dynamics did not exist before 1977 when the first paper presenting a detailed molecular-dynamics simulation of a small protein was published (71). In the next five years more than 50 theoretical papers appeared. They explored dynamic phenomena in depth for a variety of proteins (protein inhibitors, transport and storage proteins, enzymes). The magnitudes and time scales of the motions were delineated and related to a variety of experimental measurements, including NMR, X-ray diffraction, fluorescent depolarization, infra-red spectroscopy and Raman scattering. It was shown how to extend dynamical methods from the subnanosecond time range accessible to standard molecular-dynamics simulations to much longer time scales for certain processes by the use of activated, harmonic and simplified model dynamics. Further, the effect of solvent was introduced by stochastic dynamic techniques or accounted for in full dynamic simulations including also the crystal environment. Concomitantly, a wealth of experimental information on the motions appeared. The interplay between theory and experiment provides a basis for the present vitality of the field of protein dynamics.

What is known and what remains to be done? On the subnanosecond time scale our basic knowledge of protein motions is essentially complete; that is, the types of motion that occur have been clearly presented, their characteristics evaluated and the important factors determining their properties delineated. Simulation methods have shown that the structural fluctuations in proteins are sizable; particularly large fluctuations are found where steric constraints due to molecular packing are small (e.g. in the exposed side chains and external loops), but substantial mobility is also found in the protein interior. Local atomic displacements in the interior of the protein are correlated in a manner that tends to minimize disturbances of the global structure of the protein. This leads to fluctuations larger than would be permitted in a rigid polypeptide matrix.

For motions on a longer time scale, our understanding is more limited. When the motion of interest can be described in terms of a reaction path (e.g. hinge-bending, local-activated event), methods exist for determining the nature and rate of the process. However, for the motions that are slow owing to their complexity and involve large-scale structural changes, extensions of the approaches described in this review are required. Harmonic and simplified model dynamics, as well as reaction-path calculations, can provide information on slower processes, such as opening fluctuations and helix-coil transitions, but a detailed treatment of protein folding is beyond the reach of present methods.

In the theory of protein dynamics there are two directions where active study and significant progress can be expected in the near future. One concerns the more detailed examination of processes of biological interest and the other, an improvement in approaches to longer-time dynamics. As to the latter, a variety of extensions of the methodology described in this review, as well as the availability of faster computers, may yield the necessary insights. As to the former, there are many biological problems to which current dynamical methods can be applied and for which a knowledge of the dynamics is essential for a complete understanding. Some of these are listed below.

For the transport protein hemoglobin, there is more evidence concerning the role of motion than for any other protein. The tertiary and quaternary structural changes that occur on ligand binding and their relation to the allosteric mechanism are well documented. An important role of the quaternary structural change is to transmit information over a longer distance than could take place by tertiary structural changes alone; the latter are generally damped out over rather short distances unless amplified by the displacement of secondary structural elements or domains. The detailed dynamics of the allosteric mechanism has yet to be investigated; in particular, the barriers along the reaction path from the deoxy to the oxy structure have not been analyzed, nor has the importance of the fluctuations for the activated processes involved been determined. For the related storage protein, myoglobin, fluctuations in the globin are essential to the binding process; that is, the protein matrix in the X-ray structure is so tightly packed that there is no sufficiently low energy path for the ligand to enter or leave the heme pocket. Only through structural fluctuations in certain bottleneck regions can the barriers be lowered sufficiently to obtain the observed rates of ligand binding and release. Although energy minimization was used to investigate the displacements involved and the resulting barrier magnitudes, activated dynamic studies are needed to analyze the activation entropies and rates of ligand motion across the barriers.

In many proteins and peptides, the transport of substances is through the molecule rather than via overall translation as in hemoglobin. The most obvious cases are membrane systems, in which fluctuations are likely to be of great importance in determining the kinetics of transport. For channels that open and close (e.g. gramicidin) as well as for active transport involving enzymes (e.g. ATPases), fluctuations, in some cases highly correlated ones, must be involved. At present, structural details and studies of the motions are lacking, but this is an area where dynamic analyses are likely to be made in the near future.

In electron-transport proteins, such as cytochrome c, protein flexibility is likely to play two roles in the electron transfer. Evidence now favors a vibronic-coupled tunneling mechanism for transfer between cytochrome c and other proteins, although outer-sphere mechanisms are not fully excluded. In the vibronic-coupled tunneling theory, processes which would be energetically forbidden for rigid proteins become allowed if the appropriate energies for conformational distortions are available. Experimental data indicate that the important fluctuations are characterized by an average frequency on the order of 250 cm^{-1}, close to that associated with the collective modes of proteins. Also, the transfer rate is a sensitive function of donor-acceptor distance and may be greatly increased by surface side-chain displacements that allow for the closer approach of the interacting proteins.

For proteins involved in binding, flexibility and fluctuations enter into both the thermodynamics and the kinetics of the reactions. For the rate of binding of two macromolecules (protein-antigen and antibody, protein-inhibitor and enzyme), as well as for smaller multisite ligands, structural fluctuations involving side chains, hydrogen-bonding groups, etc, can lead to lowering of the free energy barriers. Dividing the binding process into successive steps for which flexibility may be needed can increase the rate. The required fluctuations are likely to be sufficiently small and local that they will be fast relative to the binding and therefore not rate limiting.

The relative flexibility of the free and bound ligand, as well as changes in the binding protein, must be considered in the overall thermodynamics of the binding reaction. If the free species have considerable flexibility and fluctuations are involved in the binding step as described above, it is likely that the bound species will be less flexible and a significant entropic destabilization will result. Thus, for strong binding in cases where the rate is not important, relatively rigid species are desirable. This would reduce the conformational entropy decrease and could lead to a very favorable enthalpy of binding if there is high complementarity in the two binding sites. However, some flexibility and an increase in the conformational space available to the bound species has a stabilizing effect that partly compensates for the loss of translational and rotational entropy on binding. Conversely, the entropy loss of binding a flexible substrate or the rigidification of a protein on substrate binding can be used to modulate the binding constant even when strong, highly specific enthalpic interactions are present. The required balance between flexibility and rigidity will be determined by the function of the binding in each case. Dynamical techniques can be employed to determine the entropy differences for such systems. Further, the mechanism and rates of the binding processes, itself, are an ideal subject for dynamical analysis.

In the function of proteins as catalysts, there is the greatest possibility of contributions from motional phenomena. The role of flexibility per se has often been discussed, particularly from the viewpoint of structural changes induced by the binding of the substrate. In addition to cooperative effects caused by quaternary alterations, a variety of results can arise from the perturbation of the tertiary structure. One example is the ordered binding of several substrates (or effectors and substrates), with the first molecule to bind altering the local conformation so as to increase or decrease the subsequent binding of other molecules. The occurrence of large-scale changes, such as the closing of active-site clefts by substrate binding, as in certain kinases, has been interpreted in terms of catalytic specificity, alteration of the solvent environment of the substrate, and exclusion of water that could compete with the enzymatic reaction. In large enzymes with more than one catalytic site or in coupled enzyme systems, conformational freedom may be important in moving the substrate along its route from one site to the next. Many of these processes are ready for the application of dynamical methods, particularly in cases where structural data are available.

The flexibility of the substrate-binding site in enzymes can result in effects corresponding to those already considered in receptor binding. In the enzyme case there exists the often-discussed possibility of enhanced binding of a substrate with its geometry and electron distribution close to the transition state; for this to occur, conformation fluctuations are essential. Entropic effects also are likely to be of significance, both with respect to solvent release on substrate binding and possible changes in vibrational frequencies that alter the vibrational entropy of the bound system in the enzyme-substrate complex or in the transition state. There are also indications that the inactivity of enzyme precursors can result from the presence of conformational freedom in residues involved in the active site. The entropic cost of constraining them in the proper geometry for interacting with the substrate may be so high that the activity is significantly reduced relative to that of the normal enzyme where the same residues are held in place more rigidly. Such a control mechanism was suggested for the trypsin, trypsinogen system, and for other proteins. As to the time dependence of fluctuations and structural alternations, there are a variety of possibilities to be considered. In the binding of reactants and release of products, the time course of fluctuations in the enzyme could interact with the motion of the substrate. The opening and closing fluctuations of active-site clefts may be modified by interactions with the substrate as it enters or leaves the binding site.

Fluctuations could play an essential role in determining the effective barriers for the catalyzed reactions. If the substrate is relatively tightly bound, local fluctuation in the enzyme could couple to the substrate in such a way as to significantly reduce the barriers. If such coupling effects exist, specific structures could have developed through evolutionary pressure to introduce directionality and enhance the required fluctuations. Frictional effects that occur in the crossing of barriers in the interior of the protein could act to increase the transition state lifetime and so alter the reaction rates relative to those predicted by conventional rate theory. Energy released locally in substrate binding may be utilized directly for catalyzing its reaction, perhaps by inducing certain fluctuations. Whether such an effect occurs would depend on the rate of dissipation of the (mainly) vibrational energy and the existence of patterns of atoms and interactions to channel the energy appropriately. It will be of great interest to determine whether any of the rather speculative possibilities outlined here for the role of the energy and directionality of structural fluctuations in enzymatic reactions can be documented theoretically or experimentally for specific systems.

A wide range of biological problems involving proteins, not to mention nucleic acids and membrane lipids, are ready for study and exciting new results can be expected as dynamical methods are applied to them. In the coming years, we shall learn how to calculate meaningful rate constants for enzymatic reactions, ligand binding and many of the other biologically important processes mentioned above. The role of flexibility and fluctuations will be understood in much greater detail. It should become possible to determine the effects upon the dynamics of changes in solvent conditions and protein amino acid sequence. As the predictive powers of the theoretical approaches increase, applications will be made to practical problems arising in areas such as genetic engineering and industrial enzyme technology.

Literature Cited

1. Phillips, D. C. 1966. *Sci. Am.* 215:78
2. Dickerson, R. E., Geis, I. 1969. *Structure and Action of Proteins.* New York: Harper & Row
3. Fermi, G., Perutz, M. F. 1981. *Haemoglobin and Myoglobin.* Oxford: Clarendon
4. Richardson, J. A. 1981. *Adv. Protein Chem.* 34:167–339
5. Tanford, C. 1980. *The Hydrophobic Effect,* p. 142. New York: Wiley. 2nd ed.
6. Phillips, D. C. 1981. *Biomolecular Stereodynamics,* ed. R. H. Sarma, pp. 497–98. New York: Adenine
7. Marquart, M., Deisenhofer, J., Huber, R., Palm, W. 1980. *J. Mol. Biol.* 141:369–91
8. Perutz, M. F., Mathews, F. S. 1966. *J. Mol. Biol.* 21:199–202
9. Watson, H. C. 1969. *Prog. Stereochem.* 4:299
10. Williams, R. J. P. 1977. *Angew. Chem. Int. Ed. Engl.* 16:766–77
11. Blundell, T., Wood, S. 1981. *Ann. Rev. Biochem.* 51:123–54
12. Citri, N. 1973. *Adv. Enzymol. Relat. Areas Mol. Biol.* 37:397–648
13. Jencks, W. P. 1975. *Adv. Enzymol. Relat. Areas Mol. Biol.* 43:219–430
14. Koshland, D. E. Jr. 1976. *FEBS Lett.* 62:E47–E52 (Suppl.)
15. Sturtevant, J. M. 1977. *Proc. Natl. Acad. Sci. USA* 74:2236–41
16. Levich, V. G. 1966. *Adv. Electrochem. Electrochem. Eng.* 4:249–57
17. Hopfield, J. J. 1974. *Proc. Natl. Acad. Sci. USA* 71:3640–44
18. Jortner, J. 1976. *J. Chem. Phys.* 64:4860–67
19. Salemme, F. R. 1977. *Ann. Rev. Biochem.* 46:299–329
20. Mendelson, R. A., Morales, M. F., Botts, J. 1973. *Biochemistry* 12:2250–55
21. Harvey, S. C., Cheung, H. C. 1977. *Biochemistry* 16:5181–87
22. Highsmith, S., Kretzschmar, K. M., O'Konski, C. T., Morales, M. F. 1977. *Proc. Natl. Acad. Sci. USA* 74:4986–90
23. Anderson, C. M., Zucker, F. H., Steitz, T. A. 1979. *Science* 204:375–80
24. Johnson, L. N. 1983. In *Inclusion Compounds,* ed. J. L. Atwood, T. E. D. Davies, D. D. MacNicol. New York: Academic
25. Janin, J., Wodak, S. J. 1983. *Prog. Biophys. Mol. Biol.* In press
26. Givol, D. 1976. *Receptors and Recognition* (Ser. A), ed. P. Cuatrecasas, M. F. Greaves. New York: Halsted
27. DeLisi, C. 1976. In *Lectures Notes in Biomathematics,* Vol. 8. New York: Springer
28. Huber, R., Deisenhofer, J., Coleman, P. M., Matsushima, M., Palm, W. 1976. *Nature* 264:415–20
29. Harrison, S. C. 1978. *Trends Biochem. Sci.* 3:3–7

30. McQuarrie, D. A. 1976. *Statistical Mechanics* New York: Harper & Row. 641 pp.
31. Hansen, J. P., McDonald, I. R. 1976. *Theory of Simple Liquids.* New York: Academic
32. Chandler, D. 1974. *Acc. Chem. Res.* 7:246–51
33. Hynes, J. T. 1977. *Ann. Rev. Phys. Chem.* 28:301–21
34. Richards, F. M. 1977. *Ann. Rev. Biophys. Bioeng.* 6:151–76
35. McCammon, J. A., Karplus, M. 1980. *Ann. Rev. Phys. Chem.* 31:29–45
36. Karplus, M., McCammon, J. A. 1981. *CRC Crit. Rev. Biochem.* 9:293–349
37. Karplus, M. 1981. In *Structural Molecular Biology,* ed. D. B. Davis, W. Saenger, S. S. Danyluk, pp. 427–53. New York: Plenum
38. Karplus, M. 1982. *Ber. Bunsenges. Phys. Chem.* 86:386–95
39. McCammon, J. A., Karplus, M. 1983. *Acc. Chem. Res.* In press
39a. Levitt, M. 1982. *Ann Rev. Biophys. Bioeng.* 11:251–71
40. Campbell, I. D., Dobson, C. M., Williams, R. J. P. 1978. *Adv. Chem. Phys.* 39:55–107
41. Peticolas, W. L. 1978. *Methods Enzymol.* 61:425–58
42. Woodward, C. K., Hilton, B. D. 1979. *Ann. Rev. Biophys. Bioeng.* 8:99–127
43. Gurd, F. R. N., Rothgeb, T. M. 1979. *Adv. Protein Chem.* 33:73–165
44. Jardetzky, O. 1981. *Acc. Chem. Res.* 14:291–98
45. Debrunner, P. G., Frauenfelder, H. 1982. *Ann. Rev. Phys. Chem.* 33:283–99
46. Careri, G., Fasella, P., Gratton, E. 1979. *Ann. Rev. Biophys. Bioeng.* 8:69–97
47. Weber, G. 1975. *Adv. Protein Chem.* 29:1–83
48. Cooper, A. 1981. *Sci. Prog. Oxford* 66:473–97
49. Williams, R. J. P. 1979. *Biol. Rev.* 54:389–420
50. Williams, R. J. P. 1980. *Chem. Soc. Rev.* 9:325–64
51. O'Connor, M., ed. 1982. Mobility and Function in Proteins and Nucleic Acids. *Ciba Found. Symp. 93.* London: Pitman
52. Rossmann, M. G., Argos, P. 1981. *Ann. Rev. Biochem.* 50:497–532
53. Kim, P. S., Baldwin, R. L. 1982. *Ann. Rev. Biochem.* 51:459–89
54. Careri, G., Fasella, P., Gratton, E. 1975. *CRC Crit. Rev. Biochem.* 3:141–64

55. Snyder, G. H., Rowan, R., Karplus, M., Sykes, B. D. 1975. *Biochemistry* 14:3765–77
56. Wagner, G., DeMarco, A., Wüthrich, K. 1976. *Biophys. Struct. Mech.* 2:139–58
57. Gelin, B. R., Karplus, M. 1975. *Proc. Natl. Acad. Sci. USA* 72:2002–06
58. Hetzel, R., Wüthrich, K., Deisenhofer, J., Huber, R. 1976. *Biophys. Struct. Mech.* 2:159–80
59. Northrup, S. H., Pear, M. R., Lee, C. Y., McCammon, J. A., Karplus, M. 1982. *Proc. Natl. Acad. Sci. USA* 79:4035–39
60. Campbell, I. D., Dobson, C. M., Moore, G. R., Perkins, S. J., Williams, R. J. P. 1976. *FEBS Lett.* 70:96–98
61. *Structure and Motion in Molecular Liquids,* 1978. Faraday Disc. Chem. Soc., Vol. 66
62. Wilson, E. B., Decius, J. C., Cross, P. C. 1955. *Molecular Vibrations* New York: McGraw-Hill
63. Levy, R. M., Karplus, M. 1979. *Biopolymers* 18:2465–95
64. Warme, P. K., Scheraga, H. A. 1974. *Biochemistry* 13:757–67
65. Levitt, M. 1974. *J. Mol. Biol.* 82:393–420
66. Gelin, B., Karplus, M. 1979. *Biochemistry* 18:1256–68
67. Lifson, S. 1982. In *Structural Molecular Biology,* ed. D. G. Davies, W. Saenger, S. S. Danyluk. New York: Plenum
68. Pauling, L., Corey, R. R., Branson, H. B. 1951. *Proc. Natl. Acad. Sci. USA* 37:205–11
69. Ramachandran, G. N., Ramakrishnan, C., Sasisekharan, V. 1963. *J. Mol. Biol.* 7:95–99
70. Brooks, B., Bruccoleri, R. E., Olafson, B. D., States, D. J., Swaminathan, S., Karplus, M. 1983. *J. Comput. Chem.* In press
71. McCammon, J. A., Gelin, B. R., Karplus, M. 1977. *Nature* 267:585–90
72. McCammon, J. A., Wolynes, P. G., Karplus, M. 1979. *Biochemistry* 18:927–42
73. van Gunsteren, W. F., Karplus, M. 1983. *Macromolecules.* In press
74. Chandrasekhar, S. 1943. *Rev. Mod. Phys.* 15:1–89
75. Ermak, D. L., McCammon, J. A. 1978. *J. Chem. Phys.* 69:1352–60
76. Levy, R. M., Karplus, M., McCammon, J. A. 1979. *Chem. Phys. Lett.* 65:4–11
77. Helfand, E., Wasserman, Z. R., Weber, T. A. 1980. *Macromolecules* 13:526–33
78. McCammon, J. A., Northrup, S. H.,

Karplus, M., Levy, R. M. 1980. *Biopolymers* 19:2033–45

79. Gō, M., Gō, N. 1976. *Biopolymers* 15:1119–27

80. Suezaki, Y., Gō, N. 1976. *Biopolymers* 15:2137–53

81. Levy, R. M., Perahia, D., Karplus, M. 1982. *Proc. Natl. Acad. Sci. USA* 79:1346–50

82. Chandler, D. 1978. *J. Chem. Phys.* 68:2959–70

83. Pechukas, P. 1976. In *Dynamics of Molecular Collisions*, Part B. ed. W. H. Miller. New York: Plenum

84. Keck, J. C. 1962. *Discuss. Faraday Soc.* 33:173–82

85. McCammon, J. A., Karplus, M. 1979. *Proc. Natl. Acad. Sci. USA* 76:3585–89

86. McCammon, J. A., Karplus, M. 1980. *Biopolymers* 19:1375–405

87. van Gunsteren, W. F., Berendsen, M. J. C. 1977. *Mol. Phys.* 34:1311–27

88. McCammon, J. A., Gelin, B. R., Karplus, M., Wolynes, P. G. 1976. *Nature* 262:325–26

89. McCammon, J. A., Wolynes, P. G. 1977. *J. Chem. Phys.* 66:1452–56

90. Karplus, M., Weaver, D. L. 1976. *Nature* 260:404–06

91. Karplus, M., Weaver, D. L. 1979. *Biopolymers* 18:1421–37

92. Miyazawa, T. 1967. In *Poly-α-Amino Acids*, ed. G. D. Fasman. New York: Marcel Dekker

93. Gō, N., Scheraga, H. A. 1976. *Macromolecules* 9:535–42

94. Bunker, D. L., Wang, F.-M. 1977. *J. Am. Chem. Soc.* 99:7457–59

95. Warshel, A., Karplus, M. 1974. *J. Am. Chem. Soc.* 96:5677–89

96. Case, D. A., Karplus, M. 1979. *J. Mol. Biol.* 132:343–68

97. Flory, P. J. 1969. *Statistical Mechanics of Chain Molecules*. New York: Wiley

98. Levitt, M. 1976. *J. Mol. Biol.* 104:59–107

99. Levitt, M. 1976. *Models for Protein Dynamics.* CECAM Workshop Report. Univ. Paris XI, Orsay, France

100. Pear, M. R., Northrup, S. H., McCammon, J. A., Karplus, M., Levy, R. M. 1981. *Biopolymers* 20:629–32

101. Northrup, S. H., Pear, M. R., McCammon, J. A., Karplus, M., Takano, T. 1980. *Nature* 287:659–60

102. Frauenfelder, H., Petsko, G. A., Tsernoglou, D. 1979. *Nature* 280:558–63

103. Sternberg, M. J. E., Grace, D. E. P., Phillips, D. C. 1979. *J. Mol. Biol.* 130:231–45

104. Artymiuk, P. J., Blake, C. C. F., Grace, D. E. P., Oatley, S. J., Phillips, D. C.,

Sternberg, M. J. E. 1979. *Nature* 280:563–68

105. Takano, T., Dickerson, R. E. 1980. *Proc. Natl. Acad. Sci. USA* 77:6371–75

106. Olafson, B., Ichiye, T., Karplus, M. 1983. *J. Mol. Biol.* Manuscript in preparation

107. van Gunsteren, W. F., Karplus, M. 1981. *Nature* 293:677–78

108. Kuryan, J., Swaminathan, S., Petsko, G., Karplus, M., Levy, R. M. 1983. *Biochemistry*. Manuscript in preparation

109. Aschaffenburg, R., Blake, C. C. F., Dickie, H. M., Gayen, S. K., Keegan, R., Sen, A. 1980. *Biochem. Biophys. Acta* 625:64–71

110. van Gunsteren, W. F., Karplus, M. 1982. *Biochemistry* 21:2259–74

111. van Gunsteren, W. F., Berendsen, M. J. C., Hermans, J., Hol, W. G. J., Postma, J. P. M. 1983. *Proc. Natl. Acad. Sci. USA* In press

112. Levitt, M. 1980. In *Protein Folding*, ed. R. Jaenicke, pp. 17–39. New York: Elsevier/North Holland

113. Watenpaugh, K. D., Margulis, T. N., Sieker, L. C., Jensen, L. H. 1978. *J. Mol. Biol.* 122:175–90

114. Karplus, M., McCammon, J. A. 1979. *Nature* 277:578

115. Northrup, S. H., Pear, M. R., Morgan, J. D., McCammon, J. A., Karplus, M. 1981. *J. Mol. Biol.* 153:1087–1109

116. Morgan, J. D., McCammon, J. A., Northrup, S. H. 1983. *Biopolymers*. In press

117. Karplus, M., Swaminathan, S., Ichiye, T., van Gunsteren, W. F. 1982. See Ref. 51

118. Swaminathan, S., Ichiye, T., van Gunsteren, W. F., Karplus, M. 1982. *Biochemistry* 21:5230–41

119. Mao, B., Pear, M. R., McCammon, J. A., Northrup, S. H. 1982. *Biopolymers.* 21:1979–89

120. Noguti, T., Gō, N. 1982. *Nature* 296:776–78

121. Levy, R. M., Karplus, M., McCammon, J. A. 1981. *J. Am. Chem. Soc.* 103:994–96

122. Levy, R. M., Karplus, M., Wolynes, P. G. 1981. *J. Am. Chem. Soc.* 103:5998–6011

123. Hoch, J. C., Dobson, C. M., Karplus, M. 1982. *Biochemistry* 21:1115–25

124. Levitt, M. 1981. *Nature* 294:379–80

125. Karplus, M., Kushick, J. N. 1981. *Macromolecules* 14:325–32

126. Parak, F., Frolov, E. N., Mössbauer, R. L., Goldanskii, V. I. 1981. *J. Mol. Biol.* 145:825–33

127. Levy, R. M., Szabo, A. 1982. *J. Am. Chem. Soc.* 104:2073–207
128. Kasprzak, A., Weber, G. 1982. *Biochemistry* 21: In press
129. Lakowicz, J. R., Maliwal, B., Cherek, H., Balter, A. 1983. *Biochemistry.* In press
130. Ichiye, T., Karplus, M. 1983. *Biochemistry.* Submitted for publication
131. Northrup, S. H., Hynes, J. T. 1980. *J. Chem. Phys.* 73:2700–14
132. Karplus, M., McCammon, J. A. 1981. *FEBS Lett.* 131:34–36
133. Takano, T. 1977. *J. Mol. Biol.* 110:569–84
134. Phillips, S. E. 1978. *Nature* 273:247–48
135. Austin, R. H., Beeson, K. W., Eisenstein, L., Frauenfelder, H., Gunsalus, I. C. 1975. *Biochemistry* 14:5355–73
136. Beece, D., Eisenstein, L., Frauenfelder, H., Good, D., Marden, M. C., et al. 1980. *Biochemistry* 19:5147–57
137. McCammon, J. A., Northrup, S. H. 1981. *Nature* 293:316–17
138. Northrup, S. H., Zarrin, F., McCammon, J. A. 1982. *J. Phys. Chem.* 86:2314–21
139. Szabo, A., Shoup, D., Northrup, S. H., McCammon, J. A. 1982. *J. Chem. Phys.* 77:4484–93
140. Hartsuck, J. A., Lipscomb, W. N. 1971. *Enzymes* 3:1–56
141. Banner, D. W., Bloomer, A. C., Petsko, G. A., Phillips, D. C., Pogson, D. I., et al. 1975. *Nature* 255:609–14
142. Pratt, L. R., Chandler, D. 1977. *J. Chem. Phys.* 67:3683–704
143. Imoto, T., Johnson, J. N., North, A. C. T., Phillips, D. C., Rupley, J. A. 1972. *Enzymes* 7:665
144. Wolynes, P. G., McCammon, J. A. 1977. *Macromolecules* 10:86–87

145. McCammon, J. A., Karplus, M. 1977. *Nature* 268:765–66
146. Mao, B., Pear, M. R., McCammon, J. A., Quiocho, F. A. 1982. *J. Biol. Chem.* 257:1131–33
147. Colonas, F., Perahia, D., Karplus, M., Ecklund, H., Bränden, C. I. 1983. *J. Mol. Biol.* In preparation
148. Newcomer, M. E., Lewis, B. A., Quiocho, F. A. 1981. *J. Biol. Chem.* 256:13218–22
149. Ecklund, M., Nordström, B., Zeppezauer, E., Söderlund, G., Ohlsson, I., et al. 1976. *J. Mol. Biol.* 102:27–59
150. Ecklund, H., Samama, J. P., Wallén, L., Bränden, C.-I., Åkeson, Å., Alwyn Jones, T. 1981. *J. Mol. Biol.* 146:561–87
151. Harrison, S. C. 1980. *Biophys. J.* 32:139–51
152. Perutz, M. F. 1979. *Ann. Rev. Biochem.* 48:327–86
153. Perutz, M. F. 1970. *Nature* 228:726–34
154. Baldwin, J., Chothia, C. 1979. *J. Mol. Biol.* 129:175–220
155. Gelin, B. R., Karplus, M. 1977. *Proc. Natl. Acad. Sci. USA* 74:801–05
156. Gelin, B. R., Lee, A. W.-M., Karplus, M. 1983. *J. Mol. Biol.* In press
157. Anderson, L. 1973. *J. Mol. Biol.* 79:495–506
158. McDonald, R. C., Steitz, T. A., Engelman, D. M. 1979. *Biochemistry* 18:338–42
159. Pickover, C. A., McKay, D. B., Engelman, D. M., Steitz, T. A. 1979. *J. Biol. Chem.* 254:11323–29
160. Hanson, D. C., Yguerabide, J., Schumaker, V. N. 1981. *Biochemistry* 20:6842–52
161. Hartmann, H., Parak, F., Steigemann, W., Petsko, G. A., Runge Ponti, D., Frauenfelder, H. 1982. *Proc. Natl. Acad. Sci. USA* 79:4967–71

Ann. Rev. Biochem. 1984. 53:595–623
Copyright © 1984 by Annual Reviews Inc. All rights reserved

7

THREE-DIMENSIONAL STRUCTURE OF MEMBRANE AND SURFACE PROTEINS

David Eisenberg

Molecular Biology Institute and Department of Chemistry and Biochemistry, University of California, Los Angeles, California 90024

CONTENTS

PERSPECTIVES AND SUMMARY

Far less is known about the structures of membrane-related proteins than of soluble proteins. In large part this is because X-ray crystallography and other biophysical tools are not so easily applied to membrane-related proteins. Because of our lack of knowledge, models for the functions of membrane proteins—active and passive transport of molecules into and out of cells and organelles; transduction of energy among light, electrical, and chemical forms; and reception and transduction of chemical signals across membranes—are not so advanced as models for the various

207

biological processes that proteins mediate in aqueous surroundings. We know the details of atomic arrangement in many enzyme-substrate complexes, but we know little of the way that membrane proteins bind and respond to ligands or even of how and why they fold into membranes.

Bacteriorhodopsin (the purple membrane protein) is the only proper membrane protein whose three-dimensional structure is known even to moderate resolution. Even for this protein, specific amino acids have not yet been assigned with certainty to features of the three-dimensional model, although several plausible proposals have been put forward. Most of these models identify seven α helices running roughly perpendicular to the plane of the membrane.

The difficulty in determining structures of membrane proteins has been in growing the crystals needed for diffraction studies. Fortunately, there has been recent progress. For at least some membrane proteins it is possible to replace membrane lipids with small detergents or other amphiphiles, and then to render the protein-amphiphile complex insoluble by addition of a precipitant. This has resulted in well-ordered three-dimensional membrane-protein crystals of porin and a photosynthetic reaction center; there is every reason to believe that the technique is quite general. Because such crystals contain aqueous channels, determination of structures by the standard method of isomorphous replacement should be possible.

Structures are known for several small peptides that bind to, or insert into, membranes. These include alamethicin, which associates with itself to form an ion-conductive channel in membranes, and melittin, which binds to and breaks up membranes. Both are bent α helices. The melittin helix is highly amphiphilic: the residues on one side are largely hydrophobic and the residues on the other are largely hydrophilic. This may suggest that this protein has evolved to bind to the surface of membranes. Several other small, lytic peptides have amino acid sequences homologous to that of melittin, suggesting that their action may be related to their tendency to form amphiphilic helices.

In the near absence of information on the three-dimensional structures of membrane-related proteins, an abundance of information on their amino acid sequences has been gained from rapid DNA-sequencing methods. This has tempted investigators to infer various aspects of three-dimensional structure from sequences. Many of the sequences contain continuous segments of relatively apolar residues. These segments are in many cases about the length (20–25 amino acid residues) that might be expected for a transmembrane α helix. It is plausible that many of these apolar sequences are in fact α helices, but this is unproven. There is no reason to believe that other types of protein secondary structure, such as barrels of β strands, are not found in some membrane proteins.

Amino acid sequences of membrane-related proteins can also be analyzed for segments of high amphiphilicity. A measure of amphiphilicity is the hydrophobic moment, and it is found that melittin and related peptides have unusually large hydrophobic moments. Some segments of membrane proteins are also capable of forming relatively amphiphilic structures, and these may be parts of aqueous membrane channels.

Many techniques have contributed to our understanding of the structure of membrane-related proteins. This review focuses on two of those techniques: (a) diffraction studies at the molecular level, and (b) correlations and predictions from amino acid sequences of aspects of the three-dimensional structure.

DIFFRACTION STUDIES OF MEMBRANE-RELATED PROTEINS

Specimen Preparation

Diffraction methods require ordered specimens—crystals. Michel (1) has suggested that three-dimensional crystals of membrane proteins are of two types (Figure 1). Type I crystals resemble stacked membranes. Lipids surround the proteins in a lipid-bilayerlike fashion. The apolar regions of the lipid bind to hydrophobic portions of the protein, stabilizing a two-dimensional network (as in a horizontal row in the top panel of Figure 1). These networks then stack on each other to form a crystal. Polar interactions are important only in the third dimension. To date several examples of Type I crystals have been grown, but all too small for high-resolution X-ray diffraction studies. They are mitochondrial cytochrome oxidase (2), light-harvesting chlorophyll a/b protein (3), and purple membrane protein (bacteriorhodopsin) (4). The difficulty in growing these crystals is presumably that they contain two types of interprotein contacts: polar in the stacking direction and hydrophobic mediated by lipid in the membranelike layers. Michel (1) notes that it is difficult to change simultaneously the strength of both types of interactions, as may be necessary for crystal growth.

In Type II crystals, the apolar ends of detergents bind to membrane proteins, compensating their hydrophobic surfaces that once bound in membranes. Then the detergent-protein complex is induced to crystallize by the addition of a precipitant, such as a salt or polyethylene glycol. In Type II crystals it is the polar parts of proteins that interact, forming a three-dimensional structure (Figure 1, lower panel).

Thus, Type II crystals are akin to crystals of globular proteins, and several of diffraction quality have been grown. Bacteriorhodopsin (5) and porin (6) have been crystallized, using β-D-octylglucopyranoside. In the

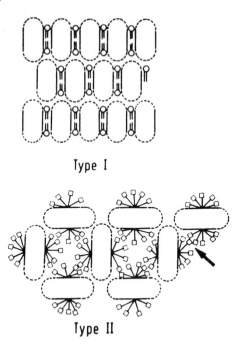

Type I

Type II

Figure 1   The two basic types of crystals of membrane proteins, according to Michel (1), with the hydrophilic surface areas of proteins shown by dashed lines. Type I crystals resemble stacked membranes (shown here stacked in horizontal layers). The contacts between the layers are polar, whereas within a layer they are mainly hydrophobic, and mediated by lipid. In Type II crystals, the apolar ends of detergents bind to the hydrophobic patches on the membrane proteins, where lipids once bound. The polar ends of the detergents form the interparticle contacts, as at the arrow. Reproduced from (1).

case of bacteriorhodopsin, the use of radiolabeled detergent indicates about 30 detergent molecules per protein (7).

Small amphiphiles may be even more effective than conventional detergents in forming Type II crystals, because they are smaller and tend less to disturb the crystalline packing (1). X-ray–grade crystals of the photosynthetic reaction center were grown using heptane-1,2,3-triol and triethylammonium phosphate (8). The proper choice of detergent, precipitant, and possible other components is still very much a matter for empirical study (9).

For structural studies by electron microscopy, two-dimensional crystals are suitable. Sometimes annealing of isolated membranes is sufficient to produce crystalline arrays, as it was for membrane-bound acetylcholine receptor (10). For mitochondrial ubiquinol : cytochrome c reductase, both single and double layer membrane crystals were grown by mixing the

enzyme-Triton complex with phospholipid-Triton micelles and subsequently removing the Triton (11).

Diffraction Studies of Membrane Proteins

PURPLE MEMBRANE PROTEIN (BACTERIORHODOPSIN)   This protein remains
the membrane protein about which we have the most detailed knowledge of
structure (12). The pioneering study of Henderson & Unwin (13) combined
electron diffraction and low-dose electron microscopy to establish the basic
architecture of the molecule in a trigonal crystal. More recently, similar
studies of an orthorhombic crystal (14) have provided an independent
confirmation of the basic structure (15). In both structures, seven regions of
density run roughly parallel to each other, in the direction perpendicular to
the plane of the membrane. These seven rods were postulated to be α helices
based on (a) the suggestion from high-angle X-ray diffraction (16) of four to
seven α helices in each bacteriorhodopsin molecule, (b) the amino acid
sequence (17, 18) that contains seven relatively hydrophobic segments that
are of the proper length to bridge the hydrophobic portion of the bilayer
(about 30–35 Å) when coiled as α helices, and (c) the shapes of the rods.

A comparison of the models of bacteriorhodopsin derived from the
trigonal and orthorhombic crystals is shown in Figure 2. Rods labeled 3, 4,
5, 6, and 7 have similar shapes in the two structures, although some of the
rods in the orthorhombic map may be slightly longer. Rods 1 and 2 of this

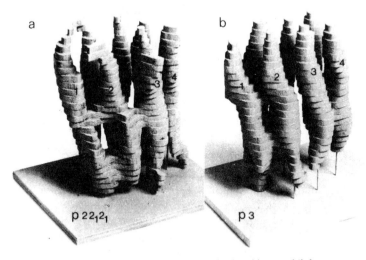

Figure 2   Bacteriorhodopsin models: Newer orthorhombic crystal (*left*, space group P22$_1$2$_1$)
and older trigonal crystal (*right*, space group P3), showing the seven bent rods. Both models
are skeletal in the sense that they contain only 36% of the volume expected for a protein of M_r
26,000. The numbering of helices is from (19). Reproduced from (15).

map are more kinked toward the center of the membrane and display two close connections. These features may reflect better phases in the ortho-rhombic map, or conceivably can be noise or differences in the specimens (15). In the orthorhombic map is the suggestion of a connection at the tops of rods 1 and 7. Other connections of the polypeptide chain are uncertain from the maps.

Electron diffraction studies are not yet of sufficient resolution to reveal which amino acids occupy each rod of density, or even the order in which the rods are linked. This tantalizing situation has led investigators to use other evidence to formulate models (19–22). For example, Engelman et al (20) used data from proteolytic cleavage and energetic considerations to identify possible helical segments in the structure. In proposing another model, Agard & Stroud (21) attempted to correct for the missing conical region of diffraction data by computational processing, and they produced a new map with longer helices (47 Å vs 35 Å), and increased connectivity among the rods. Another model (22), based in part on data from ultraviolet circular dichroism and infrared spectroscopy, suggests five α helices and four strands of β sheet, rather than the seven α helices more commonly accepted. Neutron diffraction studies have led to proposals for the angular orientation of some of the α helices (23) and for the location of the chromophore (24). Also, chemical studies (25, 26) have identified the residue (lysine 216) to which the retinylidene chromophore is bound.

OTHER PROTEINS   Table 1 summarizes several other studies of membrane proteins by diffraction methods. They provide interesting information on molecular shapes and possible subunit locations, but not as yet details of the paths of the polypeptide chains. Recent reports (6, 8) of large X-ray diffraction–grade crystals foreshadow far more detailed models for porin and the photosynthetic reaction center from *Rhodopseudomonas viridis*.

Structure of Alamethicin and a Proposal for an Ion Channel

Alamethicin is a 20-residue polypeptide antibiotic that inserts into black lipid membranes and forms voltage-gated ion channels (30). Crystals of alamethicin were grown from methanol and acetonitrile by Fox & Richards (31) and the atomic structure determined and refined at 1.5 Å resolution. The asymmetric unit of the crystal consists of three alamethicin molecules, each mainly an α helix slightly bent at proline residue 14, with small variations in the pattern of hydrogen bonding among the three helices. The crystal contains a complex array of irregular channels that run in the direction of the three parallel helices. However, the helices do not pack as a symmetric trimer, nor do their polar side chains segregate toward the center of one of the irregular channels.

Table 1 Some three-dimensional diffraction studies of membrane proteins

Protein	Method	Resolution in Å	Molecular shape	Reference
Purple membrane trigonal	Electron microscopy/ electron diffraction	7 in membrane plane, 14 perpendicular	Seven rods traversing membrane, roughly perpendicularly	Henderson & Unwin (13)
Purple membrane orthorhombic	Electron microscopy/ electron diffraction	6.5 in plane, 12 perpendicular	Confirms trigonal structure	Leifer & Henderson (15)
Cytochrome c oxidase	Electron microscopy/ image processing	20	Y-shaped molecule protruding from membrane over 50 Å into solution	Deatherage et al (27)
Acetylcholine receptor	Electron microscopy/ X-ray diffraction		Funnel-shaped molecule 110 Å long with a central channel	Kistler et al (28)
Ubiquinol: cytochrome c reductase	Electron microscopy/ image processing	25–35	H-shaped dimer with two-fold axis perpendicular to membrane	Leonard et al (29)

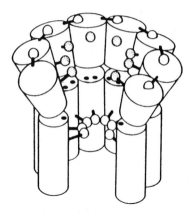

Figure 3   A model proposed by Fox & Richards (31) for a circular membrane pore formed from alamethicin molecules. Each 20-residue helix is represented as a cylinder, broken at residue Pro 14. The lower portions of the cylinders are formed from the N termini of the peptides and inserted perpendicularly into a bilayer. The spheres and dots are polar groups, expected to be in contact with water, as explained in the text. Reproduced from (31).

On the basis of the crystal structure of the monomer and plausible intermolecular contacts, Fox & Richards (31) proposed a model for an ion channel. They assume that there is *n*-fold symmetry in the channel about the normal to membrane plane, that the N termini of all alamethicin chains are on one side of the membrane and all C termini are on the other, and that interchain hydrogen bonding and side-chain packing are maximized. A portion of such a pore is shown schematically in Figure 3. Each alamethicin monomer is represented as two cylindrical α-helical segments interrupted by the proline residue 14, with the shorter C-terminal segments pointing up. The stippled spheres at the mouth of the pore represent Glu 18, the only charged side chains in the molecule, and the open spheres indicate proposed hydrogen-bonded rings formed by Gln 7 and Gln 19. Two solvent-accessible carbonyl groups are shown by black dots. The authors also proposed a structure for the oligomeric pore in the absence of an applied voltage, involving less complete penetration of the C termini of the helices into the membrane.

Structure of Melittin, a Surface-Seeking Protein

Melittin from bee venom is water soluble, yet integrates into membranes and lyses cells, and also forms surface monolayers at air-water interfaces. Each melittin chain consists of 26 amino acid residues, and in aqueous salt solutions it exists as a tetramer (32). The tetramer can be crystallized by salting out, and its structure was determined in two crystal forms to 2.0 and 2.5 Å resolution, respectively by Terwilliger, Weissman, and Eisenberg (33–

36). In both crystal forms the melittin polypeptide is a bent α-helical rod, with an "inner" surface consisting largely of hydrophobic side chains and an "outer" surface consisting of hydrophilic side chains. This asymmetry of hydrophobicity is often called *amphiphilicity*. In the tetramer, the four bent amphiphilic helices contribute their inner apolar side chains to a hydrophobic interaction in the molecular center, reminiscent of a micelle. In both crystal forms, the tetramers are packed in layers, producing alternating hydrophobic and hydrophilic levels, making the crystal a packed set of parallel protein bilayers.

The highly amphiphilic structure of the melittin helix suggests that it seeks the air-water interface or a membrane surface because at such a phase boundary it can expose its hydrophilic face to the aqueous phase and still bury its largely hydrophobic face in the apolar phase. In fact, CD and NMR experiments (37, 38) indicate that in detergent and in membranes the backbone conformation of melittin is similar to that in the tetramer. Thus, the amphiphilic helix seen in the two crystal forms may be the conformation of the protein at the membrane surface, perhaps at the air-water interface (36).

A model for melittin at the surface of a bilayer is shown in Figure 4. The melittin helix is shown at the left, oriented so that its hydrophilic side chains extend upward toward the aqueous phase, and its hydrophobic side chains

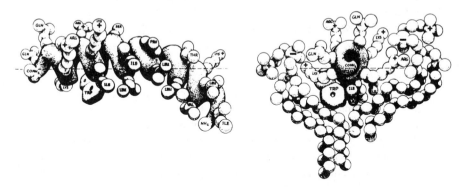

Figure 4   A model for melittin binding at a surface. On the left, the amphiphilic α helix is shown schematically at a surface (the dashed line) between an aqueous phase (above) and a hydrophobic phase (below). Melittin is oriented so that its hydrophilic side chains extend upward toward the aqueous phase, and its hydrophobic side chains extend downward into lipid. The structure shown for this monomer is identical to that found for the monomer in the crystalline tetramer, except that the side chains of Lys 23 and Gln 26 have been bent so that they do not extend into the lipid. In the view to the right, two phosphatidylcholine molecules have been added, showing polar contacts between their head groups with polar side chains on melittin and apolar contacts between the hydrocarbon tails and hydrophobic side chains of melittin. Reproduced from (30).

extend downward. Hypothetical interactions between two phosphatidylcholine molecules and the C terminus of the melittin helix are shown at the right. Conceivably the lytic effect of melittin stems from its limited penetration into the upper leaflet of the bilayer, which creates a kink as shown in the figure.

In the presence of an applied electric field, melittin has been observed to permit voltage-dependent ionic conductance across black lipid membranes (39, 40). This shows that melittin is able to rearrange, in the presence of a field, into a pore or channel or in some other way to transport charge across the membrane. Several kinks such as those of Figure 4 could join to form a pore with hydrophilic interior.

Three-Dimensional Structures of Other Membrane-Related Proteins

BACTERIOCHLOROPHYLL PROTEIN AS A MODEL FOR PROTEIN-LIPID INTERACTIONS   Matthews (41) has analyzed lipid conformation, lipid packing, and lipid-protein interactions in the structure of a bacteriochlorophyll protein, with the structure known to 2.7 Å resolution (42, 43). This molecule of M_r 150,000 consists of three identical subunits, each formed from 15 strands of twisted β sheet. This β-sheet subunit is exposed on the outside to solvent and encloses a central core of seven bacteriochlorophyll a molecules. Each bacteriochlorophyll consists of a chlorin ring and a phytol side chain.

In a detailed analysis of the conformations and interactions of the phytol side chains seen in the electron density map, Matthews (41) concludes that they adopt more or less irregular conformations. Sometimes they are extended, but by no means in an idealized all-*trans* conformation. The main influence seems to be the shape of the protein, to which the phytol chains adapt their conformations. Where the hydrocarbon tails do interact, they lie parallel to each other, but the phytol-phytol interactions do not seem to dominate over phytol-protein interactions to the extent that they determine the structure. In the most obvious example of phytol-protein interactions, three phytol chains lie next to each other and all against a β sheet, perpendicular to the polypeptide strands. While it is unknown how general such an interaction may be in membrane proteins, it certainly demonstrates that α helices are not the only type of secondary structure to be expected to interact with lipids.

PHOSPHOLIPASE A$_2$, A LIPID INTERFACIAL ENZYME   Three-dimensional structures are known for phospholipase A$_2$ from both cow pancreas (44, 45) and snake (46); a recent review (47) focuses on the pancreatic enzyme as a possible model for membrane-bound enzymes. Phospholipase A$_2$ catalyzes the hydrolysis of the 2-acyl ester bond of 3-*sn*-phosphoglycerides, and is most active when the substrate is at an organized lipid-water interface. Volwerk & de Haas (47) believe that the mechanism of action of the enzyme on lipid-water interfaces can be described in terms of a binding of the enzyme to the organized lipid-water interface, followed by binding of a single substrate at the active site and then catalysis. From chemical modification studies, the interfacial recognition is known to involve the N terminus, which is hydrogen bonded and buried in the vicinity of the active site. Further crystallographic and spectroscopic studies of the enzyme with a modified N terminus suggest that the function of the bound N terminus in the native enzyme is to constrain the movement of the N-terminal chain (48). In transaminated enzymes, lacking the free N terminus, the chain is more mobile, and the capacity to bind to micellar substrates is lost. Perhaps the inability to bind comes from the entropy decrease that must accompany the binding of the more mobile N terminus.

YOLK LIPOPROTEIN COMPLEX   Banaszak & co-workers (49, 50) have succeeded in determining a low resolution (about 25 Å) picture of this complex by coordinated electron microscopy and powder X-ray diffraction. The complex consists of eight polypeptide chains and about 104 lipid molecules, and contains a two-fold axis of symmetry. The overall shape is that of a flattened ellipsoid with approximate dimensions of 55 × 115 × 250 Å. Tentative assignments of the various polypeptide chains have been made to regions of the model, and a low-density region has been assigned to lipid in a micellar arrangement. Banaszak et al (49) note that this structure is completely different from the quasi-spherical micellar models suggested by others for plasma lipoproteins.

Other Models for Membrane Proteins

In the past, the near void of experimental information on the three-dimensional structures of membrane-related proteins has tempted authors to put forward proposals for types of polypeptide structures to be expected in membrane proteins. Aside from their intrinsic interest, such proposals may also be of use to those attempting to decipher low-to-moderate resolution electron-density maps, and to those attempting to think of amino acid sequences in terms of the three-dimensional structures they form.

From model building and spectroscopic evidence, Urry (51, 52) has proposed a so-called β helix for the structure of the transmembrane channel formed by gramicidin A. This is a twisted, internally hydrogen bonded strand of polypeptide, in which the hydrogen-bonding pattern resembles that between parallel chains of a β sheet, and which to form requires alternating D and L amino acids. Urry (53) has considered variations of this idea for other types of membrane channel, and Kennedy (54) has also discussed β structures for membrane proteins. Preliminary X-ray diffraction studies of gramicidin A crystals have been reported (55, 56) that are compatible with any of several types of β helix.

Many proposals for α-helices in membrane proteins have been formulated. Dunker & Zaleske (57) studied the stereochemical constraints in bundles of 4 to 14 coiled α helices, which they suggest as models for membrane proteins. This work is extended in (58). Engelman & Steitz (59) have focused on the initial event in the secretion of proteins across membranes, which they propose is the penetration of the membrane by a helical hairpin of the peptide chain.

ANALYSIS OF AMINO ACID SEQUENCES OF MEMBRANE-RELATED PROTEINS

Hydrophobicities

In contrast to our very limited knowledge of the three-dimensional structures of membrane proteins, knowledge of their amino acid sequences is absolutely burgeoning, thanks to gene sequencing. This situation has prompted investigators to develop hypotheses concerning which segments of the sequence penetrate the membrane, as well as other aspects of membrane protein structure. Central to these inferences is the notion of a *hydrophobicity scale* for amino acid residues—a measure of their relative affinities for hydrophobic phases.

No generally accepted method exists to measure or calculate hydrophobicities; each method used constitutes a separate operational definition. Attempts to compare and to utilize different scales can be confusing, because different scales often reflect different properties of amino acids. Of course, it would be unrealistic to expect that all aspects of the interaction of a residue with water can be summarized in a single number.

A recent discussion of hydrophobicity scales has been given by Edsall & McKenzie (60) in the context of a detailed review of the interaction of water with proteins. Table 2 lists some of the hydrophobicity scales compared in their review as well as other recent scales. The scale of Wolfenden et al (61) is the most closely linked to experiment. These authors have recorded the Gibbs free energy of transfer from dilute aqueous solution to the vapor of substances of the class RH, where R represents an amino acid side chain. For example, RH for glycine is H_2. This procedure yields a direct measure of the free energy for transfer from water to a hydrophobic phase, but of course the measure does not reflect other interactions of the side chain, including those that link it covalently to the rest of the protein. The scale correlates well with other measures of hydrophobicity, particularly the observed distribution of amino acid side chains between the surface and interior of the protein (62, 63), with the exception of glycine, which by this scale is considered the most hydrophobic of all the residues. The well-known scale of Nozaki & Tanford (64) is based on free energy of transfer of amino acids from water to ethanol, but did not include values for all amino acid residues. Values for some other residues found in transmembrane sequences were suggested by Segrest & Feldman (65). The dependence of this scale on the choice of ethanol as the hydrophobic phase may produce too hydrophobic a value for tryptophan. The scale of Janin (62) is based on the fraction of each type of residue that is found buried in globular proteins, ascribing the largest hydrophobicities to those residues that are most commonly buried away from solvent. This algorithm probably over-emphasizes the hydrophobicity of Cys, which frequently forms disulfide linkages in the protein interior. The theoretical scale of von Heijne & Blomberg (66) was designed to describe three energetic effects of the transfer of amino acid side chains to a hydrophobic from a hydrophilic phase. These are the covering of hydrophobic surface area, hydrogen-bond breakage, and charge neutralization.

The consensus scale of Eisenberg et al (67) was designed to mitigate the effects of outlying values in any one scale, produced by the peculiarities of the method, and is simply an average of the four scales to the right of it in Table 2. Another combined scale was formulated by Kyte & Doolittle (68) who amalgamated values from water-to-vapor transfers (61) and from internal-external distribution of amino acid residues (63), and who "did not hesitate to adjust the values subjectively" in the second decimal place. Kyte & Doolittle also offer some illuminating comments on values from experiments on free energy of transfer. A "membrane-buried preference" scale was derived by Argos et al (69) from the relative frequencies of 1125 amino acids found in protein segments judged to be within membranes. These 1125 residues were taken from the seven helical spans in one model (19) for bacteriorhodopsin as well as from signal sequence segments delineated by von Heijne (70), and from other models for membrane-bound proteins. The scale may depend strongly on the selections of this data base.

One other scale is in Table 3, the OMH (optimal matching hydropho-bicity) scale (71). This scale was derived on the assumption that families of proteins that fold in the same way (such as the globins), do so because they have the same pattern of residue hydrophobicities along their amino acid sequences. The OMH hydrophobicities are the set of residue values that yield the best numerical match for 1572 amino acid replacements among closely related proteins.

Comparison of the various scales in Table 2 is inconvenient, in part

Table 2 Hydrophobicity scales for amino acid residues[a]

Amino acid	Consensus (67)	von Heijne (66)	Janin (62)	Chothia (63)	Wolfenden (61)	Tanford/Segrest (64, 65)	Kyte (68)	Argos (69)
Ile	0.73	4.4	0.7	0.24	2.15	5.0	4.5	1.67
Phe	0.61	5.2	0.5	0.0	−0.76	5.0	2.8	2.03
Val	0.54	3.9	0.6	0.09	1.99	3.0	4.2	1.14
Leu	0.53	4.2	0.5	−0.12	2.28	3.5	3.8	2.93
Trp	0.37	3.9	0.3	−0.59	−5.88	6.5	−0.9	1.08
Met	0.26	2.1	0.4	−0.24	−1.48	2.5	1.9	2.96
Ala	0.25	2.9	0.3	−0.29	1.94	1.0	1.8	1.56
Gly	0.16	1.9	0.3	−0.34	2.39	0.0	−0.4	0.62
Cys	0.04	−0.08	0.9	0.0	−1.24	0.0	2.5	1.23
Tyr	0.02	3.6	−0.4	−1.02	−6.11	4.5	−1.3	0.68
Pro	−0.07	1.1	−0.3	−0.90	—	1.5	−1.6	0.76
Thr	−0.18	1.2	−0.2	−0.71	−4.88	0.5	−0.7	0.91
Ser	−0.26	0.36	−0.1	−0.75	−5.06	−0.5	−0.8	0.81
His	−0.40	−1.5	−0.1	−0.94	−10.27	1.0	−3.2	0.29
Glu	−0.62	−4.0	−0.7	−0.90	−10.20	—	−3.5	0.23
Asn	−0.64	−1.0	−0.5	−1.18	−9.68	−1.5	−3.5	0.27
Gln	−0.69	−0.52	−0.7	−1.53	−9.38	−1.0	−3.5	0.51
Asp	−0.72	−5.6	−0.6	−1.02	−10.95	—	−3.5	0.14
Lys	−1.1	−2.3	−1.8	−2.05	−9.52	—	−3.9	0.15
Arg	−1.8	−9.4	−1.4	−2.71	−19.92	—	−4.5	0.45

[a] The order is by decreasing hydrophobicity on the consensus scale. The magnitudes for all but the two scales on the right may be considered roughly in kcal mol^{-1} for transfer from a hydrophobic to a hydrophilic phase. As discussed in the text, the scales do not all measure the same property.

Table 3 Normalized hydrophobicity scales[a]

Amino acid	Consensus (67)	Janin (62)	Kyte (68)	Argos (69)	OMH (71)
Ile	1.4	1.2	1.7	0.77	1.2
Phe	1.2	0.87	1.1	1.2	1.9
Val	1.1	1.0	1.6	0.14	0.91
Leu	1.1	0.87	1.4	2.3	1.2
Trp	0.81	0.59	-0.14	0.07	0.50
Met	0.64	0.73	0.80	2.3	1.0
Ala	0.62	0.59	0.77	0.64	-0.40
Gly	0.48	0.59	0.03	-0.48	-0.67
Cys	0.29	1.4	1.0	0.25	0.17
Tyr	0.26	-0.40	-0.27	-0.41	1.7
Pro	0.12	-0.26	-0.37	-0.31	-0.49
Thr	-0.05	-0.12	-0.07	-0.13	-0.28
Ser	-0.18	0.02	-0.10	-0.25	-0.55
His	-0.40	0.02	-0.91	-0.87	-0.64
Glu	-0.74	-0.83	-1.0	-0.94	-1.2
Asn	-0.78	-0.55	-1.0	-0.89	-0.92
Gln	-0.85	-0.83	-1.0	-0.61	-0.91
Asp	-0.90	-0.69	-1.0	-1.0	-1.3
Lys	-1.5	-2.4	-1.1	-1.0	-0.67
Arg	-2.5	-1.8	-1.3	-0.68	-0.59

[a] The values are those of Table 2, but normalized to a mean of 0.0 and a standard deviation of 1.0. The OMH scale (71) has been used only in normalized form.

because they have different means and different spreads. To eliminate this factor, the five scales presented in Table 3 have been normalized to a mean of 0.0 and a standard deviation of 1.0. Normalization makes direct comparison easier, but the question remains: Which scale, if any, reflects very well the tendency of residues to associate with membranes? This is still an active area of research. In any case, the OMH scale, which was designed for matching of sequences, is not necessarily comparable to the others, which were formulated to measure hydrophobicity or affinity for membranes.

Hydrophobic Moments and the Hydrophobic Moment Plot

A second tool for analyzing amino acid sequences of membrane-related proteins is the *hydrophobic moment*, a measure of the amphiphilicity or asymmetry of hydrophobicity of a segment of a polypeptide chain (72). It was recognized from the earliest known protein structures, myoglobin and hemoglobin, that most of the α helices are amphiphilic (73); i.e. one surface of each helix projects mainly hydrophilic side chains, while the opposite surface projects mainly hydrophobic side chains. This helix amphiphilicity was represented by Schiffer & Edmundson (74) as two-dimensional "helical wheel" diagrams, a projection down the helix axis showing the relative orientations of residues. Helical wheel diagrams have been used to estimate the amphiphilicity of membrane helices (69), but it has been emphasized that this method is prone to misinterpretation (75).

A quantitative measure of the amphiphilicity perpendicular to the axis of any periodic peptide structure, such as the α helix, was proposed by Eisenberg et al (72) and called the *hydrophobic moment*. It can be calculated for an amino acid sequence of N residues and their associated hydrophobicities H_n from the definition (Equation 1):

$$\mu_H = \left\{ \left[\sum_{n=1}^{N} H_n \sin(\delta n) \right]^2 + \left[\sum_{n=1}^{N} H_n \cos(\delta n) \right]^2 \right\}^{1/2}. \qquad 1.$$

In this definition the sums are over all residues of the peptide, and δ is the angle in radians at which successive side chains emerge from the backbone when the periodic segment is viewed down its axis. For the α helix, $\delta = 100°$.

A geometric representation of this equation is given by Figure 5 for two helices, the upper one showing an 11-residue helical segment from lactate dehydrogenase and the lower one showing residues 5–22 from melittin. The vector sum of the solid arrows is μ_H, the hydrophobic moment. It is small for the upper helix where the residues are evenly distributed about the helix, and large for the lower helix, which has most of the hydrophobic residues on one side and most of the hydrophilic residues on the other. Thus, the hydrophobic moment measures the extent of amphiphilicity of a helix. The concept is not restricted to α helices (76), but more general definitions are just beginning to find applications to membrane-related proteins (90).

The hydrophobic properties of a helix can be represented as a point on a *hydrophobic moment plot*, on which the vertical axis is the hydrophobic moment per residue, and the horizontal axis is the hydrophobicity per residue. Figure 6 is an example, similar to those in (77). Helices from globular proteins generally plot in the region labeled GLOBULAR, at intermediate values of hydrophobicity and hydrophobic moment. The melittin helix, and other possible helices from surface-seeking proteins plot in the region labeled SURFACE at high hydrophobic moment. Possible

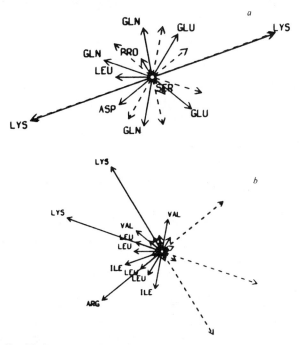

Figure 5   Graphical representations of residue contributions to the helical hydrophobic moment of two α helices. Hydrophobic residues have positive values for the hydrophobicity, H_n, and are represented as vectors extending out from the center of the axial projection of the α helix. Hydrophilic residues have negative values of H_n and their positions about the helix axis are represented by dashed vectors extending from the center. Their vector contributions are shown by solid lines 180° away. The vector sum of the H_n is μ_H. (a) The 11-residue segment of a helix in lactate dehydrogenase starting at residue 308. The value of the hydrophobic moment is small, $<\mu_H> = 0.1$/residue, as indicated by the symmetric distribution of solid vectors. (b) Residues 5–22 of melittin, a highly amphiphilic helix with $<\mu_H> = 0.4$/residue. The resulting hydrophobic moment would be off the page to the left. The hydrophobicity values used in this figure are those of Janin (62). The small residue contributions are represented by vectors but not labeled. Reproduced from (72).

helices from transmembrane proteins plot at large values of hydrophobicity. The boundary curve to the upper right of Figure 6 shows the largest possible hydrophobic moment that a helix of any given hydrophobicity can have. It represents values for hypothetical copolymers of arginine and isoleucine.

The tendency of different classes of proteins to plot in different regions of the hydrophobic moment plot is illustrated in Figure 7 for the proteins hemoglobin, melittin, and bacteriorhodopsin. A "window" of 11 residues is moved through the sequence of each, and the hydrophobicity and hydrophobic moment per residue is calculated, and this point is placed on the plot. In doing so, the implicit assumption is made that the entire protein

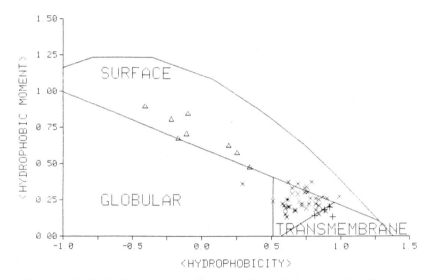

Figure 6   A hydrophobic moment plot, showing the characteristics of some 11-residue surface and membrane helices. The vertical axis gives the hydrophobic moment per residue, and the horizontal axis gives the hydrophobicity per residue, both based on the consensus hydrophobicities of Table 2. Most helices from globular proteins plot in the region labeled GLOBULAR. The eight surface protein segments from Table 4 define the region labeled SURFACE, and are plotted as Δ. Each point represents the 11-residue segment of the protein having the largest hydrophobic moment. In the region labeled TRANSMEMBRANE, 48 plausible helices from membrane proteins are plotted with + symbols for probable monomeric membrane anchors, and X for plausible helices from dimeric or channel membrane proteins. Each point represents the 11-residue segment of the plausible helix having the largest hydrophobic moment. Notice that the monomeric membrane anchors plot below the line that slopes up to the right, and accordingly, this region is called the MONOMERIC region. The other division of the transmembrane region, above and to the left of the monomeric region, is called the MULTIMERIC region. The consensus hydrophobicity scale was used in this figure, as well as in Figures 7 and 8.

is α helical, which is not strictly true for the largely α-helical proteins hemoglobin and bacteriorhodopsin. Nevertheless, many features of Figure 7 are consistent. The points for hemoglobin lie mainly in the GLOBULAR region, although some spill into the SURFACE and TRANSMEMBRANE regions. Most points for melittin lie in or near the SURFACE region, but those for its highly hydrophilic C terminus (LysArgLysArgGlnGln) lie in the GLOBULAR region. The points corresponding to the seven transmembrane helices of bacteriorhodopsin lie mainly in the TRANSMEMBRANE region, whereas the remaining points for this protein lie in the GLOBULAR region, as might be expected for these protein segments in contact with the solvent.

Surface Proteins

Some proteins, or segments of proteins, may have evolved a surface-seeking function, i.e. an affinity for the interface between an aqueous and a hydrophobic phase. Melittin, described above, is an example. In the case of melittin, it seems likely that the surface-seeking capacity is related to its high amphiphilicity when the peptide is coiled as an α helix. That is, melittin seeks surfaces because it has a large helical hydrophobic moment.

Several other small proteins are known to have surface-seeking functions and may also have large hydrophobic moments; these are listed in Table 4. They include a second form of mellitin (78); two synthetic cytotoxins that are melittin analogs (79; W. F. DeGrado, personal communication); two

Table 4 Surface-seeking proteins[a]

Protein	Starting residue	Plot coordinates $\langle H \rangle$	$\langle \mu_H \rangle$	Reference
Cecropin A	3 (LYS)	−0.22	0.80	Steiner et al (81)
Cecropin B	3 (LYS)	−0.41	0.89	
δ Hemolysin				
S. aureus	16 (ILE)	−0.11	0.70	Fitton et al (80)
S. aureus (canine strain)	16 (ILE)	−0.17	0.67	
Mellitin				
A. mellifera	12 (GLY)	0.25	0.57	Habermann (32)
A. florea	12 (GLY)	0.34	0.47	Kriel (78)
Artificial cytotoxin	12 (LEU)	0.19	0.62	DeGrado et al (79)
Artificial cytotoxin	2 (LEU)	−0.10	0.84	DeGrado (personal communication)

[a] The hydrophobic properties of these eight small proteins are represented in the hydrophobic moment plot of Figures 5, on which the 11-residue segment having the largest hydrophobic moment is plotted. The first residue of the 11 is given in the second column.

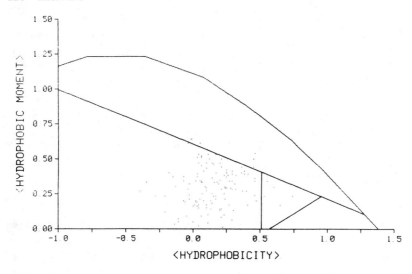

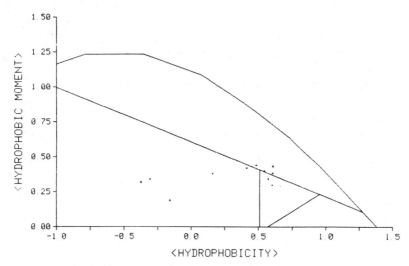

Figure 7   Hydrophobic moment plots for three largely α-helical proteins: (*upper left*) hemoglobin, human β chain [sequence from (94)]; (*lower left*) melittin (32); (*upper right*) bacteriorhodopsin (17); and (*lower right*) seven helical transmembrane segments of bacteriorhodopsin. Each point represents one 11-residue segment of the amino acid sequence, having coordinates $<H>$ and $<\mu_H>$. Every successive 11-residue segment of the proteins is plotted, making the implicit assumption that the proteins are entirely α helical. This assumption is of course not completely warranted for hemoglobin or bacteriorhodopsin. Notice that most of the points for hemoglobin fall in the GLOBULAR region, most for the bacteriorhodopsin helices fall in the TRANSMEMBRANE region, and most for melittin fall in or near the SURFACE region.

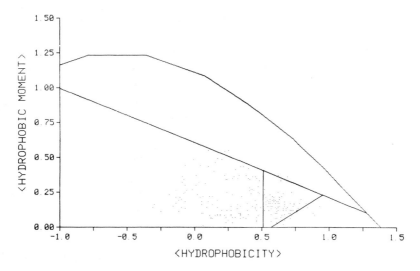

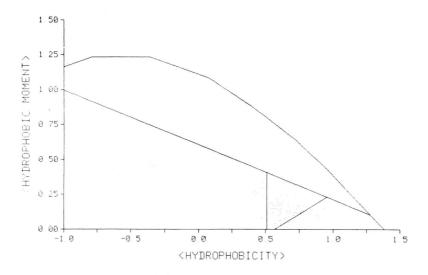

forms of δ hemolysin, a lytic protein secreted by *Staphylococcus aureus* (80);
and two cecropins, lytic antibacterial peptides from the moth *Hyalophora
cecropia* (81). Of these, the synthetic cytotoxins and cecropin A are known
to assume the α-helical conformation, at least in some solutions. The
hydrophobic properties of all eight proteins listed in Table 4 are shown by
triangles on the hydrophobic moment plot of Figure 6, and it can be seen
that all have relatively large values of hydrophobic moment. This is the

rationale for calling the high hydrophobic moment area of the plot the SURFACE region.

The observation that proteins of Table 4, all having the surface-seeking or lytic activity, also have large hydrophobic moments suggests that it may be possible to detect surface-seeking proteins from their amino acid sequences. The method is simply to ask if the amino acid sequence leads to a large helical hydrophobic moment. If so, and if the segment having the large moment is a sizable proportion of the whole protein, then the protein is likely to be surface seeking, and/or lytic. Some globular proteins, not known to have surface-related functions, contain individual helices having sizable hydrophobic moments. These are invariably helices on the protein surface, and are in all cases only a small proportion of the entire molecule. Apparently surface-seeking proteins must be highly amphiphilic over a large fraction of the molecule.

The discussion of amphiphilicity and surface proteins in this review has been from the viewpoint of hydrophobic moments. There have been discussions from other viewpoints (82), including some especially important articles on the design and synthesis of amphiphilic peptides (83, 84).

Detection of Membrane-Spanning Helices from Amino Acid Sequences

In determining amino acid sequences for membrane-related proteins (e.g. 85, 86), it has seemed reasonable to assign hydrophobic segments to transmembrane protein features. Earlier, Segrest & Feldman (65) and Rose (87) noted that numerical hydrophobicities might be effective in detecting such hydrophobic segments. More recently, several authors (68, 69, 77) have proposed and tested algorithms on membrane-related proteins and globular controls. The major obstacle to such tests is the limited amount of experimental evidence on exactly which protein segments are within the membrane.

Even ignoring this difficulty, one finds it is not easy to develop an algorithm that detects possible membrane-associated segments of amino acid sequences in membrane proteins and *also* correctly fails to identify "membrane-associated segments" in the sequences of globular proteins, used as controls. Kyte & Doolittle (68) found that averaging hydrophobicity over segments of 19 residues is most effective in distinguishing membrane-spanning segments from globular controls. Yet, they found that the most hydrophobic 19-residue segment from any protein they studied was in soluble dogfish lactate dehydrogenase. Why is this segment not membrane spanning? In a general sense, we know that the lowest free-energy state of a protein is determined by cooperative interactions among various parts of the protein and of the protein with solvent. Thus, consideration only of local interactions of a protein with lipid may be insufficient to detect membrane-spanning sequences. It may be necessary to consider cooperativity to be able to identify membrane-spanning sequences.

Cooperative effects were included in a recent study (77) of the amino acid sequences of some 46 membrane proteins and some 19 globular protein controls. The controls were selected for their binding of hydrophobic ligands, their large size (implying perhaps a small surface-to-volume ratio and thus highly hydrophobic interiors), or their exceptionally high hydrophobicity. The authors used the normalized consensus hydro-phobicities of Table 3, and averaged over a segment of 21 residues that moves through each of the sequences. A segment of 21 residues is taken because 21 residues coiled into an α helix approximate the thickness of the apolar portion of a lipid bilayer (21 residues $\times$ 1.5 Å/residue $=$ 32 Å), in similar fashion to other algorithms (20, 68). Membrane proteins were found generally to have one or two highly hydrophobic 21-residue segments: average hydrophobicity of 0.65 for one or a sum of 1.10 for two. This was termed an "initiator." Then other 21-residue segments having a mean hydrophobicity of 0.42 or greater were accepted as membrane associated. No globular protein was found with an initiator, although several such proteins contained 21-residue segments having average hydrophobicity over 0.42.

This algorithm (77) detects membrane-spanning segments in many membrane proteins, including seven such segments in bacteriorhodopsin and rhodopsin, but there are exceptions. Three proteins believed to be membrane associated contained no segments selected by the algorithm. One of the three is porin (88) believed to fold with β structure and with relatively little α helix (89). The algorithms of recent computer studies (68, 69, 77) are designed to seek membrane-spanning α helices, which are longer than possible membrane-spanning β segments, and hence easier to detect and distinguish from amino acid segments in soluble proteins.

In summary, computer algorithms can probably detect many trans-membrane α helices, but assessment of the accuracy must await further experimental studies. Other types of membrane-bound protein secondary structure, such as strands of β sheets, may presently elude detection altogether.

Classification of Membrane-Related Helices

The classification of membrane-related helices by their positions on a hydrophobic moment plot seems to be correlated with their function (77). To determine the position of a helix on the plot, a window of 11 residues is passed through the amino acid sequence of a proposed membrane segment, and the average hydrophobicity and hydrophobic moment per residue is computed. A point is plotted on the hydrophobic moment plot of Figure 6 that corresponds to the one 11-residue segment having the largest hydrophobic moment. The lower-right portion of Figure 6 contains 49 such points, each corresponding to a probable membrane-penetrating helix. (The use of the helical hydrophobic moment in Figure 6 implies that the membrane-spanning segment is an α helix.)

The membrane-related helices shown on the hydrophobic moment plot of Figure 6 cluster according to function. Surface-seeking proteins as discussed above, plot in the region of high hydrophobic moment. But even among membrane-spanning helices there are distinctions, with single-chain membrane anchors plotting in the region of highest hydrophobicity and lowest hydrophobic moment. Five HLA Class I sequences, which are believed to be single-helix membrane anchors, plot below the line sloping up to the right and are both very hydrophobic and evenly hydrophobic around the helix. These helices are indicated by + signs (two of which are superimposed). Other plausible helices from membrane proteins in Figure 6 are indicated by X. They are from bacteriorhodopsin, rhodopsin, the four protein chains of acetylcholine receptor, and membrane IgG and IgM chains. These plot at higher values of hydrophobic moment and lower values of hydrophobicity. Such values are consistent with channel-forming functions, which may involve relatively amphiphilic helices. A similar computation by Finer-Moore & Stroud (90) also suggests highly amphiphilic helices in acetylcholine receptor. The dimeric IgG and IgM proteins might be expected to pair in the membrane, with the more amphiphilic faces of the two helices together. Thus, amphiphilicity for these proteins may also reflect some function. In summary, Figure 6 suggests that the hydrophobic moment, as well as the hydrophobicity of membrane-spanning helices, can yield some information.

These classifications can be pictorially displayed. Several studies (e.g. 68) of amino acid sequences of membrane proteins have displayed results as a hydrophobicity profile along the sequence of a protein. A variation of this is shown in Figure 8, in which the sequence is shown with especially hydrophobic and surface features having distinctive symbols. The sequence of hemoglobin displays relatively few 11-residue segments having surface or membrane values. In contrast, the sequence of melittin shows several 11-residue segments of the surface or transmembrane type, and the sequence of bacteriorhodopsin shows seven clusters of transmembrane segments, perhaps associated with the seven transmembrane rods discussed above. Especially interesting is the profile for diphtheria toxin, which shares functional characteristics of globular, membrane, and surface proteins. It is a single polypeptide chain, soluble in aqueous solution, which can bind to

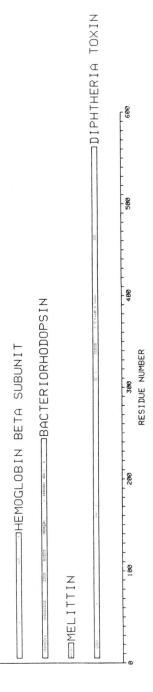

Figure 8   Membrane association profiles of four proteins, as determined by the method of (77). The amino acid sequence number is plotted along the horizontal axis. Each point is plotted at the amino acid sequence number of the central residue of an 11-residue segment. Symbols indicate the type of membrane association, in terms of the regions of Figure 6: no symbol is plotted for GLOBULAR, two vertical dots indicate the MULTIMERIC region, three indicate the MONOMERIC region, and one indicates the SURFACE region. Notice the seven regions of TRANSMEMBRANE character (MULTIMERIC or MONOMERIC symbols) in bacteriorhodopsin.

planar lipid bilayers, forming transmembrane channels (91). In addition, lipid bilayers are disorganized by a cyanogen bromide fragment of diphtheria toxin, as they are by melittin and other surface proteins (92). In view of these studies, it is intriguing that the profile for diphtheria toxin, based on the amino acid sequence of (93), shows five possible trans-membrane helices and one segment of nine consecutive 11-residue windows that plot in the SURFACE region. The "transmembrane" segment at the N terminus is the hydrophobic leader peptide. The other four segments are within the portion of the protein destined to become the membrane-inserting B fragment. Moreover, the surface region is contained within the cyanogen bromide fragment that disrupts lipid bilayers. When the three-dimensional structure of diphtheria toxin is known, it may reveal if these segments are involved in transmembrane function.

CONCLUSIONS

More than anything else, a lack of crystals has prevented molecular structures of membrane proteins from being determined by diffraction methods. However, recent successes in crystal growth, using detergents and small amphiphiles, suggest that the era of ignorance may be coming to an end.

The one shining exception has been the determination of the structure of bacteriorhodopsin at moderate resolution by a novel method of combined electron microscopy and diffraction (13). This study and coordinated work show that most transmembrane portions of the protein consist of α-helical rods running perpendicular to the membrane plane. These helical rods contain 20–25 mainly hydrophobic residues, but exactly which residues, and which residues are in given rods, are unanswered questions. Many other protein sequences contain segments of 20–25 relatively hydrophobic residues, and it is plausible but unproven that these are coiled in α helices. The tangible model of bacteriorhodopsin should not lead us to forget that other secondary structures may well exist in membrane proteins. As has often been pointed out, these can include β sheet twisted into barrels, as well as less-regular structures.

In addition to the membrane-penetrating proteins like bacteriorhodop-sin, other proteins have evolved the capacity to associate with membrane. One class is that of surface-seeking proteins like melittin. Melittin's strong amphiphilicity suggests that it may initially associate with membranes with its helical axis parallel to the plane of the membrane. Subsequently, several melittin molecules may form a pore, such as has been envisaged for the more hydrophobic helix alamethicin (31).

Computational analysis of amino acid sequences from membrane proteins can yield structural information of value. Transmembrane α-helical segments can probably be detected. Also, surface-seeking proteins can be recognized from their sequences by large hydrophobic moments. It has also been suggested that putative helices can be classified by consideration both of their hydrophobicities and hydrophobic moments. Computational analysis will be much more effective once several structures for membrane proteins are known. The code that links primary to three-dimensional structure may be more easily deciphered for membrane proteins than for soluble proteins.

ACKNOWLEDGMENTS

I am grateful to the following colleagues and co-workers for stimulating discussions on membrane-related proteins: R. J. Collier, J. A. Lake, D. C. Rees, E. M. Schwarz, R. M. Sweet, T. C. Terwilliger, P. Thornber, E. Tobin, R. M. Weiss, W. Wickner. I also thank J. Hutcheson for diligent typing, E. M. Schwarz for preparation of Figures 6, 7, and 8, and NIH and NSF for financial support of related research.

Literature Cited

1. Michel, H. 1983. *Trends Biochem. Sci.* 8:56–59
2. Vanderkooi, G., Senior, A. E., Capaldi, R. A., Hayashi, H. 1972. *Biochim. Biophys. Acta* 274:38–48
3. Li, J., Hollingshead, C. 1982. *Biophys. J.* 37:363–70
4. Henderson, R., Shotton, D. 1980. *J. Mol. Biol.* 139:99–109
5. Michel, H., Oesterhelt, D. 1980. *Proc. Natl. Acad. Sci. USA* 77:1283–85
6. Garavito, R. M., Rosenbusch, J. P. 1980. *J. Cell Biol.* 86:327–29
7. Michel, H. 1982. *EMBO J.* 1:1267–71
8. Michel, H. 1982. *J. Mol. Biol.* 158:567–72
9. Garavito, R. M., Jenkins, J. P. 1983. *The Structure and Function of Membrane Proteins*, ed. F. Palmieri. Amsterdam: Elsevier/North Holland
10. Kistler, J., Stroud, R. M. 1981. *Proc. Natl. Acad. Sci. USA* 78:3678–782
11. Leonard, K., Wingfield, P., Arad, T., Weiss, H. 1981. *J. Mol. Biol.* 149:259–74
12. Henderson, R. 1977. *Ann. Rev. Biophys. Bioeng.* 6:87–109
13. Henderson, R., Unwin, P. N. T. 1975. *Nature* 257:23–32
14. Michel, H., Oesterhelt, D., Henderson, R. 1980. *Proc. Natl. Acad. Sci. USA* 77:338–42
15. Leifer, D., Henderson, R. 1983. *J. Mol. Biol.* 163:451–66
16. Henderson, R. 1975. *J. Mol. Biol.* 93:123–38
17. Ovchinnikov, Y. A., Abdulaev, N. G., Feigina, M. Y., Kiselev, A. V., Lobanov, N. A. 1979. *FEBS Lett.* 100:219–24
18. Khorana, H. G., Gerber, G. E., Herlihy, W. C., Gray, C. P., Anderegg, R. J., Nihei, K., Biemann, K. 1979. *Proc. Natl. Acad. Sci. USA* 76:5046–50
19. Engelman, D. M., Henderson, R., McLachlan, A. D., Wallace, B. A. 1980. *Proc. Natl. Acad. Sci. USA* 77:2023–27
20. Engelman, D. M., Goldman, A., Steitz, T. A. 1981. *Methods Enzymol.* 88:81–88
21. Agard, D. A., Stroud, R. M. 1982. *Biophys. J.* 37:589–602
22. Jap, B. K., Maestre, M. F., Hayward, S. B., Glaeser, R. M., 1983. *Biophys. J.* 43:81–90
23. Engelman, D. M., Zaccai, G. 1980. *Proc. Natl. Acad. Sci. USA* 77:5894–98
24. King, G. I., Mowery, P. C., Stoeckenius, W., Crespi, H. L., Schoenborn, B. P. 1980. *Proc. Natl. Acad. Sci. USA* 77:4726–30
25. Bayley, H., Huang, K.-S., Radhakrishnan, R., Ross, A. H., Takagaki, Y., Khorana, H. G. 1981. *Proc. Natl. Acad. Sci. USA* 78:2225–29

26. Katre, N. V., Wolber, P. K., Stoeckenius, W., Stroud, R. M. 1981. *Proc. Natl. Acad. Sci. USA* 78:4068–72
27. Deatherage, J. F., Henderson, R., Capaldi, R. A. 1982. *J. Mol. Biol.* 158:501–14
28. Kistler, J., Stroud, R. M., Klymkowsky, M. W., Lalancette, R. A., Fairclough, R. H. 1982. *Biophys. J.* 37:371–83
29. Leonard, K., Wingfield, P., Arad, T., Weiss, H. 1981. *J. Mol. Biol.* 149:259–74
30. Mueller, P., Rudin, D. O. 1968. *Nature* 217:713–19
31. Fox, R. O., Richards, F. M. 1982. *Nature* 300:325–30
32. Habermann, E. 1972. *Science* 177:314–22
33. Anderson, D., Terwilliger, T. C., Wickner, W., Eisenberg, D. 1980. *J. Biol. Chem.* 255:2578–82
34. Terwilliger, T. C., Eisenberg, D. 1982. *J. Biol. Chem.* 257:6010–15
35. Terwilliger, T. C., Eisenberg, D. 1982. *J. Biol. Chem.* 257:6016–22
36. Terwilliger, T. C., Weissman, L., Eisenberg, D. 1982. *Biophys. J.* 37:353–61
37. Lauterwein, J., Bosch, C., Brown, L. R., Wuthrich, K. 1979. *Biochim. Biophys. Acta* 556:244–64
38. Brown, L. R., Lauterwein, J., Wuthrich, K. 1980. *Biochim. Biophys. Acta* 622:231–44
39. Tosteson, M. T., Tosteson, D. 1982. *Biophys. J.* 36:109–16
40. Kempf, C., Klausner, R. D., Weinstein, J. N., Renswoude, J. V., Pincus, M., Blumenthal, R. 1982. *J. Biol. Chem.* 257:2469–76
41. Matthews, B. W. 1982. In *Lipid-Protein Interactions*, ed. P. C. Jost, O. H. Griffith, 1:1–23. New York: Wiley
42. Fenna, R., Matthews, B. W. 1975. *Nature* 258:573–77
43. Matthews, B. W., Fenna, R. E., Bolognesi, M. C., Schmid, M. F., Olson, J. M. 1979. *J. Mol. Biol.* 131:259–85
44. Dijkstra, B. W., Drenth, J., Kalk, K. H. 1981. *Nature* 289:604–6
45. Dijkstra, B. W., Kalk, K. H., Hol, W. G., Drenth, J. 1981. *J. Mol. Biol.* 147:97–123
46. Keith, C., Feldman, D. S., Deganellos, S., Glick, J., Ward, K. B., Jones, E. D., Sigler, P. B. 1981. *J. Biol. Chem.* 256:8602–7
47. Volwerk, J. J., de Haas, G. H. 1982. See Ref. 41, 1:69–149
48. Dijkstra, B. W., Kalk, K. H., Drenth, J., de Haas, G. H., Egmond, M. R., Slotboom, A. J. 1983. *J. Mol. Biol.* Submitted for publication
49. Banaszak, L. J., Ross, J. M., Wrenn, R. F. 1982. See Ref. 41, pp. 233–50
50. Ohlendorf, D. H., Collins, M. L.,

Puronen, E. O., Banaszak, L. J., Harrison, S. C. 1975. *J. Mol. Biol.* 99:153–65
51. Urry, D. W. 1971. *Proc. Natl. Acad. Sci. USA* 68:672–76
52. Urry, D. W., Goodall, M. C., Glickson, J. D., Mayers, D. F. 1971. *Proc. Natl. Acad. Sci. USA* 68:1907–11
53. Urry, D. W. 1982. *Prog. Clin. Biol. Res.* 79:87–111
54. Kennedy, S. J. 1978. *J. Membr. Biol.* 42:265–79
55. Koeppe, R. E., Hodgson, K. O., Stryer, L. 1978. *J. Mol. Biol.* 121:41–54
56. Koeppe, R. E., Berg, J. M., Hodgson, K. O., Stryer, L. 1979. *Nature* 279:723–5
57. Dunker, A. K., Zaleske, D. J. 1977. *Biochem. J.* 163:45–57
58. Burres, N., Dunker, A. K. 1980. *J. Theor. Biol.* 21:723–36
59. Engelman, D. M., Steitz, T. A. 1981. *Cell* 23:411–22
60. Edsall, J. T., McKenzie, H. A. 1983. *Adv. Biophys.* 16:53–183
61. Wolfenden, R., Andersson, L., Cullis, P. M., Southgate, C. C. B. 1981. *Biochemistry* 20:849–55
62. Janin, J. 1979. *Nature* 277:491–92
63. Chothia, C. 1976. *J. Mol. Biol.* 105:1–14
64. Nozaki, Y., Tanford, C. 1971. *J. Biol. Chem.* 246:2211–17
65. Segrest, J. P., Feldman, R. J. 1974. *J. Mol. Biol.* 87:853–58
66. von Heijne, G., Blomberg, C. 1979. *Eur. J. Biochem.* 97:175–81
67. Eisenberg, D., Weiss, R. M., Terwilliger, T. C., Wilcox, W. 1982. *Faraday Symp. Chem. Soc.* 17:109–20
68. Kyte, J., Doolittle, R. F. 1982. *J. Mol. Biol.* 157:105–32
69. Argos, P., Rao, J. K. M., Hargrave, P. A. 1982. *Eur. J. Biochem.* 128:565–75
70. von Heijne, G. 1981. *Eur. J. Biochem.* 116:419–22
71. Sweet, R. M., Eisenberg, D. 1983. *J. Mol. Biol.* In press
72. Eisenberg, D., Weiss, R. M., Terwilliger, T. C. 1982. *Nature* 299:371–74
73. Perutz, M. F., Kendrew, J. C., Watson, H. C. 1965. *J. Mol. Biol.* 13:669–78
74. Schiffer, M., Edmundson, A. B. 1967. *Biophys. J.* 7:121–35
75. Flinta, C., von Heijne, G., Johansson, J. 1983. *J. Mol. Biol.* 168:193–96
76. Eisenberg, D., Weiss, R. M., Terwilliger, T. C. 1984. *Proc. Natl. Acad. Sci. USA.* 81:140–44
77. Eisenberg, D., Schwarz, E. M., Komaromy, M., Wall, R. 1984. Submitted for publication
78. Kriel, G. 1973. *FEBS Lett.* 33:241–44
79. DeGrado, W. F., Kezdy, F. J., Kaiser, E. T. 1981. *J. Am. Chem. Soc.* 103:679–81

80. Fitten, J. E., Dell, A., Show, W. V. 1980. *FEBS Lett.* 115:209–12
81. Steiner, H., Hultmark, D., Engstrom, A., Bennich, H., Boman, H. G. 1981. *Nature* 292:246–48
82. Segrest, J. P., Feldman, R. J. 1977. *Biopolymers* 16:2053–65
83. Kaiser, E. T., Kezdy, F. J. 1983. *Proc. Natl. Acad. Sci. USA* 80:1137–43
84. Segrest, J. P., Chung, B. H., Brouillette, C. G., Kanellis, P., McGahan, R. 1983. *J. Biol. Chem.* 258:2290–95
85. Tomita, M., Marchesi, V. 1975. *Proc. Natl. Acad. Sci. USA* 72:2964–68
86. Dratz, E. A., Hargrave, P. A. 1983. *Trends Biochem. Sci.* 8:128–31
87. Rose, G. D. 1978. *Nature* 272:586–90
88. Chen, R., Kramer, C., Schmidmayr, W., Henning, U. 1979. *Proc. Natl. Acad. Sci.*

89. Rosenbuch, J. P. 1974. *J. Chem. Biol.* 249:8019–29
90. Finer-Moore, J., Stroud, R. M. 1984. *Proc. Natl. Acad. Sci. USA.* 81:155–59
91. Donovan, J. J., Simon, M. I., Draper, R. K., Montal, M. 1981. *Proc. Natl. Acad. Sci. USA* 78:172–76
92. Kayser, G., Lambotte, P., Falmagne, P., Capiau, C., Zanen, J., Ruysschaert, J. M. 1981. *Biochem. Biophys. Res. Commun.* 99:358–63
93. Greenfield, L., Bjorn, M. J., Horn, G., Fong, D., Buck, G. A., et al. 1984. *Proc. Natl. Acad. Sci. USA* In press
94. Feldman, R. J. 1976. *Atlas of Macromolecular Structure on Microfiche.* Rockville, Md: Tractor, Jitco

USA 76:5014–17

Ann. Rev. Biochem. 1984. 53 : 35–73

MYOSIN

William F. Harrington and Michael E. Rodgers

Department of Biology, Johns Hopkins University, Baltimore, Maryland, 21218

PERSPECTIVES AND SUMMARY

Myosin is one of the principal protein components of numerous contractile systems and comprises almost 50% of the total protein in skeletal muscles. It is arranged almost invariably in the form of thick filaments and appears to be the major element of energy transduction and force development in these systems. Myosins from various sources have been isolated and studied. These include vertebrate skeletal muscle (mainly rabbit and chicken), cardiac muscle, several smooth muscles, moluscan, nematode, and insect flight muscles, as well as nonmuscle systems. Although each myosin has unique properties depending on the origin, certain features are common to all.

Myosin is a highly asymmetric hexameric protein. The molecule consists of two globular head regions and a rodlike tail portion that is a "coiled coil" of α-helices. The rod portion is responsible for the assembly of myosin to form the functional thick-filament structure. The globular heads contain both the enzymatic active site and the active binding region. Each head also has two noncovalently bound light chains that appear to be involved in the

237

regulation of contraction. In addition, the enzymatic hydrolysis of MgATP by all myosins is markedly enhanced in the presence of actin.

The functional form of myosin in both muscle and nonmuscle systems is the aggregated state. In muscle cells several hundred myosin molecules are systematically organized into bipolar structures, the thick filaments, whose length and diameter vary depending on the species. The heads of the myosin molecules project from the surface of these structures and undergo cyclic interaction with neighboring thin (actin) filaments to generate a relative contractile force between the thick and thin filaments. Although the systematic arrangement of myosin heads along the surfaces of several species of thick filaments has been determined, the three-dimensional packing of myosin tails within the cores is still not resolved and is currently an area of intense research activity. Nonmuscle myosins also form bipolar thick filaments under physiological conditions, and there is growing evidence that this process is intimately related to phosphorylation of their regulatory light chain.

The molecular events responsible for force generation in a cycling cross-bridge are currently a major focus of interest in the muscle field. The dominant question is this: Is a conformational transition in the myosin head, resulting in a change in the effective angle of the actin-attached cross-bridge, responsible for force generation? Evidence is accumulating from spectroscopic studies that simple head rotation is not occurring, suggesting that different models should be considered.

Since the last review of contractile proteins in this series (1) an enormous effort has been expended in trying to further characterize the structure and function of the myosin molecule. In this chapter, we first discuss some of the advances in our understanding of the chemical and physical nature of the myosin molecule in the monomeric state. We then consider the structure of the thick filament and the assembly processes that result in their formation. Finally, we discuss some aspects of the mechanism of action of this molecule, which is one of the major goals of all research on contractility.

Owing to the brevity of this review, several important aspects of myosin function have not been included. In particular, the kinetics of ATP hydrolysis by myosin and the role of phosphorylation in the regulation of contraction are noticeably absent. The reader is referred to another review (2) for a thorough discussion of the ATPase kinetics of myosin and acto-myosin systems. The role of phosphorylation in the regulation of contraction has also been considered in several recent review articles (3–5).

STRUCTURE OF MYOSIN

The principal site of energy transduction in contractile systems is generally believed to reside within the myosin molecule. Myosin is a relatively large

($M_r \sim$ 500,000), asymmetric, hexameric protein containing several structural and functional domains. Two globular heads located at one end of the molecule are joined to a very long rodlike tail segment thus endowing myosin with properties of both globular and fibrous proteins. Each molecule comprises two heavy chains ($M_r \sim$ 230,000), two essential light chains ($M_r \sim$ 16,000–20,000) and two regulatory light chains ($M_r \sim$ 16,000–20,000). Approximately 50% of the heavy chains starting at the C-terminus fold together to form a coiled coil of α-helices referred to as rod. This region contains the binding sites for assembly of myosin into thick filaments, the functional form of myosin in muscles. The remaining half of each heavy chain folds together with one essential light chain and one regulatory light chain to form the globular head regions called subfragment-1 (S-1). Each head contains a site for ATP binding and hydrolysis, the actin-binding site and several divalent cation-binding sites.

Investigations of the structure and function of myosin have involved the use of numerous organisms and a variety of tissues within these organisms. The general properties outlined above appear to be common to virtually all myosin molecules. A majority of the studies discussed in this section have been done using vertebrate skeletal myosin, principally from rabbit. Thus, the source of myosin is included only when discussing features that may not be common to most myosins. Since many studies deal with various proteolytic subfragments of myosin rather than the intact molecule, this section is divided into two parts, the first dealing with the globular heads of myosin and the second with rod.

Subfragment-1

MORPHOLOGY   The structure of myosin subfragment-1 at atomic resolution has not been obtained to date, but a number of studies have provided information about its shape and substructure. Negatively stained complexes of F-actin and S-1 exhibit a distinctive repeating arrowhead pattern when examined in the electron microscope (6). Image reconstruction of electron micrographs of S-1–decorated actin filaments by Moore et al (7) showed S-1 to be an elongated molecule about 15 nm in length having a bend near the center. The assignment of the protein densities was such that S-1 was bound somewhat tangentially to the outer surface of the actin filament and its long axis was tilted and slewed with respect to the filament axis. This arrangement was able to account for the arrowhead pattern and provided the first detailed information about the shape of the myosin head and the structure of the Acto–S-1 complex. Recently, several groups have reexamined this question and have proposed a different and more detailed structure for S-1.

Taylor & Amos (8) have obtained electron micrographs of decorated actin filaments under low-dose conditions resulting in increased resolution and better signal-to-noise ratios. The resulting reconstructions show a significant difference in the density distribution near the particle axis from that described by Moore et al (7). By interpreting the density closest to the axis as actin rather than a region of S-1, these authors obtain a somewhat different shape for S-1. The long axis of the molecule bends similarly to that proposed earlier, but the end of S-1 proximal to actin contains more mass than the distal region. The actin binding site of S-1 appears to be at the outside of the rather broad head and extends into the long-pitch helical groove of actin. The molecule bends at a point about 6.2 nm from the actin binding site, tapers down and away from the filament axis and ends in a dovetail-like structure. The dimensions of the S-1 in the region proximal to actin are 6.2–6.5 nm × 4.8–5.6 nm and the overall length is ∼ 12–13 nm.

Wakabayashi & Toyoshima (9) also published a paper on an Acto–S-1 reconstruction about the same time as did Taylor & Amos (8). They interpret their results in terms of a multidomain structure with 50% of the S-1 mass being in one domain centered about 6.2 nm from the filament axis and at the same axial level as the actin to which it is bound. One small, elongated domain is more distal to the axis and one or two domains are closer to the filament axis. The overall shape, however, is quite similar to that of Taylor & Amos (8). The angle of S-1 attachment is suggested to be almost perpendicular to the filament axis, and the arrowhead appearance of the Acto–S-1 complex is attributed to the curved nature of S-1. This orientation is different from the 55°–65° angle suggested by Taylor & Amos (8). The apparent difference seems to lie in the choice of chords representing the principal S-1 axis. Owing to the complex shape of S-1, assignment of the S-1 axis must be considered somewhat arbitrary.

The shape and size of S-1 predicted from image reconstruction (8, 9) is consistent with low-angle X-ray scattering profiles of subfragment-1 (10). These studies also reveal a 12 ± 1 nm length for S-1. In addition, comparison of the experimental data with predicted scattering patterns based on various solid models resulted in a best fit for a model based on a three-dimensional reconstructed image of S-1 (Seymour & O'Brien, unpublished). The 12–13 nm length of S-1, however, is somewhat smaller than the length of the head in intact myosin. Elliott & Offer (11) observed a 19 nm length for the myosin head based on electron micrographs of shadowed myosin molecules and a similar length was also reported by Craig et al (12). Myosin heads of this length can readily span the gap between thick and thin filaments in intact muscle. The discrepancy in the length of S-1 now appears to be related to the presence or absence of the regulatory light chain as discussed below. In the previous studies (8, 9), this light chain is known to be absent.

Craig et al (12) have examined the complex of scallop thin filaments with scallop S-1 in the presence and absence of regulatory light chain. In the presence of light chain, the arrowheads of the Acto–S-1 complex are "barbed" in appearance and the maximum width of the complex is 26 ± 2 nm. When the regulatory light chain is absent, the barbs are no longer observed and the width of the complex is significantly smaller (19 ± 2 nm). These arrowheads are termed "blunted" and are similar to those used in the reconstruction studies above (8, 9). Vibert & Craig (13) have recently reported three-dimensional image reconstruction results of the scallop Acto–S-1 complex in the presence and absence of regulatory light chain. The general features of the reconstructed image in the absence of light chain are quite similar to those of rabbit Acto–S-1 (8, 9) and the contour length of S-1 is ~ 13.5 nm. When the regulatory light chain is present, however, the results show that S-1 is about 4–5 nm longer and additional mass is located in the region of S-1 most distant from the actin binding site. It is suggested that myosin-linked regulation may involve changes in the structure of the S-1/rod junction as a result of Ca^{2+} binding to the regulatory light chains.

Katayama & Wakabayashi (14) have provided the first reconstructed image of an Acto-HMM complex. HMM (heavy meromyosin) is a proteolytic fragment of myosin consisting of both heads and part of the rod (referred to as subfragment-2 or S-2). Since HMM contains the S-2 region of myosin in addition to S-1, one might expect significant differences between the reconstructed images of Acto–S-1 and Acto-HMM. This was found to be the case. The arrowhead pattern of Acto-HMM is "barbed" in appearance unlike that of Acto–S-1. This is consistent with the results of Vibert & Craig (13), since the regulatory light chain is essentially intact in HMM. Additionally, whiskerlike filamentous structures are observed emanating from the outer regions of the complex and pointing in the same direction as the arrowhead. These have an estimated contour length of 50 nm and a width of 3 nm and have tentatively been assigned to S-2. The previous multidomain structure of S-1 could be readily superimposed on the Acto-HMM structure, but two additional domains were present at higher radii. A globular domain extends from the S-1 region most distal to the actin binding site and has been tentatively assigned to the regulatory light chain. The second domain is filamentous and appears to originate at the presumptive light-chain region and wind clockwise around the main body of the complex when viewed from the top of the arrowhead. This domain is somewhat shorter than the whiskerlike structures seen in electron micrographs and probably corresponds to a part of S-2 held rigid enough to retain the symmetry required for reconstruction.

SUBSTRUCTURE   The image reconstruction studies discussed above provide the general shape of the S-1 region of myosin but give little detailed information about its internal structure. Mapping of the substructure of S-1 has been pursued mainly through biochemical techniques. Most of this work has centered around proteolytic fragmentation of S-1, cross-linking and specific labeling of various reactive groups. The early studies of Balint et al (15, 16) demonstrated that S-1 is readily cleaved by trypsin to form three major fragments held together by noncovalent links. These have molecular weights of 27 K, 50 K and 20 K and occur in that order from the N-terminus to the C-terminus of the S-1 heavy chain. The three-dimensional arrangement of the three tryptic peptides within the head is not known; however, a number of spatial constraints have been determined. At least part of the 20-kd (kilodalton) peptide must lie in the neck region of myosin since it contains the C-terminus of S-1 (16). The 20-kd peptide also contains part of the actin binding site, the remainder of which is found on the 50-kd peptide (17–20). One antigenic site on the 27-kd peptide is located at the tip of S-1 most distal to the neck (21). Two proteolytically sensitive peptides with molecular weights of $\sim$2–2.5-kd connect these three relatively stable regions. The susceptibility of both of these segments to tryptic digestion is altered by the presence of actin. The region between the 50-kd and 20-kd peptides is protected from digestion when S-1 is bound to actin (17, 22–24). Binding to actin also alters the cleavage pathway of the segment connecting the 50-kd and 27-kd peptides (25). The essential light chains appear to bind to both the 27-kd (26) and the 20-kd peptide (27–29) while the regulatory light chain is bound at least to the 20-kd region (27, 29–31). These results provide a general outline for the arrangement of the polypeptide chains on the surface of the myosin head.

The presence of two highly reactive thiols in each myosin head was first demonstrated by Kielley & Bradley (32) and later confirmed by Sekine & Yamaguchi (33). Chemical modification of these thiols alters the ATPase activity of myosin. In addition, S-1 contains one highly reactive lysine residue that on modification also affects the ATPase activity (34, 35). Alteration of the myosin ATPase activity by modification of these residues suggests some type of linkage with the ATP binding site. Many recent studies have been aimed at mapping these groups and other structural and functional regions within S-1.

The location of the two reactive thiol groups within S-1, referred to as SH_1 and SH_2, is important to understanding the mechanism of myosin action. Elzinga & Collins (36) have shown that these thiols are separated by nine residues in the primary sequence and Balint et al (16) found that they reside in the 20-kd tryptic peptide of the S-1 heavy chain. Further studies by Sutoh (37) indicated that they lie $\sim$7 kd from the N-terminus of this fragment. Their precise location is now known to be 66 (SH_1) and 56 (SH_2) residues from the N-terminus of the 20-kd fragment (38, 39). Modification of SH_1 is associated with an increase in the Ca-ATPase activity and loss of the EDTA ATPase of myosin. Subsequent modification of SH_2 results in the loss of both ATPase activities (33, 40–42). Several studies using cross-linking reagents have shown that upon addition of ligands such as MgADP, the separation of SH_1 and SH_2 can change from > 12 Å to as little as 2 Å (42–47). Furthermore, cross-linking of SH_1 and SH_2 in the presence of nucleotide can trap this ligand in the active site (44–47). These studies suggest a marked flexibility in the polypeptide chain connecting SH_1 and SH_2 and reveal that these groups have a structural and functional role in the ATPase activity of myosin although direct involvement in the active site now seems unlikely (48, 49).

Recently, fluorescence energy transfer has been used to measure the time-average spatial separation between probes attached to SH_1 and SH_2 (50). Results indicate an interchromophoric separation of 24–29 Å. Addition of F-actin produced a 5 Å decrease in the spatial separation. MgADP, on the other hand, caused no spatial change, but other local perturbations were observed. The authors point out, however, that the distances measured include contributions from the bulky modifiers and the magnitude of such effects is currently unknown. Similar studies by Dalbey (51) indicate a comparable probe-to-probe distance but in this case, addition of MgADP produced an ~ 8 Å decrease in the spatial separation of the probes. The spatial relationship of SH_1 and SH_2 to other functional regions of S-1 is discussed below.

Both the ATP-binding site (52) and the reactive lysine residue (53–55) have been located in the 27-kd NH_2-terminal tryptic peptide of S-1. Fluorescence energy transfer experiments have been used to determine the spatial separation of these regions and the SH_1 residue. Tao & Lamkin (56) determined the most probable distance between SH_1 and the active site to be ~ 39 Å. This is consistent with earlier results that show that SH_1 does not directly interact with the active site (48, 49). The interchromophoric distance between labels on SH_1 and the reactive lysine has been reported to be ~ 26 Å based on similar experiments (57). Other workers have measured the distance between SH_1 and a site on the essential light chain (58) and between SH_1 and cysteine 373 of actin (59). These studies place many constraints on the arrangement of the peptide chains within the S-1 structure. Morales et al (60) have incorporated the results of these and other studies into a highly schematic model for the arrangement of polypeptide chains within S-1. In the absence of a high-resolution structural determination of S-1, these types of studies provide the only currently available information on the substructure of the myosin head.

Most of the known functional units within S-1 have been located in the 20-kd and 27-kd tryptic peptides. Little is known about the structure and function of the 50-kd peptide region except that it contains three of the conformationally sensitive tryptophan residues (61) and part of the actin-binding site (18, 20). It is curious that there is so little information about this region, which comprises almost half of the total mass of S-1. Future investigations into the functional role of the 50-kd peptide and its interactions with actin and the other subsections of myosin will be of considerable interest.

Recently a number of workers have referred to the three tryptic peptides of S-1 as domains. It is not clear that this designation is deserved. In a general sense, domains are regions of folded polypeptide chains in a complex structure separated by relatively well defined boundaries. Such regions have been observed in the image reconstruction studies mentioned earlier. The relationship between these and the three tryptic peptides is not known. Thermodynamically, domains are regions of structure that undergo denaturation as a cooperative unit (62, 63). It is not known whether the folded segments of S-1 behave in this manner. From the studies discussed above, it seems likely that the polypeptide chains of S-1 are intricately interwoven and the assignment of the tryptic fragments to domains should await higher resolution data. For these reasons, we have chosen to refer to the tryptic peptides only as segments of primary structure.

ACTIN BINDING SITE   The interactions between myosin and actin are of fundamental importance to the contractile process. Cyclic attachment and detachment of S-1 to actin in conjunction with ATP hydrolysis is essential for force production and muscular contraction. Although a substantial body of information on the kinetics of this interaction has been reported in the past, structural information has generally been lacking until recently. The image reconstruction studies of Moore et al (7) provided the first three-dimensional structural interpretation of this interaction. As discussed earlier, extensions of these studies have led to a somewhat more refined view of the structure of this complex. However, an unambiguous assignment of the domains in these models to actin and S-1 is still uncertain owing to the limit of resolution of this technique. An alternate approach aimed at providing more detailed chemical information about the Acto–S-1 interaction has yielded some surprising and somewhat controversial results.

Mornet et al (17–19) have utilized chemical cross-linking in conjunction with tryptic digestion of S-1 to investigate the structure of the actin-myosin interface. Cross-linking of the Acto–S-1 complex was carried out using the zero-length cross-linker EDC [1-ethyl-3(3-dimethylaminopropyl) carbodiimide]. Analysis of the cross-linked products by electrophoresis on SDS-containing[1] polyacrylamide gels revealed the presence of a protein doublet with an estimated M_r of ~ 180,000. Cross-linking fluorescent-

---

[1] SDS, sodium dodecylsulfate.

labeled S-1 with unlabeled actin or labeled actin with unmodified S-1 demonstrated that this component was an Acto–S-1 complex since fluorescent label appeared in the cross-linked doublet on SDS gels in both cases. The apparent molecular weight of the complex suggested that it consisted of two actin monomers and one S-1 heavy chain. This was verified by measuring the stoichiometry of actin:S-1 in the cross-linked species using ^{3}H-actin and ^{14}C–S-1 (18). The origin of the doublet was not explained. Further analysis of this complex was carried out using tryptically digested S-1.

As discussed earlier, tryptic hydrolysis of S-1 occurs principally at two sites and yields three peptides of M_r 27K, 50K, and 20K. Under appropriate conditions, the heavy chain of S-1 can be cleaved at either one of these sites or at both (64). Mornet et al (17, 18) were able to show that actin is cross-linked to both the 50-kd species and the 20-kd species by carrying out EDC cross-linking of actin and these various forms of S-1. Furthermore although the 20-kd and 50-kd species formed cross-links with only a single actin monomer, the 70-kd S-1 peptide (50 kd + 20 kd) formed a complex with two actins as judged by molecular weights determined from SDS gels. These results taken together implied that each actin monomer has two sites of attachment for S-1, one for the 50-kd tryptic peptide and one for the 20-kd tryptic peptide. Furthermore, each S-1 appeared to be bound to two contiguous actin monomers in the Acto–S-1 complex. Although these results do not alter the net stoichiometry of Acto–S-1 binding (1 : 1) they add new insight to our conception of this interaction.

Additional support for the idea that S-1 interacts with two sites on F-actin has subsequently been provided by Amos et al (65). These workers have extended their earlier image reconstruction analysis (8) to allow finer resolution of their averaged image. Although much of the original detail was unchanged, the division of the actin monomer into two domains was more pronounced. This is consistent with other structures proposed for F-actin (66–68). In addition, a connection between S-1 and an actin monomer on the opposite strand from the principal binding sites was detected. Until unambiguous assignment of the density peaks in the reconstructed image is possible, however, this interpretation must be viewed with caution.

Studies on the actin-myosin interaction sites have also been reported by Sutoh (20, 69, 70) using zero-length cross-linking as described above (17) and peptide mapping. The actin peptide involved in these interactions was identified and a more detailed mapping of the cross-linking sites on the S-1 peptides was obtained. The results indicate that both the 20-kd and 50-kd tryptic S-1 peptides cross-link to actin and are consistent with previous studies (17, 18, 22). Contrary to earlier interpretations, however, peptide-mapping studies show that both fragments cross-link to the 12-residue N-terminal segment of actin. An additional site of Acto–S-1 cross-linking was also found linking the C-terminal region of actin to the N-terminal segment of the essential light chain (20). Further peptide mapping experiments have localized the actin binding sites of both the 20-kd and 50-kd tryptic peptides of S-1. These are located in regions 18–20 and 27–35 kd from the C-terminus of the intact S-1 heavy chain (70).

Sutoh has also carried out zero-length cross-linking experiments between intact S-1 and actin (70). Consistent with the earlier studies, a doublet of cross-linked products was found with apparent molecular weights of 175K and 165K, but his interpretation of this result is somewhat different from that of Mornet et al (18). Since a complex of two actin monomers and S-1 requires the formation of at least two cross-links, one should expect to see intermediate species either at short times of cross-linking or with very low concentrations of reagent. Such intermediates were never observed. Unless the formation of the two cross-links is highly cooperative, which seems unlikely, this result suggests that a cross-linked complex of one actin and one S-1 may migrate with an anomolously high molecular weight. Using dye binding and fluorescent labeling to measure the stoichiometry of binding, Sutoh finds an actin:S-1 ratio close to one for several different experiments. Finally, as mentioned above, the two segments of S-1 that bind to actin cross-link to the same 12-residue peptide at the actin N-terminus. These results suggest that S-1 binds to only one actin monomer, but two types of attachment between S-1 and actin are possible. The origin of the doublet of cross-linked species observed on denaturing gels is attributed to cross-linking of actin to one or the other of the S-1 binding sites. However, the actin:S-1 stoichiometry of 2:1 obtained by Mornet et al (18) using radiolabels is not consistent with this information. Thus the question of whether S-1 binds to two actins is not yet resolved.

In all of the studies discussed above, the Acto–S-1 complex is that of the rigor conformation. This is generally believed to be the orientation of the actin-attached head at the end of the cross-bridge cycle. The conformation of the Acto–S-1 complex may well be different during other parts of the cycle (see section on Role of Myosin in Force Generation).

LIGHT CHAINS   The light chains of myosin fall into two general classes, regulatory and essential, with one of each class being bound to each head of the myosin molecule. It should be noted that the light chains of myosin are often referred to by other names depending on their source, method of removal from myosin or other factors. We shall refer to them here as regulatory or essential. The regulatory light chains are believed to play a role in the regulation of contraction by Ca^{2+}. This is well established in the case of invertebrate systems and in vertebrate smooth muscle and has recently been reviewed by Szent-Györgyi (71), Kendrick-Jones & Scholey (4), and Kendrick-Jones et al (5). Several studies also suggest that light chain–linked Ca^{2+} regulation may occur in vertebrate striated muscle (72–75) but the results are not as clear in these systems. The essential light chains were so named because they appeared to be essential to the ATPase activity of myosin. The role of this class of light chains in the contractile mechanism is somewhat more ambiguous than the regulatory class.

The disposition of the light chains within the head of the myosin molecule is of fundamental importance to their role in the contractile process and a number of recent studies are relevant to this point. Early EPR (76) and proteolytic digestion (77) studies showed that divalent cation binding to the regulatory light chain of skeletal myosin protected the S-1/rod junction from digestion. These experiments suggested that the light chain with bound Ca^{2+} may sterically block the enzymatic attack at the head-rod junction or that Ca^{2+} binding might induce a conformational change in this region. Electron microscopic studies of the Acto–S-1 complex of regulated vs desensitized scallop S-1 (12, 13) and rabbit Acto-HMM (14) reveal that a significant part of the mass of the regulatory light chain is located in the neck region of myosin. Similar results have been obtained by Flicker et al (27), who used antibodies to the regulatory light chain of scallop. Electron microscopic observation of the myosin antibody complex showed that the antibody is preferentially bound to the neck region of myosin. These findings have led to the speculation that the regulatory light chain may function by modulating the structure or mobility of the myosin heads at the S-1/rod junction.

The location of the essential light chain in S-1 has also been investigated. Antibodies to scallop essential light chains have been shown to bind near the S-1/rod junction of myosin (27). Proteolytic digestion experiments (28) show that a 3-kd peptide at the C-terminus of rabbit S-1 contains a major binding site for the essential light chains. The location of the essential light chain, however, seems to be spatially more extensive than this since the essential light chain of rabbit myosin can be cross-linked to actin in the rigor complex (20). Prince et al (78) have also found evidence for interaction between the essential light chain of rabbit myosin and actin using ^1H-NMR. Thus it seems that the essential light chain may well span the length of S-1 from the neck region to the actin binding site.

Although a fair amount of information on the function of the regulatory light chains has been obtained, the function of the essential light chain is currently unknown. It seems unlikely that this class of light chains is truly essential to the ATPase activity of myosin since the S-1 heavy chain can hydrolyze ATP in the absence of light chains (79, 80). A possible role for the essential light chains in the regulation of scallop myosin has recently been proposed by Hardwicke et al (81). Wallimann & Szent-Györgyi (82, 83) have shown that binding of antibodies to the essential light chain of scallop myosin abolishes calcium regulation both in myofibrils and in isolated myosin. In addition, the binding of these antibodies to myosin, previously freed of regulatory light chain, inhibits the rebinding of the regulatory chains. Their results suggested that the regulatory and essential light chains may be arranged in close proximity and that some interplay might occur between these chains. In an extension of these studies, Hardwicke et al (84) have now shown that the presence of the regulatory light chain inhibits chemical modification of thiols on the essential chain, thus supporting the idea that these light chains are in close proximity. A direct demonstration of this proposed interaction was provided by Wallimann et al (85) using photochemical cross-linking. The regulatory and essential light chains were readily cross-linked by a reagent with a length of 6.5 Å. Further results showed that at least 50% of the regulatory chain is in contact with the essential light chain suggesting extensive overlap between these molecules. Recently, these studies have been extended to show that although the C-terminal region of the regulatory light chain remains stationary, the N-terminal portion can move with respect to the essential light chain. The N-terminus of the regulatory chain can be cross-linked to the essential chain under rigor conditions, but not under relaxing conditions. This movement is associated with the on-off state of the regulatory switch and seems to occur on the myosin head before attachment to actin. The studies described above have provided an elegant demonstration of interplay between the light chains in the regulation of the contractile process.

SUBFRAGMENT-1/ROD JUNCTION   The head-rod junction region of myosin is a relatively short peptide connecting the C-terminus of S-1 to the N-terminus of rod. It is clear from the results discussed above that at least part of both of the regulatory and essential light chains lie in or near this region. Furthermore, binding of divalent cation to the regulatory light chain alters the conformation and proteolytic susceptibility of this region, suggesting a possible role in the regulation of contraction.

The head-rod junction is often referred to as the swivel since it is generally believed to behave like a universal joint. Approximately thirty amino acids lie between the rod and S-1 regions in the primary sequence of myosin (39), but it is not clear whether the swivel comprises all or part of this region or whether it extends into the folded S-1 structure. There is considerable evidence that this region is quite flexible both in isolated myosin and in thick filaments. This point is considered in more detail in a recent review (86).

Myosin Rod

The rod portion of the myosin molecule is a coiled coil of α-helices approximately 156 nm long (11). It can be separated from intact myosin by either papain digestion (87) or chymotryptic digestion in the absence of divalent cation (77). This fragment of myosin is insoluble at physiological ionic strength. Under appropriate conditions, the rod can be further degraded to form LMM (light meromyosin) and S-2 (87). LMM, the C-terminal half of the rod, is responsible for its solubility properties and forms the core of vertebrate skeletal thick filaments. S-2 is soluble under physiological conditions and this region of the rod is believed to be loosely bound to the thick-filament surface. This could allow it to swing out from the thick filament during the cross-bridge cycle to permit S-1 to interact with actin (88). The nature and interactions of the rod are of fundamental importance to the assembly and structure of the functionally active form of myosin, the thick filament. It may also play an important role in the mechanism of force production and shortening in muscles (89–91).

PRIMARY AND SECONDARY STRUCTURE   The information contained in the primary sequence of the rod is important to an understanding of the interactions occurring between adjacent myosin molecules in the thick filament. The sequence of two segments of rod, a 258-residue region from S-2 and a 148-residue portion of LMM, have been known for some time (92, 93). Both sequences exhibit the heptapeptide repeat (a-b-c-d-e-f-g) typical of coiled-coil structures where residues a and d are hydrophobic and form the interface between the α-helices in the folded protein. In the S-2 fragment, however, ~20% of the residues in positions a and d are polar compared to 5% for the LMM fragment and 4% for tropomyosin. This finding has led to the proposal (92) that the two chains of S-2 may be more loosely bound than those in LMM or tropomyosin. Parry (94) has further analyzed these sequences for other structural features. Positions e and g are frequently occupied by acidic and basic residues, respectively. Residues at these positions are most readily capable of forming the interchain ionic interactions required for stabilization of the coiled coil. The maximum number of such interactions is found when the two chains are arranged in axial register. It has since been demonstrated that this is the correct arrangement based on cross-linking studies (95, 96). Both chains have discontinuities in the heptapeptide repeat, but are still likely to be 90%–100% α-helical from structure predictions. Finally, analysis of these sequences using Fast Fourier Transform methods has revealed a ~28-residue repeat in the linear disposition of both acidic and basic residues with the basic and acidic periodicities out of phase by 180°. This arrangement was shown to be consistent with parallel packed myosin molecules that are staggered by odd multiples of 14.3 nm with respect to one another as in native thick filaments (97).

Although the myosin heavy chain has proved to be difficult to sequence by classical methods, the techniques of molecular cloning and DNA sequencing have successfully overcome these problems. Karn et al (39) have isolated and sequenced the unc-54 myosin heavy chain gene from the soil nematode *Caenorhabditis elegans*. The amino acid sequence of myosin rod deduced from the unc-54 gene sequence has been extensively analyzed by McLachlan & Karn (98, 99). The entire rod is 1117 residues long ending with a proline residue at the N-terminus. Comparison of this sequence with the completed portions of rabbit (92, 93), rat (100), and *Drosophila* (101) rods indicates a high degree of homology: 42.5% between nematode and rabbit, 48.1% between rat and nematode, and 75.9% between rat and rabbit. In addition, most of the substitutions are conserved at positions where hydrophobic or charged residues are preferred.

The entire segment of rod from Pro-1 at the N-terminus to Pro-1095 near the C-terminus was found to have high helix probability (70%–90% overall with 15 weak spots below 50%) the coiled-coil heptapeptide pattern was followed throughout and strong evidence for the 28-residue unit was found. In fact the entire rod sequence can be mapped as a 28-residue cyclic pattern with 4 principal exceptions where one extra residue is inserted between positions c and d of the heptapeptide repeat. It has been suggested that these "skip" residues modulate the pitch of the coiled-coil (98, 99).

Analysis of this primary sequence data in terms of the interactions occurring between myosin molecules is relevant to any model for thick-filament structure. Because of the high charge density on the coiled-coil surface, ionic interactions would seem to be the most important in filament assembly. McLachlan & Karn (98, 99) have used a linear charged cylinder model to investigate interactions between two rods with various degrees of stagger as the sum of attractions-repulsions. Staggers of 98 residues (14.3 nm) and 294 residues (43 nm) show a high degree of almost uniform attraction throughout LMM and almost no net attraction in the S-2 region. This is consistent with the idea that strong interactions between LMM regions occur in the core of the thick filament and immobilize these regions whereas S-2, because of its weak interactions, can swing away from the filament surface (88, 91). Recent data by Lu et al (102) suggest that a ~5-kd region of LMM near the C-terminus is responsible for the insolubility of rabbit LMM at low ionic strength. It will be interesting to see how this finding will correlate with the above type of analysis. Finally, McLachlan & Karn (98, 99) have discussed several possible packing schemes for coiled coils with the characteristics of myosin rod. Although no firm conclusions were drawn, this information will be useful in future tests of various models for the packing of myosin molecules in the systematic assembly of the thick filament.

PHYSICAL PROPERTIES   The question of flexibility and structural stability of the myosin rod has been a topic of considerable interest for many years. Between the 140-kd LMM segment of the rod and the 40-kd N-terminal segment (short S-2) there is a region (~ 20 kd) that is highly susceptible to proteolysis (77, 87, 103, 104). These studies have led to the suggestion (104) that the regular coiled-coil structure is disrupted in this region leading to hinging of the rod (96, 104). A number of studies have tended to support this idea. Shadowed electron micrographs of myosin by Elliott & Offer (11) show a sharp bend in the rod at ~ 43 nm from the heads (in about 25% of the molecules) and the axial extent of this hinge region is at most 9 nm. They also observed that the rods tended to show a gentle curvature throughout their length. Takahashi (105) has obtained similar results from negatively stained specimens and has additionally observed a flexible region between 63 and 77 nm from the C-terminus. Recently, Trybus et al (106) have clearly demonstrated a bent monomeric conformation of gizzard myosin occurring in $\sim 100\%$ of the molecules (see section on Vertebrate Smooth Muscle and Nonmuscle Filaments). They further showed that this conformation exists both in electron micrographs and in solution.

The early studies of Burke et al (104) using proteolytic digestion, thermal denaturation, and viscosity led these authors to suggest that rabbit myosin rod consists of two relatively stable regions separated by a flexible segment of low thermal stability. Several more recent studies on the rod and its subfragments support this conclusion. Thermal melting studies by Tsong et al (107, 108) using optical rotatory dispersion indicate that as much as 25% of long S-2 (a fragment comprised of the stable short S-2 plus the hinge) may be melted at physiological temperatures. Temperature-jump studies indicated that this helix-coil transition occurs in the submillisecond time scale and its possible role in force generation was discussed. The existence of a "hinge" region of lower thermal stability than short S-2 was also observed by Swenson & Ritchie (109) from differential scanning calorimetry experiments of long S-2 and short S-2. Recently, Potekhin et al (110) investigated the thermal denaturation of the entire rod using differential scanning calorimetry. Their results suggest that rod melts in a set of at least six quasi-independent cooperative domains with the region of lowest stability located near its center. It should be noted that the secondary-structure prediction mentioned earlier (99) suggests that although there are a few weak areas along the rod, the sequence data are compatible with a regular coiled-coil conformation throughout its length. Given the high proteolytic susceptibility of the hinge region and its relatively low thermal stability, this conclusion may be oversimplified. Recent studies on similar structures have indicated that single-residue changes can significantly alter the thermal stability of coiled coils (111). It seems that secondary-structure analyses are not yet able to predict or account for the presence of quasi-independent melting domains within these structures.

The relationship between the thermal stability and proteolytic sensitivity of the rod and its hydrodynamic properties is ambiguous. Although it seems reasonable that a region of high proteolytic sensitivity and low thermal stability in a rodlike particle might be quite flexible, it is not clear that this is the case. A number of physicochemical studies have addressed this question. In an early set of experiments, Highsmith et al (112) used transient electric birefringence to determine the rotational relaxation times for rod and its proteolytic fragments, LMM and S-2. Their results indicated that the measured decay time for LMM fit quite well with that expected for a rigid cylindrical molecule of the size of LMM (M_r 140,000). However, both S-2 and rod exhibited decay times significantly less than expected for the appropriate cylindrical models. These results suggest that myosin rod is flexible somewhere within the S-2 region. Recent modeling studies by Garcia de la Torre & Bloomfield (113) confirm that these rotational relaxation times are consistent with hinging of the myosin rod about the S-2/LMM junction.

Recently, Highsmith et al (114) used depolarized light scattering and high-resolution ^{1}H-NMR to examine internal motions in the myosin rod. Consistent with the previous results, LMM was found to behave like a rigid linear molecule and rod was observed to be somewhat flexible. Modeling of the rod data using a theory developed by Zero & Pecora (115, 116) suggests that the myosin tail is best described as a once broken rod and that the LMM and S-2 domains can diffuse freely within a conical volume having a maximum angle of $\sim 130°$ with respect to the rod axis. Complementary ^{1}H-NMR experiments suggest that the structure responsible for flexibility is not likely to be random coil since less than 4% of the total rod could be assigned as random coil. Both LMM, which is rigid, and rod, which is flexible, contained the same amount of random structure. A similar result was obtained from ^{1}H-NMR melting studies on long and short S-2 by Stewart & Roberts (117). The results, however, did not exclude the possibility that the coiled coil could "melt" to form a nonflexible conformation on the NMR time scale.

Although the above studies provide evidence that myosin rod is a flexible hinged molecule, other studies have given contradictory results. Fluorescence depolarization experiments of Harvey & Cheung (118) indicate that myosin rod is essentially rigid at neutral pH over a temperature range of 5°–50°C. Below pH 4, appreciable bending was observed consistent with the viscosity studies of Burke et al (104). It seems unlikely, however, that myosin rod behavior under these conditions is relevant to its physiological state.

Studies on the viscoelastic properties of myosin and myosin rod have also indicated that the rod is not hinged. Using the Birnboim-Schrag multiple lumped resonator apparatus, Rosser et al (119) and Hvidt et al (120) have measured the storage and loss shear moduli of myosin and rod at 7°C in the frequency range of 150–8000 Hz. Their results indicate that myosin rod is neither rigid nor hinged but behaves as a semiflexible rod at this temperature. Although the presence of a hinge was inconsistent with the viscoelastic data, the rms (root-mean-square) excursion of the N-terminus of S-2 (due to Brownian motion) was estimated to be 26.5 nm from the thick-filament surface (120). Thus the cross-bridge could easily span the gap between thick and thin filaments during the contractile cycle.

The origin of the disparity of results on the flexibility of myosin rod remains to be established. Part of the problem may be the time domain in which motions occur, since various methods used to study this problem are sensitive to motions on vastly different time scales. Additionally, many of the hydrodynamic measurements have been made at relatively low temperatures (20°C or less) and may not be relevant to the state of the rod at physiological temperature. Another possible source of variations lies in the effect of solvent since the proteolytic susceptibility of myosin rod is highly sensitive to the presence and concentration of Hofmeister series anions (121). Recently the question of flexibility within the myosin molecule has been thoroughly reviewed by Harvey & Cheung (86). Although the high flexibility of myosin heads about the S-1/rod junction is now generally accepted, a flexible joint within the myosin rod remains to be unambiguously established.

AGGREGATION PROPERTIES OF MYOSIN

Structure of Thick Filaments

THICK FILAMENTS OF VERTEBRATE STRIATED MUSCLE   The thick filaments of muscle are remarkable bipolar structures in which myosin molecules are systematically ordered with their tails in the backbone and heads arranged along the surface. The basic organization of myosin molecules in the vertebrate striated thick filament was originally deduced by Huxley (6) from electron micrographs of negatively stained filaments. Packing is antiparallel near the center of the filament (the bare central zone) and parallel throughout the remainder of the structure which has an overall length of 1.6

μm. In recent years, X-ray diffraction studies of living, relaxed muscles from a variety of animals have provided important information on filament symmetry, cross-bridge conformation, and backbone structure. X-ray patterns from resting vertebrate striated muscle (fast and slow skeletal and cardiac) show a series of layer lines that are believed to arise from the myosin cross-bridges (97, 122, 123). The pattern is consistent with an approximately helical arrangement having an axial repeat of 43 nm and a subunit repeat of 14.3 nm (97). Since the X-ray patterns contain no phase information, they do not reveal the handedness of the helical arrangement nor do they provide an unambiguous answer to the number of cross-bridges, n (the rotational symmetry), at each successive 14.3 nm level. Because the rotational symmetry is important in the architecture of the thick filament and its mode of interaction with neighboring thin filaments, a number of groups have attempted to determine this parameter by other more indirect methods. Values of $n = 3.5$ have been obtained from nucleotide binding in myofibrils (124, 125). Two reports based on quantitative studies using SDS polyacrylamide gel electrophoresis have given the value $n = 2.7$ (126, 127), and two others, $n = 3.9$ (128) and $n = 3.8$ (129). Calculations based on the myosin content of muscle coupled with the number of thick filaments per unit volume gives the value $n = 4.4$ (130). Estimation of the number of thick filaments using a particle counting technique gives the value $n = 4.3$ (128).

It seems clear from these studies that $n = 3$ or 4, but in view of the lack of agreement among the various methods a more direct approach is required to provide a definitive value for the rotational symmetry. Two recent investigations using the scanning transmission electron microscope (STEM) seem to satisfy this requirement. The STEM technique, which determines the mass per unit length of the filament by quantitative electron scattering, has given values of $n = 2.7$ (131) and $n = 2.9$ (132).

Evidence bearing on this question also comes from image processing of electron micrographs. Thin transverse sections through different parts of the A-band region of cross-linked vertebrate skeletal muscle show back-bone and subunit profiles of the thick filament consistent with three-fold rotational symmetry (133, 134) and, by inference, a three-stranded helix. The symmetric relationships possible in a sarcomere superlattice constructed around three-fold thick filaments provides additional support for this model (135, 136). Recent computer-image analysis of isolated, negatively stained frog vertebrate skeletal thick filaments are also in accord with a three-stranded structure (137). This helical arrangement of bridges is compatible with Squire's earlier (138, 139) generalized structure of the thick filament in which myosin molecules from diverse muscles (vertebrate and invertebrate) have a single basic mode of packing. In these models the rotational symmetry is proportional to filament diameter. All myosin molecules are in equivalent environments and are tilted and twisted slightly with respect to the filament axis so that the heads are at the surface while the myosin tails form the backbone of the structure. The core of these structures is hollow or contains paramyosin.

The presence of substructure in thick filaments from a variety of muscles has been recognized for some time. Early electron micrographs of transverse sections of invertebrate muscle (140, 141) indicated subunits spaced 2.5–4.5 nm apart. Optical diffraction of transverse sections of fixed vertebrate skeletal muscle has also revealed the presence of regular substructure with subunit spacings of about 4 nm (142–144). These optical diffraction patterns changed in a predictable manner when the sections were tilted, suggesting that the substructure is probably one of parallel rods. Pepe & Drucker (145) employed photographic rotational superposition (146) to enhance the substructural detail in vertebrate-skeletal thick-filament cross sections. Their results suggested the presence of 12 subfilaments, 3 centrally located and 9 peripherally located, arranged on an approximately hexagonal lattice. More recently, Stewart et al (147) have used computer-image processing of electron micrographs of transverse and tilted sections (143) to establish the arrangement of the subfilaments. This study provides supporting evidence for the presence of 12 subunits parallel to the long axis of the filament, hexagonally packed, with center-to-center spacing of about 4 nm. Subunit spacings of about 4 nm have also been observed by X-ray diffraction in vertebrate (148) as well as in crustacean muscles (149, 150).

The subunit structure described above appears to be incompatible with the original three-stranded model of Squire (139) in which individual myosin molecules are tilted with respect to the filament axis, but it does have a remarkably close correspondence to the two-stranded structure of the thick filament proposed by Pepe (151). The 12 parallel subfilaments in Pepe's model of the vertebrate thick-filament backbone form three equivalent subgroups. Within each subgroup, subfilaments have axial displacements of 14 and 43 nm, while the displacement between the subgroups is 28.6 nm. The backbone has a three-fold screw axis but no rotation axis; the axial repeat is 86 nm. The myosin cross-bridge distribution has an approximate two-fold rotation axis with four myosin molecules per 14.3 nm interval. Thus, there are four myosin heads per cross-bridge. Since the lateral spacing between each subfilament is 4 nm and the diameter of the α-helical coiled coil of the myosin rod is about 2 nm, Pepe proposes that each subfilament is constructed by assembly of myosin dimers (147, 152). A (parallel) dimer has also been proposed as the basic structural unit of the thick filament based on physicochemical studies of the self-association properties of myosin (153–155).

Most recent measurements favor a rotational symmetry of three for the vertebrate skeletal thick filament. If this value turns out to be correct, it would be incompatible with Pepe's model. An additional difficulty with Pepe's model is that, unlike the Squire model, the myosin molecules are not in equivalent local environments in the packing scheme. This arrangement is at variance with the rule of quasi equivalence of packing environments observed in other multiple subunit structures (156). Nevertheless, the recent finding (152, 157, 158) that vertebrate skeletal-myosin filaments can be made to splay into three groups of subfilaments at low ionic strengths appears to provide supporting evidence for the presence of parallel subfilaments in the intact thick-filament shaft. This observation would seem to exclude Squire's early (139) packing scheme but may not be inconsistent with his later models (159, 160), which retain the principle of equivalence of interactions but suggest alternative structures built of subfilaments of varying size.

The principal of quasi equivalence employed in the Squire models assumes that these structures are constructed from identical myosin molecules. Recently, Miller et al (161) determined the structural location of two genetically distinct myosin molecules in nematode body-wall thick filaments. They propose that one type of myosin (myosin A) interacts in antiparallel fashion to form the bare central zone of these filaments. The other type (myosin B) participates in parallel interactions either with myosin A in the overlap regions flanking the bare zone or with other myosin B molecules to form the distal, polar regions of the filaments. This structural organization may be relevant to other striated muscles that are known to produce two kinds of myosin heavy chain (162–164). Such a packing scheme could account for the puzzling observation that the M-line proteins (165) bind only in the bare central region of thick filaments. It might also explain the anomalous binding of C-protein that appears in electron micrographs of fixed, longitudinal sections to be restricted to seven stripes at intervals of 43 nm on either side of the bare central zone of vertebrate thick filaments (166, 167). Clearly, additional structural information is essential before the detailed three-dimensional packing arrangement of myosin molecules within the vertebrate skeletal filament can be convincingly established.

THICK FILAMENTS OF INVERTEBRATE STRIATED MUSCLE   Invertebrates show a striking diversity in the architecture of striated muscle structures. The thick filaments of these muscles also exhibit considerable variation in molecular structure depending on the species and the anatomical location of the muscle. X-ray patterns from three invertebrate muscles in the relaxed state, representative of three classes (168) of myosin filament structure have been reported by Wray et al (169). Diffraction patterns of a member of the first class, scallop striated adductor muscles, show the helical repeat of cross-bridges to be 48 nm and the number of myosin molecules at each 14.5 nm level to be six or seven. Analysis of the layer line profiles suggested that the bridges are densely packed at each level along the length of the filament and that the heads are tilted rather than perpendicular to the filament axis. These conclusions have been strongly supported by recent electron microscope studies (170) of negatively stained scallop thick filaments isolated according to the mild preparative procedure developed by Kensler & Levine (171). Three-dimensional image analysis reveals (170) many of the structural features found in the X-ray diffraction patterns of the living, relaxed scallop muscle. The cross-bridge array has seven-fold rotational symmetry with surface projections running almost parallel to the filament surface but are slewed to lie along right-handed helical tracks with a pitch of 48 nm.

The second class of invertebrate filament structure, defined by its X-ray diffraction pattern (172, 173, 169), is represented by *Limulus* (horseshoe crab). The leg muscles of tarantula spiders have an almost identical X-ray pattern (168). The surface lattice of cross-bridges in these structures has a helical repeat of 43.5 nm with subunit repeat of 14.5 nm. Electron microscopy and image analysis (171, 174–176) of negatively stained *Limulus* filament preparations demonstrate that the helical cross-bridge array is also right-handed with four-fold rotational symmetry. Again, the myosin heads appear to be markedly tilted and are inclined at an angle of about 30° to the helical path.

The third class of invertebrate myosin filaments comprises certain crustacean and insect flight muscles. The bridges of the relaxed muscles in this class are very disordered compared to those of scallop and *Limulus* filaments suggesting considerable movement about their average positions. In this class, the helical repeat is not constant but ranges in value: 30–32 nm for fast muscles of lobsters and crayfish, 33 nm for striated crab muscles, 35–36 nm for lobster slow muscles, 36 nm for dragonfly flight muscles, and 38.5 nm for asynchronous flight muscles (168). Unlike *Limulus* and scallop muscles, negatively stained lobster and insect flight muscle thick filaments show no well-defined surface lattice, so it has proved impossible to date to establish the rotational symmetry of the cross-bridge array by electron microscopy and image analysis. A recent determination of the thick-filament mass of insect flight muscle (*Lethocerus* and *Musca*) from STEM measurements (132) gave a value of four for the rotational symmetry.

From a comparison of X-ray diffraction data of several crustacean muscles, Wray (150) has suggested that the thick-filament backbone of these invertebrates has the architecture of a simple, multistranded cable wrapped around a hollow core. Individual strands (subfilaments) of the cable have diameters of about 4 nm consisting of a cylindrical group of three myosin rods. The myosin rods forming each subfilament twist around each other with a helical repeat of 43.5 nm in order to make all the myosin heads accessible to the filament surface. Neighboring subfilaments are staggered by 14.5 nm giving a regular helical array of cross-bridges with subunit repeat of 14.5 nm but a helical repeat that varies among muscles.

All of the invertebrate thick filaments discussed above contain varying amounts of the α-helical, coiled-coil fibrous protein paramyosin (168, 177, 178), which has been shown by hydrodynamic and light-scattering data to be about 135 nm long with a M_r of 220,000 (179). The arrangement and function of the paramyosin molecules is still not clear in these muscles but they are thought to occur in the interior of the thick filaments along part or all of their length (180–182). Electron microscopy of sectioned invertebrate muscles from several species suggest that the filaments have a hollow core (170, 183, 184). There is also considerable variation in the diameter (15–25 nm) and length of filaments among the invertebrate striated muscles (184). Scallop striated muscles, for example, have lengths of 1.8 μm (181), whereas crab eyestalk muscle thick filaments are about 7.5 μm (185).

THICK FILAMENTS OF CATCH MUSCLE   Molluscs possess a slowly acting, high-tension smooth muscle, the catch muscle, which can hold the shell of these bivalves closed for hours or even days with a very small turnover of ATP. Like vertebrate skeletal muscle, molluscan adductor muscles contain thick and thin filaments and are believed to contract by the relative sliding of these two sets of filaments through the operation of cross-bridges (186). The thick filaments of these muscles are bipolar (187), roughly circular in cross-section (188, 189) and show large variations in diameter and length. Diameters up to about 100 nm and lengths of several microns have been reported in the catch muscle of the scallop *Pecten maximus*. Thick filaments of the smooth adductor muscle of the clam *Mercenaria mercenaria* have diameters varying from 50–150 nm and lengths up to at least 15 μm (190). The greater part of the filaments (up to about 90% by weight) is not myosin but paramyosin, which can be correlated with their great lengths and diameters. The paramyosin is known to form the underlying core of these filaments, which is surrounded by a surface layer of myosin (187).

Early electron microscope studies of stained thick filaments from the adductor muscle of the clam *M. mercenaria* showed a regular two-dimensional pattern of spots (191), which is now known to come from the paramyosin backbone. This pattern, the "Bear-Selby" net, has since been observed in thick filaments isolated from a variety of catch muscles. It has a repeat distance (c) along the fiber axis of 72.5 nm (192, 193) and a transverse repeat (a) that varies substantially with water content. Two different models have been proposed for the molecular organization of the thick-filament backbone. G. F. Elliott (189) suggested a crystalline layer structure within the filaments based on X-ray diffraction and electron microscope observations. A. Elliott & Lowy (194) proposed a model in which a surface, constructed from a two-dimensional Bear-Selby net, was rolled up like a carpet to form a helicoid. Recent electron microscope studies of stained filaments and sectioned muscle (190, 195) have now provided convincing evidence that the paramyosin core resembles a single crystal in its molecular arrangement: On rotation of individual, negatively stained filaments, the Bear-Selby net disappears and is replaced by a pattern of transverse lines 14.4 nm apart, which is incompatible with a helical structure. A three-dimensional synthesis computed from electron micrographs of filaments with a range of tilts about the long axis (196, 197) confirm that the basic packing scheme is a layered structure, although the crystalline pattern is somewhat different than that proposed earlier by G. F. Elliott (189). These studies show that the dark-staining spots forming the checkerboard pattern of the Bear-Selby net are gaps between successive paramyosin molecules along the filament axis (198) that run right through the filament from one side to the other. Transverse sections of smooth adductor muscle show no internal features along the filament axis but, on tilting, striations appear on the filaments that can be correlated with the planes of the Bear-Selby net (199).

The packing arrangement of myosin molecules on the surface of the paramyosin cores is still unresolved. Squire (138) has suggested, in accord with his generalized model for muscle thick filaments, that the myosin molecules form a mantle of constant thickness (3.5 nm) around the paramyosin core. This arrangement could account for the high paramyosin:myosin ratio in very thick molluscan filaments. Recently, Cohen (200) has proposed that the myosin in the thick filaments of catch muscles comprises a single layer of molecules whose assembly and contractile activity are controlled by the underlying core of paramyosin. In her model, each myosin molecule would occupy a constant surface area of about 435 nm^2 compared with 150 nm^2 in other thick-filament models (e.g. those of Squire and Wray). She proposes that the noncatch state is a weak linkage of the myosin S-2 region with the paramyosin core, whereas catch is an unusual state resulting from strong paramyosin interactions along the entire myosin rod. These interactions are believed to be modulated by phosphorylation (201) of paramyosin.

Disassembly and Assembly of Myosin Filaments

NATIVE THICK FILAMENTS   Native thick filaments from all muscle types are dissociated in high salt ($\sim$0.6 M KCl) into their constituent protein subunits, suggesting that the packing forces holding these structures together are mainly ionic in nature. According to equilibrium sedimentation and quasi-elastic light-scattering studies, vertebrate skeletal myosin molecules still exhibit a strong tendency to associate even in high-salt solvents where they are in a rapidly reversible monomer-dimer equilibrium (153, 202–204) but this view has been challenged by Szuchet (205) and Emes & Rowe (206), who find only myosin monomers under these conditions.

Dissociation of native vertebrate filaments appears to occur in three distinct stages on increasing the salt concentration (207). Electron microscope studies show that the lengths of these filaments drop gradually and continuously with increasing amounts of salt until at a concentration of 160 mM KCl (pH 7.0) the filaments are reduced to a length of about 1.3 μm. A further elevation in salt to 170 mM KCl results in a precipitous, cooperative drop in length to 0.8 μm. Above 170 mM KCl only the shorter fragments ("stubs") are found. As the KCl concentration is elevated to 300 mM, the lengths of the stubs again decreases smoothly to a third intermediate state of length about 0.2 μm. Above this concentration complete dissociation of these fragments occurs. The three stages in the sequential dissembly process presumably result from variations in packing of myosin and the presence of other thick-filament proteins, notably the C-protein that, as mentioned earlier, is located at seven positions 43 nm apart and flanking the central M-band region of the filaments.

Dialysis of native myosin filaments to either pH 8 or 0.2 M KCl (pH 7) yields native bare-zone assemblies (BZA) of length 0.3 μm similar to the thick-filament stubs described above. Niederman & Peters (208) have now shown that these fragments can be used to nucleate the reassembly of native-sized myosin filaments. On transferring the dialyzing system to polymerizing conditions (0.1 M KCl) the BZA, and the dissociated distal myosin molecules recombine to form 1.5 μm bipolar filaments with a narrow size range. When the ratio BZA/distal-myosin is altered, the resulting filament length distribution is found to be inversely related to the amount of BZA present. Native-length filaments are no longer generated once the bare central structure is destroyed. When native thick filaments, isolated BZA fragments, or distal myosin molecules are dissolved in 0.6 M KCl, then dialyzed against low-salt polymerizing solvents, very long filaments (5 μm) are formed. The nucleating BZA fragments apparently retain (208) the M-line proteins located in the bare zone of the myosin filament (209, 210). However, the C-protein would appear to be dissociated, along with the distal myosin, since it is localized in the outer 3/4 segments of the filament. It is not clear whether BZA fragments devoid of M-line proteins can nucleate assembly of native-like filaments under these ionic conditions.

SYNTHETIC THICK FILAMENTS   The study of filaments formed in vitro from purified myosin has provided additional information on the mechanism of assembly and structure of thick filaments. Like many other self-associating reactions, the process of filament formation is critically dependent on changes in pH and ionic strength, and both the length and diameter depend on the exact ionic environment present during the polymerization process. When the ionic strength of a solution of myosin from rabbit skeletal muscle is lowered from 0.6 to 0.1 M, bipolar filaments markedly similar in morphology to those found in intact muscle are formed (6, 211). Rapid dilution at neutral pH generally gives shorter particles (0.5 μm) with a more uniform length distribution than does slow dilution or dialysis (6, 211, 212) consistent with a nucleation-controlled reaction. Slow dialysis can yield filaments between 1 and 12 μm in length (213). Early electron microscope studies of the assemblies formed by overnight dialysis of myosin vs a low-salt (0.15 M KCl) solvent also revealed striking dimensional variations depending on pH (211–214). At pH 6.2, long, threadlike particles are observed with a very broad distribution of lengths (varying from 1–12 μm) and with diameters between 30 and 50 nm. Near neutrality, particles with topological features similar to those of native thick filaments are seen. These have a mean length near 1.2 μm. Longer particles are observed on slow dialysis or dilution that show no bare central zone (215, 216). Filaments generated near pH 8 exhibit a remarkably sharp size distribution with mean length near 0.63 μm. These short particles have numerous projections, a bare central region and a diameter of 10–15 nm characteristic of native filaments (218).

Detailed centrifugation studies of this system have shown that the filament is in rapid reversible equilibrium with a more slowly sedimenting species, provisionally identified as monomer (218). It was further established that the equilibrium is displaced toward monomer with increasing pH, salt, and hydrostatic pressure (217–219). Above a critical concentration, a constant monomer concentration is in equilibrium with any concentration of polymer in these systems. The critical concentration depends on the ionic strength, pH and hydrostatic pressure showing that skeletal myosin does indeed form filaments by a nucleation-controlled polymerization mechanism. Vertebrate *smooth-muscle* myosin polymerization is sensitive to the same solution conditions, but does not exhibit a critical concentration. In this system the concentration of monomer varies with total protein concentration (220). The myosin used in these studies was largely unphosphorylated. It has since been shown (221, 222) that phosphorylation of the 20-kd light chains of this myosin has a marked effect on filament assembly (see below). This might account for the different polymerization mechanisms observed between vertebrate striated and smooth-muscle myosins [as suggested by Lowey & Megerman (220)] and further comparative studies of phosphorylated and unphosphorylated myosins are required to resolve this question.

The dependence of the myosin-filament equilibrium state on small changes in salt concentration, pH, and hydrostatic pressure supports the view that the forces holding these filaments together are mainly ionic in nature. It is also supported by the lack of a temperature dependence of the equilibrium constant (218, 220). Ionic bonding is generally characterized by negligible heat effects, whereas hydrophobic "bonding" usually exhibits small but positive enthalpies of formation. Judging from the volume change ($+300$ cm^3 per mole monomer) derived from the pressure dependence of the equilibrium constant, only about 25 bonds are formed per mole monomer incorporated suggesting a delicate ionic balance between the filament and its surrounding medium. Indeed it has been shown (223) that the binding of only one or two molecules of ATP, ADP or pyrophosphate to the rod portion of rabbit skeletal myosin (at pH 8.3) destabilizes the polymer, whereas the presence of millimolar concentrations of Mg^{2+} or Ca^{2+} in the polymerization medium effectively counteracts this effect (223, 224). Pinset-Härström & Truffy (224) recently reported that the width of the very long filaments generated by slow dilution at pH 7 is markedly reduced (to physiological diameters of 15–17 nm) from the 30–50 nm characteristic of these filaments in the presence of MgATP.

MECHANISM OF ASSEMBLY   It seems clear that the bare central zone of the thick filament is formed early in the assembly process and that this region is constructed from myosin molecules arranged in overlapping antiparallel (tail-to-tail) array to form short bipolar structures. This process is followed by addition of myosin in a parallel fashion (head-to-tail) at each end of the bipolar unit (6, 97, 151, 211–214, 216, 225). Recent pressure-jump kinetic studies (155, 226, 227) have provided important clues to an understanding of the overall self-assembly reaction. When the hydrostatic pressure of the myosin self-assembly equilibrium at pH 8.1 is increased abruptly, a biphasic dissociation curve is observed in which a linear decrease in turbidity is followed by a second pressure-insensitive phase. The reaction is rapidly reversible over the first phase; it is much slower when the filaments are reconstituted from the second phase of the dissociation curve. The results suggest that the first phase represents the effect of pressure on the propagation or growth phase of the assembly, whereas the second phase represents the effect on the more stable bare-zone region at the center of the filament. Electron micrographs of filaments chemically cross-linked at various hydrostatic pressures show a linear relationship between filament length and pressure suggesting that filaments shorten symmetrically towards the bare zone while maintaining a narrow length distribution. From an analysis of the relaxation kinetics following abrupt increases and decreases of pressure in the growth phase of the reaction, it was proposed that the building block for filament construction is a myosin dimer (227, 228). The dimer is postulated to add to the filament ends at a rate that is independent of filament length, whereas the rate constant for dissociation increases exponentially (by a factor of 500) as the filament grows from the bare zone to its full length, suggesting that the length of the filament is kinetically controlled. The presence of dimer at high concentrations in the assembly equilibrium at pH 8.1 was established by chemical cross-linking of species present at various hydrostatic pressures (155). The predominant species (73%) observed in electron micrographs of the cross-linked myosin was a parallel dimer with an axial displacement of 44 nm. A dimer of the same geometry was proposed earlier as the basic building block of the thick filament based on hydrodynamic (153, 154) and cross-linking experiments (229) of the assembly equilibrium. The results of turbidometric stop-flow experiments of Katsura & Noda (212, 214) are also consistent with a nucleation-growth scheme for filament assembly in which the myosin dimer is a basic building unit in this reaction.

The experiments discussed above suggest bilateral growth of the thick filament by addition of parallel dimers to the short bipolar bare central zone fragments. Cross-linked antiparallel dimers were also observed in these studies [amounting to about 8% (155) and 13% (229) of the species counted], which might serve as building units of the bare central zones. The processes involved in the formation of this, the nucleating region, are still obscure, but recent investigations by Reisler and colleagues (230–232) are particularly relevant to this stage of assembly. Homogeneous, bipolar particles of length 0.30 μm and diameter, 8 nm, consisting of 16–18 myosin molecules (minifilaments) have been prepared (230) by a two-stage dialysis of myosin from 0.6 M KCl into low-salt buffers [pyrophosphate (5 mM) then citrate-tris (10 mM)] at pH 8.0. These particles, which are very stable at pH 8.0, have the appearance of short bipolar filaments (6) with a bare central region of 160–180 nm. Addition of KCl to a solution of these minifilaments (at pH 8.0) results in rapid bilateral growth (231) to form the 0.63-μm-long synthetic filaments described by Josephs & Harrington (213). Small amounts of minifilaments seed the explosive growth of filaments at pH 8.0 consistent with their role as the nucleating units in assembly.

VERTEBRATE SMOOTH-MUSCLE AND NONMUSCLE MYOSIN FILAMENTS   Vertebrate smooth-muscle thick and nonmuscle (cytoplasmic) filaments show a common and remarkable dependence of their assembly properties on phosphorylation of their constituent 20-kd light chains (221, 222). At physiological ionic strength and pH, nonphosphorylated, smooth-muscle (gizzard) and nonmuscle (thymus and platelet) myosin filaments are dissociated by MgATP to form a myosin species of sedimentation coefficient $\sim 11S$ (106, 221, 233). When the 20-kd light chains are phosphorylated, the 11S species reassemble into filaments that are stable in MgATP (221, 222, 234, 235). Hydrodynamic and electron microscope studies have shown that the 11S species is a monomeric myosin molecule that has a folded, pretzel-like conformation (106, 236–238) in contrast to the extended shape characteristic of the conventional 6S monomeric myosin in high salt (11, 87). Rotary shadowed specimens reveal that the 11S molecules are folded at two hinge regions located approximately one-third and two-thirds along the length of the tail from the myosin heads (237, 238). The folded tail appears to be bound near the distal second hinge to the neck region of the myosin heads. Light-chain phosphorylation (in the presence of light-chain kinase, calmodulin and Ca^{2+}) results in a striking change in the conformation of the folded myosin molecules (238). The myosin tail unfolds to form the conventional elongated 6S structure, which then rapidly assembles into filaments. Thus phosphorylation of the light chains in these systems induces the formation of bipolar filaments that may act in combination with actin to produce force and movement.

Vertebrate smooth-muscle myosin can assemble into short (0.5 μm) bipolar filaments with a central bare zone, similar to those formed by striated muscle (216, 239–243) or it can form short bipolar segments (244). Under certain ionic conditions smooth-muscle myosin can also assemble into long "side-polar" filaments that have cross-bridges of the same polarity (14-nm spacing) along one side of the filament and the opposite polarity along the other side (243, 245). These filaments have no bare central zone. Thick filaments having an appearance similar to the synthetic side-polar form have been observed in smooth-muscle homogenates (216, 246) but it is not clear whether such structures do reflect an in vivo form of myosin. X-ray diffraction patterns of living, vertebrate smooth muscle (247, 248) sometimes show a weak 14.3-nm axial reflection similar to striated muscle, suggesting that some of the myosin must be in filament form under physiological conditions, but the type of structure possessed by these filaments (bipolar or side-polar) is still uncertain.

Cytoplasmic myosins polymerize to form bipolar filaments under physiological conditions (249–251) with a bare zone the length of the myosin tail, which is flanked by terminal regions of variable length. The myosin heads protrude from the surface of the filament at 15-nm intervals. The filaments formed by purified cytoplasmic myosins are small compared to vertebrate and invertebrate striated muscle thick filaments. Platelet myosin filaments, for example, are composed of about 30 myosin molecules and are 10–11 nm wide and 0.3 μm long (251). *Acanthamoeba* myosin II filaments form structures of two different sizes: thin bipolar filaments 7 nm wide and 200 nm long consisting of 16 myosin II molecules reminiscent of vertebrate skeletal minifilaments (230) and thick filaments of variable width (14–19 nm), which consist of 40 or more myosin II molecules (252).

Acanthamoeba myosin II has three phosphorylation sites (six sites per molecule) near the end of the tail of each of its two heavy chains (253, 254). Myosin II with four to six sites phosphorylated shows little or no actin-activated ATPase activity, while the maximally dephosphorylated enzyme has maximal actin-activated ATPase activity (253, 255, 256). Both phosphorylated and dephosphorylated myosin II form bipolar filaments (257) under conditions of the enzyme assay. Kuznicki et al (258) have now reported the startling observation that regulation of the actin-ATPase activity occurs through phosphorylation at these sites in intact bipolar filaments, suggesting some type of generalized conformational effect within the backbone of the filament. These findings may be relevant to the catch mechanism of Cohen (200) described earlier. In her model, dephosphorylation of paramyosin within the core of the catch-muscle thick filament results in relaxation of catch (release of actin-attached S-2 elements from the filament surface) and rapid cycling of cross-bridges.

The in vitro experiments described above for both cytoplasmic and smooth-muscle myosin show that filaments are formed only when the regulatory light chain is phosphorylated. Thus functional bipolar filaments may exist only during periods of contractile activity. This interpretation should be treated with caution, however. Somlyo et al (259) recently reported that in resting smooth muscle, which has been rapidly frozen to avoid the effects of fixation, thick filaments are clearly present but they are not phosphorylated.

ROLE OF MYOSIN IN FORCE GENERATION

The central problem in muscle research today is to understand the molecular mechanism of the force-generating event. There is now general agreement that contraction involves an active sliding process developed between filaments of actin and myosin (260, 261). There is also a large body of well-established evidence, based on structural, biochemical, and physiological studies, that the elements responsible for the generation of contractile force reside in the cross-bridges, the HMM regions of myosin.

When muscle is stimulated to contract, the actin and myosin filaments slide past each other by several hundreds of nanometers and ATP is cleaved at a rate that requires repetitive attachment and detachment of the bridges. The filaments themselves do not shorten in this process, but the movable elements, the cross-bridges, have a range of axial translation of about 10–12 nm per cycle (262, 263). Biochemical studies of the interaction of isolated S-1 subunits with actin filaments in vitro reveal that ATP is cleaved each time S-1 undergoes binding and release; the maximum rate of cleavage approximates the cycle time of the cross-bridge in its activated physiological state (2). There are, therefore, very good reasons to believe that the biochemical cycle observed in the test tube is closely linked to the mechanical cycle responsible for force generation in contracting muscle.

From early X-ray diffraction studies of vertebrate striated muscle, as well as observations of sectioned muscle in the electron microscope (97, 264, 265), it is clear that the actin and myosin filaments are parallel to each other and organized into a regular two-dimensional hexagonal lattice. In other muscles the same basic arrangement exists, although the details may vary, such as the number and location of the thin filaments around each thick-filament backbone. When the sarcomere shortens during contraction, the two-dimensional lattice expands laterally (266, 267) as the thin filaments are drawn into the thick-filament array (A-band region). This means that at long sarcomere lengths the filaments are closer together than at short sarcomere lengths so the tension-generating mechanism must be able to operate over a range of interfilament distances. Thus at different sarcomere lengths the cross-bridges must differ in length or in orientation to accommodate the changing lateral distance between filaments. In current cross-bridge models for force generation this is accomplished by means of two flexible joints located at either end of the S-2 region of myosin: the S-1/rod swivel and the LMM/HMM hinge. This arrangement permits the cross-bridges to move easily through all necessary interfilament distances as the fiber contracts.

In the sliding filament-rotating head model (88, 262, 268) for force generation, the cross-bridge (HMM region) swings away from the thick-filament surface and the heads attach to actin in a nearly 90° orientation. This step is followed by a rapid transition in the effective angle to a more stable 45° configuration—the rigor angle (269, 270). In principle this process could give the radial and longitudinal components deduced from observed changes in the X-ray patterns of living, activated muscle (97, 271) and could result in sliding of the actin filaments toward the center of the A-band by about 10 nm per cycle. Additional refinements and modifications of this model (272–277) have been proposed in recent years, but in all cases some type of structural transition resulting in a change in the effective angle of the actin-attached myosin head is presumed to be the force-generating event.

The original H. E. Huxley (88) model assumed that the S-2 region of the cross-bridge was a rigid, inextensible link between the filament backbone and the S-1 subunits. In the model of A. F. Huxley & Simmons (262), which was based on analysis of tension transients in isometrically contracting muscle fibers following abrupt step changes in length, S-2 was assigned a more active role. In this model S-2 contains an elastic element capable of being extended by as much as 10 nm as the head rotates on the actin filament.

A different type of force-generating mechanism in a cycling cross-bridge has been proposed (89, 90) in which a conformational change in the S-2 region rather than head rotation, is considered to be the force-generating event. According to this model S-2 is constrained in the α-helical conformation in the resting state of muscle where it is part of the thick-filament surface. When the S-1 subunit attaches to actin in a cross-bridge cycle, the S-2 link is released from this surface resulting in a rapid helix-coil transition in a part (the LMM-HMM hinge domain) of the coiled-coil structure. This process causes shortening in S-2 and produces force.

Changes in the X-ray diffraction pattern (myosin layer-line reflections) when frog sartorious muscle is stimulated to contract show that the helical arrangement of cross-bridges around the backbone of the thick filaments in the resting state undergoes disordering on a time scale (millisecond) approximating the development of tension (278). It also seems clear from changes in the equatorial patterns that the cross-bridges move away from the surface of the thick filaments on activation (279–281). Neither the equatorial nor the layer-line changes occur in muscle stretched to nonoverlap lengths between thick and thin filaments (280, 282, 283). Hence the decrease in helical ordering appears to be associated with attachment of cross-bridges to actin when they develop tension.

The evidence that the myosin heads attach to actin in a 90° orientation is less compelling. Both X-ray and electron microscope studies suggest a variety of resting configurations of the cross-bridges depending on the muscle (169, 170, 176, 284). More importantly, the S-1 subunits of resting muscle are believed to be executing large angle ($\sim 40°$) Brownian motion rotations about their mean resting orientations (285–288), so it is possible for them to attach to actin over a wide range of angles in the initial phase of the cycle (2).

Present evidence indicates that the S-1 subunits are of sufficient length (18–19 nm) to attach to actin before release of the cross-bridge from the thick-filament surface even at the rigor angle (9, 11, 13). Cross-linking experiments (289, 290) on glycerinated muscle fibers in rigor show that the myosin heads can be cross-linked to the backbone of the thick filaments while they are still attached to the thin filaments without any detectable change in lattice spacing. This point is of particular significance in the helix-coil mechanism since the S-1 subunit must bind to the thin filament *before* the α-helical S-2 region of the cross-bridge is released from the surface of the thick filament in order to generate tension between the two sets of filaments.

Despite intense effort, unambiguous and conclusive evidence for changes in the orientation of myosin heads while they are attached to actin in a contractile cycle has not been demonstrated. Several recent studies have dealt directly with this question. Huxley et al (271) have measured time-resolved changes in the low-angle X-ray diffraction pattern of contracting frog muscles during the transient response to abrupt (1 msec) changes in muscle length. They found large changes in the intensity of the 14.3 meridional reflections consistent with longitudinal movement of the actin-attached cross-bridges, but little or no change was observed in the equatorial pattern that would signal head rotation. The polarization of a fluorescence probe bound to S-1 has been reported to fluctuate during contraction, but not during relaxation or rigor (291). However, it is not clear if this change occurs on activated, actin-attached heads (292). Indeed, EPR spectral measurements (293) of a spin label attached to SH_1 on the myosin heads of activated (rabbit psoas) muscle show that $\sim 80\%$ of the probes show a random distribution (on a 10 microsecond time scale) like relaxed muscle (285) while the remaining 20% are highly ordered at the same angle found in rigor muscle (294). These findings suggest that a domain of the myosin head does not change orientation during the power stroke. Güth (292) has reported that the intrinsic tryptophan polarization of fluorescence of activated glycerinated muscle fiber remains unchanged following a quick release step (< 1 msec). His studies have been interpreted as indicating that either the polarization ($P_\perp$) degree is insensitive to the orientation of the cross-bridges, or that the bridges do not rotate. The degree of polarization of fluorescent nucleotide analogs bound in the active site of the S-1 subunits support this conclusion since little difference between the angular distribution of these probes in rigor and activated muscle is observed (295, 296).

Experiments designed to search for head rotation in *rigor fibers* under applied tensile stresses have also given negative results. Naylor & Podolsky (297) measured the intensity ratio of two equatorial X-ray reflections ($I_{1,1}/I_{1,0}$) and concluded that the angle of attachment of S-1 to actin remains unchanged under stress. No change was observed in fluorescent probes bound to S-1 on application of large forces to fibers in rigor (298). EPR spectra of spin-labeled cross-bridges are in agreement with the X-ray and fluorescence results in showing that the probe angle of rigor fibers remains unaltered on application of stress (299). One possibility to explain the EPR and fluorescence probe results is that the probed region of the myosin head remains rigid, while another region (18, 23, 300) of the S-1 subunit rotates to give the required 10 nm axial displacement (293, 301). Another possibility is that the entire S-1 subunit maintains a nearly invariant orientation (rigor angle), while some other part of myosin, for example the S-2 region, shortens (89, 90, 302).

In the Huxley-Simmons model (262), head rotation extends the spring (elastic element) in S-2 by about 10 nm. Hence the power stroke, which results from contraction of the spring, slides the actin filament on average by about this amount to bring the active tension in the bridge to zero (262, 263, 303). According to the helix-coil model, melting of ~ 165 residues per polypeptide strand would allow an equivalent shortening of the S-2 link (90). As we discussed earlier in this section, this is approximately the size (103, 108) of the proteolytically sensitive region (hinge domain) at the C-terminal end (304) of long S-2. The maximum force that can be produced by such a shortening (helix-coil) process is about 3 pico Newtons (pN) per bridge (90), which is comparable to the maximum tension estimated per bridge in a working muscle (262).

It seems likely that the proteolytically sensitive hinge region has a distinctly lower thermal stability ($T_m = 35°-40°C$) than the flanking LMM and short S-2 region of myosin (108, 109). Temperature-jump measurements of the rates of helix-coil transition of long S-2 (107) and its cyanogen bromide fragments (108) show that at $40°C$ these coiled-coil structures can undergo reversible helix-coil transitions on a time scale compatible with the cycle time of a cross-bridge. The fast (submillisecond) relaxation process observed in long S-2 at low temperature ($T_m = 35°C$) appears to be localized in the hinge (108, 302) suggesting that *both* the series-elastic and the force-generating elements (262) of a cycling cross-bridge could reside in this region of the S-2 lever arm.

The rates of cross-linking the S-1 and S-2 regions of myosin to the thick-filament backbone of synthetic filaments, as well as myofibrils and glycerinated fibers in rigor, drops sharply when the pH is elevated over a narrow range of 7.0–8.4 (96, 290, 305), suggesting that the cross-bridges are released and swing away from the filament surface. This interpretation is consistent with fluorescence depolarization (306) and EPR (307) studies that show increased rotational mobility of the heads in synthetic filaments on raising the pH from 7.0 to 8.3. The hinge domain of S-2 also appears to undergo a reversible conformational transition to a more open and proteolytically accessible state when the cross-bridges are released from the thick-filament surface of rigor myofibrils (96) or rod filaments (308) under these conditions. Thus it seems likely that relatively small changes in the local ionic environment can release the entire HMM segment from the filament backbone. Assuming a two-state transition, about two protons are dissociated when the cross-bridge is detached and swings out from the thick-filament surface (96). From circular dichroism studies (309) of rod and myosin minifilaments at 20°C, these charge-dependent changes appear to be accompanied by a partial melting (9% and 5%, respectively) of the α-helical structure in the S-2 region. The enzyme probe technique has also been used to investigate conformational changes in active contracting psoas muscle fibers held at constant length (91). When the contractile apparatus is activated, the hinge domain becomes markedly susceptible to cleavage (by chymotrypsin) and LMM is formed at an increased rate (about tenfold), which depends on the concentration of MgATP.

Literature Cited

1. Mannherz, H. G., Goody, R. S. 1976. *Ann. Rev. Biochem.* 45 : 427–65
2. Taylor, E. W. 1979. *CRC Crit. Rev. Biochem.* 6 : 103–64
3. Adelstein, R. S., Eisenberg, E. 1980. *Ann. Rev. Biochem.* 49 : 921–56
4. Kendrick-Jones, J., Scholey, J. M. 1981. *J. Muscle Res. Cell Motil.* 2 : 347–72
5. Kendrick-Jones, J., Jakes, R., Tooth, P., Craig, R., Scholey, J. 1982. In *Basic Biology of Muscles: A Comparative Approach*, ed. B. M. Twarog, R. J. C. Levine, M. M. Dewey, pp. 255–72. New York : Raven
6. Huxley, H. E. 1963. *J. Mol. Biol.* 7 : 281–308
7. Moore, P. B., Huxley, H. E., DeRosier, D. J. 1970. *J. Mol. Biol.* 50 : 279–95
8. Taylor, K. A., Amos, L. A. 1981. *J. Mol. Biol.* 147 : 297–324
9. Wakabayashi, T., Toyoshima, C. 1981. *J. Biochem.* 90 : 683–701
10. Mendelson, R., Kretzschmar, K. M. 1980. *Biochemistry* 19 : 4103–8
11. Elliott, A., Offer, G. 1978. *J. Mol. Biol.* 123 : 505–19
12. Craig, R., Szent-Györgyi, A. G., Beese, L., Flicker, P., Vibert, P., Cohen, C. 1980. *J. Mol. Biol.* 140 : 35–55
13. Vibert, P., Craig, R. 1982. *J. Mol. Biol.* 157 : 299–319
14. Katayama, E., Wakabayashi, T. 1981. *J. Biochem.* 90 : 703–14
15. Balint, M., Sréter, F. A., Gergely, J. 1975. *Arch. Biochem. Biophys.* 168 : 557–66
16. Balint, M., Wolf, I., Tarcsafalui, A., Gergely, J., Sréter, F. A. 1978. *Arch. Biochem. Biophys.* 190 : 793–99
17. Mornet, D., Bertrand, R., Pantel, P.,

Audemard, E., Kassab, R. 1981. *Biochemistry* 20 : 2110–20
18. Mornet, D., Bertrand, R., Pantel, P., Audemard, E., Kassab, R. 1981. *Nature* 292 : 301–6
19. Labbe, J.-P., Mornet, D., Roseau, G., Kassab, R. 1982. *Biochemistry* 21 : 6897–6902
20. Sutoh, K. 1982. *Biochemistry* 21 : 3654–61
21. Winkelmann, D. A., Lowey, S. 1983. *Biophys. J.* 41 : 228a
22. Mornet, D., Pantel, P., Audemard, E., Kassab, R. 1979. *Biochem. Biophys. Res. Commun.* 89 : 925–32
23. Lovell, S. J., Harrington, W. F. 1981. *J. Mol. Biol.* 149 : 659–74
24. Yamamoto, K., Sekine, T. 1979. *J. Biochem.* 86 : 1855–62
25. Muhlrad, A., Hozumi, T. 1982. *Proc. Natl. Acad. Sci. USA* 79 : 958–62
26. Labbe, J.-P., Mornet, D., Vandest, P., Kassab, R. 1981. *Biochem. Biophys. Res. Commun.* 102 : 466–75
27. Flicker, P., Wallimann, T., Vibert, P. 1981. *Biophys. J.* 33 : 279a
28. Burke, M., Sivaramakrishnan, M., Kamalakannan, V. 1983. *Biochemistry* 22 : 3046–53
29. Szentkiralyi, E. M. 1982. *Biophys. J.* 37 : 39a
30. Kuwayama, H., Yagi, K. 1980. *J. Biochem.* 87 : 1603–7
31. Mocz, G., Biro, E. N. A., Balint, M. 1982. *FEBS Lett.* 126 : 603–9
32. Kielley, W. W., Bradley, L. B. 1956. *J. Biol. Chem.* 218 : 653–59
33. Sekine, T., Yamaguchi, M. 1963. *J. Biochem.* 54 : 196–98
34. Kubo, S., Tokura, S., Tonomura, Y. 1960. *J. Biol. Chem.* 235 : 2835–39

35. Fábián, F., Mühlrad, A. 1968. *Biochem. Biophys. Acta* 162:596–603
36. Elzinga, M., Collins, J. H. 1977. *Proc. Natl. Acad. Sci. USA* 74:4281–84
37. Sutoh, K. 1981. *Biochemistry* 20:3281–85
38. Gallager, M., Elzinga, M. 1980. *Fed. Proc.* 39:2168
39. Karn, J., McLachlan, A. D., Barnett, L. 1982. In *Muscle Development: Molecular and Cellular Control*, ed. M. L. Pearson, H. F. Epstein, pp. 129–42. Cold Spring Harbor, NY: Cold Spring Lab.
40. Sekine, T., Barnett, L. M., Kielley, W. W. 1962. *J. Biol. Chem.* 237:2769–72
41. Sekine, T., Kielley, W. W. 1964. *Biochem. Biophys. Acta* 81:336–45
42. Reisler, E., Burke, M., Harrington, W. F. 1974. *Biochemistry* 13:2014–22
43. Burke, M., Reisler, E. 1977. *Biochemistry* 16:5559–63
44. Wells, J. A., Weber, M. M., Legg, J. I., Yount, R. G. 1979. *Biochemistry* 18:4793–99
45. Wells, J. A., Weber, M. M., Yount, R. G. 1979. *Biochemistry* 18:4800–5
46. Wells, J. A., Yount, R. G. 1979. *Proc. Natl. Acad. Sci. USA* 76:4966–70
47. Wells, J. A., Yount, R. G. 1980. *Biochemistry* 19:1711–17
48. Wiedner, H., Wetzel, R., Eckstein, F. 1978. *J. Biol. Chem.* 253:2763–68
49. Botts, J., Ue, K., Hozumi, T., Samet, J. 1979. *Biochemistry* 18:5157–63
50. Cheung, H. C., Gonsoulin, F., Garland, F. 1983. *J. Biol. Chem.* 258:5775–86
51. Dalbey, R. 1983. *Biophys. J.* 41:98a
52. Szilagyi, L., Balint, M., Sréter, F. A., Gergely, J. 1979. *Biochem. Biophys. Res. Commun.* 87:936–45
53. Hozumi, T., Muhlrad, A. 1981. *Biochemistry* 20:2945–50
54. Mornet, D., Pantel, P., Bertrand, R., Audemard, E., Kassab, R. 1980. *FEBS Lett.* 117:183–88
55. Miyanishi, T., Tonomura, Y. 1981. *J. Biochem.* 89:831–39
56. Tao, T., Lamkin, M. 1981. *Biochemistry* 20:5051–55
57. Takashi, R., Muhlrad, A., Botts, J. 1982. *Biochemistry* 21:5661–68
58. Marsh, D. J., Lowey, S. 1980. *Biochemistry* 19:774–84
59. Takashi, R. 1979. *Biochemistry* 18:5164–69
60. Morales, M. F., Borjedo, J., Botts, J., Cooke, R., Mendelson, R. A., Takashi, R. 1982. *Ann. Rev. Phys. Chem.* 33:319–51
61. Hozumi, T. 1981. *J. Biochem.* 90:785–88
62. Privalov, P. L. 1982. *Adv. Prot. Chem.* 35:1–104
63. Privalov, P. L. 1979. *Adv. Prot. Chem.* 33:167–241
64. Mornet, D., Pantel, P., Bertrand, R., Audemard, E., Kassab, R. 1981. *FEBS Lett.* 123:54–58
65. Amos, L. A., Huxley, H. E., Holmes, K. C., Goody, R. S., Taylor, K. A. 1982. *Nature* 299:467–69
66. Suck, D., Kabsch, W., Mannherz, H. G. 1981. *Proc. Natl. Acad. Sci. USA* 78:4319–23
67. Egelman, E. H., DeRosier, D. J. 1982. *J. Cell. Biol.* 95:288a
68. Smith, P. R., Fowler, W. E., Pollard, T. D., Aebi, U. 1983. *J. Mol. Biol.* 167:641–60
69. Sutoh, K. 1982. *Biochemistry* 21:4800–4
70. Sutoh, K. 1983. *Biochemistry* 22:1579–85
71. Szent-Györgyi, A. G. 1980. In *Muscle Contraction: Its Regulatory Mechanisms*, ed. S. Ebashi, K. Maruyama, M. Endo, pp. 375–89. Berlin: Springer-Verlag
72. Lehman, W. 1978. *Nature* 274:80–81
73. Chin, T. K., Rowe, A. J. 1982. *J. Muscle Res. Cell. Motil.* 3:118
74. Pulliam, D. L., Sawyna, V., Levine, R. J. C. 1983. *Biochemistry* 22:2324–31
75. Wagner, P., Stone, D. 1983. *Biophys. J.* 41:91a
76. Bagshaw, C. R. 1977. *Biochemistry* 16:59–67
77. Weeds, A. G., Pope, B. 1977. *J. Mol. Biol.* 111:129–57
78. Prince, H. P., Trayer, H. R., Henry, G. D., Trayer, I. P., Dalgarno, D. C., et al. 1981. *FEBS Lett.* 121:213–19
79. Wagner, P. D., Giniger, E. 1981. *Nature* 292:560–62
80. Sivaramakrishnan, M., Burke, M. 1982. *J. Biol. Chem.* 257:1102–5
81. Hardwicke, P. M. D., Wallimann, T., Szent-Györgyi, A. G. 1983. *Nature* 301:478–82
82. Wallimann, T., Szent-Györgyi, A. G. 1981. *Biochemistry* 20:1176–87
83. Wallimann, T., Szent-Györgyi, A. G. 1981. *Biochemistry* 20:1188–97
84. Hardwicke, P. M. D., Wallimann, T., Szent-Györgyi, A. G. 1982. *J. Mol. Biol.* 156:141–52
85. Wallimann, T., Hardwicke, P. M. D., Szent-Györgyi, A. D. 1982. *J. Mol. Biol.* 156:153–73
86. Harvey, S. C., Cheung, H. C. 1982. In *Cell and Muscle Motility*, ed. R. M. Dowben, J. W. Shay, 2:279–302. New York: Plenum
87. Lowey, S., Slayter, H. S., Weeds, A. G., Baker, H. 1969. *J. Mol. Biol.* 42:1–29

88. Huxley, H. E. 1969. *Science* 164:1356–66
89. Harrington, W. F. 1971. *Proc. Natl. Acad. Sci. USA* 68:685–89
90. Harrington, W. F. 1979. *Proc. Natl. Acad. Sci. USA* 76:5066–70
91. Ueno, H., Harrington, W. F. 1981. *Proc. Natl. Acad. Sci. USA* 78:6101–5
92. Capony, J. P., Elzinga, M. 1981. *Biophys. J.* 33:148a
93. Trus, B. L., Elzinga, M. 1981. In *Structural Aspects of Recognition and Assembly in Biological Macromolecules*, ed. M. Balaban, J. L. Sussman, W. Traub, A. Yonath, 1:361. Boston: Balaban Int. Sci.
94. Parry, D. A. D. 1981. *J. Mol. Biol.* 153:459–64
95. Stewart, M. 1982. *FEBS Lett.* 140:210–12
96. Ueno, H., Harrington, W. F. 1981. *J. Mol. Biol.* 149:619–40
97. Huxley, H. E., Brown, W. 1967. *J. Mol. Biol.* 30:383–434
98. McLachlan, A. D., Karn, J. 1982. *Nature* 299:226–31
99. McLachlan, A. D., Karn, J. 1983. *J. Mol. Biol.* 164:605–26
100. Mahdavi, V., Periasamy, M., Nadal-Ginard, B. 1982. *Nature* 297:659–64
101. Bernstein, S. I., Mogami, K., Donady, J. J., Emerson Jr., C. P. 1983. *Nature* 302:393–97
102. Lu, R. C., Nyitray, L., Balint, M., Gergely, J. 1983. *Biophys. J.* 41:228a
103. Sutoh, K., Karr, T., Harrington, W. F. 1978. *J. Mol. Biol.* 126:1–22
104. Burke, M., Himmelfarb, S., Harrington, W. F. 1973. *Biochemistry* 12:701–10
105. Takahashi, K. 1978. *J. Biochem.* 83:905–8
106. Trybus, K. M., Huiatt, T. W., Lowey, S. 1982. *Proc. Natl. Acad. Sci. USA* 79:6151–55
107. Tsong, T. Y., Karr, T., Harrington, W. F. 1979. *Proc. Natl. Acad. Sci. USA* 76:1109–13
108. Tsong, T. Y., Himmelfarb, S., Harrington, W. F. 1983. *J. Mol. Biol.* 164:431–50
109. Swenson, C. A., Ritchie, P. A. 1980. *Biochemistry* 19:5371–75
110. Potekhin, S. A., Trapkov, V. A., Privalov, P. L. 1979. *Biofizika* 24:46–50
111. Talbot, J. A., Hodges, R. S. 1982. *Acc. Chem. Res.* 15:224–30
112. Highsmith, S., Kretzschmar, K. M., O'Konski, C. T., Morales, M. F. 1977. *Proc. Natl. Acad. Sci. USA* 74:4986–90
113. Garcia de la Torre, J., Bloomfield, V. A. 1980. *Biochemistry* 19:5118–23
114. Highsmith, S., Wang, C.-C., Zero, K.,

Pecora, R., Jardetzky, O. 1982. *Biochemistry* 21:1192–97
115. Zero, K. M., Pecora, R. 1982. *Macromolecules* 15:87–93
116. Zero, K., Pecora, R. 1982. *Macromolecules* 15:1023–27
117. Stewart, M., Roberts, G. C. K. 1982. *FEBS Lett.* 146:293–96
118. Harvey, S. C., Cheung, H. C. 1977. *Biochemistry* 16:5181–87
119. Rosser, R. W., Nestler, F. H. M., Schrag, J. L., Ferry, J. D., Greaser, M. 1978. *Macromolecules* 11:1239–42
120. Hvidt, S., Nestler, F. H. M., Greaser, M. L., Ferry, J. D. 1982. *Biochemistry* 21:4064–73
121. Stafford, W. F., Margossian, S. S. 1982. *Biophys. J.* 37:55a
122. Matsubara, I. 1974. *J. Physiol.* 238:473–86
123. Matsubara, I., Millman, B. M. 1974. *J. Mol. Biol.* 82:527–36
124. Maruyama, K., Weber, A. 1972. *Biochemistry* 11:2990–98
125. Marston, S. B., Tregear, R. T. 1972. *Nature New Biol.* 235:23–24
126. Potter, J. D. 1974. *Arch. Biochem. Biophys.* 162:436–41
127. Treager, R. T., Squire, J. M. 1973. *J. Mol. Biol.* 77:279–90
128. Morimoto, K., Harrington, W. F. 1974. *J. Mol. Biol.* 83:83–97
129. Pepe, F. A., Drucker, B. 1979. *J. Mol. Biol.* 130:379–93
130. Huxley, H. E. 1960. In *The Cell*, ed. J. Brachet, A. E. Mirsky, Vol. 4, Chap. 7, pp. 363–481. New York: Academic
131. Lamvik, M. K. 1978. *J. Mol. Biol.* 122:55–68
132. Reedy, M. K., Leonard, K. R., Freeman, R., Arad, T. 1981. *J. Muscle Res. Cell Motil.* 2:45–64
133. Luther, P. K., Munro, P. M. G., Squire, J. M. 1981. *J. Mol. Biol.* 151:703–30
134. Freundlich, A., Luther, P. K., Squire, J. M. 1980. *J. Muscle Res. Cell Motil.* 1:321–43
135. Squire, J. M. 1974. *J. Mol. Biol.* 90:153–60
136. Luther, P. K., Squire, J. M. 1980. *J. Mol. Biol.* 141:409–39
137. Kensler, R. W., Stewart, M. 1983. *J. Cell Biol.* 96:1797–1802
138. Squire, J. M. 1971. *Nature* 233:457–62
139. Squire, J. M. 1973. *J. Mol. Biol.* 77:291–323
140. Baccetti, B. 1965. *J. Ultrastruct. Res.* 13:245–56
141. Gilev, V. P. 1966. *Biophysics* 11:312–17
142. Pepe, F. A., Dowben, P. 1977. *J. Mol. Biol.* 113:199–218
143. Pepe, F. A., Ashton, F. T., Dowben, P.,

Stewart, M. 1981. *J. Mol. Biol.* 145: 421–40

144. Ashton, F. T., Pepe, F. A. 1981. *J. Microsc.* 123: 93–108

145. Pepe, F. A., Drucker, B. 1972. *J. Cell. Biol.* 52: 255–60

146. Markham, R., Frey, S., Hills, G. J. 1963. *Virology* 20: 88–102

147. Stewart, M., Ashton, F. T., Lieberson, R., Pepe, F. A. 1981. *J. Mol. Biol.* 153: 381–92

148. Millman, B. M. 1979. In *Motility in Cell Function*, ed. F. A. Pepe, J. W. Sanger, V. T. Nachmias, pp. 351–54. New York: Academic

149. Wray, J. 1979. See Ref. 148, pp. 347–50

150. Wray, J. 1979. *Nature* 277: 37–40

151. Pepe, F. A. 1967. *J. Mol. Biol.* 27: 203–25

152. Pepe, F. A. 1982. See Ref. 86, pp. 141–71

153. Harrington, W. F., Burke, M. 1972. *Biochemistry* 11: 1448–55

154. Burke, M., Harrington, W. F. 1972. *Biochemistry* 11: 1456–62

155. Davis, J. S., Buck, J., Greene, E. P. 1982. *FEBS Lett.* 140: 293–97

156. Caspar, D. L. D., Klug, A. 1962. *Cold Spring Harbor Symp. Quant. Biol.* 27: 1–24

157. Maw, M. C., Rowe, A. J. 1980. *Nature* 286: 412–14

158. Trinick, J. A. 1981. *J. Mol. Biol.* 151: 309–14

159. Squire, J. M. 1975. *Ann. Rev. Biophys. Bioeng.* 4: 137–63

160. Squire, J. M. 1979. In *Fibrous Proteins: Scientific, Industrial and Medical Aspects*, ed. D. A. D. Parry, L. K. Creamer, 1: 27–70. London: Academic

161. Miller, D. M. III, Ortiz, I., Berliner, G. C., Epstein, H. F. 1983. *Cell* 34: 477–90

162. Starr, R., Offer, G. 1973. *J. Mol. Biol.* 81: 17–31

163. Chizzonite, R. A., Everett, A. W., Clark, W. A., Jakovcic, S., Rabinowitz, M., Zak, R. 1982. *J. Biol. Chem.* 257: 2056–65

164. Whalen, R. G., Bugaisky, L. B., Butler-Browne, G. S., Sell, S. M., Schwartz, K., Pinset-Härström, I. 1982. See Ref. 39, pp. 25–33

165. Harrington, W. F. 1979. In *The Proteins*, ed. H. Neurath, R. L. Hill, 4: 245–409. New York: Academic. 3rd ed.

166. Pepe, F. A., Drucker, B. 1975. *J. Mol. Biol.* 99: 609–17

167. Craig, R., Offer, G. 1976. *Proc. R. Soc. London Ser. B* 192: 451–61

168. Wray, J. S. 1982. See Ref. 5, pp. 29–36

169. Wray, J. S., Vibert, P. J., Cohen, C. 1975. *Nature* 257: 561–64

170. Vibert, P., Craig, R. 1983. *J. Mol. Biol.* 165: 303–20

171. Kensler, R. W., Levine, R. J. C. 1982. *J. Cell. Biol.* 92: 443–51

172. Millman, B. M., Warden, W. J., Colflesh, D. E., Dewey, M. M. 1974. *Fed. Proc. Am. Soc. Exp. Biol.* 33: 1333 (abstr.)

173. Wray, J. S., Vibert, P. J., Cohen, C. 1974. *J. Mol. Biol.* 88: 343–48

174. Kensler, R. W., Levine, R. J. C. 1981. *Biophys. J.* 33: 242a

175. Kensler, R. W., Levine, R. J. C. 1982. *J. Muscle Res. Cell Motil.* 3: 349–61

176. Stewart, M., Kensler, R. W., Levine, R. J. C. 1981. *J. Mol. Biol.* 153: 781–90

177. Winkelman, L. 1976. *Comp. Biochem. Physiol. B* 55: 391–97

178. Levine, R. J. C., Elfvin, M., Dewey, M. M., Walcott, B. 1976. *J. Cell. Biol.* 71: 273–79

179. Lowey, S., Kucera, J., Holtzer, A. 1963. *J. Mol. Biol.* 7: 234–44

180. Elfvin, M., Levine, R. J. C., Dewey, M. M. 1976. *J. Cell. Biol.* 71: 261–72

181. Millman, B. M., Bennett, P. M. 1976. *J. Mol. Biol.* 103: 439–67

182. Bullard, B., Hammond, K. S., Luke, B. M. 1977. *J. Mol. Biol.* 115: 417–40

183. Jahromi, S. S., Atwood, H. L. 1969. *J. Exp. Zool.* 171: 25–38

184. Pringle, J. W. S. 1972. In *The Structure and Function of Muscle*, ed. G. H. Bourne, 1: 10. New York: Academic

185. Hoyle, G., McNeill, P. A. 1968. *J. Exp. Zool.* 167: 487–522

186. Hanson, J., Lowy, J., Huxley, H. E., Bailey, K., Kay, C. M., Rüegg, J. C. 1957. *Nature* 180: 1134–35

187. Szent-Györgyi, A. G., Cohen, C., Kendrick-Jones, J. 1971. *J. Mol. Biol.* 56: 239–58

188. Lowy, J., Hanson, J. 1962. *Physiol. Rev.* 42 (Suppl. 5): 34–47

189. Elliott, G. F. 1964. *J. Mol. Biol.* 10: 89–104

190. Elliott, A., Bennett, P. M. 1982. See Ref. 168, pp. 11–28

191. Hall, C. E., Jakus, M. A., Schmidt, F. O. 1945. *J. Appl. Physics* 16: 459–65

192. Bear, R., Selby, C. C. 1956. *J. Biophys. Biochem. Cytol.* 2: 55–69

193. Cohen, C., Szent-Györgyi, A. G. 1971. In *Contractility of Muscle Cells and Related Processes*, ed. R. J. Podolsky, pp. 23–36. New York: Prentice Hall

194. Elliott, A., Lowy, J. 1970. *J. Mol. Biol.* 53: 181–203

195. Elliott, A. 1979. *J. Mol. Biol.* 132: 323–41

196. Dover, S. D., Elliott, A. 1979. *J. Mol. Biol.* 132: 340–41

197. Dover, S. D., Elliott, A., Kernaghan, A. K. 1981. *J. Microsc.* 122: 23–33

198. Cohen, C., Szent-Györgyi, A. G., Kendrick-Jones, J. 1971. *J. Mol. Biol.* 56:223–37
199. Bennett, P. M., Elliott, A. 1981. *J. Muscle Res. Cell Motil.* 2:65–81
200. Cohen, C. 1982. *Proc. Natl. Acad. Sci. USA* 79:3176–78
201. Achazi, R. K. 1979. *Pflügers Arch.* 379:197–201
202. Godfrey, J., Harrington, W. F. 1970. *Biochemistry* 9:886–93
203. Godfrey, J., Harrington, W. F. 1970. *Biochemistry* 9:894–907
204. Herbert, T. J., Carlson, F. D. 1971. *Biopolymers* 10:2231–52
205. Szuchet, S. 1977. *Arch. Biochem. Biophys.* 180:493–503
206. Emes, C. H., Rowe, A. J. 1978. *Biochim. Biophys. Acta* 537:110–24
207. Trinick, J. A., Cooper, J. 1980. *J. Mol. Biol.* 141:315–21
208. Niederman, R., Peters, L. K. 1982. *J. Mol. Biol.* 161:505–17
209. Kundrat, E., Pepe, F. A. 1971. *J. Cell. Biol.* 48:340–47
210. Eaton, B. L., Pepe, F. A. 1972. *J. Cell. Biol.* 55:681–95
211. Kaminer, B., Bell, A. L. 1966. *J. Mol. Biol.* 20:391–401
212. Katsura, I., Noda, H. 1971. *J. Biochem.* 69:219–29
213. Josephs, R., Harrington, W. F. 1966. *Biochemistry* 5:3474–87
214. Katsura, I., Noda, H. 1973. *J. Biochem.* 73:245–56
215. Moos, C., Offer, G., Starr, R., Bennett, P. M. 1975. *J. Mol. Biol.* 97:1–9
216. Hinssen, H., D'Haese, J., Small, J. V., Sobieszek, A. 1978. *J. Ultrastruct. Res.* 64:282–302
217. Josephs, R., Harrington, W. F. 1967. *Proc. Natl. Acad. Sci. USA* 58:1587–94
218. Josephs, R., Harrington, W. F. 1968. *Biochemistry* 7:2834–47
219. Harrington, W. F., Josephs, R. 1968. *Dev. Biol. Suppl.* 2:21–62
220. Megerman, J., Lowey, S. 1981. *Biochemistry* 20:2099–110
221. Suzuki, H., Onishi, H., Takahashi, K., Watanabe, S. 1978. *J. Biochem.* 84:1529–42
222. Scholey, J. M., Taylor, K. A., Kendrick-Jones, J. 1980. *Nature* 287:233–35
223. Harrington, W. F., Himmelfarb, S. 1972. *Biochemistry* 11:2945–52
224. Pinset-Härström, I., Truffy, J. 1979. *J. Mol. Biol.* 134:173–88
225. Harrington, W. F. 1972. In *Current Topics in Biochemistry*, ed. C. B. Anfinsen, R. F. Goldberger, A. N. Schechter, pp. 135–85. New York: Academic
226. Davis, J. S. 1981. *Biochem. J.* 197:301–8

227. Davis, J. S. 1981. *Biochem. J.* 197:309–14
228. Davis, J. S., Gutfreund, H. 1976. *FEBS Lett.* 72:199–207
229. Reisler, E., Burke, M., Josephs, R., Harrington, W. F. 1973. *J. Mechanochem. Cell Motil.* 2:163–79
230. Reisler, E., Smith, C., Seegan, G. 1980. *J. Mol. Biol.* 143:129–45
231. Reisler, E., Cheung, P., Oriol-Audit, C., Lake, J. A. 1982. *Biochemistry* 21:701–7
232. Cheung, P., Reisler, E. 1982. *Biochemistry* 21:6906–10
233. Kendrick-Jones, J., Tooth, P., Taylor, K. A., Scholey, J. M. 1982. *Cold Spring Harbor Symp. Quant. Biol.* 46:929–38
234. Scholey, J. M., Taylor, K. A., Kendrick-Jones, J. 1981. *Biochimie* 63:255–71
235. Scholey, J. M., Smith, R. C., Drenckhahn, D., Gröschel-Stewart, U. Kendrick-Jones, J. 1983. *J. Biol. Chem.* 257:7737–45
236. Suzuki, H., Kamata, T., Onishi, H., Watanabe, S. 1982. *J. Biochem.* 91:1699–1705
237. Onishi, H., Wakabayashi, T. 1982. *J. Biochem.* 92:871–79
238. Craig, R., Smith, R., Kendrick-Jones, J. 1983. *Nature* 302:436–39
239. Hanson, J., Lowy, J. 1964. *Proc. R. Soc. London Ser. B* 160:523–24
240. Kaminer, B. 1969. *J. Mol. Biol.* 39:257–64
241. Sobieszek, A. 1972. *J. Mol. Biol.* 70:741–44
242. Wachsberger, P. R., Pepe, F. A. 1974. *J. Mol. Biol.* 88:385–91
243. Craig, R., Megerman, J. 1977. *J. Cell. Biol.* 75:990–96
244. Kendrick-Jones, J., Szent-Györgyi, A. G., Cohen, C. 1971. *J. Mol. Biol.* 59:527–29
245. Craig, R., Megerman, J. 1979. See Ref. 148, pp. 91–102
246. Small, J. V. 1977. *J. Cell. Sci.* 24:329–49
247. Lowy, J., Poulsen, F. R., Vibert, P. J. 1970. *Nature* 225:1053–54
248. Shoenberg, C. F., Haselgrove, J. C. 1974. *Nature* 249:152–54
249. Pollard, T. D., Stafford, W. F., Porter, M. E. 1978. *J. Biol. Chem.* 253:4798–4808
250. Hinssen, H. 1970. *Cytobiologie* 2:326–31
251. Niederman, R., Pollard, T. D. 1975. *J. Cell. Biol.* 67:72–92
252. Pollard, T. D. 1982. *J. Cell. Biol.* 95:816–25
253. Coté, G. P., Collins, J. H., Korn, E. D. 1981. *J. Biol. Chem.* 256:12811–16
254. Collins, J. H., Coté, G. P., Korn, E. D. 1982. *J. Biol. Chem.* 257:4529–34
255. Collins, J. H., Korn, E. D. 1980. *J. Biol. Chem.* 255:8011–14

256. Collins, J. H., Korn, E. D. 1981. *J. Biol. Chem.* 256 : 2586–95
257. Collins, J. H., Kuznicki, J., Bowers, B., Korn, E. D. 1982. *Biochemistry* 21 : 6910–15
258. Kuznicki, J., Albanesi, J. P., Côte, G. P., Korn, E. D. 1983. *J. Biol. Chem.* 258 : 6011–15
259. Somlyo, A. V., Butler, T. M., Bond, M., Somlyo, A. P. 1981. *Nature* 294 : 567–69
260. Huxley, H. E., Hanson, J. 1954. *Nature* 173 : 973–76
261. Huxley, A. F., Niedergerke, R. 1954. *Nature* 173 : 971–73
262. Huxley, A. F., Simmons, R. M. 1971. *Nature* 233 : 533–38
263. Barden, J. A., Mason, P. 1978. *Science* 199 : 1212–13
264. Huxley, H. E. 1953. *Proc. R. Soc. London Ser. B* 141 : 59–62
265. Huxley, H. E. 1953. *Biochim. Biophys. Acta* 12 : 387–94
266. Elliott, G. F., Lowy, J., Worthington, C. R. 1963. *J. Mol. Biol.* 6 : 295–305
267. Elliott, G. F., Lowy, J., Millman, B. M. 1967. *J. Mol. Biol.* 25 : 31–45
268. Huxley, H. E. 1972. In *The Structure and Function of Muscle*, ed. G. H. Bourne, 1 : 301–87. New York : Academic
269. Reedy, M. K., Holmes, K. C., Tregear, R. T. 1965. *Nature* 207 : 1276–80
270. Reedy, M. K. 1968. *J. Mol. Biol.* 31 : 155–76
271. Huxley, H. E., Simmons, R. M., Faruqi, A. R., Kress, M., Bordas, J., Koch, M. H. J. 1981. *Proc. Natl. Acad. Sci. USA* 78 : 2297–2301
272. Marston, S. B., Tregear, R. T., Rodger, C. D., Clarke, M. L. 1979. *J. Mol. Biol.* 128 : 111–26
273. Morales, M. F., Botts, J. 1979. *Proc. Natl. Acad. Sci. USA* 76 : 3857–59
274. Eisenberg, E., Hill, T. L. 1978. *Prog. Biophys. Mol. Biol.* 33 : 55–82
275. Eisenberg, E., Hill, T. L., Chen, Y. D. 1980. *Biophys. J.* 29 : 195–227
276. Schoenberg, M. 1980. *Biophys. J.* 30 : 51–67
277. Schoenberg, M. 1980. *Biophys. J.* 30 : 69–77
278. Huxley, H. E., Faruqi, A. R., Kress, M., Bordas, J., Koch, M. H. J. 1982. *J. Mol. Biol.* 158 : 637–84
279. Huxley, H. E. 1968. *J. Mol. Biol.* 37 : 507–20
280. Huxley, H. E. 1979. In *Crossbridge Mechanism in Muscle Contraction*, ed. H. Sugi, G. H. Pollack, pp. 391–401. Tokyo : Univ. Tokyo Press
281. Haselgrove, J. C., Huxley, H. E. 1973. *J. Mol. Biol.* 77 : 549–68
282. Yagi, N., Matsubara, I. 1980. *Science* 207 : 307–8
283. Huxley, H. E., Faruqi, A. R., Bordas, J., Koch, M. H. J., Milch, J. R. 1980. *Nature* 284 : 140–43
284. Haselgrove, J. C. 1980. *J. Muscle Res. Cell. Motil.* 1 : 177–91
285. Thomas, D. D., Ishiwata, S., Seidel, J. C., Gergely, J. 1980. *Biophys. J.* 32 : 873–90
286. Barnett, V. A., Thomas, D. D. 1983. *J. Mol. Biol.* Submitted for publication
287. Mendelson, R. A., Wilson, M. G. A. 1982. *Biophys. J.* 39 : 221–27
288. Poulsen, F. R., Lowy, J. 1983. *Nature* 303 : 146–52
289. Sutoh, K., Harrington, W. F. 1977. *Biochemistry* 16 : 2441–49
290. Chiao, Y. C., Harrington, W. F. 1979. *Biochemistry* 18 : 959–63
291. Borejdo, J., Putnam, S., Morales, M. F. 1979. *Proc. Natl. Acad. Sci. USA* 76 : 6346–50
292. Güth, K. 1980. *Biophys. Struct. Mech.* 6 : 81–93
293. Cooke, R., Crowder, M. S., Thomas, D. D. 1982. *Nature* 300 : 776–78
294. Thomas, D. D., Cooke, R. 1980. *Biophys. J.* 32 : 891–906
295. Yanagida, T. 1981. *J. Mol. Biol.* 146 : 539–60
296. Yanagida, T. 1983. In *Contractile Mechanisms in Muscle*, ed. G. H. Pollack, H. Sugi. New York : Plenum. In press
297. Naylor, G. R. S., Podolsky, R. J. 1981. *Proc. Natl. Acad. Sci. USA* 78 : 5559–63
298. Dos Remedios, C. G., Millikan, R. G. C., Morales, M. F. 1972. *J. Gen. Physiol.* 59 : 103–20
299. Cooke, R. 1981. *Nature* 294 : 570–71
300. Yamamoto, K., Sekine, T. 1979. *J. Biochem.* 86 : 1855–62
301. Holmes, K. C., Goody, R. S. 1983. In *Mobility and Function in Proteins and Nucleic Acids, Ciba Found. Symp.* 93 : 139–53. London : Pitman
302. Harrington, W. F., Ueno, H., Tsong, T. Y. 1983. See Ref. 301, pp. 186–203
303. Ford, L. E., Huxley, A. F., Simmons, R. M. 1977. *J. Physiol.* 269 : 441–515
304. Lu, R. C. 1980. *Proc. Natl. Acad. Sci. USA* 77 : 2010–13
305. Sutoh, K., Chiao, Y. C., Harrington, W. F. 1978. *Biochemistry* 17 : 1234–39
306. Mendelson, R. A., Cheung, P. H. 1976. *Science* 194 : 190–92
307. Thomas, D. D., Seidel, J. C., Gergely, J., Hyde, J. S. 1975. *J. Supramol. Struct.* 3 : 376–90
308. Reisler, E., Liu, J. 1982. *J. Mol. Biol.* 157 : 659–69
309. Applegate, D., Reisler, E. 1983. *Biophys. J.* 41 : 229a (Abstr.)

II
Nucleic Acid Structure

Ann. Rev. Biochem. 1982.51:395–427

9

THE THREE-DIMENSIONAL STRUCTURE OF DNA[1]

Steven B. Zimmerman

Laboratory of Molecular Biology, National Institute of Arthritis, Diabetes, and Digestive and Kidney Diseases, National Institutes of Health, Bethesda, Maryland 20205

CONTENTS

Perspectives and Summary

This is a particularly interesting time to review the three-dimensional structures of DNA. Among recent dramatic events are the demonstration of a radically new type of DNA structure—the left-handed Z-helix—as well as the determination of the detailed structure of a segment of B form DNA. The latter is now especially relevant, as an increasing number of heretical candidates are being proposed to replace several familiar models for DNA, including the Watson-Crick model for the B form. Accordingly, the emphasis of this review is on the regular structures that have been proposed for double-stranded DNA.

The review starts by considering two of the major conformations (or "forms") that have been proposed for DNA, namely the B and Z forms,

---

[1]The US Government has the right to retain a nonexclusive, royalty-free license in and to any copyright covering this paper.

with a brief mention of several others, including the A, C, and D forms. Although this survey contains some recent information on the interconversions between certain of these forms and a mention of studies on their dynamic behavior, no attempt is made to be comprehensive in these areas. The status of the structure of DNA-RNA hybrids is briefly reviewed. Obviously, a detailed review of all of these areas is impossible and considerable selection has been exercised. Studies of transfer RNA or of oligonucleotide or polynucleotide complexes with drugs or proteins are only discussed as they seem directly pertinent to DNA structure per se. Theoretical studies are deemphasized relative to experimental studies. The literature covered by this review extends through the middle of 1981.

DNA structure was last reviewed in this series by Jovin (1). A number of other reviews have appeared on nucleic acid structure (2–8) and physical properties (9) as well as on the properties of DNA-protein complexes (10, 11), chromatin (12), and topoisomerases (13–15), and complexes of nucleic acids or their components with metals (16, 17), with drugs (18), or with water (19). Related reviews have centered upon the data obtained from optical techniques (20) or NMR (21–25) (24 reviews nucleosides and nucleotides). Superhelical DNA has been also been reviewed (26–29).

Structural Models for DNA

ORIGINS OF POLYNUCLEOTIDE MODELS With the exception of the recently discovered Z form, the various conformations of DNA were all originally distinguished and defined by their X-ray fiber diffraction patterns. In all cases, this type of diffraction record has provided the richest source of data against which detailed structural proposals for polymeric DNA can be tested. Such fiber diffraction patterns provide two major types of information (30): the helical parameters (pitch, residue repeat distance, and residues/turn), which are often readily obtainable from spacings of the reflections in the X-ray patterns, and the intensity distribution pattern itself. The helical parameters are used to set limits on the type of detailed structural proposals ("models") that need to be considered, and trial structures are built within these constraints. Such provisional models are further subjected to the laws of structural chemistry: covalent bond angles or distances must be within the ranges determined by accurate structure determinations on small molecules. The coordinates of stereochemically acceptable models are used to calculate predicted diffraction patterns, which are then compared with the intensity distribution in the observed diffraction pattern. The model typically goes through a number of cycles of adjustment to improve the fit of its calculated diffraction pattern to the one observed.

It is in the nature of this train of argument that the best model so obtained is not necessarily unique. The coordinates are presented at atomic resolution because they must be so specified for testing of the model, not because they are necessarily a unique solution dictated by the diffraction data. Because of the shortage of diffraction data, assumptions are incorporated into the model building. For example, base-pairing schemes are commonly assumed from physicochemical or other sources. Sugar puckers or backbone dihedral angles may be fixed to certain values or constrained to be within certain ranges. Indeed, even the hand of the basic helix must be assumed, at least initially, although subsequently right- vs left-handed models can be tested against each other.

These limitations on DNA models derived from fiber data have stimulated major reinterpretations and have led to proposals for left-handed and "side-by-side" models. These proposals are discussed in some detail. In the case of Z DNA, the original structure was obtained from single crystal X-ray diffraction analysis of short oligomers. This approach does yield a unique structure; however, it must then be shown that polymers can assume the same structure as did the oligomers. This stage of the demonstration once again has rested heavily on fiber diffraction studies, in conjunction with other physical chemical evidence. Several models are compared in Figure 1 and their parameters are summarized in Table 1. Finally, it should be noted that the original structural proposals for the various forms of DNA stem from data on solid samples, i.e. fibers or crystals. A number of recent studies have concluded that the original models must be modified if they are to portray realistically the structures present in solution.

TERMINOLOGY AND CONVENTIONS   For those interested in the definitions of specific conformational features such as the various sugar puckers and the ranges of dihedral angles the descriptions of Saenger (47) or of Arnott (48) are suggested.

A designation such as "the B form" of DNA is commonly used in the literature to refer to both the actual structure that gives rise to the characteristic diffraction pattern and to the hypothetical structural model that is proposed to rationalize the diffraction pattern. In this review, the former meaning is employed. Specific structural proposals such as types of sugar puckers and sets of atomic coordinates are considered to be attributes of models for the actual structure.

B FORM DNA   The Watson-Crick proposal (49) provided the basis for the familiar model of B form DNA. Elegant studies (50) led to a stereochemically acceptable version of this model that was consistent with the observed

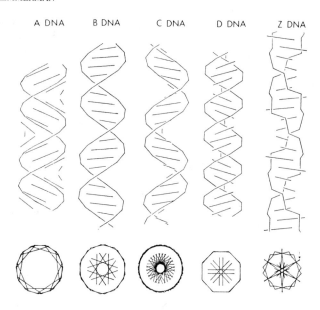

Figure 1   Models for various conformations of DNA. Segments containing 20 base pairs are shown for right-handed models of A (31), B (31), C (32) or D DNA (33) and for the left-handed Z_I form of DNA (34). The upper views are perpendicular to the helical axes and the lower views look along the helical axes. The continuous helical lines are formed by linking the phosphorus atoms along each strand. The line segments indicate the positions of the base pairs and are formed by joining the C1' atoms of each base pair. This simplified mode of representation emphasizes the differences in helical parameters and in positions of the base pairs among these models. The arrow in the lower view of Z DNA indicates the C1' of the deoxyguanosine residue, the base of which is relatively exposed in this structure (see text); the deoxycytidine residue is more centrally located.

diffraction pattern. Arnott & Hukins (31) refined this structure and provided a set of atomic coordinates that have been widely used in discussions of B DNA. Certain aspects of the B form are not presently controversial: all current models assume a double helix with antiparallel strands and with Watson-Crick base pairs oriented roughly at right angles to the helix axis (Figure 1). Controversy exists as to whether the helix is right-or left-handed or both, as to the exact number of base pairs/turn, and as to the disposition of the bases (coplanarity of bases within the pairs and their orientation with respect to the helix axis).

Most DNA can adopt the B form as defined by its characteristic X-ray fiber diffraction pattern: The bulk of the sequences of natural DNA over a wide range of base compositions (51, 52) as well as synthetic DNA of several simple base sequences (35) yields the B pattern under appropriate

DNA conformation	Occurrence	Axial rise per base pair (Å)	Base pairs per turn	Base pairs per repeating unit	Ref
A	Most natural and synthetic DNA	2.6	11.0	1	31, 35, 36
B	Most natural and synthetic DNA	3.4	10.0[a]	1	31, 35, 37
Alternating B	Several alternating purine-pyrimidine DNA	3.4	10.0	2	38
C	Most natural and synthetic DNA	3.3	7.9–9.6	1	33, 35, 39, 40
D	Several synthetic DNA	3.0	8.0	1	33, 35, 41
T	Glucosylated bacteriophage DNA	3.3–3.4	8.0–8.4	1	42, 43
Z	Several alternating purine-pyrimidine DNA	3.6–3.8	12.0	2	34, 44–46

[a] Alternative models with different parameters are discussed in text.

conditions. Further, the patterns are similar to each other in detail, which suggests a narrow range of variability in B form structure. This apparent homogeneity is in contrast to recent suggestions from nondiffraction techniques and from single-crystal diffraction studies on oligonucleotides, which indicate either static or dynamic structural heterogeneity (see below). Early indications of deviant structures in DNA of high AT-content have been correlated with the presence of non-DNA material (53). There are a few species of synthetic DNA, namely poly(dA)·poly(dT), poly(dI)·poly(dC), and poly(dA-dI)·poly(dC-dT), that have a B form that is significantly different from that of the other complementary deoxypolymers. These polymers have different intermolecular packing arrangements and a slightly changed value for the rise per residue (35, 54). A striking indication of altered structure or structures for the conformations of poly(dA)·poly(dT) and poly(dI)·poly(dC) is their refusal to undergo a transition from the B form to the A form (35, 54–56). These unusual properties assume particular interest given the existence of relatively long dA·dT sequences in vivo (57).

B form adopted by self-complementary oligodeoxynucleotides The structure of a long self-complementary oligodeoxynucleotide, d(CGCGAATTCGCG), has recently been solved by single-crystal diffraction techniques (58–61). The dodecanucleotide forms more than a full turn of a helix whose overall structure is notably like that proposed for the canonical model for B DNA. This study therefore allows an unambiguous examination at high resolution of a prototype for the B form of DNA, and is certainly a most important addition to our knowledge.

The structure adopted by the dodecanucleotide is a right-handed antiparallel double-stranded helix with the bases essentially perpendicular to the helix axis (58). While these major attributes of the B form model are present, the detailed structure of the oligomer departs from that of the familiar model in a number of significant ways.

The well-defined molecular axis is not straight; rather the axis traces a smooth bend of significant curvature (a total of 19° bend distributed over the entire length, which corresponds to a radius of curvature of 112 Å; for reference, the diameter of the B helix itself is $\sim$ 21 Å.) A clear basis for the bend is seen in the intermolecular contacts within the crystal (58). The terminal three G-C base pairs of each duplex form five H bonds with atoms of the next duplex "above" themselves in a manner that requires a bend in the helix axis. There are two formal possibilities: either the DNA is intrinsically bent and the crystal packing simply accommodates this innate tendency or, alternatively, the packing causes the bending. The authors provide several reasonable arguments for the latter interpretation, but ultimately we must await the results of similar studies with other oligomers. It is the terminal alternating G-C sequences that are involved in the canted interactions between duplexes, so that it will be important to see what structures form in oligomers lacking these terminal sequences or, alternatively, containing longer internal sequences. The bend in the dodecamer clearly poses a problem if we wish to know the structure of unbent DNA, since deconvoluting the bent structure requires some arbitrary decisions. Notwithstanding, it is most useful to examine this remarkable structure.

The crystal structure differs from the usual models derived from fiber diffraction in a fundamental way: The dodecamer duplex is not simply a set of 24-nucleotidyl residues of essentially identical conformation joined monotonously into 12 base pairs of essentially identical structure. Rather there are enormous variations in the helical relationships between successive base pairs, and often significant conformational differences between the residues (59, 60). The helical variations span the full range observed by fiber diffraction for structures as disparate as those of the A, B, and D forms of DNA. For example, the eleven local helical steps correspond to 9.4, 9.1, 10.8, 9.6, 9.6, 11.2, 10.0, 8.7, 11.1, 8.0, and 9.7 residues/turn (59). These values may be compared to the values from fiber diffraction of $\sim$ 11, 10, and 8 residues/turn for the A, B, and D helices, respectively (Table 1). Hence, although the average (local) helical parameters in the dodecanucleotide (9.65 residues/turn and 3.33 Å rise/residue) are within a few percent of those inferred from the B form diffraction, the structure has tremendous local variation. Indeed, while the base pair tilt and position relative to the helix axis for the central six base pairs is similar to that in the B DNA models from fiber diffraction studies (31), the adjacent base pair on either side has a distinctly A DNA conformation, and another peripheral base pair is like that of the D form. The other backbone conformational angles are generally similar to those inferred from earlier fiber diffraction or theoretical studies, with a few values in unusual ranges.

The bases of a given base pair are not coplanar in the dodecanucleotide (59, 60). The average propellor twist (i.e. total dihedral angle between base planes in one H-bonded base pair) is 17.3° ± 0.4° for the central four A-T residues and 11.5° ± 5.1° for the G-C pairs at the ends. [In contrast, an earlier model for B DNA from fiber diffraction studies (31) has an almost coplanar arrangement of the bases within a base pair (twist = 4°).]

The variation in sugar conformations within the dodecamer has provided several insights (59, 59a). First, the conformations span essentially the whole range possible for deoxyfuranose rings. They show a correlation between the glycosidic torsion angle and the sugar ring conformation. The purine residues tend toward the C2'-*endo* sugar pucker and a high value for their glycosidic angle, while the pyrimidine residues tend toward a lower glycosidic angle and their sugars are more like C3'-*endo* puckered sugars. This behavior was rationalized in terms of steric contacts between the sugars and the O2 of the pyrimidines. The authors note a further striking fact, which they formalize as the "principle of anticorrelation:" the two sugars of a given base pair tend to have values of their internal torsion angles (specifically about C4'–C3') that are equidistant from that of a central value corresponding to the C1'-*exo* conformation. In other words, if one sugar in a given base pair is C3'-*endo,* the other sugar in that pair tends to have a C2'-*endo* conformation. Several deoxyribosyl structures and a recent DNA-RNA hybrid structure seem to be consistent with this principle, whereas the sugars in yeast phenylalanine tRNA are not.

In addition to the striking static heterogeneity in local structure, the dodecanucleotide provides evidence of dynamic heterogeneity. The thermal vibrations of individual atoms were inferred from their temperature factors as obtained in the X-ray structure determination. Drew et al (59) indicate that the relatively larger vibrations of peripheral atoms are consistent with the rapid intramolecular motions ascribed to DNA from NMR and other measurements (see below)

DNA in solution is often presumed to be in the B form (see section on the helical repeat of DNA). While modeling studies of solvent accessibility to the surfaces of the B form and of other polynucleotide structures have appeared (62, 63), experimental knowledge of the organization of water about duplex DNA has been sparse. [The single-crystal structure of the proflavine complex with dCpdG is a notable exception (64).] The dodecanucleotide structure provides a unique opportunity to visualize water structure in a sizeable unit of B form DNA (61). There are 72 ordered molecules of water per duplex. Of these, 50 are either in the grooves or closely associated with the phosphate groups. The most striking arrangement of water molecules is in the region of the minor groove associated with the central AATT sequence. Solvent is organized in layers up to three or four molecules deep. The innermost layer forms a regular "spine" of water bridging the hydrophilic groups of alternate bases. This backbone of water is compatible with any sequence of A-T or I-C base pairs, but *not* with G-C residues, due to the disrupting influence of the 2-amino group of guanine. This pattern suggests that guanine residues will destabilize the B form, an inference that is in general concurrence with the fiber diffraction survey of Leslie et al (35).

Finally, we note a study that suggests that the dodecanucleotide structure may represent a reasonable model for B DNA in solution. Lomonosoff et al [(65) and further discussion in (60)] have found a good correlation between the rates of nuclease cleavage of internucleotide bonds of the dodecamer in solution and their "exposure" (magnitude of the local helical rotation between base pairs) in the crystal.

A complex of a second sizable oligodeoxynucleotide, d(CGTACG), with daunomycin has been shown by single-crystal diffraction analysis to contain a short segment that adopts a structure similar to that of the canonical B form. This complex forms a self-complementary right-handed anti parallel mini-DNA helix, with a molecule of drug intercalated between the C-G base pairs at each end (66). The structure of the ends of the molecules is modified for intercalation, but the central A and T residues share many characteristics with the familiar B DNA model. The sugars of the central A and T residues conform to the principle of anticorrelation above (59). The authors indicate however that the backbone conformations of the central base pairs of the hexanucleotide have significant departures from those in the usual A or B form models.

B form model for alternating copolymers   A single-crystal X-ray structure determination (67, 67a) on the tetranucleotide d(ATAT) has led to interesting proposals, (38, 67a) for a modified B conformation suitable for a regularly alternating copolymer. In the crystal, the two base pairs at each end of the tetranucleotide form H bonds to two different adjacent tetranucleotide molecules, which yield short segments of right-handed antiparallel double helix. The most striking feature of the crystal structure is the regular alternation in sugar pucker and glycosidic dihedral angle, χ. The adenosine residues both have a C3'-*endo* sugar conformation, while the thymidine residues have a C2'-*endo* pucker. This alternation has been incorporated into a model for the polymer poly(dA-dT)·poly(dA-dT), which is generally rather similar to the canonical B form model [cf Figure 10 in (68)]. This model is proposed as one of a family of basically similar alternating structures that could be built to maximize base-stacking interactions (38; see also 67a, 69). Such alternating B models are consistent with a number of the properties of poly(dA-dT)·poly(dA-dT) previously observed in solution [see (38) for references]. Shindo et al (70) have independently suggested a model with an alternating backbone conformation for this polymer in solution based on two resonance peaks in the ^{31}P NMR spectrum. Recent ^{31}P NMR studies have also detected the two resonance peaks expected for the alternating B model in fibers of poly(dA-dT)·poly(dA-dT) (71), as well as in solutions of poly(dA-dbr^{5}U)·poly(dA-dbr^{5}U) and poly(dI-dC)·poly(dI-dC) (68). The tetramethylammonium ion has been suggested to favor an alternating conformation for poly(dA-dT)·poly(dA-dT) (72).

Lomonosoff et al (65) have provided evidence that poly(dG-dC)·poly(dG-dC) in low-salt solutions can also adopt an alternating type of conformation, presumably of the B form, based upon pancreatic DNase I digestion patterns of the polymer. [An alternating B model was proposed (73) for the high-salt form of (dG-dC)-oligomers in an earlier study; in hindsight, it seems as likely that the dinucleotide repeat inferred is related to the more recently delineated Z form (see below).] Less detailed models for alternating structures have been suggested from theoretical studies (74, 75).

Bent models for B DNA   Structural models for DNA have generally been built as unbent double helices. There have, however, been several suggestions of smoothly bent or discontinuously bent ("kinked") models. Smoothly bent DNA has actually been observed, in the form of the dodecanucleotide duplex (58) discussed in a previous section. There have been a number of theoretical studies of smoothly bent models for DNA wound around the histone cores of chromatin (76–79).

Various kinked models feature straight segments of B form DNA interspersed at regular intervals with residues having alterations in certain torsional angles of their sugar-phosphate backbones (80–83). A variety of superhelical arrangements can result, depending upon both the frequency of kinking and its detailed mechanism. Such kinked models form a conceptual framework for observations as diverse as the compaction of chromatin and its regular patterns of nuclease digestion, the "breathing" of DNA, or the mechanism of drug intercalation. Solvent bombardment or thermal sources have been suggested as possible origins for kinks and other types of localized or traveling fluctuations in DNA structure (84–86). There is no direct evidence for kinking at present. ^{31}P NMR studies of DNA in either chromatin or free in solution have not indicated the presence of species with an altered chemical shift as might be expected if kinking occurred and if the phosphodiester geometry was markedly changed at the kinked site (87–90).

Left-handed models for B DNA   There has in the past generally been a consensus that B form DNA is a right-handed helix. As particularly emphasized by Sasisekharan and his collaborators, this assumption does not have a secure experimental basis. Let us review the arguments for left-handed DNA, reserving for the moment a discussion of mixtures of left- and right-handed DNA in the form of the side-by-side DNA model. It has long been known that DNA in the solid state can rapidly and reversibly be interconverted between the A, B, and C forms, simply by varying the relative humidity and salts present (37). The ease of these transitions in fibers was interpreted to mean that all of these forms are of the same hand. Since the original model building for the A form appeared to rule out a left-handed structure (cf section on A DNA below), the A, B, and C forms were all inferred to be right handed. This reasoning is subject to dispute at several points: First, it is only an intuitive conclusion at present that transitions in fibers can not change the hand of the structures involved. Second, even if this is accepted as a working hypothesis, several groups have shown that the B form of DNA interconverts in fibers with the Z form (46, 91), a presumably left-handed structure; hence, if anything, this line of argument suggests that the B form could be left handed. Third, satisfactory left-handed models apparently can be built for the A form (and also for the B and D forms) (69, 92). Detailed sets of coordinates have been supplied by Gupta et al (69) that are stated to fulfill the usual stereochemical criteria. Several groups have noted that, surprisingly enough, only minor differences in the values of backbone dihedral angles need occur between left- and right-handed conformations, although the orientation of the base undergoes a significant shift (2, 7, 75, 92–94). Energy calculations have indicated to some a preference for right-handed conformations (2, 7, 94), although base stacking energies seem to be similar (95). A major criterion for an acceptable model is that it yield a predicted diffraction pattern consistent with the observed diffraction pattern. The left-handed B DNA model appears satisfactory in this respect (69). Hence, from the results of the fiber diffraction approach, there is no clear basis for preferring left- to right-handed DNA.

It may be noted that there are several unambiguous observations of right-handed helices in oligodeoxynucleotides of mixed base sequence in both the A and B forms and in RNA [the short A form–like helices of yeast phenylalanine tRNA (96)]. As yet, the only examples of a left-handed helix are seen in certain oligodeoxynucleotides and polydeoxynucleotides, which have a regularly alternating purine-pyrimidine sequence and a structure that is presumably not of general occurrence (see section on Z form DNA), or in short oligoribonucleotides that are constrained by a second covalent link between base and sugar (97). In sum, while there are indications that natural DNA is right handed, there seems no obvious reason why it should not be able to adopt a left-handed conformation.

Side-by-side models for DNA   Two groups have independently proposed a new class of model for the B form of DNA (98–101). These so-called side-by-side or SBS models are basically different from either the uniformly right- or uniformly left-handed structures discussed above. All of the SBS models feature a regular alternation of short segments of left- and right-handed double-stranded DNA. The versions proposed by Sasisekharan and collaborators (98, 100) have alternating segments of five base pairs, so that the two strands do not undergo a net winding around themselves. The most recent SBS model of Millane & Rodley (101) has a slight right-handed bias so that the strands wind around each other, i.e. they are "linked" about once for every 77 base pairs. In contrast, the uniform left- or right-handed models previously discussed are linked once for every pitch length of approximately ten base pairs. A debate has developed in the literature between the proponents of the double helix and those favoring the SBS conformation. We first consider the arguments based on model building and X-ray diffraction studies and then consider topological and other types of evidence.

Models of SBS DNA have many more degrees of freedom available than do models of uniform helices, since the repeating unit of SBS DNA is long (ten base pairs rather than one or two base pairs) and heterogeneous (left-handed, right-handed, and junction regions). Consequently, while modeling of such a large unit is a technically difficult problem, there seems little doubt that stereochemically acceptable SBS models can be built (101, 102). Detailed studies of the conformation and base stacking have appeared (95). It may be surprising to the reader that the expected diffraction patterns for such models are not totally different from those of the regularly helical models. Similarities arise because the fiber diffraction patterns are dominated by two features that appear similarly in SBS and uniform models. The first feature is the apparent helical repeat: canonical models have a pitch of ~ 34 Å, consistent with a tenfold helix and implying diffraction layer lines at spacings of $N/34$ Å (where N is an integer). Similarly spaced layer lines may be generated by SBS models with an exact or approximate structural repeat distance of 34 Å, which corresponds to a unit of five left-handed and five right-handed base pairs. The second common feature of SBS and double-helical models is the centrally located Watson-Crick base pairs, which are situated approximately at right angles to the helix axis and spaced at an average of ~ 3.4 Å. This 3.4-Å average spacing between base pairs results in the strong meridional reflection that is observed at $N/3.4$ Å and that also dominates the form of an interatomic scattering function called the axial Patterson function, which has been applied to this problem (103). The general similarities in this function for the SBS and canonical B DNA models therefore do not form a basis for distinguishing between these structures as candidates for the B form. A more incisive test is to calculate the actual pattern expected from the atomic coordinates of the models in question. This has been done by two groups for several versions of SBS models (102, 104). In both studies, the authors concluded that the predicted diffraction patterns of the particular SBS models they examined do not fit the observed diffraction data nearly as well as do the predicted patterns of the best double-helical structures. In particular, SBS models predict several meridional reflections that are not observed. These studies have dealt with diffraction from fibers of noncrystalline DNA, the so-called continuous transform. There is also considerable diffraction data from fibers of semicrystalline B DNA. Arnott (105) notes that the intensity distribution of the crystalline data agrees better with the double-helical rather than the SBS model. Further, a point originally made by Dover (106) is applied (105). The crystal lattice adopted by the lithium salt of DNA generates precisely regular intermolecular contacts between neighboring molecules for base pairs related by steps of $36.15° \pm 0.25°$ of helical rotation, i.e. for a helix with precisely ten base pairs per turn of the helix. This relationship makes immediate sense in that it optimizes favorable lattice interactions for each base pair if the structure involved is that of a regular tenfold helix; however, it would be essentially a fortuitous result for an SBS type of structure, since SBS structures do not have a regular repeat at 36° intervals about the molecular axis.

As mentioned, the two strands of SBS DNA are relatively unlinked because of the alternation between left- and right-handed segments. A possible consequence of this feature, as noted a number of years ago (107), and presumably one of the motive forces behind SBS DNA, is that the relative lack of linking of the strands might facilitate the separation of the two strands during replication or other processes. However, the cell has developed elegant enzymatic mechanisms for winding and unwinding DNA

(13–15), so that the ease of strand separation in SBS models is not a compelling argument in their favor (cf 108). Further, there is topological evidence that the two strands of closed-circular DNA are indeed linked by intertwining about each other once every 10 base pairs (109). In contrast, under special circumstances, several studies have observed coiling of DNA strands without linking, specifically when closed-circular single-stranded DNA interacts with closed-circular single- or double-stranded DNA of complementary sequence (110, 111). The regularity and conformations involved in these interactions are unknown, but such situations are possible candidates for mixtures of left- and right-handed helical segments. Finally, Greenall et al (104) raise the question of the conformations involved in the A ⇌ B interconversion if the B form is indeed in SBS structure, since an SBS version of the A form has apparently not yet been described.

In sum, there seems to be little experimental basis for SBS DNA, while there is evidence for uniform double helices. Until some more convincing evidence is brought forth, SBS DNA seems to be an unlikely possibility.

z FORM DNA   Some years ago, Pohl & Jovin observed that the circular dichroism (CD) spectrum of the alternating copolymer, (poly(dG-dC)·poly(dG-dC), underwent a novel inversion when the polymer was exposed to high salt or ethanol concentrations (112, 113). Recent demonstrations (34, 44, 45, 114) that oligomers of this material can assume a left-handed conformation, the Z form, have suggested that this structure is the basis for the inverted CD spectrum (see below).

Z Form in crystalline oligonucleotides of d(C-G)·d(C-G)   The structures of several oligomers of alternating d(C-G)·d(C-G) have been determined within the last few years using single-crystal diffraction techniques. Given the large number of departures of these oligomer structures from those of canonical polynucleotides, it is fortunate that this methodology yields structures that are free of the kinds of assumptions that typically are made in fiber diffraction analysis. The initial structure in this series was derived from crystals of the hexamer, d(CGCGCG) (44), and was followed shortly by determinations on the tetramer, d(CGCG) (45, 114). There are now a number of examples of these oligomers crystallized from a variety of media. The major features of their structures are similar. I first discuss the hexamer (44) as a prototype, and then summarize the differences among the various oligomer structures.

In the crystals of the hexamer, two molecules of d(CGCGCG) join in an antiparallel fashion to form a left-handed minihelix. The C and G residues on one strand form Watson-Crick hydrogen bonds to the G and C residues on the other strand. The six base pairs so formed comprise one half of a left-handed helical turn.

The conformations of the residues also alternate in this structure of alternating base sequence. In general, the dC residues are similar to the residues in the canonical models for B DNA, while the dG residues are very different. For example, the sugars of the dC and dG residues display a regular alternation between the C2'-*endo* and C3'-*endo* conformations, respectively. Further, the dihedral angle around the C4'–C5' bond alternates markedly (between a *gauche-gauche* conformation in dC residues, as in B DNA, and a *gauche-trans* conformation in dG residues). Since the C and G residues have different conformations, the repeating unit on a given strand is a *dinucleotide*. These dinucleotide units are very regularly arrayed in the minihelix, although the symmetry of the lattice does not demand this regularity. The regularity even extends across the gap where duplexes stack upon each other except, of course, for the missing phosphate groups. The result is that the crystal structure approximates that of a continuous polymer, with 6 dinucleotides or 12 bases for each helical turn of a given strand. These alternating conformations produce an irregular zig-zag course of the sugar-phosphate backbone, hence the Z DNA designation for this structure (Figure 1). The residues also alternate in the relationship of the base to its sugar. The dC residue has an *anti* relationship, which seems to be generally needed to relieve steric interactions between pyrimidine bases and their sugars (115). The guanine residues in contrast assume a *syn* conformation wherein the base closely approaches its sugar. This is the first example of a residue of a natural nucleotide that adopts the *syn* position in an oligomeric or larger system, although the bases in polynucleotides can be constrained toward this position by a bulky base substituent (116) or by a second covalent linkage between base and sugar (97).

As might be expected from the unique alternation of backbone and base-sugar conformations, the positioning of base pairs within the helix and the base-stacking relationships are remarkable. The base pairs are relatively peripheral in location in Z DNA. Because of the *syn* conformation of the guanosine, the reactive N7 and C8 positions of guanine are located at the margins of the structure and thereby made accessible to environmental insults (see below). The base pairs at the d(CpG) sequences are displaced laterally by 7 Å relative to each other, so that ordinary base overlaps do not occur. Rather, the cytosine residue of one base pair stacks with a neighboring cytosine residue on the opposite strand. The guanine residues at this sequence do not stack with the adjacent bases, but rather overlap the furanose ring oxygens of the adjacent sugars. In contrast the stacks at the d(GpC) sequences are relatively similar to those of B DNA with overlap of the bases that adjoin on the same strand. The Z form has a single very deep helical "groove" that corresponds in location to the minor groove (i.e. the groove between the sugar-phosphate backbones) in the A or B forms.

Limited but significant variation has been observed in the structures solved to date for d(CGCG) or d(CGCGCG) crystallized under various conditions (34, 44, 45, 114, 117). The principal variation occurs in the orientation of the phosphate group of the GpC sequence. In the various crystals that have been solved, one or more of the phosphate residues has rotated from its position in the predominant conformation so that it lies further outwards and away from the minor groove (34, 45, 114, 117). Models for continuous polymers based upon each of these two phosphate conformations, labelled Z_I and Z_{II}, have been described (see below). A related change in the conformation of the phosphate group occurs in the crystalline tetramer solved by Drew et al (45). In this structure, labeled Z', the phosphate group of the GpC sequence again rotates outward, as in going from Z_I to Z_{II}. The reasons for the rotations in Z' and Z_{II} are apparently different, being correlated with repulsion by a neighboring chloride ion for Z' (45) or, in some cases at least, with binding to a nearby hydrated magnesium ion for Z_{II} (34, 114). The altered phosphate positioning in Z' is correlated with a change of sugar pucker of the internal deoxyguanosine residues, so that those sugars adopt a conformation (Cl'-exo) similar to that in the deoxycytidine residues. The result is that the Z' backbone has an almost uniform sugar pucker, which demonstrates that the characteristic dinucleotide repeat of Z DNA does not necessarily need to extend to a major involvement of the sugar conformation (45).

Z form in fibers of alternating polymers   The demonstration of the Z structure in crystals of oligomers has led to attempts to identify such a structure in fibers or in solutions of polymeric DNA. X-ray fiber diffraction has provided relatively unambiguous evidence in the case of fibers. Arguments for the Z form in solution are summarized in the next section.

Fibers of certain alternating deoxypolymers yield a distinctive X-ray diffraction pattern identified with the Z form [which has also been called the S form in polymers (35)]. The original observation of Z form diffraction *from fibers* was made with poly(dG-dC)·poly(dG-dC) (46). Basically similar patterns have been collected from fibers of poly(dA-dC)·poly(dG-dT) (35) and poly(dG-dm⁵C)·poly(dG-dm⁵C) (118). A related pattern from poly(dA-ds⁴T)·poly(dA-ds⁴T) (119) has been reinterpreted in terms of the Z structure (46).

Models play an important role in fiber diffraction in validating proposed structures, as outlined earlier. In the case of polymeric DNA, the models for the Z form have followed the structures that were determined by single-crystal diffraction of the oligomers. The initial model for poly(dG-dC)·poly(dG-dC) (46) differs significantly in backbone position and base-stacking arrangements from the oligomer structures; a more recent model (120) approaches the oligomer structure more closely. In neither case were predicted diffraction patterns presented for the models. The discussion of Wang et al (34) contains the most detailed polymeric models based on the oligomers. These authors have generated full sets of atomic coordinates for each of the two variations on the Z form described above, i.e. Z_I and Z_{II}, which they observed in their oligomer studies. The predicted diffraction pattern for their model of the predominant conformation (Z_I) (34) fits the observed pattern (46), which provides strong evidence for the Z conformation in fibers of this polymer.

Is an RNA backbone compatible with the Z form? In the initial description of the Z form, the authors (44) indicated that the 2'-hydroxyl position of the sugar would point outwards from the helix and so would not necessarily be sterically encumbered. There are, however, no spectral indications that the ribopolymer, poly(rG-rC)·poly(rG-rC), enters a high-salt form under conditions where the deoxypolymer does (113). The relationship between such high-salt conformations and the Z form is considered next.

The high-salt form of alternating polymers and its relationship to the Z form

The studies of Pohl & Jovin on the high-salt form of poly(dG-dC)·poly(dG-dC) were clearly an important factor in the choice of d(CGCG) and d(CGCGCG) for crystallization, which led to the elucidation of the Z structure. The original studies, which showed a requirement for high-salt concentrations (113) or intermediate levels of ethanol (112), have been extended to include a variety of cations that are able to elicit this form (based on spectral criteria) at much lower concentrations (121). For example, poly(dG-dm⁵C)·poly(dG-dm⁵C) undergoes an inversion in its circular dichroism spectrum in the presence of 5 μM hexamine cobalt or 2 μM spermine. This study also demonstrates a dramatic influence of polymer composition: the methylated polymer undergoes the spectral transition at ~ 1 mM Mg^{2+}, a concentration about 1000-fold lower than that needed for the unmethylated polymer (121). Although I continue to use the high-salt designation for the conformation with inverted CD spectrum described by Pohl & Jovin (113), these results clearly show it can be induced under specific conditions at low ionic strength. As is described in the next section, binding of certain substituents also favors the occurrence of the high-salt form.

Other properties besides circular dichroism and ultraviolet absorbance (112, 113) have been used to measure the transition. The ³H-exchange properties of the high-salt form are markedly different from those of the low-salt form. Two protons have half-times at least 50-fold longer than those in more usual polynucleotide conformations. Ramstein & Leng (122) note their potential use to assay for the high-salt conformation in natural DNA. The transition is also accompanied by marked changes in Raman scattering spectra (123).

There is considerable evidence that the high-salt form corresponds to the Z form. First, the laser Raman spectrum of crystalline d(CGCGCG) is similar that of poly(dG-dC)·poly(dG-dC) in high-salt solution and different from that of the polymer in low-salt solution (123a). Second, the increase in molecular length going through the transition at intermediate ethanol concentrations corresponds to that expected for the conversion from the B to the Z form (123b). Third, the high-salt form (as evidenced by its CD spectrum) is observed for polymers with regularly alternating purine-pyrimidine sequences, i.e. polymers of alternating (dG-dC) (113), (dG-dm ^{5}C) (121), or (dI-dbr^{5}C) (73), but not for the nonalternating polymer, poly(dG)·poly(dC) (113). This specificity is consistent with the basic structure of the Z form, which requires alternation of purines and pyrimidines. Incidentally, the inversion of the CD spectrum might seem to be evidence for the presumed conversion from a right-handed B form to a left-handed Z form. However, theoretical considerations indicate that inversion does not necessarily correspond to a change in helical sense (124). In addition, there are instances of similar inverted CD spectra for which there are viable alternative explanations. For example, the product of the annealing of complementary closed-circular DNA, Form V DNA, has a related spectrum (111), although its irregular base sequence seems incompatible with a regular Z conformation. A second inverted spectrum is that of poly(dI-dC)·poly(dI-dC) in low-salt media. As described in the section on D form DNA, it has been variously interpreted by different groups as due to left- or right-handed helices quite different from the Z form.

Further evidence for correspondence between the high-salt and Z forms comes from NMR experiments. These studies suggest an alternating conformation for the high-salt form of alternating (dG-dC). The ^{1}H NMR spectrum of (dG-dC)$_8$ (73) indicates that one but not both glycosidic torsion angles and one but not both sugar puckers change in the Pohl-Jovin transition. The ^{31}P NMR spectrum of this material (73) as well as that of a 145 base-pair length of alternating (dC-dG) (68, 125) splits into two resonances of approximately equal intensity, which indicates an alternating conformation about the phosphodiester linkage. Magnetic shielding constants calculated for poly(dG-dC)·poly(dG-dC) in high salt were in agreement with the presence of the Z form but not of the B form (126). These various NMR results are consistent with but not uniquely diagnostic for the Z form. Adoption of the Z structure is also suggested by studies of derivatives of alternating sequence polymers (see next section).

There are two studies that raise the possibility that the high-salt form may not be the Z form. First, determination of the helical periodicity of either poly(dG-dC)·poly(dG-dC) or poly(dG-dm^{5}C)·poly(dG-dm^{5}C) in the high-salt form by nuclease digestion while the polymers are bound to absorbents (see section on the helical repeat of DNA) gives $\geq$ 13 base pairs/ turn rather than the value of 12 expected for the Z structure. It is not clear whether the discrepancy reflects the effects of adsorption or whether a different structure is present. Second, the calculated CD spectrum for the Z form does not agree with that observed for the high-salt form of poly(dG-dC)·poly(dG-dC) (127). Whether the disagreement arises from assumptions in the calculations or in the choice of structure has not been resolved.

The structural transition of alternating (dG-dC) in solution generally occurs at a relatively high-salt concentration (113). Crawford et al (114) note that, despite this, the various hexamer and tetramer crystals containing segments of Z DNA were generally obtained from low-salt solutions. They therefore suggest that there is an equilibrium between left- and right-handed conformations, which shifts toward the Z form as crystallization proceeds. The crystals themselves nominally qualify as high-salt environments, having concentrations of charged groups of the order of several molar (44, 45). A form of d(CGCG) crystallized at lower salt concentrations has been described but not solved in detail; this form undergoes a reversible salt-dependent transition to the Z' structure while still in crystalline form (128).

Reaction of alternating (dG-dC) with ligands   Several ligands favor the transition of poly(dG-dC)·poly(dG-dC) to a form with an inverted CD spectrum similar to that induced by high-salt levels. Mitomycin C induces such a change with a polymer specifity like that for the salt-induced conversion: alternating deoxypolymers of (dG-dC) are affected, but not the corresponding ribopolymer or the nonalternating deoxypolymer. The drug also can cause a qualitatively similar but much less extensive change in the CD spectrum of natural DNA (129).

Extensive reaction of the carcinogen, N-acetoxy-N-acetyl-2-aminofluorene, can cause poly(dG-dC)·poly(dG-dC) to undergo an inversion in its CD spectrum even at low ionic strengths (130–132). At relatively low levels of derivatization, the polymer remains more prone to undergo the transition, as judged by reduced levels of ethanol needed for the reaction. The major site of covalent attachment of the bulky fluorene derivative is at the

C8 position of guanine, which leads to very interesting speculation in terms of the Z structure. (The usual caveat applies: we do not know that the inverted spectrum of the fluorene derivative corresponds to that of the Z form, although it is certainly worth making this working hypothesis to entertain the implications.) Reactive sites on guanine (N7 and C8) are particularly exposed to the medium in the Z structure, and the potential importance of the Z form in reactions with chemicals in the environment was noted by Wang et al (44). Several consequences have been suggested. First, the guanine residues may be more reactive in the Z form. While this suggestion was difficult to test in high-salt or ethanolic media (131), it may be approachable in the aqueous media recently described (121). Second, once reacted, the altered residues may lock the sequence into the Z form.

Related studies have been performed with a second carcinogen, N-hydroxy-N-2-aminofluorene (130) or with dimethyl sulfate (132a). Also the intercalating dye, ethidium bromide, was early noted to be an effective inhibitor of the transition to the high-salt form, presumably due to a preference for binding to the low-salt form (133).

Z form in natural DNA? The unusual properties associated with Z DNA and with the high-salt form of DNA of alternating purine-pyrimidine sequence have prompted experiments that seek evidence for the occurrence of either or both states in DNA of heterogeneous sequence. As just outlined, ligands may induce limited amounts of the high-salt form in natural DNA. Several studies that do not depend on such ligands are available at the moment; more are expected.

Supercoiled phage PM2 DNA, a closed-circular DNA of heterogeneous sequence, undergoes a salt-induced cooperative transition to a form adsorbed by nitrocellulose filters (134). The basis of the adsorption is unknown; formation of a Z structure is proposed, with either the Z form itself or single-stranded regions between B and Z regions binding to the filter. (The degree of retention of poly(dG-dC)·poly(dG-dC) under these conditions would be of interest.) The NaCl concentration required for the transition is similar to that for the high-salt transition of alternating d(G-C). A higher salt level is required for *linear* PM2 DNA, in accord with earlier suggestions (135, 136) that the excess energy of negative supercoiling would aid the transition from a right-handed to a left-handed structure. Prior treatment of the DNA with N-acetoxy-N-acetyl-2-aminofluorene (see preceding section) or with ultraviolet irradiation lowered the salt concentration needed to induce the transition. The authors speculate that an increased tendency to enter the proposed Z conformation as a result of DNA damage may provide a signal to DNA repair systems.

Klysik et al (137) have cloned plasmids containing inserts of various lengths of alternating d(G-C). Inserts of longer than 40 base pairs of d(G-C) were unstable and suffered deletions. The authors generated a short segment (138–176 base pairs) of heterogeneous sequence DNA containing near its center about a third of its length as alternating d(G-C) residues [which were in turn interrupted by a single d(GATC) sequence]. This DNA segment underwent a salt-induced transformation to a form with a partially inverted CD spectrum. The high-salt spectrum was equivalent to that of a mixture of heterogeneous sequence DNA with an amount of alternating d(G-C) lower than that which was actually present, which lead the authors to suggest that d(G-C) residues near the edges of the insert might not be in a Z type of conformation. As they note, this argument makes several assumptions, including a lack of signal from the proposed junction regions. In contrast, the ^{31}P NMR spectrum in high-salt media showed the appearance of a second resonance in about the amount expected for total conversion of the insert. Klysik et al further applied an important test of the proposed conversion from right- to left-handed regions. A DNA segment with alternating d(G-C) ends was inserted into a plasmid; the linking number of the DNA was estimated from the band pattern in gels run at salt concentrations that spanned the transition. A high-salt conversion clearly occurred that was formally equivalent to the unwinding of about half of the expected number of supercoils for a B to Z transition. The authors suggest that the relatively small change reflects a partial conversion. An alternative interpretation may be considered: Because of their self-complementary sequences, the d(G-C) inserts could "loop out" to make a cruciform structure, with an expected change in linking number about equal to that actually observed. This ambiguity can be avoided in principle by using inserts of a sequence that is not self-complementary but that forms the Z structure. Poly(dA-dC)·poly(dG-dT) appears to fit these requirements (35). It may be noted that a stretch of 62 base pairs of this latter sequence has been found in vivo (138), while comparably long sequences of alternating d(G-C) have apparently not been described. Dickerson & Drew (60) suggest that the tendency of alternating (dC-dG) sequences to adopt the Z form may indeed be small, given the absence of Z structure in their crystals of d(CGCGAATTCGCG). Those crystals were comparable in salt level to those of d(CGCG) or d(CGCGCG) containing the Z form.

Antibodies specific for the high-salt form have been elicited in several animals in response to injections of either brominated or unbrominated poly (dG-dC)·poly(dG-dC) (140a). The fluorescence staining patterns of these antibodies on the polytene chromosomes of Drosophila have provided the first evidence for the Z form in natural DNA (140b). A regulatory role has been suggested for such Z form regions (121, 140b).

Transition between B and Z forms   Arnott et al (35, 46) observed the Z form in fibers of alternating d(G-C) that earlier had given the B form diffraction pattern. Sasisekharan & Brahmachari (91) showed a relatively rapid conversion between these forms that was controlled by changes in the ambient relative humidity. The two groups reached opposite conclusions: one argued that B and Z forms are of different hand (46), while the other concluded that a ready transition under mild conditions ruled out a change in handedness of the helix (91). Whatever the outcome in fibers, Simpson & Shindo (125) note that the transition in solution between the high- and low-salt forms of the 145 base pair pieces of alternating d(G-C) can occur without total strand dissociation; it is, of course, not clear that the same forms are involved as in fibers.

The Z form (74, 139, 140) and the B ⇌ Z transition (136) have also been considered from a theoretical point of view. Possible model structures for the interface between B and Z helices have been discussed, both with unstacked (44) or completely stacked junctions (141).

OTHER FORMS OF DNA   Studies of polymers of simple repeated sequence have been crucial in delimiting the variation expected from base sequences. A recent survey of many complementary deoxypolymers of defined repeating sequences, by Leslie et al (35), based on X-ray fiber diffraction methods should be noted.

A form DNA   Most DNA will enter the A form. Notable exceptions are poly(dA)·poly(dT) (see above), poly(dA-dG)·poly(dC-dT) (35), and the glucosylated DNA from bacteriophage T2 (142). The A form of poly(dA-dT)·poly(dA-dT) was originally reported to be unstable relative to the D form (described in the next section) (41), although exceptions to this behavior have been noted (35, 143). Fibers of different materials in the A form tend to be quite crystalline and to share the same space group and the same lattice parameters (35, 36), which suggests that the A conformation may be favored by specific packing interactions. The A form of DNA can apparently, however, occur in solution in ethanolic solvents (144, 145).

The original model for A DNA (36) and a refined model (31) are right-handed helices with 11 base pairs per turn (Figure 1). The structures of two self-complementary oligodeoxynucleotides, d(GGTATACC)(145a) and d(iodo-CCGG) (145b), have recently been determined by single-crystal methods to be of the A form. It is very significant that both form right-handed helices. Left-handed models for polymeric A form DNA which are stated to be stereochemically acceptable have been reported, but the details are not yet available (69, 92). Given the differences in the molecular transforms of left- and right-handed versions (69), a quantitative comparison with the semi-crystalline fiber diffraction data must be awaited. NMR results indicate that the A and B conformations can exist in apposition in a single RNA-DNA duplex (146). A model-building study of this duplex has appeared (147).

DNA in solution can be induced to undergo a reversible cooperative transition in secondary structure over a small range of ethanol concentrations. The species involved have distinctive CD spectra that were early correlated with the presence of the B form at low ethanol levels and the A form at high ethanol levels (148). This assignment has been corroborated by a direct demonstration of the species involved in the ethanol-induced transition by diffraction techniques (149, 150).

C form DNA   The characteristic C form X-ray diffraction pattern can be obtained from fibers of a wide variety of natural and synthetic DNAs held under relatively dehydrated conditions (35, 39, 40). The original model for the C form (40) as well as a subsequent refinement (32) correspond to a right-handed structure with 9.3 base pairs/turn (Figure 1). There is a widespread tendency in the literature to equate the C form with these models. However, unlike the forms so far discussed, the C form diffraction pattern is obtained from a family of structures possessing a wide range of helical parameters [7.9 to 9.6 residues/turn (39, 40), a range wider than that which separates the A and B forms].

DNA in concentrated salt solutions or in certain organic solvent-water mixtures adopts a distinctive CD spectrum. A similar spectrum is assumed by DNA in chromatin and in certain viruses. This spectrum had been correlated with the presence of the C form (148). However, a number of studies indicate that the conformation actually present is close to that of the B form (39, 151–157), which leaves the CD spectrum of the C form presently undefined.

D form DNA   The characteristic X-ray diffraction pattern of the D form of DNA was originally obtained by Davies & Baldwin from fibers of poly (dA-dT)·poly(dA-dT) (41). The closely related pattern from poly(dI-dC) ·poly(dI-dC) (158) corresponded to an eightfold helix. In addition to those just mentioned, several other nonguanine-containing deoxypolymers have been found to adopt the D form (35, cf 33): poly(dA-dA-dT)·poly(dA-dT-dT), poly(dA-dI-dT)·poly(dA-dC-dT), poly(dA-dI-dC)·poly(dI-dC-dT), and poly(dA-dC)·poly(dI-dT). In general, the D form seems to occur in fibers under relatively dehydrated conditions, more like those favoring the A or C forms than those yielding the B form.

Mitsui et al (158) were unable to build a fully satisfactory *right-handed* model consistent with the D form pattern from poly(dI-dC)·poly(dI-dC). They made the then startling suggestion that a left-handed model (with an unusual sugar conformation) was at least as likely as a right-handed model. Details of the models and their predicted diffraction patterns were not presented. The inverted CD spectrum of the polymer (which occurs at low ionic strengths) was taken as encouragement for the left-handed model. Whatever the handedness of the structure in solution, it is a substrate for a variety of enzymes that act on natural DNA (159). Such enzymatic action could be used to argue for a similar helical sense in solution for natural DNA and poly(dI-dC)·poly(dI-dC) if we knew more about the binding sites of the enzymes. Arnott et al (33) subsequently rejected the left-handed model of Mitsui et al on the grounds of its unusual nature and sugar conformation, and suggested a more usual right-handed model (Figure 1) to explain the diffraction pattern. Very recently, several residues of the dodecamer structure of Drew et al (59) (see section on B form DNA) have provided instances of the O'-*endo* sugar pucker suggested by Mitsui et al, and certainly the left-handedness of their model cannot currently be used as a basis to reject their structure. What does prevent serious consideration of their proposal for D DNA is the lack of a detailed model to provide a basis for judging stereochemical acceptability and fit to the diffraction pattern. However, as part of their survey of left-handed structures, Sasisek-haran and collaborators (69) have built a left-handed model for D DNA with an unexceptional sugar conformation (C2'-*endo*), as well as a right-handed version. They find their left-handed model agrees with the crystalline diffraction data about as well as does the previous right-handed model (33).

Extended DNA   Arnott et al (160) have recently interpreted the fiber diffraction pattern from a complex of DNA with a platinum-containing intercalator in terms of an unusual linear (i.e. nonhelical) model, which they call L DNA. The proposed structure is stabilized by intercalation between every second base pair. In general terms, the model is obtained by unwinding the helical turns of a double helix without unpairing the base pairs. The result is a ladder-like structure with the base pairs as rungs, generally much like the "*cis*-ladder" proposed by Cyriax & Gäth (161) from a modeling study of nonintercalated DNA. These developments recall the venerable observation of Wilkins et al (162) of a reversibly altered phase of DNA formed by the mechanical extension of DNA fibers. The extension of the fibers was tentatively suggested to result from the actual extension of the molecules.

Miscellaneous models Two novel types of models have been proposed on theoretical grounds. In the "vertical helix" a relatively open duplex with its bases oriented parallel to the helical axis is generated by using the high *anti* range of the glycosidic torsion angle (163). In another series of models, a reversed backbone polarity has been suggested, which leads to the *syn* conformation for right-handed duplexes or the *anti* conformation for left-handed duplexes (164).

RNA-DNA hybrids The early fiber diffraction studies of RNA-DNA hybrids (165–167) yielded RNA-like patterns with 11 or 12 base pairs/turn. These results are often cited to show that hybrids adopt only such RNA-like conformations. The NMR spectrum of a model system containing hybrid sequences $[(rC)_{11}-(dC)_{16}$ annealed to poly(dG)] is also consistent with an A form for the hybrid sequences (146).

A number of observations, however, indicate that at least some hybrids may rival DNA-DNA duplexes in structural capabilities. For example, Gray & Ratliff (168) have shown that certain synthetic hybrids undergo an ethanol-induced transition that is similar to the transition between the A and B forms of natural DNA (see section on A form DNA). Further, a diffraction study of highly solvated fibers of poly(rA)·poly(dT) yielded a pattern similar to that of B DNA (169). A structural model has been proposed that has major similarities to the canonical B form model except that the backbone conformations of the two strands differ from each other and from those in the usual B form model (31, 50). Poly(rA)·poly(dT) can also adopt A- or A'-like forms (6, 169) with 11 or 12 residues/turn, respectively, which emphasizes that at least some hybrid sequences are polymorphic. The hybrid poly(dI)·poly(rC) also forms a tenfold helix in fibers (6); its structure has been suggested to be similar to that of the original Watson-Crick model for B DNA (49), which had the sugar pucker (C3'-*endo*) most often associated with polyribonucleotide models.

Helical Repeat of DNA

The X-ray techniques that have had such a dominant role in deriving structural models for DNA require samples with considerable local order. The ordering may be as low as that in a rudimentary fiber where elongated molecules tend to pack with their axes in the same direction, or the ordering may be as high as that in a crystal with its precisely repeated units. In either case, order is usually obtained by repetitive intermolecular interactions in solid samples. Such interactions can influence the structure. Hence, studies in solution assume a double importance, being useful not only in themselves but also to help in evaluation of the effect of the solid nature of the samples on the characteristics of the structure. Recently, two new techniques and a variation on an old technique have been used to determine one of the most basic characteristics of a model, its helical repeat, under conditions relatively free from intermolecular interactions (Table 2).

Wang and his collaborators developed an elegant technique employing dilute aqueous solutions of DNA. Sequences of known length were inserted into closed circular DNA. The resulting changes in linking number were evaluated from gel electrophoretic patterns and used to deduce the number of base-pairs/turn. Insets of heterogeneous sequences had a repeat of 10.6 $\pm$ 0.1 residues/turn, while inserts of the homopolymer poly(dA)·poly(dT) were distinctly different (10.1 $\pm$ 0.1 residues/turn) (171, 171a). Values for inserts of other sequences are summarized in Table 2.

A second approach is based on the observation of Liu & Wang (175) that when DNA that has been adsorbed to crystallites of calcium phosphate is digested with pancreatic DNase I, a regular pattern of single-strand cuts is found, presumably due to limited steric accessibility. The periodicity in lengths of the DNase products is equated by the authors with the helical repeat. Rhodes & Klug (172, 173) have extensively characterized this technique and shown the periodicity to be independent of the particular choice of adsorbent and nuclease. They argue that the helical repeats so obtained are indicative of solution values based on the indifference of the repeat to details of the adsorption or digestion conditions. Although the technique is likely to be free of DNA-DNA intermolecular interactions, the possibility of changes induced by surface adsorption must be kept in mind. There is good correspondence in the results of this and the preceding technique for the materials so far tested by both approaches. The adsorption technique gave helical repeats of 10.6 $\pm$ 0.1 and 10.0 $\pm$ 0.1 for DNA of heterogeneous sequence and for poly(dA)·poly(dT), respectively, in agreement with the insertion technique (Table 2). Values for several alternating copolymers, namely poly(dA-dT)·poly(dA-dT), poly(dG-dC)·poly(dG-dC) and poly(dG-dm^{5}C)·poly(dG-dm^{5}C) are summarized in Table 2.

In comparing the helical repeats obtained by these techniques to those from fiber data, we must decide which fiber conformation is most appropriate. The canonical B form is an obvious, but not necessarily correct, choice to correspond to DNA in solution. The poly(dA)·poly(dT) periodicity does indeed agree exactly with that estimated from fiber data for the B form, perhaps reflecting the previous discussed reluctance of this polymer to leave the B form. In contrast, poly(dA-dT)·poly(dA-dT), poly(dG)·poly(dC), and DNA of heterogeneous sequence all had helical repeats between those of the classical A and B forms, i.e. between 11 and 10 residues/turn,

Table 2   Experimental estimates of the residues per turn of DNA duplexes

DNA sample	Technique	Base pairs per helical turn	Ref
Natural DNA			
B form fibers (high humidity)	X-ray	10.0 ± 0.15	170
Wetted fibers	X-ray	9.9 ± 0.14	170
Solution	Linking changes	10.6 ± 0.1	171, 171a
Adsorbed on surface	Nuclease digestion	10.6 ± 0.1	172, 173
d (CGCGAATTCGCG)			
Crystal	X-ray	9.65[a], 10.1[b]	60
Poly (dA) · poly (dT)			
B form fibers	X-ray	10.0	54
Solution	Linking changes	10.1 ± 0.1	171, 171a
Adsorbed to surface	Nuclease digestion	10.0 ± 0.1	173
Poly (dA–dT) · poly (dA–dT)			
B form fibers	X-ray	10.0	41
Solution	Linking changes	10.7 ± 0.1	171a
Adsorbed to surface	Nuclease digestion	10.5 ± < 0.1	173
Poly (dG) · poly (dC)			
B form fibers	X-ray	10.0	174
Solution	Linking changes	10.7 ± 0.1	171
Poly (dG–dC) · poly (dG–dC)			
B form fibers	X-ray	10.0	35
Adsorbed to surface[c]	Nuclease digestion	10 ± 1	118
Z form fibers	X-ray	12.0	46
Adsorbed to surface[d]	Nuclease digestion	13 ± 1	118
Poly (dG–dm⁵C) · poly (dG–dm⁵C)			
Adsorbed to surface[c]	Nuclease digestion	10.4 ± 0.4	118
Z form fibers	X-ray	12.0	118
Adsorbed to surface[d]	Nuclease digestion	13.6 ± 0.4	118

[a] Average value using local helical axes.
[b] Using a single helical axis for all residues.
[c] Adsorbed under conditions favoring low-salt form.
[d] Adsorbed under conditions favoring high-salt form.

respectively, while the repeats of the alternating d(G-C) and d(G-m⁵C) polymers were significantly greater than the fiber repeats for any duplex DNA, including Z DNA. Generally, then, the surface adsorption technique and the insertion technique give values for the residues per turn that are greater than those expected from the fiber or crystal structures. A model for DNA of heterogeneous sequence with 10.6 residues/turn has also been proposed based on energy calculations for DNA in the absence of water (76).

An alternative approach (170) attempted to bridge the gap between fiber and solution structures. Fibers were swollen with up to three volumes of water, and their X-ray fiber diffraction patterns were collected. Molecular orientation was sufficiently retained in this range to obtain helical parameters. The diffraction pattern remained of the B form, and the helical parameters were not significantly altered from those in fibers. There is therefore disagreement between the helical repeat of wetted DNA fibers (9.9 ± 0.14 residues/turn) and that from the insertion or adsorption techniques (10.6 ± 0.1 residues/turn). The behavior of the wetted fibers has been suggested to be influenced by remaining intermolecular interactions which either directly (172) or through multivalent cations (172a) might constrain the helical repeat to stay near 10.0 residues/turn as in the fibers. Such interactions would have to extend through water layers which average at least several molecules in thickness. There is no evidence that multivalent ions occur in significant amounts in the wetted fiber samples.

Heterogeneity in DNA Structure

STATIC HETEROGENEITY   DNA is highly polymorphic. Both natural DNA of heterogenous sequence and a variety of synthetic DNAs of simple repetitive sequence are capable of adopting more than one conformation. Base sequence rather than overall base composition often clearly influences the preferred conformation under a given set of conditions. Among many examples studied by fiber diffraction are the ability of poly(dA-dt)·poly(dA-dT) to adopt the A,B,C, and D forms (41), while poly(dA)·poly(dT) has only been observed in a B form (54), or the ability of poly(dG-dC)·poly(dG-dC) to adopt the A,B, or Z forms (46), whereas poly(dG)·poly(dC) is apparently restricted to the A and B forms (174). In addition, there are numerous instances of variation in physical, chemical, or other properties depending on base sequence, which undoubtedly involve conformational changes (3).

Examples of restricted variation in local nucleotide conformation are those observed in Z DNA or suggested for the alternating B DNA model; in each case two different nucleotidyl conformations alternate regularly. At a more complex level, the dodecanucleotide structure from Dickerson's group is a marvelous example of local structural variation, as discussed in the section on B form DNA. Considerable heterogeneity in backbone conformation is also suggested by the ^{31}P NMR spectra of an oligodeoxynucleotide (176) and of fibers of natural DNA in the B form (177); such heterogeneity is not ascribed to the A form of natural DNA (177, 178).

Natural DNA sequences are not random. Analyses of the deviations from randomness have revealed patterns that raise the possibility of more subtle influences of sequence upon structure (179–182).

DYNAMIC HETEROGENEITY   The heterogeneity so far discussed is static in nature, i.e. under a given set of conditions a particular conformation is assumed to accompany a particular base sequence. An alternative source of heterogeneity is a dynamic variation of structure, not necessarily associated with particular base sequences. DNA in solution or in B form DNA fibers [but not in the A form (177, 178)] has often been suggested to undergo internal motions over a wide range of time constants and with significant amplitudes [however, note (183)]. A detailed review of dynamic structural heterogeneity is beyond the scope of this review, but a few references are cited in this rapidly expanding field. Evidence concerning dynamic heterogeneity in DNA or in DNA-ligand complexes has been obtained by fluorescence depolarization (184, 185), by electron paramagnetic resonance (186), by NMR (87, 177, 183, 187–192); for review see 21), and, as mentioned in the section on B form DNA, by crystallography (59).

Concluding Remarks

A number of detailed models for DNA have been discussed. To the extent to which the models conform to reality, they provide snapshots of likely conformations for biological contexts. However, much is missing from the pictures, particularly the selection and distortion enforced by interaction with proteins or other molecules. Only recently have detailed structures for the classes of protein of interest begun to become available from single-crystal diffraction results. The structures of two regulatory proteins that interact with DNA have been solved at high resolution (193). Neither crystal structure includes oligodeoxynucleotides or other representatives of DNA, so that the interactions of the proteins with DNA had to be inferred from modeling studies. In both cases, it was proposed that exposed α-helices protruding from the protein surface interact with the major grooves of a basically B form model for DNA. However, there is a dramatic difference between the two models: the catabolite gene activator protein of *E. coli* (194) is proposed to interact with *left*-handed B form DNA, while a *right*-handed B form DNA gives a neat fit for the cro repressor of bacteriophage λ (195). These models of the DNA-protein complexes must be regarded as hypothetical, and may not prove to be representative. Nevertheless, these studies suggest the beginning of an exciting period wherein the functional importance of a variety of DNA structures can be judged not only by detailed studies of uncomplexed DNA but also by detailed studies of complexes with biologically important proteins or other molecules.

ACKNOWLEDGMENTS

The many editorial and logistic contributions of Barbara H. Pheiffer have been invaluable and are acknowledged with appreciation. I also thank the authors who sent preprints of unpublished work. Critical comments from my colleagues at the National Institutes of Health have been most valuable.

Literature Cited

1. Jovin, T. M. 1976. *Ann. Rev. Biochem.* 45:889–920
2. Sundaralingam, M. 1979. In *Symposium on Biomolecular Structure, Conformation, Function and Evolution*, ed. R. Srinivasan, pp. 259–83. New York: Pergamon
3. Wells, R. D., Goodman, T. C., Hillen, W., Horn, G. T., Klein, R. D., Larson, J. E., Müller, U. R., Neuendorf, S. K., Panayotatos, N., Stirdivant, S. M. 1980. *Prog. Nucleic Acid Res. Mol. Biol.* 24:167–267
4. Wells, R. D., Blakesley, R. W., Hardies, S. C., Horn, G. T., Larson, J. E., Selsing, E., Burd, J. F., Chan, H. W., Dodgson, J. B., Jensen, K. F., Nes, I. F., Wartell, R. M. 1977. *Crit. Rev. Biochem.* 4:305–40
5. Arnott, S., Chandrasekaran, R., Bond, P. J., Birdsall, D. L., Leslie, A. G. W., Puigjaner, L. C. 1980. *7th Ann. Katzir Katchalsky Conf. Struct. Aspects Recognition Assem. Biolog. Macromol.* Glenside, Pa: Int. Sci. Serv.
6. Chandrasekaran, R., Arnott, S., Banerjee, A., Campbell-Smith, S., Leslie, A. G. W., Puigjaner, L. 1980. *ACS Symp. Ser.* 141:483–502
7. Sundaralingam, M., Westhof, E. 1979. *Int. J. Quantum Chem. Symp.* 6:115–30
8. Jack, A. 1979. *Int. Rev. Biochem: Chem. Macromolecules II A* 24:211–56
9. Record, M. T. Jr., Mazur, S. J., Melançon, P., Roe, J.-H., Shaner, S. L., Unger, L. 1981. *Ann. Rev. Biochem.* 50:997–1024
10. Champoux, J. J. 1978. *Ann. Rev. Biochem.* 47:449–79
11. von Hippel, P. 1979. *Biol. Regul. Develop.* 1:279–347
12. McGhee, J. D., Felsenfeld, G. 1980. *Ann. Rev. Biochem.* 49:1115–56
13. Cozzarelli, N. R. 1980. *Cell* 22:327–28
14. Cozzarelli, N. R. 1980. *Science* 207:953–60
15. Gellert, M. 1981. *Ann. Rev. Biochem.* 50:879–910
16. Swaminathan, V., Sundaralingam, M. 1979. *Crit. Rev. Biochem.* 6:245–336

17. Lippard, S. J. 1978. *Acc. Chem. Res.* 11:211–17
18. Patel, D. J. 1979. *Acc. Chem. Res.* 12:118–25
19. Texter, J. 1978. *Prog. Biophys. Mol. Biol.* 33:83–97
20. Tinoco, I. Jr., Bustamante, C., Maestre, M. F. 1980. *Ann. Rev. Biophys. Bioeng.* 9:107–41
21. Shindo, H. 1982. In *Magnetic Resonance in Biology*, Vol. 2, ed. J. Cohen, New York: Wiley. In press
22. Kearns, D. R. 1977. *Ann. Rev. Biophys. Bioeng.* 6:477–523
23. Cohen, J. S. 1980. *Trends Biochem. Sci.* 5:58–60
24. Davies, D. B. 1978. *Prog. NMR Spectrosc.* 12:135–225
25. Schweizer, M. P. 1980. In *Magnetic Resonance in Biology*, ed. J. Cohen, 1:259–302. New York: Wiley
26. Bauer, W. R. 1978. *Ann. Rev. Biophys. Bioeng.* 7:287–313
27. Crick, F. H. C. 1976. *Proc. Natl. Acad. Sci. USA* 73:2639–43
28. Wang, J. C. 1980. *Trends Biol. Sci.* 5:219–21
29. Bauer, W. R., Crick, F. H. C., White, J. H. 1980. *Sci. Am.* 243:118–33
30. Wilson, H. R. 1966. *Diffraction of X-rays by Proteins, Nucleic Acids and Viruses.* London: Arnold. 137 pp.
31. Arnott, S., Hukins, D. W. L. 1972. *Biochem. Biophys. Res. Commun.* 47:1504–9
32. Arnott, S., Selsing, E. 1975. *J. Mol. Biol.* 98:265–69
33. Arnott, S., Chandrasekaran, R., Hukins, D. W. L., Smith, P. J. C., Watts, L. 1974. *J. Mol. Biol.* 88:523–33
34. Wang, A. H.-J., Quigley, G. J., Kolpak, F. J., van der Marel, G., van Boom, J. H., Rich, A. 1981. *Science* 211:171–76
35. Leslie, A. G. W., Arnott, S., Chandrasekaran, R., Ratliff, R. L. 1980. *J. Mol. Biol.* 143:49–72
36. Fuller, W., Wilkins, M. H. F., Wilson, H. R., Hamilton, L. D., Arnott, S. 1965. *J. Mol. Biol.* 12:60–80
37. Langridge, R., Wilson, H. R., Hooper,

C. W., Wilkins, M. H. F., Hamilton, L. D. 1960. *J. Mol. Biol.* 2:19–37

38. Klug, A., Jack, A., Viswamitra, M. A., Kennard, O., Shakked, Z., Steitz, T. A. 1979. *J. Mol. Biol.* 131:669–80

39. Zimmerman, S. B., Pheiffer, B. H. 1980. *J. Mol. Biol.* 142:315–30

40. Marvin, D. A., Spencer, M., Wilkins, M. H. F., Hamilton, L. D. 1961. *J. Mol. Biol.* 3:547–65

41. Davies, D. R., Baldwin, R. L. 1963. *J. Mol. Biol.* 6:251–55

42. Mokul'skii, M. A., Kapitonova, K. A., Mokul'skaya, T. D. 1972. *Mol. Biol.* 6:883–901

43. Mokul'skaya, T. D., Smetanina, E. P., Myshko, G. E., Mokul'skii, M. A. 1975. *Mol. Biol.* 9:552–55

44. Wang, A. H.-J., Quigley, G. J., Kolpak, F. J., Crawford, J. L., van Boom, J. H., van der Marel, G., Rich, A. 1979. *Nature* 282:680–86

45. Drew, H., Takano, T., Tanaka, S., Itakura, K., Dickerson, R. E. 1980. *Nature* 286:567–73

46. Arnott, S., Chandrasekaran, R., Birdsall, D. L., Leslie, A. G. W., Ratliff, R. L. 1980. *Nature* 283:743–45

47. Saenger, W. 1973. *Angew. Chem.* 12:591–601

48. Arnott, S. 1970. *Prog. Biophys. Molec. Biol.* 21:265–319

49. Crick, F. H. C., Watson, J. D. 1954. *Proc. R. Soc.* 223A:80–96

50. Langridge, R., Marvin, D. A., Seeds, W. E., Wilson, H. R., Hooper, C. W., Wilkins, M. H. F., Hamilton, L. D. 1960. *J. Mol. Biol.* 2:38–64

51. Hamilton, L. D., Barclay, R. K., Wilkins, M. H. F., Brown, G. L., Wilson, H. R., Marvin, D. A., Ephrussi-Taylor, H., Simmons, N. S. 1959. *J. Biophys. Biochem. Cytol.* 5:397–403

52. Prémilat, S., Albiser, G. 1975. *J. Mol. Biol.* 99:27–36

53. Selsing, E., Arnott, S. 1976. *Nucleic Acids Res.* 3:2443–50

54. Arnott, S., Selsing, E. 1974. *J. Mol. Biol.* 88:509–21

55. Pilet, J., Blicharski, J., Brahms, J. 1975. *Biochemistry* 14:1869–76

56. Langridge, R. 1969. *J. Cell Physiol.* 74: Suppl. 1, pp. 1–20

57. Baralle, F. E., Shoulders, C. C., Goodbourn, S., Jeffreys, A., Proudfoot, N. J. 1980. *Nucleic Acids Res.* 8:4393–404

58. Wing, R., Drew, H., Takano, T., Broka, C., Tanaka, S., Itakura, K., Dickerson, R. E. 1980. *Nature* 287:755–58

59. Drew, H. R., Wing, R. M., Takano, T., Broka, C., Tanaka, S., Itakura, K.,

Dickerson, R. E. 1981. *Proc. Natl. Acad. Sci. USA* 78:2179–83

59a. Dickerson, R. E., Drew, H. R. 1981. *Proc. Natl. Acad. Sci. USA* 78:7318–22

60. Dickerson, R. E., Drew, H. R. 1981. *J. Mol. Biol.* 149:761–86

61. Drew, H. R., Dickerson, R. E. 1981. *J. Mol. Biol.* 151:535–56

62. Alden, C. J., Kim, S.-H. 1979. *J. Mol. Biol.* 132:411–34

63. Thiyagarajan, P., Ponnuswamy, P. K. 1979. *Biopolymers* 18:2233–47

64. Neidle, S., Berman, H. M., Shieh, H. S. 1980. *Nature* 288:129–33

65. Lomonossoff, G. P., Butler, P. J. G., Klug, A. 1981. *J. Mol. Biol.* 149:745–60

66. Quigley, G. J., Wang, A. H.-J., Ughetto, G., van der Marel, G., van Boom, J. H., Rich, A. 1980. *Proc. Natl. Acad. Sci. USA* 77:7204–8

67. Viswamitra, M. A., Kennard, O., Jones, P. G., Sheldrick, G. M., Salisbury, S., Falvello, L. 1978. *Nature* 273:687–88

67a. Viswamitra, M. A., Shakked, Z., Jones, P. G., Sheldrick, G. M., Salisbury, S. A., Kennard, O. 1981. *Biopolymers.* In press

68. Conner, J. S., Wooten, J. B., Chatterjee, C. L. 1981. *Biochemistry* 20:3049–55

69. Gupta, G., Bansal, M., Sasisekharan, V. 1980. *Int. J. Biol. Macromol.* 2:368–80

70. Shindo, H., Simpson, R. T., Cohen, J. S. 1979. *J. Biol. Chem.* 254:8125–28

71. Shindo, H., Zimmerman, S. B. 1980. *Nature* 283:690–91

72. Marky, L. A., Patel, D., Breslauer, K. J. 1981. *Biochemistry* 20:1427–31

73. Patel, D. J., Canuel, L. L., Pohl, F. M. 1979. *Proc. Natl. Acad. Sci. USA* 76:2508–11

74. Sasisekharan, V., Gupta, G., Bansal, M. 1981. *Int. J. Biol. Macromol.* 3:2–8

75. Yathindra, N., Jayaraman, S. 1980. *Current Sci.* 49:167–71

76. Levitt, M. 1978. *Proc. Natl. Acad. Sci. USA* 75:640–44

77. Sussman, J. L., Trifonov, E. N. 1978. *Proc. Natl. Acad. Sci. USA* 75:103–7

78. Camerini-Otero, R. D., Felsenfeld, G. 1978. *Proc. Natl. Acad. Sci. USA* 75:1708–12

79. Olson, W. K. 1979. *Biopolymers* 18: 1213–33

80. Sobell, H. M., Tsai, C.-C., Gilbert, S. G., Jain, S. C., Sakore, T. D. 1976. *Proc. Natl. Acad. Sci. USA* 73:3068–72

81. Sobell, H. M., Tsai, C.-C., Jain, S. C., Sakore, T. D. 1978. *Philos. Trans. R. Soc. London Ser. B* 283:295–98

82. Crick, F. H. C., Klug, A. 1975. *Nature* 255:530–33

83. Yathindra, N. 1979. *Current Sci.* 48:753–56
84. Sobell, H. M., Lozansky, E. D., Lessen, M. 1978. *Cold Spring Harbor Symp. Quant. Biol.* 43:11–19
85. Lozansky, E. D., Sobell, H. M., Lessen, M. 1979. In *Stereodynamics of Molecular Systems*, ed. R. Sarma, pp. 265–82. New York: Pergamon
86. Englander, S. W., Kallenbach, N. R., Heeger, A. J., Krumhansl, J. A., Litwin, S. 1980. *Proc. Natl. Acad. Sci. USA* 77:7222–26
87. Klevan, L., Armitage, I. M., Crothers, D. M. 1979. *Nucleic Acids Res.* 6:1607–16
88. Simpson, R. T., Shindo, H. 1979. *Nucleic Acids Res.* 7:481–92
89. Shindo, H., McGhee, J. D., Cohen, J. S. 1980. *Biopolymers* 19:523–37
90. Kallenbach, N. R., Appleby, D. W., Bradley, C. H. 1978. *Nature* 272:134–38
91. Sasisekharan, V., Brahmachari, S. K. 1981. *Curr. Sci.* 50:10–13
92. Gupta, G., Bansal, M., Sasisekharan, V. 1980. *Proc. Natl. Acad. Sci. USA* 77:6486–90
93. Sasisekharan, V., Pattabiraman, N. 1978. *Nature* 275:159–62
94. Yathindra, N., Sundaralingam, M. 1976. *Nucleic Acids Res.* 3:729–47
95. Gupta, G., Sasisekharan, V. 1978. *Nucleic Acids Res.* 5:1655–73
96. Rich, A., RajBhandary, U. L. 1976. *Ann. Rev. Biochem.* 45:805–60
97. Ikehara, M., Uesugi, S., Yano, J. 1972. *Nature New Biol.* 240:16–17
98. Sasisekharan, V., Pattabiraman, N., Gupta, G. 1978. *Proc. Natl. Acad. Sci. USA* 75:4092–96
99. Rodley, G. A., Scobie, R. S., Bates, R. H. T., Lewitt, R. M. 1976. *Proc. Natl. Acad. Sci. USA* 73:2959–63
100. Sasisekharan, V. 1981. *Curr. Sci.* 50:107–11
101. Millane, R. P., Rodley, G. A. 1981. *Nucleic Acids Res.* 9:1765–73
102. Albiser, G., Prémilat, S. 1980. *Biochem. Biophys. Res. Commun.* 95:1231–37
103. Bates, R. H. T., McKinnon, G. C., Millane, R. P., Rodley, G. A. 1980. *Pramaña* 14:233–52
104. Greenall, R. J., Pigram, W. J., Fuller, W. 1979. *Nature* 282:880–82
105. Arnott, S. 1979. *Nature* 278:780–81
106. Dover, S. D. 1977. *J. Mol. Biol.* 110:699–700
107. Pohl, F. M. 1967. *Naturwissenschaften* 54:616
108. Pohl, W. F., Roberts, G. W. 1978. *J. Math. Biol.* 6:383–402
109. Crick, F. H. C., Wang, J. C., Bauer, W. R. 1979. *J. Mol. Biol.* 129:449–61
110. DasGupta, C., Takehiko, S., Cunningham, R. P., Radding, C. M. 1980. *Cell* 22:437–46
111. Stettler, U. H., Weber, H., Koller, T., Weissmann, C. 1979. *J. Mol. Biol.* 131:21–40
112. Pohl, F. M. 1976. *Nature* 260:365–66
113. Pohl, F. M., Jovin, T. M. 1972. *J. Mol. Biol.* 67:375–96
114. Crawford, J. L., Kolpak, F. J., Wang, A. H.-J., Quigley, G. J., van Boom, J. H., van der Marel, G., Rich, A. 1980. *Proc. Natl. Acad. Sci. USA* 77:4016–20
115. Haschemeyer, A. E. V., Rich, A. 1967. *J. Mol. Biol.* 27:369–84
116. Govil, G., Fisk, C. L., Howard, F. B., Miles, H. T. 1981. *Biopolymers* 20:573–603
117. Drew, H. R., Dickerson, R. E. 1981. *J. Mol. Biol.* 152:723–36
118. Behe, M., Zimmerman, S., Felsenfeld, G. 1981. *Nature* 293:233–35
119. Saenger, W., Landmann, H., Lezius, A. G. 1973. *Jerusalem Symp. Quantum Chem. Biochem.* 5:457–66
120. Gupta, G., Bansal, M., Sasisekharan, V. 1980. *Biochem. Biophys. Res. Commun.* 95:728–33
121. Behe, M., Felsenfeld, G. 1981. *Proc. Natl. Acad. Sci. USA* 78:1619–23
122. Ramstein, J., Leng, M. 1980. *Nature* 288:413–14
123. Pohl, F. M., Ranade, A., Stockburger, M. 1973. *Biochim. Biophys. Acta* 335:85–92
123a. Thamann, T. J., Lord, R. C., Wang, A. H.-J., Rich, A. 1981. *Nucleic Acids Res.* 9:5443–57
123b. Wu, H. M., Dattagupta, N., Crothers, D. M. 1981. *Proc. Natl. Acad. Sci. USA* 78:6808–11
124. Bayley, P. M. 1973. *Prog. Biophys.* 27:1–76
125. Simpson, R. T., Shindo, H. 1980. *Nucleic Acids Res.* 8:2093–2103
126. Mitra, C. K., Sarma, M. H., Sarma, R. H. 1981. *Biochemistry* 20:2036–41
127. Vasmel, H., Greve, J. 1981. *Biopolymers* 20:1329–32
128. Drew, H. R., Dickerson, R. E., Itakura, K. 1978. *J. Mol. Biol.* 125:535–543
129. Mercado, C. M., Tomasz, M. 1977. *Biochemistry* 16:2040–46
130. Sage, E., Leng, M. 1980. *Proc. Natl. Acad. Sci. USA* 77:4597–601
131. Santella, R. M., Grunberger, D., Weinstein, I. B., Rich, A. 1981. *Proc. Natl. Acad. Sci. USA* 78:1451–55
132. Sage, E., Leng, M. 1981. *Nucleic Acids Res.* 9:1241–50

132a. Möller, A., Nordheim, A., Nichols, S. R., Rich, A. 1981. *Proc. Natl. Acad. Sci. USA* 78:4777–81

133. Pohl, F. M., Jovin, T. M., Baehr, W., Holbrook, J. J. 1972. *Proc. Natl. Acad. Sci. USA* 69:3805–9

134. Kuhnlein, U., Tsang, S. S., Edwards, J. 1980. *Nature* 287:363–64

135. Gellert, M., Mizuuchi, K. 1980. As cited in Davies, D. R., Zimmerman, S. B. *Nature* 283:11–12

136. Benham, C. J. 1980. *Nature* 286:637–38

137. Klysik, J., Stirdivant, S. M., Larson, J. E., Hart, P. A., Wells, R. D. 1981. *Nature* 290:672–77

138. Nishioka, Y., Leder, P. 1980. *J. Biol. Chem.* 255:3691–94

139. Zakrzewska, K., Lavery, R., Pullman, A., Pullman, B. 1980. *Nucleic Acids Res.* 8:3917–32

140. Jayaraman, S., Yathindra, N. 1980. *Biochem. Biophys. Res. Commun.* 97:1407–19

140a. Lafer, E. M., Möller, A., Nordheim, A., Stollar, B. D., Rich, A. 1981. *Proc. Natl. Acad. Sci. USA* 78:3546–50

140b. Nordheim, A., Pardue, M. L., Lafer, E. M., Möller, A., Stollar, B. D., Rich, A. 1981. *Nature* 294:417–22

141. Gupta, G., Bansal, M., Sasisekharan, V. 1980. *Biochem. Biophys. Res. Commun.* 97:1258–67

142. Skuratovskii, I. Ya., Bartenev, V. N. 1978. *Mol. Biol.* 12:1359–76

143. Shindo, H., Wooten, J. B., Zimmerman, S. B. 1981. *Biochemistry* 20:745–50

144. Potaman, V. N., Bannikov, Yu. A., Shlyachtenko, L. S. 1980. *Nucleic Acids Res.* 8:635–42

145. Zavriev, S. K., Minchenkova, L. E., Frank-Kamenetskii, M. D., Ivanov, V. I. 1978. *Nucleic Acids Res.* 5:2657–63

145a. Shakked, Z., Rabinovich, D., Cruse, W. B. T., Egert, E., Kennard, O., Sala, G., Salisbury, S. A., Viswamitra, M. A. 1981. *Proc. R. Soc. Ser. B.* 213:479–87

145b. Conner, B. N., Takano, T., Tanaka, S., Itakura, K., Dickerson, R. E. 1981. *Nature.* 295:294–99

146. Selsing, E., Wells, R. D., Early, T. A., Kearns, D. R. 1978. *Nature* 275:249–50

147. Selsing, E., Wells, R. D., Alden, C. J., Arnott, S. 1979. *J. Biol. Chem.* 254:5417–22

148. Tunis-Schneider, M. J. B., Maestre, M. F. 1970. *J. Mol. Biol.* 52:521–41

149. Zimmerman, S. B., Pheiffer, B. H. 1979. *J. Mol. Biol.* 135:1023–27

150. Gray, D. M., Edmondson, S. P., Lang, D., Vaughan, M. 1979. *Nucleic Acids Res.* 6:2089–2107

151. Wang, J. C. 1969. *J. Mol. Biol.* 43:25–39

152. Maniatis, T., Venable, J. H. Jr., Lerman, L. S. 1974. *J. Mol. Biol.* 84:37–64

153. Goodwin, D. C., Brahms, J. 1978. *Nucleic Acids Res.* 5:835–50

154. Gray, D. M., Taylor, T. N., Lang, D. 1978. *Biopolymers* 17:145–57

155. Baase, W. A., Johnson, W. C. Jr. 1979. *Nucleic Acids Res.* 6:797–814

156. Sprecher, C. A., Baase, W. A., Johnson, W. C. Jr. 1979. *Biopolymers* 18:1009–19

157. Lee, C.-H., Mizusawa, H., Kakefuda, T. 1981. *Proc. Natl. Acad. Sci. USA* 78:2838–42

158. Mitsui, Y., Langridge, R., Shortle, B. E., Cantor, C. R., Grant, R. C., Kodama, M., Wells, R. D. 1970. *Nature* 228:1166–69

159. Grant, R. C., Kodama, M., Wells, R. D. 1972. *Biochemistry* 11:805–15

160. Arnott, S., Bond, P. J., Chandrasekaran, R. 1980. *Nature* 287:561–63

161. Cyriax, B., Gäth, R. 1978. *Naturwissenschaften* 65:106–8

162. Wilkins, M. H. F., Gosling, R. G., Seeds, W. E. 1951. *Nature* 167:759–60

163. Olson, W. K. 1977. *Proc. Natl. Acad. Sci. USA* 74:1775–79

164. Hopkins, R. C. 1981. *Science* 211:289–91

165. Milman, G., Langridge, R., Chamberlin, M. J. 1967. *Proc. Natl. Acad. Sci. USA* 57:1804–10

166. O'Brien, E. J., MacEwan, A. W. 1970. *J. Mol. Biol.* 48:243–61

167. Higuchi, S., Tsuboi, M., Iitaka, Y. 1969. *Biopolymers* 7:909–16

168. Gray, D. M., Ratliff, R. L. 1975. *Biopolymers* 14:487–98

169. Zimmerman, S. B., Pheiffer, B. H. 1981. *Proc. Natl. Acad. Sci. USA* 78:78–82

170. Zimmerman, S. B., Pheiffer, B. H. 1979. *Proc. Natl. Acad. Sci. USA* 76:2703–7

171. Peck, L. J., Wang, J. C. 1981. *Nature* 292:375–78

171a. Strauss, F., Gaillard, C., Prunell, A. 1981. *Eur. J. Biochem.* 118:215–22

172. Rhodes, D., Klug, A. 1980. *Nature* 286:573–78

172a. Mandelkern, M., Dattagupta, N., Crothers, D. M. 1981. *Proc. Natl. Acad. Sci. USA* 78:4294–98

173. Rhodes, D., Klug, A. 1981. *Nature* 292:378–80

174. Arnott, S., Selsing, E. 1974. *J. Mol. Biol.* 88:551–52

175. Liu, L. F., Wang, J. C. 1978. *Cell* 15:979–84

176. Patel, D. J., Canuel, L. L. 1979. *Eur. J. Biochem.* 96:267–76

177. Shindo, H., Wooten, J. B., Pheiffer, B. H., Zimmerman, S. B. 1980. *Biochemistry* 19:518–26
178. Nall, B. T., Rothwell, W. P., Waugh, J. S., Rupprecht, A. 1981. *Biochemistry* 20:1881–87
179. Nussinov, R. 1980. *Nucleic Acids Res.* 8:4545–62
180. Zhurkin, V. B. 1981. *Nucleic Acids Res.* 9:1963–71
181. Trifonov, E. N. 1980. *Nucleic Acids Res.* 8:4041–53
182. Grantham, R. 1980. *FEBS Lett.* 121:193–99
183. DiVerdi, J. A., Opella, S. J. 1981. *J. Mol. Biol.* 149:307–11
184. Wahl, Ph., Paoletti, J., LePecq, J.-B. 1970. *Proc. Natl. Acad. Sci. USA* 65: 417–21
185. Barkley, M. D., Zimm, B. H. 1979. *J. Chem. Phys.* 70:2991–3007
186. Robinson, B. H., Lerman, L. S., Beth, A. H., Frisch, H. L., Dalton, L. R., Auer, C. 1980. *J. Mol. Biol.* 139:19–44
187. Hogan, M. E., Jardetzky, O. 1980. *Biochemistry* 19:2079–85
188. Hogan, M. E., Jardetzky, O. 1980. *Biochemistry* 19:3460–68
189. Bolton, P. H., James, T. L. 1980. *J. Am. Chem. Soc.* 102:25–31
190. Rill, R. L., Hilliard, P. R., Jr., Bailey, J. T., Levy, G. C. 1980. *J. Am. Chem. Soc.* 102:418–20
191. Hogan, M. E., Jardetzky, O. 1979. *Proc. Natl. Acad. Sci. USA* 76:6341–45
192. Early, T. A., Kearns, D. R. 1979. *Proc. Natl. Acad. Sci. USA* 76:4165–69
193. Davies, D. R. 1981. *Nature* 290:736–37
194. McKay, D. B., Steitz, T. A. 1981. *Nature* 290:744–49
195. Anderson, W. F., Ohlendorf, D. H., Takeda, Y., Matthews, B. W. 1981. *Nature* 290:754–58

Ann. Rev. Biochem. 1981. 50:997–1024

10

DOUBLE HELICAL DNA: CONFORMATIONS, PHYSICAL PROPERTIES, AND INTERACTIONS WITH LIGANDS[1]

M. T. Record, Jr., S. J. Mazur, P. Melançon, J.-H. Roe, S. L. Shaner, and L. Unger

Department of Chemistry, University of Wisconsin, Madison, Wisconsin 53706

CONTENTS

Perspectives and Summary

Substantial progress has been made recently in understanding the unique physical features of double helical DNA and their implications for its interactions with ligands and for packaging DNA in vivo. These unique features of DNA include: 1. its lateral and torsional stiffness; 2. its high axial

[1]Abbreviations used are: bp, base pair; deg, angular degree; e.u., entropy units (cal mol^{-1} K^{-1}); mbp, mole base pairs.

313

charge density (two phosphate groups per 3.4 Å) and moderate surface charge density (two phosphate groups per 210 Å^2); and 3. its superposition of two binding lattices, one a relatively homogeneous array of phosphate charges that will interact electrostatically and nonspecifically with cationic ligands, the other a heterogeneous array of stacked base pairs that will interact with ligands nonelectrostatically and specifically. We examine the molecular basis of these physical features and discuss their implications for a variety of thermodynamic properties of DNA, including the secondary and higher order structure of the molecule as a function of solution conditions, the interaction of DNA with ligands, and the ligand or solvent induced collapse of DNA into a folded state. Our intent is to provide a basic quantitative as well as qualitative description and a critical analysis of these topics. We have attempted to be comprehensible, if not comprehensive. Fortunately a number of general (1, 2) and specific (3–8) reviews of topics in DNA physical chemistry have recently appeared.

DNA Conformation

The double helical nature of DNA has now been convincingly established (9), and except for minor details the experimental evidence confirms the canonical B-DNA structure: an antiparallel, right-handed double stranded helix with internal Watson-Crick base pairing. The refined structure was shown by model building to fit the X-ray diffraction pattern of DNA fibers at high humidity. This approach and the results have been discussed extensively elsewhere (1–3, 10, 11); we only list here, in Table 1, some useful molecular parameters. Of particular interest for modeling DNA as a polyelectrolyte or as a lattice of binding sites are the actual helical charge distribution and geometric quantities such as groove depth and width. (To date, it has been difficult to incorporate real geometry into polyelectrolyte theories, and with a few exceptions (12–14; LeBret, M., personal communication), DNA is usually modeled as a line charge or cylinder.) The major portion of this review is concerned with the effect of environmental factors on the thermodynamic properties of DNA. It therefore is important to

Table 1 Structural parameters of DNA duplex helices (10)

Sample	Humidity (%)	Winding angle (deg)	Rise per residue (Å)	Width of grooves (Å)		Depth of grooves (Å)	
				minor	major	minor	major
DNA·A,Na$^+$	75	32.7	2.56	11.0	2.7	2.8	13.5
DNA·B,Na$^+$	92	36	3.46	—	—	—	—
DNA·B,Li$^+$	66	36	3.37	5.7	11.7	7.5	8.5
DNA·C,Li$^+$	66	38.6	3.31	4.8	10.5	7.9	7.5

understand the extent to which these factors affect DNA conformation. The present discussion focuses on what the preferred conformation of DNA in solution is, as a function of environmental variables, and how much the structure can vary about that average conformation.

SOLUTION CONFORMATION   There are several reasons to suspect that DNA conformation in solution differs somewhat from the structure in the fiber. First, the conformation in the fiber (see Table 1) and in solution (see below) depends on the environment. Second, it has been suggested that intermolecular packing forces will affect the observed structural parameters (15, 16), although an increase in the distance between molecules through extreme hydration does not affect the values obtained from fibers (17). The lack of precise information on DNA conformation in solution has motivated much research on the subject. Some controversy exists with regard to the exact number of base pairs per turn (helical repeat), the tilt of the base pairs relative to the helix axis, and the sequence dependence of the structure. Information on the helical repeat of DNA under physiological conditions has come mainly from two different approaches. Using the bandshift method, which exploits the topological properties of closed circular DNA, Wang measured an average helical repeat of 10.6 ± 0.1 (18; J. Wang, personal communication). The interpretation of the precise size distribution of DNA fragments produced by nuclease digestion of DNA immobilized on different surfaces gave a similar nonintegral value for the helical repeat of 10.6 ± 0.1 (19). Related results from the nuclease digestion of nucleosome cores (20, 21) can only yield model-dependent values (22). Neither the bandshift nor the nuclease digestion techniques can distinguish between the reported smaller helical winding and a combination of right-handed helical winding and left-handed intrinsic writhe of the helix axis (23).

Electric dichroism measurements on DNA fragments indicated that the tilt of the bases was not 83 deg as in the B-DNA structure but closer to 73 deg (16, 24), a value similar to that predicted by the conformational energy minimization technique of Levitt (25). However a tilt of 83 deg was observed if measurements were done under conditions at which aggregation of DNA fragments occurs, which suggests that intermolecular interactions may affect structure (16).

Investigation of the sequence dependence of the conformation with the bandshift method has so far yielded preliminary results that indicate that nonalternating AT has 10.1 ± 0.1 bp per turn whereas nonalternating GC gives 10.7 ± 0.1 bp per turn (Wang, J. C., personal communication). Also, the very recent determination of the atomic coordinates for the dodecamer d(CpGpCpGpApApTpTpGpCpGpC) allows a direct observation of variation of the winding angle along the sequence (23). An extreme case of a sequence dependent effect is given by the alternating B-structure proposed for poly(dA-dT) (27) and by the left-handed helices observed at high salt for alternating GC oligomers (28, 29) and polymers (30, 31).

ENVIRONMENTAL EFFECTS   As shown in Table 1, DNA takes on different conformations in different environments in the fiber. For example, depending on the type and amount of salt or the amount of hydration, a B, C, or A structure is observed. This polymorphism has been recently discussed (32). Circular dichroism (CD) spectra of isotropic DNA films prepared under conditions similar to those of the fibers have been measured (33). Correlation between optical properties and geometrical structure suggests that transitions of one form into another can be observed directly. Changes in the near UV CD spectrum of DNA induced by variations in salt concentration, type of salt, solvent composition, temperature, or in the DNA itself (superhelicity, collapse) have been reported and interpreted by many authors (34, 35). Although useful as a characterization of the different transitions, the approach is limited by the fact that interpretation in terms of defined conformational changes is rather difficult and controversial (34, 36, 36a, 37). [An example is the decrease in the 275 nm CD band observed at high salt, and attributed to a B-form to C-form transition. This interpretation would require a change in the winding angle of up to 2.6 deg (see Table 1), whereas a direct determination of that change by ethidium bromide sedimentation velocity titration gives a maximum value of 0.8 deg (36).]

A more precise estimate of the effect of salt and temperature on linear DNA conformation has come from studies that take advantage of the topological properties of closed circular DNA molecules. The average change in the winding of DNA upon transfer from an incubation solution to a fixed electrophoresis condition has been measured as a function of such parameters as temperature, type of salt, and salt concentration. The winding angle shows a significant temperature dependence equivalent to $-13 \pm 2 \times 10^{-3}$ deg bp^{-1} K^{-1}, which implies that the helical structure unwinds continuously as the temperature is increased (38). The winding angle also varies with counterion type at constant salt concentration, and increases in the order $Na^+ < K^+ < Li^+ < Rb^+ < Cs^+ < NH_4^+$. In all cases the winding angle increases linearly with the logarithm of the salt concentration over the range studied (39). No effect of anion substitution was observed. The slopes are such that at 20°C, one observes a decrease in winding angle of about 0.1 deg per base pair in going from 0.2 M NaCl to 0.05 M NaCl, or 0.2 deg per base pair in changing the salt from NH_4Cl to NaCl at 0.2 M. It is plausible to interpret a decrease in the winding angle as corresponding to a decrease in the axial charge density of the helix (10). Consequently, the observed linear dependence of the winding angle on the logarithm of the salt concentration can be understood as a general polyelectrolyte effect. A decrease in axial charge density reduces the requirement for the local accumulation of counterions near the polyion [thermodynamic counterion binding (40)]. Consequently counterions are released to the bulk solution as the axial charge density is reduced; the entropic contribution from counterion release (a free energy of dilution) varies as the logarithm of the bulk salt concentration. Manning is developing a more quantitative description of these observations (G. S. Manning, personal communication).

The effect of cation type on the winding angle could be related to a more specific interaction of the counterion with the DNA. It is interesting to note that there is a good correlation between the ordering of counterions derived from their effect on DNA conformation (41–45) and their relative affinity for DNA as determined by NMR from their ability to displace ^{23}Na from the surface region (46). The origin of these cation-specific effects is unclear. Site-binding of cations to either phosphates or bases, driven by the high local ion concentration, is a possible explanation. However, the formation of inner sphere complexes, as observed (47, 48) in crystals of the dinucleotides ApU and GpC (under low hydration conditions), is unlikely in solution. Neither field jump kinetic measurements on the interaction of Na^+ with poly (dA) (49), nor ^{23}Na NMR relaxation measurements on DNA (46, 50, 51), detected the degree of dehydration that accompanies inner sphere site binding. In addition, differential equilibrium dialysis experiments failed to show any base-specific interactions of these counterions with DNA (52).

Thermodynamic Analysis of DNA Stability

HELIX-COIL EQUILIBRIA   The stability of native DNA as a function of base composition and solution conditions is characterized by the denaturation temperature T_m. From a calorimetric determination of the enthalpy of denaturation ($\Delta H^0_{T_m}$), the entropy of denaturation ($\Delta S^0_{T_m}$) is obtained as $\Delta S^0_{T_m} = \Delta H^0_{T_m}/T_m$. Since in general the heat capacity difference between denatured and native states (ΔC^0_p) will be nonzero, $\Delta H^0_{T_m}$ and $\Delta S^0_{T_m}$ will be functions of absolute temperature as well as of the variables used to affect T_m. Consequently it is important to obtain accurate values of ΔC^0_p and use them to extrapolate thermodynamic functions to a standard reference temperature (e.g. 298 K). Such standard enthalpies, entropies, and free energies, determined as a function of base composition and solution variables, will provide information about the contributions of various noncovalent interactions to helix stability. (Stability is of course a relative term, and the extent of residual noncovalent interactions in the denatured state must be known in such an analysis.) The development of high precision differential adiabatic scanning microcalorimeters has allowed the determination of ΔC_p^0 and $\Delta H^0{}_{T_m}$ with sufficient accuracy to perform the above analysis [see e.g. (53, 54)]. To date, only the transition of poly(rA)·poly(rU) has been investigated (55, 56). Interpretation of the calorimetric results is complicated by the existence of residual base stacking in rA above the transition temperature of poly(rA)·poly(rU) (57, 58). Though different thermodynamic parameters for poly(rA) stacking have been obtained by different groups, and therefore different corrections of the poly(rA)· poly(rU) calorimetric data to obtain thermodynamic functions for the transition to unstructured single strands have been applied, the potential of the method is such that we cover one plausible treatment of the data here.

Filimonov & Privalov (56) measured ΔC_p^0 and $\Delta H^0{}_{T_m}$ for the denaturation of poly(rA)·poly(rU) as a function of [NaCl] (0.01 – 0.1 M). An apparent ΔC_p^0 of approximately 80 cal K^{-1} mol^{-1} bp^{-1} was obtained; calorimetric enthalpies ranged from ~6.8 kcal mbp^{-1} (0.01 M NaCl; $T_m = 310.5$ K) to 8.6 kcal mbp^{-1} (0.1 M NaCl; $T_m = 331$ K). Single stranded poly(rA) shows a broad, salt independent, noncooperative transition (from a stacked to an unstacked state) centered at 313 K, for which a calorimetric enthalpy of 3.0 kcal per mole base and entropy of 9.8 cal K^{-1} per mole base were obtained. Correction for the contribution of residual stacking in poly(rA) to the observed ΔC_p^0 gave a heat capacity difference between unstructured strands and native helix of 30 ± 4 cal K^{-1} mol^{-1} bp^{-1}, and increased the enthalpies of denaturation to 8.5 kcal mol^{-1} bp^{-1} (at 0.01 M NaCl) and to 9.2 kcal mbp^{-1} (at 0.1 M NaCl). Filimonov & Privalov observed that the variation of the corrected $\Delta H^0{}_{T_m}$ with T_m is exactly that predicted from the value of ΔC_p^0 for the effect of temperature itself on the enthalpy change. Consequently, upon converting the thermodynamic functions to a standard temperature of 298 K, the authors found that the enthalpy difference between denatured and native states of poly(rA)·poly(rU) was independent of [NaCl]; the salt effect on helix stability has an entropic origin. Values of ΔS^0 at 298 K range from 26.3 cal K^{-1} mol^{-1} bp^{-1} at 0.01 M NaCl to 24.3 cal K^{-1} mol^{-1} bp^{-1} at 0.1 M NaCl. The corresponding free energy differences are 0.3 kcal mol^{-1} bp^{-1} and 0.8 kcal mol^{-1} bp^{-1}, respectively. Earlier calorimetric measurements of the enthalpy of denaturation of T2 phage DNA (59) and calf thymus DNA (60) as a function of NaCl concentration showed a similar dependence of $\Delta H^0{}_{T_m}$ on T_m to that observed for poly(rA)· poly(rU), which suggests that in general the effect of electrolyte concentration on the stability of the double helix is primarily entropic.

A molecular basis for this result is provided by polyelectrolyte theory (7, 8, 40). The high structural charge density on a polyion such as DNA generates locally steep gradients in the concentrations of counterions and coions surrounding it. (Even at low concentrations of added salt, the local counterion concentration at the DNA surface is predicted to be in the molar range; the concentration of coions near the surface is essentially zero.) This effect is analogous to, but much more dramatic than, the Debye-Hückel screening effect in an ordinary electrolyte solution. Because the local counterion concentration is so high, a significant amount of binding of counterions to the polyion charged groups will occur if there is any chemical (as distinct from electrostatic) affinity between these species. The fraction of a counterion physically bound per polyion monomer is denoted $1 - \alpha$. Both of the above effects (ion gradients, counterion binding) contribute to a reduction in the chemical potential of the electrolyte component as a result of the presence of the polyelectrolyte, and may be considered together to define the extent of thermodynamic binding of counterions. The fraction of a counterion thermodynamically bound per polyion monomer is denoted $1 - i$ or ψ (40, 61, 62). Such thermodynamic quantities as the Donnan coefficient, osmotic coefficient, and electrolyte activity coefficient are directly related to the extent of thermodynamic binding of counterions (40); effects of electrolyte concentration on conformational equilibria or ligand binding reactions of nucleic acids result from differences in the extent of thermodynamic binding of counterions between product and reactant states of the nucleic acid (8, 40). At low salt concentrations, ψ is entirely determined by the structural charge density of the polyion; in the experimental range of salt concentrations, ψ is *reduced* from its low salt value to an extent that depends on the amount of physical counterion binding and the salt concentration (40). ^{23}Na NMR experiments are consistent with an extent of physical binding of Na$^+$ ions to helical DNA that is relatively independent of salt, and in the range $0.25 \leq 1 - \alpha \leq 0.75$ (41, 51).

The denaturation of a nucleic acid reduces its structural charge density, which reduces the steepness of the local ion concentration gradients and therefore reduces ψ. Consequently, denaturation is accompanied by the release of $\psi_h - \psi_c$ thermodynamically bound counterions per nucleotide transformed from the helix (h) to the coil (c) state; $\psi_h - \psi_c$ appears from both theory and experiment to be relatively independent of salt concentration (7, 8, 40, 59, 63). As the salt concentration (m_3) is reduced, the contribution from the release of these ions to the observed free energy difference (at a specified temperature) between denatured and native states of a base pair (ΔG°_{obs}) becomes more important:

$$\frac{d\Delta G^{\circ}_{obs}}{d \ln m_3} = 2RT \, (\psi_h - \psi_c)$$

To the extent that ψ is independent of temperature, this effect is entropic in origin (a free energy of dilution); that is

$$\frac{d\Delta S^\circ_{obs}}{d \ln m_3} \cong 2R(\psi_h - \psi_c) = \frac{\Delta H^\circ_{T_m}}{T^2_m}\frac{dT_m}{d \ln m_3}$$

and

$$\frac{d\Delta H^\circ_{obs}}{d \ln m_3} \cong 0$$

where $\Delta H^0_{T_m}$ is the enthalpy of denaturation at T_m, and ΔS^0_{obs} and ΔH^0_{obs} are the entropy and enthalpy differences between denatured and native states of a base pair at a fixed reference temperature. From either the calorimetric or T_m data on poly(rA)·poly(rU) denaturation to unstacked coils, one finds that 0.19 ions are released (in the thermodynamic sense) per nucleotide denatured ($\psi_h - \psi_c = 0.19$).

Values of ΔC^0_p as a function of base composition and solution conditions for natural DNA molecules are not available, and only limited calorimetric data is available on the variation of the denaturation enthalpy with T_m at constant electrolyte concentration and variable base composition (64). It is interesting to note that the dependence of $\Delta H^0_{T_m}$ on T_m is in close agreement with that predicted from the ΔC^0_p of poly(rA)·poly(rU) denaturation (56, 60, 64), which suggests (as in the case where T_m was varied by changing the salt concentration) that much of the apparent variation of $\Delta H^0_{T_m}$ with base composition may be a heat capacity effect and not directly an effect of base composition. This would imply that the additional thermodynamic stability of the G·C base pair as compared to the A·T base pair may have a primarily entropic origin. One effect that may contribute to a variation in ΔS^0_{obs} with base composition is the apparent variation in the thermodynamic extent of ion release ($\psi_h - \psi_c$) with base composition. Evidence has accumulated to indicate that the dependence of T_m on the fraction of G·C base pairs (X_{GC}) is a function of salt concentration. At low salt concentrations dT_m/X_{GC} is larger than the canonical Marmur-Doty value ($dT_m/dX_{GC} = 41$ deg) (65) applicable at 0.2 M salt; dT_m/dX_{GC} decreases dramatically in the molar salt concentration range (66–68). If dT_m/dX_{GC} is a function of salt concentration (m_3), then $dT_m/d \ln m_3$ must be a function of X_{GC}. An empirical equation that represents one set of denaturation data in NaCl is T_m (°C) = 176 − (2.6 − X_{GC}) (36 − 3.06 ln m_3) (68), from which one obtains $\psi_h − \psi_c$ = 0.16 − 0.06 X_{GC}. A similar result has recently been derived by Blake & Haydock (69) from an elegant study of the salt dependence of the subtransitions of 34 cooperatively melting regions of phage λ DNA, using a high resolution differential melting technique. As these authors point out, ~40% more ions are released (in the thermodynamic sense) per nucleotide denatured in an AT-rich region than in a GC-rich region. Consequently AT-rich regions are predicted to be preferentially destabilized by a reduction in salt concentration; the effect is predicted to be primarily entropic. Small changes in the structural charge density or in the chemical affinity for counterions in either the denatured and/or the native form as a function of G·C content could account for this result. If the G·C dependence of thermodynamic counterion binding resides in part in a greater extent of thermodynamic binding of electrolyte ions to AT-rich regions in the helix (which would be of great importance physiologically), this effect must not be ion-specific, since difference equilibrium dialysis experiments using alkali metal cations did not detect any ion-specific effects of base composition (52).

CYCLIZATION   There has been renewed interest in the cyclization and oligomerization reactions of DNA molecules or fragments with complementary single-stranded termini, both as a result of the ability to generate such fragments by digestion of DNA with a variety of restriction endonucleases, and because of the importance of such end-joining reactions in molecular cloning procedures. The definitive work on the kinetics and equilibria of the end-joining reactions is that of Wang & Davidson (70–72) with λ DNA and sheared half-molecules thereof. The process is adequately represented as a two-state equilibrium between cohered and free ends, which proceeds through a transient intermediate in which the cohesive ends are aligned but not yet base paired. Comparison of the van't Hoff enthalpy obtained from the temperature dependence of the end-joining equilibrium constant with the calorimetric enthalpy of base pair formation provided an accurate estimate of the number of nucleotides in the single-stranded ends. Recently, the dependence of the end-joining equilibrium constant on salt concentration (72) has been analyzed by the method described in the previous section to obtain the total increase in thermodynamic binding of counterions that accompanies the reaction (73). From this, the number of nucleotides in the single-stranded ends could be determined; again agreement with the chemically determined value is excellent.

Equilibrium constants (and also association rate constants) for cyclization and linear dimer formation are related by the Jacobson-Stockmayer factor $j = (3/2\pi L b_e)^{3/2}$, where L is the contour length and b_e is the statistical segment length or Kuhn length of the DNA under the ionic conditions of the experiment (see below); j is the concentration of one end of a randomly coiling polymer chain in the vicinity of the other (70). In predicting the distribution of cyclic and linear products from an end-joining reaction involving molecules with $L \gg b_e$ so that Gaussian chain statistics apply, the significant quantity is the ratio j/i, where i is the bulk concentration of cohesive ends (75). For large DNA molecules, j will accurately represent the concentration of proximate ends. As the fragment contour length decreases, the applicability of Gaussian statistics breaks down and j will increasingly overestimate the concentration of ends sufficiently close to cyclize. The ratio j/i can still be used as a semiempirical measure of concentrations that promote cyclization. The extent of cyclization may be maximized for short fragments by selecting incubation conditions (salt concentrations and/or temperature) that stabilize the closed form (74, 75). Recent computations of the radial distribution functions of short, ideal DNA chains predict that fragments as small as 256 bp may cyclize, although those with only 128 bp should not (75a).

Flexibility of DNA

LATERAL RIGIDITY   Unstructured single-stranded polynucleotides possess a limited extent of stiffness (statistical segment length of ~40 Å) (76). The statistical segment length does increase (by a factor of 2–3) under conditions favoring a large amount of base stacking, but the single-stranded molecules fail to achieve the rigidity displayed by double helical DNA.

Stiff macromolecules such as DNA are conveniently modeled as worm-like coils (2). This representation treats the polymer as a continuously curving chain for which the direction of curvature at any point along the chain is random. In this model, the stiffness of the molecule is characterized by its persistence length a which represents, in the limit of infinite contour length, the average projection of its end-to-end distance vector along a z-axis defined by the orientation of the first segment of the chain. Larger values of a correspond to greater local rigidity. For contour lengths of the order of a, the chain approximates rodlike behavior. For very long contour lengths, the chain approaches the behavior of a Gaussian coil. The persistence length and the statistical segment length b_e (used in the Gaussian chain model of the polymer as a series of segments connected by universal joints) are simply related: $a = b_e/2$.

Long-range excluded volume effects in addition to short-range stiffness perturb the dimensions and properties of real polymer chains from those of the ideal flexible random coil. Excluded volume effects arise both because the segments occupy a finite volume and because of differences in segment-

segment and segment-solvent interactions. Obtaining a rigorous theoretical description of excluded volume effects on chain statistics has been a difficult problem (77). One commonly used method for introducing these effects is by introducing a parameter ϵ into the equation for the mean square end-to-end distance: $\langle h^2 \rangle = N_e^{1+\epsilon} b_e^2$ where N_e is the number of statistical segments in the total contour length.

Most of the determinations of the persistence length of DNA are at 0.2 M NaCl, neutral pH, and 25°C. Within the last decade, a substantial degree of consensus on the value of a has finally appeared. When excluded volume has been neglected, the persistence length has been found to be 600 ± 100 Å (78–84). Consideration of excluded volume in determining a has produced values of 400 ± 100 Å with $\epsilon \approx 0.1 \pm 0.02$ (85–87a).

Whether excluded volume need be considered has been much debated (88–91). The sedimentation coefficient and intrinsic viscosity of native DNA vary with powers of the molecular weight (0.445 and 0.665, respectively), which differ from the 0.5 power dependence expected for an unperturbed random coil. This deviation was postulated to be due to excluded volume. However, it has also been argued, based on measured values of the second virial coefficient, that the effects of excluded volume on chain dimensions are small at 0.2 M salt (a 10% increase in the radius of gyration over its unperturbed value for molecules with $M = 30 \times 10^6$) (88). Recent measurements of the second virial coefficient as a function of salt indicates that it increases from 2.5×10^{-4} mL mol g^{-2} to 5.0×10^{-4} mL mol g^{-2} between 4.0 and 0.2 M NaCl. A reduction of the salt concentration to 0.005 M increases the second virial coefficient to 22.2×10^{-4} mL mol g^{-2} (86). This suggests that excluded volume effects are more significant at lower salt, where long-range polyelectrolyte effects are important.

Several studies of the salt dependence of the persistence length exist. The results are summarized in Table 2. All of the available data agree that as the salt concentration is lowered, a increases. A large dependence of the persistence length upon salt concentration was obtained from a statistical analysis of local curvature of T2 DNA fragments examined by electron microscopy (93). Fragments were adsorbed on to a cytochrome c coated grid from solutions containing ammonium acetate. Evaluation of these results is impeded by lack of knowledge of the effect of salt upon the cytochrome c-DNA interaction. Moreover, the effects of excluded volume on the conformation of these fragments, stated to be a few microns in length and therefore possibly subject to such constraints in solution, were not taken into account (93a). Harrington (91) combined flow birefringence and intrinsic viscosity data on T2 DNA to obtain a steep dependence of a on salt when the experimental data was calibrated with $a = 660$ Å at 0.2 M NaCl. A recent light-scattering study on the linear form of the ColE1 plasmid ($M = 4.4 \times 10^6$) (86) determined a with and without an excluded volume correction (see Table 2). When excluded volume is considered, $a = 370$ Å at 0.2 M NaCl, and the increase in a with decreasing salt is reduced. Recalibration of Harrington's data (R. E. Harrington, personal communication) using this value of a results in a less dramatic salt dependence that is in reasonable agreement with the light-scattering data, as well as with that obtained in an earlier analysis (87) of salt dependent hydrodynamic data.

Analysis of the rotational relaxation times of the transient electric bire-fringence of short (587 bp) blunt-ended restriction fragments indicates that the persistence length is 500 ± 50 Å at or above 1 mM NaCl and that electrostatic contributions to the persistence length are negligible for these fragments above 1 mM NaCl. Similar experiments in $MgCl_2$ show that the asymptotic value (530 ± 50 Å) is obtained at concentrations above 0.1 mM (93a).

Two similar theoretical treatments have derived an expression for the electrostatic component of the persistence length (94–96). The polyelectrolyte is modeled as a structureless space curve and electrostatic interactions are assumed to occur via screened Coulombic potentials. The electrostatic contribution is predicted to be negligible at 1 M NaCl and to increase to 65 Å at 0.005 M NaCl (97). This predicted dependence is not in quantitative agreement with any of the available experimental data.

The total free energy of bending has been considered for a model equivalent to the wormlike coil (2, 98). It has been found that $\Delta G° = \beta\theta^2/2L$, where β, the bending force constant of the chain of contour length L, is related to the persistence length ($\beta = akT$) and θ is the angle between the tangent vectors of the origin and the end of the chain. A similar result has been obtained by considering the discrete analogue of this model (99). Thus by studying the variation of the persistence length as a function of temperature, the thermodynamic parameters for bending may be obtained.

Table 2 Salt dependence of the persistence length of DNA

[NaCl] M	Hearst et al (87) a ($\epsilon = 0.072$) Å	Frontali et al (93)[a] a ($\epsilon = 0$) Å	Borochov et al (86) a ($\epsilon = 0$) Å	Borochov et al (86)		Harrington (91)		
				a (Å)	ϵ	a^b Å	a^c Å	ϵ
0.005	660	2100	910	530	0.200	1540	530	—
0.1	400	800	550	400	0.118	740	370	0.102
0.2	400	560	500	370	0.100	660	370	0.085
1.0	330	540	360	290	0.090	460	280	0.070

[a] Values interpolated from the available salt data.
[b] Values obtained using $a = 660$ Å at 0.2 M salt as calibration.
[c] Values obtained using $a = 370$ Å at 0.2 M salt as calibration (R. E. Harrington, personal communication).

Only two such studies exist. Gray & Hearst (100) obtained the temperature dependence (5–49°C) of a from a sedimentation study on several phage DNAs at 0.2 M salt. The enthalpy and entropy of bending were determined to be 6.9 ± 2.2 cal bp mol^{-1} deg^{-2} and (–1.5 ± .8) × 10^{-2} e.u. bp deg^{-2}. A recent flow birefringence study by Harrington (101) on T2 DNA in 2.0 M NaCl determined that ΔH = 12.4 ± .9 cal bp mol^{-1} deg^{-2} and ΔS = (1.8 ± 3.2) × 10^{-3} e.u. bp deg^{-2}. These values for the enthalpy and entropy were obtained by assuming a value of 660 Å for the persistence length at 25° C. If the flow birefringence data is calibrated with a value of 370 Å for a at 25°C, the values for the enthalpy (6.7 ± .5 cal bp mol^{-1} deg^{-2}) and entropy (0.002 ± 0.004 e.u. bp deg^{-2}) are in agreement with the results of Gray & Hearst.

TORSIONAL RIGIDITY   Several studies concerned with the torsional rigidity of DNA have appeared recently. These studies arrive at the common conclusion that DNA is much less resistant to twisting than was previously supposed (102–106). The torsional rigidity determines the energy required to change the angle between base pairs by a given amount and is related to the extent to which the conformation of DNA may be perturbed by solution conditions or deformed by the binding of a protein.

The fluorescence depolarization of intercalated ethidium bromide demonstrated the existence of twisting and bending motions of the double helix occurring on a time scale of 10^{-8} seconds (107). The dynamics of a bead-spring model (102, 103) and an elastic continuum model (102) have been developed in order to gain an understanding of the motions that contribute to the depolarization of fluorescence. For an ideal elastic cylinder, the energy required for deformation by twisting is given by $U = \frac{1}{2}C\theta^2$, where C is the torsional force constant and θ is the angular displacement per unit length. The torsional force constants found in these studies were in the range C = 23–53 cal bp mol^{-1} deg^{-2}. Recently, the fluorescence depolarization has been measured with increased accuracy and subnanosecond time resolution (103a, 104). The observed polarization of the fluorescence decays as an exponential in (time)$^{\frac{1}{2}}$. This is the time dependence predicted for twisting motions by the elastic continuum model and also by the discrete model after an initial period. Thus the torsional motions of the DNA helix occurring in a time range of 10^{-8} to 10^{-7} seconds can be described by simple models and are characterized by a torsional force constant C = 17 cal bp mol^{-1} deg^{-2}. In addition, these results have been interpreted as evidence against the existence of widely spaced torsion joints in an otherwise stiff molecule (103a).

A similar model has also been developed to describe the contributions of twisting and bending motions to the electron spin resonance (ESR) correlation time of an intercalated spin labeled propidium derivative (105). The ESR measurements can be described by a model in which the intercalated compound undergoes an average torsional motion of 4 deg (at 20°C). This average amplitude corresponds to a lower limit of $C = 40$ cal bp mol^{-1} deg^{-2}. This model approximately predicts the observed correlation times and their dependence on temperature, segment length, and magnetic relaxation rates.

Another estimate of the torsional force constant has been obtained from a combination of experimental and theoretical work on supercoiled DNA (106, 106a). The resulting torsional force constant is $C = 21$ cal bp mol^{-1} deg^{-2}.

For an isolated torsion spring, the root-mean-squared amplitude of a fluctuation in the displacement is given by $\theta_{rms} = (RT/C)^{1/2}$. The force constants given here correspond to a base pair motion of $\theta_{rms} = 3$–6 deg at 25°C, which is substantially larger than expected.

Thermodynamic and Molecular Description of Supercoiling

Circular, double-stranded DNA isolated from a variety of sources has been found to have a convoluted tertiary structure and altered chemical and biochemical reactivities. DNA in this state, known as supercoiled or superhelical DNA, has been the subject of a great deal of research and several reviews (4, 108). The emphasis in this section is on the recent advances toward a thermodynamic and molecular description of DNA in the supercoiled state.

The physical constraint that causes superhelicity is that the number of times one strand encircles the other remains invariant as long as the strands are not broken. This number is called the linking number or the topological winding number and is designated by Lk, L, or α (4, 109, 110). The linking number is the sum of the twisting of one strand about the other in the double helix, which is referred to as the twist or Tw, and the three dimensional disposition of the helix axis, which is called the writhe or Wr. This is described by the equation $Lk = Tw + Wr$. For a closed, double-stranded molecule, a change in the twist results in a compensating change in writhe. A useful quantity is the number of titratable superhelical turns, τ, which may be thought of as the writhe of a molecule of a given linking number when Tw is fixed at its value for a linear molecule. The form of τ that is independent of molecular weight is called the superhelix density and is defined by $\sigma = 10\tau/N$, where N is the number of base pairs. The properties of these quantities and methods for the measurement of τ are discussed in detail in the references cited above and in several texts (1, 2).

The evidence from sedimentation (91, 111–113), intrinsic viscosity (113), light scattering (114, 115), and electron microscopy studies (112, 116) suggests that the shape of a closed-circular DNA changes as the superhelix density is increased. At low superhelix densities ($|\sigma| \gtrsim 0.05$), the DNA remains a random coil that becomes more compact as the superhelix density increases. Molecules with higher superhelix densities ($0.06 < |\sigma| < 0.19$) have been characterized as interwound or branched interwound superhelices. This behavior is approximately symmetric for positive and negative superhelix densities (112, 117). The details of molecular structure in these limiting regions, and the transition between them are influenced by temperature and ionic environment (111, 112) as well as by the origin of the DNA (115). In addition to these physical studies that characterize superhelical DNA by shape, and therefore indirectly by writhe, there are spectroscopic studies that bear indirectly on twist. The circular dichroism of superhelical DNA differs significantly from that of linear or nicked DNA (43, 118, 119). There is evidence that this effect may either decrease (120) or increase (120–122) with increasing salt concentration, depending on the nature of the DNA. The present lack of understanding of the effect of tertiary structure on CD (37) forestalls its use in a quantitative description of superhelical DNA.

The thermodynamic description of supercoiling is comparatively simple. In terms of intensive quantities, the free energy of supercoiling is given by $\Delta G^0 = uRT\sigma^2$, where u is a reduced force constant. From buoyant density sedimentation (123) and gel electrophoresis experiments (38, 124, 125), it was found that $u = 10.2 \pm 0.6$. Little variation in u was observed for different salt types and concentrations (0.002 M $MgCl_2$, 0.2 M NaCl, 5.8 M CsCl), temperatures (2–28°C), superhelix densities ($|\sigma| < 0.06$), and for DNA from various sources. For higher superhelix densities, a cubic term is included in the free energy expression (123). The value of u found by spectrophotometric titration with ethidium bromide was significantly lower: $u = 5.4 \pm 0.4$ (3M CsCl, 2.5–20°C) (126). These authors also report a slight temperature dependence above 20°C. The insensitivity of u to temperature suggests that the free energy of supercoiling is largely entropic. The precision of the available data, however, does not permit a more quantitative statement (126).

Theoretical descriptions of highly supercoiled molecules have developed from an elastic model of DNA. The molecule is described as an isotropic rod characterized by a force constant for bending, A, and a force constant for twisting, C (see section on flexibility). The elastic model is expected to be an appropriate model for highly supercoiled molecules (127). The equilibrium shapes of an elastic rod subject to torsional stress have been obtained by solution of the equations of linear elasticity (128, 129). These results have not been usefully applied to supercoiled, circular molecules because the constraint of ring closure is not explicitly included. The general solutions of the equilibrium shapes of twisted rings are apparently not known at present. Another approach has been to assume that a highly supercoiled DNA takes the shape of an interwound superhelix (130). Subject to the constraint of a distance of closest approach of the arms of the superhelix, r_o, the free energy of twisting and bending is minimized with respect to the number of physical superhelical turns. By specifying either a torsional force constant or a value for r_o, these authors have explained the observed insensitivity of the sedimentation coefficient to the superhelix density in the range $0.06 < |\sigma| < 0.09$. Introduction of recent estimates of the torsional force constant (102–106) results in a solution only when r_o is large.

Molecules with low superhelix density have been the subject of recent theoretical developments based on ideal models of polymer chains. In this formalism, supercoiling is a constraint that reduces the configurational entropy of the closed random coil. By calculating the writhes of many computer generated random walks, it was found that a population of circular molecules equilibrated by strand scission and closure will have a Gaussian distribution of writhes (106, 106a). The variance of the writhe is proportional to the number of statistical segments in the random walk (see section on flexibility). When combined with the experimental observation that the equilibrium distribution in linking number is Gaussian (38, 124), this result predicts that the distribution in twist is also Gaussian, and gives an independent estimate of the torsional force constant. When the statistical segment length is taken as 1150 Å, the resulting theoretical variance in the writhe accounts for about half of the observed variance in the linking number. In this case, a change in the linking number of 1 causes a change in the writhe of ½ and a change in the twist of ½. If the more recently devalued statistical segment lengths are used instead (see section on flexibility), the variance in the writhe is increased. The resulting, larger torsional force constant is in better agreement with other estimates (see section on torsional rigidity).

The investigation of an interesting form of DNA, produced by annealing complementary closed circular single strands, suggests that there are limits to the variance in writhe (119). Analysis by spectroscopy and electron microscopy show that this DNA is 60–90% double stranded. The linking number of this structure must be zero, but the sedimentation and gel electrophoresis characterization is not consistent with the compensation of all of the right-handed double-helix turns by left-handed writhing. The authors propose that a large fraction of the compensating turns takes the form of left-handed duplex turns.

The expression of the free energy of supercoiling in terms of fluctuations in secondary structure or altered chemical and biochemical reactivity has been recently reviewed (4). Briefly, DNA with a negative superhelix density is characterized by an enhanced reactivity toward ligands or proteins that unwind the double helix and toward reagents or enzymes that react specifically with single-stranded DNA. This is generally attributed to a superhelix density dependent increase in either transient fluctuations in secondary structure or stable alternate structures such as cruciforms and melted regions. Some sequences in negatively supercoiled DNA are reactive with single-strand specific reagents (131). Recent work identifies sites of nuclease sensitivity as inverted-repeated sequences and argues for the existence of cruciform structures (131a). A closely related phenomenon is the uptake of homologous single strands by supercoiled DNA (132–134). This reaction is driven by the free energy of supercoiling. A careful study analysis of the kinetics of this reaction has resulted in a model in which the rate limiting step is the opening or unstacking of a small number of bases in the double helix (133). This is taken as evidence against the existence of stable denatured regions.

Helix-coil transition theories have been applied to the problem of local melting induced by the free energy of supercoiling (135–139). The theories predict a substantial destabilization of the double helix at high superhelix densities, although there is some disagreement about the probability of stable melted regions and cruciform structures. An interesting prediction is that significant fluctuations disrupting base pairing will appear at positive ($\sigma > 0.2$) as well as negative superhelix densities ($\sigma < -0.02$) (136). Recent experimental work has shown that PM2 DNA becomes sensitive to *Alteromonas* nuclease at positive and negative superhelix densities very close to those predicted (117).

Thermodynamic Analysis of Ligand Binding

Thermodynamic studies complement structural studies by revealing how much each type of interaction contributes to the free energy of complex formation under specified conditions, and the extent to which each contribution is enthalpic or entropic. Moreover, a molecular interpretation of thermodynamic data can be used to obtain estimates of the number of each type of interaction (see section on cyclization). Here we review some elements of the thermodynamic analysis of the noncovalent interactions of proteins and other large cationic ligands with double helical DNA. No attempt is made to review the totality of thermodynamic information available on ligand-DNA interactions. Comprehensive reviews of the interactions of DNA with cationic ligands (7, 8), oligopeptides (140), intercalators (141), and various proteins (3, 5, 6, 142–144) are available.

THE DNA DOUBLE HELIX FROM A LIGAND'S PERSPECTIVE   Double helical DNA may be viewed as the superposition of two lattices: the regular exterior array of phosphate charges, and the interior array of stacked base pairs, accessible to ligand binding from either the major or minor groove.

To an approaching ligand the phosphate lattice appears relatively homogeneous, affording sites for nonspecific electrostatic interactions. The lattice(s) of base pairs, which exhibit both sequence heterogeneity and, consequently, local conformational heterogeneity, may bind ligands either through intercalation or by specific interactions with accessible functional groups on the bases. Both lattices are involved in the formation of high affinity, site-specific complexes between proteins such as *lac* repressor or RNA polymerase and DNA (145–147). These proteins, like any ligand having positively charged groups, also interact in a weaker, nonspecific electrostatic mode with the lattice of phosphates (148–152).

ANALYSIS OF NONSPECIFIC BINDING EQUILIBRIA

McGhee & von Hippel (153) have derived a modified version of the Scatchard binding isotherm that has proved to be widely applicable in the analysis of both electrostatic and intercalative types of binding to nucleic acid lattices [see e.g. (141, 148)]. In a relatively simple and powerful way their probabilistic derivation incorporates the consequences of nearest-neighbor cooperativity and overlap. The latter effect is characterized by n, the number of contiguous sites (nucleotides) rendered inaccessible to further binding by the binding of one ligand. A potential ambiguity in the analysis of ligand-nucleic acid binding equilibria arises in the physical interpretation of n: does it represent a number of nucleotides or half that number of nucleotide pairs? For example, the value of n deduced for the nonspecific binding of *lac* repressor is approximately 24 nucleotides or (equivalently) 12 base pairs. The latter alternative suggests that the ligand occupies (sterically blocks) all radial access to the helix for ~ 1.2 helical turns. The former number indicates that over a span of ~ 2.4 helical turns only half of the cylindrical surface is inaccessible to further binding; this interpretation appears to be favored by evidence from electron microscopy on the nonspecific DNA-repressor complexes (154). Thus, a second repressor molecule could bind nonspecifically to the helical region opposite to a bound ligand. Electron microscopy (155) and chemical modification experiments (156) both indicate that the specific binding of *lac* repressor and RNA polymerase involves contacts on only one side of the DNA helix.

DEPENDENCE OF K_{obs} ON SOLUTION CONDITIONS   A unique and often dramatic feature of the interaction of cationic ligands (or proteins with a cationic binding site) with nucleic acids is the dependence of the binding constant K_{obs} on salt concentration (m_3) in the solution. The reaction of a protein with a nucleic acid in solution is a multiple equilibrium, involving a variety of different microscopic states of binding of small ions (e.g. protons or electrolyte ions) or other solvent components to both protein, nucleic acid, and complex. These small ions are consequently net participants in the equilibrium, and the observed equilibrium constant becomes a function of their concentrations. (For example, in the nonspecific interaction of *lac* repressor with DNA, $-d\ln K_{obs}/d\ln m_3 = 11 \pm 1$ (148, 149, 151).) This effect of small ions is *not* simply an ionic strength effect but rather is ion specific. Ionic strength (I) is not a useful quantity for interpreting competitive binding equilibria, such as the effects of Mg^{2+} and Na^+ on the nonspecific binding constant of lac repressor to double stranded DNA (145, 148). The value of K_{obs} is essentially the same ($\sim 3 \times 10^5$ M^{-1}) in 0.12 M NaCl ($I = 0.12$), in 0.09 M NaCl, 0.003 M $MgCl_2$ ($I = 0.099$), and in 0.01 M NaCl, 0.01 M $MgCl_2$ ($I = 0.04$). Since, in the absence of $MgCl_2$, the nonspecific binding affinity of *lac* repressor increases dramatically as the NaCl concentration is reduced, the Mg^{2+} ion clearly functions as a direct competitor with the protein for DNA sites; this effect of Mg^{2+} on K_{obs} is much larger than that which might be expected from the contribution of $MgCl_2$ to the ionic strength.

A second source of the dependence of K_{obs} on electrolyte concentration is the extreme deviation from uniformity of electrolyte concentrations near the DNA polyanion. As a result of this screening effect, the activity coefficient of the DNA is a strong function of salt concentration. The neutralization of phosphate charges by the ligand may be expected to reduce the local nonuniformity of ion concentrations. An approximate theory that interprets salt effects on K_{obs} in terms of the release of thermodynamically bound (by binding and screening interactions) counterions from the DNA upon the neutralization of phosphate charges by cationic groups on the ligand has been developed by Record and co-workers (8, 40, 61). Based originally on Manning's limiting-law (low salt) polyelectrolyte theory (61, 157), the binding theory has recently been extended to higher (physiologically relevant) salt concentrations using thermodynamic results obtained from the cylindrical Poisson-Boltzmann model under conditions of excess salt (40, 158). Alternative molecular thermodynamic models to describe the effects of salt on K_{obs} have been developed by Manning (7) and by Bloomfield and co-workers (159).

Since the thermodynamic ion binding theory neglects end effects in the region of the bound ligand and cannot treat the interactions of electrolyte ions with the (unknown) distribution of charges on the oligocationic binding site of the ligand, it is not expected to be as successful in this context as it has proven to be in the analysis of helix-coil equilibria of nucleic acids (see section on helix-coil equilibria), where both reactant and product species are polyelectrolytes, and where the binding of anions need not be considered. Use of the thermodynamic ion binding parameter ψ gives the result that

$$-\frac{d \ln K_{obs}}{d \ln m_3} = Z\psi + k$$

where Z is the net cationic valence of the binding site of the ligand (or equivalently the number of DNA phosphates neutralized by the ligand), ψ is the fraction of an electrolyte ion thermodynamically bound per DNA phosphate (and therefore released from each phosphate neutralized), and k is the unknown net amount of ion release from the ligand upon complex formation (8, 40). At low electrolyte concentrations, ψ approaches its limiting-law value (0.88 for double helical DNA); at higher salt concentrations ψ is reduced by an amount that depends on the salt concentration and the amount of physical counterion binding. At 0.2 M NaCl, the Poisson-Boltzmann thermodynamic theory (40) predicts a range of values of ψ from 0.53 in the absence of any physical counterion binding, to 0.80 if the extent of counterion binding is as great as that hypothesized by Manning (157). In the absence of information about the anion release term k, the logarithmic derivative $-d \ln K_{obs}/d \ln m_3$ may be considered to provide a maximum estimate of the amount of thermodynamic ion release from the nucleic acid $(Z\psi)_{max}$, from which a maximum value of Z can be obtained.

The dependence of K_{obs} on m_3 has been measured for a large number of oligocationic ligands of known charge, including Mg^{2+}, oligopeptides, and various intercalators (see e.g. 7, 8, 160–164). Without exception, $\ln K_{obs}$ is found to be a linear function of $\ln m_3$, in the absence of competition by other cationic species; moreover, values of Z estimated from the slopes of such plots are in good agreement ($\pm$ 10–20%) with the charge on the ligand, *using the limiting-law value of ψ* ($\psi = 0.88$) and neglecting possible effects of anion release from the ligand. These studies may be considered to calibrate the binding theory, and may further indicate a substantial amount of physical counterion binding to the DNA, since this is required by the Poisson-Boltzmann analysis in order for ψ to remain in the vicinity of the low salt limiting value at higher electrolyte concentrations. The above results can also be interpreted using the molecular thermodynamic model of Manning (7) (which assumes that the ligand replaces counterions in the delocalized surface (condensed) layer surrounding the polyion, and predicts that $-d \ln K_{obs}/d \ln m_3 = Z + k$) or by the numerical Poisson-Boltzmann calculations of Bloomfield and co-workers (159) (which evaluate the actual amount of counterion release from a cylindrical shell around the DNA upon introducing an oligocationic ligand). Each theory has its own limitations and approximations. At present it is difficult to distinguish between these approaches since all provide a reasonable fit to experimental results.

The strong dependences of K_{obs} on solution conditions make it possible to obtain a wide range of binding constants for the interaction of a cationic ligand with DNA by proper choice of pH and ion concentration. Consequently it is useful to define a standard condition for comparison of binding constants. Since the thermodynamic equivalent of the binding reaction is

$$\text{ligand} + \text{DNA} \rightleftharpoons \text{complex} + (Z\psi + k) \text{ electrolyte ions,}$$

an approximate standard condition is at an electrolyte concentration of 1 M. [The questions of whether the standard state should be at unit activity or unit concentration, and the choice of concentration scale, are as yet unresolved (8, 40).] Such standard state binding constants for the interactions of oligolysines with nucleic acid helices are in the range $0.1-1 M^{-1}$, and provide an estimate of the standard free energy of formation of a single lysine-phosphate interaction at 1 M NaCl: $\Delta G° \simeq +0.2 \pm 0.1$ kcal. The large increase in K_{obs} obtained at lower electrolyte concentration results from the favorable free energy of dilution of the $(Z\psi + k)$ ions from the standard state to the salt concentration of the experiment. In favorable cases, where protonation and anion effects on the binding constant of a ligand-nucleic acid interaction have been quantified, then extrapolation of K_{obs} to the 1 M salt standard state provides an estimate of the nonelectrostatic contribution to the binding free energy, by comparison with the binding free energy expected for an oligolysine with the same valence as that of the binding site on the ligand (61).

For *lac* repressor, assuming no anion contribution to the salt dependence of K_{obs}, analysis of nonspecific binding data indicated that $Z = 12 \pm 2$ (148, 149, 151) and that the nonspecific complex probably involved only electrostatic interactions. Estimates of Z ranging from 5 to 11 have been obtained for the specific interaction of repressor with the operator site on λ p*lac* DNA (145, 147, 165); there is direct evidence that the amount of ion release accompanying formation of the specific complex is less than for the nonspecific complex (166, 167). A probable value of Z for the specific interaction is 8 ± 1 (145); this is in general agreement with the number of phosphates in the operator region that cannot be chemically modified by an ethylating agent without blocking specific binding (5). [Curiously, the amount of ion release (~ 1.5) accompanying the interaction of repressor with a synthetic operator fragment is much less than that observed with λ p*lac* DNA (165); a further decrease can be caused by 5-bromouracil substitution in the fragment (168).] Estimates of the nonelectrostatic component of the binding free energy of the repressor-operator interaction from the extrapolated binding constant at 1 M salt are in the range 9–12 kcal/mol (145, 147); these values agree with that obtained from binding studies (165) using a modified (core) repressor lacking the 59-residue NH_2 termini that contribute the ionic interactions with operator. Under "physiological" ionic conditions (0.2 M NaCl, 0.003 M MgCl$_2$), approximately 40% of the binding free energy is contributed by electrostatic interactions and counterion release (145). Further work should lead to a detailed molecular and thermodynamic picture of the specific and nonspecific interactions.

Investigation of the kinetics of ligand-DNA interactions in vitro provides information about the reaction mechanism. The kinetics of these binding processes exhibit unusual features that are traceable to the polymeric and polyelectrolyte characteristics of DNA (147, 169, 170). Thus in addition to defining the time scale of the reaction (and therefore indicating what time dependent conformational events might be involved), kinetic studies provide insights into mechanisms not encountered in nonpolymeric systems. Although the polymeric and polyelectrolyte character of DNA introduces mechanistic complexities, it also suggests logical variables (chain length and ion concentrations) to use in distinguishing between alternative mechanisms.

Intramolecular Folding of DNA

The in vitro monomolecular collapse (condensation) of DNA into highly compacted structures may be caused by the addition of neutral or anionic (171, 172) polymers, or a variety of cationic species [polyamines (173–175), histones (176), poly-L-lysine (177) etc] to very dilute solutions of DNA. The structure of the DNA in the collapsed state is not well characterized (178–181). The experimental evidence available suggests that the structure the compacted form assumes may depend upon the species inducing collapse, and/or upon the solution conditions.

Intramolecular segment-segment interactions are normally highly unfavorable for DNA because of the strong repulsive forces between the charges of the backbone. To produce a compacted molecule one must find a means of either increasing the favorability of the segment-segment interactions and/or of making segment-solvent interactions even more unfavorable than the intramolecular interactions. The cation induced collapse occurs because neutralization of a large fraction [$\gtrsim$90% has been observed experimentally (182)] of the phosphate charge by the cationic species provides the necessary reduction in the electrostatic free energy of the compacted form (183). The collapse caused by addition of neutral or anionic polymers and salt is an example of the alternate strategy. These polymers exclude DNA from the solution volume. Collapse of the DNA reduces this entropically unfavorable effect on free energy. [To obtain collapse in the presence of a polymer, the salt concentration must be sufficiently high ($>$.3 M) to stabilize the collapsed DNA.] In both ways of inducing compaction, monomolecular collapse occurs rather than aggregation and/or precipitation simply because the DNA is at high dilution, and intramolecular interactions are favored over intermolecular ones (171, 174, 180, 184, 185).

A theoretical description of the monomolecular collapse of DNA has appeared (184). The expression for the free energy of mixing a polymer and solvent, obtained from the lattice theory of Flory and Huggins, was extended to include third virial coefficient effects. Calculation of the free energy of mixing as a function of polymer chain expansion indicated that when segment-segment interactions are sufficiently favorable, the preferred polymer configuration is highly compacted. For comparatively stiff polymers such as DNA, the transition from the extended coil to the collapsed form is predicted to occur abruptly as segment-solvent interactions become increasingly unfavorable compared to segment-segment interactions. This theoretical prediction is in agreement with the experimental observation that collapse occurs over a narrow concentration range of the species inducing the effect (173–175, 180, 182, 185).

It was suggested by Manning (7) that DNA may spontaneously fold upon the neutralization of some critical fraction of the backbone charge. Wilson & Bloomfield (182) used light scattering to study the effects of cation concentration and valence on DNA collapse, and then applied Manning's theory (7) for a mixed ion system to calculate that about 90% of the phosphate charge was neutralized under the solution conditions at which collapse occurred. According to Manning's theory, the divalent ions putrescine and Mg^{2+} neutralize only 88% of the charge on DNA (7) and consequently are unable to produce collapse. A study of several inert trivalent metal ion complexes (predicted by Manning to neutralize 92% of the phosphate charge) showed that each was capable of inducing collapse in aqueous solution (175).

The finding that a high-degree of charge neutralization is required for collapse is not surprising in view of the analysis of Bloomfield and coworkers (183, 186) of the energetic factors for the packaging of T4 DNA into its phage head. Order of magnitude estimates for the various possible sources of unfavorable contributions to the free energy of packaging were obtained. Polyelectrolyte repulsions are found to be the major thermodynamic obstacle to the process of packaging T4 viral DNA (10^5 kcal of free energy per mole of virus), although the free energy of bending (10^3 kcal per mole of virus) and that due to the loss of configurational entropy upon collapse (10^2 kcal per mole of virus) were also found to be nonnegligible. Similar rankings for T5 and λ have been calculated (186). Riemer & Bloomfield (183) estimated that polyamine interactions with DNA could be sufficiently favorable to cancel the unfavorable electrostatic repulsions. Another calculation (186), based on a Poisson-Boltzmann analysis of the repulsive potential generated in one model for the collapsed state, indicates that with the reduction in surface charge density produced by multivalent counterions, dispersion forces may be sufficient to drive compaction. Ion-dipole interactions (187), and the cationic cross-bridging of DNA segments (175) have also been proposed as possible attractive forces that could drive collapse.

ACKNOWLEDGMENTS

We acknowledge with thanks the preprints and discussions with many colleagues that contributed to this review. Dr. Charles F. Anderson provided helpful comments on the manuscript. Work from this laboratory was supported by NSF grant PCM79-04607 and NIH grant GM 23467. The assistance of Mary Ehren in preparing the manuscript is gratefully acknowledged.

Literature Cited

1. Cantor, C. R., Schimmel, P. R. 1980. *Biophysical Chemistry*. San Francisco: Freeman. 1371 pp.
2. Bloomfield, V. A., Crothers, D. M., Tinoco, I. Jr. 1974. *Physical Chemistry of Nucleic Acids.* New York: Harper & Row. 517 pp.
3. Wells, R. D., Goodman, T. C., Hillen, W., Horn, G. T., Klein, R. D., Larson, J. E., Müller, U. R., Neuendorf, S. K., Panayotatos, N., Stirdivant, S. M. 1980. *Prog. Nucl. Acids Res. Mol. Biol.* 24:167–267
4. Bauer, W. R. 1978. *Ann. Rev. Biophys. Bioeng.* 7:287–313
5. von Hippel, P. H. 1979. In *Biological Regulation and Development*, Vol. 1, ed. R. F. Goldberger, pp. 279–347. New York: Plenum
6. von Hippel, P. H., McGhee, J. D. 1972. *Ann. Rev. Biochem.* 41:231–300
7. Manning, G. S. 1978. *Q. Rev. Biophys.* 11:179–246
8. Record, M. T. Jr., Anderson, C. F., Lohman, T. M. 1978. *Q. Rev. Biophys.* 11:103–78
9. Crick, F. H. C., Wang, J. C., Bauer, W. R. 1979. *J. Mol. Biol.* 129:449–61
10. Arnott, S. 1977. In Proc. *Cleveland Symp. Macromolecules 1st*, ed. A. G. Walton, pp. 87–104. Amsterdam: Elsevier
11. Davies, D. R. 1967. *Ann. Rev. Biochem.* 36:321–64
12. Deleted in proof
13. Skolnick, J. 1979. *Macromolecules* 12:515–21
14. Soumpasis, D. 1978. *J. Chem. Phys.* 69:3190–96
15. Dover, S. D. 1977. *J. Mol. Biol.* 110:699–700
16. Mandelkern, M., Dattagupta, N., Crothers, D. M. 1981. *Nature*. In press
17. Zimmerman, S. B., Pheiffer, B. H. 1979. *Proc. Natl. Acad. Sci. USA* 76:2703–7
18. Wang, J. C. 1979. *Proc. Natl. Acad. Sci. USA* 76:200–3
19. Rhodes, D., Klug, A. 1980. *Nature* 286:573–78
20. Trifonov, E. N., Bettecken, T. 1979. *Biochemistry* 18:454–56
21. Prunnel, A., Kornberg, R. D., Lutter, L., Klug, A., Levitt, M., Crick, F. H. C. 1979. *Science* 204:855–58
22. McGhee, J. D., Felsenfeld, G. 1980. *Ann. Rev. Biochem.* 49:1115–56
23. Drew, H. R., Wing, R. M., Takano, T., Broka, C., Tanaka, S., Itakura, K., Dickerson, R. E. 1981. *Proc. Natl. Acad. Sci. USA.* In press
24. Hogan, M., Dattagupta, N., Crothers, D. M. 1978. *Proc. Natl. Acad. Sci. USA* 75:195–99
25. Levitt, M. 1978. *Proc. Natl. Acad. Sci. USA* 75:640–44
26. Deleted in proof
27. Klug, A., Jack, A., Viswamitra, M. A., Kennard, O., Shakked, Z., Steitz, T. A. 1979. *J. Mol. Biol.* 131:669–80
28. Wang, A. H. J., Quigley, G. J., Kolpak, F. J., Crawford, J. L., van Boom, J. H., Marel, G., Rich, A. 1979. *Nature* 282:680–86
29. Drew, H. R., Takano, T., Tanaka, S., Itakura, K., Dickerson, R. E. 1980. *Nature* 286:567–73

30. Pohl, F. M., Jovin, T. M. 1972. *J. Mol. Biol.* 67:375–96
31. Arnott, S., Chandrasekaran, R., Birdsall, D. L., Wheslie, A. G., Ratliff, R. L. 1980. *Nature* 283:743–45
32. Arnott, S. 1980. *Trends Biochem. Sci.* 5:231–34
33. Tunis-Schneider, M. J., Maestre, M. F. 1970. *J. Mol. Biol.* 52:521–41
34. Tinoco, I. Jr., Bustamante, C., Maestre, M. F. 1980. *Ann. Rev. Biophys. Bioeng.* 9:107–41
35. Woody, R. W. 1977. *J. Poly. Sci. Macromol. Rev.* 12:181–321
36. Baase, W. A., Johnson, W. C. Jr. 1979. *Nucleic Acids Res.* 6:797–814
36a. Zimmerman, S. B., Pheiffer, B. H. 1980. *J. Mol. Biol.* 142:315–30
37. Parthasarathy, N., Schmitz, K. 1980. *Biopolymers* 19:1137–1151
38. Depew, R. E., Wang, J. C. 1975. *Proc. Natl. Acad. Sci. USA* 72:4275–79
39. Anderson, P., Bauer, W. 1978. *Biochemistry* 17:594–601
40. Klein, B. J. 1980. *Thermodynamic studies of the polyelectrolyte behavior of nucleic acids.* PhD thesis. Univ. Wisconsin-Madison. 198 pp.
41. Anderson, C. F., Record, M. T. Jr., Hart, P. A. 1978. *Biophys. Chem.* 7:301–16
42. Ivanov, V. I., Minchenkow, L. E., Schyolkina, A. K., Poletayev, A. I. 1973. *Biopolymers* 12:89–110
43. Zimmer, C., Luck, G. 1973. *Biochem. Biophys. Acta* 312:215–27
44. Hanlon, S., Brudno, S., Wu, T. T., Wolf, B. 1975. *Biochemistry* 14:1648–60
45. Chan, A., Kilkuskie, R., Hanlon, S. 1979. *Biochemistry* 18:84–91
46. Bleam, M. L., Anderson, C. F., Record, M. T. Jr. 1980. *Proc. Natl. Acad. Sci. USA* 77:3085–89
47. Seeman, N. C., Rosenberg, J. M., Suddath, F. L., Parkkim, J. J., Rich, A. 1976. *J. Mol. Biol.* 104:105–44
48. Rosenberg, J. M., Seeman, N. C., Day, R. O., Rich, A. 1976. *J. Mol. Biol.* 104:145–67
49. Pörschke, D. 1976. *Biophys. Chem.* 4:383–94
50. Bleam, M. L. 1980. *NMR Studies of the interactions of small cations with DNA.* PhD thesis. Univ. Wisconsin-Madison. 211 pp.
51. Reuben, J., Shporer, M., Gabbay, E. 1975. *Proc. Natl. Acad. Sci. USA* 72:245–47
52. Shapiro, J. T., Stannard, B. S., Felsenfeld, G. 1969. *Biochemistry* 8:3232–41
53. Privalov, P. L., Khechinashvili, N. N. 1974. *J. Mol. Biol.* 86:665–84
54. Privalov, P. L. 1979. *Adv. Protein Chem.* 33:167–241
55. Suurkuusk, J., Alvarez, J., Freire, E., Biltonen, R. 1977. *Biopolymers* 16:2641–52
56. Filimonov, V. V., Privalov, P. L. 1978. *J. Mol. Biol.* 122:465–70
57. Breslauer, K. J., Sturtevant, J. M. 1977. *Biophys. Chem.* 7:205–9
58. Dewey, T. G., Turner, D. H. 1979. *Biochemistry* 18:5757–62
59. Privalov, P. L., Ptitsyn, O. B., Birshtein, T. M. 1969. *Biopolymers* 8:559–71
60. Shiao, D. D. F., Sturtevant, J. M. 1973. *Biopolymers* 12:1829–36
61. Record, M. T. Jr., Lohman, T. M., deHaseth, P. L. 1976. *J. Mol. Biol.* 107:145–58
62. Gross, L. M., Strauss, U. P. 1966. In *Chemical Physics of Ionic Solutions,* ed. B. E., Conway, R. G. Baradas. pp. 361–89. New York: Wiley
63. Krakauer, H., Sturtevant, J. M. 1968. *Biopolymers* 6:491–512
64. Klump, H., Ackermann, T. 1971. *Biopolymers* 10:513–22
65. Marmur, J., Doty, P. 1962. *J. Mol. Biol.* 5:109–31
66. Owen, R. J., Hill, L. R., LaPage, S. P. 1969. *Biopolymers* 7:503–16
67. Gruenwedel, D. W., Han, C. H. 1969. *Biopolymers* 7:557–70
68. Frank-Kamenetskii, M. D. 1971. *Biopolymers* 10:2623–24
69. Blake, R. D., Haydock, P. V. 1979. *Biopolymers* 18:3089–3109
70. Wang, J. C., Davidson, N. 1966. *J. Mol. Biol.* 15:111–23
71. Wang, J. C., Davidson, N. 1966. *J. Mol. Biol.* 19:469–82
72. Wang, J. C., Davidson, N. 1968. *Cold Spring Harbor Symp. Quant. Biol.* 33:409–15
73. Record, M. T. Jr., Lohman, T. M. 1978. *Biopolymers* 17:159–66
74. Mertz, J. E., Davis, R. W. 1972. *Proc. Natl. Acad. Sci. USA* 69:3370–74
75. Dugaiczyk, A., Boyer, H. W., Goodman, H. M. 1975. *J. Mol. Biol.* 96:174–84
75a. Olson, W. K. 1979. *Biopolymers* 18:1213–33
76. Inners, L. D., Felsenfeld, G. 1970. *J. Mol. Biol.* 50:373–89
77. Yamakawa, H. 1971. *Modern Theory of Polymer Solutions.* New York: Harper & Row. 419 pp.
78. Godfrey, J. E., Eisenberg, H. 1976. *Biophys. Chem.* 5:301–18

79. Jolly, D., Eisenberg, H. 1976. *Biopolymers* 15:61–95
80. Voordouw, G., Kam, Z., Borochov, N., Eisenberg, H. 1978. *Biophys. Chem.* 8:171–89
81. Yamakawa, H., Fujii, M. 1974. *Macromolecules* 7:128–35
82. Yamakawa, H., Fujii, M. 1974. *Macromolecules* 7:649–54
83. Kovacic, R. T., van Holde, K. E. 1977. *Biochemistry* 16:1490–98
84. Record, M. T. Jr., Woodbury, C. P., Inman, R. B. 1975. *Biopolymers* 14:393–408
85. Sharp, P., Bloomfield, V. A. 1968. *Biopolymers* 6:1201–11
86. Borochov, N., Eisenberg, H., Kam, Z. 1981. *Biopolymers* In press
87. Hearst, J. E., Schmid, C. W., Rinehart, F. P. 1968. *Macromolecules* 1:491–94
87a. Harpst, J. A. 1980. *Biophys. Chem.* 11:295–302
88. Hays, J. B., Magar, M. E., Zimm, B. H. 1969. *Biopolymers* 8:531–36
89. Sharp, P., Bloomfield, V. A. 1968. *J. Chem. Phys.* 48:2149–55
90. Schmid, C. W., Rinehart, F. P., Hearst, J. E. 1971. *Biopolymers* 10:883–93
91. Harrington, R. E. 1978. *Biopolymers* 17:919–36
92. Crothers, D. M., Zimm, B. H. 1965. *J. Mol. Biol.* 12:525–36
93. Frontali, C., Dore, E., Ferrauto, A., Gratton, E., Bettini, A., Pozzan, M. R., Valdevit, E. 1979. *Biopolymers* 18:1353–73
93a. Hagerman, P. J. 1981. *Biopolymers.* In press
94. Skolnick, J., Fixman, M. 1977. *Macromolecules* 10:944–48
95. Odijk, T. 1977. *J. Polymer Sci. Polymer Phys. Ed.* 15:477–83
96. Odijk, T., Houwaart, A. C. 1978. *J. Polymer Sci. Polymer Phys. Ed.* 16:627–39
97. Odijk, T. 1979. *Biopolymers* 18:3111–13
98. Landau, L., Lifshitz, E. 1958. *Statistical Physics*, pp. 478–82. London: Pergamon
99. Schellman, J. A. 1974. *Biopolymers* 13:217–26
100. Gray, H. B. Jr., Hearst, J. E. 1968. *J. Mol. Biol.* 35:111–29
101. Harrington, R. E. 1977. *Nucleic Acids Res.* 4:3519–35
102. Barkley, M. D., Zimm, B. H. 1979. *J. Chem. Phys.* 70:2991–3007
103. Allison, S. A., Schurr, J. M. 1979. *Chem. Phys.* 41:35–59
103a. Thomas, J. C., Allison, S. A., Appel-

lof, C. J., Schurr, J. M. 1980. *Biophys. Chem.* 12:177–88
104. Millar, D. P., Robbins, R. J., Zewail, A. H. 1980. *Proc. Natl. Acad. Sci. USA* 77:5593–97
105. Robinson, B. H., Lerman, L. S., Beth, A. H., Frisch, H. L., Dalton, L. R., Aver, C. 1980. *J. Mol. Biol.* 139:19–44
106. Vologodskii, A. V., Anshelevich, V. V., Lukashin, A. V., Frank-Kamenetskii, M. D. 1979. *Nature* 280:294–98
106a. LeBret, M. 1980. *Biopolymers* 19:619–37
107. Wahl, Ph., Paoletti, J., LePecq, J. B. 1970. *Proc. Natl. Acad. Sci. USA* 65:417–21
108. Bauer, W., Vinograd, J. 1974. In *Basic Principles in Nucleic Acid Chemistry*, ed. P. O. P. Ts'o, 2:262–303. New York: Academic
109. Crick, F. H. C. 1976. *Proc. Natl. Acad. Sci. USA* 73:2639–43
110. Fuller, F. B. 1978. *Proc. Natl. Acad. Sci. USA* 75:3557–61
111. Wang, J. C. 1969. *J. Mol. Biol.* 43:25–39
112. Upholt, W. B., Gray, H. B. Jr., Vinograd, J. 1971. *J. Mol. Biol.* 62:21–38
113. Ostrander, D. A., Gray, H. B. Jr. 1973. *Biopolymers* 12:1387–1419
114. Campbell, A. M., Jolly, D. J. 1973. *Biochem J.* 133:209–66
115. Campbell, A. M., Eason, R. 1975. *FEBS Lett.* 55:212
116. Campbell, A. M. 1976. *Biochem. J.* 155:101–5
117. Lau, P. P., Gray, H. B. Jr. 1979. *Nucleic Acids Res.* 6:331–57
118. Bram, S. 1971. *J. Mol. Biol.* 58:277–88
119. Stettler, U. H., Weber, H., Koller, T., Weissman, C. 1979. *J. Mol. Biol.* 131:21–40
120. Maestre, M. F., Wang, J. C. 1971. *Biopolymers* 10:1021–30
121. Campbell, A. M., Lochhead, D. S. 1971. *Biochem. J.* 123:661–63
122. Belintsev, B. N., Gagua, A. V., Nedospasov, S. A. 1979. *Nucleic Acids Res.* 6:983–92
123. Bauer, W., Vinograd, J. 1970. *J. Mol. Biol.* 47:419–35
124. Pulleyblank, D. E., Shure, M., Tang, D., Vinograd, J., Vosberg, H. P. 1975. *Proc. Natl. Acad. Sci. USA* 72:4280–84
125. Shure, M., Pulleyblank, D. E., Vinograd, J. 1977. *Nucleic Acids Res.* 4:1183–1205
126. Hsieh, T. S., Wang, J. C. 1975. *Biochemistry* 14:527–35
127. Fuller, F. B. 1971. *Proc. Natl. Acad. Sci. USA* 68:815–19

128. Benham, C. J. 1977. *Proc. Natl. Acad. Sci. USA* 74:2397–41
129. Benham, C. J. 1979. *Biopolymers* 18:609–23
130. Camerini-Otero, R. D., Felsenfeld, G. 1978. *Proc. Natl. Acad. Sci. USA* 75:1708–12
131. Hale, P., Woodward, R. S., Lebowitz, J. 1980. *Nature* 284:640–44
131a. Panayotatos, N., Wells, R. D. 1981. *Nature* 289:466–70
132. Holloman, W. K., Wiegand, R., Hoessli, C., Radding, C. M. 1975. *Proc. Natl. Acad. Sci. USA* 72:2394–98
133. Beattie, K. L., Wiegand, R. C., Radding, C. M. 1977. *J. Mol. Biol.* 116:783–803
134. Wiegand, R. C., Beattie, K. L., Holloman, W. K., Radding, C. M. 1977. *J. Mol. Biol.* 116:805–24
135. Laiken, N. 1973. *Biopolymers* 12:11–26
136. Benham, C. J. 1979. *Proc. Natl. Acad. Sci. USA* 76:3870–74
137. Benham, C. J. 1980. *J. Chem. Phys.* 77:3633–39
138. Anshelevich, V. V., Vologodskii, A. V., Lukashin, A. V., Frank-Kamenetskii, M. D. 1979. *Biopolymers* 18:2733–44
139. Vologodskii, A. V., Lukashin, A. V., Anshelevich, V. V., Frank-Kamenetskii, M. D. 1979. *Nucleic Acids Res.* 6:967–82
140. Helene, C. 1981. *Crit. Rev. Biochem.* In press
141. Wilson, W. D., Jones, R. L. 1981. In *Intercalation Chemistry,* ed. M. S. Whittingham, A. J. Jacobson. New York: Academic: In press
142. Barkley, M. D., Bourgeois, S. 1980. In *The Operon,* ed. W. S. Reznikoff, J. H. Miller, pp. 171–220. Cold Spring Harbor, New York: Cold Spring Harbor Lab.
143. Chamberlin, M. J. 1976. In *RNA Polymerase,* ed. R. Losick, M. Chamberlin. pp. 159–92. Cold Spring Harbor, New York: Cold Spring Harbor Lab.
144. Jovin, T. M. 1976. *Ann. Rev. Biochem.* 45:889–920
145. Record, M. T. Jr., deHaseth, P. L., Lohman, T. M. 1977. *Biochemistry* 16:4791–95
146. Strauss, H. S., Burgess, R. R., Record, M. T. Jr. 1980. *Biochemistry* 19:3504–15
147. Barkley, M. D., Lewis, P. A., Sullivan, G. E. 1981. *Biochemistry.* In press
148. Revzin, A., von Hippel, P. H. 1977. *Biochemistry* 16:4769–76
149. deHaseth, P. L., Lohman, T. M., Record, M. T. Jr. 1977. *Biochemistry* 16:4783–90
150. deHaseth, P. L., Lohman, T. M., Burgess, R. R., Record, M. T. Jr. 1978. *Biochemistry* 17:1612–22
151. Lohman, T. M., Wensley, C. G., Cina, J., Burgess, R. R., Record, M. T. Jr. 1980. *Biochemistry* 18:3516–22
152. Revzin, A., Woychik, R. P. 1981. *Biochemistry* 30:251–55
153. McGhee, J. D., von Hippel, P. H. 1974. *J. Mol. Biol.* 86:469–89
154. Zingsheim, H. P., Geisler, N., Weber, K., Mayer, F. 1977. *J. Mol. Biol.* 115:565–70
155. Hirsh, J., Schleif, R. 1976. *J. Mol. Biol.* 108:471–90
156. Goeddel, D. V., Yansura, D. G., Caruthers, M. H. 1978. *Proc. Natl. Acad. Sci. USA* 75:3578–82
157. Manning, G. S. 1969. *J. Chem. Phys.* 51:924–33
158. Anderson, C. F., Record, M. T. Jr. 1980. *Biophys. Chem.* 11:353–60
159. Wilson, R. W., Rau, D. C., Bloomfield, V. A. 1980. *Biophys. J.* 30:317–26
160. Lohman, T. M., deHaseth, P. L., Record, M. T. Jr. 1980. *Biochemistry* 19:3522–30
161a. Howe-Grant, M., Lippard, S. J. 1979. *Biochemistry* 18:5762–69
161b. Becker, M. M., Dervan, P. B. 1979. *J. Am. Chem. Soc.* 101:3664–66
162. Saucier, J. M. 1977. *Biochemistry* 16:5879–89
163. Capelle, N., Barbet, J., Dessen, P., Blanquet, S., Roques, B. P., LePecq, J. B. 1979. *Biochemistry* 18:3354–62
164. Wilson, W. D., Lopp, I. G. 1979. *Biopolymers* 18:3025–41
165. O'Gorman, R. B., Dunaway, M., Matthews, K. S. 1980. *J. Biol. Chem.* 255:10100–10106
166. Riggs, A. D., Suzuki, H., Bourgeois, S. 1970. *J. Mol. Biol.* 48:67–83
167. Lin, S.-Y., Riggs, A. D. 1975. *Cell* 4:107–11
168. Goeddel, D. V., Yansura, D. G., Winston, C., Caruthers, M. H. 1978. *J. Mol. Biol.* 123:661–87
169. Lohman, T. M., deHaseth, P. L., Record, M. T. Jr. 1978. *Biophys. Chem.* 8:281–94
170. Belintsev, B. N., Zauriev, S. K., Shemyakin, M. F. 1980. *Nucleic Acids Res.* 8:1391–404
171. Lerman, L. 1971. *Proc. Natl. Acad. Sci. USA* 68:1886–90
172. Evdokimov, Y. M., Platonov, A. L., Tikhonenko, A. S., Varshavsky, Ya. M. 1972. *FEBS Lett.* 23:180–84
173. Gosule, L. C., Schellman, J. A. 1976. *Nature* 259:333–35

174. Gosule, L. C., Schellman, J. A. 1978. *J. Mol. Biol.* 121:311–26
175. Widom, J., Baldwin, R. L. 1980. *J. Mol. Biol.* 144:431–53
176. Olins, D. E., Olins, A. L. 1971. *J. Mol. Biol.* 57:437–55
177. Haynes, M., Garrett, R. A., Gratzer, W. B. 1970. *Biochemistry* 9:4410–16
178. Chattoraj, D. K., Gosule, L. C., Schellman, J. A. 1978. *J. Mol. Biol.* 121:327–37
179. Laemmli, U. K. 1975. *Proc. Natl. Acad. Sci. USA* 72:4288–92
180. Jordan, C. F., Lerman, L. S., Venable, J. H. Jr. 1972. *Nature New Biol.* 236:67–70
181. Maniatis, T., Venable, J. H. Jr., Lerman, L. S. 1974. *J. Mol. Biol.* 84:37–64
182. Wilson, R. W., Bloomfield, V. A. 1979. *Biochemistry* 18:2192–96
183. Riemer, S. C., Bloomfield, V. A. 1978. *Biopolymers* 17:785–94
184. Post, C. B., Zimm, B. 1979. *Biopolymers* 18:1487–501
185. Dore, E., Frontali, C., Gratton, E. 1972. *Biopolymers* 11:443–59
186. Bloomfield, V. A., Wilson, R. W., Rau, D. C. 1980. *Biophys. Chem.* 11:1339–43
187. Oosawa, F. 1971. *Polyelectrolytes.* New York: Dekker

Ann. Rev. Biochem. 1984. 53:791–846

11

THE CHEMISTRY AND BIOLOGY OF LEFT-HANDED Z-DNA

Alexander Rich, Alfred Nordheim, and Andrew H.-J. Wang

Department of Biology, Massachusetts Institute of Technology, Cambridge, Massachusetts 02139

CONTENTS

341

PERSPECTIVE AND SUMMARY

One of the significant developments in our understanding of the biochemistry of DNA in recent years is our awareness that the double helix has considerable conformational flexibility. DNA is no longer looked upon as a static molecule but rather a dynamic structure in which different conformations are in equilibrium with each other. The concept of structural flexibility in DNA was sharply illustrated by the discovery of left-handed Z-DNA. We have considerable information about the structural details of this particular DNA conformation as derived from high-resolution X-ray diffraction studies of single crystals containing double-helical DNA fragments.

Both right-handed B-DNA and left-handed Z-DNA are double-helical conformations with antiparallel chains that are held together by Watson-Crick hydrogen bonding between the bases. All of the nucleotides along B-DNA have the same conformation. However, nucleotides along the left-handed double helix alternate in *syn* and *anti* conformations of the bases. Since the *syn* conformation is more stable for purines than for pyrimidines, Z-DNA is favored in nucleotide sequences that have alternations of purines and pyrimidines. Not all sequences are equally favorable. Segments of DNA with alternating d(CG) sequences are the most favored for forming Z-DNA. The next most effective sequences are d(CA) or d(TG). Long sequences of d(CA/TG)$_n$ represent a class of middle repetitive elements in eukaryotic genomes. These sequences form Z-DNA readily and their widespread distribution may indicate a significant biological function. Finally, d(AT) sequences favor Z-DNA least of all. We know that Z-DNA can also form in sequences in which some base pairs are found out of purine-pyrimidine alternation.

An equilibrium exists between B-DNA and Z-DNA. In physiological solutions, Z-DNA is less stable than B-DNA largely owing to electrostatic repulsion between the negatively charged phosphates on opposite strands that are closer together in Z-DNA than in B-DNA. Z-DNA may be present only in low concentrations in solution. However, Z-DNA can be stabilized most effectively by negative supercoiling of DNA. In vivo DNA appears negatively supercoiled or unwound, and the free energy of supercoiling represents a powerful force for stabilization of Z-DNA.

The molecular architecture of the Z-DNA double helix is considerably different from that of B-DNA. This leads to a different reactivity with other molecules, and there is a class of proteins found in eukaryotes that bind to Z-DNA but not to B-DNA. Z-DNA binding proteins appear to be of many types and they have yet to be fully characterized in terms of the sequence specificity of their binding and their specific roles in the organization and activity of chromatin. Unlike B-DNA, Z-DNA is very immunogenic and this has led to the production of highly specific polyclonal and monoclonal antibodies. These antibodies have been very useful in identifying Z-DNA in many systems.

Z-DNA has attracted the interest of a wide number of biochemists and physical chemists. Many studies have been carried out describing its altered reactivity, including the manner in which it reacts with carcinogens. A number of chemical and environmental modifications have been shown to influence the B-Z equilibrium in solution.

Only recently have we started to accumulate information about the biological roles of Z-DNA. At present, the data suggests that Z-DNA may play a role in the transcriptional enhancer of the DNA tumor virus SV40, acting as a positive regulatory transcription signal. Work in other systems indicates that Z-DNA may also act as a negative regulatory signal in decreasing the level of transcription. As is the case with B-DNA, it is likely that positive or negative regulatory roles of Z-DNA depend upon the location and sequence composition of the left-handed segments. It is likely that most of these activities will be found to operate through specific interactions with Z-DNA binding proteins.

In eukaryotes, methylation of cytosine C5 in d(CG) sequences in vivo is associated with gene inactivation. Methylation of d(CG) sequences also strongly favors Z-DNA formation. Whether these two phenomena are related has yet to be discovered.

In its nucleotide sequence, DNA contains coding information that directs the synthesis of RNA. DNA also contains conformational information in that certain nucleotide sequences can adopt altered double-helical structures. The conformational information in DNA is expressed by its ability to facilitate the formation of altered conformations that may be used in biological systems in various ways. The formation of Z-DNA is a striking case of conformational information in DNA but it is not likely to be the sole example.

INTRODUCTION

For the past 30 years, scientists have been working with the biochemical and genetic consequences of the double-helical structure of DNA formulated by Watson & Crick in 1953 (1). Their model stimulated a large number of investigations into the chemistry and biology of DNA and this knowledge is now at the core of biochemistry. Although there was some concern about the handedness of DNA at the time of its formulation, great interest in the possibility of alternative conformations for DNA has developed only recently. There were some early attempts to build left-handed double-helical structures, but this was not an extensive activity. The left-handed Z-DNA structure was not discovered by thinking through the possibilities of left-handed double-helical models; rather, it emerged somewhat unexpectedly on solving the structure of a crystalline fragment of double-helical DNA (2). The earlier structural work on DNA involved mostly X-ray diffraction studies of fibers. These have only limited resolution and are generally disordered to varying extents. Only in the past few years has it been possible to study DNA molecular structure by single crystal X-ray analysis. This development is an outgrowth of the advances in organic synthetic nucleotide chemistry that can produce oligonucleotides of defined sequence in quantities sufficient for crystallization experiments. Single crystals diffract X rays often at or near atomic resolution (~ 1 Å). Solution of these crystal structures provides a wealth of detail in contrast to the results of DNA fiber X-ray diffraction analyses. The first visualization of a double helix at atomic resolution was seen in this laboratory in 1973 with dinucleoside monophosphates showing two right-handed base-paired RNA fragments (3, 4). The structure of left-handed Z-DNA was solved here in 1979 as a double helix containing a crystalline hexanucleoside pentaphosphate with the sequence d(CpGpCpGpCpG) (2). The crystal diffracted to 0.9 Å and its solution produced a large amount of experimental detail. The structure explained a polymorphism that had been observed in solution earlier with the synthetic DNA polymer, poly(dG-dC) (5, 6).

This left-handed structure posed an interesting challenge to biochemists. Was there biological relevance to this alternative conformation? In a sense, the investigations stemming from these initial studies are the inverse of biochemical studies generally. Usually factors or substances are isolated and we learn something about their chemistry and biology. Only at a later stage is the three-dimensional structure available, which serves to pull together aspects of the chemistry and the biology. In the present case, a DNA structure was available. It presented a challenge to develop aspects of its chemistry and its relevance, if any, to biological systems.

In this review, we survey the efforts that have been made on this problem in the ensuing four years. We review the salient structural features of Z-DNA and compare them to the more familiar features of right-handed B-DNA.

MOLECULAR STRUCTURE OF Z-DNA

The Z-DNA structure was unanticipated because it has many conformational features that distinguish it from B-DNA. In the crystal, the double-helical hexanucleoside pentaphosphate molecules are aligned along the crystal c axis in a manner that makes it appear as if there is a continuous helix running through the crystal. Figure 1 shows a van der Waals diagram

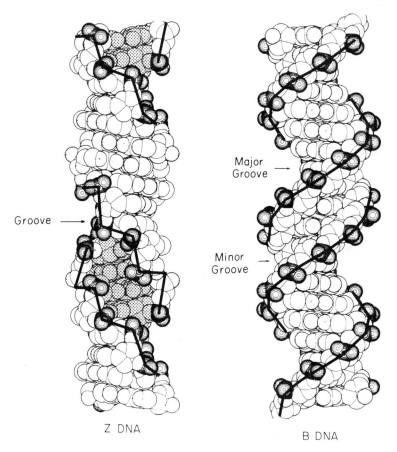

Figure 1   van der Waals models of Z-DNA and B-DNA. The irregularity of the Z-DNA backbone is illustrated by the heavy lines that go from phosphate to phosphate residue along the chain. The Z-DNA diagram shows the molecules as they appear in the hexamer crystal. The groove in Z-DNA is quite deep, extending to the axis of the double helix. In contrast, B-DNA has a smooth line connecting the phosphate groups and two grooves, neither one of which extends to the helix axis of the molecule (2).

of Z-DNA as it appeared in the crystal lattice. Three molecules are shown in the diagram and there is continuity of base-pair stacking along the helical axis. A stereo diagram of Z-DNA is shown in Figure 2. In contrast to the right-handed B-DNA, Z-DNA has one deep helical groove that is formally analogous to the minor groove of B-DNA. The concave major groove of B-DNA forms the convex outer surface of Z-DNA. In Z-DNA the asymmetric

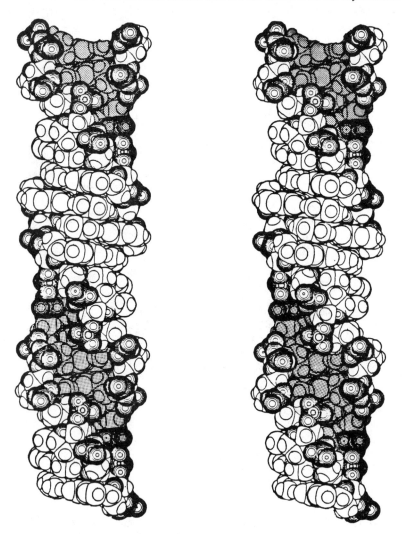

Figure 2   Stereodiagram of Z-DNA. Three dimensionality can be seen either with stereo viewers or simply by looking at the diagram and relaxing the eye muscles until the two figures merge. The zig-zag array of phosphate groups shows up clearly (14).

unit is a dinucleotide compared to the mononucleotide found in B-DNA. A heavy line is drawn in Figure 1 between adjacent phosphate groups. The zig-zag organization of the phosphates is a direct consequence of the two different conformations found in the nucleotides of Z-DNA. Figure 3 compares the conformation of deoxyguanosine in Z-DNA and B-DNA. In B-DNA, all of the nucleotides have the *anti* conformation and a C2' *endo* pucker of the deoxyribose ring. In the Z-DNA crystal containing alternating cytosine and guanine residues, the deoxycytidines all have the *anti* conformation but the deoxyguanosines all have the *syn* conformation. Figure 4 is a schematic diagram that shows the conformation of nucleotide hexamers in the B and Z forms. The dinucleotide repeat in Z-DNA is seen in the alternations of *anti* and *syn* conformation. A Z-DNA crystal structure with this hexamer sequence d(CpGpTpApCpG) has been solved by Wang et al (7) to 1.2 Å resolution.

In B-DNA there are typically 10.5 bp (base pairs) per helical turn, a helical pitch of 34 Å and a diameter of near 20 Å. Z-DNA has 12 bp per helical turn with a pitch of 44.6 Å and a diameter of approximately 18 Å. Thus Z-DNA is somewhat slimmer than B-DNA and has a larger number of base pairs per helical turn.

Both Z-DNA and B-DNA have two antiparallel polynucleotide chains

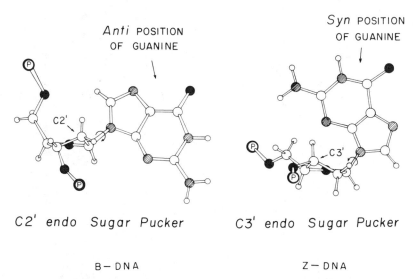

Figure 3 Conformation of deoxyguanosine in B-DNA and Z-DNA. The sugar is oriented so that the plane defined by C1'-O1'-C4' is horizontal. Atoms lying above this plane are in the *endo* conformation. In Z-DNA the C3' is *endo* while in B-DNA the C2' is *endo*. In addition, Z-DNA has guanine in the *syn* position, in contrast to the *anti* position in B-DNA. A curved arrow around the glycosyl carbon-nitrogen linkage indicates the site of rotation.

held together by Watson-Crick base pairs. However, the base pairs have a different relationship to the sugar phosphate chains, as shown in Figure 5. The B-DNA molecule is drawn in the familiar ladder representation in which the planar bases are shown as flat plates. In converting a section of B-DNA to Z-DNA, the base pairs must flip so that they are now upside down relative to the orientation that they had in B-DNA. This is brought about by rotating the purine residue about its glycosyl bond, from *anti* to *syn*. In the case of the pyrimidines, both the base as well as the sugar rotate about. It is this rotation of the sugar that produces the zig-zag backbone conformation of the Z-DNA.

Figure 6 illustrates the stacking of successive base pairs in Z-DNA and B-DNA as viewed down the helix axis, which is marked by a solid dot. The base pair drawn with the heavier lines are closer to the reader. The stacking of base pairs in Z-DNA is quite different in the sequence d(CpG) and in d(GpC), while in B-DNA the stacking is not very different. Base pairs in the d(CpG) sequence in Z-DNA are considerably sheared so that the cytosine residues on opposite strands are stacking one over the other in the center of the molecule; however, the guanine residues are no longer stacking on

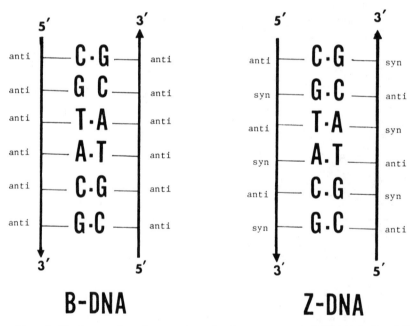

Figure 4   The base conformations are shown for the same segment of DNA in the B and Z conformations. Alternating residues in Z-DNA adopt *syn* and *anti* conformations. The three-dimensional structure of a hexamer with this sequence has been found to adopt the Z-DNA conformation (7).

bases but stack upon the O1′ oxygen atoms of the sugar residue below. For the d(CpG) sequence there is only a small rotation (−9°) between successive base pairs, while the sequence d(GpC) has a much larger rotation (−51°). The sugar-phosphate chains are drawn in the same orientation for all four diagrams in Figure 6. In B-DNA, the minor groove is at the top of the diagram. In Z-DNA, what corresponds to the minor groove is now at the bottom of the figure close to the helical axis. It is the difference in the position of the helical axis that accounts for the major change in the organization of the molecule. The helical axis passes through the center of the base pairs in B-DNA, producing a molecule with two grooves. In contrast, the helix axis in Z-DNA falls outside the base pairs, producing a molecule in which there is only one deep groove. Furthermore, there is a considerable change in the relationship of the bases to the sugar-phosphate backbone in comparing the two molecules. The imidazole ring of guanine is found prominently on the outer part of the Z-DNA molecule with con-

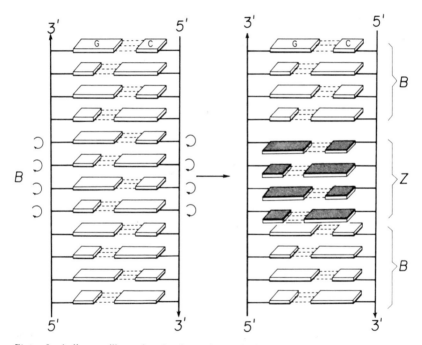

Figure 5   A diagram illustrating the change in topological relationship if a four–base-pair segment of B-DNA were converted into Z-DNA. The conversion can be accomplished by rotation or flipping of the base pairs as indicated. The turning is indicated by the curved arrows. Rotation of the guanine residues about the glycosylic bond yields deoxyguanosine in the *syn* conformation, while for deoxycytidine residues, both cytosine and deoxyribose are rotated (2).

siderable exposure of guanine N7 and C8. These atoms are shielded in B-DNA. An arrow in Figure 6 points to the C8 residues of guanine in both conformations. In Z-DNA, this atom is at the periphery of the molecule while in B-DNA the hydrogen atom attached to guanine C8 is in van der Waal's contact with the sugar-phosphate chain on the outside of the molecule. This difference in accessibility explains some of the considerable differences in the chemical reactivity of the two molecules.

Figure 7 compares end views of Z-DNA and B-DNA in which a guanine-cytosine base pair has been emphasized by shading. In B-DNA, the base pair sits in the center of the molecule surrounded by the symmetrically disposed sugar-phosphate chains. In Z-DNA, the base pair is moved away from the center so that the guanine imidazole ring is found near the periphery. Both of these diagrams are idealized representations that assume that the molecules are completely regular (8). This degree of regularity does

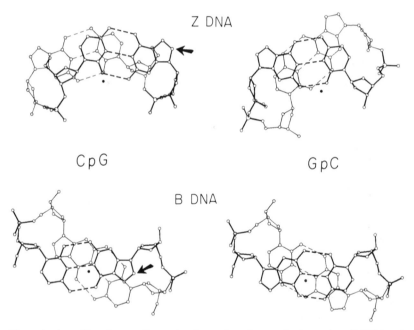

Figure 6   A stacking diagram illustrating the overlap of successive bases along Z-DNA and B-DNA helix. The base pair drawn with heavier lines is stacked above the pair drawn with lighter lines. The left-hand column represents d(CpG) sequences of both Z-DNA and B-DNA, while d(GpC) sequences are on the right. The direction of the deoxyribose-phosphate chains is the same in all these diagrams. Note that the minor groove in B-DNA is found at the top of the B-DNA diagrams while the analogous side of the base pairs is found at the bottom of the Z-DNA diagrams. The solid black dot indicates the helical axis and the arrow pointing to guanine C8 shows their different environment in B and Z-DNA (2).

not exist in real molecules nor in real crystals; they are averages that are generated for reasons of convenience. For B-DNA our perception of its regularity was more apparent than real, since the initial structures were based on fiber diffraction studies in which it was assumed that the molecule was regular. In contrast, crystallized fragments of B-DNA do not show this type of regularity as shown by Dickerson and colleagues (9). Likewise, all of the crystals of Z-DNA show a variety of slightly different conformations (2, 8, 10–15). In the case of Z-DNA at least three major conformations have been seen, two of which are associated with changes in the orientation of the phosphate groups (8). The conformation found most frequently (called Z_I) is illustrated in Figure 7. The conformation of the base pair is modified slightly when negatively charged chloride ions are found close to the base pairs (10).

A van der Waals diagram of an end view of Z-DNA is shown in Figure 8. The depth of the groove can be seen in this diagram, which has three base pairs. The Z-DNA structure was the first in which the *syn* conformation was used systematically in the formation of a polynucleotide structure. For many years it has been known that both purines and pyrimidines can rotate about their glycosyl bonds and these two conformations have been seen in a variety of crystallographic as well as solution studies. One of the earliest deoxynucleoside complexes studied was deoxycytidine hydrogen-bonded

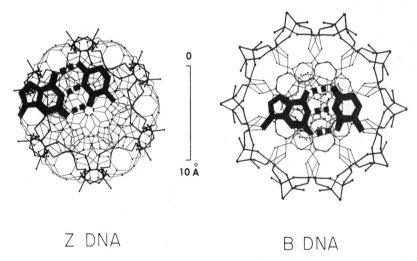

0

10 Å

Z DNA   B DNA

Figure 7   End views of regular, idealized helical skeletal diagrams of Z-DNA and B-DNA. Heavier lines are used for the phosphate-sugar backbone. A guanine-cytosine base pair is shown by shading and the difference in the positions of the base pairs is quite striking; they are near the center of B-DNA but at the periphery of Z-DNA (8).

to deoxyguanosine, in which the deoxyguanosine residues were in the *syn* conformation (16). An early theoretical study by Haschemeyer & Rich (17) of the relative stability of purines and pyrimidines in *syn* and *anti* conformations suggested that although purines could form the *syn* conformation without loss of energy, there was a small amount of steric hindrance to the formation of pyrimidine residues in *syn* conformations. These theoretical studies were reinforced by experimental studies in solution that generally indicate that purine residues can form *syn* conformations relatively easily but are less common for pyrimidines (18). In Z-DNA, the organization of the base pairs is such that every other residue along the chain is in the *syn* conformation. This implies that Z-DNA is more likely to be found in sequences that have alternations of purines and pyrimidines, so

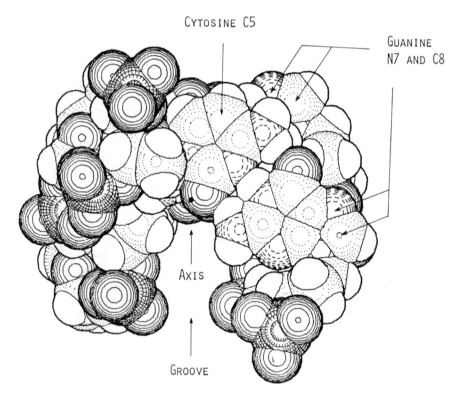

Figure 8   End view of Z-DNA in which three base pairs are shown in a van der Waals diagram. The groove extends almost down to the helical axis, which passes near the cytosine O2 atom. As the base pairs come toward the reader, they rotate in a clockwise direction, thereby exposing three phosphate groups on the left. Oxygen atoms are shaded with circles, phosphorus atoms have spiked circles, and nitrogen atoms are indicated by dashed circles. The position of the methylation site, cytosine C5, is shown as well as the guanine imidazole ring N7 and C8 atoms (8).

that the purines can exist in the *syn* conformations. It should be emphasized that although there is some energy loss due to close van der Waals crowding when pyrimidines are in the *syn* conformation, this energy loss is not very large. Deviations from strict alternations of purines and pyrimidines have been found in segments of Z-DNA (19) and have also been observed in crystalline Z-DNA structures (20).

The Z-DNA conformation has been visualized in crystal structures containing four to eight base pairs and the molecules have crystallized in orthorhombic and hexagonal lattices (2, 7, 8, 10–15, 20). The sequences have involved guanine-cytosine as well as adenine-thymine base pairs (7). Some crystal structures have modifications such as cytosine residues with methyl groups on the 5 position (14). A self-complementary Z-DNA structure with the sequence d(CpGpTpApCpG) (Figure 4) has been solved in which the cytosine residues have either methyl or bromine atoms attached to the C5 position (7). The geometry of the AT base pairs in Z-DNA is similar to that of the CG base pairs with the adenine residues in the *syn* conformation and stacking interactions similar to those seen in Figure 6. A significant difference, however, is that the water molecules in the helical groove of Z-DNA are disordered near the AT base pairs in contrast to the high level of order found in the solvent of the segments containing CG base pairs. The ordering is largely due to the presence of the amino group on the 2 position of guanine, which hydrogen bonds to a water molecule and helps to organize others in the groove. Its absence in the AT base pair is undoubtedly associated with the solvent disordering which may contribute to the observation that AT base pairs form Z-DNA less readily than CG.

Z-DNA crystal structures are no longer confined to alternations of purines and pyrimidines. A Z-DNA structure has been solved recently with the sequence d(CGATCG) in which the cytosine residues are brominated on C5 (20). The crystal is stabilized in the Z conformation through the use of a $Co(NH_3)_6^{3+}$ cation. Fiber diffraction studies have been carried out by Arnott and others (21–24) on poly(dG-dC) which revealed it could form the Z-DNA conformation; these studies suggested that B- to Z-DNA conversions could take place in the fiber itself. The conversion depends upon relative humidity (213) and solvent composition (214). The diffraction pattern produced by fibers of poly(dG-dC) in the Z conformation has been quantitatively interpreted using data from the crystal structure analysis (8). Several articles have reviewed aspects of Z-DNA structure (13, 25–35).

METHODS FOR DETECTING Z-DNA

In 1971 Pohl (5) and in 1972 Pohl & Jovin (6) reported that poly(dG-dC) underwent a cooperative conformational change when the concentration of salt in the medium was raised. In 4 M NaCl, the circular dichroism of the solution was nearly inverted compared to the spectrum in 0.1 M NaCl. Pohl and colleagues (36, 37) carried out a number of additional studies of this conformational change, which showed that the high-salt form of poly(dG-dC) could be induced by a variety of other changes including added alcohol. The interpretation of this as a conformational change was reinforced by the observation that the Raman spectra of the high and low salt forms of poly(dG-dC) differed considerably (38). Raman spectra measure molecular vibrations and many of them are conformation dependent. Raman spectra can be obtained both in solution and in the solid state. Accordingly, an analysis of the Z-DNA crystals by Thamann et al (39) indicated that the crystals had a Raman spectrum that was virtually identical to that of the high-salt form of poly(dG-dC) and quite different from that of the low-salt form. From this it was inferred that the low-salt form represented right-handed B-DNA while the high-salt form was Z-DNA. In solution, there is an equilibrium between right-handed and left-handed forms of DNA. The actual distribution between these two states is strongly influenced by environmental conditions as well as by the sequence of nucleotides in the double helix. Several investigations have supported the concept of a dynamic equilibrium between these two states.

A variety of methods have been developed for detecting and measuring Z-DNA and differentiating it from B-DNA. They are listed here in two categories.

Physical-Chemical Methods

Table 1 lists physical and chemical techniques for identifying Z-DNA, while the information obtained from the use of these techniques is discussed in a

Table 1   Physical and chemical methods for detecting Z-DNA

Method	Comments	References
1) Circular dichroism (UV and vacuum UV)	Near inversion of CD spectrum between B- and Z-DNA; especially useful in polymer studies.	(6, 40, 41, 42–44)
2) X-ray crystallography (single crystal)	Defines the conformation; can be at atomic resolution.	(2, 7, 8, 10–12, 14, 20)
3) X-ray crystallography (fiber)	Largely useful after the conformation is defined as by single crystal analysis; usually does not yield enough data to solve unknown structure.	(21, 22, 34, 213, 214)
4) UV absorbance changes	Small but easily measured, especially with polymers; A_{260}/A_{295} ratio is useful.	

Table 1   (*continued*)

Method	Comments	References
5) Raman spectroscopy	Z-DNA specific vibrational modes identified; useful for both solution and solid-state studies.	(38, 39, 45, 46, 47–50)
6) ^{31}P NMR	Detects differences in phosphate conformation between B- and Z-DNA.	(51–54, 55–57)
7) ^{1}H NMR	Nuclear Overhauser Effect is valuable for measuring close contacts and conformational changes.	(52, 56–59)
8) Hydrogen-deuterium exchange	Differential exchange rates are conformationally dependent; hydrogen bonding protons are released more slowly from Z-DNA.	(60, 61)
9) Gel electrophoresis (one- and two-dimensional)	Topoisomer distribution of plasmids modified by Z-DNA formation due to changes in supercoiling	(62–66, 67–70)
10) Sedimentation	Z-DNA changes supercoiling of plasmids and hence their sedimentation rates.	(65)
11) Nitrocellulose filtration	Z-DNA elution retarded by salt	(71)
12) Light scattering	Decreased flexibility seen in Z-DNA	(72, 73)
13) Nuclease sensitivity	Decreased sensitivity of Z-DNA except an increase at B-Z junction for S1 nuclease	(74–76)
14) Binding of chiral molecules	Phenanthroline ruthenium complexes are chiral and have differential binding to B- and Z-DNA.	(77, 78)
15) Anti–Z-DNA antibody binding	See Table 2.	

section below. The DNA molecule undergoes a distinct structural reorganization in converting from B-DNA to Z-DNA; it is therefore not surprising that there are many different methods that are useful for detecting these changes. Since Z-DNA was first detected in solution with poly(dG-dC) and since this polymer forms Z-DNA readily, most of the physical-chemical methods use this polymer in describing the modified properties.

Use of Specific Anti–Z-DNA Antibodies

B-DNA and Z-DNA are in equilibrium with each other. Under most physiological conditions, Z-DNA is less stable than B-DNA, especially with linear DNA fragments. Synthetic poly(dG-dC) is stable as Z-DNA in 4 M NaCl, but it was necessary to stabilize it in order to have it remain as Z-DNA in the lower ionic environment of a physiological medium. The arrows in Figure 4 point to the relative accessibility of the imidazole ring of guanine especially C8 in Z-DNA compared to B-DNA. One of the first methods for stabilizing Z-DNA in a low-salt solution was through chemical bromination (75, 79). Poly(dG-dC) was placed in a 4 M salt solution where it assumed the Z-DNA conformation and Br_2 was added. The bromine reacted largely with the C8 position of guanine and to a lesser extent with the C5 position of cytosine. The bromine atom at C8 sterically prevents guanine from adopting the *anti* conformation (80, 81). When the salt is dialyzed away from brominated poly(dG-dC), it remains as Z-DNA. Only one third of the guanine C8H atoms are replaced by bromine and one cytosine in six has reacted to stabilize Z-DNA. Immunological experiments were then carried out on this material by Lafer et al (79) who demonstrated that Z-DNA is a strong immunogen. In contrast, B-DNA is a very poor immunogen. The organism probably becomes tolerant to B-DNA during the early stages of embryological development. It is likely that Z-DNA is not seen by the cells of the immune system during early embryogenesis. When DNA is released from cells that have broken down, nuclease cleavage probably converts any Z-DNA to B-DNA by releasing torsional strain. Leng and others (82–86) have shown that Z-DNA can also be stabilized in poly(dG-dC) by other chemical modifications including platinum complexes of guanine and bromination of cytosine C5 alone. All of these Z-DNA forms are powerful immunogens that elicit the production of a high titre of antibodies. The antibodies have a high degree of specificity for Z-DNA and do not interact with B-DNA or with any other DNA or RNA poly- or oligonucleotide. Antibodies specific for Z-DNA arise in certain autoimmune disease states and are found in both murine and human systemic lupus erythematosus (79, 87).

Monoclonal antibodies have been produced against Z-DNA by Moller et al (88) and by Pohl and associates (89, 90). They have been shown to combine with different parts of the surface of Z-DNA. Some antibodies bind to the base pairs on the surface of the molecule while others have a preference for the sugars and negatively charged phosphate groups. These differences are uncovered by studying the reactivity of monoclonal antibodies with a variety of chemically modified polynucleotides that can form Z-DNA. For example, methylation of the C5 position of cytosine in poly(dG-m^{5}dC) blocks some monoclonal antibodies raised against Z-DNA without the methyl group on the cytosine, whereas others that bind on the sugar-phosphate chains are not blocked (88).

One of the interesting properties of polyclonal and some monoclonal antibodies against Z-DNA is that they combine with Z-DNA even in the presence of high concentrations of NaCl (84, 91). This is particularly useful as it allows one to work with polymers as well as plasmids in different salt concentrations. Varying the salt concentration can be used to control the degree of Z-DNA formation and the specificity and reactivity of the antibodies can be measured.

The choice of Z-DNA modification can influence both the immunogenicity of the Z-DNA polymer and the specificity of the antibodies produced by it. The chemically brominated poly(dG-dC) produces the highest level of antibody, while the enzymatically brominated polymer poly(dG-br^{5}dC) produces significantly less antibody (79, 84, 86, 92). In addition, more of the antibodies produced by the latter polymer are needed to react with unmodified poly(dG-dC) in a high-salt solution. This may be so because the systematic presence of a bromine atom in the antigen may be recognized by a number of the polyclonal antibodies. This points out that antibody preparations should be carefully screened for their specificities before carrying out analyses with them.

A number of methods for detecting Z-DNA employ antibodies specific for this molecule as outlined in Table 2. One of the first methods used was indirect immune fluorescence on *Drosophila* polytene chromosomes (54, 91–96). In these studies an anti–Z-DNA antibody raised in the rabbit, for example, is added to a fixed cytological chromosome preparation; this is followed by the addition of a second goat antibody raised against rabbit antibodies. The second antibody has a fluorescent chromophore conjugated to it. This is visualized by illuminating the preparation at a wavelength that excites fluorescence. Photographs are then taken at the emitting wavelength so the position of the initial antibodies can be seen.

Another method for using the antibody involves the retention of DNA on nitrocellulose filters. Naked DNA molecules can pass through the pores in nitrocellulose filters while protein-DNA complexes are retained on the filters. Nordheim et al (19) used retention on nitrocellulose filters to detect the presence of Z-DNA when anti–Z-DNA antibodies are present in the solution. The process can be further modified by cross-linking the antibody to the DNA through the use of glutaraldehyde or other cross-linking agents. The DNA is then digested with restriction endonucleases and the digest is passed through a nitrocellulose filter. The filtrate contains all of the fragments that are not attached to the antibody, while those bound to the antibody are retained on the nitrocellulose. This makes it possible to identify specific sequences that are forming Z-DNA by using a variety of restriction endonucleases (19).

Antibodies bound to Z-DNA can also be visualized in the electron microscope. Antibody visualization can be carried out by shadowing or through the use of ferritin coupled to the antibody (44, 68, 105, 106). This makes it possible to identify particular regions of a genome or of plasmids that form Z-DNA by measuring the position at which antibody molecules are found. Antibodies have been covalently attached to columns by Thomae et al (90) and they are used as a method for separating plasmids that contain Z-DNA from those that do not contain Z-DNA. This method

Table 2 Methods using specific antibodies for detection of Z-DNA (both polyclonal and monoclonal antibodies are available)

Method	Comments	References
1) Indirect immune fluorescence	Useful for cytological preparations, defines regions with Z-DNA	(54, 91, 93–100)
2) Plasmid retention on nitrocellulose filters	DNA does not pass through filters if complexed to protein; useful for detecting Z-DNA in linear or negatively supercoiled molecules	(19, 101, 102)
3) Antibody cross-linking to Z-DNA	Fixes antibody on Z-DNA; restriction cleavage of DNA is useful for identifying specific Z-DNA nucleotide sequences.	(19, 102–104)
4) Immuno-electron microscopy	Visualization of antibodies attached to Z-DNA, useful for identifying this in a larger genome, ferritin labeling can be used.	(19, 31, 68, 105, 106)
5) Antibodies attached to columns for Z-DNA–plasmid isolation	Useful for screening large numbers of plasmids to isolate those containing Z-DNA segments.	(90)
6) Plasmid retardation in agarose gel electrophoresis	Useful for detecting Z-DNA segments in supercoiled plasmids	(44)
7) Competition assay	Useful for detecting Z-DNA in unknown samples by competing with binding of known Z-DNA to antibodies	(79, 84, 87, 107)

is particularly valuable for identifying the occurrence of Z-DNA in genomic libraries so that particular sequences can be isolated.

Because of their high degree of specificity, the use of antibodies specific for Z-DNA represent a valuable method for identifying Z-DNA in a variety of different systems. Such identification is often unambiguous: this is in contrast to some physical methods that rely on changes in properties that are consistent with Z-DNA but not necessarily specific.

PHYSICAL-CHEMICAL STUDIES OF Z-DNA

Ultraviolet Circular Dichroism and Absorbance

Raising the salt concentration in a solution of poly(dG-dC) produces a near-inversion of the ultraviolet (UV) circular dichroism spectrum. In a low-salt solution there is a positive band at 280 mm; it is converted to a more intense negative band with a minimum at 290 nm in 4 M NaCl (6, 40, 108, 109, 110). Similarly, there is an inversion of the negative bands at 253 to a positive band at 265 nm. This circular dichroism inversion reported by Pohl in 1971 (5) and Pohl & Jovin in 1972 (6) is among the simplest methods used to study Z-DNA formation in polymers. However, one must be cautious. Some chemical modifications introduce chiral centers, as shown by Tomasz et al (111) where reaction of poly(dG-dC) with mitomycin yields an inversion of the UV circular dichroism but without formation of Z-DNA.

A number of attempts have been made to detect changes in the circular dichroism of other synthetic polymers with alternations of purines and pyrimidines. Raising the salt concentration, and using a variety of salts, often produces dichroic changes that are highly suggestive but usually do not have the decisive character of those seen with poly(dG-dC) (112–114).

The inversion of the circular dichroism also occurs in the vacuum ultraviolet region between 180 and 230 nm (42, 43). B-DNA has a large positive peak at 187 nm while the Z-DNA form has a large negative peak at 194 nm and a positive band below 186 nm. The magnitude of the differences observed between B- and Z-DNA below 200 nm are about ten times greater than those observed between 230 and 300 nm. This sensitive region of the spectrum has also been used in demonstrating the existence of Z-DNA in Form V DNA as discussed below, as the spectrum of that molecule appears to be a near 1:1 mixture of right-handed B and left-handed Z spectra (49, 50). The vacuum UV spectrum has also been used to address the question of whether poly(dI-dC) forms a right-handed or a left-handed double helix. Poly(dI-dC) in a low-salt solution has a circular dichroism spectrum in the range 210–300 nm similar to the high-salt spectrum of poly(dG-dC) (115). There was some question, however, as to whether this corresponds to a left-handed or a right-handed helix (115). Examination of the vacuum UV CD spectrum of poly(dI-dC) in the region 180 to 220 nm shows that it is totally different from that of the Z form of poly(dG-dC) and it has been interpreted as a right-handed helix (43). The vacuum UV CD spectrum may be a more reliable index for the handedness of a helix as the absorption bands responsible for the spectrum arise largely from the backbone (43).

In a high-salt solution, poly(dG-dC) shows a decrease in absorbance at A_{260} and an increase at A_{295}, compared to the low-salt spectrum. Although the effect is small, it is a useful index for Z-DNA formation, especially in polymeric solutions (36).

Raman Spectra

Raman spectroscopy of biological molecules has become a powerful tool in recent years once it was wedded to the technology of laser excitation. It is now possible to obtain detailed laser Raman spectra from very small samples. Raman spectra measure the different types of vibrations of the components of macromolecules including bond stretching and bending. Some of these vibrations are sensitive to conformational change. To a first approximation the Raman spectrum of a molecule is independent of whether it is fixed in a crystal lattice or tumbling free in solution. Raman spectroscopy is thus a uniquely powerful tool for asking questions about the identity of a molecule in two different physical states. Poly(dG-dC) has significant differences in its Raman spectrum in high (4M) and low (0.1 M) NaCl solutions (38), and the high-salt form has a spectrum identical to that produced by the crystal (39). The crystal analysis covered the range from 400 to 1800 cm^{-1} and over 14 different bands were identified that were sensitive to conformation. Further analysis by Benevides & Thomas (47) characterized many of these vibrations. A number of them arise from changes in the vibrational frequency of the phosphate group, as might be expected since the phosphate groups have quite different conformations in Z-DNA compared to B-DNA. Other vibrations that are conformationally sensitive arise from the guanine residues and the backbone.

By comparing the Raman spectra of poly(dG-dC) with a restriction fragment containing regions with alternating CG residues, Wartell and associates (45, 46) showed that essentially all of the dC-dG regions are in the Z conformation in 4.5 M NaCl. Moreover, this conversion also perturbed the B-DNA segment to a considerable extent. This conclusion was different from that obtained by a study of circular dichroism of the same restriction fragments at low and high salt (116). The circular dichroism results suggested that not all of the dG-dC inserts converted to Z-DNA. This is perhaps a good example of the relative usefulness of the two different techniques. The Raman spectra produce bands that are quantitative in nature and can be used to measure the fraction of a molecule in one conformation or another. However, there is no adequate theory for circular dichroism and the spectra cannot be interpreted quantitatively at present.

Raman spectra have also been used to analyze the vibrations of poly(dA-br^{5}dC)·poly(dG-dT) (54). The difference spectra from this polymer in high and low salt solutions have some similarities to the difference spectra of poly(dG-dC) in high and low salt solutions, which provides physical evidence for the formation of Z-DNA in the brominated polymer.

Laser Raman spectra have been used in the analysis of form V DNA to demonstrate the existence of Z-DNA (49, 50). By comparing the Raman spectrum of form V DNA as well as the native plasmids from which the form V plasmid was obtained, significant spectral differences could be seen that were interpreted as Z-DNA.

A recent analysis has been carried out on the crystals of d(CGCATGCG) and d(m^{5}CGTAm^{5}CG) (48). In both of these crystal structures the molecules exist as Z-DNA in the crystal lattice. The Raman spectra show features associated with the vibrations of guanine and cytosine residues in the Z form, but also show vibrations attributed to Z-form adenine-thymine base pairs. The advantage of analyzing spectra from crystals is that the molecules are completely in a known conformation. In addition, the solvent environment of the vibrating groups is also known and the extent to which it perturbs or modifies the vibrations can be assessed.

Raman spectral analysis is of great value in identifying the Z-DNA conformation in a number of environments. Further analysis is likely to lead to a greater understanding of the vibrational signatures associated with Z-DNA.

Nuclear Magnetic Resonance Studies

NMR spectroscopy provides a powerful method for studying the conformation of molecules in solution. The first NMR evidence of an unusual conformation for poly(dG-dC) came from the observation by Patel et al (51) that the ^{31}P NMR spectra of the polymer in a high-salt solution differed in a striking fashion from that obtained in a low-salt solution. Instead of exhibiting one resonance peak, a splitting was seen in high-salt solution. Later investigations suggested that this was due to the different environment of the phosphate groups in either the d(GpC) or d(CpG) sequences (2, 53, 55). A downfield shift of one peak of approximately 1.4 ppm is found in poly(dG-dC) in 4 M NaCl, while poly(dG-m^{5}dC), which forms Z-DNA in 1.5 M NaCl, has a downfield shift of slightly over 1.2 ppm. The identity of the shifted peak has been established by Jovin et al (54, 96) through studying poly(dG-dC) synthesized with a phosphorothioate linkage in either the

GpC or CpG positions. This led to the assignment of the lower field resonance to the phosphate in the GpC sequence. Studies on hexamers of $(m^5dC\text{-}dG)_3$ that form Z-DNA either in solutions of $NaClO_4$ or with added methanol also made it possible to make assignments because of the presence of three phosphates from CpG sequences and only two from GpC (56, 57). These assignments supported those obtained from the thiol-substituted polymers. The observed chemical shift difference of 1.1 ppm in the oligomers is very similar to that seen in the polymer.

There have been several proton NMR studies of Z-DNA polymers (52, 58, 117, 118). These include studies that employ the Nuclear Overhauser Effect (NOE), i.e. the change in the intensity of a given nuclear spin on saturation of a nearby dipolar coupled spin. In terms of protons, this means that if one irradiates at the absorbing frequency of one proton, nearby protons are perturbed in a manner proportional to the inverse sixth power of the distance between the protons. The perturbation can be used as a measure of distance between atoms. When guanosine is in the *syn* conformation, the C8 proton of guanine is fairly close to the C1' proton of the deoxyribose (Figure 3). However, when guanine is in the *anti* conformation, these two protons are farther apart. NOE spectra can thus be used to determine the distribution of *syn* and *anti* nucleotides. Studies of this type have been carried out on individual nucleotides showing that guanosine phosphates are flexible in solution and adopt the *syn* conformation a large part of the time (18). In studies on poly(dG-dC), a strong NOE effect was found between the deoxyguanosine H8 and deoxyribose H1' protons in poly(dG-dC) in a 4 M salt solution but not in a 0.1 M salt solution (52, 117). These transient NOE measurements demonstrated a *syn* glycosyl torsion angle at guanosine residues in contrast to the *anti* conformations found in cytidine. Similar measurements were also carried out on poly(dG-m^5dC) in 1.5 m NaCl (52). These studies represent the solution analog to the X-ray crystallographic studies that demonstrated the two different types of conformations (2).

NOE studies are also carried out in a continuous fashion in a so-called two-dimensional NMR study. These experiments allow one to determine all interactions within the molecule in one experiment. In this way, spatial relations of all atoms in a particular conformation can be mapped. A two-dimensional NMR study has been carried out by Feigon et al (57) on the hexamer $(m^5dC\text{-}dG)_3$ in deuterated solutions. In 0.1 M NaCl, the proton NMR spectrum is largely due to B-DNA with 2% to 4% of Z-DNA visible. This is a graphic demonstration of the equilibrium between the two conformations. The NMR spectrum changes when methanol is added. Analysis of these spectra has led to the assignment of most of the proton resonances in both the B-DNA and Z-DNA conformations. These assignments can be made without reference to a specific structure. However, the conclusions are in accord with a Z-DNA conformation in a 40% methanol solution. One of the interactions that shows up under these conditions is a methyl-methyl cross peak between the two methyl groups on the 5 position of cytosine on opposite strands (see Figures 6 and 9). Because of the great insight this method provides in ascertaining aspects of structure, it is likely that it will be used significantly in the future to monitor the effects of a variety of ligands and other interacting substances on the equilibrium between B-DNA and Z-DNA.

Other Methods

Light-scattering studies on poly(dG-dC) in the Z form or the B form reveal that the Z-DNA molecule is stiffer and has a greater persistence length in solution (73). Sedimentation studies have also been used in examining the B to Z conversion. This has been most useful in cases in which Z-DNA formation in plasmids is associated with a change in the level of supercoiling, which influences sedimentation rate. Using an insert of alternating guanine and cytosine residues, the change in sedimentation constant has been quantitatively interpreted to indicate that the entire segment forms Z-DNA (65).

In a tritium-exchange experiment, the exchange of solvent protons with those involved in base-pair hydrogen bonding is measured. Studies by Leng and associates (60, 108) reveal that the exchange half-time for some protons in Z-DNA is much longer than in B-DNA. By comparing different polymers, it has been shown that the slowly exchanging protons in Z-DNA are those on the cytosine amino group. Presumably the slower exchange rate implies that the Z-DNA helix opens less rapidly than B-DNA in agreement with the increased stiffness of the Z-DNA molecule as detected by light scattering.

Electron-microscopic (EM) studies have been carried out on Z-DNA. It has been noted that Z-DNA formation is frequently associated with extensive self-association (119). Furthermore, self-association can take place in an ordered manner giving rise to toroidal or ropelike structures. A variety of entangled and branched polymer filaments are seen that depend on the size of the aggregates (120). Structures are seen that look like four-chain hairpins aggregated together.

Theoretical Analyses: Energetics and Mechanics

Z-DNA represents an alternative conformation of DNA molecules and it has become an attractive target for a number of theoretical studies with the aim of understanding the differences in the chemical properties of Z-DNA as compared to B-DNA (121).

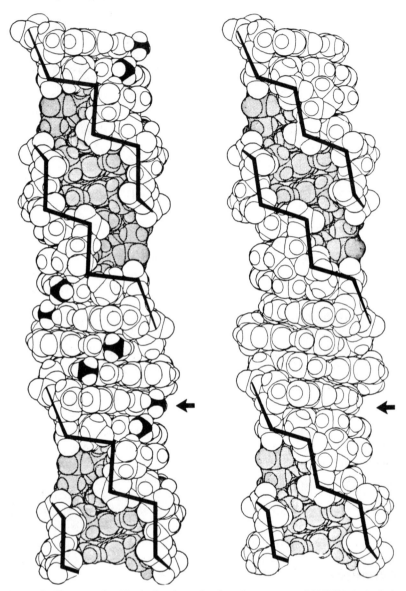

Figure 9 Two van der Waals drawings showing the structure of Z-DNA in both its unmethylated and methylated forms as determined in single-crystal structures of (dC-dG)₃ and (m⁵dC-dG)₃, respectively. The groove in the molecule is shown by the shading. The black zig-zag lines go from phosphate group to phosphate group to show the arrangement of the sugar-phosphate backbone. The methyl groups on the C5 position of cytosine in the methylated molecule (*left*) are drawn in black. The arrow shows that a depression on the surface of the unmethylated polymer (*right*) is filled by methyl groups (*left*). The methyl group is in close contact with the imidazole ring of guanosine and the C1′ and C2′ atoms of the sugar ring (14).

Harvey (122) points out that the longitudinal breathing motion of Z-DNA may be a partial mechanism for the B to Z transition. This is because base pairs must flip over in order to go from B-DNA to Z-DNA and he suggests that they flip over with breaking hydrogen bonds. Olson et al (123) have generated other transition mechanisms for flipping the base pairs that involve more than one base pair at a time. Tidor et al (124) have applied the techniques of molecular and harmonic dynamics to the internal mobility of double-stranded DNA hexamers in both the Z and the B conformations. They and Kollman et al (125) find that the atomic fluctuations depend upon the conformation to a significant extent. Hopkins (126–128) has pointed out that Z-DNA is one of four possible models, two right-handed and two left-handed. Sasisekharan et al (129–131) have considered several left-handed helices including Z-DNA. B-Z transitions have also been considered by Benham (132, 133) in torsionally stressed DNA.

There are significant differences in the distribution of the phosphate groups and in the electrostatic potential in B- and Z-DNA. Pullman and colleagues (134–137) have calculated the electrostatic molecular potentials of Z- vs B-DNA and pointed out the differences in the electric charges of the bases. This is likely to affect the manner in which these different conformations interact with other molecules. Van Lier et al (138) have calculated the difference in the *syn* and *anti* energetic barriers in native DNA and in methylated DNA. They find high activation energies for the pyrimidine nucleosides and only moderate ones for the purines. They also carried out theoretical work involving the mechanism for B-Z transitions.

The use of plasmids with segments of alternating purine-pyrimidine residues has made it possible to carry out a number of calculations dealing with the energy of B-Z transitions. Studies by Wang and associates (19, 62) suggested that there were two significant numbers to be considered for inserts with alternating CG. One is the free energy per base pair in the Z conformation and the other is the energy of the B-Z junction. These can be calculated in a semiempirical manner using the different negative super-helical energies at which levels various segments form Z-DNA (19). Using two-dimensional gel electrophoresis, it was shown that the B- to Z-DNA conversion within supercoiled plasmids is highly cooperative and the free energy per base pair is shown to be close to 0.3 Kcal per mole, while the free energy of the junction is +5 Kcal per mole. With these parameters, it is possible to calculate the energy of the B-Z transition for varying lengths of the insert as driven by negative supercoiling. The calculations also suggest that there is likely to be some unwinding of the B-DNA near the B-Z junction (62). Most of the models used at present consist of alternating CG sequences. It will be of interest in the future when these theoretical studies are extended to include a variety of additional Z-forming sequences including base pairs that are out of purine-pyrimidine alternation.

CHEMICAL FACTORS INFLUENCING THE
EQUILIBRIUM BETWEEN B- AND Z-DNA

All sequences of DNA have an equilibrium between right-handed B-DNA and left-handed Z-DNA. The position of that equilibrium is determined by the sequence of nucleotides, and alternations of purines and pyrimidines favor the Z-DNA conformation, especially CG sequences. For this reason, poly(dG-dC) has been used most widely in searching for factors that influence the equilibrium.

In this equilibrium B-DNA is usually the lower energy state and Z-DNA can become the lower energy state only when the system is perturbed in some way to stabilize it. The relative instability of Z-DNA relative to B-DNA is partly associated with the fact that the phosphate groups on opposite strands come closer together in Z-DNA than in B-DNA, as can be seen in Figure 1. The distance of closest approach of the phosphate groups across the groove in Z-DNA is 7.7 Å compared to 11.7 Å in B-DNA (8). Because this relative instability of Z-DNA has a large electrostatic component, it is not surprising that the initial observations concerning Z-DNA were found in solutions with high concentrations of salt, which reduce the phosphate-phosphate repulsion (6). Many factors are now known that stabilize Z-DNA or lower its energy so that the equilibrium shifts in its favor. With the development of structural knowledge concerning Z-DNA a large number of chemical modifications have been studied and different environmental conditions have been discovered that can influence the equilibrium. These are listed in Table 3. One of the most important modifications from a physiological point of view is the methylation of cytosine C5 in CpG sequences. That is discussed separately below. A basic understanding of the effect of chemical modifications on the B-Z equilibrium is not well developed, even though it is often possible to make qualitative statements to estimate the direction in which a particular perturbation will affect the equilibrium.

Covalent DNA Modifications

A number of covalent modifications are listed in the first section of Table 3. The first covalent modification made on poly(dG-dC) was guided by structural information on Z-DNA. Bromination of poly(dG-dC) in the Z-DNA form occurs largely at the C8 position of guanine and to a lesser extent on the C5 position of cytosine (91, 75). The bromination reaction on the C8 position of guanine was well known, and structural studies on C8-brominated purines showed that they were all confined to the *syn* conformation for steric reasons (80, 81). After brominating poly(dG-dC) in 4 M NaCl so that 35% of the guanine residues had reacted in the C8

Table 3 Chemical factors influencing B-Z equilibrium [usually measured with poly(dG-dC)]

		References
A. Covalent modifications		
Changes at	1) Methylation	(14, 40, 139)
cytosine C5	2) Halogenation—Br, I	(140)
	3) 5-Azacytosine	(54)
Changes at	1) Bromination	(75, 141)
guanine C8	2) Acetylaminofluorene	(142–151)
Changes at	1) Methylation	(152)
guanine N7	2) Aflatoxin	(153)
	3) Platinum complexes	(154, 155)
Phosphate group changes		
Thio substitution		(54, 96)
Others	4-Thio uridine	(54)
	Chemical Cross-linking	(156)
B. Ionic changes in solution		
Effect of cations	1) Monovalent—Na^+, K^+, Li^+	(6, 89, 157)
	2) Divalent—Mg^{2+}, Ca^{2+}, Mn^{2+}	(6, 40, 158, 159)
	3) Higher valencies—$Co(NH_3)_3^{3+}$, polyamines	(40)
Effect of anions	$NaClO_4$, sodium acetate	(160, 6)
C. Solvent modifications		
Alcohols:	Ethanol	(1, 37)
	Methanol	(37, 54)
	Ethylene glycol	(161)
	Trifluoroethane	(37, 57, 114, 158, 159)
D. Small molecule effectors		
Intercalating	Ethidium	(38, 162)
agents	Daunomycin	(163)
	Adriamycin	(163a)
	Actinomycin D	(162)
Groove binder	Netropsin	(164)

position and 17% of the cytosine residues at C5, the molecule remained in the Z-DNA conformation after the salt was removed by dialysis (75). The brominated polymer was in the Z-DNA form as judged by several different physical and chemical criteria, including changes in the circular dichroism, the UV absorption spectrum, the ^{31}P NMR spectrum and modifications in its filtration through nitrocellulose filters in high salt.

Halogenation of cytosine at the C5 position of poly(dG-dC) leads to considerable stabilization of Z-DNA. This is especially true for the fully brominated polymer that appears to exist as Z-DNA exclusively almost independent of salt concentration (83, 54). Greater stabilization of Z-DNA has also been reported by the substitution of an iodine atom at the same position (54).

Covalent modifications of the guanine residue in poly(dG-dC) has also led to stabilization of Z-DNA. As mentioned above, bromination of guanine C8 stabilizes the *syn* conformation (75, 91, 141). Likewise, reaction of the C8 position of guanine with the carcinogen N-acetoxyaminofluorene also stabilizes the Z-DNA conformation of this polymer as shown by the lower amount of NaCl required to produce an inversion of the circular dichroism after this reaction (142, 143, 145–151, 165). In this case the presence of a bulky substituent at the 8 position of guanine prevents the assumption of the *anti* conformation.

A number of reactions have also been found to occur at the N7 position. Methylation at N7 leads to stabilization of Z-DNA (152). When poly(dG-dC) is fully methylated in this position the polymer is stable as Z-DNA in a physiological salt solution. Methylation of N7 is associated with introduction of a formal positive charge in this position, which may contribute to the stabilization of Z-DNA by electrostatic means similar to concentrated NaCl solutions. Another class of compounds that stabilizes the Z form of poly(dG-dC) is platinum complexes, including chloro(diethylenetriamine) platinum (II) chloride (154, 155). They are probably coordinated onto the N7 of guanine and are likely to act in large part through the electrostatic contribution of the positively charged cation. The carcinogen aflatoxin reacts with the N7 of guanine (153) and it also induces the formation of a positive charge at that position. However, its effect in the B-Z equilibrium is just the opposite of N7 methylation (152). It stabilizes the B conformation and prevents its conversion to Z-DNA when salt is added to the medium. The large aflatoxin molecule probably binds to B-DNA and prevents the conversion to Z-DNA.

Other covalent changes that have been shown to influence the B-Z equilibrium are the substitution of sulphur atoms for the oxygen atoms in the phosphate group. There are two types of phosphate groups and the thio phosphate is more effective in one position than the other in Z-DNA stabilization (54, 96).

Ions, Solvents, and Small Molecules

The most widely studied changes are those introduced into the solution. Ions, especially cations, strongly influence the equilibrium of Z-DNA and B-DNA (6, 160). It is likely that the predominant interaction in modifying the equilibrium is due to the cations clustering around the negatively charged phosphates and reducing the phosphate-phosphate repulsive interactions. The monovalent cations, sodium, potassium, and lithium, have all been shown to influence this equilibrium. The midpoint for the sodium conversion as judged from the change in the circular dichroism is near 2.7 M (6). For divalent cations lower concentrations are required; for magnesium the concentration is near 0.7 M for the midpoint of the conversion. It is not surprising that ions with higher valencies are even more effective, such as the polyamines with charges of $+3$ or $+4$ (2, 40, 166). The cobalt hexamine complex has been noted to be especially useful since it stabilizes the Z-DNA form at millimolar concentrations (40). The exact position of the cobalt-hexamine cation in Z-DNA is likely to be identified soon as Z-DNA crystals stabilized by cobalt hexamine are now available.

Anions can also influence the distribution. Sodium perchlorate stabilizes Z-DNA beyond the contribution of its cation (6). Sodium acetate has been shown to induce an intermediate form between B- and Z-DNA as demonstrated by circular dichroism and laser Raman spectroscopy (160). A variety of agents that change the dielectric constant of water has been found to stabilize Z-DNA. This includes the alcohols, ethanol, methanol, and ethylene glycol; experiments have also been reported with trifluoroethane (37, 57, 114, 158, 159). Z-DNA is stabilized in 44% ethanol while 60% of trifluoroethane is required. The mechanism for this stabilization is not obvious. Lowering of the dielectric constant means that ionic interactions will be felt more strongly; this will result in a closer clustering of cations near the negatively charged phosphate group. It is possible that this results in an effective screening and reduction of the phosphate-phosphate electrostatic repulsion. However, other elements may also play a role, including features that are specific to the structure of the individual alcohol.

Another type of interaction that influences the equilibrium between B- and Z-DNA is the presence of small molecules that interact with one or the other sequence. An example of this is seen in the intercalators (36, 162). Addition of ethidium to poly(dG-dC) stabilized in the Z-DNA form by either salt or ethanol effectively converted the polymer back to intercalated B-DNA as shown by the circular dichroic spectrum. B-DNA has a more flexible backbone and is able to form an intercalated complex with ethidium; Z-DNA is a more rigid molecule and may not be able to accommodate an intercalative agent between its base pairs (73, 167). The antitumor agents daunomycin and adriamycin facilitate the formation of B-DNA when added to poly(dG-dC) or poly(dG-m^{5}dC) in the Z-DNA form (163, 168). Molecules that bind to DNA also influence the distribution. For example, addition of netrospin or distamycin to Z-DNA reverses it to B-DNA (164).

Cytosine Methylation in CG Sequences Stabilizes Z-DNA

One of the most common modifications of eukaryotic DNA is methylation of cytosine C5 in CG sequences. The CG sequence is sharply reduced in eukaryotic DNA (169), yet a considerable body of evidence suggests that methylation of CG residues is associated with gene inactivation; subsequent removal of the methyl group is associated with gene activation (170).

The CG dinucleotide sequence in Z-DNA has a conformation quite distinct from that of the CG sequence in B-DNA (Figure 6). In the initial structural discovery, Wang et al (2) suggested that this may have relevance in terms of the role of methylation in modifying gene expression. The effect of methylation on the B-Z equilibrium was shown dramatically by the work of Behe & Felsenfeld (40), who compared the B-Z equilibrium in poly(dG-dC) and poly(dG-m^{5}dC). Whereas the midpoint of the transition from B- to Z-DNA required 0.7 M Mg^{2+} in a solution with 50 mM NaCl, a lowering of the magnesium requirement by three orders of magnitude is seen for the methylated polymer. Z-DNA is the stable conformation of poly(dG-m^{5}dC) in a physiological salt solution. Furthermore, much less Na^+ or K^+ was required to stabilize Z-DNA in the methylated polymer. The polyamines spermine and spermidine are very effective in stabilizing Z-DNA in the methylated polymer; the midpoint in the B-Z equilibrium is 2 μM in spermine. These experiments showed that methylation of cytosine residues had a dramatic effect in modifying the equilibrium between B- and Z-DNA, strongly stabilizing the Z-DNA form. Methylation is also effective in stabilizing Z-DNA in inserts of alternating CG residues in plasmids. Methylation reduces the amount of negative supercoiling required to stabilize the Z conformation (139).

The structure of the methylated molecule was revealed when the hexamer (m^{5}dC-dG)$_3$ was solved by Fujii et al (14) at 1.3 Å resolution. The overall form of the molecule is quite similar to that of the unmethylated Z-DNA structure as seen in the (dC-dG)$_3$ crystal (2). The cytosine residues on opposite chains of adjacent base pairs are stacked above each other (Figure 6) and their methyl groups are close to each other, separated by 4.6Å. A diagram of the structure is shown in Figure 9. The methyl groups are found in pairs on the surface and they fill a slight hydrophobic depression formed by the imidazole group of guanine of the next base pair and the C1' and C2' hydrogen atoms of the sugar. In the absence of the methyl group, the depression is normally filled by water molecules. The methyl group acts to form a small hydrophobic patch on the surface of the molecule that stabilizes it by excluding water from the hydrophobic pocket (Figure 9). The position of the methyl group in Z-DNA is in marked contrast to its position in B-DNA, where it projects into the major groove of the double helix and is surrounded by water molecules. The difference in the environment of the methyl group of cytosine in Z-DNA and B-DNA is one of the major factors in leading to its strong stabilization of Z-DNA (14). X-ray diffraction studies have also been carried out on fibers of poly(dG-m^{5}dC) that form

Z-DNA. Their diffraction patterns are similar to those produced by the unmethylated polymer in the Z-DNA form (22, 34).

The methylated hexamer $(m^5dC-dG)_3$ has also been used in other physical studies (14, 29, 56, 171). In a low-salt aqueous solution, this oligomer is predominantly in the B-DNA form but 2% to 4% of the Z-DNA can be detected in its NMR spectrum (57). Addition of either salt or methanol facilitates the conversion of this hexamer into Z-DNA. The addition of $NaClO_4$ also produces the Z-DNA form as shown by changes in UV absorbance, circular dichroism, and ^{31}P NMR (56). The methylated polymer poly(dG-m^{5}dC) has also been studied by NMR investigations with respect to its B-Z transition (52, 55, 56).

Methylated poly(dG-dC) has been used to study nucleosome formation. Under some conditions nucleosomes reconstitute in a B-DNA form (172), while in other experiments Z-DNA is found in the nucleosomes (173, 174). The nucleosomal particle prepared from Z-DNA was found to be more resistant to DNase I than the particle formed from B-DNA. Further work will have to be done to understand the implications of nucleosomes that can apparently form with either B-DNA or Z-DNA.

The prokaryotic Hha I methylase adds a methyl group to the CG sequence in GCGC. B-DNA has been shown to be its substrate, but not Z-DNA (76). Murine and human methylases show no preference for Z-DNA (175, 176). At the present time, it appears that methylation of CG sequences in the 5′-flanking promoter region prevents transcription of the gene (170). The unanswered question is whether this effect is carried out via Z-DNA formation. Nickol & Felsenfeld (170a) have methylated the CCGG sequences in the 5′-flanking sequences of the chicken β-globin gene and found that it does not induce the formation of Z-DNA in a supercoiled plasmid as judged by changes in negative supercoiling. As yet there is no in vivo bridge connecting the stabilization of Z-DNA by methylation with gene inactivation. However, Z-DNA binding proteins exist and if there are any that are specific for methylated CG sequences, these could be a bridge. The involvement of Z-DNA in gene inactivation can be regarded only as an attractive hypothesis at present.

NEGATIVE SUPERCOILING STABILIZES Z-DNA

Topological Constraints in DNA

In a double-stranded DNA circle the two DNA strands are coiled around each other to form a double helix with a right-handed helical repeat of approximately 10.5 bp. Consequently, in a relaxed state the two circular DNA strands will be intertwined (or linked) as many times as double-helical turns have been formed. This can be described in terms of a linking number

α. In the relaxed state the linking number α is a measure of the number of helical turns and is written as α_0. However, if the *actual* linking number α deviates from the linking number of the *relaxed* state, α_0, then the DNA circle assumes the shape of a supercoiled molecule in which the topological linkage of the two DNA strands is described by the equation: $\tau = \alpha - \alpha_0$. In this equation τ defines the linking difference $\alpha - \alpha_0$ and it is a measure of the number of superhelical turns. The linking difference $\alpha - \alpha_0$ can also be normalized over α_0. The specific linking difference σ (or superhelical density) is written as $\sigma = \tau/\alpha_0 = [\alpha - \alpha_0]/\alpha_0$. All circular DNA isolated from natural sources exists in the underwound, negatively supercoiled state, i.e. $\tau < 0$ or $\sigma < 0$. The supercoiled state is associated with an unfavorable free energy relative to the relaxed state, whereby the excess free energy, ΔG, is proportional to the square of the number of superhelical turns: $\Delta G \sim |\alpha - \alpha_0|^2$ or the square of the superhelical density (177). Consequently, processes that reduce the number of superhelical turns are energetically favored. In negatively supercoiled circles, these processes include unpairing of bases, strand separation, cruciform formation, unwinding of the double helix, and protein binding as in nucleosome formation. Another process that is facilitated in a negatively supercoiled plasmid is the stabilization of Z-DNA. This was demonstrated independently by Singleton et al (63) and Peck et al (65). Negative supercoiling that exists in virtually all naturally occurring DNA has been found to be one of the most important factors in the stabilization of Z-DNA (35, 44, 63–65, 67, 70, 102, 103, 106, 178–180).

Changes in superhelical density associated with Z-DNA formation affect hydrodynamic properties of circular plasmids, e.g. sedimentation rate or migration velocity in electrophoretic separations, etc. Figure 10 shows a diagram of a negatively supercoiled plasmid in equilibrium with a less supercoiled plasmid that has a segment of Z-DNA. The energy of supercoiling is used to stabilize a segment of Z-DNA. B-Z transitions can be easily detected and measured in supercoiled circles because of the considerable change in hydrodynamic features and conformation. The initial experimental studies dealt with forming Z-DNA in plasmids with inserts of uninterrupted alternating CG residues up to 42 bp in length (19, 63–65). B- to Z-DNA transitions began to occur at the relatively low negative superhelical density of $\sigma = -0.03$ (65) as judged by agarose gel electrophoretic separation of topological isomers. An abrupt change in migration rate could also be brought about by other factors which were known to induce the B- to Z-DNA transitions, including increasing concentrations of NaCl or cobalt hexamine. An inverse correlation could be found between the superhelical energy requirements and the lengths of DNA segments to be stabilized in the Z-DNA form (19, 64).

Antibodies specific for Z-DNA were used by Nordheim et al (19) to detect the conversion to Z-DNA upon negative supercoiling. Nitrocellulose filters allow protein-free DNA to pass through them but DNA complexed to antibodies is retained on the filter. In these experiments, plasmids were adjusted to increasing superhelical densities by topoisomerase I relaxation in the presence of ethidium bromide. Antibody binding curves could be obtained that reflected the conversion of B-DNA to Z-DNA at various levels of negative superhelical density. These experiments made it possible to make a quantitative estimate of the superhelical energies at which B- to Z-DNA conversion occurs. The Z-DNA–forming segments could be identified by mapping the antibody binding sites on the plasmid genomes (19). This mapping technique employs covalent cross-linking of the antibodies to their DNA binding site, restriction endonucleolytic fragmentation of the DNA, and separation of protein-free fragments from antibody-bound fragments by nitrocellulose filtration. This technique led to the first identification of a natural sequence that formed Z-DNA in plasmid pBR322 (19) (Figure 11). The sequence consists of 14 bp of alternating purine pyrimidine residues with one base pair out of alternation. Further evidence that Z-DNA formation occurred here was shown by the inhibition by antibody binding of the activity of three restriction endonucleases that have recognition sequences in that DNA segment (Figure 11) (103).

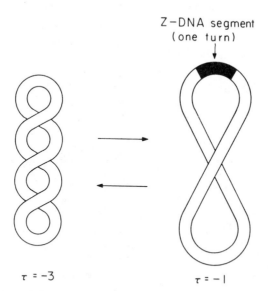

Z–DNA segment
(one turn)

$\tau = -3$

$\tau = -1$

Figure 10   A schematic diagram illustrating the effect on supercoiling of converting a segment of plasmid from B-DNA (no shading) to Z-DNA (shaded black). In this diagram, a 12-bp segment of B-DNA has converted to Z-DNA, resulting in a loss of two negative supercoils. The letter τ is the number of supercoils.

At higher superhelical densities additional Z-DNA segments are formed in pBR322 and they can be identified by antibody cross-linking or visualized by using immuno-electron microscopy (44). The latter technique is not as precise a method for identifying Z-DNA segments as is the separation of restriction fragments by nitrocellulose filtration, but it yields a comprehensive overview of large plasmid structures. This makes it possible to estimate the total extent of Z-DNA formation in negatively supercoiled plasmids. It is also possible to make measurements of the changes in the circular dichroism of plasmids with different levels of negative supercoiling and from this to estimate the amount of Z-DNA formed (44).

The negative superhelical density of plasmids in bacterial systems is generally in the range of 0.05–0.07, although higher levels of negative supercoiling have been reported. For example, bacteriophage PM2 exists as a covalently closed circular molecule with a superhelical density of $\sigma = -0.12$. Kuhnlein et al (71) postulated the occurrence of B- to Z-DNA transitions within the PM2 genome that were induced by salt and negative supercoiling. Antibody binding studies by van de Sande and colleagues (68, 106) confirmed this.

Negative DNA supercoiling has been shown to induce the formation of Z-DNA in DNA segments with the repeating sequence $d(CA/TG)_n$ (67, 102). These segments are naturally occurring in the eukaryotic genome. Z-DNA formation was shown by antibody binding studies (102) as well as by changes in electrophoretic mobility of negatively supercoiled plasmids as determined in a two-dimensional gel analysis (67). These two methods of measurement are complementary and it is noteworthy that comparable negative superhelical densities were found to be required to induce the formation of Z-DNA in $d(CA/TG)_n$ sequences using these two different systems. It is interesting that small changes in salt concentration can greatly influence the energy requirements for B- to Z-DNA transitions induced by

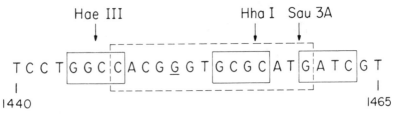

Figure 11   The nucleotide sequences of plasmid pBR322 from residues 1440 to 1465. The solid boxes correspond to recognition sites for the restriction endonucleases Hae III, Hha I, and Sau 3A, which are blocked by the antibody attached to the negatively supercoiled plasmid. Enclosed in the dashed box is a 14-bp sequence of alternating purine and pyrimidine residues with one base pair (*G*) out of alternation. This box corresponds to the proposed attachment site of the anti–Z-DNA–specific antibody (103).

supercoiling. In the concentration range 50 mM–300 mM, NaCl clearly inhibits the B to Z-DNA conversion as the concentration increases (62, 63, 102, 103). This phenomenon has been postulated to be due to salt effects on either the formation of B-Z junctions (103) or the free energy change for the transition itself (62).

These supercoiling experiments were important for demonstrating that Z-DNA segments can be formed in plasmids even though the bulk of the molecule is in the form of right-handed B-DNA. It also provided a way of analyzing junction regions between B-DNA and Z-DNA. Interestingly, Wells and colleagues (63, 70, 74, 181) showed that these junctions displayed increased activity as substrates for the single strand–specific nucleases Sl or Bal 31. The supercoiling experiments also demonstrated that fairly short segments can form Z-DNA, even though energetic considerations suggest that short segments will be unfavorable. In addition, natural sequences were identified that form Z-DNA, and it was shown that bases can be out of purine-pyrimidine alternation and still form Z-DNA (19, 103). The formation of Z-DNA was first demonstrated in linear polymers of poly(dG-dC) in which salt was the driving force for carrying out this conversion (6). The studies on supercoiling demonstrated that it is a stronger factor in facilitating Z-DNA formation than is salt. This is supported by the fact that some segments that do not form Z-DNA fully in chemically unmodified form (140) at high salt concentrations (113, 114, 140, 149) can be made to form Z-DNA readily at modest negative superhelical densities (67, 102). Studies of negatively supercoiled systems also allow careful analysis of free-energy parameters of B- to Z-DNA transitions (62). Finally these systems make it possible to study the kinetics of Z-DNA formation in natural sequences as distinct from the kinetics of forming sequences in Z-DNA polynucleotides.

Form V DNA

Form V DNA is prepared by annealing complementary single-stranded circles of DNA. The annealed circles reassociate to form a molecule with well-defined physicochemical characteristics that suggest the existence of an ordered, hydrogen-bonded, double-helical structure (182). If the plasmid reanneals to form right-handed helical segments, it is topologically constrained also to form a similar number of left-handed helices. In the original description, Stettler et al (182) showed that form V plasmids had a circular dichroism that resembled an intermediate state between that seen in the high-salt form of poly(dG-dC) (6) and regular B-DNA. The form V system, in which two single-stranded circles are forced into double-helical coiling, may be looked upon as one which is negatively supercoiled to the limit, formally equivalent to a superhelical density that approaches -1.0.

The left-handed segments have been shown by several groups (49, 105, 107) to contain Z-DNA.

Brahms and colleagues (49, 50) showed that the circular dichroism of form V DNA in both the ultraviolet and vacuum ultraviolet regions was quantitatively the sum of the circular dichroism of right-handed B-DNA and left-handed Z-DNA with about 10% nonpaired bases. They also carried out a Raman spectrum analysis of form V DNA and compared it with the Raman spectra of the B- and Z-DNA forms of poly(dG-dC) obtained by Pohl et al (38). This Raman study also supports the existence of Z-DNA segments in form V DNA. Some 80% to 90% of the residues in these plasmids were in an ordered and base-paired structure, and the proportion of the left-handed Z-DNA was estimated to be as high as 35% to 40%. Since these plasmids do not have this high a percentage of alternating purine-pyrimidine residues, it strongly indicates that sequences far removed from alternation are adopting the Z-DNA conformation.

Pohl, Leng, and their associates (105, 107) used anti–Z-DNA antibodies to study the disposition of Z-DNA in form V plasmids. A considerable part of form V DNA interacted with the antibodies. Furthermore, the electron microscopic studies revealed that widespread segments of the form V plasmid were covered by antibodies specific against Z-DNA. Forty percent of the residues were estimated to be in the Z-DNA form, as judged by antibody binding.

These results suggest that Z-DNA is the major left-handed form of DNA. It would be of great interest to know if there exist alternative forms of left-handed DNA other than Z-DNA; these might be detected as segments of left-handed DNA that are shown not to be Z-DNA. Further quantitative studies of form V plasmid may make it possible to state if any other left-handed forms of DNA can be stabilized by topological constraint.

PROTEINS THAT BIND TO Z-DNA

An understanding of the biological significance of Z-DNA requires insight into specific Z-DNA–protein interactions. Initial studies have concentrated on analyzing differential protein binding to B- vs Z-DNA forms as well as attempts to purify eukaryotic nuclear Z-DNA binding proteins. Addition of poly-L-arginine but not poly-L-lysine to poly(dG-dC) in high-salt solution allows the Z conformation to be retained upon dialysis to lower salt concentration (183). This suggests that the stabilization of the Z-DNA form is not due to simple electrostatic factors but may also involve some specificity in the interaction of the arginine side chain with the Z-DNA helix. Using a different methodology, Russell et al (166) studied poly(dG-dC) in the presence of 25% ethylene glycol where it can be readily converted to the Z conformation by the addition of micromolar $MnCl_2$ (161). In this environment, the transition to B-DNA can be reversed by chelating the Mn^{2+} ions with EDTA or by diluting them out with the addition of NaCl. The DNA interconversions were monitored by circular dichroism measurements. In this test system, poly-L-arginine could suppress the Z-to-B transition while poly-L-lysine was ineffective. Different effects were found on examining a number of purified histones. Nucleosome core histones H3, H4, H2a, and H2b, as well as protamine, stabilized the Z-DNA form induced by Mn^{2+} and prevented its transition to B-DNA after removing Mn^{2+} or salt. In contrast, histones H1 and H5 promoted the transition from Z- to B-DNA as detected by a change in the circular dichroism. These experiments show differential effects among the histones and may support the observations by Miller et al (173) that it is possible to form nucleosomes with Z-DNA since all four of the core histones appear to support Z-DNA stabilization.

Nordheim et al (101) used the method of Z-DNA affinity chromatography to isolate proteins from the nuclei of *Drosophila* Schneider tissue culture cells that have the ability to bind to the Br-poly(dG-dC) Z-DNA form but not the B-DNA form of poly(dG-dC) or to B-DNA with naturally occurring sequences. By this method, five major and several minor proteins were isolated using the criterion of strict Z-DNA binding capability without any B-DNA binding activity. Many of the proteins had molecular weights above 70,000. None of the histones were isolated by this procedure. In the isolation protocol, competitor B-DNA in the form of sonicated *E. coli* DNA was added to the extract, and proteins with any B-DNA binding activity were eliminated in the column fractionation. The naturally occurring *Drosophila* Z-DNA binding proteins isolated also appeared to have the ability to influence the B-Z equilibrium of poly(dG-m^{5}dC) in 0.1 M NaCl as assayed in filter-binding experiments. The proteins also bind to negatively supercoiled plasmids containing inserts of either alternating $d(CG)_n$ or alternating $d(CA/GT)_n$ in the Z form, but not to the relaxed plasmids. These nuclear proteins are heterogeneous and they have to be purified further so that they can be characterized more fully.

As further work is carried out on Z-DNA binding proteins, it will be useful to distinguish between those that bind Z-DNA with sequence specificity and those that bind independent of sequence. The same dichotomy exists in B-DNA binding proteins, and there is considerable functional differentiation among these two classes. Finally, some proteins may be discovered that have the property of binding to both B-DNA and Z-DNA, and some of them may act to shift the B-Z equilibrium. It is also possible that proteins may be found with enzymatic activities that facilitate establishing the B-Z equilibrium, and they may be designated as "DNA conformases."

NUCLEOTIDE SEQUENCES THAT FAVOR Z-DNA FORMATION

Polymers and Repetitive Sequences, e.g. d(CA/GT)ₙ

Z-DNA formation is favored in nucleotide sequences containing alternations of purines and pyrimidines because of the ease with which purines can adopt the *syn* conformation in comparison to the pyrimidines (17). There are three kinds of regular polymers with simple alternations of purine and pyrimidine sequences: poly(dG-dC), poly(dC-dA)·poly(dG-dT), and poly(dA-dT). Poly(dG-dC) forms Z-DNA readily under various ionic conditions (6). Attempts were made to look for a Z-DNA form of poly(dA-dC)·poly(dG-dT) in solutions with elevated salt concentrations. Suggestive changes in the circular dichroism spectrum were found in high concentrations of CsF as well as in ethanolic solutions (113, 114). Treatment of poly(dC-dA)·poly(dG-dT) with N-acetylaminofluorene accompanied by elevated salt concentration produced a near-inversion of the circular dichroism that suggested even more strongly that Z-DNA was forming under these conditions (149). Addition of a methyl group to the cytosine C5 position facilitated the salt-induced Z-DNA formation in poly(dA-m^{5}dC)·poly(dG-dT) (140). A Z-DNA–type fiber diffraction pattern has been reported for poly(dC-dA)·poly(dG-dT) (21).

As outlined above, DNA supercoiling provides a powerful force for Z-DNA formation. On this basis, chemically unmodified d(CA/TG)ₙ segments could be stabilized as Z-DNA when cloned into circular plasmids. Binding of anti–Z-DNA antibodies to a 64-bp fragment of d(CA/TG)ₙ (102) as well as two-dimensional gel electrophoresis on plasmids carrying 52–120 bp of d(CA/TG)ₙ (67) demonstrated Z-DNA formation in these inserts. This suggested that the driving force of negative supercoiling is a more effective way of promoting Z-DNA formation than alteration of the environment through added salts or even chemical modification of DNA. Methylation of cytosine residues in a d(CG)ₙ plasmid insert also stabilizes Z-DNA as it does in the poly(dG-dC) polymer (40, 139).

At present, attempts to demonstrate the formation of Z-DNA in poly(dA-dT) have not yielded spectroscopic changes that indicate the formation of Z-DNA. However, substitution of 2-amino purine in place of adenine leads to a copolymer that yields spectroscopic data (UV, CD, NMR), suggesting Z-DNA formation in a 2 M NaCl solution (54).

We can summarize the tendency of the pyrimidine-purine dinucleotides to form Z-DNA in polymers as follows: m^{5}CG > CG > TG = CA > TA. Some indication of the structural basis for these differences has emerged from analysis of the Z-DNA crystal structure of d(m^{5}CGTAm^{5}CG) by Wang et al (7). This crystal diffracts to a resolution of 1.2 Å and its structure showed that the AT base pairs form Z-DNA with a geometry rather similar to the CG base pairs. However, there was a significant difference in the ordering of the water molecules in the deep helical groove. Near CG sequences there are well-ordered molecules, while these are disordered at the TA base pairs. The ordered solvent binding is probably associated with Z helix stabilization in CG sequences.

Although the repeating sequence $d(CG/GC)_n$ is not found widely in biological systems, the sequence $d(CA/GT)_n$ is widely distributed in eukaryotic genomes where it may be considered a form of middle repetitive DNA. These tracts are found in all gene systems that have been extensively sequenced, including globins, immunoglobulins, actins, etc (184–86). They have even been found in an unusual class of *alu* sequences where they are 18 bp in length (187). Generally, these sequences are found in intergenic segments or in the intervening sequences of genes. Hamada et al (185) have carried out a hybridization analysis and surveyed a number of eukaryotic genomes. The hybridization stringency used would detect a $d(CA/GT)_n$ segment of 50 bp or greater. The human genome was found to have 50,000 copies of these, salmon, 20,000; *Drosophila*, 2,000; and yeast, 100. In all they have been detected in 10 different organisms, including slime mold, *Xenopus*, and several birds (184, 186). In the case of yeast, these segments have been found near the telomere ends of the yeast chromosome (188). A survey of neighboring DNA sequences near these segments does not reveal any special base arrangements (102). The widespread occurrence of these $d(CA/TG)_n$ sequences has posed an interesting challenge to molecular biologists since the sequences obviously can form Z-DNA quite readily at negative superhelical densities that are in the physiological range. The extent to which this Z-DNA potential is used in chromatin of eukaryotic cells awaits further investigation.

Z-DNA in Other Naturally Occurring Sequences

The potency of negative supercoiling to induce Z-DNA formation combined with the use of anti–Z-DNA–specific antibodies provide the most sensitive technique to date for the identification of short segments with potential for Z-DNA formation. Antibody binding sites are either located by restriction endonuclease fragmentation followed by fragment separation on nitrocellulose filters (19) or visualized by electron microscopy (44, 105). Experiments carried out on negatively supercoiled pBR322 revealed the greatest potential for Z-DNA formation in a segment of the plasmid containing 14 bp of alternating purines and pyrimidines with one base pair out of alternation (19) (Figure 11). At higher levels of negative supercoiling, other segments form Z-DNA as well. The segment contained one base pair out of alternation, suggesting that the energy penalty paid for one pyrimidine in the *syn* conformation is outweighed by the energy gain in forming a longer segment of Z-DNA.

An immuno-electron microscopic analysis of the entire pBR322 plasmid revealed four Z-DNA segments, one of which was identified as the sequence indicated in Figure 11 (44). This method has also been applied to the bacteriophage PM2, which is known to have a high negative superhelical density of -0.11 (68, 106, 189). The molecule was found to have eight different segments binding anti–Z-DNA antibodies in regions that had high concentrations of alternating purine-pyrimidine sequences by EM analysis.

Z-DNA formation has also been detected in negatively supercoiled circular DNA obtained from the simian virus 40 (SV40) (104). Z-DNA formed in the transcriptional enhancer region where three segments of eight alternating purine-pyrimidine residues were identified as major antibody binding sites. These are the longest stretches with alternating purines and pyrimidines found in the SV40 virus, except for one nine base-pair segment that contains several AT base pairs.

One can anticipate that in the near future many more Z-DNA–forming sequences will be detected. The antibody cross-linking methods will undoubtedly be helpful, and it is reasonable to look forward to the development of chemical methods to allow detection of Z-DNA directly without the use of antibodies. Such methods would have great utility in exploring the existence of Z-DNA in vivo without the need to extract DNA and run the risk of perturbing the physical state of chromatin in that process.

CYTOLOGICAL STUDIES OF Z-DNA IN CHROMOSOMES

Drosophila and *Chironomus*

The availability of antibodies that react specifically with Z-DNA initiated the use of cytological techniques for Z-DNA detection in eukaryotic nuclei and chromosomes. At first, the polytene chromosomes of *Drosophila* salivary glands were studied using indirect immune fluorescence. Polytene chromosomes have 1,000–2,000 copies of sister chromatids lying in register generating a complex system of densely staining bands and less densely staining interbands. Fixation of the polytene chromosomes is necessary for staining reactions and it is generally carried out through the use of acetic acid, which results in removal of many chromosomal proteins. The initial investigation by Nordheim et al (91) revealed the widespread appearance of anti–Z-DNA fluorescence located largely in the interband region but also in some selected bands as well as transcriptionally active puff areas. The staining pattern did not follow the distribution of DNA concentrations along the chromosome (91, 95). Modifications of the fixing technique were carried out in which the proteins were initially cross-linked by the use of glutaraldehyde. These produced a greatly reduced but otherwise largely unchanged fluorescent staining pattern. Leng and collaborators (85, 97) obtained *Drosophila* staining patterns similar to those described by Nordheim et al (91) and furthermore observed anti–Z-DNA fluorescence in band regions of *Chironomus*. In the studies by Jovin and colleagues (54, 93, 96) the *Chironomus* and *Drosophila* polytene chromosomes were analyzed and staining was revealed in the band regions. Hill & Stollar (94) micro-dissected individual polytene chromosomes from *Drosophila* and then tried to stain them before exposure to an acid fixative. They found no immunofluorescence for the unfixed chromosome. However, when acetic acid was added to the chromosome interband staining was observed first and upon prolonged fixation intense fluorescence was found in the band region. It was suggested that the removal of chromosomal proteins by acid fixation was generating Z-DNA through the release of torsional strain, which is normally absorbed in the nucleosomes. After acid fixation, relaxation of the isolated chromosomes with topoisomerase I abolished the immune fluorescence.

In general, the cytological staining patterns can be regarded as revealing the presence of segments with potential for Z-DNA formation; independent experiments will be necessary to ascertain to what extent this Z-DNA potential is realized at any time in the polytene chromatin of the living cell. The interpretation of this cytological data is complex because of a class of selective Z-DNA binding proteins that have been isolated from *Drosophila* nuclei (101). Some of these proteins are present in substantial quantities; they are bound to chromatin and are likely attached to segments of Z-DNA. Furthermore, these proteins are removed more readily than B-DNA binding proteins. Possible removal of such Z-DNA stabilizing proteins in the unfixed chromosomes of Hill & Stollar (94) could account for the lack of anti–Z-DNA staining before nucleosome disruption by acetic acid. The production of antibodies against Z-DNA binding proteins may make it possible to address this problem more fully in the future.

Other Species

Lipps et al (98) reported an immuno-fluorescence study of the nuclei in the ciliated protozoan *Stylonichia*. *Stylonichia* has a very convenient control since it contains two types of nuclei in the cell. The larger macronucleus is transcriptionally active, while the smaller micronucleus is transcriptionally inactive but is used for sexual reproduction. The DNA concentration in these two nuclei are similar and both stain with an antibody against

B-DNA showing that they are both equally accessible. However, only the macronucleus stains with an antibody against Z-DNA, while the micronucleus does not stain at all. This is found when the nuclei are fixed with 45% acetic acid, as in the classical technique, as well as with 95% ethanol, which removes fewer proteins. DNA relaxation upon cleavage by DNAse I abolishes the fluorescent signal. Although differential nucleosome destruction in the two types of nuclei remains a possibility, the data suggest that in *Stylonichia*, Z-DNA is found associated with transcriptionally active nuclei but not with the transcriptionally inactive nucleus.

In the study of the metaphase chromosomes of both man and the primate *Cetus*, heavy staining was found in the R band–positive heterochromatin, which is believed to be rich in CG base pairs (99). Weaker staining was found in the euchromatin. Similar results were found in gerbil nuclei (190). A similar staining has been observed in the metaphase chromosomes of plant tissue (191). Rye metaphase chromosomes have been examined and show a banding pattern that has increased staining in the heterochromatin regions. Very strong staining is also observed in the telomere regions of many of the chromosomes. Interphase nuclei of several different varieties of plants show a more diffuse staining by Z-DNA antibodies.

An interesting immuno-histochemical study of rat tissues was carried out by Morgenegg et al (100), who were looking for Z-DNA immunoreactivity in several different rat cell types. In examination of tissues from brain, kidney, liver, and testes it was found that some nuclei produce intense staining against the Z-DNA antibody while other nuclei do not. In the case of the seminiferous tubules of the testes, where one sees a progression of differentiation states leading to the formation of spermatazoa, it is apparent that only certain stages in this developmental process are actively staining, while others are not staining. A number of different fixation procedures were studied and all gave similar results. Since all of these nuclei are fixed under similar conditions, it is difficult to suggest that these differences are artifacts due to the extraction of protein from some nuclei and not others.

It is premature to say exactly what the staining patterns indicate, but the data clearly suggest that the Z-DNA antibody is detecting a feature that is apparent in some nuclei and not in the others. Further work will have to be carried out in a variety of carefully controlled systems in order to assess fully the significance of anti–Z-DNA antibody staining in chromosomal material. It is anticipated that studies using antibodies against Z-DNA binding proteins will be of great importance.

SUGGESTED BIOLOGICAL ROLES FOR Z-DNA

Although it seemed reasonable on general grounds to anticipate that the formation of Z-DNA in biological systems would have biological effects, it is only relatively recently that specific examples have developed. Twisting the double helix in the reverse direction requires energy, and untwisting it is likely to be spontaneous if it is allowed to occur. Since this process affects the level of negative supercoiling, it could lead to effects that may be close or distant (33, 35). The sequence of nucleotides determines the position and extent of Z-DNA formation. These sequences may thus be regarded as a form of conformational information in DNA, in contrast to the more familiar coding information.

Z-DNA and Transcriptional Enhancers

As described above, a method exists for identifying Z-DNA–forming nucleotide sequences in negatively supercoiled plasmids using anti–Z-DNA antibodies (19, 192). The first eukaryotic DNA investigated by this technique was the circular genome of the simian DNA tumor virus (SV40). Mapping of Z-DNA sequences in deproteinated, negatively supercoiled SV40 DNA revealed three segments of eight base pairs of alternating purines and pyrimidines that represent three major sites that form Z-DNA (Figure 12)(104). Two of these segments are identical and are found within a 72-bp repeated segment, while one is just outside this region. A computer scan of the 5243-bp SV40 DNA reveals that there is one segment of nine base pairs of alternating purines and pyrimidines that contains the sequence ATAT, making it a less likely candidate for forming Z-DNA. The only three sequences containing eight base pairs of alternating purines and pyrimidines are clustered as shown in Figure 12.

This study identified a potential for Z-DNA formation within the region of SV40 that regulates transcription and replication. More precisely, the

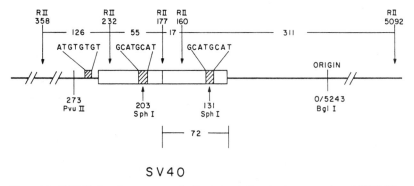

S V 4 0

Figure 12 Z-DNA–forming sequences in the transcriptional enhancer region of SV40. The two 72-bp repeats are shown in boxes and a number of restriction sites are indicated. The EcoRII sites were used to help identify the three Z-DNA–forming regions shown in cross-hatched boxes. The eight–base pair segments with purine-pyrimidine alternation are indicated. The Sph I sites lie in two of these regions (104).

Z-DNA segments are part of the transcriptional enhancer element of the viral early promoter (193–195). Enhancers are *cis*-acting components of eukaryotic promoters that stimulate transcription in a manner relatively independent of position and orientation. Enhancers may represent entry sites for transcriptional factors into the chromatin, possibly RNA polymerase. The chromatin structure of the SV40 minichromosome displays a nucleosome-free region over the viral control segments (196, 197). Additionally, two segments of hypersensitivity for DNAse I cleavage have been identified within this region, covering the origin of replication and the transcriptional enhancer, respectively. DNase I hypersensitivity is usually associated with transcriptionally active chromatin. Fine structure mapping of DNAse I hypersensitive sites in the enhancer region (198) located DNAse I cleavage approximately 25 bp on either side of the segments forming Z-DNA. It was suggested that these hypersensitive sites on either side of the Z-DNA–forming regions had their positions determined in part by the presence of a Z-DNA binding protein (104). A Z-DNA binding protein has been isolated from SV40 minichromosomes and is likely to be attached to the Z-DNA regions (Figure 12) (199).

Extensive mutational analyses have been carried out by several groups to characterize the boundaries of the SV40 enhancer element. The data derived from these studies suggest that two of the three identified Z-DNA segments might be important for the physiological activity of the SV40 enhancer. A survey of other transcriptional enhancers revealed that pairs of segments with the potential of forming Z-DNA are found in a number of transcriptional enhancers with a distance of 50 to 80 base pairs between the two Z-DNA–forming segments (104). The distance between Z-DNA–forming regions may be of importance, as it might reflect the formation of a structure in which the DNA is wrapped around a dimer of Z-DNA binding proteins so that both Z-DNA–forming regions are in contact with the two dimer subunits. Formation of such a particle might be recognized by RNA polymerase or other transcriptional factors that could dislodge the Z-DNA binding proteins; the subsequent spontaneous conversion of the Z-DNA segments to B-DNA could produce a local increase in negative supercoiling that may alter local chromatin structure and facilitate entry of RNA polymerase onto the molecule. However, the speculative nature of this hypothesis underscores the need for more detailed experimental analysis of the mechanism of transcriptional enhancement. It is possible that involvement of Z-DNA is utilized in one class of enhancers, whereas other classes may achieve equivalent effects (i.e. postulated generation of local superhelical stress) by different mechanisms.

Some variants of SV40 found in nature contain only one copy of the 72-bp repeat. This variant of SV40 has been used to examine the effect of transitions and transversions in the Z-DNA–forming regions on transcriptional enhancement and viral activity (200). The experiments are outlined in Figure 13. The results reveal that the introduction of transitions that maintain the alternations of purines and pyrimidines and the potential for forming Z-DNA do not result in a significant impairment of viral activity as

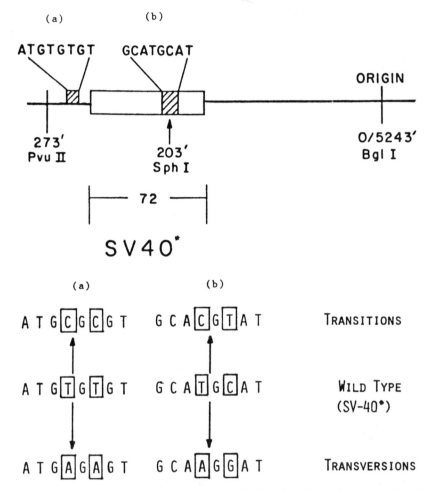

Figure 13 Diagram of mutagenesis experiments to test the role of Z-DNA in a transcriptional enhancer. An SV40 wild type is used with one copy of the 72-bp repeat (SV40*) and mutations are introduced into the two Z-DNA–forming segments, *a* and *b*. Two pyrimidines are mutated in each Z segment. In the transitions, the thymines are changed to cytosine, or cytosine to thymine using site-directed mutagenesis. In the transversions, thymine is changed to adenine and cytosine to guanine. The transversions no longer have the alternating purine-pyrimidine sequence that is favored for Z-DNA formation (200).

measured by plaque formation. They also do not significantly impair transient expression of RNA synthesis employing the enhancer-dependent (201) β-globin promoter. However, the introduction of transversions has a marked effect on all of these activities. When transversions are introduced into both of the Z-DNA–forming segments, the virus is inactivated. Further, the transcriptional enhancer shows greatly reduced ability to activate the β-globin promoter. The transversions in Figure 13 break up the alternation of purines and pyrimidines, which is optimal for Z-DNA formation.

Karin et al (202) have studied the transcriptional effects of deletions in the 5′-flanking region of the human metallothionein gene. They have discovered the position of a glucocorticoid receptor and two metallothionein apoprotein binding sites. In addition, they identified a "Z" region that was essential for transcription, containing the sequence GCGTGTGCA. There are similar Z sequences with 8 or 9 bp of purine-pyrimidine alternation 43 and 73 bp upstream. Karin and colleagues (personal comm.) have found that this DNA region is associated with transcriptional enhancer activity. It is interesting that one of the metallothionein binding sites is found between Z sequences. There is no induction of the gene until a heavy metal removes the metallothionein apoprotein binding to the DNA. Perhaps this is a mechanism for preventing attachment of a Z-DNA binding protein by the metallothionein apoprotein. Its removal may then lead to enhancer regulated transcription.

The experiments on the SV40 transcriptional enhancer provide the most detailed example to date of a specific biological role that may be carried out by Z-DNA. In particular, the use of site-directed mutagenesis in comparing the effects of either transitions or transversion on Z-DNA formation and resulting phenotypic consequences may prove valuable in a number of other systems in which the biological role of Z-DNA is being investigated.

Other Potential Effects on Transcription

A provocative role for Z-DNA as a negative effector of RNA polymerase III transcription has been demonstrated in the work of Hipskind, Clarkson, and colleagues (203, 204), who investigated in vitro transcription of two *Xenopus* met-tRNA genes. One of these genes is fully active for transcription, while the other is largely inactivated. Comparison of the 5′-flanking sequences revealed that the inactivated gene had a nine–base pair TGCGCGTGC segment five base pairs upstream from the site of initiation of transcription. Gradual removal of this segment with its potential of forming Z-DNA resulted in transcriptional reactivation. In this system, the observed transcriptional effects require a supercoiled template. Further experiments will have to be carried out to ascertain with certainty that the inhibition of pol III transcription is related to Z-DNA formation itself and not to some other peculiarity of the system.

Several types of transcriptional in vitro studies have been carried out with Z-DNA. Jovin and colleagues (54, 96, 158, 205) have studied an alcohol-containing solution that stabilized Z-DNA formation in poly(dG-dC). These experiments suggested that the aggregated Z-DNA formed in this solvent (termed Z*) was capable of acting as a substrate for RNA polymerase II, although the rate of production of transcripts was slower than that found when the same polymer was in the B-DNA form.

In other experiments, Peck & Wang (206) inserted 32 bp of alternating CG downstream of a prokaryotic promoter. They found that supercoil-induced formation of Z-DNA in the inserted segment resulted in termination of transcription at the B-Z junction. Likewise, Butzow et al (205a) found that converting poly(dG-dC) to Z-DNA resulted in a marked decrease of transcriptional activity using E. coli RNA polymerase. This raises the possibility that Z-DNA formation in vivo may result in aborting the activity of RNA polymerase and as such may have the potential of a regulatory system that is responsive to the level of negative supercoiling in the template DNA.

Recombination

In a characterization of human fetal γ-globin genes, Smithies and co-workers (206a) identified nearby segments of $d(CA/GT)_{20}$ and suggested the possibility that these might act to facilitate the recombinational gene conversion events that apparently led to the expansion in the number of globin genes. Work carried out on plasmids containing segments of $d(CA/GT)_n$ showed that these segments formed Z-DNA quite readily at intermediate superhelical densities (67, 102). A segment of DNA in which Z-DNA is more heavily represented might also transiently have some regions that are neither B-DNA nor Z-DNA and be relatively unwound. Such segments may be able to undergo strand separation and as such may provide a focus for recombination with a similar segment nearby (102). Recent investigations on synapsis formation between homologous circular plasmids as promoted by Rec I protein from Ustilago demonstrated the formation of left-handed DNA in the region of the paranemic joint (207). This leads to the general question of whether recombination is intrinsically associated with generation of negative superhelical strain and, consequently, with Z-DNA formation. If that is the case, then correlations may be found between the formation of Z-DNA and the recombinational efficiency of different segments of DNA.

Chromatin Structure

Since left-handed Z-DNA can represent unique recognition signals for specific DNA-protein interactions, it may participate in the organization of the nuclear chromatin structure of eukaryotic chromosomes. As mentioned above, there is some evidence that Z-DNA can form nucleosomes and that B-DNA nucleosomes are able to convert into Z-DNA nucleosomes (173). The properties of these two types of structures are likely to be unique. One of the problems frequently discussed is that of nucleosome phasing, i.e. the positioning of nucleosomes on DNA relative to some functional sequence. It is possible that Z-DNA formation plays a role in this process and may be involved in generation of the nucleosome gap in SV40 minichromosomes (196, 197).

The DNA in eukaryotes is organized in domains so that every 50–100 kb (kilobase-pairs) is attached to the scaffold protein of the nuclear matrix. As pointed out above, there is roughly one 50-bp or longer sequence of $d(CA/GT)_n$ every 50–100 kb in eukaryotes (185). Since this sequence can form Z-DNA readily at physiological negative superhelical densities (67, 102), such conversion could play a role in domain activation (33, 102). The Z-forming segment is long enough so that a B-to-Z (or Z-to-B) conversion would change the superhelical density of the entire domain. Other Z-forming sequences might also act in this manner. A change in superhelical density can act to regulate gene expression (208) over a large region.

The minichromosomes of SV40 have DNAse I–hypersensitive sites (198) that are usually found in active chromatin. Most of these sites are found in the control region of SV40, which is free of nucleosomes in the activated minichromosome. Three segments of the transcriptional enhancer part of the control region form Z-DNA in negatively supercoiled SV40 DNA and there is evidence that these regions are covered with Z-DNA binding proteins (104, 199). Some DNase I–hypersensitive sites were located about 25 bp on either side of the Z-DNA–forming regions (104). This suggests that a subclass of the hypersensitive sites may be associated with Z-DNA and its binding proteins. This would represent an example of local chromatin organization that is influenced by Z-DNA in a regulatory region, in contrast to the larger level of domain activation discussed above.

Z-DNA and Carcinogens

The conformations of B- and Z-DNA differ substantially and it is not surprising that interesting effects are found among the class of carcinogens that react with DNA (209). One of the principal targets of carcinogen attack is the guanine residue, and it plays a major role in Z-DNA stabilization. The reactivity of B- and Z-DNA differ from each other (135) and this distinction could be important (210). The guanine N7 and C8 atoms are completely accessible in Z-DNA (Figure 6) but are much less accessible in B-DNA. This is consistent with stabilizing the Z conformation by N-acetoxyaminofluorene, which reacts covalently with guanine C8 (142, 143, 145, 147–149, 150). Methylating carcinogens that add methyl groups to guanine N7 also stabilize Z-DNA (152), although less effectively than N-acetoxyaminofluorene. Aflatoxin that is bulky reacts with guanine N7 but stabilizes the B-DNA form of poly(dG-dC), possibly owing to its interaction with other substituents in the major groove (153).

Other substances, such as adriamycin (162) and daunomycin (163), which are active as mutagens and carcinogens convert the Z-DNA form of poly(dG-dC) to an intercalated B-DNA form.

Is it possible that some of the carcinogenic effect of these agents involves interfering with or modifying the B-Z equilibrium? If a regulatory region that must undergo B-to-Z (or Z-to-B) conversion is restrained by these compounds, could this produce the inheritable defect that we recognize as oncogenic conversion? The answer is not known.

Other Suggestive Roles

Since Z-DNA formation is dependent upon alternations of purines and pyrimidines, the occurrence of these segments in genomic DNA is of interest. For example, surveys of the 5'-flanking sequences of yeast genes have noted frequent stretches of alternating purines and pyrimidines in this region. Such examples include the genes for alcohol dehydrogenase I and cytochrome C (211). Similar sequences are found in the promoter regions of other eukaryotic genes. It is important to know if these segments are subjected in vivo to negative superhelical stresses that induce Z-DNA formation. Their presence in promoter regions indicates the possibility that they play a regulatory role. Such regulatory roles may be either positive or negative in nature depending on the site where it occurs, as stressed above. It is not surprising that Z-DNA can have both positive and negative effects as we are quite familiar with the ability of B-DNA itself to act both in a positive and negative regulatory role depending upon the particular sequences and their location.

An intriguing phenomenon is seen in a study of the distribution of d(CA/GT)$_n$ sequences in the *Drosophila* polytene chromosome. Hybridization experiments have revealed about 2,000 copies of these sequences containing 50 bp or more in the *Drosophila* chromosome (185). Through in situ hybridization it has been demonstrated that these are found on all of the *Drosophila* chromosomes; however, a considerably higher concentration of these sequences is found on the X chromosome and a lower concentration on the autosomes (212). The biological significance of this observation is not clear; however, it is known that the genes on the X chromosomes of *Drosophila* are dosage compensated in the case of the male, which has only one X chromosome. These genes are expressed to a higher level than is found in the case of the female, in which two X chromosomes are present. We do not know whether or not the presence of these sequences with their potential for forming Z-DNA is at all related to the phenomenon of dosage compensation in *Drosophila*. However, the observation is intriguing, as it clearly differentiates the *Drosophila* X chromosomes from autosomes on the molecular level.

Van de Sande, Jovin, and co-workers (54, 96, 158) have pointed out Z-DNA's proclivity for self-aggregation. This is the basis for the formation of what is termed Z*-DNA, an aggregated form of Z-DNA seen in a solution containing $MgCl_2$ and ethanol. It was suggested that such aggregation may be important in the condensation of chromosomes that occurs regularly in the mitotic cycle. The wide distribution of long segments containing $d(CA/GT)_n$ makes this an interesting possibility.

In both yeast and tetrahymena, the $d(CA/GT)_n$ sequences are positioned near the telomeres (188). The telomeres are unique places in the eukaryotic chromosome and their presence in this region may be associated with the special topological properties of chromosomal ends.

As mentioned above, m^5CG sequences stabilize Z-DNA in a powerful manner (40). These methylated sequences are also associated with gene inactivation (170). New discoveries are needed before we can relate these two phenomena.

In describing the potential role of Z-DNA, one can in principle think of two different effects, distal and proximal. The proximal effects are those associated with interactions in the immediate vicinity of the segment forming Z-DNA. Distal effects are those that may be distributed over long stretches of the genome in which the role of Z-DNA in regulating the negative superhelical density of a DNA domain may be important in modifying its biological activity. In the consideration of potential biological roles, it may be useful to recognize these two alternative modes of action (33).

CONCLUDING REMARKS

Ultimately, the biological roles of Z-DNA will be understood in the context of identifying and characterizing the various proteins that interact with this left-handed form of the double helix. The external form of Z-DNA differs significantly from B-DNA so that these proteins are likely to be distinct from B-DNA binding proteins.

Too little is known to allow observations about the relative extent to which Z-DNA is used in prokaryotic vs eukaryotic systems. It is interesting, however, that there are significant differences already observable between the relative amounts of Z-DNA that form in negatively supercoiled eukaryotic SV40 DNA on the one hand and prokaryotic plasmid pBR322 on the other. In SV40, three short segments of Z-DNA formation have been identified as the major sites in the genome (104). On the other hand, negatively supercoiled plasmid pBR322 has a much longer identified sequence forming Z-DNA, and a number of additional segments have been visualized in EM investigations (44, 103). The qualitative impression from these experiments is that the prokaryotic DNA seems to form Z-DNA more readily than the eukaryotic DNA. This may be a consequence of the markedly reduced frequency of CG sequences in eukaryotic DNA compared to prokaryotic (169). This reduction may be related to the methylation of CG sequences in the 5 position of cytosine found in eukaryotes but not in prokaryotes. The spontaneous deamination of methylated cytosine residues leads to thymine, which would cause considerable biological confusion. The reduction of CG sequences in eukaryotic DNA down to approximately 10% of what is observed in prokaryotic DNA (169) may lead to Z-DNA being found less frequently in eukaryotic DNA. However, this may make it possible for such sequences to be used more readily for regulatory roles in eukaryotic systems because of their relative infrequency. A notable difference between these two systems is the presence of the middle repetitive DNA sequences $d(CA/GT)_n$, $n > 50$, which occur frequently in eukaryotic systems and are entirely absent in prokaryotic systems (185).

It is likely that many of these issues will be resolved soon. Our ability to modify the genome through recombinant DNA techniques makes it possible either to insert or to remove Z-DNA–forming sequences. In the near future it will be possible for us to learn a great deal more about the roles that Z-DNA plays in biological systems.

ACKNOWLEDGMENTS

We wish to thank many colleagues who sent us preprints before publication. We thank Eileen Lafer for assisting with the text and Pamela Alexander for patience and expert preparation of the document. The work was supported by grants from the American Cancer Society, the National Institutes of Health, and the National Aeronautics and Space Administration. A. N. was supported by the Charles A. King Trust of Boston, Massachusetts.

Literature Cited

1. Watson, J. D., Crick, F. H. C. 1953. *Nature* 171:737
2. Wang, A. H.-J., Quigley, G. J., Kolpak, F. J., Crawford, J. L., van Boom, J. H., van der Marel, G., Rich, A. 1979. *Nature*

282:680–86
3. Rosenberg, J. M., Seeman, N. C., Kim, J. J. P., Suddath, F. L., Nicholas, H. B., Rich, A. 1973. *Nature* 243:150–54
4. Day, R. O., Seeman, N. C., Rosenberg,

J. M., Rich, A. 1973. *Proc. Natl. Acad. Sci. USA* 70:849–53

5. Pohl, F. M. 1971. In *1st Eur. Biophys. Cong.*, ed. E. Broda, A. Locker, H. Springer-Lederer, pp. 343–47. Vienna: Academy of Medicine

6. Pohl, F. M., Jovin, T. M. 1972. *J. Mol. Biol.* 67:375–96

7. Wang, A. H.-J., Hakoshima, T., van der Marel, G., van Boom, J. H., Rich, A. 1984. *Cell.* In press

8. Wang, A. H.-J., Quigley, G. J., Kolpak, F. J., van der Marel, G., van Boom, J. H., Rich, A. 1981. *Science* 211:171–76

9. Wing, R., Drew, H., Takano, T., Broka, C., Tanaka, S., Itakura, K., Dickerson, R. E. 1980. *Nature* 287:755–58

10. Drew, H., Takano, T., Tanaka, S., Itakura, K., Dickerson, R. E. 1980. *Nature* 286:567–73

11. Crawford, J. L., Kolpak, F. J., Wang, A. H.-J., Quigley, G. J., van Boom, J. H., et al. 1981. *Proc. Natl. Acad. Sci. USA* 77:4016–20

12. Drew, H. R., Dickerson, R. E. 1981. *J. Mol. Biol.* 152:723–36

13. Rich, A., Quigley, G. J., Wang, A. H.-J. 1982. In *Biomolecular Stereodynamics*, ed. R. H. Sarma, pp. 35–52. Guilderland, NY: Adenine

14. Fujii, S., Wang, A. H.-J., van der Marel, G., van Boom, J. H., Rich, A. 1982. *Nucleic Acids Res.* 10:7879–92

15. Wang, A. H.-J., Fujii, S., van Boom, J. H., Rich, A. 1983. *Cold Spring Harbor Symp. Quant. Biol.* 47:33–44

16. Haschemeyer, A. E. V., Sobell, H. M. 1964. *Nature* 202:969–71

17. Haschemeyer, A. E. V., Rich, A. 1967. *J. Mol. Biol.* 27:369–84

18. Davies, D. B. 1978. *Prog. NMR Spectrosc.* 12:135–86

19. Nordheim, A., Lafer, E. M., Peck, L. J., Wang, J. C., Stollar, B. D., Rich, A. 1982. *Cell* 31:309–18

20. Wang, A. H.-J., Gessner, R., van der Marel, G., van Boom, J. H., Rich, A. 1984. In preparation

21. Arnott, S., Chandrasekaran, R., Birdsall, D. L., Leslie, A. G. W., Ratliff, R. L. 1980. *Nature* 283:743–45

22. Behe, M., Zimmerman, S., Felsenfeld, G. 1981. *Nature* 293:233–35

23. Arnott, S., Chandrasekaran, R., Banerjee, A. K., He, R., Walker, J. K. 1983. *J. Biomol. Struct. Dynam.* 1:437–51

24. Arnott, S., Chandrasekaran, R., Hall, I. H., Puigjaner, L. C., Walker, J. K., Wang, M. 1983. *Cold Spring Harbor Symp. Quant. Biol.* 47:53–66

25. Davies, D. R., Zimmerman, S. 1980. *Nature* 283:11–12

26. Rich, A. 1982. In *Primary and Tertiary Structure of Nucleic Acids and Cancer Research*, ed. M. Miwa, et al, pp. 153–64. Tokyo: Japan Sci. Soc.

27. Dickerson, R. E., Drew, H. R., Conner, B. N., Wing, R. M., Fratini, A. V., Kopka, M. L. 1982. *Science* 216:475–78

28. Rich, A. 1982. In *Molecular Genetic Neuroscience*, ed. F. O. Schmitt, S. J. Bird, F. Bloom, pp. 13–22. New York: Raven

29. Fujii, S., Wang, A. H.-J., van Boom, J. H., Rich, A. 1982. *Nucleic Acids Res.* 11:109–12

30. Rich, A. 1982. In *The Synthesis, Structure and Function of Biochemical Molecules*, Welch Found. Conf. Chem. Res., ed. W. O. Milligan, pp. 95–117. Houston: Welch Found.

31. Dickerson, R. E., Drew, H. R., Conner, B. N., Kopka, M. L., Pjura, P. E. 1983. *Cold Spring Harbor Symp. Quant. Biol.* 47:1–12

32. Rich, A. 1983. In *Structure, Dynamics, Interactions and Evolution of Biological Macromolecules*, ed. C. Helene, pp. 3–21. Dordrecht, Holland: Reidel

33. Rich, A. 1983. *Cold Spring Harbor Symp. Quant. Biol.* 47:1–12

34. Zimmerman, S. B. 1982. *Ann. Rev. Biochem.* 51:395–427

35. Rich, A., Wang, A. H.-J., Nordheim, A. 1983. In *Nucleic Acid Research*, ed. K. Mizobuchi, I. Watanabe, J. D. Watson, pp. 11–35. New York: Academic

36. Pohl, F. M., Jovin, T. M., Baehr, W., Holbrook, J. J. 1972. *Proc. Natl. Acad. Sci. USA* 69:3805–9

37. Pohl, F. M. 1976. *Nature* 260:365–66

38. Pohl, F. M., Ranade, A., Stockburger, M. 1973. *Biochim. Biophys. Acta* 335:85–92

39. Thamann, T. J., Lord, R. C., Wang, A. H.-J., Rich, A. 1981. *Nucleic Acids Res.* 9:5443–57

40. Behe, M., Felsenfeld, G. 1981. *Proc. Natl. Acad. Sci. USA* 78:1619–23

41. Quadrifoglio, F., Manzini, G., Vasser, M., Dinkelspiel, K., Crea, R. 1981. *Nucleic Acids Res.* 9:2195–206

42. Sutherland, J. C., Griffin, K. P., Keck, P. C., Takacs, P. Z. 1981. *Proc. Natl. Acad. Sci. USA* 78:4801–4

43. Sutherland, J. C., Griffin, K. P. 1983. *Biopolymers* 22:1445–48

44. DiCapua, E., Stasiak, A., Koller, T., Brahms, S., Thomae, R., Pohl, F. M. 1983. *EMBO J.* 2:1531–35

45. Wartell, R. M., Klysik, J., Hillen, W., Wells, R. D. 1982. *Proc. Natl. Acad. Sci. USA* 79:2549–53

46. Wartell, R. M., Harrell, J. T., Zacharias,

W., Wells, R. D. 1983. *J. Biomol. Struct. Dynam.* 1:83–95

47. Benevides, J. M., Thomas, G. J. Jr. 1983. *Nucleic Acids Res.* 11:5747–61

48. Benevides, J. M., Thomas, G. J. Jr., van der Marel, G., van Boom, J. H., Wang, A. H.-J., Rich, A. 1984. *Nucleic Acids Res.* In press

49. Brahms, S., Vergne, J., Brahms, J. G., DiCapua, E., Bucher, P., Koller, T. 1982. *J. Mol. Biol.* 162:473–93

50. Brahms, S., Vergne, J., Brahms, J. G., DiCapua, E., Bucher, P., Koller, T. 1983. *Cold Spring Harbor Symp. Quant. Biol.* 47:119–24

51. Patel, D. J., Canuel, L. L., Pohl, F. M. 1979. *Proc. Natl. Acad. Sci. USA* 76:2508–11

52. Patel, D. J., Kozlowski, S. A., Nordheim, A., Rich, A. 1981. *Proc. Natl. Acad. Sci. USA* 79:1413–17

53. Cohen, J. S., Wooten, J. B., Chatterjee, C. L. 1981. *Biochemistry* 20:3049–55

54. Jovin, T. M., McIntosh, L. P., Arndt-Jovin, D. J., Zarling, D. A., Robert-Nicoud, M., van de Sande, J. H., et al. 1983. *J. Biomol. Struct. Dynamics* 1:21–57

55. Chen, C.-W., Cohen, J. S., Behe, M. 1983. *Biochemistry* 22:2136–42

56. Hartmann, B., Thuong, N. T., Pouyet, J., Ptak, M., Leng, M. 1983. *Nucleic Acids Res.* 11:4453–66

57. Feigon, J., Wang, A. H.-J., van der Marel, G. A., van Boom, J. H., Rich, A. 1983. *Nucleic Acids Res.* 12:1243–63

58. Mitra, C. K., Sarma, M. H., Sarma, R. H. 1981. *Biochemistry* 20:2036–41

59. Tran-Dinh, S., Taboury, J., Neumann, J.-M., Huynh-Dinh, T., Genissel, B., et al. 1983. *FEBS Lett.* 154:407–10

60. Hartmann, B., Pilet, J., Ptak, M., Ramstein, J., Malfoy, B., Leng, M. 1982. *Nucleic Acids Res.* 10:3261–77

61. Pilet, J., Leng, M. 1982. *Proc. Natl. Acad. Sci. USA* 79:26–30

62. Peck, L., Wang, J. 1983. *Proc. Natl. Acad. Sci. USA* 80:6206–10

63. Singleton, C. K., Klysik, J., Stirdivant, S. M., Wells, R. D. 1982. *Nature* 299:312–16

64. Stirdivant, S. M., Klysik, J., Wells, R. D. 1982. *J. Biol. Chem.* 257:10159–65

65. Peck, L. J., Nordheim, A., Rich, A., Wang, J. C. 1982. *Proc. Natl. Acad. Sci. USA* 79:4560–64

66. Haniford, D. B., Pulleyblank, D. E. 1983. *J. Biomol. Struct. Dynam.* 1:593–609

67. Haniford, D. B., Pulleyblank, D. E. 1983. *Nature* 302:632–34

68. Miller, F. D., Jorgenson, K. F., Winkfein, R. J., van de Sande, J. H., Zarling,

D. A., et al. 1983. *J. Biomol. Struct. Dynam.* 1:611–20

69. Wang, J. C., Peck, L. J., Becherer, K. 1983. *Cold Spring Harbor Symp. Quant. Biol.* 47:85–92

70. Wells, R. D., Brennan, R., Chapman, K. A., Goodman, T. C., Hart, P. A., et al. 1983. *Cold Spring Harbor Symp. Quant. Biol.* 47:77–84

71. Kuhnlein, U., Tsang, S. S., Edwards, J. 1980. *Nature* 287:363–64

72. Wu, H. M., Dattagupta, N., Crothers, D. M. 1981. *Proc. Natl. Acad. Sci. USA* 78:6808–11

73. Thomas, T. J., Bloomfield, V. A. 1983. *Nucleic Acids Res.* 11:1919–30

74. Kilpatrick, M. W., Wei, C. F., Gray, H. B. Jr., Wells, R. D. 1983. *Nucleic Acids Res.* 11:3811–22

75. Möller, A., Nordheim, A., Kozlowski, S. A., Patel, D., Rich, A. 1984. *Biochemistry* 23:54–62

76. Vardimon, L., Rich, A. 1984. *Proc. Natl. Acad. Sci. USA.* In press

77. Barton, J. K., Basile, L. A., Danishefsky, A., Alexandrescu, A. 1984. *Proc. Natl. Acad. Sci. USA.* In press

78. Barton, J. K. 1983. *J. Biomol. Struct. Dynam.* 1:621–32

79. Lafer, E. M., Möller, A., Nordheim, A., Stollar, B. D., Rich, A. 1981. *Proc. Natl. Acad. Sci. USA* 78:3546–50

80. Bugg, C. E., Thewalt, U. 1969. *Biochem. Biophys. Res. Commun.* 37:623–28

81. Tavale, S., Sobell, H. M. 1970. *J. Mol. Biol.* 48:109–23

82. Malfoy, B., Leng, M. 1981. *FEBS Lett.* 132:45–48

83. Malfoy, B., Rousseau, N., Leng, M. 1982. *Biochemistry* 21:5463–67

84. Lafer, E. M., Möller, A., Valle, R. P. C., Nordheim, A., Rich, A., Stollar, B. D. 1983. *Cold Spring Harbor Symp. Quant. Biol.* 47:155–62

85. Leng, M., Harmann, B., Malfoy, B., Pilet, J., Ramstein, J., Sage, E. 1983. *Cold Spring Harbor Symp. Quant. Biol.* 47:163–70

86. Zarling, D. A., McIntosh, L. P., Arndt-Jovin, D. J., Robert-Nicoud, M., Jovin, T. M. 1984. *J. Biomol. Struct. Dynam.* In press

87. Lafer, E. M., Valle, R. P. C., Möller, A., Nordheim, A., Schur, P. H., et al. 1983. *J. Clin. Invest.* 71:314–21

88. Möller, A., Gabriels, J. E., Lafer, E. M., Nordheim, A., Rich, A., Stollar, B. D. 1982. *J. Biol. Chem.* 257:12081–85

89. Pohl, F. M. 1983. *Cold Spring Harbor Symp. Quant. Biol.* 47:113–18

90. Thomae, R., Beck, S., Pohl, F. M. 1983. *Proc. Natl. Acad. Sci. USA* 80:5550–53

91. Nordheim, A., Pardue, M. L., Lafer, E. M., Möller, A., Stollar, B. D., Rich, A. 1981. *Nature* 294:417–22
92. Zarling, D. A., Arndt-Jovin, D. J., Robert-Nicoud, M., McIntosh, L. P., Thomae, R., Jovin, T. M. 1984. *J. Mol. Biol.* In press
93. Arndt-Jovin, D. J., Robert-Nicoud, M., Zarling, D. A., Greider, C., Weimer, E., Jovin, T. M. 1983. *Proc. Natl. Acad. Sci. USA* 80:4344–48
94. Hill, R. J., Stollar, B. D. 1983. *Nature* 305:338–40
95. Pardue, M. L., Nordheim, A., Lafer, E. M., Stollar, B. D., Rich, A. 1983. *Cold Spring Harbor Symp. Quant. Biol.* 47:171–76
96. Jovin, T. M., van de Sande, J. H., Zarling, D. A., Arndt-Jovin, D. J., Eckstein, F., et al. 1983. *Cold Spring Harbor Symp. Quant. Biol.* 47:153–54
97. Lemeunier, F., Derbin, C., Malfoy, B., Leng, M., Taillandier, E. 1982. *Exp. Cell Res.* 141:508–13
98. Lipps, H. J., Nordheim, A., Lafer, E. M., Ammermann, D., Stollar, B. D., Rich, A. 1983. *Cell* 32:435–41
99. Viegas, E. 1983. *Proc. Natl. Acad. Sci. USA* 80:5890–94
100. Morgenegg, G., Celio, M. R., Malfoy, B., Leng, M., Kuenzle, C. C. 1983. *Nature* 303:540–43
101. Nordheim, A., Tesser, P., Azorin, F., Kwon, Y. H., Möller, A., Rich, A. 1982. *Proc. Natl. Acad. Sci. USA* 79:7729–33
102. Nordheim, A., Rich, A. 1983. *Proc. Natl. Acad. Sci. USA* 80:1821–25
103. Azorin, F., Nordheim, A., Rich, A. 1983. *EMBO J.* 2:649–55
104. Nordheim, A., Rich, A. 1983. *Nature* 303:674–79
105. Lang, M. C., Malfoy, B., Freund, A. M., Daune, M., Leng, M. 1982. *EMBO J.* 1:1149–53
106. Stockton, J. F., Miller, F. D., Jorgenson, K. F., Zarling, D. A., Morgan, A. R., Rattner, J. B., van de Sande, J. H. 1983. *EMBO J.* 2:2123–28
107. Pohl, F. M., Thomae, R., DiCapua, E. 1982. *Nature* 300:545–46
108. Ramstein, J., Leng, M. 1980. *Nature* 288:413–14
109. Greve, J., Vasmel, H. 1981. *Biopolymers* 20:1329–32
110. Ivanov, V. I., Minyat, E. E. 1981. *Nucleic Acids Res.* 9:4783–98
111. Tomasz, M., Barton, J. K., Magliozzo, C. C., Tucker, D., Lafer, E. M., Stollar, B. D. 1983. *Proc. Natl. Acad. Sci. USA* 80:2874–78
112. Vorlickova, M., Kypr, J., Kleinwachter, V., Palecek, E. 1980. *Nucleic Acids Res.* 8:3965–71
113. Vorlickova, M., Kypr, J., Stokrova, S., Sponar, J. 1982. *Nucleic Acids Res.* 10:1071–80
114. Zimmer, C., Tymen, S., Marck, C., Guschlbauer, W. 1982. *Nucleic Acids Res.* 10:1081–91
115. Mitsui, Y., Langridge, R., Shortle, B. E., Cantor, C. R., Grant, R. C., et al. 1970. *Nature* 228:1166–69
116. Klysik, J., Stirdivant, S. M., Larson, J. E., Hart, P. A., Wells, R. D. 1981. *Nature* 290:672–76
117. Dhingra, M. M., Sarma, M. H., Gupta, G., Sarma, R. H. 1983. *J. Biomol. Struct. Dynam.* 1:417–27
118. Sarma, M. H., Gupta, G., Dhingra, M. M., Sarma, R. H. 1983. *J. Biomol. Struct. Dynam.* 1:59–81
119. Castleman, H., Erlanger, B. F. 1983. *Cold Spring Harbor Symp. Quant. Biol.* 47:133–42
120. Revet, B., Delain, E., Dante, R., Niveleau, A. 1983. *J. Biomol. Struct. Dynam.* 1:857–71
121. Jayaraman, S., Yathindra, N. 1980. *Biochem. Biophys. Res. Commun.* 97:1407–19
122. Harvey, S. C. 1983. *Nucleic Acids Res.* 11:4867–78
123. Olson, W. K., Srinivasan, A. R., Marky, N. L., Balaji, V. N. 1983. *Cold Spring Harbor Symp. Quant. Biol.* 47:229–42
124. Tidor, B., Irikura, K. K., Brooks, B. R., Karplus, M. 1983. *J. Biomol. Struct. Dynam.* 1:231–51
125. Kollman, P., Weiner, P., Quigley, G., Wang, A. 1982. *Biopolymers* 21:1945–69
126. Hopkins, R. C. 1981. *Science* 211:289–91
127. Hopkins, R. C. 1983. *Cold Spring Harbor Symp. Quant. Biol.* 47:129–32
128. Hopkins, R. C. 1983. *J. Theor. Biol.* 101:327–33
129. Sasisekharan, V., Gupta, G. 1980. *Curr. Sci.* 49:43–48
130. Gupta, G., Bansal, M., Sasisekharan, V. 1980. *Proc. Natl. Acad. Sci. USA* 77:6486–90
131. Sasisekharan, V. 1983. *Cold Spring Harbor Symp. Quant. Biol.* 47:45–52
132. Benham, C. J. 1980. *Nature* 286:637–38
133. Benham, C. J. 1983. *Cold Spring Harbor Symp. Quant. Biol.* 47:219–28
134. Lavery, R., Pullman, B. 1981. *Nucleic Acids Res.* 9:4677–88
135. Zakrzewska, K., Lavery, R., Pullman, A., Pullman, B. 1980. *Nucleic Acids Res.* 8:3917–32
136. Pullman, B., Lavery, R., Pullman, A. 1982. *Eur. J. Biochem.* 124:229–38
137. Pullman, B. 1983. *J. Biomol. Struct. Dynam.* 1:773–94

138. Van Lier, J. J. C., Smits, M. T., Buck, H. M. 1983. *Eur. J. Biochem.* 132:55–62
139. Klysik, J., Stirdivant, S. M., Singleton, C. K., Zacharias, W., Wells, R. D. 1983. *J. Mol. Biol.* 168:51–71
140. McIntosh, L. P., Greiger, I., Eckstein, F., Zarling, D. A., van de Sande, J. H., Jovin, T. M. 1983. *Nature* 294:83–86
141. Uesugi, S., Shida, T., Ikehara, M. 1982. *Biochemistry* 21:3400–8
142. Sage, E., Leng, M. 1981. *Nucleic Acids Res.* 9:1241–50
143. Sage, E., Leng, M. 1980. *Proc. Natl. Acad. Sci. USA* 77:4597–601
144. Grunberger, D., Santella, R. M. 1981. *J. Supramol. Struct. Cell Biochem.* 17:231–44
145. Santella, R. M., Grunberger, D., Broyde, S., Hingerty, B. E. 1981. *Nucleic Acids Res.* 9:5459–67
146. Santella, R. M., Grunberger, D., Weinstein, I. B., Rich, A. 1981. *Proc. Natl. Acad. Sci. USA* 78:1451–55
147. Santella, R. M., Grunberger, D., Nordheim, A., Rich, A. 1982. *Biochem. Biophys. Res. Commun.* 106:1226–32
148. Spodheim-Maurizot, M., Malfoy, B., Saint-Ruf, G. 1982. *Nucleic Acids Res.* 10:4423–30
149. Wells, R. D., Miglietta, J. J., Klysik, J., Larson, J. E., Stirdivant, S. M., Zacharias, W. 1982. *J. Biol. Chem.* 257:10166–70
150. Rio, P., Leng, M. 1983. *Nucleic Acids Res.* 11:4947–56
151. Santella, R. M., Grunberger, D. 1983. *Environ. Health Perspect.* 49:107–15
152. Möller, A., Nordheim, A., Nichols, S. R., Rich, A. 1981. *Proc. Natl. Acad. Sci. USA* 78:4777–81
153. Nordheim, A., Hao, W. M., Wogan, G. N., Rich, A. 1983. *Science* 219:1434–36
154. Malfoy, B., Hartmann, B., Leng, M. 1981. *Nucleic Acids Res.* 9:5659–69
155. Ushay, H. M., Santella, R. M., Caradonna, J. P., Grunberger, D., Lippard, S. J. 1982. *Nucleic Acids Res.* 10:3573–88
156. Castleman, H., Hanau, L. H., Erlanger, B. F. 1983. *Nucleic Acids Res.* 11:8421–29
157. Behe, M. J., Szu, S. C., Charney, E., Felsenfeld, G. 1983. *Biopolymers.* In press
158. van de Sande, J. H., Jovin, T. M. 1982. *EMBO J.* 1:115–20
159. van de Sande, J. H., McIntosh, L. P., Jovin, T. M. 1982. *EMBO J.* 1:777–82
160. Zacharias, W., Martin, J. C., Wells, R. D. 1983. *Biochemistry* 22:2398–405
161. Zacharias, W., Larson, J. E., Klysik, J.,

Stirdivant, S. M., Wells, R. D. 1982. *J. Biol. Chem.* 257:2775–82
162. Mirau, P. A., Kearns, D. R. 1983. *Nucleic Acids Res.* 11:1931–41
163. Chaires, J. B. 1984. *Nucleic Acids Res.* 11:8485–94
163a. Van Helden, P. D. 1983. *Nucleic Acids Res.* 11:8415–20
164. Zimmer, C., Marck, C., Guschlbauer, W. 1983. *FEBS Lett.* 154:156–60
165. Hanau, L. H., Santella, R. M., Grunberger, D., Erlanger, B. F. 1984. *J. Biol. Chem.* 259: In press
166. Russell, W. C., Precious, B., Martin, S. R., Bayley, P. M. 1983. *EMBO J.* 2:1647–53
167. Gupta, G., Dhingra, M. M., Sarma, R. H. 1983. *J. Biomol. Struct. Dynam.* 1:97–113
168. Chen, C. W., Knop, R. H., Cohen, J. S. 1983. *Biochemistry* 22:5468–71
169. Bird, A. P. 1980. *Nucleic Acids Res.* 8:1499–504
170. Doerfler, W. 1983. *Ann. Rev. Biochem.* 52:93–124
170a. Nickol, J. M., Felsenfeld, G. 1983. *Cell* 35:467–77
171. van der Marel, G. A., Wille, G., Westerink, H., Wang, A. H.-J., Rich, A., et al. 1982. *J. R. Netherlands Chem. Soc.* 101:77–78
172. Nickol, J., Behe, M., Felsenfeld, G. 1982. *Proc. Natl. Acad. Sci. USA* 79:1771–75
173. Miller, F. D., Rattner, J. B., van de Sande, J. H. 1983. *Cold Spring Harbor Symp. Quant. Biol.* 47:571–76
174. Prevelige, P. E. Jr., Fasman, G. D. 1983. *Biochim. Biophys. Acta* 739:85–96
175. Bestor, T. H., Ingram, V. M. 1983. *Proc. Natl. Acad. Sci. USA* 80:5559–63
176. Pfeifer, G. P., Grunwald, S., Boehm, T. L. J., Drahovsky, D. 1983. *Biochem. Biophys. Acta* 740:323–31
177. Wang, J. C. 1980. *Trends Biochem. Sci.* 5:219–21
178. Cantor, C. R. 1981. *Cell* 25:293–95
179. Nordheim, A., Peck, L. J., Lafer, E. M., Stollar, B. D., Wang, J. C., Rich, A. 1983. *Cold Spring Harbor Symp. Quant. Biol.* 47:93–100
180. O'Connor, T., Kilpatrick, M. W., Klysik, J., Larson, J. E., Martin, J. C., et al. 1983. *J. Biomol. Struct. Dynam.* 1:999–1009
181. Singleton, C. K., Klysik, J., Wells, R. D. 1983. *Proc. Natl. Acad. Sci. USA* 80:2447–51
182. Stettler, U. H., Weber, H., Koller, T., Weissman, C. 1979. *J. Mol. Biol.* 131:21–40
183. Klevan, L., Schumaker, V. N. 1982. *Nucleic Acids Res.* 21:6809–17

184. Miesfeld, R., Krystal, M., Arnheim, N. 1981. *Nucleic Acids Res.* 9:5931–38
185. Hamada, H., Petrino, M. G., Kakunaga, T. 1982. *Proc. Natl. Acad. Sci. USA* 79:6465–69
186. Hamada, H., Kakunaga, T. 1982. *Nature* 298:396–98
187. Saffer, J. D., Lerman, M. I. 1983. *Mol. Cell. Biol.* 3:960–64
188. Walmsley, R. M., Szostak, J. W., Petes, T. D. 1983. *Nature* 302:84–86
189. Miller, F. D., Winkfein, R., Stockton, J. F., Rattner, J. B., van de Sande, J. H. 1984. In preparation
190. Viegas-Pequignot, E., Derbin, C., Lemeunier, F., Taillandier, E. 1982. *Ann. Genet.* 25:218–22
191. Jones, J., Nordheim, A., Rich, A. 1984. *Nature.* In press
192. Rich, A., Nordheim, A., Azorin, F. 1983. *J. Biomol. Struct. Dynam.* 1:1–19
193. Gruss, P., Dhar, R., Khoury, G. 1981. *Proc. Natl. Acad. Sci. USA* 78:943–47
194. Benoist, C., Chambon, P. 1981. *Nature* 290:304–10
195. Khoury, G., Gruss, P. 1983. *Cell* 33:313–14
196. Varshavsky, A. J., Sundin, O., Bohn, M. 1981. *Nucleic Acids Res.* 5:5931–38
197. Scott, W. A., Wigmore, D. J. 1978. *Cell* 15:1511–18
198. Cereghini, S., Herbomel, P., Jounneau, J., Saragosti, S., Katinka, M., et al. 1983. *Cold Spring Harbor Symp. Quant. Biol.* 47:935–44
199. Azorin, F., Rich, A. 1984. In preparation
200. Herr, W., Gluzman, Y., Nordheim, A., Rich, A. 1984. In preparation
201. Banerji, J., Rusconi, S., Schaffner, W. 1981. *Cell* 27:299–308
202. Karin, M., Haslinger, A., Holtgreve, H., Richards, R. I., Krantes, P., et al. 1984. *Nature.* In press
203. Hipskind, R. A., Mazabraud, A., Corlet, J., Clarkson, S. G. 1983. *Cold Spring Harbor Symp. Quant. Biol.* 47:873–78
204. Hipskind, R. A., Clarkson, S. G. 1983. *Cell* 34:881–90
205. Durand, R., Job, C., Zarling, D. A., Teissere, M., Jovin, T. M., Job, D. 1983. *EMBO J.* 2:1707–14
205a. Butzow, J. J., Shin, Y. A., Eichhorn, G. L. 1984. *Biochemistry.* In press
206. Peck, L., Wang, J. 1984. *Cell.* In press
206a. Slightom, J. L., Blechl, A. E., Smithies, O. 1980. *Cell* 21:627–38
207. Kmiec, E. B., Holloman, W. K. 1984. *Cell.* 36:593–98
208. Smith, G. R. 1981. *Cell* 24:191–93
209. Neidle, S. 1981. *Nature* 292:292–93
210. Gosselin, G., Imbach, J.-L. 1983. *Eur. J. Med. Chem.-Clin. Ther.* 18:393–400
211. Smith, M., Leung, D. W., Gillam, S., Astell, C. R., Montgomery, D. L., Hall, B. D. 1979. *Cell* 16:753–61
212. Pardue, M. L., Nordheim, A., Rich, A. 1984. *Cell.* In press
213. Mahendrasingam, A., Pigram, W. J., Fuller, W., Brahms, J., Vergne, J. 1983. *J. Mol. Biol.* 168:897–901
214. Sasisekharan, V., Brahmachari, S. K. 1981. *Curr. Sci.* 50:10–13

Ann. Rev. Biochem. 1984. 53: 119–62

12

STRUCTURE OF RIBOSOMAL RNA

Harry F. Noller

Thimann Laboratories, University of California, Santa Cruz, California, 95064

CONTENTS

PERSPECTIVES AND SUMMARY

Ribosomes are ribonucleoprotein particles that are responsible for translation of the genetic code. Unlike other cellular polymerases, ribosomes typically contain 50 to 60 percent RNA as an integral part of their structures. It is an intriguing problem to understand why this is so, and its elucidation is likely to be inseparable from an understanding of the structure, function, assembly, and evolution of these particles. This review is an attempt to summarize recent progress in our understanding of ribosomal RNA (rRNA) structure, and to point out some of the implica-

397

tions of these findings as they relate to broader biological questions. The reader is referred to previous reviews on the topics of ribosome structure and function (1–6) and ribosomal RNA (7–14). For the most part, emphasis is on the large (16S- and 23S-like) rRNAs; consideration of 5S rRNA is confined to a few salient topics. Detailed discussions of 5S rRNA and 5.8S rRNA may be found in several recent review articles (15–20). A recent compilation of 5S rRNA sequences is in (21).

Many complete sequences are now available for the large rRNAs from organisms representing a wide phylogenetic spectrum. This has made possible elucidation of the secondary structures of the rRNAs by comparative sequence analysis and identification of highly conserved as well as phylogenetically variable regions of these molecules. The secondary structures begin to suggest details of the molecular organization of the rRNAs, outlining major structural domains and placing certain constraints on tertiary folding. Clues to the three-dimensional structure of rRNA in the ribosome are beginning to emerge, but we are not yet at the stage where a detailed model can be built.

The question of the direct involvement of rRNA in translation has become an increasingly interesting one in recent years. Affinity labeling experiments and specific protection of rRNA from chemical and enzymatic probes by functional ligands imply participation of RNA in a variety of ribosomal functions. Genetic studies, particularly in the yeast mitochondrial system, have shown that point mutations in rRNA genes can confer functional alterations upon ribosomes. The suggestion that ribosomal RNA plays a fundamental and direct role in translation is supported by the available experimental evidence and is entirely consistent with the very highly conserved nucleotide sequences and secondary structure elements that are found in these molecules.

rRNA Species—Nomenclature

In prokaryotic cells (in both the eubacterial and archaebacterial lines), ribosomes contain three ribosomal RNA molecules, usually called 5S, 16S, and 23S rRNA. These molecules contain about 120, 1540, and 2900 nucleotides, respectively (23–30). This is true for chloroplast ribosomes as well (31–32, 35–36), although mitochondria show a wide range of variability in the sizes of their rRNAs (37–49), and appear to lack 5S rRNA (50). Many chloroplasts contain a 4.5S rRNA corresponding to the 3' terminal 96 nucleotides of 23S rRNA (51, 52), and some contain a 7S rRNA, corresponding to the 5' 291 nucleotides of 23S rRNA (53). The 23S-like rRNA from *Paramecium* mitochondria also contains a 7S rRNA corresponding to the 5' ~310 nucleotides of 23S rRNA (44). In eukaryotic cytoplasmic ribosomes, the three main classes of rRNA are usually called 5S, 17S or 18S, and 26S or 28S rRNA. Most eukaryotic cytoplasmic ribosomes also contain a 5.8S rRNA, which corresponds to the 160 nucleotides at the 5'-terminus of prokaryotic 23S rRNA (29, 54–56); some insects also have a 2S rRNA corresponding to the 3' ~ 25 nucleotides of 5.8S rRNA (57).

For the sake of clarity, I often refer to rRNAs from small ribosomal subunits as "16S-like" rRNAs, or "small subunit" rRNAs; similarly, the large rRNAs from the large subunit are referred to as "23S-like" or "large subunit" rRNAs. To avoid confusion, I always refer to 5S rRNA as such. Other nomenclature (e.g., 12S, 18S, etc) is used where specifically required. In referring to specific positions in rRNA structures, I use here the numbering of *Escherichia coli* 16S, 23S, and 5S rRNA (22–23, 29, 58) unless otherwise noted.

PRIMARY STRUCTURE

At this writing, nucleotide sequences for over twenty 16S-like rRNAs, and over a dozen 23S-like rRNAs have been completed (Table 1). For the most part, these have been deduced by DNA sequencing of the corresponding cloned rRNA genes. The DNA sequencing approach has been favored for several reasons: Ribosomal RNA genes are easily cloned, since it is usually relatively easy to obtain rRNA for use as a hybridization probe, and rRNA genes are present in multiple copies in most organisms; restriction enzymes provide a wide variety of precise fragments for sequencing; and both strands of the DNA can be sequenced, providing independent verification of the sequence. Drawbacks to DNA sequencing are that location of the ends of the mature rRNAs and information about post-transcriptionally modified nucleotides must be obtained independently. Also, the extent of sequence heterogeneity in the rRNA population is not apparent using this approach. In any case, the accuracy of the resulting sequence is by far the most crucial factor in any consideration of sequencing methodology, and in this regard DNA methods are presently superior.

Examination of Table 1 shows that the size of the 16S-like and 23S-like rRNAs varies over a two to three fold range. The smallest sequenced 16S-like rRNAs are the trypanosome mitochondrial 9S rRNA (597 residues) (37; Table 1) and the mammalian mitochondrial 12S rRNAs (~ 950 residues). The largest examples are the eukaryotic cytoplasmic 18S rRNAs (59–62; Table 1), which can comprise over 1800 nucleotides. This extreme variation in size is not to be taken as an absence of structural constraints, however. As is discussed below, there is clear homology between the higher-order structures of rRNAs from all sources. Furthermore, certain regions of rRNA show extremely high sequence conservation. It is convenient to think of "typical" 16S-like and 23S-like rRNAs as having about 1500 and 2900 nucleotides, respectively, and the other types as being variants of these. Typical rRNAs would comprise those of eubacteria, archaebacteria, chloroplasts, and plant mitochondria; the variant types would include eukaryotic cytoplasmic and the remaining mitochondrial rRNAs.

The content of post-transcriptionally modified nucleotides in rRNA is much less than that of tRNA, and is typically less than 1% in prokaryotes. In the cases of E. coli 16S rRNA and Xenopus laevis 18S rRNA, all or most of the modified nucleotides have been identified (60, 67–69) and placed in the RNA chain (23–24, 60). In eubacteria, the majority of modifications are base methylations (67–69). In E. coli 16S rRNA, these are clustered in six regions (Figure 1): $^7_mG_{527}$; $^2_mG_{966}$ and $^5_mC_{967}$; $^2_mG_{1207}$; $^4_mC_{m1402}$ and

Table 1 Sequenced 16S-like and 23S-like rRNAs

Organism	Residues[g]	Reference
16S-like		
Eubacteria		
E. coli	1542	23, 24
B. brevis	(1540)	25
P. vulgaris	1544	26
A. nidulans	(1487)	27
Chloroplasts		
Maize chloroplast	1490	31
Tobacco chloroplast	1485	32
Euglena chloroplast	1491	33
Chlamydomonas chloroplast	1475	34
Archaebacteria		
H. volcanii	1469	28
Eukaryotes		
S. cerevisiae cytoplasmic (18S)	1799	59
X. laevis (18S)	1825	60
Rat liver (18S)	1874	61, 62
Dictyostelium (17S)	1873	a
Maize (18S)	—	b
Mitochondria		
Trypanosome	597[c]	37
Human	954	38
Bovine	958	39
Mouse	956	40
Rat	953	41
Yeast	1686	42, 43
Aspergillus	1437	44
Paramecium	(1607)	45
Wheat	1957	d
Maize	(1962)	e

Table 1 (*continued*)

Organism	Residues[g]	Reference
23S-like		
Eubacteria		
E. coli	2904	29
B. stearothermophilus	2931	30
Chloroplasts		
Maize chloroplast	2903	35
Tobacco chloroplast	2904	36
Eukaryotes		
S. cerevisiae (26S)	3392	65
S. carlsbergensis (26S)	3393	64
Physarum (26S)	3788	63
X. laevis (28S)	(4110)	f
Rat (28S)	4718	66
Mitochondria		
Trypanosome[c]	1152	37
Human	1559	38
Bovine	1571	39
Mouse	1582	40
Rat	1584	41
Paramecium	(2380+)	47
Yeast	3273	48
Aspergillus	2768	49

[a] M. Sogin, personal communication.
[b] J. Messing, personal communication.
[c] P. Sloof, personal communication.
[d] M. Gray, personal communication.
[e] R. Sederoff, personal communication.
[f] S. Gerbi, personal communication.
[g] In cases where there is some uncertainty about the actual length of the mature rRNA chain, the best estimate for the number of residues is given in parentheses.

$_m^5C_{1407}$; $_mU_{1498}$; $_m^2G_{1516}$, $_{m2}^6A_{1518}$ and $_{m2}^6A_{1519}$ (23–24). In *E. coli* 23S rRNA, pseudouridines and ribothymidines are found, in addition to base and ribose methylations (67–68). There is also evidence for the existence of other kinds of base modifications in *E. coli* 23S rRNA, as yet uncharacterized (S. Turner, unpublished). In *X. laevis* 18S rRNA, there are numerous ribose methylations in addition to base modifications (60), as appears to be generally true for eukaryotic cytoplasmic rRNAs (70–71). The modified bases tend to occur in positions in the sequence analogous to those in eubacteria, while the ribose methylations are more widely distributed. In most cases, the post-transcriptional modifications in *E. coli* 16S rRNA have been found to be located in regions that are accessible to chemical probes in the intact 30S ribosomal subunit (72). These regions furthermore tend to be highly conserved in primary and secondary structure (10, 13) (see Structure and Function section).

Most intriguing are the regions of very high conservation of primary structure in the large rRNAs. In the 16S-like rRNAs, positions 322–329, 515–533, 691–699, 1047–1061, 1390–1407, and 1492–1506 are nearly universal (13; Figure 1). Only the mitochondria show significant variation within these sequences. The three sequences around positions 530, 1400, and 1500, each containing about twenty conserved nucleotides, are particularly striking. The loop around position 530 maintains its conserved sequence for the most part even in the most extreme mitochondrial examples, and is virtually unchanged across the three major phylogenetic lines. One can only conclude that these nucleotide sequences are among the most ancient of those presently in existence, and that their persistence in the face of evolutionary pressures is a measure of the crucial role of ribosomal RNA in the translation process. A detailed analysis of the phylogenetic conservation of each section of the 16S-like rRNAs may be found in (13). In the 23S-like rRNAs, sequences that are universal, or nearly so, include those at the approximate positions 671–677, 803–811, 1664–1677, 1833–1838, 1900–1905, 1927–1943, 1959–1973, 1990–1996, 2059–2065, 2445–2454, 2467–2483, 2500–2506, 2552–2556, 2583–2589, and 2654–2665.

The high homology among the sequences of rRNAs from phylogenetically diverse sources is by itself persuasive evidence that all ribosomes are related through a common ancestral ribosome. Based on this notion, Woese and colleagues have used rRNA sequences, or RNase T_1 oligomers thereof, to measure evolutionary distances between organisms (73). The power of this approach is documented by their discovery of the third major line of descent, the archaebacteria (74); sequences of 16S rRNAs from these organisms were shown to be no more closely related to those of eubacteria than to those of eukaryotes. Still another application is the use of comparative sequence analysis in deducing secondary structure, as described in the following section.

SECONDARY STRUCTURE

General Approaches

COMPARATIVE SEQUENCE ANALYSIS Determination of secondary structure, even in a relatively small RNA molecule, can be surprisingly difficult, as demonstrated in the case of E. coli 5S rRNA. Although only 120 nucleotides in length, eight years elapsed between elucidation of its primary (22) and secondary (75–76) structures. The large number of spurious secondary structures for this molecule is well documented (77). Because the complexity of the problem of secondary-structure determinations increases approximately as the second power of the length of the RNA chain in question, one can readily sense the need for extremely reliable criteria with which to test the validity of each base-pairing possibility for the large rRNAs. As clearly demonstrated in the cases of tRNA and 5S rRNA, the most convincing proof for the existence of biologically significant pairing (short of X-ray crystallographic evidence, perhaps) comes from comparative sequence analysis.

Briefly, the comparative approach begins with the assumption that the molecules under comparison have essentially the same secondary structure. The primary structures are then aligned, using positions of high sequence homology as a guide. Success in this approach is crucially dependent upon the precision of the sequence alignment; if there is uncertainty at this stage, any conclusions about secondary structure can be taken to be equivocal. Next, base changes between the compared sequences are noted, and where compensating base changes maintain complementarity between two potential pairing regions, this is taken as evidence for the existence of a true helix at that position. Conversely, uncompensated changes, leading to mismatches (except in the case of G-C to G-U changes), are evidence against the helix. Clearly, the more independent examples of compensating changes that can be found for a given helix, the more likely it is to be correct. In studies on the large rRNAs, a useful criterion has been to consider two independent sets of compensating base changes within a helix as proof for its existence (13, 78). In principle, comparative sequence analysis amounts to assessing the data from a kind of preexisting genetic experiment in which all existing organisms can be considered pseudo-revertants of the various mutations in rRNA genes that have occurred during evolution.

Most investigators now use the comparative approach in studies on the secondary structures of the large rRNAs (10, 13, 78–82a). Discrepancies among structural models proposed by different laboratories can usually be attributed to the degree of emphasis placed on the comparative approach relative to other criteria, or to lack of rigor in its use, most commonly in the sequence-alignment step. The more sequences that become available, the more straightforward the alignment process becomes, as patterns of conservation vs variation become established. Furthermore, as in the case of the tRNA cloverleaf, the initial secondary-structure model becomes a prediction that must be fulfilled by a precise fit to each newly derived nucleotide sequence. That a 16S rRNA secondary structure derived on the basis of only three complete sequences and RNase T_1 oligomer catalogs has survived testing by all of the currently available 24 complete 16S-like rRNA sequences (which represent a very broad sampling of the phylogenetic spectrum; Table 1) is in itself convincing evidence for its validity.

COMPUTER ANALYSIS   Because of the size of the large rRNA molecules, it is natural to employ computer methods at various levels of secondary-structure analysis. These approaches can be distinguished by the extent to which the problem is handled in the computer, and by which aspects of the problem are addressed by this method. At one extreme, the computer may be used simply to store sequence information; at the other extreme, the output may take the form of a fully developed secondary structure. Most of these methods center on free-energy minimization algorithms (83–88), using empirically determined values for stability parameters (89). Some approaches also allow incorporation of additional biological information, such as comparative sequence information (85), and positions that are susceptible to single or double strand–specific probes (88). Thus far, satisfactory secondary structures for the large rRNAs have not been obtained by use of purely computational approaches, although this appears to have been achieved in the case of 5S rRNA (85). A major difficulty in applying computer-based methods to the large rRNAs is that the requirement for computing time and/or working memory usually increases as the third or higher power of the length of the polynucleotide chain. This should become a less serious obstacle as faster processors with much larger memories become available and as more efficient algorithms are developed for making use of comparative sequence information.

FREE-ENERGY PREDICTION   Another problem that continues to be explored is the estimation of free energies of potential RNA secondary structures. If the true biological structure(s) of an RNA molecule is that which has the lowest global free energy, then it is important to determine which aspects of the structure contribute significantly to this number. Values for nearest-neighbor stacking energies and costs of unpaired loops, bulges and mismatches have been empirically estimated (89) and refined (90–91; I. Tinoco, personal communication) for helical RNA oligomers. These are potentially powerful tools, but alone do not presently appear to suffice for prediction of secondary structure. For example, in certain regions of 16S rRNA, the biologically significant structure (as judged by all other criteria, including comparative sequence analysis) appears to comprise helices that are predicted to be significantly less stable than some that can be rigorously excluded by comparative analysis (79). It may well be that when one compares the total free energies of all helices in competing structures, the *global* free energy is in fact lower in the biologically significant structure, even though individual helices may not be the most stable ones on a helix-by-helix basis. A practical difficulty arises in attempting to test this possibility, since the total number of potential secondary models to be tested for a molecule the size of 16S rRNA is greater than 10^{100} (259) (for comparison, the number of fundamental particles in the universe is estimated to be about 10^{80}).

Another important consideration is that structure other than secondary may contribute significantly to the stability of the folded RNA molecule. There are indications that certain tertiary features of tRNA are in fact more stable than some secondary interactions (92). Tight binding of multivalent cations may also make important contributions. Finally, ribosomal RNAs exist as ribonucleoprotein particles, interacting strongly and specifically with many ribosomal proteins. Binding of r-proteins to rRNA is known to alter the thermal stability of the higher-order structure of the RNA, usually decreasing or eliminating low-temperature "premelting" effects that are observed in melting curves of the naked rRNAs, and often increasing the cooperativity or sharpness of the melting transition (93). The influence of ribosomal proteins on the secondary structure of rRNA is probably limited, however, judging by the close agreement of biochemical probe experiments on naked rRNA with the phylogenetically derived secondary-structure models (see below). Very likely, all of these factors ultimately play a part in determining and stabilizing the structure of rRNA, but quantitative estimation of their respective contributions to the conformational free energy is not presently within reach.

EXPERIMENTAL METHODS   A wide variety of experimental approaches has been employed in the study of rRNA structure. It is probably fair to say that experimental methods alone are not yet sufficient to deduce a correct secondary structure but are an extremely important test of the structural features derived by comparative sequence analysis. The most successful of these approaches include (a) the use of chemical and enzymatic probes, both of which are represented by single strand– and double strand–specific classes (72, 95–111); (b) two-dimensional gel systems, in which RNA fragments from partial nuclease digests, associated by base pairing in the first dimension, are resolved under dissociative conditions in the second dimension (25, 108, 141); (c) oligonucleotide probes, which presumably bind to free, unpaired regions of the rRNA (112–115); (d) direct observation of structural features in partially unfolded rRNA by electron microscopy (120, 260); and nuclear magnetic resonance techniques (116–119).

Among the more useful single strand–specific chemical probes are kethoxal (G-specific) (72, 95), diethyl pyrocarbonate (A- and G-specific) (96), bisulfite (C- and U-specific) (79), dimethyl sulfate (C-specific as a single-strand probe; G-specific as a "major groove" or tertiary-structure probe at ^7N) (96–97), and m-chloroperbenzoate (79) and monoperphthalate (98) (both A-specific). The reliability of chemical modification in obtaining detailed structural information has been verified in several cases by studies on tRNA, where results can be compared with a known structure (96, 99). Examples of chemical reagents that have been used as double strand–specific probes are the psoralen derivatives aminomethyltrioxsalen (AMT) and hydroxymethyltrimethylpsoralen (HMT) (100–105). Psoralens inter-calate between stacked base pairs in nucleic acids and undergo photo-chemically induced cycloaddition to pyrimidines on opposite strands (106). A number of ribonucleases have been used as structural probes, including RNase T_1 (24), RNase A (24), RNase T_2 (107), S_1 nuclease (108) (all single strand specific) and cobra venom RNase (109–111) (double strand specific). A potential danger in the use of nucleases, particularly single strand–specific ones, as structural probes is that cleavage of the sugar-phosphate chain may lead to significant unfolding, or other rearrangement of the RNA structure. Thus, it is important to distinguish between primary and secondary cleavage events, and to rely only on primary cleavage sites as indicators of structure; in smaller RNAs, this has been achieved by an elegant method utilizing alternate 5' and 3' end labeling (111). Oligo-nucleotides complementary to specific sequences in rRNA have been used to probe single-stranded regions (112–115). With the ready availability of synthetic DNA oligomers of predetermined sequence, this general ap-proach should become increasingly useful. A potential hazard is that the oligomer may itself perturb the RNA structure, in cases where the stability of an oligomer-rRNA duplex is significantly more stable than an existing structural feature of the rRNA. Also, care must be taken to ensure (e.g. by the use of RNase H) that the mode of binding of the oligomer is by conventional base pairing (114).

Nuclear magnetic resonance has been used with considerable success in probing the structure of 5S rRNA (116–118), and the colicin fragment of 16S rRNA (119), comprising the 3'-terminal 49 nucleotides of the latter molecule. Application of the Nuclear Overhauser Effect (NOE) (92) allows unambiguous assignment of individual proton resonance to specific nucleotide residues. One is then permitted to make measurements in solution to study the dynamic behavior of individual nucleotides, as well as to establish or confirm the existence of base pairing and, in principle, other higher-order structure. Unfortunately, this method appears to be limited to the study of small RNA molecules or RNA fragments. Another physical approach that has been employed with some success is electron microscopy (120, 260). RNA molecules are spread on grids under partial unfolding conditions, and the sizes of resulting loops and tails are measured. In this approach, choice of unfolding conditions is crucial, so that some features are maintained while others are disrupted, without at the same time introducing structural rearrangements. The resolution of this method is somewhat limited, not only by the resolution of the electron microscope, but also because small hairpins or other local structure, too small to be resolved, nevertheless contribute to variable shortening of the apparent length of the RNA chain.

Description of Secondary-Structure Models

General secondary-structure models for 16S and 23S rRNAs have been proposed by three different groups of investigators (7, 10, 13, 78–82) and numerous proposals for the secondary structure of 5S rRNA have been made (15–20, 75–76). Here, discussion will center on the models presented in Figures 1–3 (13; R. Gutell, H. Noller, C. Woese, unpublished); 16S rRNA models proposed by the other two groups (80–81), are in general agreement with the model shown in Figure 1. A detailed discussion of specific differences between the three 16S rRNA models may be found elsewhere (13). The model for the secondary structure of 23S rRNA shown in Figure 2 differs in several significant aspects from those proposed by the other two groups (82–82a), summarized below. As in the case of 16S rRNA, there is better agreement with the Strasbourg-Freiburg model (82a) than with the Berlin-Freiburg model (82).

16S rRNA   The secondary-structure model for 16S rRNA presented in Figure 1 (13) contains a number of refinements that have been added since the last published descriptions (10). Significantly, every one of the helices that was considered proven by comparative analysis in the first published version of the structure (7) has survived to the present version, having been tested by some twenty additional complete 16S-like RNA sequences and an abundance of experimental information. Refinements have been made possible mainly by the number and phylogenetic variety of newly completed sequences (Table 1). The helices considered to be proven by comparative sequence analysis (13, 79) (two or more phylogenetically independent base-pair changes occurring in a given helix) are indicated in the figures by shading. Few possibilities for additional base pairing remain, and nearly all of the helices depicted in Figure 1 are supported by comparative analysis. Helices not shown as shaded are either so constant in sequence that base changes have not yet been identified (e.g. 564–570/880–886) or are so variable that strictly analogous structures have not been found in two or more organisms for a given region (e.g. 1435–1445/1457–1466). Nonshaded helices should therefore be taken as tentative.

The 16S rRNA molecule is subdivided into three major structural domains, and one minor domain, by three sets of long-range base-paired interactions (Figure 1). The 5' domain (residues $\sim$ 26–557) is defined by the helix 27–37/547–556; the central domain (residues $\sim$ 564–912) by the helix 564–570/880–886; and the 3' major (residues $\sim$926–1391) and 3' minor (residues ~1392–1542) domains by the 926–933/1384–1391 helix. The extent to which these domains, defined here only in terms of secondary structure, correspond to true structural domains, is relevant to our understanding of ribosomal architecture. Fragments of 16S rRNA, corresponding closely to the three major domains, have actually been isolated in connection with studies on ribosomal protein binding sites (see section on Quaternary Structure). The ability of certain r-proteins to remain bound to these fragments, or even to rebind to RNA fragments from which proteins have been removed (121–124), supports the suggestion that these domains are true structural entities. This is not to imply that there may not be extensive interactions among domains, however.

Within each domain, the structure is organized into series of simple and compound helices, separated by various interior loops and bulges. Many of these are structural types not found in tRNA. Single-bulged nucleotides are found at positions 31, 55, 94, 397, 746, 1042, 1049, 1227, and 1441. The bulged adenosine at position 1227 is an extremely conserved feature, found in all 16S-like rRNAs thus far sequenced. The bulged nucleotides at positions 31, 397, and 746 are nearly always present, although their precise positions may shift, and in some cases they are replaced by mismatches or other irregularities. Multiple G·U pairs are almost always found in the 829–840/846–857 helix, although their positions are variable.

Frequent juxtaposition of A and G, most often at the ends of helices, but sometimes at internal positions (e.g. positions 148/174, 321/332, 663/742, 665/741, 1413/1487, 1417/1483, and 1418/1482) suggested that actual *anti-anti* A-G pairing may occur in rRNA (7, 78–79). This possibility is supported by several lines of evidence. First, comparative sequence analysis shows that in the archaebacterium *Halobacterium volcanii*, yeast mitochondria, and all eukaryotic cytoplasmic 18S rRNAs sequenced to date, the 321/332 A-G pair is replaced by a normal Watson-Crick pair (13). In the eukaryotes, an A-G juxtaposition appears instead at positions 318/335. Conversely, the two G·U pairs at positions 1425/1475 and 1426/1474 become A-G juxtapositions in *Bacillus brevis*. Residues involved in the above-mentioned A-G pairs are furthermore generally resistant to single strand–specific reagents, both chemical and enzymatic. Digestion of *B. brevis* 30S ribosomal subunits with RNase A under mild conditions releases pairs of fragments from the 1410–1430 and 1470–1490 regions, which remain stably associated under gel electrophoresis conditions in spite of no less than 5 A-G juxtapositions and several G·U pairs (25). Surprisingly, the double strand–specific cobra venom RNase cleaves *between* the 1417/1483 and 1418/1482 A-G pairs in *E. coli* and *Bacillus stearothermophilis* 16S rRNAs and in yeast 18S rRNA (126). A-G pairing has been observed in the

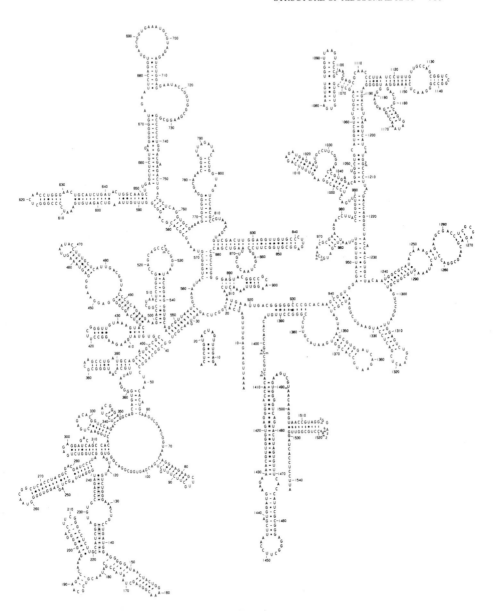

Figure 1   Secondary-structure model for eubacterial 16S rRNA (*E. coli*) (7, 10, 13, 79). Sequence and numbering are from (23, 58). Helices considered proven by comparative sequence analysis (two or more pairs of compensating base changes in the helix proper) are shaded.

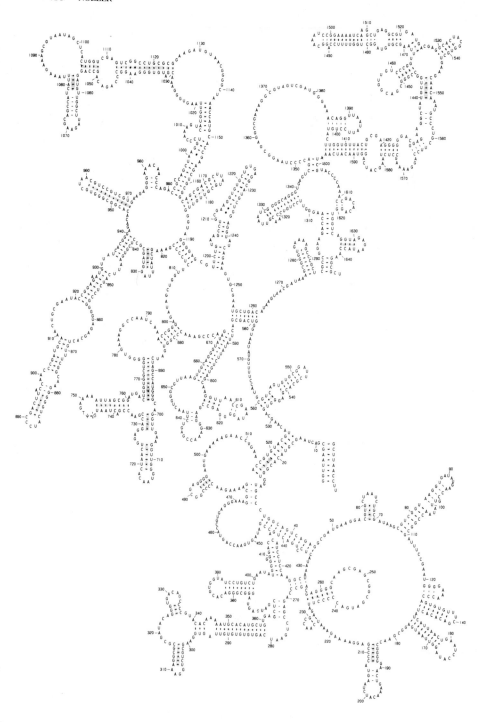

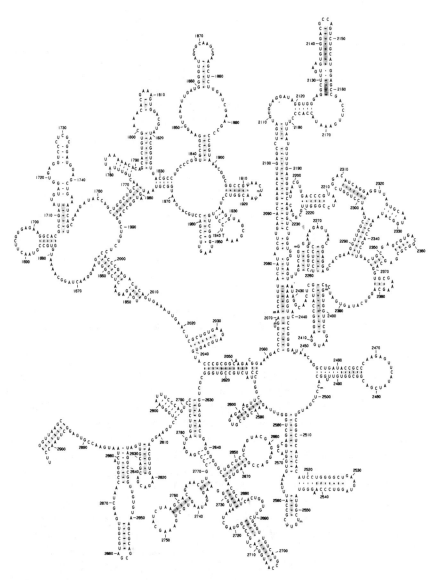

Figure 2 Secondary-structure model for eubacterial 23S rRNA (*E. coli*) (78; R. Gutell, H. Noller, C. Woese, unpublished). Helices considered proven (cf Figure 1) are shaded. Sequence and numbering are from (29). (Other half of figure on facing page.)

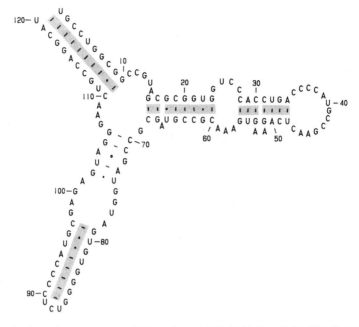

Figure 3   Secondary-structure model for eubacterial 5S rRNA (*E. coli*) (15, 75, 242). Helices considered proven (cf Figure 1) are shaded. Sequence and numbering are from (22).

structure of tRNA^{Phe}, although it involves m_2^2G (127–128). An early objection to *anti-anti* A-G pairing was that the exocyclic 2-amino group of guanine would not be fully able to hydrogen-bond with water (129). Thus, the A-G pair in tRNA^{Phe} could be rationalized, in that methylation would obviate that theoretical objection. More recently, direct evidence for formation of A-G pairs has come from NMR studies of DNA oligomers (130; D. Patel, unpublished); duplexes containing A-G pairs appear to have stabilities comparable to those that have G·U pairs at the corresponding positions. Another group of investigators has noted the high proportion of A-G juxtapositions flanking helices in rRNA, and has suggested that *anti-syn* pairing may be involved (125). In conclusion, the existence of bona fide A-G pairs in rRNA seems to be a very real possibility, and they may even be relatively commonplace structural features, along with G·U pairs; in retrospect, our understanding of base pairing in RNA appears to have been skewed by DNA pairing rules, which are turning out to be a special case.

Somewhat surprising is the absence of long, regular helices in rRNA. Instead, the structure is formed by joining of many short helices, the junctions of which tend to create a variety of structural irregularities. The reason for this kind of architecture has been rationalized in two ways (10): Multiple small helices allow a much more complex three-dimensional structure, which could approach that of a globular protein. Another reason may be that long, stable helices could create thermodynamic "traps," preventing unfolding of nonproductive intermediates during assembly. Significantly, no "knots" (131) are found in the secondary structures of the rRNAs (apart from the 9–13/21–25 vs 17–20/915–918 helices), i.e. where the loop contained by a helix is involved in pairing with a sequence outside the helix. This may also be part of a strategy to avoid dead ends in assembly.

23S rRNA   Figure 2 shows a current model for the secondary structure of 23S rRNA (78; R. Gutell, H. Noller, C. Woese, unpublished), again derived principally on the basis of comparative sequence analysis. Over one hundred individual helices are distributed among six domains, defined, as in 16S rRNA, by long-range base-paired interactions. These are referred to as domains I (residues 16–524), II (579–1261), III (1295–1645), IV (1648–2009), V (2043–2625) and VI (2630–2882). Structural organization within domains is analogous to that of 16S rRNA; short helices are connected by interior loops and bulges. Again, evidence for A-G pairing is found (78). A notable example of the latter is the tandem pairs 1039/1116 and 1040/1115, which are replaced by standard G-C and A-U pairs in *B. Stearothermophilus* and *Zea mays* chloroplast, respectively. Many single-base bulges are found, at about the same frequency as for 16S rRNA, as well as some double-base bulges (e.g. residues 1321 and 1322).

5S rRNA   Numerous groups have presented detailed discussions of 5S rRNA secondary structure recently (15–20), and the reader is referred to these for a more complete treatment. The first models derived on the basis of comparative sequence analysis for eubacterial (75) and eukaryotic (76) 5S rRNA many years ago are currently accepted as generally correct, although some new elements have been added to the basic models by various authors. Figure 3 shows a model that incorporates most of the currently accepted refinements to the original Fox-Woese proposal for *E. coli* 5S rRNA. Its five helices exemplify the kinds of local features seen in the large rRNAs. There are examples of single- and double-base bulges, an A-G pair (in this case unsubstantiated by comparative evidence) and helices connected by internal loops. The small size of 5S rRNA makes it amenable to computer analysis, and one such method (85) (which, significantly, incorporates a comparative algorithm) has produced a credible secondary structure for this molecule.

Phylogenetic Comparison

Comparative sequence analysis has allowed derivation of specific secondary-structure models for each of the rRNAs whose complete sequence is known. Again, models for the secondary structures of the various mitochondrial chloroplast, and eukaryotic cytoplasmic species have been presented by several groups. Here, the discussion centers on structures based on (13). Figure 4 shows some representative examples of 16S-like secondary structures (13). There is striking overall similarity between structures of 16S-like RNAs from the three major evolutionary lines, although the eukaryotic example shows considerable variation from the other two types. The central framework of the structure, including the long-range interactions that define the domains, is conserved even in mammalian mitochondria rRNAs, where 40% of the structure (relative to *E. coli*) is absent (37–40). Also readily identifiable in all cases are the parts of the structure containing the most highly conserved sequences (see also below). Similar conclusions can be drawn from comparison of 23S rRNA structures, which also share the same fundamental architecture.

Variation in size among the different 16S-like and 23S-like rRNAs is accounted for by insertions and deletions, which tend to be constrained to a few specific regions of these molecules (13). Thus, deletions in the 16S-like rRNAs tend to occur around positions 80, 200, 400–500, 590–650, 840, and 1450, which can be characterized as the most phylogenetically variable regions of the sequence in those organisms that retain these regions (13). Insertions in mitochondrial 16S-like rRNAs tend to occur around positions 1140 and 1450 while insertions in the eukaryotic cytoplasmic rRNAs are in the 200, 395, 580, 840, 1140, and 1450 regions (60–61, 132, 133); most of the increased size of the eukaryotic 18S rRNAs can be accounted for by the insertion of ~ 200 nucleotides around position 580. Similarly, the greater size of eukaryotic cytoplasmic 23S-like rRNAs is mainly due to insertions around positions 270, 550, 650, 1400, 1730, 2200, and 2800 (64–66, 134). The inserted sequences in both cases tend to be highly variable, and are much less conserved phylogenetically than the sequences in the parts of the molecule for which there is corresponding structure in the prokaryotic examples.

Experimental Tests

SINGLE OR DOUBLE STRAND–SPECIFIC PROBES   It was mentioned earlier that large fragments of the 16S and 23S rRNAs obtained by partial nuclease digestion correspond well to domains (or subdomains) generated by secondary structure, providing experimental support for the overall structural organization implicit in their secondary-structure models. At a

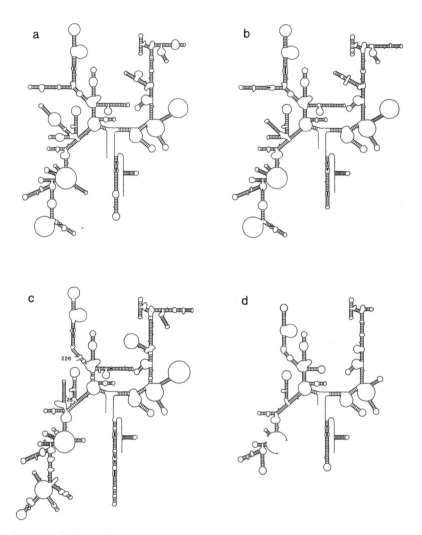

Figure 4 Phylogenetic comparison of secondary structures of 16S-like rRNAs from
(a) eubacteria (*E. coli*), (b) archaebacteria (*H. volcanii*), (c) eukaryote (*S. cerevisiae*), and
(d) a "minimal" structure showing those features common to all sequenced 16S-like rRNAs,
including mitochondrial types. Positions of insertion of additional nucleotides in eukaryotic
rRNA with respect to bacterial rRNA are indicated by 25 and 226 in (c).

detailed level, each nucleotide can, in principle, be tested experimentally by the chemical and enzymatic probes listed above. In general, attack by a single strand–specific probe is taken as evidence for the residue at that position existing in single-stranded conformation (or conversely, for a double-strand probe). In some cases, resistance to attack has been well documented; here, of course, it should be kept in mind that tertiary or quaternary structure as well as base pairing may account for protection. About 85%–90% of these kinds of probe data (summarized in 10, 13, 79) are in agreement with the secondary-structure models described here; this level of accord is probably an underestimate, because the consistency of results between any two experimental approaches is also usually only on the order of 85%–90%. The lack of complete agreement between different experiments can possibly be attributed to whether the RNA molecule is probed as part of the ribosome or in the free state, differences in ionic strength, magnesium ion concentration, methods of RNA isolation, and, most importantly, whether (and how) the RNA was ascertained to be in its "native" conformation. In the case of enzymatic probing, the problem of secondary cuts can be a significant one, as discussed above. Finally, correct identification of modified residues in a large RNA molecule is often technically difficult, and the reliability of such data are often difficult to assess from literature accounts.

Oligonucleotides can, in principle, be used as probes of single-stranded regions of rRNA. An important question to be asked is, how much of the rRNA molecule exists as open, single-stranded structure, capable of base pairing with such oligomers? Detailed studies with chemical probes suggest that 16S rRNA, even in naked form is surprisingly highly structured, and has very few regions, if any, containing many consecutive unstructured nucleotides. In one type of approach, an oligodeoxynucleotide (AAGGAGGT) complementary to the 3'-terminal region of 16S rRNA (residues 1534–1541) was synthesized and used as a probe of the accessibility of its cognate sequence in naked RNA, 30S subunits and 70S ribosomes (115). It was found that naked 16S rRNA or 30S ribosomes in the salt-depleted "inactive" form do not bind the oligomer, whereas active 30S subunits are capable of binding it; this finding in itself was unanticipated, since the RNA of the inactive form of 30S subunits is generally much more reactive toward kethoxal (135). Furthermore, protein S21 is required for oligomer binding. These results are presumably indicative of the exposure of residues 1534–1541 to solvent, and dramatically demonstrate the care that must be taken in preparing ribosomes or rRNA for such probing experiments. Most significantly, the effect of a specific ribosomal protein on rRNA topography is clear, and the likely conclusion is that S21 induces a conformational change in the 30S subunit that in some way exposes the 3'-terminal sequence.

In another approach (113–114), a random mixture of DNA hexamers is hybridized with 16S rRNA in 30S ribosomal subunits and the accessible positions are then determined by cleavage with RNase H. In these experiments, nucleotides around positions 8–13, 996–998, 1408–1410, 1495, 1500–1506, and 1531–1532 showed accessibility to the oligomers, consistent with the secondary-structure model. However, nucleotides around positions 80, 1044–1046, 1423, and 1484–1485 also bound oligomers; binding of a hexanucleotide to any of these regions requires disruption of helices considered to be proven by comparative sequence analysis. To account for this discrepancy, it can be argued that the regions of 16S rRNA in question can exist, at least transiently, in unpaired conformation while in a 30S ribosomal subunit. It must remain a possibility that the oligomers themselves may help to convert the rRNA to the open form, by virtue of their ability to bind the single-stranded form.

The latter experiments are examples of a few specific instances where phylogenetically established helices are reproducibly found to be susceptible to single strand–specific probes under certain conditions. These include the 9–13/21–25 and 17–20/915–918 helices and the 1050–1067/1189–1204 and 1410–1490 regions of the 16S-like rRNAs (72, 79, 97, 133, 135), and the 2063–2103 region of 23S-like rRNAs (78, 134, 136). Whether these results are mere reflections of the inherent instability of these structures under experimental conditions or clues to structural dynamics of rRNA is an open question at the present time.

DIRECT EVIDENCE FOR INTRAMOLECULAR INTERACTIONS   Use of various psoralen derivatives has provided confirmation of base-paired features of rRNA (summarized in Table 2) and, in addition, has suggested RNA-RNA interactions that are not readily accounted for by secondary structure. Lack of agreement between early secondary-structure models (137) based on psoralen cross-linking (101–102) with those derived from comparative evidence can possibly be ascribed to a combination of factors. Some cross-links were obtained using naked rRNA at very low ionic strength (101), where the danger of disrupting conformation is significant. Identification of cross-linked regions was in some cases incorrect because of difficulties in determining the polarity of the RNA chain in electron micrographs; this problem has now been overcome (138). Another factor is the possibility that psoralen may cross-link nucleotides that are involved in tertiary, as well as secondary, interactions (see Tertiary Structure section).

Nitrogen mustards (139) and irradiation with ultraviolet light (140) have also been used to produce RNA-RNA cross-links. The former react readily with nucleophiles, primarily ^7N of guanine, in the case of nucleic acids, and this reaction proceeds with base-paired as well as unpaired guanines. Cross-links induced by UV irradiation usually involve cycloaddition

Table 2 Experimental evidence for 16S rRNA secondary structure from isolation of noncovalently associated or cross-linked pairs of RNA fragments

Helix	Noncovalent association[a]	Psoralen cross-linked[b]	UV cross-linked[c]
17–20/915–918		+	
39–42/394–403	+		
73–82/87–97	+		
122–128/233/239	+		
136–142/221–227	+		
240–245/281–286	+	⎫	
247–249/279–277	+	⎬ +	
252–259/267–274	+	⎭	
289–295/302–311	+		
437–440/494–497	+		
442–446/488–492	+		
455–462/470–477	+	+	
564–570/880–886		+	
576–580/761–765	+		
584–587/754–757	+	⎫	+
588–595/644–651	+	⎬ + ⎫	+
597–606/633–641	+	⎭ ⎬	+
612–617/623–629	+	⎭	
655–662/743–751	+		
666–672/734–740	+		
821–828/872–879	+		+
829–840/846–857	+		
926–933/1384–1391	+	+	
938–943/1340–1345	+		
984–990/1215–1221	+		(+)[d]
997–1003/1037–1044	+		
1006–1012/1017–1023	+	+	
1046–1053/1205–1211	+		(+)
1055–1060/1197–1202	+		
1063–1067/1189–1193	+		
1068–1071/1104–1107	+		
1113–1117/1183–1187	+	+	
1239–1247/1290–1298	+	+	
1308–1314/1323–1329		+	
1350–1356/1366–1372	+	+	+
1409–1430/1470–1491	+		
1435–1445/1457–1466	+		

[a] Sources: (25, 108, 141).
[b] Sources: (101–103, 105, 155).
[c] Source: (140).
[d] In cases where there is some uncertainty in the identification of cross-linked nucleotides, the probable assignment is given in parentheses.

between the 5 and 6 positions of stacked pyrimidines. Thus, cross-linking of residues that are nonadjacent in the RNA primary structure cannot occur if they are both involved in the same helix. This is consistent with the finding that UV-induced cross-links occur in regions of the structure containing internal loops (Tables 2 and 3) where bases on opposite chains may be able to stack on one another; results of this kind may shed light on the details of base stacking in such areas of structural irregularity.

Another method that has been used to identify base-paired regions employs two-dimensional gel electrophoresis of partial nuclease digests of non-cross-linked rRNA (25, 108, 141). Pairs of fragments that remain associated in the nondenaturing first dimension are dissociated and resolved in the second dimension under denaturing conditions. This

Table 3 Experimental evidence for 23S rRNA secondary structure from RNA-RNA cross-linking and electron microscopy

Helix	Electron microscopy[a]	Psoralen[b]	Nitrogen mustard[c]	UV[d]
		Cross-linkage with		
16–25/515–524	+			
281–293/347–359				
295–297/341–343				
588–601/656–669				
812–817/1190–1195	+			
1013–1019/1143–1149	+			+
868–870/907–909		+		
1206–1209/1237–1240			} +	
1213–1222/1227–1236			}	
1405–1415/1587–1597	+	+		
1435–1437/1555–1557				+
1481–1492/1498–1508			+	
1656–1663/1997–2004	+			
1710–1750 region		+		
1841–1846/1894–1899				+
2043–2050/2618–2625	+			
2063–2070/2441–2447	+			
2630–2637/2781–2788	+			
2836–2843/2874–2882				(+)[e]
2851–2856/2861–2866				(+)

[a] Source: (260).
[b] Source: (104).
[c] Source: (139).
[d] Source: (158).
[e] In cases where there is some uncertainty in the identification of cross-linked nucleotides, the probable assignment is given in parentheses.

method has the advantage that no chemical modification is employed, but anomalous reassociation of fragments following nuclease digestion may be a potential hazard. Evidence for many helices has emerged from these studies (summarized in Table 2), as well as some interactions that do not correspond to features of the secondary-structure models (discussed below).

There is striking agreement between results obtained by electron microscopy of partially unfolded rRNAs and the proposed secondary structures. Large loops seen in such preparations of 23S rRNA correspond closely to long-range base-paired interactions in the secondary-structure model (Table 3; Figure 2; 260). The critical factor required for reproducible observation of these features appears to be the presence of adequate magnesium ion concentrations during preparation. The fact that the long-range interactions are detected at all suggests that they are more stable than many of the other interactions under these conditions. Inspection of the helices involved shows that they are, almost without exception, predicted to be quite stable from standard free-energy calculations. This may have important implications for ribosome assembly, possibly ensuring that each domain is maintained as a separate structural entity during early stages of folding.

Results of the above studies, in which direct experimental evidence has been obtained for intramolecular association of specific regions of 16S and 23S rRNA, are summarized in Tables 2 and 3. Data are much more abundant for 16S rRNA, as is the case for the structure-specific probe experiments. There is clear evidence for association or close juxtaposition of sequences making up 37 individual helices in 16S rRNA. The results listed in Tables 2 and 3 include only those that correspond to helices whose existence is compatible with comparative sequence analysis and represent most of the available data. Base pairing consistent with the comparative sequence criteria, outlined above, has so far not been found for the remaining examples of experimentally observed intramolecular association. This can be explained in one of three ways: (a) phylogenetically consistent helices exist for those interactions, but have not yet been found; (b) experimental conditions have caused rearrangement of rRNA conformation; or (c) the interactions are tertiary in nature. None of these possibilities can yet be rigorously ruled out. Experimental conditions in some cases involve very low ionic strength (101) and/or chelation of magnesium (108), and even use of 2-M urea (141), all of which are known to perturb rRNA conformation. The noncovalent association experiments involve cleaving the RNA chain with nucleases, and the possibility of subsequent structural rearrangement must be considered. Results from electron microscopy suffer from low resolution, usually on the order of ± 20 nucleotides at best, and determination of the polarity of the RNA chain has often been a problem.

Finally, many of these studies are performed on naked rRNA, and while it is becoming clear that its structure must be very much like that found in the ribosomes, significant differences may exist; in any case, such studies depend crucially on the methods of isolation and treatment of the RNA before the actual experiment. Most intriguing, however, is the possibility that some of these interactions may be clues to tertiary folding of the rRNA, to be discussed below (see Tertiary Structure section).

Recently, detailed investigations of rRNA conformation using high-resolution NMR methods have directly confirmed some secondary-structure features, and have established a basis for future studies on rRNA such as solution dynamics and RNA-protein interaction. Analysis of 5S rRNA has directly identified features of the 1–10/110–119 and 79–86/92–97 helices, using the NOE to assign imino proton resonances to nucleotides in adjacent base pairs (116–118). Similarly, the 1506–1515/1520–1529 helix of 16S rRNA has been studied using the 3'-terminal colicin fragment (119). Unfortunately, this approach is limited to the study of relatively small rRNA molecules or fragments, up to about the size of 5S rRNA.

TERTIARY STRUCTURE

Most important for understanding the biological role of rRNA will be a description of its three-dimensional structure with its associated proteins. We are clearly far from the realization of this goal, which, as for other large biological structures, will likely require X-ray crystallographic analysis. Nevertheless, it is possible that a general scheme for the three-dimensional folding of the large rRNAs can be achieved in the absence of any crystallographic information. Relevant data are emerging from two general approaches. In the first approach, the electron microscope (e.m.) model for the structure of the ribosome and its subunits (142–143) is taken as a starting point, and attempts are made to assign specific features of rRNA to positions in the e.m. model. In the second kind of approach, the secondary-structure models for the rRNAs are taken as a starting point, and attempts are made to fold them into three dimensions. The two approaches are formally independent, and so there is the possibility to test or confirm the predictions of one approach using the other.

In order to assign a site in rRNA to a location in the e.m. model, there must be some recognizable feature in the electron micrograph that can be related, directly or indirectly, to the site in the RNA. Many individual ribosomal proteins have been localized by immuno-electron microscopy (142–143) and by neutron diffraction studies (144); since some of these have known binding sites on the rRNA, it can be inferred that the latter must be located reasonably near the positions of their cognate proteins. More recently, antibodies have been raised to methylated nucleotides located at unique positions in the 16S rRNA chain, and to haptens coupled to the 5' or 3' termini, permitting direct immuno-e.m. localization of these sites (145–152). By these general approaches, the sites in 16S rRNA identified with proteins S7 (U_{1240} and C_{1378} or G_{1379}), S8 (residues ~ 590–650), S15 (residues ~ 650–750) and S20 (residues ~ 250–280) can be located according to the positions of the corresponding proteins. Similarly, $^{7}_{m}G_{527}$ (146), $_{m_2}^{6}A_{1518}$ and $_{m_2}^{6}A_{1519}$ (145), and the 5' (150) and 3' (147–149) termini of 16S rRNA have been placed directly. Furthermore, the proteins found associated with the 3' major domain (124) (residues ~ 930–1390) are all located in the "head" of the e.m. model. Less direct assignments for proteins S4, S6, and S18 can also be made (10). These placements are summarized in Figure 5.

In 23S rRNA, similar references can be made although there is presently much less information available pertaining to its structure in the large subunits. Its central lobe is proximal to 5S rRNA, from studies using antibodies to end-labeled 5S rRNA (151–152), and from the location of its

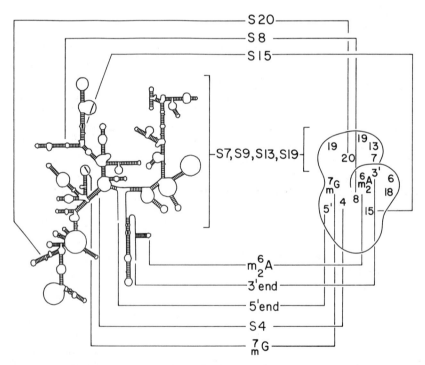

Figure 5 Placement of specific features of 16S rRNA structure in the electron microscopy model (142) of the 30S ribosomal subunit. Positions of ribosomal proteins (142–144) and sites in the RNA (145–152) are indicated.

cognate proteins, L18 and L25 (151). The peptidyl transferase center (see below) has also been placed near the central lobe, from the positions of proteins affinity-labeled by various tRNA derivatives, and from the site of attachment of the antibiotic analog iodoamphenicol (153). The 2100–2200 region of 23S rRNA can be placed on the "left-hand" lobe of the 50S subunit, because of its association with protein L1 (142–143). Finally, the 5' and 3' termini have been placed on the lower rear of the particle, by end-labeling studies (154).

In the second approach, folding the molecule into three dimensions, is based on information establishing proximity between regions of the RNA that are nonproximal in the secondary-structure model. As discussed above, some of these data may reflect tertiary contacts. Relevant findings include cross-links, induced by psoralen, nitrogen mustard, and UV irradiation, and noncovalent association of RNA fragments for which no helices supported by comparative analysis are known.

From studies employing psoralen, cross-links were found between the following regions of 16S rRNA (to be taken approximately, as the precise nucleotides involved have not yet been determined): 360/1330, 620/1420, and 960/1510 (105) (all determined by sequencing cross-linked fragments); and 1/680, 10/1180, 490/980, 610/1320, and 930/1540 (155) (all determined by electron microscopy). All of these results imply interaction between domains, and several are found between the 5' and 3' major domains. Evidence for interaction between the latter two domains comes also from studies in which large fragments from these two regions co-migrate under conditions of gel electrophoresis in the absence of proteins (108, 141, 156–157). Interaction between the 5' and 3'-major domains is also implied by the location of S20, which binds to the 5' domain (8), in the "head" region of the e.m. model, surrounded by proteins that bind to the 3' major domain (143).

Cross-linked sites in 23S rRNA that represent potential tertiary contacts are 570/2030, 740/2610, and 1780/2570 (158) (from uv irradiation) and 760/1570 (139) (from nitrogen mustard). These results imply contact between domains II and III, II and V, IV and V.

Singlet-singlet energy transfer has been used to measure inter- and intramolecular distances between the ends of the rRNAs (159). Distances between fluorophores attached to the 3' ends of 16S rRNA and 5S rRNA or 23S rRNA were estimated to be about 55 and 71 Å, respectively. The corresponding distance between 5S rRNA and 23S rRNA was too large to be measured accurately with the available probes but was estimated to be greater than 65 Å.

Even if all the results described here are taken as evidence for tertiary structure, far more information of this kind will be required to construct low resolution three-dimensional models for rRNA. Nevertheless, the apparent spatial proximity of sites that are distant from each other in the primary and secondary structure, for example the 960 and 1500 regions of 16S rRNA (both of which are implicated in ribosomal function; see Structure and Function section), is valuable information that will be of use in formulating models of both structure and function.

Cross-linking of G_{41} and G_{72} in 5S rRNA (160) has stimulated three specific proposals for "tertiary" interactions in this molecule (16, 160–161). Two of thse involve loop-loop Watson-Crick base pairing, of residues 37–40/73–76 in one model (160), and 41–44/74–77 in another (161). Besides the high susceptibility of both U_{40} and G_{41} to single strand–specific probes (162–163), even in intact ribosomes (164–165), which prevents their involvement in any permanent base-paired interactions, comparative sequence analysis does not support proposed pairings. Several phylogenetically independent disproofs can be found for both models. The third model proposes *parallel* base pairing of 38–40 with 76–78 (16). In this case, the sequences are more highly conserved, but changes in *Photobacterium* and *Aspergillus*, for example, produce U-U and A-C mismatches, respectively. Thus, the available evidence, both experimental and phylogenetic, argues against these proposals. Furthermore, it remains to be ruled out that the cross-link itself, obtained in comparatively low yield (160), is not due to the B conformer of 5S rRNA, in which G_{41} and G_{72} are believed to be brought into juxtaposition (166).

QUATERNARY STRUCTURE

r-Protein–rRNA Interactions

A complete discussion of the subject of interaction of ribosomal proteins with ribosomal RNA is beyond the scope of this article. The reader is referred to other specific review articles on the subject (8). Here, the questions to be addressed are: What are the specific recognition signals embodied in the structure of rRNA by which ribosomal proteins discern their correct binding sites? And, do the proteins significantly alter rRNA conformation, either locally or globally?

RECOGNITION SIGNALS The answer to the first question depends on a detailed description of ribosomal protein-binding sites. Information bearing on this has been obtained principally by methods that include (*a*) the classic "bind and chew" approach (in which a protein is bound to rRNA, extraneous rRNA removed by RNase treatment, and the remaining bound RNA characterized), (*b*) chemical or UV-induced cross-linking, (*c*) protection of the rRNA from chemical modification by the bound protein, (*d*) modification/selection experiments (in which the RNA is partially chemically modified, and the fraction of rRNA molecules retaining their ability to bind the protein are analyzed to see which sites are *not* modified), and (*e*) electron microscopy of protein-rRNA complexes.

Most informative are those protein binding sites that have been localized to a relatively concise region of the rRNA. Some ribosomal proteins yield very large RNA fragments in nuclease protection experiments; the "binding-site fragment" for protein S4, for example, contains about 500 nucleotides, about seven times the mass of the protein itself. It is unlikely that the protein recognizes more than a fraction of this fragment, but protein-RNA interactions may stabilize the structure and thereby enhance the RNA-RNA interactions. Unfortunately, it is often difficult to discern which features of these large fragments are the actual protein contact sites. Proteins that have been found to yield RNA binding-site fragments of more manageable size (mass on the order of that of the protein, or smaller) include S8, S15, L1, L11, L17, L18, and L25. Binding sites for these proteins have been localized approximately within 16S rRNA for S8 (residues 585–610, 630–654) and S15 (residues 655–675, 730–755) (122, 167), 23S rRNA for L1 (residues 2090–2200) (168), and L11 (residues 1050–1110) (169) and 5S rRNA for L18 (residues 15–35, 45–69) (170–172) and L25 (residues 70–110) (11, 172–73). Most of these fragments contain on the order of 40–60 nucleotides, and probably far fewer nucleotides actually make contact with protein in each case. It is important to keep in mind that these proteins may well bind RNA sites outside their respective "binding fragments," but that the other contact sites could escape detection if the binding constants are low, or if the RNA moiety is unusualy unstable to nuclease treatment.

It is reasonable to ask whether there are any structural features common to these various protein-binding sites. Interestingly, the majority would appear to contain two helices connected by an internal loop. Because of the abundance of such structures in rRNA, it is difficult to judge the significance of this. Binding-site fragments for S15, L18, and L17 all contain single-bulged adenosine residues, and the L18 site appears to contain, in addition, a pair of bulged adenosines (Figures 1–3). The S8 and L1 binding sites tend to be unusually rich in G · U pairs (6 and 4, respectively). The best-studied protein-binding sites are those in 5S rRNA, largely due to the ease of preparation of this experimental system. For both L18 and L25, there is good evidence that the proteins bind to double-helical structures. In both cases helices are protected from attack by cobra venom RNase by the bound proteins (172), and L18 protects several base-paired guanines from reaction with dimethyl sulfate (171). Furthermore, spectroscopic studies with L18 and L25 (94, 174) and NMR studies with L25 (118) indicate that the RNA helices are not disrupted when the proteins bind. Single-stranded residues flanking the binding-site helices have also been shown to be protected from chemical and enzymatic probes by the proteins in the cases of L18 and S15 (122, 170–172). Protein S8 has been cross-linked to single-stranded nucleotides (approximate positions 593–597, 629–633, and 651–654) on either side of, and between, its two binding-site helices (175). These results seem to imply that the ribosomal proteins recognize both helical and nonhelical elements. The possible role of irregularities such as bulged nucleotides, multiple G·U pairs and A·G pairs is at present an open question. The existence of bulged nucleotides in protein-binding sites from systems other than ribosomal suggests at least some connection between them; whether the bulge is a recognition signal for protein binding or is stabilized by the protein for some other purpose is not clear (171). What is clear is that the bulged nucleotides are often highly conserved phylogenetically (171), and in at least two well-documented cases (S15 and L18) are located well within protein-binding sites. It is not unreasonable to conclude that much of the architecture of the ribosomal RNAs has evolved to accommodate interactions with ribosomal proteins and, very likely, translational factors. The extent to which this is the case will not be clear without a much more complete description of the protein-binding sites in rRNA.

Studies on the translational regulation of ribosomal protein biosynthesis indicate that the structures of their mRNAs have similarly evolved to accommodate RNA-protein interactions. Certain ribosomal proteins that are known to bind directly to rRNA act as translational repressors by binding to polycistronic mRNAs, usually encoding several r-proteins, and in some cases other proteins crucial for transcriptional and translational processes (176). The regulatory sites on these mRNAs have been proposed to involve structures that mimic the rRNA binding sites for the regulatory r-proteins (177–180).

PROTEIN-INDUCED CONFORMATIONAL CHANGES Is the conformation of rRNA significantly affected by interaction with ribosomal proteins? There is controversy currently concerning the overall size and shape of naked 16S rRNA compared with 30S ribosomal subunits. There are studies using electron microscopy (181) and hydrodynamic methods (182) that show the naked 16S rRNA as having essentially the same size and shape as the 30S subunit. Other studies, however, also based on electron microscopy (183) and hydrodynamic approaches (184), have reported very different shapes and dimensions for the two cases, and describe the naked rRNA as unfolded or somewhat amorphous compared with the intact subunit. Here, the underlying question is whether the rRNA itself contains the requisite information for folding, which would merely be stabilized by the proteins, or whether some of the large-scale structuring of the molecule is completely dependent on the proteins. Thermodynamic studies on the process of in vitro assembly of 30S (185) and 50S (186) subunits show that a conformational rearrangement necessary for correct assembly [the "RI → RI*" conversion] depends on prior binding of several ribosomal proteins to the rRNA in both cases. The sedimentation constant of the 30S assembly intermediate particle changes from 21–22S to 25–26S as a result of the RI → RI* conversion (187), indicating a substantial change in global conformation (i.e. most likely a compaction of the particle). Little is known, however, concerning the specific structural changes that occur during these events.

Studies on specific rRNA-protein interactions show that little change in rRNA conformation results in most cases. An exception is the binding of L18 to 5S rRNA, which causes an increase of about 20% in the magnitude of the 263-nm circular dichroism band of the RNA (99, 174). Since this parameter is usually taken as a direct measure of helicity, the result has been interpreted as an increase in base pairing in 5S rRNA induced by the binding of L18 (94); alternatively, it has been interpreted as a change in the structural regularity of existing double-stranded segments of the molecule (174). It is interesting to consider that, if L18 were to stabilize the irregular extensions to its two binding-site helices beyond their respective bulged nucleotides (i.e. stabilize the pairing of 16–17/67–68 and 28–30/54–56) (Figure 3), this would amount to an increase in helicity of approximately the observed magnitude.

High-resolution NMR studies on 5S rRNA, its L25 binding fragment (residues 1–11, 71–120), and their respective interactions with L25 show effects of protein binding on the RNA structure (116–118). All of the assignable perturbations involve the 79–85/91–97 helix, and include the disappearance of the NOE linking base pairs U_{82}/A_{94} with G_{83}/C_{93}. In addition, several new unidentified low field resonances appear; a tantalizing possibility is that some of these may represent hydrogen bonds involved in RNA-protein contacts. Clearly, this approach is a potentially powerful tool for elucidation of specific details of protein-RNA interaction.

In conclusion, perturbation of rRNA structure by r-proteins is poorly understood. A reasonable scenario for ribosome assembly is that proteins capable of binding independently to rRNA recognize, for the most part, preexisting structural features (most likely a combination of primary, secondary, and tertiary) or in some cases may trap conformers that exist only transiently in the absence of bound proteins; this may then permit the binding of proteins that make protein-protein contacts, or whose RNA binding sites are stabilized by attachment of the primary binding proteins. Early steps may involve stabilization of structure within each domain, followed by domain-domain interactions. The lack of solid information pertinent to the structure of rRNA during assembly is a measure of how little we understand of this process, in spite of the considerable information available regarding the assembly of ribosomal proteins.

rRNA-rRNA Interactions

Thus far, no convincing evidence for direct interactions between the ribosomal RNAs has come to light. Suggestive complementarities between 5S rRNA and the large rRNAs in *E. coli* (188–189) do not withstand comparative analysis. Nor have phylogenetically consistent base-pairing schemes for 16S–23S rRNA interactions been identified, although direct cross-linking between the large rRNAs in 70S ribosomes has been reported (130).

STRUCTURE AND FUNCTION

Early studies tended to emphasize ribosomal proteins as the components likely to be directly involved in ribosomal function. This can probably be attributed to a variety of circumstances. Enzymes are made of protein, so it was anticipated that ribosomal proteins would be responsible for the various translational functions. Ribosomal mutants, most notably those carrying antibiotic resistance alleles, were found to be due to altered proteins at the outset (190); this was probably because most genetic studies of this kind were carried out on *E. coli* and other eubacteria, which carry multiple copies of the rRNA genes in their genomes, masking potential rRNA-related phenotypes. In vitro reconstitution studies again drew attention to the function of proteins (191), although the requirement for 5S rRNA became apparent (186, 192). Finally, the greater inherent chemical reactivity of proteins tended to deemphasize rRNA in affinity labeling and chemical modification studies.

It can now be said that there is at least a comparable body of evidence to support the reverse paradigm, that the essence of ribosomal function is embodied in its RNA, and that ribosomal proteins serve to enhance its function. A first consideration is the evolutionary argument, the chicken-and-egg problem, that the original ribosome cannot have used proteins for any crucial purpose, since the ribosome had yet to make them. It is also more plausible to imagine a ribosome originally constructed from RNA, to which proteins have been added during the course of evolution, than the converse. Furthermore, both in vitro and in vivo studies have demonstrated that ribosomes lacking individual ribosomal proteins are functionally active (193–194). Finally, the concept of a "functional nucleic acid," one that is capable of directing the making and breaking of covalent bonds, for example, is no longer confined to the realm of speculation, since the discovery of a self-splicing RNA (195). The chemical nature of RNA would appear to have been greatly underestimated.

rRNA Genetics

Genetic studies have now amply documented the role of rRNA in several functional contexts. Methylation of rRNA, or the lack of it, has been shown to affect resistance to the antibiotics kasugamycin (196), erythromycin (197), viomycin (199), and thiostrepton (169). Mitochondria usually have only a single set of rRNA genes, in contrast to most eubacteria. Strikingly, antibiotic resistance in mitochondrial ribosomes is almost always attributable to mutations in the rRNA. Thus, resistance to chloramphenicol (201–204), erythromycin (205), and paromomycin (43) have been found to be due to point mutations in rRNA genes in yeast and mammalian mitochondria. Cold-sensitive mutations mapping in mitochondrial rRNA genes have also been found (206). More recently, erythromycin resistance in *E. coli* has also been attributed to a mutation in a plasmid-encoded rRNA gene (207). Use of site-directed mutagenesis of cloned *E. coli* rRNA genes has produced a number of mutants in 16S and 23S rRNA (208–209). Surprisingly, most of these mutations affect the growth rate of the organism significantly. A most interesting finding is an ochre suppressor in yeast mitochondria that maps in or near its 15S rRNA gene (210). In view of these results, it can be said that genetic evidence alone provides a compelling case for a direct role for rRNA in translation.

Dramatic examples of the effect of small alterations of the covalent structure of rRNA are provided by colicin E3 and α-sarcin. These toxic proteins completely inactivate protein biosynthesis by cleaving single phosphodiester bonds in rRNA; the A_{1439}–G_{1494} linkage of 16S rRNA by colicin E3 (211–212), and the G_{2661}–A_{2662} linkage of 23S-like rRNAs by α-sarcin. Both scissions occur in universally conserved sequences (13).

Specific Structure-Function Correlations

In the past few years, evidence from a wide variety of experimental approaches has emerged to create a first impression of many of the functional properties of ribosomal RNA, and in many cases specific sites in the RNA have been related to specific functions. It is useful to review these findings briefly, and to sense the extent to which the structures of the ribosomal RNAs are beginning to take on biological meaning.

mRNA INTERACTIONS AND INITIATION   That mRNA initiation sites are selected by direct interaction with 16S rRNA as proposed by Shine & Dalgarno (215) is now well documented, and has been reviewed in detail by Steitz (216) and Gold et al (198). Comparative analysis of a large number of mRNA sequences shows a varying degree of complementarity between mRNA at a position about 10 nucleotides distal to the 5′ side of the initiator codon and the conserved CCUCC sequence at position 1535–1539 of

16S rRNA. This has now been amply substantiated by a number of elegant biochemical and genetic studies (217–219). It is likely that such a mechanism operates in eubacteria and archaebacteria, but not in eukaryotic, cytoplasmic ribosomes nor in most mitochondria.

The initiation factor IF3, involved in promoting dissociation of vacant 70S ribosomes and binding mRNA during initiation of translation (220), has long been suspected of binding to 16S rRNA. Cross-linking of IF3 to 16S rRNA has been achieved by photolysis using near-UV irradiation (221). At least one site of cross-linking is reported to be in the 3'-terminal region; removal of the 3'-terminal colicin fragment decreases the binding of IF3 to 30S subunits. Recently, protection of specific residues in the 3'-terminal region of 16S rRNA has been demonstrated in studies of the interaction of IF3 with the 49 nucleotide colicin fragment (222). The 3'-terminal region is also implicated in subunit association (see below), consistent with the possibility that IF3 exerts its "dissociation factor" activity by binding to a region of 16S rRNA that contributes significantly to the strength of the 30S–50S interaction.

Kasugamycin inhibits binding of the initiator tRNA to ribosomes, and as such appears to be a specific inhibitor of translational initiation (223). Resistance to this antibiotic has been shown to be conferred by nonmethylation of the two adenosines (normally N-6-dimethylated) at positions 1518 and 1519 of 16S rRNA (196). Thus, by several independent lines of evidence, the 3'-terminal domain of 16S rRNA appears to be directly implicated in the process of translational initiation.

In an affinity labeling study in which a derivatized trinucleotide mRNA analog was used to probe the rRNA environment around the presumed mRNA binding site, residues 462 and 474 of 16S rRNA were shown to be labeled (224). The possible generality of this result is diminished by the clear absence of the region encompassing residues 455–477 in 16S-like rRNAs from chloroplasts (31–34), in at least one archaebacterium (*H. volcanii*) (28), in mitochondria (37–44), and in eukaryotic cytoplasmic ribosomes (59–62); furthermore, the sequence in this region is phylogenetically variable among eubacteria (13). It may be relevant that the 455–477 region has been found to be a prominent site for psoralen intercalation (103), in that the probe used in the above studies contained a phenylgyloxal moiety, itself a potential intercalator. Several groups have used other derivatized oligonucleotides, as well as poly(U) and poly-(4-thio-U) as affinity probes of the mRNA binding site; these have all been shown to react to some extent with 16S rRNA, but the sites of reaction have not yet been found (225–227).

ASSOCIATION OF RIBOSOMAL SUBUNITS When 30S and 50S ribosomal subunits associate to form 70S monosomes, a number of sites that are accessible to single and double strand–specific probes (both chemical and enzymatic) become shielded (228–230). The protected regions of 16S rRNA are mainly, but not exclusively, in the central and 3′ minor domains, and include positions 337, 674, 703, 705, 773, 791, 803, 818, 1064, 1405, 1408, 1409, 1490, 1497, and 1516. Protection results must be interpreted with the caveat that an induced conformational change can, in principle, give rise to protection that would otherwise be most simply interpreted as the result of intermolecular interaction. Following the simpler interpretation, these results place specific sites in 16S rRNA (mainly in the central and 3′ minor domains) in the region of contact between the 30S and 50S ribosomal subunits, often referred to as the subunit interface.

Whether a particular site is crucial for intermolecular interaction can be tested by modification/selection experiments. Modification of 30S subunits with kethoxal destroys their ability to interact with 50S subunits (231). To test which kethoxal-reactive sites were responsible, researchers partially inactivated a population of 30S subunits by limited reaction with kethoxal. Association-competent 30S subunits were then selected from the modified population by virtue of their ability to form 70S monosomes. The modified sites in the association-competent subunits were compared with those of the total modified population. Several sites were found not to be modified, or much less extensively modified, in the competent population. These sites are residues 674, 703, 705, 791, 818, 1064, 1497, and 1516 (231). These results provide evidence that several sites in 16S rRNA that are protected from kethoxal, RNase T_1, and cobra venom RNase in 70S ribosomes are also crucial for subunit association; 50S subunits strongly select against 30S subunits in which these sites are kethoxal modified. Electron microscopic analysis of crystalline arrays of eukaryotic ribosomes, in which rRNA could be selectively visualized by contrast matching, show the subunit interface to be rich in RNA (232). Placement of rRNA in this region of the ribosome, commonly thought to be the center of the translational process, again points to a direct functional role for rRNA.

tRNA-RIBOSOME INTERACTIONS   Certain tRNAs contain modified bases in the wobble position of their anticodons rendering them capable of near-UV photochemistry. When in the ribosomal P site, they react to form a cycloaddition product with C_{1400} of 16S rRNA, in high yield (233–234). The significance of this finding is that it places the site of codon-anticodon interaction in extremely close proximity (~ 4 Å) to a nucleotide in the middle of one of the most highly conserved sequences known to biologists, the 1390–1410 region of 16S rRNA. This is unlikely to be coincidental. This region of 16S rRNA has already been placed at the subunit interface by the studies described above.

Paromomycin-resistant mutants have been shown to have alterations at the base of the stem flanking this region of 16S rRNA. In yeast mitochondria C_{1409} is replaced by G (43), and in *Tetrahymena* 18S rRNA, G_{1491} becomes A (E. Blackburn, personal communication); in both cases the 1409/1491 base pair is disrupted (Figure 1). The proximity of the site of base changes conferring resistance to paromomycin, an antibiotic known to cause misreading, to the region of 16S rRNA cross-linked to the anticodons of several tRNAs is striking.

Several sites in 16S rRNA remain kethoxal-reactive in vacant 70S ribosomes. However, when polysomes are formed, four of these sites, at positions 530, 966, 1338, and 1517, become substantially protected, in proportion to the tRNA occupancy of the polysomes (235). These protected sites may correspond to the sites of inactivation of tRNA binding by kethoxal, reported in earlier studies (236). The very highly conserved sequence in the loop at position 530 has already been noted. All four of the protected sites are in conserved sequences and are associated with secondary-structure features that are common to all 16S-like rRNAs. It is significant that three of the clusters of methylated bases in 16S rRNA are found among these four sites.

A great deal has been said concerning the postulated interaction of the GTψC sequence of tRNA with the GAAC sequence (residues 44–47) of 5S rRNA, both of which are quite conserved (77). Evidence for this is based on experiments using various oligonucleotide analogs of the GTψC sequence or tRNA fragments which inhibit aminoacyl tRNA binding, presumably by acting on the 50S subunit (237–239). Less convincing are studies placing the binding site for such fragments on 5S rRNA (240). Recent studies by Pace and colleagues (241, 242) have tested this hypothesis directly. Residues 42–46 or 42–52 were excised from 5S rRNA, and 50S ribosomal subunits were reconstituted using the deleted RNA. Ribosomes utilizing the reconstituted 50S subunits were tested for their activity in protein synthesis, programmed either by poly U (241) or by phage MS2 RNA (242). In addition, activity of these ribosomes in f·Met-tRNA binding, EF-T-dependent aminoacyl tRNA binding, ppGpp synthesis, synthesis of phage proteins of the correct size and misreading under various conditions were studied. Ribosomes carrying deleted 5S rRNA were found to be active in all of these assays, although some diminution of activity was noted in AUG- [but not poly (A, U, G) or MS2 RNA] directed fMet-tRNA binding and ppGpp synthesis, especially in the case of the larger (residues 42–52) deletion. In particular, EF-Tu-dependent aminoacyl-tRNA binding was affected very little by deletion of three of the four nucleotides in the GAAC sequence (242). Comparative analysis of the relevant sequences, in fact, predicts that this should be so. In most eukaryotic 5S rRNAs, the analog of the GAAC sequence is GAUC, yet the sequence in eukaryotic noninitiator tRNAs remains GTψC. Although eukaryotic initiator tRNAs have an altered sequence, GAψC, an apparent compensation to restore complementarity to 5S rRNA, plant 5S rRNAs retain the GAAC sequence in their 5S rRNAs, arguing against possible interaction of 5S rRNA with initiator tRNA in this fashion. In conclusion, the possibility of specific interaction(s) between the GTψC sequence of tRNA and some site in the ribosome is still alive, but that any exists between tRNA and the GAAC sequence of 5S rRNA seems finally to have been ruled out.

Possible contact between the 3' terminus (acceptor end) of tRNA and rRNA has been explored extensively. However, because of its necessary proximity to the peptidyl transferase region, these studies are discussed in the following section.

PEPTIDYL TRANSFERASE   The ribosomal activity responsible for catalysis of peptide bond formation is called peptidyl transferase. Affinity-labeling experiments as well as isolation of mutants resistant to peptidyl transferase-related antibiotics both indicate that the structure responsible for this activity is at, or near, a region in domain V of 23S rRNA.

An antibiotic closely identified with the peptidyl transferase function is chloramphenicol (243). Chloramphenicol-resistant mutations mapping in the 23S-like RNA genes have been found in several types of mitochondria. The sites of mutation have been found to be located at positions corresponding to residues 2447, 2451, 2452, 2503, and 2504 (201–204). In the secondary-structure model (Figure 2) these sites are all clustered around the very highly conserved central loop in domain V. Erythromycin is also a peptidyl transferase inhibitor, although its mode of action is quite different from that of chloramphenicol (243). An erythromycin-resistant mutation in yeast mitochondria has been shown to result from a base substitution at a position corresponding to 2058 of 23S rRNA (205). This site is also found on the central loop in domain V (Figure 2). Erythromycin resistance in *Staphylococcus* (197) has been shown to be due to dimethylation of adenine in a GAAAG sequence somewhere in 23S rRNA. The presence of a GAAAG sequence in 23S rRNA at positions 2057–2061, precisely at the site of the mitochondrial mutation discussed above, suggests that this is also a sequence crucial for sensitivity to erythromycin in bacteria.

A wide range of affinity probes of the peptidyl transferase site have been constructed, usually by attachment of reactive groups to the aminoacyl end of tRNA or to antibiotics implicated in this function (153). In several studies, these probes have been shown to react with 23S rRNA (244–251). In a few cases, oligonucleotide sequences from the sites of reaction have been reported (252–253), but only in one case has the precise position of reaction with 23S rRNA been determined (A. Barta, personal communication). This elegant study utilizes a photoreactive benzophenone derivative of yeast

Phe-tRNA to label *E. coli* ribosomes. The position of the labeled site was first localized to the 2500–2600 region of 23S rRNA by hybridization of 23S rRNA containing the covalently bound radioactive benzophenone-Phe moiety with defined restriction fragments of the 23S rRNA gene. The precise position of attachment was identified by DNA-primed reverse transcription of the modified RNA, in which the covalently modified residue causes the reverse transcriptase to "pause," generating a band on the gel pattern corresponding to cDNA chains terminating at that position. Remarkably, the main site of modification is U_{2584}, with some modification of U_{2585}, once again placing the peptidyl transferase site in the proximity of the central loop in domain V.

ELONGATION-GTPase ACTIVITIES   Thiostrepton appears to directly inhibit ribosomal GTPase, whether EFG, EFT or IF2 dependent (254). Thiostepton resistance has been localized to methylation of A_{1067} of 23S rRNA in *Streptomyces azureus* (200), implicating this region of domain II of 23S-like rRNAs in ribosomal GTPase functions. This is the same region of the 23S rRNA that binds protein L11, itself strongly implicated in GTPase-related functions (255). Recently EFG has been covalently cross-linked to 23S rRNA within the region 1055–1081, possibly to A_{1067}, further substantiating the above hypothesis (256).

"SWITCHES"   Ribosomes move along the mRNA chain during translation; other kinds of possible inter- and intramolecular movement can be imagined to take place during translation, and raise the question of whether ribosomes have "moving parts." If this turns out to be the case, and their movement is fundamental to the process of translation, then very likely they will be found to be constructed, at least in part, from RNA (by the same evolutionary arguments summarized above).

It has been suggested previously that movement of rRNA structure could be generated by the making and breaking of helices (7, 12, 75). Early proposals (7) have been ruled out by comparative sequence analysis since the elucidation of many complete rRNA sequences, nor do more recent models (12) survive comparative analysis when rigorous alignment criteria are invoked. Only one pair of helices has been identified that could be considered to contain mutally exclusive structures. These are the 9–13/21–25 and 17–20/915–918 helices of 16S rRNA (Figure 1), which could, in principle, coexist. They are, however, the only two helices thus far confirmed by comparative analysis where the loop contained by a helix is involved in pairing with a sequence outside the helix, and is thus formally analogous to a "knot." Biochemical probe experiments, summarized above, often show these sequences to be susceptible to single strand–specific agents. Whether or not the two helices actually coexist in vivo is presently an open question.

Another attractive hypothetical switch mechanism involves coaxial stacking and unstacking of helices. This possibility has been discussed previously in connection with 5S (257), 16S (13), and 23S (78) rRNA. Such a mechanism could bring about precise, large-scale conformational changes without necessarily involving disruption of base pairing or other hydrogen bonded interactions. To my knowledge, direct evidence for the existence of such a mechanism is also lacking.

Yet another possibility would be switches that involve tertiary structure. Direct evidence for a tertiary structural change in isolated 5S rRNA has been found (258). A conformational change triggered by increase in pH, temperature or cation concentration, all within physiological range, produces a structure that is more compact, has greater cation-binding capacity but less proton binding. It has been suggested that this conformational switch may be biologically significant, and could provide the necessary movement for translocational events in the ribosome (258). The parts of the 5S rRNA involved in the observed transition are unknown.

PROSPECTS

Studies of ribosome structure and function seem to be about to enter a new phase. Structural studies are poised at the threshold of the identification of elements of tertiary structure. Evidence from biochemical and genetic studies is converging on the identification of functional sites in ribosomal RNA. Specific models for protein-RNA recognition seem imminent. Supporting all of this are the broad underpinnings of phylogenetic comparison, bringing the evolutionary perspective to bear not only on questions of evolution per se, but on the actual details of the structure and mechanism of this ancient machine.

ACKNOWLEDGMENTS

I wish to thank R. Gutell and B. Weiser for computer-generated drawings, and R. Gutell, J. Prince, K. Triman, S. Turner, and L. Zagorska for reading the manuscript. Secondary-structure models have been developed in collaboration with C. Woese. This work was supported by NIH grant No. GM-17129.

Literature Cited

1. Chambliss, G., Craven, G. R., Davies, J., Davis, K., Kahan, L., Nomura, M. ed. 1979. *Ribosomes: Structure, Function and Genetics.* Baltimore, Md: Univ. Park.
2. Garrett, R. A. 1979. In *Chemistry of Macromolecules IIB,* 25:127–77. Baltimore, Md: Univ. Park
3. Brimacombe, R., Stöffler, G., Wittmann, H. G. 1978. *Ann. Rev. Biochem.* 47:217–49
4. Wittmann, H. G. 1982. *Ann. Rev. Biochem.* 51:155–83
5. Liljas, A. 1982. *Prog. Biophys. Mol. Biol.* 40:161–228
6. Nierhaus, K. H. 1982. *Curr. Top. Microbiol. Immunol.* 97:81–155
7. Noller, H. F. 1979. See Ref. 1, pp. 3–22

8. Zimmermann, R. A. 1979. See Ref. 1, pp. 135–70
9. Bogdanov, A. A., Kopylov, A. M., Shatsky, I. N. 1980. *Sub-Cell. Biochem.* 7:81–116
10. Noller, H. F., Woese, C. R. 1981. *Science* 212:403–11
11. Noller, H. F., van Knippenberg, P. H. 1983. *Horizons in Biochemistry and Biophysics*, ed. F. Palmieri, 7: In press. London: Wiley
12. Brimacombe, R., Maly, P., Zwieb, C. 1983. *Prog. Nucleic Acid Res. Mol. Biol.* 28:1–48
13. Woese, C. R., Gutell, R. R., Gupta, R., Noller, H. F. 1983. *Microbiol. Rev.* In press
14. Thompson, J. F., Hearst, J. E. 1983. *Cell* 33:19–24
15. Garrett, R. A., Douthwaite, S., Noller, H. F. 1981. *Trends Biochem. Sci.* 6:137–39
16. Böhm, S., Fabian, H., Welfle, H. 1982. *Acta Biol. Med. Ger.* 41:1–16
17. de Wachter, R., Chen., M.-W., Vandenberghe, A. 1982. *Biochimie* 64:311–29
18. Nazar, R. N. 1982. *Cell Nucleus* 10:1–28
19. MacKay, R. M., Spencer, D. F., Schnare, M. N., Doolittle, W. F., Gray, M. W. 1982. *Can. J. Biochem.* 60:480–89
20. Delihas, N., Andersen, J. 1982. *Nucleic Acids Res.* 10:7323–44
21. Erdmann, V. A., Huysmans, E., Vandenberghe, A., De Wachter, R. 1983. *Nucleic Acids Res.* 11:r105–33
22. Brownlee, G. G., Sanger, F., Barrell, B. G. 1967. *Nature* 215:735–36
23. Brosius, J., Palmer, M. L., Kennedy, P. J., Noller, H. F. 1978. *Proc. Natl. Acad. Sci. USA* 75:4801–5
24. Carbon, P., Ehresmann, C., Ehresmann, B., Ebel, J.-P. 1979. *Eur. J. Biochem.* 100:399–410
25. Kop, J., Kopylov, A. M., Noller, H. F., Siegel, R., Gupta, R., Woese, C. R. 1983. *Nucleic Acids Res.* Submitted for publication
26. Carbon, P., Ebel, J.-P., Ehresmann, C. 1981. *Nucleic Acids Res.* 9:2325–33
27. Tomioka, N., Sugiura, M. 1983. *Mol. Gen. Genet.* 191:46–50
28. Gupta, R., Lanter, J., Woese, C. R. 1983. *Science* 221:656–59
29. Brosius, J., Dull, T. J., Noller, H. F. 1980. *Proc. Natl. Acad. Sci. USA* 77:201–4
30. Kop, J., Wheaton, V., Gupta, R., Woese, C. R., Noller, H. F. 1983. *Nucleic Acids Res.* Submitted for publication
31. Schwartz, Z., Kössel, H. 1980. *Nature* 283:739–42
32. Tohdoh, N., Sugiura, M. 1982. *Gene* 17:213–18
33. Graf, L., Roux, E., Stutz, E., Kössel, H. 1982. *Nucleic Acids Res.* 10:6369–81
34. Dron, M., Rahire, M., Rochaix, J.-D. 1982. *Nucleic Acids Res.* 10:7609–20
35. Edwards, J., Kössel, H. 1981. *Nucleic Acids Res.* 9:2853–69
36. Takaiwa, F., Sugiura, M. 1982. *Eur. J. Biochem.* 124:13–19
37. Eperon, I. C., Janssen, J. W. G., Hoeijmakers, J. H. J., Borst, P. 1983. *Nucleic Acids Res.* 11:105–25
38. Eperon, I. C., Anderson, S., Nierlich, D. P. 1980. *Nature* 286:460–67
39. Anderson, S., de Bruijn, M. H. L., Coulson, A. R., Eperon, I. C., Sanger, F., Young, I. G. 1982. *J. Mol. Biol.* 156:683–717
40. van Etten, R. A., Walberg, M. W., Clayton, D. A. 1980. *Cell* 22:157–70
41. Kobayashi, M., Seki, T., Katsuyuki, Y., Koike, K. 1981. *Gene* 16:297–307
42. Sor, F., Fukuhara, H. 1980. *C.R. Acad. Sci. Ser. D* 291:933–36
43. Li, M., Tzagoloff, A., Underbrink-Lyon, K., Martin, N. C. 1982. *J. Biol. Chem.* 257:5921–28
44. Köchel, H.-G., Küntzel, H. 1981. *Nucleic Acids Res.* 9:5689–96
45. Seilhammer, J. J., Olsen, G. J., Cummings, D. G. 1983. *J. Biol. Chem.* Submitted for publication
46. Saccone, C., Cantatore, P., Gadaleta, G., Gallerani, R., Lanave, C., Pepe, G., Kroon, A. M. 1981. *Nucleic Acids Res.* 9:4139–48
47. Seilhammer, J. J., Gutell, R. R., Cummings, D. J. 1983. *J. Biol. Chem.* Submitted for publication
48. Sor, F., Fukuhara, H. 1983. *Nucleic Acids Res.* 11:339–48
49. Köchel, H. G., Küntzel, H. 1982. *Nucleic Acids Res.* 10:4795–801
50. Buetow, D. E., Wood, W. M. 1978. *Sub-Cell. Biochem.* 5:1–85
51. MacKay, R. M. 1981. *FEBS Lett.* 123:17–18
52. Edwards, K., Bedbrook, J., Dyer, T., Kössel, H. 1981. *Biochem. Int.* 2:533–38
53. Rochaix, J.-D., Darlix, J.-L. 1982. *J. Mol. Biol.* 159:383–95
54. Nazar, R. N. 1980. *FEBS Lett.* 119:212–14
55. Jacq, B. 1981. *Nucleic Acids Res.* 9:2913–32
56. Walker, T. A., Johnson, K. D., Olsen, G. J., Peters, M. A., Pace, N. R. 1982. *Biochemistry* 21:2320–29
57. Jordan, B. R., Jourdan, R., Jacq, B. 1976. *J. Mol. Biol.* 101:85–105
58. Brosius, J., Dull, T. J., Sleeter, D. D., Noller, H. F. 1981. *J. Mol. Biol.* 148:107–27
59. Rubtsov, P. M., Musakhanov, M. M., Zakharyev, V. M., Krayev, A. S.,

Skryabin, K. G., Bayev, A. A. 1980. *Nucleic Acids Res.* 8 : 5779–94
60. Salim, M., Maden, B. E. H. 1981. *Nature* 291 : 205–8
61. Chan, Y.-L., Gutell, R. R., Noller, H. F., Wool, I. G. 1984. *J. Biol. Chem.* In press
62. Torczynski, R., Bollon, A. P., Fuke, M. 1983. *Nucleic Acids Res.* 11 : 4879–90
63. Otsuka, T., Nomiyama, H., Yoshida, H., Kukita, T., Kuhara, S., Sakaki, Y. 1983. *Proc. Natl. Acad. Sci. USA* 80 : 3163–67
64. Veldman, G. M., Klootwijk, J., de Regt, V., Planta, R., Branlant, C., et al. 1981. *Nucleic Acids Res.* 9 : 6935–52
65. Georgiev, O. I., Nikolaev, N., Hadjiolov, A. A., Skryabin, K. G., Zakharyev, V. M., Bayev, A. A. 1981. *Nucleic Acids Res.* 9 : 6953–58
66. Chan, Y.-L., Wool, I. G. 1983. *Nucleic Acids Res.* In press
67. Fellner, P., Sanger, F. 1968. *Nature* 219 : 236–38
68. Fellner, P. 1969. *Eur. J. Biochem.* 11 : 12–27
69. Van Charldorp, R., Heus, H. A., Van Knippenberg, P. H. 1981. *Nucleic Acids Res.* 9 : 2717–25
70. Chai, Y. C., Busch, H. 1978. *Biochemistry* 17 : 2551–60
71. Klootwijk, J., Planta, R. J. 1973. *Eur. J. Biochem.* 39 : 325–33
72. Noller, H. F. 1974. *Biochemistry* 13 : 4694–703
73. Fox, G. E., Stackebrandt, E., Hespell, R. B., Gibson, J., Woese, C. R., et al. 1980. *Science* 209 : 457–63
74. Woese, C. R., Fox, G. E. 1977. *Proc. Natl. Acad. Sci. USA* 74 : 5088–90
75. Fox, G. E., Woese, C. R. 1975. *Nature* 256 : 505–7
76. Nishikawa, K., Takemura, S. 1974. *J. Biochem.* 76 : 935–47
77. Erdmann, V. A. 1976. *Prog. Nucleic Acids Res. Mol. Biol.* 18 : 45–90
78. Noller, H. F., Kop, J., Wheaton, V., Brosius, J., Gutell, R. R., et al. 1981. *Nucleic Acids Res.* 9 : 6167–89
79. Woese, C. R., Magrum, L. J., Gupta, R., Siegel, R. B., Stahl, D. A., et al. 1980. *Nucleic Acids Res.* 8 : 2275–93
80. Stiegler, P., Carbon, P., Zuker, M., Ebel, J. P., Ehresmann, C. 1981. *Nucleic Acids Res.* 9 : 2153–72
81. Zwieb, C., Glotz, C., Brimacombe, R. 1981. *Nucleic Acids Res.* 9 : 3621–40
82. Glotz, C., Zwieb, C., Brimacombe, R., Edwards, K., Kössel, H. 1981. *Nucleic Acids Res.* 9 : 3287–306
82a. Branlant, C., Krol, A., Machatt, M. A., Pouyet, J., Ebel, J. P., Edwards, K., Kössel, H. 1981. *Nucleic Acids Res.* 9 : 4303–24
83. Tinoco, I., Uhlenbeck, O. C., Levine, M.

D. 1971. *Nature* 230 : 362–67
84. Pipas, J. M., McMahon, J. E. 1975. *Proc. Natl. Acad. Sci. USA* 72 : 2017–21
85. Studnicka, G. M., Rahn, G. M., Cummings, I. W., Salser, W. 1978. *Nucleic Acids Res.* 9 : 3365–87
86. Waterman, M. S., Smith, T. F. 1978. *Math. Biosci.* 42 : 257–66
87. Nussinov, R., Jacobson, A. 1980. *Proc. Natl. Acad. Sci. USA* 77 : 6309–13
88. Zuker, M., Stiegler, P. 1981. *Nucleic Acids Res.* 9 : 133–47
89. Tinoco, I., Borer, P. N., Dengler, B., Levine, M. D., Uhlenbeck, O. C., et al. 1973. *Nature New Biol.* 246 : 40–41
90. Salser, W. 1977. *Cold Spring Harbor Symp. Quant. Biol.* 42 : 985–1002
91. Ninio, J. 1979. *Biochimie* 61 : 1133–50
92. Johnston, P. D., Redfield, A. G. 1979. In *Transfer RNA : Structure, Properties and Recognition*, ed. P. R. Schimmel, D. Söll, J. N. Abelson, pp. 191–205. Cold Spring Harbor, NY : Cold Spring Harbor Lab.
93. Bear, D. G., Schleich, T., Noller, H. F., Douthwaite, S., Garrett, R. A. 1979. *FEBS Lett.* 100 : 99–102
94. Bear, D. G., Schleich, T., Noller, H. F., Garrett, R. A. 1977. *Nucleic Acids Res.* 4 : 2511–26
95. Litt, M. 1969. *Biochemistry* 8 : 3249–53
96. Peattie, D. A., Gilbert, W. 1980. *Proc. Natl. Acad. Sci. USA* 77 : 4679–82
97. Mankin, A. S., Kopylov, A. M., Bogdanov, A. A. 1981. *FEBS Lett.* 134 : 11–14
98. Silberkang, M., RajBhandary, U. L., Lück, A., Erdmann, V. A. 1983. *Nucleic Acids Res.* 11 : 605–17
99. Rhodes, D. 1975. *J. Mol. Biol.* 94 : 449–60
100. Rabin, D., Crothers, D. M. 1979. *Nucleic Acids Res.* 7 : 689–703
101. Wollenzien, P. L., Hearst, J. E., Thammana, P., Cantor, C. R. 1979. *J. Mol. Biol.* 135 : 255–69
102. Thammana, P., Cantor, C. R., Wollenzien, P. L., Hearst, J. E. 1979. *J. Mol. Biol.* 135 : 271–83
103. Turner, S., Thompson, J. F., Hearst, J. E., Noller, H. F. 1982. *Nucleic Acids Res.* 10 : 2839–49
104. Turner, S., Noller, H. F. 1983. *Biochemistry* 22 : 4159–65
105. Thompson, J. F., Hearst, J. E. 1983. *Cell* 32 : 1355–65
106. Hearst, J. E. 1981. *Ann. Rev. Biophys. Bioeng.* 10 : 69–86
107. Vigne, R., Jordan, B. R., Monier, R. 1973. *J. Mol. Biol.* 76 : 303–11
108. Ross, A., Brimacombe, R. 1979. *Nature* 281 : 271–76
109. Vassilenko, S. K., Ryte, V. C. 1975. *Biokhimya* 40 : 578–83

110. Vassilenko, S. K., Carbon, P., Ebel, J. P., Ehresmann, C. 1981. *J. Mol. Biol.* 152:699–721
111. Douthwaite, S., Garrett, R. A. 1981. *Biochemistry* 20:7301–7
112. Lewis, J. B., Doty, P. 1977. *Biochemistry* 16:5016–25
113. Kopylov, A. M., Chichkova, N. V., Bogdanov, A. A., Vasilenko, S. K. 1975. *Mol, Biol. Rep.* 2:95–100
114. Mankin, A. S., Skripkin, E. A., Chichkova, N. V., Kopylov, A. M., Bogdanov, A. A. 1981. *FEBS Lett.* 131:253–56
115. Backendorf, C., Ravensbergen, C. J. C., Van der Plas, J., van Boom, J. H., Veeneman, G., Van Duin, J. 1981. *Nucleic Acids Res.* 9:1425–44
116. Kime, M. J., Moore, P. B. 1983. *FEBS Lett.* 153:199–203
117. Kime, M. J., Moore, P. B. 1983. *Biochemistry* 22:2615–22
118. Kime, M. J., Moore, P. B. 1983. *Biochemistry* 22:2622–29
119. Baan, R. A., Hilbers, C. W., van Charldorp, R., Van Leerdam, E., Van Knippenberg, P. H., Bosch, L. 1977. *Proc. Natl. Acad. Sci. USA* 74:1028–31
120. Edlind, T. D., Bassel, A. R. 1980. *J. Bacteriol.* 141:365–73
121. Ehresmann, C., Stiegler, P., Carbon, P., Ungewickell, E., Garrett, R. A. 1980. *Eur. J. Biochem.* 103:439–46
122. Müller, R., Garrett, R. A., Noller, H. F. 1979. *J. Biol. Chem.* 254:3873–78
123. Zimmermann, R. A., Mackie, G. A., Muto, A., Garrett, R. A., Ungewickell, E., et al. 1975. *Nucleic Acids Res.* 2:279–302
124. Yuki, A., Brimacombe, R. 1975. *Eur. J. Biochem.* 56:23–34
125. Traub, W., Sussman, J. L. 1982. *Nucleic Acids Res.* 10:2701–8
126. Douthwaite, S., Garrett, R. A. 1983. *J. Mol. Biol.* In press
127. Rich, A., RajBhandary, U. L. 1976. *Ann. Rev. Biochem.* 45:805–60
128. Clark, B. F. C. 1977. *Proc. Nucleic Acid Res. Mol. Biol.* 20:1–19
129. Crick, F. H. C. 1966. *J. Mol. Biol.* 19:548–55
130. Kan, L.-S., Chandrasegaran, S., Pulford, S. M., Miller, P. S. 1983. *Proc. Natl. Acad. Sci. USA* 80:4263–265
131. Cantor, C. R. 1979. See Ref. 1, pp. 23–49
132. Stiegler, P., Carbon, P., Ebel, J. P., Ehresmann, C. 1981. *Eur. J. Biochem.* 120:487–95.
133. Hogan, J. J., Gutell, R. R., Noller, H. F. 1983. *Biochemistry.* Submitted for publication
134. Hogan, J. J., Gutell, R. R., Noller, H. F. 1983. *Biochemistry.* Submitted for publication
135. Hogan, J. J., Noller, H. F. 1978. *Biochemistry* 17:587–93
136. Herr, W., Noller, H. F. 1978. *Biochemistry* 17:307–15
137. Cantor, C. R., Wollenzien, P. L., Hearst, J. E. 1980. *Nucleic Acids Res.* 8:1855–72
138. Wollenzien, P. L., Cantor, C. R. 1982. *Proc. Natl. Acad. Sci. USA* 79:3940–44
139. Stiege, W., Zwieb, C., Brimacombe, R. 1982. *Nucleic Acids Res.* 10:7211–99
140. Zwieb, C., Brimacombe, R. 1980. *Nucleic Acids Res.* 8:2397–11
141. Glotz, C., Brimacombe, R. 1980. *Nucleic Acids Res.* 8:2377–95
142. Lake, J. A. 1979. See Ref. 1, pp. 207–36
143. Stöffler, G., Bald, R., Kastner, B., Lührmann, R., Stöffler-Meilicke, M., Tischendorf, G. 1979. See Ref. 1, pp. 171–205
144. Moore, P. B. 1979. See Ref. 1, pp. 111–33
145. Politz, S. M., Glitz, D. G. 1977. *Proc. Natl. Acad. Sci. USA* 74:1468–72
146. Trempe, M. R., Ohgi, K., Glitz, D. G. 1982. *J. Biol. Chem.* 257:9822–29
147. Olson, H. M., Glitz, D. G. 1979. *Proc. Natl. Acad. Sci. USA* 76:3769–73
148. Shatsky, I. N., Mochalova, L. V., Kojouharova, M. S., Bogdanov, A. A., Vasiliev, V. D. 1979. *J. Mol. Biol.* 133:501–15
149. Lührmann, R., Stöffler-Meilicke, M., Stöffler, G. 1981. *Mol. Gen. Genet.* 182:369–76
150. Mochalova, L. V., Shatsky, I. N., Bogdanov, A. A., Vasiliev, V. D. 1982. *J. Mol. Biol.* 159:637–50.
151. Stöffler-Meilicke, M., Stöffler, G., Odom, O. W., Zinn, A., Kramer, G., Hardesty, B. 1981. *Proc. Natl. Acad. Sci. USA* 78:5538–42
152. Shatsky, I. N., Evstafieva, A. G., Bystrova, T. F., Bogdanov, A. A., Vasiliev, V. D. 1980. *FEBS Lett.* 121:97–100
153. Cooperman, B. S. 1979. See Ref. 1, pp. 531–54
154. Shatsky, I. N., Evstafieva, A. G., Bystrova, T. F., Bogdanov, A. A., Vasiliev, V. D. 1980. *FEBS Lett.* 122:251–55
155. Wollenzien, P. L., Cantor, C. R. 1982. *J. Mol. Biol.* 159:151–66
156. Spitnik-Elson, P., Elson, D., Avital, S., Abramowitz, R. 1982. *Nucleic Acids Res.* 10:1995–2006
157. Spitnik-Elson, P., Elson, D., Avital, S., Abramowitz, R. 1982. *Nucleic Acids Res.* 10:4483–92
158. Stiege, W., Glotz, C., Brimacombe, R. 1983. *Nucleic Acids Res.* 11:1687–1706
159. Odom, O. W., Robbins, D. J., Dottavia-Martin, D., Kramer, G., Hardesty, B. 1980. *Biochemistry* 19:5947–54

160. Hancock, J., Wagner, R. 1982. *Nucleic Acids Res.* 10:1257–69
161. Pieler, T., Erdmann, V. A. 1982. *Proc. Natl. Acad. Sci. USA* 79:4599–603
162. Monier, R. 1974. In *Ribosomes*, ed. M. Nomura, A. Tissières, P. Lengyel, pp. 141–68. Cold Spring Harbor, NY: Cold Spring Harbor Lab.
163. Noller, H. F., Garrett, R. A. 1979. *J. Mol. Biol.* 132:637–48
164. Noller, H. F., Herr, W. 1974. *J. Mol. Biol.* 90:181–184.
165. Delihas, N., Dunn, J. J., Erdmann, V. A. 1975. *FEBS Lett.* 58:76–80
166. Weidner, H., Yuan, R., Crothers, D. M. 1977. *Nature* 266:193–94
167. Zimmermann, R. A., Singh-Bermann, K. 1979. *Biochim. Biophys. Acta* 563:422–31
168. Branlant, C., Krol, A., SriWidada, J., Ebel, J. P., Sloof, P., Garrett, R. A. 1976. *Eur. J. Biochem.* 70:457–69
169. Thompson, J., Schmidt, F., Cundliffe, E. 1982. *J. Biol. Chem.* 257:7915–17
170. Garrett, R. A., Noller, H. F. 1979. *J. Mol. Biol.* 132:637–48
171. Peattie, D. A., Douthwaite, S., Garrett, R. A., Noller, H. F. 1981. *Proc. Natl. Acad. Aci. USA* 78:7331–35
172. Douthwaite, S., Garrett, R. A. 1982. *Biochemistry* 21:2313–20
173. Douthwaite, S., Garrett, R. A., Wagner, R., Feunteun, J. 1979. *Nucleic Acids Res.* 6:2453–70
174. Spierer, P., Bogdanov, A. A., Zimmermann, R. A. 1978. *Biochemistry* 17:5394–42
175. Wower, I., Brimacombe, R. 1983. *Nucleic Acids Res.* 11:1419–37
176. Nomura, M., Dean, D., Yates, J. L. 1982. *Trends Biochem. Sci.* 7:92–95
177. Nomura, M., Yates, J. L., Dean, D., Post, L. E. 1980. *Proc. Natl. Acad. Sci. USA* 77:7084–88
178. Olins, P. O., Nomura, M. 1981. *Nucleic Acids Res.* 9:1757–64
179. Gourse, R. L., Thurlow, D. L., Gerbi, S. A., Zimmermann, R. A. 1981. *Proc. Natl. Acad. Sci. USA* 78:2722–26
180. Branlant, C., Krol, A., Machatt, M. A., Ebel, J.-P. 1981. *Nucleic Acids Res.* 9:293–307
181. Vasiliev, V. D., Selivanova, O. M., Koteliansky, V. E. 1978. *FEBS Lett.* 95:273–76
182. Allen, S. H., Wong, K.-P. 1978. *Biochemisry* 253:8759–66
183. Boublik, M., Robakis, N., Hellmann, W. 1982. *Eur. J. Cell. Biol.* 27:177–84
184. Tam, M. F., Dodd, J. A., Hill, W. E. 1981. *J. Biol. Chem.* 256:6430–34
185. Nomura, M., Held, W. A. 1974. See Ref. 162, pp. 193–223
186. Nierhaus, K. 1979. See Ref. 1, pp. 267–94
187. Held, W. A., Nomura, M. 1973. *Biochemistry* 12:3273–81
188. Herr, W., Noller, H. F. 1975. *FEBS Lett.* 53:248–52
189. Azad, A. A. 1979. *Nucleic Acids Res.* 7:1913–29
190. Davies, J., Nomura, M. 1972. *Ann. Rev. Genet.* 6:203–34
191. Nomura, M., Mizushima, S., Ozaki, M., Traub, P., Lowry, C. V. 1969. *Cold Spring Harbor Symp. Quant. Biol.* 34:49–61
192. Erdmann, V. A., Fahnestock, S., Higo, K., Nomura, M. 1971. *Proc. Natl. Acad. Sci. USA* 68:2932–36
193. Dobbs, E. R. 1979. *J. Bacteriol.* 140:734–37
194. Garrett, R. A. 1983. *Trends Biochem. Sci.* 8:75–76
195. Kruger, K., Grabowski, P. J., Zaug, A. J., Sands, J., Gottschling, D. E., Cech, T. R. 1982. *Cell* 31:147–57
196. Helser, T. L., Davies, J. E., Dahlberg, J. E. 1972. *Nature New Biol.* 235:6–8
197. Lai, C.-J., Dahlberg, J. E., Weisblum, B. 1973. *Biochemistry* 12:457–60
198. Gold, L., Pribnow, D., Schneider, T., Shinedling, S., Singer, B. S., Stormo, G. 1981. *Ann. Rev. Microbiol.* 35:365–403
199. Yamada, T., Mizuguchi, Y., Nierhaus, K. H., Wittmann, H. G. 1978. *Nature* 275:460–61
200. Thompson, J., Schmidt, F., Cundliffe, E. 1982. *J. Biol. Chem.* 257:7915–17
201. Dujon, B. 1980. *Cell* 20:185–97
202. Kearsey, S. E., Craig, I. W. 1981. *Nature* 290:607–8
203. Blanc, H., Adams, C. W., Wallace, D. C. 1981. *Nucleic Acids Res.* 9:5785–95
204. Blanc, H., Wright, C. T., Bibb, M. J., Wallace, D. C., Clayton, D. A. 1981. *Proc. Natl. Acad. Sci. USA* 78:3789–93
205. Sor, F., Fukuhara, H. 1982. *Nucleic Acids Res.* 10:6571–77
206. Bolotin-Fukuhara, M. 1979. *Mol. Gen. Genet.* 177:39–46
207. Sigmund, C. D., Morgan, E. A. 1982. *Proc. Natl. Acad. Sci. USA* 79:5602–6
208. Gourse, R. L., Stark, M. J. R., Dahlberg, A. E. 1982. *J. Mol. Biol.* 159:397–416
209. Stark, M. J. R., Gourse, R. L., Dahlberg, A. E. 1982. *J. Mol. Biol.* 159:417–39
210. Fox, T. D., Staempfli, S. 1982. *Proc. Natl. Acad. Sci. USA* 79:1583–87
211. Bowman, C. M., Dahlberg, J. E., Ikemura, T., Konisky, J., Nomura, M. 1971. *Proc. Natl. Acad. Sci. USA* 68:964–68
212. Senior, B. W., Holland, I. B. 1971. *Proc. Natl. Acad. Sci. USA* 68:959–63

213. Schindler, D. G., Davies, J. E. 1977. *Nucleic Acids Res.* 4:1097–110
214. Endo, V., Wool, I. G. 1982. *J. Biol. Chem.* 257:9054–60
215. Shine, J., Dalgarno, L. 1974. *Proc. Natl. Acad. Sci. USA* 71:1342–46
216. Steitz, J. A. 1979. See Ref. 1, pp. 479–95
217. Steitz, J. A., Jakes, K. 1975. *Proc. Natl. Acad. Sci. USA* 72:4734–38
218. Taniguchi, T., Weissman, C. 1978. *J. Mol. Biol.* 118:533–65
219. Dunn, J. J., Buzash-Pollert, E., Studier, F. W. 1978. *Proc. Natl. Acad. Sci. USA* 75:2741–45
220. Grunberg-Manago, M. 1979. See Ref. 1, pp. 445–77
221. Cooperman, B. S., Dondon, J., Finelli, J., Grunberg-Manago, M., Michelson, A. M. 1977. *FEBS Lett.* 76:59–63
222. Wickstrom, E. 1983. *Nucleic Acids Res.* 11:2035–52
223. Okuyama, A., Machiyama, N., Kinoshita, T., Tanaka, N. 1971. *Biochem. Biophys. Res. Commun.* 43:196–99
224. Wagner, R., Gassen, H. G., Ehresmann, C., Stiegler, P., Ebel, J. P. 1976. *FEBS Lett.* 67:312–15
225. Fiser, I., Scheit, K. H., Kuechler, E. 1977. *Eur. J. Biochem.* 74:447–56
226. Budker, V. G., Girshovich, A. S., Grineva, N. I., Karpova, G. G., Knorre, D. G., Kobets, N. D. 1973. *Dokl. Akad. Nauk SSSR* 211:725–28
227. Towbin, H., Elson, D. 1978. *Nucleic Acids Res.* 5:3389–407
228. Santer, M., Shane, S. 1977. *J. Bacteriol.* 130:900–10
229. Chapman, N, M., Noller, H. F. 1979. *J. Mol. Biol.* 109:131–49
230. Vassilenko, S. K., Carbon, P., Ebel, J.-P., Ehresmann, C. 1981. *J. Mol. Biol.* 152:699–721
231. Herr, W., Chapman, N. M., Noller, H. F. 1979. *J. Mol. Biol.* 130:433–49
232. Kühlbrandt, W., Unwin, P. N. T. 1982. *J. Mol. Biol.* 156:431–48
233. Schwartz, I., Ofengand, J. 1978. *Biochemistry* 17:2524–30
234. Prince, J. B., Taylor, B. H., Thurlow, D. L., Ofengand, J., Zimmermann, R. A. 1982. *Proc. Natl. Acad. Sci. USA* 79:5450–54
235. Brow, D. A., Noller, H. F. 1983. *J. Mol. Biol.* 163:27–46
236. Noller, H. F., Chaires, J. B. 1972. *Proc. Natl. Acad. Sci. USA* 69:3115–118
237. Ofengand, J., Henes, C. 1969. *J. Biol. Chem.* 244:6241–53
238. Shimizu, N., Hayashi, H., Miura, K.-I. 1970. *J. Biochem.* 67:373–87
239. Richter, D., Erdmann, V. A., Sprinzl, M. 1973. *Nature New Biol.* 246:132–35
240. Erdmann, V. A., Sprinzl, M., Pongs, O. 1973. *Biochem. Biophys. Res. Commun.* 54:942–48
241. Pace, B., Matthews, E. A., Johnson, K. D., Cantor, C. R., Pace, N. R. 1982. *Proc. Natl. Acad. Sci. USA* 79:36–40
242. Zagórska, L., Van Duin, J., Noller, H. F., Pace, B., Johnson, K. D., Pace, N. R. 1983. *J. Biol. Chem.* In press
243. Vazquez, D. 1979. *Antibiotic Inhibitors of Protein Biosynthesis.* Berlin Heidelberg New York: Springer-Verlag
244. Bispink, L., Matthaei, H. 1973. *FEBS Lett.* 37:291–94
245. Girshovich, A. S., Bochkareva, E. S., Kramarov, V. M., Ovchinnikov, Y. A. 1974. *FEBS Lett.* 45:213–17
246. Breitmeyer, J. B., Noller, H. F. 1976. *J. Mol. Biol.* 101:297–306
247. Barta, A., Kuechler, E., Branlant, C., SriWidada, J., Krol, A., Ebel, J. P. 1975. *FEBS Lett.* 56:170–74
248. Sonenberg, N., Wilchek, M., Zamir, A. 1977. *Eur. J. Biochem.* 77:217–22
249. Yukioka, M., Hatayama, T., Morisawa, S. 1975. *Biochim. Biophys. Acta* 390:192–308
250. Johnson, A. 1979. See Ref. 92, pp. 487–99
251. Greenwall, P., Harris, R. J., Symons, R. H. 1974. *Eur. J. Biochem.* 82:225–34
252. Eckermann, D. J., Symons, R. H. 1978. *Eur. J. Biochem.* 82:225–34
253. Yukioka, M., Hatayama, T., Omori, K. 1977. *Eur. J. Biochem.* 73:449–59
254. Cundliffe, E. 1979. See Ref. 1, pp. 555–81
255. Möller, W. 1974. See Ref. 162, pp. 711–31
256. Sköld, S.-E. 1983. *Nucleic Acids Res.* 11:4923–32
257. Stahl, D. A., Luehrsen, K. R., Woese, C. R., Pace, N. R. 1981. *Nucleic Acids Res.* 9:6129–37
258. Kao, T. H., Crothers, D. M. 1980. *Proc. Natl. Acad. Sci. USA* 77:3360–64
259. Stein, P. R., Waterman, M. S. 1978. *Discrete Math.* 26:261–72
260. Klein, B. K., King, T. C., Schlessinger, D. 1983. *J. Mol. Biol.* 168:809–30

Ann. Rev. Biochem. 1981. 50:969–96

13

NMR STUDIES ON RNA STRUCTURE AND DYNAMICS[1]

Brian R. Reid

Department of Chemistry, University of Washington and Department
of Biochemistry, University of Washington Medical School, Seattle,
Washington 98195

CONTENTS

Perspectives and Summary

Over the last five to ten years high-resolution nuclear magnetic resonance (NMR) has increasingly been used to study the solution properties of relatively small cellular RNA molecules such as tRNA and, to a lesser extent, ribosomal 5S RNA. The two natural isotopes in RNA that are

---

[1]The following abbreviations are used: FTNMR, Fourier transform nuclear magnetic resonance; D or DHU, dihydrouridine; T or rT, ribothymidine; NOE, nuclear Overhauser effect; FID, free induction decay: CW, continuous wave.

amenable to NMR studies are ^1H and ^{31}P, and the vast majority of NMR studies on RNA to date involve proton NMR. Isotope enrichment methods are currently being developed in order to study other NMR nuclei e.g. ^{13}C, ^{15}N, and ^{19}F.

RNA molecules of biochemical interest range in size from a few dozen nucleotides to ribosomal and viral RNAs with molecular weights in the millions, but not all of these are amenable to structural investigation by NMR. Each nucleotide in RNA contains 10–12 protons, and proton NMR linewidths increase with polymer molecular weight; the large number of resonances and their broad linewidths lead to overlapping unresolved NMR spectra for molecules larger than 100–200 nucleotides. Spectral resolution is increased by the high magnetic field strengths of modern very-high-resolution NMR spectrometers operating around 500 MHz, but even with the resolution afforded by the highest available field strengths (11–12 Tesla) only selected regions at the extremities of the spectrum are sufficiently resolved to attempt resonance assignments and begin structural interpretation.

The most intensively studied regions of tRNA NMR spectra are the extreme low field region (–15 to –11 ppm), which contains ring NH or imino protons that are hydrogen bonded between complementary base pairs, and the high field region (–4 to 0 ppm), which contains methyl protons from modified bases. The methyl resonance positions are often shifted by ring-current effects from neighboring stacked bases in the native structure and can be used as local reporter groups, since they move back to their characteristic unshifted resonance positions when that region of the molecule unfolds. The low field spectrum is potentially more informative, since it contains a reporter resonance from each base pair; several resonances, including some from tertiary folding, have been assigned to specific base pairs by a variety of methods, but some assignments remain controversial.

A particularly undesirable aspect of low field spectroscopy of exchangeable protons is the enormous proton resonance of the solvent, since such studies must be carried out in H_2O (rather than D_2O). Recently, Redfield has introduced pulsed FTNMR methods that greatly suppress the H_2O resonance and allow the application of double-resonance and time-resolved techniques. Double-resonance NOE measurements have unambiguously identified guanine-uracil (GU) base pair resonances and have also assigned other low field protons. Time-resolved FTNMR techniques have been used to measure the helix-coil breathing rates of individual base pairs in isolated helices and in intact tRNA. Short helices are not fully cooperative; they are more stable in the middle and relatively labile at the ends. Time-resolved studies on selected tertiary base pairs in intact tRNA are beginning to reveal the overall dynamics of the folded structure. These fundamental investigations on assignments and dynamics are opening the way to future experiments in which the effects of other molecules such as codons, cations, drugs, enzymes etc on tRNA structure can be studied.

Background Introduction

NMR spectroscopy involves inherently weak signals; compared to other forms of spectroscopy it requires relatively high sample concentrations ($\sim$ 1 mM or higher). Also, the sensitivity to detection by NMR methods is not the same for all atomic nuclei. By far the most easily detected nucleus is ^{1}H, which is why most biochemical NMR investigations of proteins and nucleic acids involve proton NMR, although ^{31}P NMR has been used in some nucleic acid studies. The natural isotopes of carbon and oxygen do not possess nuclear spin, but prior substitution of the sample with ^{13}C (and ^{19}F via fluorouracil) has been used recently in RNA structural analysis by NMR and is discussed briefly at the end of this review.

In proton NMR, the resonance linewidth is governed by interproton dipolar relaxation, the rate of which usually depends on the rotational correlation time, or tumbling time, of the entire molecule. In general, this tumbling time is directly related to the molecular weight, thus the linewidth is a linear function of polymer size. (This relationship breaks down for elongated rod-like molecules or for very flexible polymers with rapid local motion that is independent of the global tumbling.) There are several general textbooks on biochemical applications of NMR that discuss linewidths and relaxation (82, 83). Hence for high-molecular-weight samples the proton resonances are generally too broad to be resolved. The tumbling time and linewidth can be reduced by heating the sample, but biochemists like to work close to physiological temperature where the molecule carries out its biological function. The spectral resolution can also be increased by higher magnetic field strength; however, even at very-high-resolution (300–500 MHz) the maximal tractable polymer size is $\sim$ 40,000 daltons where linewidths of carbon protons are around 20 Hz at 37°C. The major RNA molecules of biological interest in this size range are tRNA and 5S RNA, and RNA NMR studies have focused principally on these molecules, especially the former.

The field of RNA proton NMR is approximately ten years old and was the topic of several reviews, now somewhat outdated, in the mid-1970s (1–4). In addition, several more general articles on nucleic acids have been written that include sections on NMR of RNA (5–10). There are some relatively recent research articles and reviews in the specific area of NMR studies on transfer RNA (11–13), and a recent review by Schweizer (14) updates the general area of RNA NMR, including studies on oligonucleotides. Many of the above articles are directed more toward a spectroscopic audience than a biochemical one; in contrast this review is aimed more at the biochemist already familiar with many aspects of tRNA, but perhaps not so familiar with the techniques and advantages of modern high-resolution NMR, and it will of necessity be somewhat restricted. The most significant development in the last two to three years has been the ability to carry out double resonance and time-resolved FTNMR experiments in H_2O solvents, and the second half of this review focuses on the application of these more recent techniques and the results obtained from them.

Types of Protons and Resonance Positions

An average tRNA or 5S RNA molecule contains approximately 1000 protons including ribose CH, ribose OH, aromatic base CH, base exocyclic NH_2, base ring NH, and CH_3 groups from methylated nucleosides. The range of chemical shifts for these protons covers the spectral region from −15 to 0 ppm, but they are far from evenly distributed. The low field region from −15 to −9 ppm contains the base ring NHs of G and U, which generally resonate below −11 ppm when they are hydrogen bonded to other bases (there are typically about 30 protons in this region in a tRNA spectrum). The high field region from −4 to 0 ppm contains methyl and aliphatic resonances from modified nucleosides and usually has an intensity corresponding to 6–30 protons. The midfield region in between contains the remaining several hundred protons of which the ribose resonances are up around −5 ppm. The spectrum can be simplified somewhat by working in D_2O, thus substituting all the exchangeable NH_2, OH, and ring NH protons for deuterons (15, 16); this eliminates the low field spectrum, reduces the midfield spectrum, and leaves the high field "methyl" region unaffected. Even in D_2O the midfield region is too crowded with an excessive number of overlapping nonexchangeable protons to be amenable to detailed interpretation, although Schmidt & Kastrup have succeeded in resolving 15 single protons out of the 89 aromatic CH resonances (−9 to −7 ppm) in *Escherichia coli* valine tRNA (17).

THE HIGH FIELD METHYL SPECTRUM (−4 TO 0 PPM)   The alkyl and methyl protons of modified bases are relatively straightforward to assign since their resonance positions in the random coil state are the same (or very similar) as the chemical shifts in the model methylated nucleosides, which are usually known. Hence a common strategy has been to monitor the high field spectrum of tRNA at several temperatures up to ∼ 90°C; the assigned peaks are then traced back to their native chemical shifts by analyzing the spectra in reverse order. Once assigned, these resonances can be used as reporters to monitor intermediate structures and local unfolding by plotting resonance position (which changes due to loss of ring-current effects during unstacking) as a function of temperature. Kastrup & Schmidt have used such an approach to monitor conformational transitions in the DHU loop (D17 methylene), the anticodon loop (m^6A37 methyl), and the ribothymidine loop (rT54 methyl) of *E. coli* valine tRNA (18, 19). The former two signals are from unpaired single-stranded residues in loop regions, whereas the latter residue is involved in a tertiary base pair in the crystal structure of yeast phenylalanine tRNA (20, 37, 38) and probably all other tRNAs as well (21, 22). Temperature-dependent chemical shift plots carried out in low salt (0.045 M free Na^+) revealed transitions centered at 55° (D17), 58° (m^6A37) and 67°, (rT54), which were interpreted to reflect unfolding of the DHU stem loop, anticodon stem loop, and rT stem loop, respectively (18). However, these chemical shift changes actually monitor events only in the loop region and reveal no direct information about the helical stem. The rT resonance and the DHU resonance indicated a slow exchange conformational equilibrium (long lifetime in each state compared to the chemical shift difference, which results in distinct resonances from each state), whereas the anticodon loop signal reflected a fast exchange two-state transition. Furthermore, the rT signal exhibited interesting multiple states at 1.9, 1.8, and 1.25 ppm; an additional peak at 1.0 ppm was initially attributed to contaminant (18), but subsequently proved to be a fourth state (actually the native state) of rT54 (19). The presence of multiple states at physiological temperature was attributed to "non-native" conformers peculiar to the low salt conditions, because they largely disappeared above 0.2 M Na^+, under which conditions a single rT54 methyl resonance was found at −1.0 ppm (19). At 0.2 M Na^+ the alternate conformations can be made to reappear, but only at higher temperature; however, they can be trapped by slowly lowering the temperature back to room temperature, whereas rapid quenching in the presence of magnesium largely abolishes these alternate states (19). This hysteresis emphasizes the importance of the past history of the sample and starting conformation(s) in analyzing unfolding.

The above approach to conformational analysis of tRNA suffers somewhat from not having enough "handles" reporting events from the many regions of interest in tRNA; this is especially true for tRNA species containing only one or two methylated nucleosides. Furthermore, some methyl resonances are not useful reporters in that they do not shift during the unfolding process or are buried downfield among the ribose protons. In general, eukaryotic tRNAs contain a much higher level of methylation and other modifications than prokaryotic tRNAs, thus they present more high field spectroscopic handles with which to monitor conformational events. For instance, yeast phenylalanine tRNA contains resolved aliphatic resonances from modified nucleosides in the D stem, D loop, anticodon stem, anticodon loop, extra loop, rT stem, and rT helix. The thermal unfolding of this molecule has been investigated via the chemical shifts of these methyl groups (23–25). The added complexity of eukaryotic tRNA high field spectra, while permitting observation of more regions of the molecule, also increases the possibility of misassigning transitions. For instance the D16,D17 resonance (C5H) and the m_2^2G26 resonance coalesce at –2.6 ppm at intermediate temperature and separate out at lower temperature. The component that moves upfield to –2.4 ppm was attributed to D16,D17 by Kan et al (23, 24), and the transition was therefore interpreted to reflect events occurring in the DHU loop. However, by means of decoupling the D16,D17 C6H, Robillard et al (25) were able to show that the D16,D17 resonance hardly moves from –2.6 ppm at all and that it is the $_2^2$G26 resonance that crosses over to –2.4 ppm; hence the observed transition is reporting events at the DHU stem-anticodon stem junction rather than in the DHU loop. The NMR transitions observed for the nonexchangeable methyl protons can be directly related to independently determined thermodynamic parameters derived from optical studies in order to cross-check the assignments and interpretations. Kan et al (23, 24) carried out parallel UV hyperchromicity studies in the same buffer as their NMR studies, whereas Robillard et al (25) adjusted their buffers to the same conditions used in the optical studies of Romer et al (26, 27) so that these T_m values and temperature-jump relaxation times could be directly related to the NMR results. In the absence of magnesium, the first unfolding events reflect changes in the acceptor helix, tertiary structure, and anticodon helix (20–40°C), followed by the rT helix transition (45–55°C); the most pronounced transitions are those of m_2^2G26 and D16,D17 at both ends of the DHU helix, which reflect the unfolding of this helix at the surprising temperature of 60–65°C (25). In the presence of magnesium the unfolding of secondary and tertiary structure becomes unresolved in a cooperative transition at ~ 74°C (24). Although much important information is beginning to emerge from high field NMR spectroscopy this spectral region contains only a few signals that directly report events in the base paired helical regions.

RESONANCES FROM BASE-PAIR HYDROGEN BONDS   The hydrogen bonded protons of complementary base pairs are, of course, solvent exchangeable and cannot be observed in D_2O where most "aqueous" NMR is carried out. Hence one must work in H_2O to detect these resonances, and this causes special problems in detection and instrumentation that are discussed later. Since solvent exchange from the exposed coil state is usually rapid, a further restriction is that these base pairing resonances will only be observed when their helix lifetimes are long compared to the chemical shift difference between the bonded and nonbonded states (for modern high-resolution spectrometers this corresponds to base pair lifetimes greater than 2–3 msec). However, compensating advantages are that the ring NH hydrogen bonds resonate in the uncrowded extreme low field region of the spectrum between –11 and –15 ppm and there is only one such signal per base pair (U N3H in an AU pair and G N1H in a GC pair). Furthermore, these low field resonances are derived from the double helical parts of the RNA and can, under appropriate conditions, reveal dynamic lifetime information about these helices.

The hydrogen bonded ring NH resonances of tRNA were first observed in the early 1970s by Kearns, Shulman, and collaborators (28, 29), and the early results have been reviewed by Kearns & Shulman (1) and by Kearns (2). The majority of tRNA species studied by NMR (class I) contain 20 secondary base pairs, and the integrated low field intensity was initially interpreted to contain 20 protons; however these interpretations are undoubtedly wrong. The first evidence that the spectra actually indicated the presence of approximately seven additional resonances from tertiary base pairs was presented by Reid et al (30), Daniel & Cohn (31), and Reid & Robillard (32). Although this point remained the subject of some controversy (33–35), the analysis of extremely pure tRNA samples at higher field strengths, combined with computer analysis and simulation of the resolved spectra, has established beyond doubt the existence of low field resonances between –15 and –11 ppm from approximately seven extra base pairs in several class I tRNAs (36). As examples Figure 1 shows the low field spectra of *E. coli* $tRNA_1^{Val}$ and $tRNA^{Ile}$, both of which contain 20 base pairs in their secondary cloverleaf sequence. The isoleucine tRNA spectrum contains 23 partially resolved peaks between –15 and –11 ppm, but the peaks at –11.8 and –13.2 ppm obviously contain 2 protons, and the peak at –12.8 ppm contains at least 3 protons. Both spectra contain 4 peaks between –11 and –9 ppm including the narrow aromatic resonance from C8H of m^7G46 at $\sim$–9.1 ppm (see below). Perhaps the best evidence for the existence of the controversial tertiary resonances is that several of them have recently been assigned (a spectacular feat if they do not exist!).

ORIGIN OF LOW FIELD RESONANCES   Among the $\sim$ 27 low field resonances observed in most class I tRNA spectra the first 20 or so are obviously derived from the hydrogen bonded ring NHs of the 19–21 Watson-Crick base pairs in the cloverleaf secondary structure of such tRNAs. The origin of the extra seven resonances between –15 and –11 ppm is difficult to determine by NMR methods alone, and progress in this area has been greatly facilitated by X-ray crystallographic structure determination. The three-dimensional structure of yeast $tRNA^{Phe}$ has been determined in two crystal forms by four laboratories, and the structural aspects of tertiary base pairing in this molecule have been reviewed recently (20, 37, 38). Although

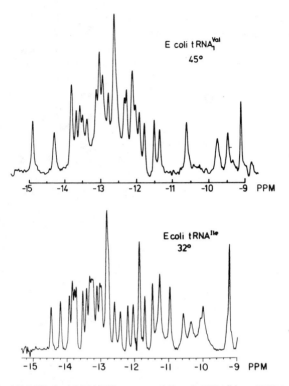

Figure 1   The 360-MHz low field NMR spectra of *E. coli* tRNA$^{Val}_1$ at 45°C (*upper*) and *E. coli* tRNA^{Ile} at 32°C (*lower*). The region between –11 and –15 ppm contains hydrogen bonded ring NHs; each proton resonance is from a different base pair in the tRNA except for the two resonances from the GU pair (–11.35 and –11.95 ppm in tRNA^{Val}; –10.9 and –11.8 ppm in tRNA^{Ile}). Both samples contained 6 mg of tRNA in 0.2 ml of buffer. The upper spectrum was obtained using rapid sweep correlation spectroscopy (2500 Hz in 0.8 sec) after 17 min of signal-averaging; the lower spectrum was obtained using Redfield 214 FTNMR with a 0.37 msec pulse and 10 min of signal averaging. The lower spectrum has been resolution-enhanced by Gaussian multiplication. The tRNA$^{Val}_1$ sample was fully thiolated at uridine 8 (note the full proton at –14.9 ppm) whereas tRNA^{Ile} contains no s^{4}U.

the structures of other tRNA species have been solved at lower resolution, the yeast tRNA^{Phe} molecule remains the only refined high-resolution structure available and hence assumes the role of a "reference tRNA structure." Klug et al (21) and Kim et al (22) have pointed out that most other class I tRNAs contain either identical base pairs or coordinated base pair changes at the strategic tertiary base pairing positions, which indicates that the crystal structure is probably a good generalized model for other tRNA species. Only the tertiary base pairs involving a hydrogen bonded ring NH are expected to generate low field NMR resonances. These hydrogen bonds involve:

1. 8–14 (reverse Hoogsteen pair involving U8 in eukaryotic tRNAs and s⁴U8 in most bacterial tRNAs);
2. 54–58 (reverse Hoogsteen pair involving the unique T54 common to most tRNAs);
3. 46–22 (GGC triple connecting the variable loop to the DHU stem via m⁷G46);
4. 19–56 (a Watson-Crick GC pair connecting the DHU loop to the rT loop);
5. 15–48 (a reverse Watson-Crick GC connecting the DHU loop and variable loop);
6. 26–44 (a twisted AG "base pair" at the junction of the anticodon helix).

Additional tertiary interactions that might possibly generate extra low field resonances are Ψ55–P58 (a ring NH-phosphate hydrogen bond) and possibly G18–Ψ55 (20, 37, 38). Furthermore, most tRNAs contain a GU pair in their secondary structure; in wobble geometry this might be expected to generate two low field resonances. Thus there is no dearth of potential candidates for the extra resonances observed in well-resolved low field spectra.

ASSIGNMENT OF SECONDARY RESONANCES   The complexity of intact tRNA spectra can be simplified by analysis of isolated helical hairpin fragments produced by controlled chemical or enzymatic cleavage, and this technique has been applied to several tRNA species (39–41). Ring current shifts from neighboring bases are undoubtedly responsible for the spectral resolution of resonances in fragments and intact tRNA, but the resulting resonance position depends on two unknowns, namely the starting inherent resonance position of an AU or GC pair and the net upfield shift from stacked neighbors. Thus there are several solutions for the two unknowns in this equation each of which can empirically rationalize the position of a given base pair resonance, but each rationalization gives rise to a different set of values for the neighboring ring current shift (2, 43, 44). Ring current shifts are both distance- and angle-dependent, and Reid et al (41) have attempted to lower the ambiguity by *assuming* that helical fragments and intact tRNA in solution maintain the 11-fold RNA geometry observed in the crystal structure. With this assumption the net shifts predicted for 11-fold geometry by Arter & Schmidt (42), using Pullman ring current values, leads to starting resonance positions of −14.35 ppm for AU pairs and −13.45 ppm for GC pairs, and accounts for the observed fragment spectra reasonably well (41). Analysis of the intact tRNA spectrum involves the assumption that it can be approximated by the sum of the component helical parts; this has been disputed by P. D. Johnston and A. G. Redfield (data submitted for publication) and by Sanchez et al (73) who have claimed that, in at least one case involving base pair 6, tertiary folding in intact tRNA causes variations in donor-acceptor geometry, which modify the starting resonance position of this base pair. The discrepancy between different sets of theoretically predicted assignments (see next paragraph) and between theoretical and experimentally calibrated ring current shift "rules" may be as great as 0.3 ppm; if the error level is this high it is, unfortunately, large enough to transpose base pair assignments between adjacent multiple-proton peaks, especially in the more crowded region between −14 and −12 ppm.

In a completely different approach Robillard et al (45) have carried out a computer calculation of the ring current effects from all nucleotides on each ring NH, using the phenylalanine tRNA X-ray coordinates. A good fit to the experimental spectrum is obtained only when unshifted AU and GC offsets of −14.35 and −13.54 ppm were used, and the resulting computer spectrum indicated no secondary base pairs below −14 ppm (45). Geerdes & Hilbers (46) and Kan & Ts'o (47) have used the same approach, but their assignments differ from those of Robillard et al (45). However, impressive support was given to the Robillard calculations by the fact that they generated a remarkably close facsimile of the observed spectrum of *E. coli* tRNA[Val] when this sequence was substituted into the program (48). Nevertheless, there is obviously room for new experimental approaches to assigning secondary base pair resonances.

ASSIGNMENT OF TERTIARY RESONANCES The tertiary folding of tRNA involves many nonstandard interactions e.g. base triples, reverse Hoogsteen, and reverse Watson-Crick pairs etc (20, 37, 38). This creates special problems in calculating their resonance positions; e.g. the two protons below −14 ppm in yeast tRNA[Phe] were calculated to be AU6 and AU12 by Kan & Ts'o (47), whereas the Robillard calculations assign them to the tertiary Hoogsteen pairs 8–14 and 54–58 (45). The assignments based on calculated spectra from three different laboratories have been compared and discussed in detail (49). Experimental assignment of tertiary resonances has been attempted by several laboratories and agreement has been reached in some cases.

The resonance at −14.8 to −14.9 ppm has been confidently assigned to the s[4]U8–A14 reverse Hoogsteen pair in species that are thiolated at residue 8. This assignment is based on the observation that this resonance moves upfield to ∼ −14.3 ppm upon dethiolating s[4]U8 (compare the valine tRNA spectra in Figures 1 and 2), and also leads to the assignment of the 8–14

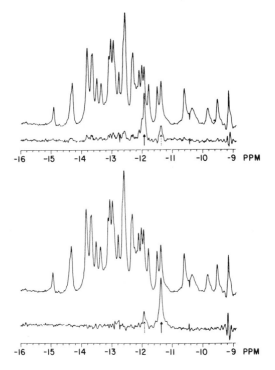

Figure 2   Identification of the resonances from the GU base pair in *E. coli* tRNA$^{Val}_1$ via the ring NH-ring NH NOE. The upper spectra are controls that were preirradiated in the valley at −11.6 ppm. The difference spectrum below each control was obtained by subtracting the spectrum after preirradiating at −11.95 ppm (*upper*) or −11.35 ppm (*lower*) from the control spectrum. Preirradiation was carried out for 0.1 sec with a 1 msec delay before the 0.37 msec observation pulse. The tRNA$^{Val}_1$ sample was only 50% thiolated at residue 8 (note the half-proton resonance at −14.9 ppm with the other half-resonance appearing as a shoulder at −14.3 ppm). [Figure taken from (55).]

resonance at −14.3 ppm in species containing unmodified uridine at position 8 (30, 31, 50, 51). This is perhaps the most definitive and unanimous tertiary assignment.

Hurd & Reid used chemical removal of the unique m^{7}G residue to assign the m^{7}G46–G22 tertiary interaction at −13.35 ppm in yeast tRNA^{Phe} and *E. coli* tRNA^{Val} (55). In the course of these studies they also established that the narrow aromatic resonance at ∼ −9.1 ppm is the C8H of m^{7}G46, and thus created a built-in measure of the extent of m^{7}G removal, which was corroborated by chemical analysis. This assignment was disputed by Salemink et al who used a similar approach but claim the m^{7}G46 imino resonance is at −12.5 ppm in yeast tRNA^{Phe} (56). However their difference spectra appear to be mis-scaled by 10–15%, which led to an apparent one proton difference spectrum in the seven-proton peak at −12.5 ppm; also the extent of m^7G removal was not documented by chemical analysis, and the spectra were unfortunately not extended far enough upfield to monitor the extent of removal via the m^7G C8H resonance at −9.1 ppm. In *E. coli* tRNA^{fMet} the m^7G is in a markedly different environment, and Hurd & Reid showed that in this case the removal of m^7G was correlated with the loss of the resonance at −14.55 ppm (55). This is in contrast to the −13.4 ppm assignment of Daniel & Cohn (57) based on s^4U8 spin labeling and comparison with fMet-3 tRNA (which contains no m^7G); however the fMet-3 sample used was not purified by the authors themselves and was not characterized with respect to m^7G content (57). Bolton & Kearns (2, 35) claim, on very weak evidence, that the m^7G tertiary resonance is not present in the spectrum. In the light of the above conflicts the assignment of this tertiary resonance remains controversial.

The G15–C48 reverse Watson-Crick tertiary base pair has been problematical and is not yet satisfactorily assigned. From the relaxation effects of paramagnetic ions including Co^{2+}, which binds to G15 in the crystal structure (79), Hurd et al conclude that the resonance at −12.25 ppm is the G15–C48 imino proton in *E. coli* tRNA^{Val} (59). This assignment has neither been disputed nor corroborated.

The G19–C56 Watson-Crick tertiary resonance is generally assigned at, or close to, −12.9 ppm in most tRNAs, and has not been disputed despite (or perhaps because of) the lack of direct evidence. This is the only tertiary interaction involving a normal Watson-Crick pair, and its deductive assignment is based on the calculated upfield shift of ∼ 0.6 ppm (45, 49) combined with the standard GC offset of ∼ −13.5 ppm, and the observation that an extra (nonsecondary) resonance is always observed at this position (41). Some support has been lent by thermal unfolding studies (25) and by exchange dynamics (52, 70, 71). Although highly reasonable, this assignment cannot be regarded as definitive in the absence of direct evidence.

The T54–A58 tertiary interaction is a reverse Hoogsteen pair and its assignment is quite controversial. The immediate environment surrounding T54–A58 is identical in yeast tRNA^{Phe} and *E. coli* tRNA^{Val}, so that its resonance position should be the same in both spectra. Reid et al (41) have argued that the five AU pairs in *E. coli* tRNA^{Val} should have similar resonance positions around −13.7 ppm; since the −14.9 ppm resonance has already been assigned to the 8–14 Hoogsteen, and the lone resonance at −14.3 ppm is too low field to be a GC pair, the only remaining candidate is T54–A58, since there are only two UA-type tertiary interactions in the crystal structure. The early loss of the −14.3 ppm resonance (and the 8–14 Hoogsteen) as the temperature is raised (41, 51), and its rapid exchange dynamics in both yeast tRNA^{Phe} (52, 70, 71) and *E. coli* tRNA^{Val} (R. E. Hurd and B. R. Reid, unpublished observations; P. D. Johnston and A. G. Redfield, unpublished observations) are consistent with this assignment; the Robillard calculations of the total ring current shift on this proton place it at −14.4 ppm when the reverse Hoogsteen unshifted offset of −14.9 ppm is used (45, 48, 49). In addition, Hurd & Reid (51) have examined the fragment encompassing residues 47–76, which contains T54 and A58 in an intact rT loop and stem; the spectrum contains a resonance at −14.35 ppm that cannot be assigned to any secondary base pairs in the stem (which is devoid of AU pairs).

However, there is evidence against the −14.3 ppm assignment for T54–A58. Kearns & Bolton (53) have observed that ethidium bromide abolishes the −14.3 ppm resonance in *E. coli* tRNA^{Val} without shifting the T54 methyl resonance at −1 ppm, and Reid has observed that the *E. coli* tRNA species Ala-1B and Phe (which contain the same rT loop as tRNA^{Val} and yeast tRNA^{Phe}) contain no resonances near −14.3 ppm (unpublished observations). The ethidium result might be due to breaking the T54–A58 hydrogen bond without directly stacking on T54, but the comparative data are difficult to rationalize. Very recently Sanchez et al (73) and P. D. Johnston and A. G. Redfield (manuscript submitted for publication) have presented additional evidence on this point; they have used double-resonance NOE techniques (described in the next section on FTNMR) that indicate that the −14.3 ppm resonance is a secondary Watson-Crick AU pair rather than a reverse Hoogsteen tertiary pair. Thus the T54–A58 assignment remains controversial and further work is needed to resolve this problem.

The best estimates of the total proton intensity between −11 and −15 ppm are 27 or 28 for *E. coli* tRNA^{Val} and 26 or 27 for yeast tRNA^{Phe} (36, 41). Both tRNAs contain a secondary GU pair, and there is now incontrovertible NOE evidence that the −11 to −15 ppm region contains two protons from the tRNA^{Val} GU pair (55, 71) and one proton from the yeast tRNA^{Phe} GU pair (70, 71). In combination with the 20 secondary resonances and the 5 tertiary resonances discussed above it now appears that, at most, only 1 other proton may be present in this spectral region. Of the three remaining crystallographic candidates the 24–44 "base pair" has been experimentally displaced upfield to the −9 to −11 ppm region (P. D. Johnston and A. G. Redfield, data submitted for publication), and this spectral region probably also contains the Ψ55 ring NH hydrogen bonded to P58 that may even have exchanged out of the spectrum as a result of its accessibility to solvent; if there is one more resonance to be accounted for in the −11 to −15 ppm region it is probably G18-Ψ55, and almost all tRNAs contain an unassigned resonance at ∼ −11.5 ppm, which is a likely position for such a "GU-type" interaction. From the foregoing it is apparent that reasonable progress in assignment has been made by extensive use of a variety of chemical and biochemical approaches, but that ambiguities still remain. Future progress in this area will probably involve spectroscopic methods using FTNMR techniques that have recently been developed for the special case of H_2O solutions.

FTNMR in H_2O Solution

Fourier transform NMR, with its advantages in sensitivity, time resolution, and double-resonance capabilities has not, until recently, been applicable to exchangeable proton studies in H_2O because of computer-digitizer dynamic range limitations. In FTNMR all resonances are excited simultaneously by a short radio frequency pulse encompassing all proton frequencies in the sample. The resulting time domain signal is a free induction decay (FID) containing the mixed frequencies of the sample resonances, which is Fourier transformed in a computer to reveal each resonance in the more normal frequency domain (60, 61). For dilute samples around 1 mM the pulse-FID sequence is repeated hundreds of times to generate reasonable signal-to-noise ratios in the final spectrum (signal averaging). The signal must first be digitized to carry out these computer manipulations, and the H_2O proton signal (110 M) obviously dominates the FID. In fact a normal 12-bit digitizer (4096 points in the vertical axis), when filled with a 110 M signal, will not register signals much below about 25 mM (i.e. 1 mM solute signals are not digitized and are lost). Even with a 16-bit digitizer, a computer with a word size of 20 bits will overflow memory after only 16 (2^4) pulses, which is insufficient signal averaging. For these reasons, until about 1977, spectra in H_2O were usually collected by CW frequency-sweep or rapid scan correlation spectroscopy methods (62), which avoided sweeping through the H_2O resonance. Such methods do not easily lend themselves to double-resonance techniques by which assignments can often be made, or to time-resolved experiments in which the dynamic aspects of the molecule can be studied.

This problem has been tackled and elegantly solved by the introduction of a new phase-shifted pulse (the 214 pulse) by Redfield and colleagues (63–67). The Redfield 214 pulse is a selective excitation pulse the most important aspect of which is that its frequency transform falls to zero amplitude at the H_2O frequency thus exciting the water resonance < 1% of a normal FT pulse. More recent "magic knob" modifications of the 214 pulse (67) lead to reductions of over 1000-fold for the H_2O amplitude in the FID, so that the effective size of the solvent signal can be reduced to ~ 100 mM, thus placing 1 mM solute signals well within the dynamic range of 12-bit digitizers and allowing at least 256 (2^8) pulses to be signal averaged in a 20-bit word computer. The length of the 214 pulse is tailored to fit the offset between the pulse carrier frequency and the water frequency, and is usually around 0.4 msec, which permits time resolution of processes in the millisecond range.

Assignments from Double-Resonance: The NOE

The nuclear Overhauser effect is a particularly useful double-resonance technique for biological polymers in which the magnetic saturation of a selectively irradiated proton resonance is transferred, via cross-relaxation, to other resonances from protons in the immediate neighborhood (68, 69). The marked distance dependence (r^6) limits the first-order NOE (in the absence of further spin diffusion) to distances in the 1–4 Å range, and it is mainly used to relate pairs of resonances that may be greatly separated in the spectrum but are very close to each other in space.

GU ASSIGNMENT VIA NOE   The vertical distance of 3.4 Å combined with the lateral displacement of a normal RNA helical pitch is such that the ring NHs of adjacent base pairs are separated by more than 4 Å and, as expected, do not exhibit ring NH-ring NH NOEs. Johnston & Redfield (70) noted that two low field protons in yeast tRNA[Phe], located at −11.8 ppm and −10.4 ppm, nevertheless did show mutual NOEs, and concluded that both ring NHs must be in the same base pair, i.e. the GU pair at position 4. E. coli tRNA[Val] also contains a secondary GU pair (at position 50), which also has been studied by Johnston & Redfield (71) and by Hurd & Reid (55); in this case the NOE-related pair of imino protons are located at −11.95 ppm and −11.35 ppm. Figure 2 shows the results from such an experiment presented as difference spectra in which the preirradiated spectrum is subtracted from the control so that the selectively saturated line appears as a positive full proton resonance. No other low field ring NHs show mutual NOEs, but it is obvious that the −11.95 ppm and −11.35 ppm resonances exhibit partial cross-relaxation (NOE).

Further proof that the two imino resonances related by mutual NOE are in fact derived from the wobble GU pair comes from the observation that two tRNAs of different sequence and different low field spectra, but with their GU pair in identical nearest neighbor environments, exhibit NOE-paired resonances at identical spectral positions (K. A. Jones and B. R. Reid, unpublished observations). Although their presence in the spectrum was not even acknowledged a few years ago (2, 4), the recent use of the NOE has now turned the GU resonances into perhaps the most reliable base pair assignment. Despite the ease of experimentally identifying GU resonances via the NOE, Geerdes & Hilbers have published a set of ring current shift rules for theoretically predicting GU resonance positions based on nearest neighbor sequence (72). However in cases such as E. coli tRNA[Phe] and tRNA $^{Ala}_{1B}$ the observed GU resonances differ by 0.5–0.7 ppm from their predicted positions (K. A. Jones and B. R. Reid, unpublished observations); Geerdes & Hilbers assumed that the helical distortion required to form a wobble pair is the same for all GC pairs regardless of nearest neighbors, but this is apparently not the case (i.e. either the G or the U, or perhaps both, can move laterally to assume wobble geometry depending perhaps on the particular stacking sequence).

SECONDARY RESONANCE ASSIGNMENT VIA NOE   Although the ring NHs of adjacent base pairs are normally too far apart to exhibit mutual NOEs, other protons located vertically above (3.4Å) and methyl protons on the same nucleoside are within NOE range of some ring NHs (see Figure 3). P. D. Johnston and A. G. Redfield (manuscript submitted for publication) have used the NOE from the m^2G10 methyl resonance at –2.75 ppm to assign the ring NH of base pair 10 at –12.63 ppm in yeast tRNA^{Phe}. The fact that Robillard et al (45) previously calculated from the crystal coordinates that this resonance should be at –12.55 ppm lends credence to the theoretical calculation approach to assignments when carried out properly. Also, the experimental approach using a fragment containing intact anticodon and DHU helices as well as fragment-calibrated ring current shifts assigned base pair 10 at –12.7 ppm in yeast tRNA^{Phe} and at –12.6 ppm in *E. coli* tRNA^{Val} (41).

Sanchez et al (73) have used the fact that the NOE acceptor from an AU ring NH donor is the adenine C2H (narrow line), whereas the acceptor from a GC ring NH donor is the guanosine amino group (broad line), to discriminate between AU and GC resonances.

TERTIARY RESONANCE ASSIGNMENT VIA NOE   As mentioned earlier, the reverse Hoogsteen T54–A58 ring NH and the m^7G46–G22 ring NH have been assigned by Hurd & Reid (51, 55) at –14.3 ppm and –13.35 ppm, respectively, in both *E. coli* tRNA^{Val} and yeast tRNA^{Phe}, but these assignments have been disputed (53, 56). Sanchez et al have used double-resonance methods to address the 54–58 assignment (73). The observed sharp NOE at –7.8 ppm upon irradiating yeast tRNA^{Phe} at –14.3 ppm should be the C8H of m^1A58 *if* the –14.3 ppm resonance is in fact the 54–58 reverse Hoogsteen (see Figure 3). However substitution of purine C8 protons with deuterium by heating at 90°C in D_2O (or more recently by growth of an auxotroph on C8 deuteroadenine), and subsequent reisolation in H_2O, did not abolish the –14.3 to –7.8 ppm NOE as expected (73). The most reasonable conclusion is that the –7.8 ppm NOE is actually an adenine C2 proton, which indicates that the –14.3 ppm resonance is a standard Watson-Crick AU pair rather than the 54–58 Hoogsteen pair. Sanchez et al (73) thus

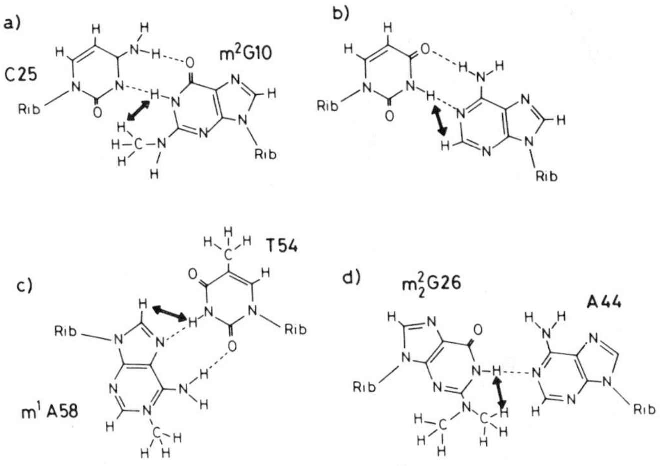

Figure 3   Diagrammatic representation of secondary and tertiary base pairs observed in the crystal structure of yeast phenylalanine tRNA: (*a*) m²G10-C25, (*b*) standard Watson-Crick AU pair, (*c*) T54-m¹A58, and (*d*) m²₂G26-A44. In each case the proximity of the hydrogen bonded ring NH to adjacent aromatic or methyl protons is shown by a double-headed arrow. (See text concerning assignment via the NOE.)

tentatively conclude that the −14.3 ppm resonance must be UA6 with the 54–58 tertiary resonance further upfield in the group of protons around −13.8 ppm, where it was assigned by Kearns (53). However, they have not been able to observe a demonstrable NOE to C8H from the low field spectrum, and it is unfortunate that yeast tRNA^Phe contains m¹A at position 58 instead of A; m¹A has peculiar chemical properties and may exhibit anomalous exchange behavior (54). Although the evidence of Sanchez et al points against assigning T54–A58 at −14.3 ppm it must remain tentative until it is found elsewhere by H/D elimination of the NOE, and analysis of the final sample proves that m¹A58–C8H has in fact been replaced by deuterium and retained the isotope. The NOE results on m⁷G by P. D. Johnston and A. G. Redfield (data submitted for publication) are consistent with, but do not prove, the m⁷G46–G22 assignment of Hurd & Reid (55).

The NOE approach has been very informative concerning the m²₂G26–A44 tertiary interaction, which appears, from the crystal structure, to involve a ring NH-ring N hydrogen bond between propeller-twisted purines (20, 37, 38; see Figure 3). Using subtractive inference and comparative spectra of different tRNAs, the extra resonance observed near −12.3 ppm was initially deduced by Reid & Hurd to be base pair 26–44 (3). However it was subsequently displaced from this position by later evidence from Hurd el al, using paramagnetic ions, which indicated that the −12.3 ppm tertiary resonance belonged to base pair 15–48 (59); this left no place for the hydrogen bonded ring NH of m_2^2G26 which, based on the crystal structure, was generally assumed to be somewhere in the –11 to –15 ppm spectral region (49). This paradox was recently solved by the NOE studies of P. D. Johnston and A. G. Redfield (manuscript submitted for publication) who realized that the methyl groups of m_2^2G26 were very close to the ring NH in question; irradiation of the m_2^2G26 methyl resonance at –2.45 ppm produced a very strong NOE at –10.4 ppm, which must be the m_2^2G26 ring NH. Hence the 26–44 tertiary imino proton does not resonate in the –11 to –15 ppm low field spectral window as previously assumed, and its unexpectedly upfield position suggests that it is probably not hydrogen bonded to A44, perhaps because of the distortion of the propeller twist. By making use of the relatively unambiguous assignments of methyl resonances, it is apparent that the future application of double-resonance NOE methods should go a long way toward eliminating much of the assignment ambiguity in the potentially more informative ring NH spectral region; the use of these methods in H_2O was made possible by the introduction of the Redfield 214 selective observation pulse (67).

RNA Dynamics Using FTNMR

Helix-coil dynamics can, in principle, be studied using steady-state CW-sweep NMR by analysis of the exchange contribution to the ring NH linewidth (74, 75). However such studies are only applicable in the restricted temperature ranges where broadening can be observed, and they are inherently less accurate than time-resolved FTNMR methods. The ring NH protons in a RNA double helix are not directly accessible to water, and double resonance FTNMR methods are beginning to reveal a great deal about helix-coil kinetics.

The helix-coil equilibrium can be described by the two rate constants k_{open} and k_{close} (the latter is obviously the greater if the structure is in the native state). The helix-coil transition is obligatory before exchange with water can occur; coil-water exchange is buffer catalyzed and can be made very rapid ($\sim 10^6$ sec^{-1}) by addition of small amounts (10–100 mM) of buffer (76, 77). If the coil lifetime is long compared to exchange from the coil state ($k_{close} << k_{cw} = k$ [cat] $= 10^5$–10^6 sec^{-1} in 10 mM buffer) then every opening event results in exchange, and the observed helix-water exchange rate is limited by, and equal to, k_{open} [For a more detailed discussion of exchange see Teitelbaum & Englander (80), Kallenbach & Berman (81), and Hilbers (75).]

Double-resonance FTNMR methods are particularly well-suited to monitoring exchange processes in the 1–1000 sec^{-1} rate range, and hence offer an exciting new approach to directly measuring k_{open} for individual base pairs with resolved ring NH resonances. After selective preirradiation of a single resonance to saturation (destruction of net magnetization) that proton will recover relatively slowly via interaction with fluctuating megnetic fields (T_1 or spin-lattice relaxation processes) in its environment (82, 83). However if, during the recovery period, that proton population is being exchanged with magnetized water protons (which are not perturbed magnetically by the 214 observation pulse or the preirradiation), recovery will be greatly accelerated; the increase directly reflects the rate of exchange with water and can be equated with k_{open} if the conditions described in the previous paragraph are met. Such experiments have been carried out on the acceptor helix base pairs of *E. coli* tRNA[Phe] by Hurd & Reid (78), and an example is shown in Figure 4. After saturation of GC5 at −12.4 ppm the spectrum is collected at various time intervals during the recovery period, using a 0.3-msec 214 observation pulse, with the results shown. The recovery rate was observed to be 28 sec^{-1} at 62°C of which 5–6 sec^{-1} is the (temperature-independent) magnetic recovery rate, which leads to a value of 22 sec^{-1} for the helix-water exchange rate. The lack of any stimulation of this rate by increasing the buffer concentration tenfold proved that exchange was open limited, and hence k_{open} is therefore 22 sec^{-1} (78). This experiment can be extended to all six base pairs in the helix and such studies reveal a positional dependence on the opening rate; for instance CG4 opens at only 5 sec^{-1} at 58°C, whereas GC6 opens at 67 sec^{-1} (78). The whole cycle of experiments can be repeated at various temperatures, thus revealing, at least in theory, the activation energy for opening each base pair; in practice such a goal for even a small six-base pair helix would require $\sim$ 150 NMR spectra, each signal-averaged for about 15 min, and would yield only a moderately accurate result (four or five points on an Arrhenius plot). Nevertheless semiquantitative interpretation of the very large activation energies in the helix interior ($\sim$ 70 kcal mol^{-1}) are consistent with no independent internal opening i.e. exchange only via sequential peeling back from the termini (78).

Saturation recovery determination of k_{open} would obviously be extremely useful in probing base pair dynamics in various regions of intact tRNA under physiological conditions. Unfortunately the interpretations of such analyses are not straightforward due to the previously discussed ambiguities in assigning the individual ring NH resonances. Undaunted by this, Redfield's group has carried out a detailed exchange analysis of all peaks in the low field spectrum of yeast tRNA[Phe] (52, 70, 71) and, more recently, the more resolved *E. coli* tRNA[Val] (P. D. Johnston et al, manuscript in preparation).

Interestingly, the technique can be used as a two-edged sword in that the results can be turned around and used to address the problem of identifying,

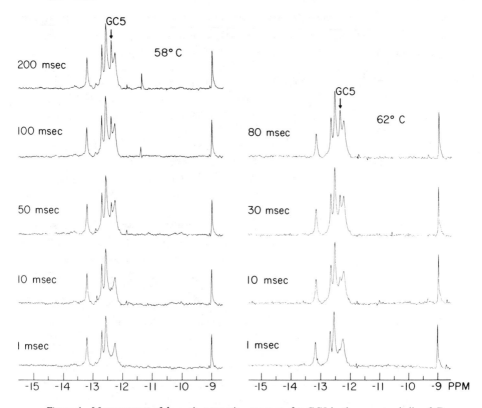

Figure 4   Measurement of k_{open} via saturation recovery for GC5 in the acceptor helix of *E. coli* tRNA[Phe]. The intact tRNA was partially unfolded in the absence of magnesium at intermediate temperature leaving only the six GC resonances of the acceptor stem, as described previously (3). The GC5 resonance was then saturated by selective preirradiation at −12.45 ppm for 0.1 sec, after which the extent of recovery was monitored at various time intervals using a 0.37 msec Redfield 214 observation pulse. The exchange rate (after correction for magnetic recovery) was found to be 6 sec^{-1} at 58°C and 22 sec^{-1} at 62°C; these values can be directly equated with k_{open} since exchange is open limited under these conditions (78).

without necessarily assigning, tertiary resonances based on their atypical dynamics. In the presence of magnesium, exchange was too slow to be measured accurately, but at moderately low ionic strength in the absence of magnesium the saturation recovery rates were exchange-dominated at around 40°C which allows the individual opening rates of some resonances to be measured (52, 70, 71). In the case of yeast tRNA[Phe] the large peaks at −13.25 and −12.5 ppm contain five base pairs and seven base pairs, respectively, and are too complex to extract rates for single base pairs from their multiphasic recovery behavior. However, the single resonances at −14.4 and −14.2 ppm both exhibit anomalously rapid (>60 sec^{-1}) water exchange at 42°C; the latter is the 8–14 tertiary pair, but the former, although originally assigned to the tertiary 54–58 pair (49, 51, 71) has been reassigned to UA6 by Sanchez et al (73), based on its NOE behavior. In fully thiolated *E. coli* tRNA^{Val} there is only one resonance between −14.5 and −14.0 ppm (see Figure 1), and it also exhibits rapid solvent exchange (R. E. Hurd and B. R. Reid, unpublished observations; P. D. Johnston et al, manuscript in preparation): however its assignment to T54–A58 (49, 51) has been weakened by the yeast tRNA^{Phe} NOE results of Sanchez et al (73). Interestingly, *E. coli* tRNA^{Phe} only contains one proton between −15 and −14 ppm, and this is the s^{4}U8–A14 tertiary resonance at −14.9 ppm (3). In samples that are only 50% thiolated at residue 8, this peak is reduced to half-intensity, and a new half-proton resonance appears at −14.3 ppm from the population of molecules containing U8–A14; this latter peak exhibits more rapid water exchange than the half proton at −14.9 ppm, which indicates that s^{4}U8–A14 is a more stable tertiary base pair than U8–A14, and perhaps suggests a functional reason for the common thiol modification at residue 8 in bacterial tRNAs (R. E. Hurd and B. R. Reid, unpublished observations). It is obvious that time-resolved helix-coil rate studies will reveal a great deal about RNA dynamics in the future. Although there are several unambiguous ring NH assignments, dynamic studies are presently somewhat limited by the absence of a complete set of thoroughly proven and incontrovertible assignments, but the information from dynamic studies is helping to bring this goal closer.

Interactions of tRNA

With reliable assignments for the high field methyl resonances, and the emergence of assignments for at least some resonances in the more complicated low field spectrum, the interaction of tRNA with cations, drugs, enzymes etc can now begin to be studied and interpreted by NMR. From the pronounced shifting of some low field resonances it is apparent that the solution structure of tRNA is different in the presence and absence of magnesium (25, 41, 70), although the spectral changes have not been interpreted in terms of detailed structural changes. From the most up-to-date assignments the changes appear to involve the D-loop interactions (15–48, 14–8, and 13–22), and perhaps the acceptor stem-rT stem junction (base pairs 7 and 49). The binding of Mn^{2+} and Co^{2+} has been studied via their specific paramagnetic relaxation effects. Chao & Kearns (84) have interpreted the selective broadening by Mn^{2+} on unfractionated tRNA and yeast tRNA^{Phe} to reflect binding at several sites in the order 8–14, U33, 54–58 and 19–56. Hurd et al (59) investigated several tRNAs at slightly lower Mn^{2+} levels and found a single site close to tertiary base pair 8–14 consistent with the P8–P9 crystallographic Mg^{2+} site. Divalent cobalt also relaxed the

8–14 resonance but occupied a different site, since it relaxed several additional resonances not relaxed by Mn^{2+} (59); the Co^{2+} effects were consistent with occupation of the crystallographic cobalt site on G15 (79).

Although all of the foregoing studies assumed the existence of discrete cation sites, a rigorous ESR and ^{31}P NMR study of tRNA-Mn^{2+} complexes by Gueron & Leroy indicates very similar Mn^{2+} lifetimes on *all* phosphates with no specific sites (85). It might be possible to explain the selective effects of Mn^{2+} on the basis of greater access of solvent ions to ring NHs in the region of the D-stem "backbone turn," but further work is needed to rationalize this point. Several tRNAs contain a single binding site for ethidium bromide, and this interaction has been studied by Kearns and colleagues (11, 86). They report that binding involves intercalation at the sixth base pair in the acceptor stem; however the interpretation is based upon the effect on the −14.3 ppm resonance, the assignment of which is controversial as discussed earlier. X-ray studies on ethidium bromide–soaked crystals by Liebman et al (87) indicate binding at the "D turn" without intercalation, but crystals of the solution complex have not been obtained.

The effect of binding the complementary codon to yeast tRNA^{Phe} has been monitored via the high field methyl resonances, by Davanloo et al (88) and also via the ring NH resonances, by Geerdes et al (89); in neither case could any conformational change be detected.

One of the ultimate goals of NMR is the analysis of tRNA recognition by specific enzymes of protein biosynthesis, although serious broadening of resonances in these high-molecular-weight complexes can be expected. Shulman and co-workers have investigated the effects on *E. coli* tRNA^{Glu} of binding glutamyl-tRNA synthetase and the elongation factor EFTu (90). These studies could only be performed at lower temperature because of the lability of the enzymes, and very broad line unresolved spectra were obtained; the authors concluded that no tRNA base pairs were broken in the complexes. More recently, advantage has been taken of thermophilic synthetase enzymes that form functional complexes with tRNA up to 65°C, and the rapid tumbling at these elevated temperatures leads to greatly reduced linewidths. In the cognate complex between *E. coli* tRNA$_1$^{Val} and *B. stearothermophilus* valyl-tRNA synthetase, at least two base pairs in the tRNA are broken (R. E. Hurd, M. LaBelle and B. R. Reid, unpublished observations); however the ring NH intensity losses are in multiple proton peaks, and cannot, as yet, be assigned to specific base pairs. The development of higher resolving spectrometers (500–600 MHz) combined with recent resolution enhancement software should make this an exciting area for further study.

NMR Studies on 5S RNA

Investigation of 5S RNA structure by NMR is in a much more primitive stage of development than the corresponding tRNA studies. There are no methyl resonances to study in the high field spectrum and as yet there is no uniform agreement on the intensity and number of base pairs in the low field spectrum; no assignments have been reported. Kearns & Wong (91) studied *E. coli* 5S RNA at temperatures between 40 and 63°C and interpreted the intensity of the spectrum at 63°C to reflect about 19 base pairs from three helices in a thermally resistant core. At lower temperatures the intensity increased by about 50% which led these authors to claim 28 ± 3 base pairs at 40°C; using −13.2 ppm as a somewhat arbitary dividing line for AU resonances they deduced the existence of 4 AU pairs and 24 GC pairs and proposed a model (4, 91). Burns et al (92), taking care to prevent conversion to the "A form," studied the "B form" of *E. coli* 5S RNA and found evidence for 33 base pairs in the low field spectrum with resonance positions that were consistent with their structural data from Raman spectra (93). Smith & Marshall (94) used ^{19}F-^{1}H NOE methods to show that *all* of the fluorouracil (FU) residues in FU-substituted *E. coli* 5S RNA had long rotational correlation times ($>10^{-8}$ sec), which indicates a highly rigid molecular framework for the entire molecule in solution. Luoma et al (95) investigated yeast 5S RNA by NMR, UV, and CD methods and concluded that it contains at least 40 base pairs of which 40% were AU pairs. Salemink et al (96) carried out similar studies on *B. licheniformis* 5S RNA and found ~ 36 base pairs, which, from their chemical shifts, were largely GC pairs. In general, 5S RNA proton spectra are not as well resolved as tRNA spectra, perhaps due to slower molecular tumbling because of their larger size or extended shape, or both; the spectra may well prove to be very difficult to assign and very little progress has been made to date.

NMR Studies with Other Nuclei

^{31}PHOSPHORUS   As mentioned in the introduction the natural isotope ^{31}P has spin ½ and is amenable to NMR, although the sensitivity to detection is only one fifteenth that of protons. In RNA we should expect one resonance from each nucleotide, and a few studies on tRNA have been reported using this nucleus. Gueron & Shulman (97) investigated *E. coli* tRNA^{Glu} and yeast tRNA^{Phe} at 40 MHz and 109 MHz (the phosphorus resonance frequency is about 40% that of protons in a given magnetic field) and observed much broader lines at the higher frequency, perhaps indicating relaxation via chemical shift anisotropy. The total range of chemical shifts encompassed almost 8 ppm, but the spectra contained a large central peak of about 65 unresolved resonances with 5 or 6 peaks on the low field side and 2 or 3 peaks at higher field. The extreme low field resonance was assigned to the 5'-terminal phosphate, since it was the only peak whose position titrated with the secondary pK of a phosphomonoester (97); and several peaks were affected by Mg^{2+}, which suggests they may be from nucleotides involved in tertiary folding. Salemink et al (98) have extended these findings in a more detailed study of yeast tRNA^{Phe} using chemical and biochemical modification; using selective cleavage at U33 with pancreatic RNase they assigned the resolved resonances c (lowfield) and j$_2$ (the highest field peak) to the anticodon loop. Although a few assignments are beginning to emerge, the tRNA ^{31}P NMR spectrum has not been particularly informative; its disadvantages are poor resolution and low sensitivity ($\sim$ 20 hr of signal averaging for 1 mM samples).

^{13}CARBON   The wide range of chemical shifts makes ^{13}C NMR a promising approach to studying RNA molecules, especially their complexes with enzymes. Although Komoroski & Allerhand (99, 100) succeeded in obtaining natural abundance ^{13}C spectra of unfractionated tRNA, the prohibitively long signal-averaging times make this impractical for most biochemical applications. Agris and co-workers (101, 102) have used the *E. coli rel met cys* mutant together with ^{13}C-methyl methionine to produce bulk tRNA with ^{13}C-enriched methyl groups. Using mononucleoside reference resonance positions and biochemical correlation of the extent of incorporation they assigned the ^{13}C-methyl resonances of rT, m^{6}A, and m^{7}G at 12 ppm, 29.1 ppm, and 36.6 ppm, respectively (101); more recently the signals from ms^{2}i^{6}A, m^{2}A, m^{1}G, mam^{5}s^{2}U, mo^{5}U, and Gm have been assigned (103). Temperature-dependent studies in unbuffered water revealed changes in the resonance positions of rT (tertiary folding) as well as m^{2}A and m^{1}G (anticodon loop) between 9 and 30°C (103). However, all the above studies were carried out with unfractionated tRNA containing $\sim$ 60 molecular species, and it is likely that individual tRNAs will show different thermal unfolding behavior. Along similar lines Hamill et al (104, 105) used an auxotroph to incorporate ^{13}C-4 uracil into tRNA, and Tompson & Agris (106) used controlled growth conditions to produce ^{13}C-2 adenine-substituted tRNA and ^{13}C-2 U/C-substituted tRNA. NMR analysis of the ^{13}C-2-substituted tRNA indicated that the broader than expected linewidth was due to chemical shift nonequivalence for the various environments of the ^{13}C in the unfractionated tRNA (107). Very recently investigations have been carried out on purified single species of tRNA that have been selectively labeled with ^{13}C. Agris & Schmidt (108) have studied *E. coli* tRNA

Cys, tRNA^{Tyr}, and tRNA^{Phe} after ^{13}C-methyl enrichment. The much simplified spectra contain resolved peaks from rT, ms^{2}i^{6}A, and m^{7}G (present only in tRNA^{Phe}). In the presence of magnesium the m^{7}G shows intermediate exchange between environments, whereas in the absence of magnesium the rT of tRNA^{Phe} shows separate resonances (slow exchange) corresponding to at least two conformations, as was shown earlier for the proton resonances of the rT methyl group (18, 19) and the low field resonances of tRNA^{Val} (41). Schweizer et al (109) have examined the spectrum of purified *E. coli* tRNA$_1$^{Val} substituted with ^{13}C-4 uracil. The spectrum revealed 11 peaks for the 14 uracils (or uracil derivatives) in this tRNA, several of which were assigned and observed to undergo differential shifts reporting local unfolding as the temperature was raised. Although the isotope substitution is somewhat laborious, and the lower sensitivity of ^{13}C necessitates a few hours of signal averaging even with enriched samples, there are several biochemical applications in which ^{13}C NMR, with its approximately ten times wider range of chemical shifts, offers distinct advantages over proton NMR, and this is obviously a fruitful avenue for further research.

^{19}FLUORINE   Although fluorine does not naturally occur in RNA, the high sensitivity and reasonably large chemical shift range combined with the incorporation of fluoro-uracil (5FU) for U in bacterial systems (110) has led to RNA NMR studies using this nucleus. In all biochemical assays, fully substituted FU-tRNA has been found to be functional (100–112). Horowitz et al have examined the ^{19}F spectrum of FU-substituted *E. coli* tRNA$_1$^{Val} (113) and succeeded in resolving 13 peaks for the 14 FU residues in the molecule; however individual assignments have not been reported. The spectrum of FU-substituted 5S RNA is disappointingly unresolved (114), but Smith & Marshall have used ^{19}F-^{1}H NOE (94) to show that there are no flexible FU residues in the rigid 5S RNA framework.

ACKNOWLEDGMENTS

This review was written in mid-1980 while on sabbatical leave at the University of Oxford. Thanks are due to the Inorganic Chemistry Laboratory and Professor R. J. P. Williams for hospitality at Oxford, and to the Fogarty Center of the NIH for the award of a Senior International Fellowship. I would like to thank all those who sent their unpublished observations and manuscripts, and also my own students, especially Ralph Hurd, Ed Azhderian, and Kathy Jones, for the use of their research data.

Literature Cited

1. Kearns, D. R., Shulman, R. G. 1974. *Acc. Chem. Res.* 7:33–39
2. Kearns, D. R. 1976. Prog. *Nucleic Acids Res. Mol. Biol.* 18:91–149
3. Reid, B. R., Hurd, R. E. 1977. *Acc. Chem. Res.* 10:396–402
4. Kearns, D. R. 1977. *Ann. Rev. Biophys. Bioeng.* 6:477–523
5. Kallenbach, N. R., Berman, H. M. 1977. *Q. Rev. Biophys.* 10:138–236
6. Crothers, D. M., Cole, P. E. 1978. In *Transfer RNA,* ed. S. Altman, pp. 196–247. Cambridge, Mass: MIT Press. 356 pp
7. Opella, S. J., Lu, P., eds. 1979. *NMR and Biochemistry.* New York: Dekker. 434 pp
8. Shulman, R. G., ed. 1979. *Biological Applications of Magnetic Resonance.* New York: Academic. 595 pp
9. Abelson, J., Schimmel, P. R., Soll, D. eds. 1979. *Transfer RNA: Structure, Properties and Recognition,* Cold Spring Harbor New York: Cold Spring Harbor Labs.
10. Schimmel, P. R., Redfield, A. G. 1980. *Ann. Rev. Biophys. Bioeng.* 9:181–221
11. Kearns, D. R., Bolton, P. H. 1978. In *Biomolecular Structure and Function,* ed. P. F. Agris, pp. 493–516. New York: Academic
12. Patel, D. J. 1978. *Ann. Rev. Phys. Chem.* 29:337–62
13. Reid, B. R. 1980. In *Topics in Nucleic Acid Structure,* ed. S. Neidle, pp. 113–39. London:Macmillan
14. Schweizer, M. 1980. In *Magnetic Resonance in Biology,* ed. J. S. Cohen. New York: Wiley. In press
15. MacDonald, C. C., Phillips, W. D., Penman, S. 1964. *Science* 144:1234–38
16. Smith, I. C. P., Yamane, T., Shulman, R. G. 1969. *Can. J. Biochem.* 47:480–84
17. Schmidt, P. G., Kastrup, R. V. 1978. In *Biomolecular Structure and Function,* ed. P. F. Agris, pp. 517–25. New York: Academic
18. Kastrup, R. V., Schmidt, P. G. 1975. *Biochemistry* 14: 3612–18
19. Kastrup, R. V., Schmidt, P. G. 1978. *Nucleic Acids Res.* 5:257–64
20. Rich, A., RajBhandary, U. L. 1976. *Ann. Rev. Biochem.* 45:805–60.
21. Klug, A., Ladner, J., Robertus, J. D. 1974. *J. Mol. Biol.* 89:511–20
22. Kim, S. H., Sussman, J. L., Suddath, F. L., Quigley, G. J., McPherson, A., Wang, A., Seeman, N. C., Rich, A. 1974. *Proc. Natl. Acad. Sci. USA* 71: 4970–75
23. Kan, L. S., Ts'o, P. O. P., von der Haar, F., Sprinzl, M., Cramer, F. 1974. *Biochem. Biophys. Res. Commun.* 59: 22–29
24. Kan, L. S., Ts'o, P. O. P., Sprinzl, M., von der Haar, F., Cramer, F. 1977. *Biochemistry* 16: 3143–52
25. Robillard, G. T., Tarr, C. E., Vosman, F., Reid, B. R. 1977. *Biochemistry* 16:5261–73
26. Romer, R., Riesner, D., Maass, G., Wintermeyer, W., Thiebe, R., Zachau, H. G. 1969. *FEBS Lett.* 5:15–19
27. Romer, R., Riesner, D., Maass, G. 1970. *FEBS Lett.* 10:352–57
28. Kearns, D. R., Patel, D., Shulman, R. G. 1971. *Nature* 229:338–39
29. Kearns, D. R., Patel, D., Shulman, R. G., Yamane, T. 1971. *J. Mol. Biol.* 61:265–72
30. Reid, B. R., Ribeiro, N. S., Gould, G., Robillard, G., Hilbers, C. W., Shulman, R. G. 1975. *Proc. Natl. Acad. Sci. USA* 72: 2049–55
31. Daniel, W. E., Cohn, M. 1975. *Proc. Natl. Acad. Sci. USA* 72:2582–88
32. Reid, B. R., Robillard, G. T. 1975. *Nature* 257: 287–91
33. Bolton, P. H., Kearns, D. R. 1976. *Nature* 262:423–24
34. Reid, B. R. 1976. *Nature* 262:424
35. Bolton, P. H., Jones, C. R., Bastedo-Lerner, D., Wong, K. L. Kearns, D. R. 1976. *Biochemistry* 15:4370–16
36. Reid, B. R., Ribeiro, N. S., McCollum, L., Abbate, J., Hurd, R. E. 1977. *Biochemistry* 16:2086–94
37. Kim, S. H. 1976. Prog. *Nucleic Acid Res. Mol. Biol.* 17: 181–216
38. Kim, S. H. 1978. In *Transfer RNA,* ed. S. Altman, pp. 248–93. Cambridge, Mass: MIT Press. 356 pp
39. Lightfoot, D. R., Wong, K. L., Kearns, D. R., Reid, B. R., Shulman, R. G. 1973. *J. Mol. Biol.* 78:71–82
40. Wong, K. L., Wong, Y. P., Kearns, D. R. 1975. *Biopolymers* 14:749–757
41. Reid, B. R., McCollum, L., Ribeiro, N. S., Abbate, J., Hurd, R. E. 1979. *Biochemistry* 18:3996–4005
42. Arter, D. B., Schmidt, P. G. 1976. *Nucleic Acids Res.* 3:1437–46
43. Shulman, R. G., Hilbers, C. W., Kearns, D. R., Reid, B. R., Wong, Y. P. 1973. *J. Mol. Biol.* 78:57–69
44. Reid, B. R., Azhderian, E., Hurd, R. E. 1979. See Ref. 7, pp. 91–115
45. Robillard, G. T., Tarr, C. E., Vosman, F., Berendsen, H. J. C. 1976. *Nature* 262:363–68

46. Geerdes, H. A. M., Hilbers, C. W. 1977. *Nucleic Acids Res.* 4:207–12
47. Kan, L. S., Ts'o, P. O. P. 1977. *Nucleic Acids Res.* 4:1633–38
48. Robillard, G. T., Tarr, C. E., Vosman, F., Sussman, J. L. 1977. *Biophys. Chem.* 6:291–97
49. Robillard, G. T., Reid, B. R. 1979. See Ref. 8, pp. 45–112
50. Wong, K. L., Bolton, P. H., Kearns, D. R. 1975. *Biochim. Biophys. Acta* 382: 446–51
51. Hurd, R. E., Reid, B. R. 1979. *Biochemistry* 18:4005–11
52. Johnston, P. D., Redfield, A. G. 1977. *Nucleic Acids Res.* 4:3599–15
53. Kearns, D. R., Bolton, P. H. 1978. In *Biomolecular Structure and Function,* ed. P. F. Agris, pp. 493–516. New York: Academic
54. Macon, J. B., Wolfenden, R. 1968. *Biochemistry* 7:3453–58
55. Hurd, R. E., Reid, B. R. 1979. *Biochemistry* 18:4017–24
56. Salemink, P. J. M., Yamane, T., Hilbers, C. W. 1977. *Nucleic Acids Res.* 4:3737–38
57. Daniel, W. E., Cohn, M. 1976. *Biochemistry* 15:3917–24
58. Romer, R., Varadi, V. 1977. *Proc. Natl. Acad. Sci. USA* 74:1561–66
59. Hurd, R. E., Azhderian, E., Reid, B. R. 1979. *Biochemistry* 18:4012–17
60. Farrar, T. C., Becker, E. D. 1971. *Pulse and Fourier Transform NMR.* New York: Academic. 115 pp.
61. Mullen, K., Pregosin, P. S. 1976. *Fourier Transform NMR Techniques.* New York: Academic. 149 pp.
62. Dadok, J., Sprecher, R. F. 1974. *J. Magn. Reson.* 13:243–53
63. Redfield, A. G. 1976. In *NMR: Basic Principles and Progress,* eds. P. Diehl, E. Fluck, R. Kosfeld, 13:137–52. Berling: Springer-Verlag
64. Redfield, A. G. 1978. *Methods Enzymol.* 49:253–70
65. Redfield, A. G., Kunz, S. D. 1975. *J. Magn. Reson.* 19:250–54
66. Redfield, A. G., Kunz, S. D., Ralph, E. K. 1975. *J. Magn. Reson.* 19:114–17
67. Redfield, A. G., Kinz, S. D. 1979. See Ref. 7, pp. 225–39
68. Noggle, J. H., Schirmer, R. E. 1971. *The Nuclear Overhauser Effect: Chemical Applications.* New York: Academic
69. Bothner-By, A. A. 1979. See Ref. 8, pp. 177–219
70. Johnston, P. D., Redfield, A. G. 1978. *Nucleic Acids Res.* 5:3913–27
71. Johnston, P. D., Redfield, A. G. 1979. See Ref. 9, pp. 191–206
72. Geerdes, H. A. M., Hilbers, C. W. 1979. *FEBS Lett.* 107:125–31
73. Sanchez, V., Redfield, A. G., Johnston, P. D., Tropp, J. 1980. *Proc. Natl. Acad. Sci. USA.* 77:5659–62
74. Crothers, D. M., Cole, P. E., Hilbers, C. W., Shulman, R. G. 1974. *J. Mol. Biol.* 87:63–88
75. Hilbers, C. W. 1979. See Ref. 8, pp. 1–43
76. Eigen, M. 1964. *Angew. Chem. Int. Ed.* 13:1–19
77. Kallenbach, N. R., Daniel, W. E., Kaminker, M. A. 1976. *Biochemistry* 15:1218–28
78. Hurd, R. E., Reid, B. R. 1980. *J. Mol. Biol..* In press
79. Jack, A., Ladner, J. E., Rhodes, D., Brown, R. S., Klug, A. 1977. *J. Mol. Biol.* 111:315–28
80. Teitelbaum, H., Englander, S. W. 1975. *J. Mol. Biol.* 92:79–92
81. Kallenbach, N. R., Berman, H. M. 1977. *Q. Rev. Biophys.* 10:138–236
82. Dwek, R. A. 1973. *Nuclear Magnetic Resonance in Biochemistry.* London: Oxford Univ. Press. 395 pp
83. James, T. L. 1975. *Nuclear Magnetic Resonance in Biochemistry.* New York: Academic. 413 pp
84. Chao, Y. H., Kearns, D. R. 1977. *Biochim. Biophys. Acta* 477:20–27
85. Gueron, M., Leroy, J. L. 1979. In *ESR and NMR of Paramagnetic Species in Biological and Related Systems,* ed. I. Bertini, R. S. Drago, pp. 327–67. Dordrecht, Netherlands: Reidel
86. Jones, C. R., Kearns, D. R. 1975. *Biochemistry* 14:2660–65
87. Liebman, M., Rubin, J., Sundaralingam, M. 1977. *Proc. Natl. Acad. Sci. USA* 74:4821–25
88. Davanloo, P., Sprinzl, M., Cramer, F. 1979. *Biochemistry* 15:3189–99
89. Geerdes, H. A. M., Van Boom, J. H., Hilbers, C. W. 1978. *FEBS Lett.* 88:27–32
90. Shulman, R. G., Hilbers, C. W., Miller, D. L., Yang, S. K., Soll, D. 1974. In *Structure and Conformation of Nucleic Acids and Protein-Nucleic Acid Interactions,* ed. M. Sundaralingam, S. T. Rao, pp. 149–70. Baltimore: Univ. Park Press
91. Kearns, D. R., Wong, Y. P. 1974 *J. Mol. Biol.* 87:755–74
92. Burns, P. D., Luoma, G. A., Marshall, A. G. 1980. *Nucleic Acids Res.* In press
93. Luoma, G. A., Marshall, A. G. 1978. *Proc. Natl. Acad. Sci. USA* 75:4901–4
94. Smith, J. L., Marshall, A. G. 1980. *Biochemistry.* In press

95. Luoma, G. A., Burns, P. D., Bruce, R. E., Marshall, A. G. 1980. *Biochemistry.* In press
96. Salemink, P. J. M., Raue, H. A., Heerschap, A., Planta, R. J., Hilbers, C. W. 1980. *Biochemistry.* In press
97. Gueron, M. and Shulman, R. G. 1975. *Proc. Natl. Acad. Sci. USA* 72:3482–85
98. Salemink, P. J. M., Swarthof, T., Hilbers, C. W. 1979. *Biochemistry* 18: 3477–85
99. Komoroski, R. A., Allerhand, A. 1972. *Proc. Natl. Acad. Sci. USA* 69:1804–8
100. Komoroski, R. A., Allerhand, A. 1974. *Biochemistry* 13:369–72
101. Agris, P. F., Fujiwara, F. G., Schmidt, C. F., Loeppky, R. N. 1975. *Nucleic Acids Res.* 2:1503–12
102. Fujiwara, F. G., Tompson, J., Loeppky, R. N., Agris, P. F. 1978. In *Biomolecular Structure and Function,* ed. P. F. Agris, pp. 527–33. New York: Academic
103. Tompson, J. G., Hayashi, F., Paukstelis, J. V., Loeppky, R. N., Agris, P. F. 1979. *Biochemistry* 10:2079–85
104. Hamill, W. D., Grant, D. M., Horton, W. J., Lundquist, R., Dickman, S. 1976. *J. Am. Chem. Soc.* 98:1276–78
105. Hamill, W. D., Horton, W. J., Grant, D. M. 1980. *J. Am. Chem. Soc.* In press
106. Tompson, J. G., Agris, P. F. 1979. *Nucleic Acids Res.* 7:765–79
107. Schmidt, P. G., Tompson, J. G., Agris, P. F. 1980. *Nucleic Acids Res.* 8:643–56
108. Agris, P. F., Schmidt, P. G. 1980. *Nucleic Acids Res.* In press
109. Schweizer, M. P., Hamill, W. D., Walkiw, I. J., Horton, W. J., Grant, D. M. 1980. *Nucleic Acids Res.* In press
110. Kaiser, I. I. 1969. *Biochemistry* 8: 231–38
111. Horowitz, J., Ou, C. N., Ishaq, M., Ofengand, J., Bierbaum, J. 1974. *J. Mol. Biol.* 88:301–12
112. Ofengand, J., Bierbaum, J., Horowitz, J., Ou, C. N., Ishaq, M. 1974. *J. Mol. Biol.* 88:313–25
113. Horowitz, J., Ofengand, J., Daniel, W. E., Cohn, M. 1977. *J. Biol. Chem.* 252:4418–20
114. Marshall, A. G., Smith, J. L. 1977. *J. Am. Chem. Soc.* 99:635–36

III
Protein-Nucleic Acid Interactions

Ann. Rev. Biochem. 1984. 53: 293–321

14

PROTEIN-DNA RECOGNITION

Carl O. Pabo

Department of Biophysics, The Johns Hopkins University
School of Medicine, Baltimore, Maryland, 21205

Robert T. Sauer

Department of Biology, Massachusetts Institute of Technology,
Cambridge, Massachusetts, 02139

CONTENTS

INTRODUCTION

Sequence-specific DNA-binding proteins regulate gene expression and also serve structural and catalytic roles in other cellular processes. How do these proteins bind to double-helical DNA and how do they recognize a particular base sequence? Here we review recent crystallographic, biochemical, and genetic studies that address these questions. For the most part, we emphasize work published between 1980 and 1983, since the first

471

three-dimensional structures of site-specific DNA-binding proteins were reported during this period. Crystal structures are now available for the Cro and cI repressors of bacteriophage lambda and for the CAP protein of *Escherichia coli* (1–3). Each of these three proteins can turn off expression of specific genes by preventing initiation of transcription; CAP and lambda repressor can also enhance gene expression by stimulating transcription from certain promoters. Other recent reviews discuss these proteins and their physiological actions (4–10), the mechanism and control of prokaryotic transcription (11), and general aspects of protein-DNA interactions (12, 13).

Cro, lambda repressor, and CAP interact with DNA in a basically similar manner. Despite differences in size, domain organization and tertiary structure, each of these proteins binds to operator DNA as a dimer and uses α-helices to contact adjacent major grooves along one face of the double helix. Moreover, sequence homologies suggest that many other DNA-binding proteins use similar α-helical regions for DNA recognition. How does each of these proteins recognize its proper binding site? What forces are involved? Is there a "recognition code"? We consider these questions after discussing the structures of Cro, lambda repressor, and CAP and describing the models proposed for the respective protein-DNA complexes.

LAMBDA CRO

Lambda Cro binds to six operator sites in the double-stranded phage DNA (14, 15). These sites are clustered in two operator regions, and each region contains three 17-bp (base pair) sites. The DNA sequences of the six sites are similar but not identical, and Cro's affinity for the different sites varies over a tenfold range (15–17). The sequence of each operator site has approximate two-fold symmetry, and the consensus sequence, shown in Figure 1, is symmetric. The Cro monomer contains 66 amino acid residues (18, 19). Cro exists as a dimer in solution (20) and this is the form active in DNA binding.

A crystallographic study at 2.8 Å resolution by Anderson, Ohlendorf, Takeda, & Matthews (1) showed that the Cro monomer contains three

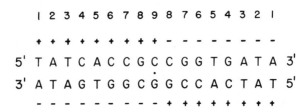

Figure 1   Consensus Operator Sequence for Lambda Cro and Lambda Repressor.

strands of antiparallel β-sheet (residues 2–6, 39–45, and 48–55) and three α-helices (residues 7–14, 15–23, and 27–36). The Cro dimer is stabilized by a region of antiparallel β-sheet that is formed by pairing Glu 54–Val 55–Lys 56 from each monomer. The four C-terminal residues of Cro (residues 63–66) are poorly represented in the electron density map and are probably somewhat disordered both in the crystal and in solution.

A Model for the Cro-Operator Complex

The structure of the Cro dimer immediately suggested its basic mechanism of DNA binding (1). In the dimer, the two copies of α-helix 3 form protruding ridges that are separated by the same center-to-center distance, 34 Å, that separates successive major grooves of B-DNA (Figure 2) (21, 22). The angle between the two Cro helices allows them to fit neatly into successive major grooves of the operator. This arrangement provides an excellent fit between the surface of the protein and the surface of the DNA, and accounts nicely for the observed DNA modification and protection data. This data, which is shown in schematic form in Figure 2, had suggested that Cro bound in a symmetric manner and that it contacted

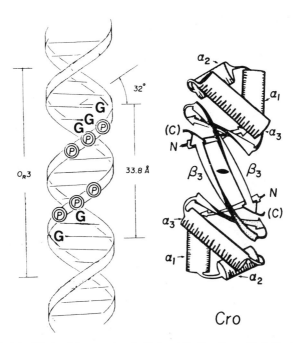

Cro

Figure 2   Sketch of the lambda operator site O_R3 and the Cro dimer. P's indicate phosphates that have been implicated as Cro contacts. G's indicate guanines implicated as contacts. Adapted with permission from (1).

adjacent major grooves along one face of the double helix (Figure 3) (15–17). The proposed complex also seems chemically reasonable, since a number of hydrophilic and charged residues can interact with exposed hydrogen-bonding groups in the major groove and with the negatively charged phosphates.

Refinement of the Cro structure at 2.2 Å and further model building by Ohlendorf et al (23) have allowed a detailed analysis of possible Cro-operator interactions. During this modeling of hydrogen-bonding interactions, energy minimization (24) was used to ensure that the stereochemistry of the proposed complex was reasonable. Model building started with a B-form operator (21, 22), since DNA in solution adopts this conformation (25). However, minor adjustments in the DNA structure were allowed, and in the best model for the complex the operator DNA was smoothly bent with a radius of curvature of 75 Å. Thus, the DNA is bent around Cro so that each end of the operator is 5 Å closer to the protein than it would be if the DNA were straight. This bending seems plausible since it should require only a few kcal of energy (26), but Ohlendorf et al (23) point out that a

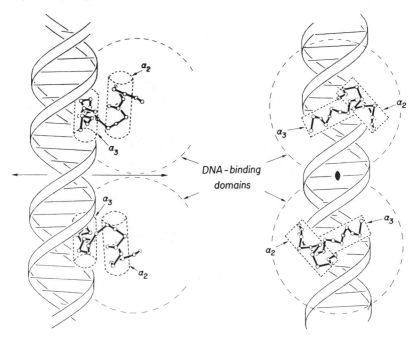

Figure 3   Alpha carbons from helices two and three of the proposed Cro-operator complex. Although the rest of the protein structures are quite different, the corresponding helical regions of repressor and CAP are quite similar and may contact the DNA in a similar manner. Used with permission from (8).

similar fit could also be obtained by a "hinge-bending" motion of the protein dimer that would allow it to contact a straight operator site.

The Cro-operator model predicts several specific contacts between each Cro monomer and the edges of base pairs in the major groove. These contacts, which all involve side chains in or near α-helix 3, are listed below. The base-numbering scheme is that for the consensus operator site shown in Figure 1:

1. The hydroxyl group of Tyr26 donates a hydrogen bond to 04 of the thymine at +1.
2. The side-chain amide of Gln27 donates a hydrogen bond to N7 of the adenine at +2, and accepts a hydrogen bond from N6 of the same adenine. This interaction is illustrated in Figure 4 (top).
3. The hydroxyl group of Ser28 forms two hydrogen bonds with N6 and N7 of adenine − 3, as shown in Figure 4 (center).
4. The amino group of Lys32 donates one hydrogen bond to 04 of the thymine at − 5, donates a second hydrogen bond to 06 of guanine − 4, and may donate a third hydrogen bond to N7 of guanine − 4.
5. The guanidinium group of Arg38 donates two hydrogen bonds to 06 and N7 of guanine − 6, as shown in Figure 4 (bottom).

In addition, Gln27, Asn31, and Lys32 seem to make a few van der Waals contacts in the major groove.

The proposed Cro-operator complex also places a large number of residues near the sugar-phosphate DNA backbone. Those with hydrogen-bonding potential include Gln16, Thr17, and Lys21 in helix 2, which is just above the major groove. They also include Asn31, His35, and Lys39, which are in or beyond helix 3, and Glu54 and Lys56 in the C-terminal β-region (23). Several additional polar interactions might also be made by residues Asn61-Lys62-Lys63-Thr64-Thr65 in Cro's flexible C-terminal region.

Evidence Supporting the Cro-Operator Model

The overall fit of Cro against the operator site is supported by a calculation of the electrostatic potential around the Cro dimer (27, 28). There is a weak negative potential on the far side of the dimer, but the overall potential is dominated by a positive region that straddles the two-fold axis. This region of positive charge coincides remarkably well with the presumed DNA-binding site.

Protein modification studies provide general support for the model of the Cro-operator complex (Y. Takeda, J. Kim, C. Caday, D. Davis, E. Steer, D. Ohlendorf, B. Matthews, W. Anderson, manuscript in preparation). As the model predicts, Lys21, Lys32, Lys56, and Lys62/63 are protected from chemical modification when Cro is bound to DNA, but Tyr26 and Lys39,

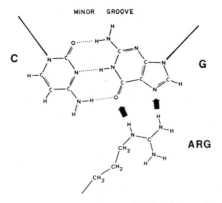

Figure 4 (*top*) Sketch indicating possible hydrogen bonds between glutamine and an A:T base pair (86). (*Center*) Sketch indicating how serine could form a pair of hydrogen bonds with an A:T base pair (23). (*bottom*) Sketch indicating the hydrogen bonds that arginine could form with a G:C base pair (86).

which are also predicted to be contact residues, are not protected. However, detailed interpretation of these results is complicated by the fact that the Cro is largely bound to nonoperator DNA in these studies, and it is likely that the structures of operator and nonoperator complexes may be somewhat different. Other experiments show that the affinity of Cro for operator DNA is reduced by carboxypeptidase digestion, and this result, together with the protection of Lys62/63, suggests that the flexible C-terminal region of Cro makes some DNA contacts.

Genetic studies also support the model proposed for the Cro-operator complex. Most of the proposed DNA contact residues are represented in a collection of mutations that are phenotypically defective in operator binding (A. Pakula, R. Sauer, manuscript in preparation). These include each of the five residues implicated in specific base contacts (Tyr26 → Asp; Gln27 → His; Ser28 → Arg/Asn; Lys32 → Thr/Gln; Arg38 → Gln) and many of the residues implicated in backbone contacts (Gln16 → His; Lys39 → Thr; Glu54 → Lys/Ala; Lys56 → Asn/Gln/Thr). In addition, three of the proposed specific contact residues have been altered by oligonucleotide replacement of the appropriate region of the *cro* gene (M. Nasoff, S. Noble, M. Caruthers, manuscript in preparation). The mutations introduced by this procedure (Tyr26 → Phe/Leu/Asp; Gln27 → Leu/Cys/Arg; Ser28 → Ala) all reduce the operator affinity of Cro. Although further analysis of all the *cro* mutations is needed to show that they do not disrupt Cro folding, it is likely that most of them owe their reduced operator binding to a defect in DNA binding. The correspondence between the positions of the mutations and the proposed DNA contact residues supports the model for the Cro-operator complex.

LAMBDA REPRESSOR

Lambda repressor recognizes the same six operator sites (14) that lambda Cro recognizes, and repressor also binds to each 17-bp operator site as a dimer (29, 30). The affinity of repressor for the six sites varies over a 50-fold range, but the sites for which repressor has highest affinity are not the sites for which Cro has highest affinity (15–17, 31). For example, the site called O_R3 is one of the weakest binding sites for repressor but is the strongest binding site for Cro. The different affinities of repressor and Cro for the six operator sites help explain the contrasting physiological roles of these two proteins (4–6). Chemical protection and modification studies show that Cro and repressor contact many of the same functional groups in the operators, but the Cro contacts seem to be a subset of the repressor contacts (compare Figures 2, 5, and 6) (15–17, 32). The repressor monomer contains 236 amino acids (33) and is thus considerably larger than Cro.

The repressor monomer folds into two domains of similar size, which can be separated by cleavage with papain or other proteases (34). The N-terminal domain, as an isolated proteolytic fragment of residues 1–92, binds specifically to the lambda operator sites and mediates both positive and negative control of transcription (35). Thus the regulatory activities of the intact protein are retained by the N-terminal fragment. However, the N-terminal fragment binds to the operator less tightly than intact repressor because the fragment dimer readily dissociates in solution (34). Intact repressor binds to the operator more tightly because the C-terminal domain stabilizes the dimer and thereby stabilizes the protein-DNA complex.

Intact lambda repressor has not been crystallized, but the structure of the N-terminal operator-binding domain has been solved at 3.2 Å resolution by Pabo & Lewis (3). The N-terminal domain consists of an N-terminal arm and five α-helices, and is a dimer in the crystal. The first eight residues of the domain form an arm that extends away from the globular region. Most of this arm packs against another molecule in the protein crystal, but residues 1–3 are disordered and not visible in the electron density map. The α-helices of the domain include residues 9–23 (helix 1), 33–39 (helix 2), 44–52 (helix 3), 61–69 (helix 4), and 79–92 (helix 5). The first four helices, along with the irregular regions of chain that connect them, form a compact, globular

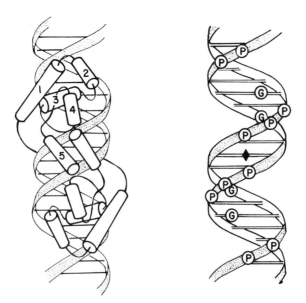

Figure 5   The proposed lambda repressor-operator complex. Panel on right summarizes the results of chemical protection experiments at the site of O_R1.

domain. The fifth helix extends off to one side of the molecule and folds against helix 5 of a neighboring monomer. This helix-helix contact seems to be the major interaction stabilizing the N-terminal dimer.

A Model for the Repressor-Operator Complex

Possible structures for the complex of the N-terminal dimer with operator DNA were evaluated by model building (3). B-form operator DNA was used for these studies because DNA in solution is B-form (25), and repressor does not significantly wind or unwind the operator DNA (36). NMR studies have subsequently shown that the operator conformation in the repressor-operator complex is similar to the expected B-conformation (M. Weiss, D. Patel, R. Sauer, M. Karplus, manuscript in preparation). The model building was also guided by chemical modification experiments that suggested that repressor contacted the major groove and that most contacts were on one side of the double-helical site (16, 17, 31). With these constraints, model building yielded only one arrangement that gave a good fit between the surface of the protein and the surface of the DNA. This complex, which is shown schematically in Figure 5, allows each subunit of the dimer to contact one half of the operator site. In each half site, the N-terminal portion of helix 3 fits directly into the major groove. Helix 2 is just above the major groove, and its N-terminal region is next to the sugar phosphate backbone of the DNA. Figure 5 also summarizes the chemical protection data, showing the guanine N7 and phosphate groups implicated as repressor contacts on the front side of the DNA helix (16, 17, 32).

In the proposed complex, the two N-terminal arms of repressor are set slightly to the sides of the operator helix and extend towards the "back" of the DNA near the center of the 17-bp site. Biochemical studies show that these arms actually make major groove contacts on the back of the operator site (37). The 92-residue N-terminal domain, like intact repressor, protects several operator guanines from chemical methylation; four of the protected sites are visible in the major groove on the "front" of the operator site, and two are visible on the "back" (Figures 5 and 6). However, a shorter N-terminal fragment, containing residues 4–92 and thus missing the first three residues of the arm, protects only the guanines on the front of the operator site. Since NMR studies show that the 1–92 and 4–92 fragments have the same conformation (M. Weiss et al, manuscript in preparation), the different protection patterns imply that the first three residues of the arm must contact the major groove on the back of the operator site. Model building indicates that repressor's arms are long enough to encircle the double helix, possibly by wrapping around the DNA in the major groove, as shown in Figure 6. The first five residues of the arm, Ser1-Thr2-Lys3-Lys4-Lys5, are polar and could readily make hydrogen bonds to bases in the major groove or interact with the sugar-phosphate backbone.

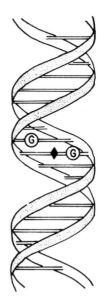

Figure 6   Proposed interaction between lambda repressor's N-terminal arm and the back of the operator site. Panel on right shows guanines that are protected on the back of operator site O_R1.

Except for adjustments in the position of the flexible N-terminal arm, the initial model building (3) simply used the crystallographic coordinates of the N-terminal dimer and searched for an overall fit of the repressor against the DNA. In a subsequent study, Lewis et al (38) made more detailed predictions about the contacts between repressor and the consensus operator site. This phase of model building started with the previous complex but then allowed surface side chains to move and allowed minor adjustments of the relative orientation of the subunits within the dimer. After these adjustments, it appeared that four side chains from each subunit of the dimer could make specific major groove contacts on the front of the operator site. Three of these side chains, Gln44, Ser45, and Ala49, are on α-helix 3, and the fourth, Asn55, is in the irregular region of protein chain just beyond helix 3. The proposed contacts with the consensus site (Figure 1) are summarized below:

1. The side-chain amide of Gln44 makes two contacts with adenine + 2. It accepts a hydrogen bond from the N6 of adenine and donates a hydrogen bond to the N7 [Figure 4 *(top)*].
2. The hydroxyl of Ser45 donates a hydrogen bond to the N7 position of guanine − 4.

3. The methyl group of Ala49 makes van der Waals contacts with the methyl groups of thymine $+3$ and thymine -5.
4. The side-chain amide of Asn55 donates two hydrogen bonds to the O6 and N7 positions of guanine -6.

Detailed predictions of contacts that involve repressor's N-terminal arm were not attempted. The conformation of the arm in the crystal seems to be determined by crystal packing forces and there is no reason to believe that it adopts a similar structure in solution or in the repressor-operator complex. In fact, NMR studies show that the arm is flexible in solution (M. Weiss et al, manuscript in preparation). Without reliable structural constraints, detailed model building gives too many possibilities to be useful.

In addition to the major groove contacts, several contacts between repressor and the sugar phosphate backbone of the DNA seem possible (38). Ethylation of any of ten phosphates in the operator interferes with repressor binding (16, 17) and, as shown in Figure 5, these phosphates are symmetrically disposed on the front face of the operator site. Six are clustered near the center of the site and two are near each end of the site. In the proposed complex, Gln33, which is the first residue in helix 2, and Asn52, which is the last residue in helix 3, contact the two phosphates near the outer edge of the operator site. Residues Asn58, Tyr60, and Asn61, in the irregular region between helices 3 and 4, appear to contact the phosphates near the center of the operator site.

None of the proposed contacts between phosphates and amino acids involve ion pairs. However, the Lys24-Lys25-Lys26 sequence, which is part of the loop of irregular chain between helices 1 and 2 (see Figure 5), is near the DNA. These residues could interact with the phosphates on the outer edge of the site if the operator DNA were allowed to partially bend around the protein, in the manner proposed for the Cro-operator complex (23). The salt-dependence of binding (30, 39) suggests that a few ion pairs are formed between the repressor dimer and operator DNA, and residues 24–26 may be responsible for these ion pairs.

Repressor Mutants Defective in DNA Binding

Genetic and biochemical studies of repressor mutants provide strong support for the fundamental features of the proposed repressor-operator complex (40, 41). Twelve mutations, affecting eight residue positions in the N-terminal domain, decrease the operator affinity of repressor but do not disrupt the structure of the mutant N-terminal domain. As shown in Figure 7, seven of the residue positions affected by these "DNA-binding" mutations cluster in the α2-α3 region of the N-terminal domain. Here, the mutations affect each of the four residues predicted to make specific major groove contacts (Gln44 → Leu/Ser/Tyr; Ser45 → Leu; Ala49 → Val/Asp; Asn55 → Lys); two of the residues predicted to make phosphate contacts (Gln33 → Ser/Tyr; Asn52 → Asp); and one residue that is close to the operator in the proposed complex (Gly43 → Glu). The only "DNA-binding" mutation outside the α2-α3 region alters a residue in the N-terminal arm (Lys4 → Gln).

In Figure 8, the positions of the "DNA-binding" mutations are shown on a space-filling model of the N-terminal dimer. This figure also shows the position of the "DNA contact" residues predicted from model building (3, 38). There is clearly a striking correspondence between the genetic and model-building results. The only significant differences involve residues 58, 60, and 61, which are near the center of the repressor dimer (Figure 8) and which were implicated as backbone contacts by model building. Although mutations at these positions were not obtained in the study cited above, two of these sites were identified by mutations (Asn58 → Ile; Asn61 → Ser) in a study where the mutant proteins were not characterized (42, 43). Thus, almost every residue that has been proposed as a DNA contact residue is altered by one or more repressor mutations. This excellent overall correspondence between the genetic results and the model building would not be expected if the model were seriously wrong, and thus the mutations provide strong experimental support for the model of the repressor-operator complex.

Thus far, we have referred to protein-DNA "contacts" without explicit reference to the energy provided by each contact. This issue has been

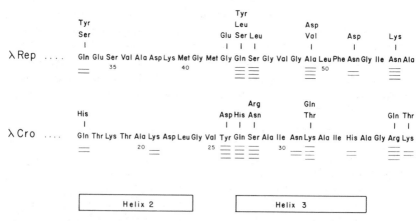

Figure 7   Sequence of the Helix 2/3 Regions from Repressor and Cro. Residues underlined twice are predicted to make contacts with the DNA backbone (23, 38). Those underlined four times are predicted to make major groove contacts. The positions of surface mutations that decrease operator binding are shown.

addressed by studying the affinities of the purified mutant repressors for operator DNA. Mutants may have reduced affinity either because favorable contacts are removed or because unfavorable steric or electrostatic contacts are introduced. Both effects may contribute when the wild-type side chain is replaced by a larger side chain in the mutant (e.g. Ser45 → Leu). However, the Lys4 → Gln, Gln33 → Ser and Gln44 → Ser mutations replace the wild-type side chain with a smaller side chain and thus their decreased operator affinity is likely to reflect the loss of favorable interactions. The operator affinities of these three mutants are about 100-fold less than the wild-type affinity (H. Nelson, R. Sauer, manuscript in preparation). This suggests that the Lys4, Gln33, and Gln44 side chains each contribute about 2.7 kcal/mole of free energy to the interaction between the dimer and the operator. If these energies are additive, then these three side chains contribute a total of about 8 kcal/mole, or half of the 16 kcal/mole free-energy change that occurs upon binding (30). Even if the Lys4, Gln33, and Gln44 contacts are somewhat stronger than average, it seems that the total repressor-operator binding energy can be reasonably explained by the contacts proposed in the model.

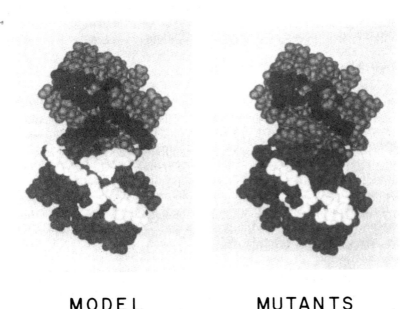

MODEL   MUTANTS

Figure 8   Space-filling models of lambda repressor's N-terminal domain. Residues predicted to contact the operator DNA are highlighted in left panel. Positions of "DNA-binding" mutations are highlighted in right panel. Computer graphics were provided by Richard J. Feldman.

Contacts between a protein side chain and the DNA sugar phosphate backbone are often referred to as "nonspecific" contacts. However, such contacts could readily contribute to the specificity of binding. This point is illustrated by the Gln33 → Ser mutant of lambda repressor. In the proposed complex, Gln33 makes a contact with one of the phosphates near the outer edge of the operator site (3, 38), and substitution of serine at this position reduces the operator affinity of repressor about 100-fold. However, wild-type repressor and the mutant Gln33 → Ser repressor have the same affinity for nonoperator DNA (H. Nelson, R. Sauer, manuscript in preparation). Since specificity depends on the ratio of operator to nonoperator binding (12), Gln33 increases the specificity of operator recognition. How can we rationalize this in molecular terms? One possibility is that Gln33 only contacts the DNA backbone in the specific repressor-operator complex. In complexes of repressor with nonoperator DNA, steric interference in the major groove might prevent Gln33 from approaching the backbone closely enough to make a contact. The Asn52 → Asp repressor mutant also affects a proposed phosphate contact but in this case the mutant has reduced affinity for both operator and nonoperator DNA. Presumably the reduced affinity is caused by electrostatic repulsion between the mutant side chain and the phosphate backbone, and this suggests that Asn52 is close to the DNA in both the operator and nonoperator complexes.

CAP PROTEIN

The catabolite gene activator protein (CAP), also called the cyclic AMP receptor protein (CRP), regulates several catabolite-sensitive gene operons in *E. coli* (7, 44). When cyclic AMP is present at a sufficient concentration, it forms a complex with CAP, and this complex is active in specific DNA binding (45). A consensus sequence has been suggested for the CAP binding sites (46), but many of the individual sites differ considerably from this sequence. The CAP protein contains 209 residues (47, 48) and has two domains (49). The C-terminal domain binds DNA, while the N-terminal domain binds cyclic AMP and provides most of the dimer contacts. CAP is a stable dimer in solution and this dimer is the active DNA binding species (50).

Crystallographic studies by McKay, Weber, & Steitz (2, 51) have determined the structure of the intact CAP dimer in a complex with cAMP. A sketch of the CAP monomer is shown in Figure 9. The N-terminal domain contains 135 residues and consists of a pair of short helices (A and B), an eight-stranded antiparallel β-roll, and a long α-helix (C). The C-terminal domain includes residues 136–209 and contains three α-helices (D, E, and F) and two pairs of short antiparallel β-strands. The CAP

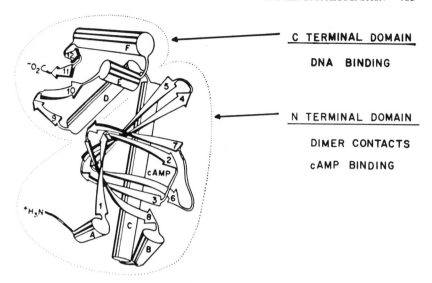

C TERMINAL DOMAIN

DNA BINDING

N TERMINAL DOMAIN

DIMER CONTACTS

cAMP BINDING

Figure 9   Sketch of the CAP monomer. The approximate extent of each domain is outlined, and the primary functions of each domain are listed. Adapted with permission from (51).

dimer contacts, which are in the N-terminal domain, involve a pairing of the C-helices and some additional contacts between the C-helix of one subunit and the β-roll of the other subunit. The cyclic AMP also occupies part of this dimer interface. It is completely buried within the interior of the N-terminal domain, but it forms hydrogen bonds that bridge the dimer interface. In the crystal form studied, the CAP dimer is somewhat asymmetric, since there are different orientations of the N-terminal and C-terminal domains in the two subunits. One subunit has an "open" conformation with a cleft between the domains, while the other subunit has a "closed" conformation.

Models for CAP-DNA Interactions

Several different models have been proposed for the interaction of CAP with DNA (2, 3, 51–54) but current results suggest that CAP binds to right-handed B-DNA and uses the N-terminal portions of its F-helices to contact the major groove as shown in Figure 10 (3, 54). Calculation of the electrostatic potential at the surface of the CAP dimer provides some support for this model (55), since the only regions of net positive charge are near the amino-terminal portions of the F-helices. Originally, McKay & Steitz (2) had proposed that CAP binds to left-handed B-form DNA. This conformation of DNA (which is quite distinct from Z-DNA) has never been observed. However, it seems conformationally plausible (56) and in model-

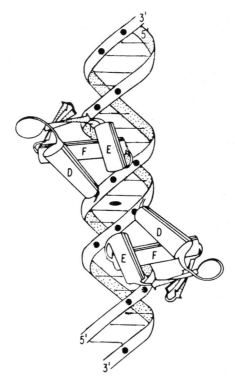

Figure 10   The proposed interaction between the C-terminal domain of CAP and the CAP binding site in the *lac* operon. Black dots indicate contacts with phosphates (45). Adapted with permission from (54).

building experiments, the two F-helices of the CAP dimer can fit neatly into successive major grooves of this left-handed DNA. However, biochemical experiments rule out this model. If CAP were to bind to left-handed DNA, then it should unwind the right-handed DNA found in solution by four turns (1440 degrees). This does not occur. In fact, Kolb & Buc (57) have shown that CAP binding unwinds DNA by no more than 30 degrees.

Several specific CAP-operator interactions have recently been proposed on the basis of model building with right-handed B-DNA (I. Weber, T. Steitz, personal communication). In this model the DNA is bent around the protein with a radius of curvature similar to that predicted for the Cro-operator complex. The specific contacts are listed below using the base numbering scheme of Figure 11:

1. The guanidinium group of Arg180 donates two hydrogen bonds to O6 and N7 of guanine 3 [(Figure 4 (*bottom*)] and donates two hydrogen bonds to the symmetrically related guanine 16.
2. The side chain of Glu181 in one monomer accepts one hydrogen bond

from N4 of cytosine 5 and accepts a second hydrogen bond from N6 of adenine 4. The side chain of Glu181 in the other monomer accepts one hydrogen bond from N6 of adenine 15 and may accept a second hydrogen bond from the N4 of cytosine 14.

3. The amino group of Lys188 donates two hydrogen bonds to O6 and N7 of guanine 5, and two to the symmetrically related guanine 14.

4. Arg185 donates one hydrogen bond to N7 of adenine 6, and donates one hydrogen bond to the symmetrically related adenine 13.

A genetic study of CAP mutants with altered DNA-binding specificity (R. Ebright, J. Beckwith, P. Cossart, B. Gicquel-Sanzey, manuscript in preparation) has suggested specific interactions between CAP and its binding site and also supports the model in which CAP binds to right-handed B-DNA. Figure 11 shows the CAP binding site in the lac operon, and also shows the symmetrically related L8 and L29 mutations in this site. CAP mutants with increased affinity for the L8 site or the L29 site, but with reduced affinity for the wild-type site, were selected and three different mutations were obtained. All three mutations, Glu181 → Leu, Glu181 → Val and Glu181 → Lys, change the same residue in helix F. Since model building suggests that helix F makes major groove contacts, it is likely that Glu181 normally recognizes the G : C base pairs at 5 and 14, while Leu181, Val181, and Lys181 recognize the mutant A : T base pairs at these positions. These contacts can be accommodated in the complex of CAP with right-handed DNA shown in Figure 10, but they would be difficult, if not impossible, to make if CAP bound to left-handed DNA (I. Weber, T. Steitz, personal communications).

Analysis of these CAP mutations also indicates that CAP interacts with its binding site in a symmetric fashion. Two of the CAP mutations were selected using the L8 mutation and one was selected using the L29 mutation. Nevertheless, the mutants bind equally well to the L8 site and the L29 site, which have symmetrically related base changes.

It is not known how cAMP increases the affinity of CAP for its specific DNA sites, but the crystal structure suggests some possibilities and rules

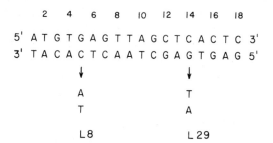

Figure 11   Sequence of the CAP binding site in the *lac* operon.

out others (51). It had been proposed, from cAMP analog studies, that the adenosine ring of cAMP interacts directly with the DNA (52). The structure of the CAP-cAMP complex shows that this proposal must be incorrect, since the cyclic AMP is buried in the interior of the large domain. However, since the buried cAMP interacts with both N-terminal subunits of the dimer, it might affect DNA binding by changing the relative orientation of the two subunits or by changing the orientation of the domains within a subunit.

A CONSERVED α-HELICAL STRUCTURE FOUND IN MANY DNA-BINDING PROTEINS

In the proposed complexes of Cro, repressor, and CAP with DNA, many of the DNA contacts are made by two α-helices that are linked by a tight turn. In both Cro and repressor, these are helices 2 and 3, and in CAP these are designated helices E and F. In each of the three models, the first helix (2 or E) sits above the groove near the DNA backbone while the second helix (3 or F) fits partly or completely into the major groove. This is illustrated in Figure 3.

The structures of these α-helical units within the three proteins are nearly identical. For CAP and Cro, 24 α-carbons from the αE-αF region and 24 α-carbons from the α2-α3 unit can be superimposed with a deviation of only 1.1 Å per α-carbon (58). The agreement between lambda repressor and Cro is slightly better. Here, 20 α-carbons from the two α2-α3 units superimpose with an average deviation of only 0.7 Å (59). Alpha-helices arranged in this way may be unique to DNA-binding proteins as they have not been found in any other protein structures (58, 59).

Superimposing the strictly conserved bihelical unit also reveals a limited structural homology among parts of helix 1 from Cro and repressor and parts of helix D from CAP (58, 59). However, this homology is not extensive and is far less precise than the other homology. In all other regions, the tertiary folds of the three proteins are completely different. It should also be noted that the arrangements of the conserved helical units with respect to the dimer axes are not identical in the three proteins. Thus, the helical units of the protein dimers cannot be superimposed as precisely as the helical units of the monomers, and the different orientations of these regions imply that the proteins could not be docked with their bihelical units contacting the DNA in precisely the same manner.

Sequence Homologies

A number of DNA-binding proteins share sequence homologies with Cro, repressor, and CAP, and several research groups have predicted that these proteins also use helix-turn-helix structures for DNA interactions (60–62). In some cases, entire protein sequences are homologous. For example, the lambda, P22, 434, and LexA repressor sequences are significantly related (61), and in these cases, the homology almost certainly implies an evolutionary relationship. In other cases, there are only limited regions of homology, but many DNA-binding proteins have regions that are homologous to the α2-α3 sequences of Cro and repressor and the αE-αF sequence of CAP. A set of such sequences is shown in Figure 12.

The pattern of conserved residues and residue types, shown in the alignment of Figure 12, suggests that the homologous proteins also form similar helix-turn-helix structures. Alpha-helices on the surface of a protein often have a characteristic pattern of nonpolar residues that face the hydrophobic core of the protein. Because the helical repeat is 3.6 residues/turn, these residues usually occur at relative helical positions 1-4-5 or 1-2-5 (63). The bihelical units of Cro, repressor and CAP contain such triplets, and in the numbering scheme of Figure 12, these triplets occupy positions 4, 5, and 8 in the first helix and positions 15, 18, and 19 in the second helix. The homologous DNA-binding proteins also tend to have nonpolar residues at these positions and, in addition, have nonpolar residues at position 10, which is part of the hydrophobic core of Cro, repressor and CAP. This means that the homologous proteins could form similar bihelical units and have predominantly nonpolar side chains facing the hydrophobic core. In the proteins of known structure, the residues at positions 1–3, 6–7, 11–14, and 16–17 are solvent exposed and hydrophilic, and the homologous proteins also tend to have hydrophilic residues at these positions. Thus, bihelical units in the homologous proteins would have a number of exposed polar residues that might be used for DNA interactions.

In the alignment of Figure 12, positions 5, 9, and 15 are among the most highly conserved. Alanine is favored at position 5, glycine predominates at 9, and either valine or isoleucine usually occupies 15. Each of these residues seems to have an important role in maintaining the structure of the bihelical unit. In Cro and repressor, the side chains at 5 and 15 in the helix 2/3 unit are in van der Waal's contact and probably help to maintain the proper angle between the two helices. As discussed below, position 9 forms part of the tight turn between the helices. If the homologous sequences also form bihelical structures, then the strong conservation at positions 5, 9, and 15 could be rationalized in structural terms.

The highly conserved glycine at position 9 of the alignment (Figure 12), illustrates an interesting problem in trying to predict structural homology from sequence homology. Originally, it was thought that glycine was required at this position, and most listings of homologies excluded

```
        1    2    3    4    5    6    7    8    9   10   11   12   13   14   15   16   17   18   19   20

...Gln-Glu-Ser-Val-Ala-Asp-Lys-Met-Gly-Met-Gly-Gln-Ser-Gly-Val-Gly-Ala-Leu-Phe-Asn...   λ Rep
    33   34   35   36   37   38   39   40   41   42   43   44   45   46   47   48   49   50   51   52
                        ***                      ***                      ***            ***
...Gln-Thr-Lys-Thr-Ala-Lys-Asp-Leu-Gly-Val-Tyr-Gln-Ser-Ala-Ile-Asn-Lys-Ala-Ile-His...   λ Cro
    16   17   18   19   20   21   22   23   24   25   26   27   28   29   30   31   32   33   34   35
                        ***                      ***                      ***            ***
...Gln-Ala-Ala-Leu-Gly-Lys-Met-Val-Gly-Val-Ser-Asn-Val-Ala-Ile-Ser-Gln-Trp-Gln-Arg...   P22 Rep
    21   22   23   24   25   26   27   28   29   30   31   32   33   34   35   36   37   38   39   40
                        ***                      ***                      ***            ***
...Gln-Arg-Ala-Val-Ala-Lys-Ala-Leu-Gly-Ile-Ser-Asp-Ala-Ala-Val-Ser-Gln-Trp-Lys-Glu...   P22 Cro
    13   14   15   16   17   18   19   20   21   22   23   24   25   26   27   28   29   30   31   32
                        ***                      ***                      ***            ***
...Gln-Ala-Glu-Leu-Ala-Gln-Lys-Val-Gly-Thr-Thr-Gln-Gln-Ser-Ile-Glu-Gln-Leu-Glu-Asn...   434 Rep
    17   18   19   20   21   22   23   24   25   26   27   28   29   30   31   32   33   34   35   36
                        ***                      ***                      ***            ***
...Gln-Thr-Glu-Leu-Ala-Thr-Lys-Ala-Gly-Val-Lys-Gln-Gln-Ser-Ile-Gln-Leu-Ile-Glu-Ala...   434 Cro
    19   20   21   22   23   24   25   26   27   28   29   30   31   32   33   34   35   36   37   38
                        ***                      ***                      ***            ***
...Arg-Gln-Glu-Ile-Gly-Gln-Ile-Val-Gly-Cys-Ser-Arg-Glu-Thr-Val-Gly-Arg-Ile-Leu-Lys...   CAP
   169  170  171  172  173  174  175  176  177  178  179  180  181  182  183  184  185  186  187  188
                        ***                      ***                      ***            ***
...Arg-Gly-Asp-Ile-Gly-Asn-Tyr-Leu-Gly-Leu-Thr-Val-Glu-Thr-Ile-Ser-Arg-Leu-Leu-Gly...   Fnr
   187  188  189  190  191  192  193  194  195  196  197  198  199  200  201  202  203  204  205  206
                        ***                      ***                      ***            ***
...Leu-Tyr-Asp-Val-Ala-Glu-Tyr-Ala-Gly-Val-Ser-Tyr-Gln-Thr-Val-Ser-Arg-Val-Val-Asn...   Lac R
     6    7    8    9   10   11   12   13   14   15   16   17   18   19   20   21   22   23   24   25
                        ***                      ***                      ***            ***
...Ile-Lys-Asp-Val-Ala-Arg-Leu-Ala-Gly-Val-Ser-Val-Ala-Thr-Val-Ser-Arg-Val-Ile-Asn...   Gal R
     4    5    6    7    8    9   10   11   12   13   14   15   16   17   18   19   20   21   22   23
                        ***                      ***                      ***            ***
...Thr-Glu-Lys-Thr-Ala-Glu-Ala-Val-Gly-Val-Asp-Lys-Ser-Gln-Ile-Ser-Arg-Trp-Lys-Arg...   λ cII
    24   25   26   27   28   29   30   31   32   33   34   35   36   37   38   39   40   41   42   43
                        ***                      ***                      ***            ***
...Gln-Arg-Lys-Val-Ala-Asp-Ala-Leu-Gly-Ile-Asn-Glu-Ser-Gln-Ile-Ser-Arg-Trp-Lys-Gly...   P22 cI
    26   27   28   29   30   31   32   33   34   35   36   37   38   39   40   41   42   43   44   45
                        ***                      ***                      ***            ***
...Lys-Glu-Glu-Val-Ala-Lys-Lys-Cys-Gly-Ile-Thr-Pro-Leu-Gln-Val-Arg-Val-Trp-Cys-Asn...   Mat α
   117  118  119  120  121  122  123  124  125  126  127  128  129  130  131  132  133  134  135  136
                        ***                      ***                      ***            ***
...Thr-Arg-Lys-Leu-Ala-Gln-Lys-Leu-Gly-Val-Glu-Gln-Pro-Thr-Leu-Tyr-Trp-His-Val-Lys...   Tet R Tn10
    27   28   29   30   31   32   33   34   35   36   37   38   39   40   41   42   43   44   45   46
                        ***                      ***                      ***            ***
...Thr-Arg-Arg-Leu-Ala-Glu-Arg-Leu-Gly-Val-Gln-Gln-Pro-Ala-Leu-Tyr-Trp-His-Phe-Lys...   Tet R pSC101
    27   28   29   30   31   32   33   34   35   36   37   38   39   40   41   42   43   44   45   46
                        ***                      ***                      ***            ***
...Gln-Arg-Glu-Leu-Lys-Asn-Glu-Leu-Gly-Ala-Gly-Ile-Ala-Thr-Ile-Thr-Arg-Gly-Ser-Asn...   Trp Rep
    66   67   68   69   70   71   72   73   74   75   76   77   78   79   80   81   82   83   84   85
                        ***                      ***                      ***            ***
...Arg-Gln-Gln-Leu-Ala-Ile-Ile-Phe-Gly-Ile-Gly-Val-Ser-Thr-Leu-Tyr-Arg-Tyr-Phe-Pro...   H-inversion
   162  163  164  165  166  167  168  169  170  171  172  173  174  175  176  177  178  179  180  181
                        ***                      ***                      ***            ***
...Ala-Thr-Glu-Ile-Ala-His-Gln-Leu-Ser-Ile-Ala-Arg-Ser-Thr-Val-Tyr-Lys-Ile-Leu-Glu...   Tn3 Resolvase
   161  162  163  164  165  166  167  168  169  170  171  172  173  174  175  176  177  178  179  180
                        ***                      ***                      ***            ***
...Ala-Ser-His-Ile-Ser-Lys-Thr-Met-Asn-Ile-Ala-Arg-Ser-Thr-Val-Tyr-Lys-Val-Ile-Asn...   γδ Resolvase
   161  162  163  164  165  166  167  168  169  170  171  172  173  174  175  176  177  178  179  180
                        ***                      ***                      ***            ***
...Ile-Ala-Ser-Val-Ala-Gln-His-Val-Cys-Leu-Ser-Pro-Ser-Arg-Leu-Ser-His-Leu-Phe-Arg...   Ara C
   196  197  198  199  200  201  202  203  204  205  206  207  208  209  210  211  212  213  214  215
                        ***                      ***                      ***            ***
...Arg-Ala-Glu-Ile-Ala-Gln-Arg-Leu-Gly-Phe-Arg-Ser-Pro-Asn-Ala-Ala-Glu-Glu-His-Leu...   Lex R
    28   29   30   31   32   33   34   35   36   37   38   39   40   41   42   43   44   45   46   47
                        ***                      ***                      ***            ***
```

Helix 2   Helix 3

Figure 12 Sequences homologous to the α2-α3 sequences of lambda Cro and lambda repressor, and the αE-αF sequence of CAP. Sequence references: Fnr protein (92); tetracyline repressors from Tn10 and pSC101 (T. Nguyen, K. Postle, K. Bertrand, manuscript in preparation); H-inversion protein (93); transposon Tn3 resolvase (94); transposon gamma-delta resolvase (95); and arabinose C protein (96). Citations for other sequences are listed in (60–62).

sequences like AraC and Tn3 repressor for this reason. However, it has been shown that a double mutant of lambda repressor, containing Asp38 → Asn (position 6) and Gly41 → Glu (position 9) is functional (64). This proves that formation of the proper turn between helices 2 and 3 does not require glycine. What then do we make of the strong conservation of glycine at position 9? In Cro, this glycine assumes a backbone conformation (phi = 60, psi = 44) commonly observed for glycine but only rarely observed for all other amino acids (65). The side-chain hydrogen atom of glycine allows it to readily assume this conformation, whereas the larger side chains of other residues cause some steric hindrance. This suggests that residues other than glycine might be accommodated at position 9 but at the cost of some modest conformational strain.

For some of the homologous proteins, further evidence suggests that they actually do form bihelical units like those of Cro, repressor, and CAP. Circular dichroism shows that lac repressor, lambda cII protein, and P22 repressor contain substantial regions of α-helix in their DNA-binding domains (66–68) and there are mutants of each protein whose properties can be explained by the proposed helix-turn-helix model (64, 69, 70; Y. Ho, M. Rosenberg, D. Wulff, unpublished). The strongest physical evidence in support of a bihelical unit is for lac repressor. Here, NMR studies have identified tertiary interactions that are predicted by the model (71) and have identified two linked α-helices in the predicted positions (72).

It seems very likely that many DNA-binding proteins use helix-turn-helix units and the question even arises whether there are any specific DNA-binding proteins that do not use bihelical units, or at least α-helical regions, for recognition. There are many DNA-binding proteins that lack obvious homology with the sequences shown in Figure 12. For example, the Mnt and Arc repressors of bacteriophage P22 are two such proteins (73). However, structural homology can be present in proteins that lack sequence homology (74) and circular dichroism studies show that both of these small DNA-binding proteins are substantially α-helical (A. Vershon, P. Youderian, M. Susskind, R. Sauer, manuscript in preparation). Structural studies will be required to determine whether the α-helices of these proteins are used for DNA binding and, if so, whether the helical regions resemble those of CAP, Cro, and lambda repressor.

USE OF α-HELICES IN DNA RECOGNITION

Several early model-building studies predicted that α-helices could fit into the major groove of B-form DNA (75–78), and the structural information now available shows that this is a common mode of DNA recognition. An α-helix with side chains has a diameter of about 12 Å while the major groove of B-DNA is about 12 Å wide and 6–8 Å deep. Thus, one side of an α-helix can fit snuggly into the major groove. The α-helical backbone can be viewed as a scaffold from which side chains can contact the edges of the base pairs in the major groove. The number of base pairs that could be contacted by a single α-helix depends on the orientation of the α-helix with respect to the groove and on the length of the side chains. Maximum contact is obtained if the helix is parallel to the local direction of the major groove (Figure 3), and with this arrangement side chains on the helix could contact four to six base pairs. (More extensive contacts are not possible because the α-helix is straight whereas the major groove curves away in both directions.) However, model building suggests that the α-helix does not need to be precisely parallel to the major groove. A variety of different orientations would still be sterically reasonable and would allow extensive contacts with the DNA. For example, an α-helix can be positioned with one of its ends, rather than its center, closest to the double-helical axis of the DNA and the helix can be tilted by 15–20 degrees with respect to the major groove. This arrangement, which is similar to the one proposed for lambda repressor, still allows the helix to contact four or five base pairs. An arrangement with the N-terminal rather than the C-terminal end closest to the groove is probably preferred, since an α-helix has a partial positive charge at its N-terminal end, and the major groove carries a partial negative charge (3, 79, 80). Moreover, since the side chains of an α-helix point toward the N-terminal end, this arrangement should help orient the side chains for interactions with the major groove.

Since lambda repressor and Cro recognize the same operator sites, it is instructive to compare the way in which they use their α-helices for DNA binding. Alpha-helix 3 in the Cro-operator complex is almost parallel to the major groove, but its N-terminal end is somewhat closer to the DNA than its C-terminal end (23). This is clear from the pattern of side-chain contacts. Gln27, the first residue in helix 3, fits directly into the major groove and contacts the edge of a base pair. However His35, near the C-terminus of helix 3, is farther from the groove and appears to contact a phosphate. In the repressor-operator complex (38), helix 3 is not as closely parallel to the major groove, but the overall arrangement, including the pattern of side-chain contacts, is similar. For example, as shown in Figure 7, each residue position in the α2-α3 region of repressor that is proposed as a specific or backbone contact is also proposed to make a similar type of contact in Cro. However, the individual side chain contacts made by the two proteins are actually quite different. Half of the common contact positions have different residues and even where the contact residues are identical there are important differences in the proposed complexes. Consider the contacts made by the Gln27-Ser28 side chains of Cro and the homologous Gln44-Ser45 side chains of repressor (Figure 7). The glutamines are both predicted to make the same contacts with adenine +2 [Figure 4 (*top*); (23, 38)]. However the contacts predicted for the serines are different. Ser28 of Cro appears to contact adenine −3 [Figure 4 (*center*)], whereas Ser45 of respressor appears to donate a single hydrogen bond to the N7 of guanine −4. These different predictions arise from differences in the structures of the two protein dimers. In the Cro dimer the α-carbons of Ser28 are some 6 Å farther apart than are the α-carbons of Ser45 in the repressor dimer (59). The α-carbons of Gln27 (in Cro) and Gln44 (in repressor) are also separated by different distances in the two proteins, but these residues can still make the same contacts with adenine +2 because the glutamine side chains, being longer, can reach this base.

Although helix 3 of Cro, helix 3 of repressor, and helix F of CAP are clearly used for recognition of sites in the major groove, it is less clear what role the preceding helices (2 or E) serve and why the two helices are so strictly conserved as a bihelical unit. In the proposed complexes, the axis of helix 2 or helix E is almost perpendicular to the sugar phosphate backbone and the partial positive charges at the N-terminal ends of these helices are close to the phosphates. This should provide a favorable electrostatic interaction. In addition, Gln16 at the N-terminal end of Cro's helix 2 and the corresponding Gln33 at the N-terminal end of repressor's helix 2 appear to hydrogen bond to the phosphates. As discussed with respect to the Gln33 → Ser mutant of lambda repressor, such backbone contacts appear to be directly responsible for some binding specificity. From a structural point of view, these contacts may serve as "clamps" that keep helix 3 from rolling in the groove. By correctly orienting the helices and side chains in the major groove, the backbone contacts could increase the specificity of the interactions with the base pairs.

OTHER MODES OF INTERACTION

Although the recent structural and genetic studies suggest that most of the site-specific contacts are made by residues from the α-helical regions of Cro, repressor and CAP, some contacts seem to be made by regions with an irregular or extended structure. For example, Arg38 of Cro and Asn55 of repressor are thought to make specific major groove contacts and both lie beyond the C-terminal end of helix 3. Since the Cro and repressor α-helices contact only four to five adjacent bases within the major groove, the use of a contact from a nonhelical region allows each protein to contact an additional contiguous base-pair.

Lambda repressor's flexible N-terminal arm provides another example of an extended structure that is used in protein-DNA recognition. The use of such a structure was first suggested by Feughelman et al (81), who proposed that an extended polypeptide chain could wrap around a double helix and bind in one of the grooves. As previously noted (37), this use of a flexible arm allows repressor to contact both sides of the DNA helix without creating a large kinetic barrier to association, as the arm can wrap around the DNA after the globular portion of the protein has bound to the front of the operator site. However, the use of flexible binding regions may limit specificity by allowing alternative contacts with different sequences, and may also limit interaction energies (82). For example, since repressor's N-terminal arms are disordered in solution but adopt a specific conformation upon binding the operator, their net binding energies will be reduced by the entropic cost of fixing their positions in the complex. The C-terminal residues of Cro apparently provide a second example of a flexible region of protein that is used to make DNA contacts.

Model-building studies have also suggested that β-sheets might be used to bind double-stranded DNA. The right-handed twist of a β-sheet should allow a pair of antiparallel β-strands to fit into the minor (83, 84) or the major groove (85) of B-form DNA. Although these proposals seem plausible, there is no structural evidence to indicate that β-sheets are actually used in site-specific recognition. Initial inspection of the Cro structure (1) had suggested that the antiparallel β-ribbon that is involved in the Cro dimer contacts might bind to the minor groove. However, more detailed model building studies (23) suggest that the β-ribbon does not lie in the minor groove and suggest that contacts from this region are limited to interactions between the side chains and the phosphate backbone.

PROSPECTS FOR A "RECOGNITION CODE"

Even in the absence of high-resolution information from co-crystals it is possible to make a number of reasonable guesses about the general nature of any "recognition code." The structural information, model-building studies, and genetic data make it almost certain that hydrogen bonds between side chains and the edges of base pairs are responsible for much of the specificity in protein-DNA interactions. This has always seemed reasonable, since hydrogen bonds are highly dependent on the position and orientation of the donor and acceptor groups, and since hydrogen bonds are responsible for specificity in so many other biological interactions.

In principle, it could have been possible that site-specific binding proteins used a simple "recognition code," involving a one-to-one correspondence between the amino acid side chains and the bases in the DNA. For example, since a glutamine side chain in the major groove can make two hydrogen bonds to adenine, and arginine can make two hydrogen bonds to guanine [see Figure 4 (*top* and *bottom*); and (86)], it was conceivable that glutamine would always be used to recognize an A : T base pair and that arginine would always be used to recognize a G : C pair. However, no one-to-one recognition code is consistent with the current data, and it seems inconceivable that any simple code could have escaped notice with all of the structural and genetic information that is now available.

The proposed hydrogen-bonding interactions for Cro, repressor, and CAP seem to indicate that the "recognition code" is degenerate; i.e. it seems that each base pair can be recognized by several different amino acids and that each amino acid can bind to several different bases. The repressor-operator and Cro-operator models certainly support this idea, since adenine is recognized by glutamine (in both complexes), but it is also recognized by serine (in the Cro complex). Serine binds to adenine (in the Cro model), but serine also binds to guanine (in the repressor model). If specificity is still to be maintained, "degeneracy" of this type implies that the "meaning" of a particular amino acid will depend on the conformation and orientation of the protein backbone. This is not surprising. It actually would be impossible to have any simple repeating pattern of contacts made by one α-helix, since the periodicity of an α-helix has no simple relationship to the periodicity of B-DNA. Thus far, in the three proposed complexes a variety of side chains including those of Gln, Asn, Ser, Tyr, Arg, Lys, Glu, Thr, and His have been used to make hydrogen bonds in the major groove or with the DNA backbone. It is likely that these residues will be commonly used for DNA recognition.

However, van der Waals interactions also seem to be an important part of the recognition process. In repressor, the methyl group of Ala49 makes one of the specific major groove interactions and, in CAP, replacement of Glu181 by Leu or Val changes the specificity of DNA binding. Several of the hydrogen-bonding side chains in Cro also seem to make significant van der Waals contacts with the operator. Although the favorable energies obtained from van der Waals interactions, hydrogen bonding, or electro-static interactions are important aspects of "recognition," the overall fit of the protein and DNA surfaces is also extremely important. For example, Gly48 in repressor's helix 3 seems to play a passive role in recognition since a larger side chain at this position would cause an unfavorable steric contact between the protein and DNA. Thus the "lock and key" analogy that describes the fit of substrates to enzymes also seems to apply to protein-DNA interactions.

At this stage, it is difficult to guess how many different bonding patterns will be used in recognition, and thus we cannot know how "degenerate" the "recognition code" actually is. However, it is still conceivable that the list of possible interactions will be small enough so that the "code" will have a predictive value and could be used to locate DNA-binding regions within a protein sequence or to predict the preferred DNA sites to which a particular protein binds.

SUMMARY

Several general principles emerge from the studies of Cro, lambda repressor, and CAP.

1. The DNA-binding sites are recognized in a form similar to B-DNA. They do not form cruciforms or other novel DNA structures. There seem to be proteins that bind left-handed Z-DNA (87) and DNA in other conformations, but it remains to be seen how these structures are recognized or how proteins recognize specific sequences in single-stranded DNA.

2. Cro, repressor, and CAP use symmetrically related subunits to interact with two-fold related sites in the operator sequences. Many other DNA-binding proteins are dimers or tetramers and their operator sequences have approximate two-fold symmetry. It seems likely that these proteins will, like Cro, repressor, and CAP, form symmetric complexes. However, there is no requirement for symmetry in protein-DNA interactions. Some sequence-specific DNA-binding proteins, like RNA polymerase, do not have symmetrically related subunits and do not bind to symmetric recognition sequences.

3. Cro, repressor, and CAP use α-helices for many of the contacts between side chains and bases in the major groove. An adjacent α-helical region contacts the DNA backbone and may help to orient the "recognition" helices. This use of α-helical regions for DNA binding appears to be a common mode of recognition.

4. Most of the contacts made by Cro, repressor, and CAP occur on one side of the double helix. However, lambda repressor contacts both sides of the double helix by using a flexible region of protein to wrap around the DNA.

5. Recognition of specific base sequences involves hydrogen bonds and van der Waals interactions between side chains and the edges of base pairs. These specific interactions, together with backbone interactions and electrostatic interactions, stabilize the protein-DNA complexes.

The current models for the complexes of Cro, repressor, and CAP with operator DNA are probably fundamentally correct, but it should be emphasized that model building alone, even when coupled with genetic and biochemical studies, cannot be expected to provide a completely reliable "high-resolution" view of the protein-DNA complex. For example, the use of standard B-DNA geometry for the operator is clearly an approximation. Recent studies of B-DNA duplexes have revealed sequence-dependent variations in local structure that could affect protein recognition (88–90). Small changes in protein structure, which may occur upon binding to the DNA, could also affect the detailed structure of the complex. Crystallographic studies of the Cro-operator (91) and repressor-operator complexes (38), which are now in progress, should be extremely helpful in evaluating and refining the current models.

ACKNOWLEDGMENTS

We thank our many colleagues for advice, helpful discussions, and unpublished information. Work performed in our laboratories was supported by grants to C. O. P. from the American Cancer Society and the NIH (GM-31471) and to R. T. S. from the NIH (AI-16892, AI-15706).

Literature Cited

1. Anderson, W. F., Ohlendorf, D. H., Takeda, Y., Matthews, B. W. 1981. *Nature* 290:754–58
2. McKay, D. B., Steitz, T. A. 1981. *Nature* 290:744–49
3. Pabo, C. O., Lewis, M. 1982. *Nature* 298:443–47
4. Ptashne, M., Jeffrey, A., Johnson, A. D., Maurer, R., Meyer, B. J., et al. 1980. *Cell* 19:1–11
5. Johnson, A. D., Poteete, A. R., Lauer, G., Sauer, R. T., Ackers, G., Ptashne, M. 1981. *Nature* 294:217–23
6. Ptashne, M., Johnson, A. D., Pabo, C. O. 1982. *Sci. Am.* 247:128–40
7. Adhya, S., Garges, S. 1982. *Cell* 29:287–89
8. Ohlendorf, D. H., Matthews, B. W. 1983. *Ann. Rev. Biophys. Bioeng.* 12:259–84
9. Gussin, G., Johnson, A. D., Pabo, C. O., Sauer, R. T. 1983. In *Lambda II*, ed. R. W. Hendrix, J. W. Roberts, F. W. Stahl, R. A. Weisberg, pp. 93–121. Cold Spring Harbor, NY: Cold Spring Harbor Press
10. Takeda, Y., Ohlendorf, D. H., Anderson, W. F., Matthews, B. W. 1983. *Science* 221:1020–26
11. von Hippel, P. H., Bear, D. G., Morgan, W. D., McSwiggen, J. A. 1984. *Ann. Rev. Biochem.* 53:389–446
12. von Hippel, P. H. 1979. In *Biological Regulation and Development*, ed. R. F. Goldberger, 1:279–347. New York: Plenum
13. Helene, C., Lancelot, G. 1982. *Prog. Biophys. Mol. Biol.* 39:1–68
14. Maniatis, T., Ptashne, M., Backman, K.,
C., Kleid, D., Flashman, S., Jeffrey, A., Maurer, R. 1975. *Cell* 5:109–13
15. Johnson, A. D., Meyer, B. J., Ptashne, M. 1978. *Proc. Natl. Acad. Sci. USA* 75:1783–87
16. Johnson, A. D. 1980. PhD thesis. Harvard Univ., Cambridge, Mass.
17. Johnson, A. D. 1983. *J. Mol. Biol.* In press
18. Hsiang, M. W., Cole, R. D., Takeda, Y., Echols, H. 1977. *Nature* 270:275–77
19. Roberts, T. M., Shimatake, H., Brady, C., Rosenberg, M. 1977. *Nature* 270:274–75
20. Takeda, Y., Folkmanis, A., Echols, H. 1977. *J. Biol. Chem.* 252:6177–83
21. Watson, J. D., Crick, F. H. C. 1953. *Nature* 171:736–38
22. Arnott, S., Hukins, D. W. L. 1972. *Biochem. Biophys. Res. Commun.* 47:1504–9
23. Ohlendorf, D. H., Anderson, W. F., Fisher, R. G., Takeda, Y., Matthews, B. W. 1982. *Nature* 298:718–23
24. Jack, A., Levitt, M. 1971. *Acta Cryst. A* 34:931–35
25. Wang, J. 1979. *Proc. Natl. Acad. Sci. USA* 76:200–3
26. Bloomfield, V. A., Crothers, D. M., Tinoco, I. 1974. *Physical Chemistry of Nucleic Acids.* New York: Harper & Row
27. Matthew, J. B., Richards, F. M. 1982. *Biochemistry* 21:4989–99
28. Ohlendorf, D. H., Anderson, W. F., Takeda, Y., Matthews, B. W. 1983. *J. Biomol. Struct. Dynam.* In press
29. Chadwick, P., Pirrotta, V., Steinberg, R., Hopkins, N., Ptashne, M. 1970. *Cold*

Spring Harbor Symp. Quant. Biol. 35: 283–94

30. Sauer, R. T. 1979. *Molecular Characterization of the Lambda Repressor and its Gene cI.* PhD thesis. Harvard Univ., Cambridge, Mass.

31. Johnson, A. D., Meyer, B. J., Ptashne, M. 1979. *Proc. Natl. Acad. Sci. USA* 76: 5061–5

32. Humayun, Z., Kleid, D., Ptashne, M. 1977. *Nucleic Acids. Res.* 4: 1595–607

33. Sauer, R. T., Anderegg, R. 1978. *Biochemistry* 17: 1092–1100

34. Pabo, C. O., Sauer, R. T., Sturtevant, J. M., Ptashne, M. 1979. *Proc. Natl. Acad. Sci. USA* 76: 1608–12

35. Sauer, R. T., Pabo, C. O., Meyer, B. J., Ptashne, M., Backman, K. C. 1979. *Nature* 279: 396–400

36. Maniatis, T., Ptashne, M. 1973. *Proc. Natl. Acad. Sci. USA* 70: 1531–35

37. Pabo, C. O., Krovatin, W., Jeffrey, A., Sauer, R. T. 1982. *Nature* 298: 441–43

38. Lewis, M., Jeffrey, A., Wang, J., Ladner, R., Ptashne, M., Pabo, C. O. 1983. *Cold Spring Harbor Symp. Quant. Biol.* 47: 435–40

39. Record, M. T., Lohman, T. M., deHaseth, P. L. 1976. *J. Mol. Biol.* 107: 145–58

40. Nelson, H. C. M., Hecht, M. H., Sauer, R. T. 1983. *Cold Spring Harbor Symp. Quant. Biol.* 47: 441–49

41. Hecht, M. H., Nelson, H. C. M., Sauer, R. T. 1983. *Proc. Natl. Acad. Sci. USA* 80: 2676–80

42. Skopek, T. R., Hutchinson, F. 1982. *J. Mol. Biol.* 159: 19–33

43. Wood, R. D., Skopek, T. R., Hutchinson, F. 1983. *J. Mol. Biol.* In press

44. de Crombrugghe, B., Busby, S., Buc, H. 1983. In *Biological Regulation and Development,* ed. K. Yamomoto, Vol. 36. New York: Plenum. In press

45. Majors, J. 1978. PhD thesis. Harvard Univ., Cambridge, Mass.

46. Ebright, R. H. 1982. In *Molecular Structure and Biological Function,* ed. J. Griffen, W. Duax, p. 91. New York: Elsevier/North Holland

47. Aiba, H., Fujimoto, S., Ozaki, N. 1982. *Nucleic Acids Res.* 10: 1345–61

48. Cossart, P., Gicquel-Sanzey, B. 1982. *Nucleic Acids Res.* 10: 1363–68

49. Aiba, H., Krakow, J. S. 1981. *Biochemistry* 20: 4774–80

50. Fried, M. G., Crothers, D. M. 1983. *Nucleic Acids Res.* 11: 141–58

51. McKay, D. B., Weber, I. T., Steitz, T. A. 1982. *J. Biol. Chem.* 257: 9518–24

52. Ebright, R. H., Wong, J. R. 1981. *Proc. Natl. Acad. Sci. USA* 78: 4011–15

53. Salemme, F. R. 1982. *Proc. Natl. Acad. Sci. USA* 79: 5263–67

54. Steitz, T. A., Weber, I. T. 1983. In *Structures of Biological Molecules and Assemblies,* ed. F. Jurnak, A. McPherson. New York: Wiley. In press

55. Steitz, T. A., Weber, I. T., Matthew, J. B. 1983. *Cold Spring Harbor Symp. Quant. Biol.* 47: 419–26

56. Gupta, G., Bansal, M., Sasisekharan, V. 1980. *Proc. Natl. Acad. Sci. USA* 77: 6486–90

57. Kolb, A., Buc, H. 1982. *Nucleic Acids Res.* 10: 473–85

58. Steitz, T. A., Ohlendorf, D. H., McKay, D. B., Anderson, W. F., Matthews, B. W. 1982. *Proc. Natl. Acad. Sci. USA* 79: 3097–3100

59. Ohlendorf, D. H., Anderson, W., Lewis, M., Pabo, C. O., Matthews, B. W. 1983. *J. Mol. Biol.* 169: 757–69

60. Matthews, B. W., Ohlendorf, D. H., Anderson, W. F., Takeda, Y. 1982. *Proc. Natl. Acad. Sci. USA* 79: 1428–32

61. Sauer, R. T., Yocum, R. R., Doolittle, R. F., Lewis, M., Pabo, C. O. *Nature* 298: 447–451

62. Weber, I. T., McKay, D. B., Steitz, T. A. 1982. *Nucleic Acids Res.* 10: 5085–5102

63. Schiffer, M., Edmundson, A. B. 1967. *Biophys. J.* 7: 121–35

64. Hochschild, A., Irwin, N., Ptashne, M. 1983. *Cell* 32: 319–25

65. Ohlendorf, D. H., Anderson, W. F., Matthews, B. W. 1983. *J. Mol. Evol.* 19: 109–14

66. Geisler, N., Weber, K. 1977. *Biochemistry* 16: 938–43

67. Ho, Y., Lewis, M., Rosenberg, M. 1982. *J. Biol. Chem.* 257: 9128–34

68. Sauer, R. T., Nelson, H. C. M., Hehir, K., Hecht, M., Gimble, F. S., DeAnda, J., Poteete, A. R. 1983. *J. Biomol. Struct. Dynam.* In press

69. Miller, J. H. 1978. In *The Operon,* ed. J. H. Miller, W. Reznikoff, pp. 31–88. Cold Spring Harbor, NY: Cold Spring Harbor Lab.

70. Beyreuther, K. 1978. See Ref. 69, pp. 123–54

71. Arndt, K. T., Boschelli, F., Lu, P., Miller, J. H. 1981. *Biochemistry* 20: 6109–18

72. Zuiderweg, E. R. P., Kaptein, R., Wuthrich, K. 1983. *Proc. Natl. Acad. Sci. USA* 80: 5837–41

73. Sauer, R. T., Krovatin, W., DeAnda, J., Youderian, P., Susskind, M. M. 1983. *J. Mol. Biol.* 168: 699–713

74. Matthews, B. W., Grutter, M. G., Anderson, W. F., Remington, S. J. 1981. *Nature* 290: 334–35

75. Zubay, G., Doty, P. 1959. *J. Mol. Biol.* 1: 1–20

76. Sung, M. T., Dixon, G. H. 1970. *Proc. Natl. Acad. Sci. USA* 67: 1616–23

77. Adler, K., Beyreuther, K., Fanning, E.,

Geisler, N., Gronenborn, B., et al. 1972. *Nature* 237: 322–27

78. Warrant, R. W., Kim, S.-H. 1978. *Nature* 271: 130–35
79. Hol, W. G. J., van Duijnen, P. T., Berendsen, H. J. C. 1978. *Nature* 273: 443–47
80. Pullman, A., Pullman, B. 1981. *Quart. Rev. Biophys.* 14: 289–380
81. Feughelman, M., Langridge, R., Seeds, W. E., Stokes, A. R., Wilson, H. R., et al. 1955. *Nature* 175: 38
82. Huber, R. 1979. *Trends Biochem. Sci.* 4: 271–76
83. Carter, C. W., Kraut, J. 1974. *Proc. Natl. Acad. Sci. USA* 71: 283–87
84. Church, G., Sussman, J. L., Kim, S.-H. 1977. *Proc. Natl. Acad. Sci. USA* 74: 1458–62
85. Blake, C. C. F., Oatley, S. J. 1977. *Nature* 268: 115–20
86. Seeman, N. C., Rosenberg, J. M., Rich, A. 1976. *Proc. Natl. Acad. Sci. USA* 73: 804–8
87. Nordheim, A., Tesser, P., Azorin, F.,

Kwon, Y. H., Moller, A., Rich, A. 1982. *Proc. Natl. Acad. Sci. USA* 79: 7729–33
88. Dickerson, R. E., Drew, H. R. 1981. *J. Mol. Biol.* 149: 761–86
89. Dickerson, R. E. 1983. *J. Mol. Biol.* 166: 419–41
90. Patel, D. J., Ikuta, S., Kozlowski, S., Itakura, K. 1983. *Proc. Natl. Acad. Sci. USA* 80: 2184–88
91. Anderson, W. F., Cygler, M., Vandonfelaar, M., Ohlendorf, D., Matthews, B. W., et al. 1983. *J. Mol. Biol.* 168: In press
92. Shaw, D. J., Guest, J. R. 1982. *Nucleic Acids Res.* 10: 6119–30
93. Simon, M., Zieg, J., Silverman, M., Mandel, G., Doolittle, R. F. 1981. *Science* 209: 1370–74
94. Heffron, F., McCarthy, B. J., Ohtsubo, H., Ohtsubo, E. 1979. *Cell* 18: 1153–63
95. Reed, R. R., Shibuya, G. I., Steitz, J. A. 1982. *Nature* 300: 381–83
96. Miyada, C. G., Horwitz, A. H., Cass, L., Timko, J., Wilcox, G. 1980. *Nucleic Acids Res.* 8: 5267–65

Ann. Rev. Biochem. 1982.51:617–54

15

SnRNAs, SnRNPs, AND RNA PROCESSING

Harris Busch, Ramachandra Reddy, Lawrence Rothblum, and Yong C. Choi[1]

Department of Pharmacology, Baylor College of Medicine, Houston, Texas 77030

CONTENTS

[1]We are grateful to all those who sent us preprints of their work, and to our colleagues in this and other laboratories for suggestions and comments on the manuscript. Work in our laboratory has been supported by grants from Cancer Center Program Grant 10893, awarded by DHEW; the Bristol-Myers Fund; the Michael E. DeBakey Medical Foundation; the Pauline Sterne Wolff Memorial Foundation; the Taub Foundation; and The William S. Farish Fund.

PERSPECTIVES AND SUMMARY

On rare occasions, convergences occur in science that lead to a clarification of several seemingly unrelated lines of investigation. Such a series of events has linked the small nuclear RNAs (snRNAs) and small nuclear ribonucleoproteins (snRNPs) (reviewed in 1–10) to splicing of premessenger RNA (pre-mRNA) transcripts (11–18) and to autoimmune diseases (19–23). From these different avenues of research, the hypothesis emerged that one function of the U-snRNP class of snRNP particles is to guide the elimination of intervening sequences or introns (IVS) of pre-mRNA by binding to splice sites of the IVS in heterogeneous nuclear RNP (hnRNP) particles. The mechanism of excision of the IVS, and ligation of the useful portions of the mRNA are not yet known, nor is it clear if the snRNPs are involved in transport of mRNA to the cytoplasm and incorporation into polysomes (12) as their primary function.

Several groups (11–13) working on viral and eukaryotic mRNA have suggested such a role in splicing for snRNP particles (Figure 1). Lerner et al (12) evaluated the hypothesis based upon calculations of numbers of hydrogen bonds between U1 RNA and IVS cleavage sites, common features of the splice junctions, and details of interactions of small and large nuclear particles. The rigorous testing of the hypothesis should provide a satisfactory evaluation of the concept and the biochemical mechanisms involved in splicing. There are serious technical problems in developing assay systems for bona fide processing of hnRNA. Now that sequencing and characterization of the U-snRNAs are essentially complete, the U-snRNPs are being considered as possible macromolecular aggregates with specific cleavage and ligase functions. When these particles are available in native states, it will be essential to devise processing reactions utilizing pre-mRNA to test their potential functions.

SMALL RNAs

Historical Aspects

In early studies (24) on the nucleolar U-snRNA species an RNA fraction was found in the 4-8S region of a sucrose gradient that had different base composition from that of the rRNA or tRNA. Due to its high content of

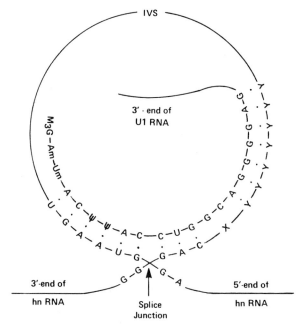

Figure 1 A proposed model for splicing of hnRNA, involving U1 snRNA. The model is derived from others (12, 13). A similar model involving viral precursor RNA and viral snRNA has been proposed (11). The consensus sequences for the splice junctions are from (12, 15) and the sequence of U1 RNA (8, 63, 118) were reported earlier.

uridylic acid it was designated U-RNA (24–29). There was initial concern that such small RNA species could be degradation products of higher-molecular-weight RNA as had been the case for "chromosomal RNA" and other small RNA products (30, 31). To control nuclear RNases is difficult partly because a number of them have specific processing functions; they differ from cytoplasmic contaminants in that they are released during isolation of nuclear subfractions. Some of these problems have been overcome using improved methods (32, 33) and RNase inhibitors.

By 1966, snRNA had been observed (34–36), but specific RNA species had not been characterized although Reich et al (37) had identified a small viral RNA. Subsequently, Pene et al (38) isolated 5.8S RNA, and other snRNAs were identified (39–51).

After studies were made on RNase contents of nuclear preparations, we chose the Novikoff hepatoma as an optimal tissue for study (32). When "RNase free" citric acid nuclei were employed (33), single species of highly purified U1 and U2 snRNAs could be isolated (25). The purification, terminal nucleoside analysis, high uridylic acid content and evidence that U-snRNA's are specifically localized were reported (25–29, 49). U3 RNA was found only in the nucleolus and not in the nucleoplasm (25–29). These results were consistent with those using gel electrophoresis (49, 51). In the absence of chemical evidence, it was not clear whether the gel bands were the same products we had isolated or were degradation products of other RNA species. It was also not apparent how many were isomeric structures like U1A and U1B RNA (4,8).

In other studies, the U3 RNA was purified by a number of techniques. Nine lines of evidence proved that the RNAs were not breakdown products of other RNA species. One line of evidence (49, 24) was that the turnover rates were very low by comparison to other species, particularly rRNA.

Despite the reproducibility of the isolation procedures, the only satisfactory way to show that U-snRNAs are unique species, and not specific cleavage products or artifacts of degradation has been to determine their sequences.

Conservation Through Evolution

Small RNAs are present in viruses and in prokaryotic and eukaryotic cells, but their gel electrophoretic patterns are different. After snRNAs were found in human cells (40, 41, 43–45, 49–54); rat tissues (24–29, 54, 55); Chinese hamster (50); and mouse (41, 42, 56, 57), systematic studies concluded that snRNAs are present in all vertebrates (49–59). snRNAs are independent of cell type and malignancy (49, 61). U1, U2, and U3 RNAs appear to be highly conserved (6, 41, 42, 62, 63). Sequence analysis of U1 RNA structures of chicken, rat, and human tissues show it to be over 95% conserved. There are only two nucleotide differences between rat and human U1 RNAs (63). Notable similarities have been found for the sequence of U1 RNA of *Drosophila* (63a) with other U1 RNA sequences.

Detailed studies (39, 41, 42) on nonvertebrates, e.g. cockroach, meal worm, blowfly, sea urchins, *Tetrahymena,* and *Chironymus tentans* revealed that they all contain small RNAs, but that their mobility on gels is different from vertebrate snRNAs. U1, U2, U3 RNAs of Dictyostelium contain a cap structure similar to that found in rat U-snRNAs (64, 64a).

The U3 RNA sequence of rat (65) is homologous to that of U3 RNA of *Dictyostelium* (66), which indicates the similarity in structure of U-snRNAs of vertebrates and invertebrates. The characteristics of small RNAs are shown in Table 1. Table 2 lists the major small RNAs, their 5'-termini, chain lengths, and their alternate nomenclature.

Most recently, we have found that dinoflagellates, which are considered to be eukaryotes that evolved 3 billion years ago, also contain U1-U6 snRNAs or very similar capped molecular species (R. Reddy, D. Spector, D. Henning, and H. Busch, unpublished results). Diener (66a) has suggested that base-pairing of U1 RNA to intron "ancestors" may be part of a mechanism for viroid formation.

Table 1 General characteristics of U SnRNAs

Size range 90–400 nucleotides

Copies per cell 1 × 10^6 molecules for the most abundant RNAs (e.g. U1 RNA);
1 × 10^4 to 5 × 10^5 for other RNAs

Half-lives U snRNAs are stable; half-lives of up to one cell cycle

Specific localization localized to specific subcellular compartments (e.g. U3 RNA
in nucleolus; U1, U2, U4, U5, U6, and La 4.5 RNAs in nucleoplasm; and Y
RNAs in cytoplasm)

Polymerases capped snRNAs U1 to U6 synthesized by pol II, and other small
RNAs synthesized by pol III

Cap structure U1 to U6 snRNAs are capped, U1 to U5 RNAs contain a trimethylguanosine cap, and U6 RNA has a different cap

RNP particles all U snRNAs exist in RNP particles

Associated with precursor RNAs The U snRNAs are associated with precursor
hnRNAs and pre-rRNA

Synthesis of Small RNAs

Transfer RNA (tRNA) and 5S RNA are synthesized by RNA polymerase III (67–69); 5.8S RNA is synthesized as part of 45S ribosomal RNA precursor by RNA polymerase I (70–72). RNA polymerase II, which synthesizes hnRNA (73), also synthesizes U-snRNA (74–78). The sensitivity of U1 and U2 RNA biosynthesis to α-amanitin in whole cel systems (73a–76) and in cell-free systems (78), and to 5-6-dichloro-1-β-D-ribofuranosyl benzimidazole (77) also indicates that polymerase II is responsible for U1 and U2 RNA biosynthesis. snRNAs may be synthesized by RNA polymerase I (52), but there is no confirmation of this. 7S RNA, 7-3 RNA (52, 79), La 4.5 RNA (80) and Y RNAs (81) have been reported to be synthesized by polymerase III. Thus it appears that capped small U1–U6 RNAs are synthesized by polymerase II, and other small RNAs are synthesized by polymerase III.

Precursors of snRNAs

There are precursors of U1 and U2 RNA that are slightly larger than the mature U-snRNA's. They are detected in the cytoplasmic fraction within 10 min of [^{3}H]uridine labeling (51, 82, 82a). Precursors of U1 and U2 RNAs have been analyzed by fingerprinting (83). Since these RNAs are capped, the larger size of the precursors is presumed to reflect elongation at 3' ends. Salditt-Georgieff et al (84) found $m^{2,2,7}_3G$ in the cap I structures isolated from the <750-base long nuclear RNA fraction of Chinese hamster ovary cells labeled briefly with [methyl-^{3}H]methionine. These data suggest that longer precursor molecules for snRNAs exist than were previously reported. Tamm et al (77) also found evidence for long precursors for

Table 2   Termini and chain lengths of snRNAs

RNA	5' terminus	Chain length	Alternative nomenclature
U snRNA			
U1	m₃GpppAmUmA	165	D
U2	m₃GpppAmUmC	188–189	C
U3	m₃GpppAmAG	210–214	A, D2
U4	m₃GpppAmGmC	142–146	F
U5	m₃GpppAmUmA	116–118	G', 5S III
U6	XpppGUG	107–108	H1, 4.5 III
Other snRNA			
tRNA	pN	~80	
5S	pppN	121	G
5.8S	pN	158	E
La 4.5	pppG	90–94	
La 4.5 I	pppG	98–99	H2
7S	pppG	294–295	L
VAI, II	pppG	157–160	
7–1	pppN	~260	M
7–2	pppN	~290	M
7–3	pppN	~300	K
8S	pppN	~400	
Y1–Y3	pppN	~100	

snRNAs but did not find trimethylguanosine in hnRNAs longer than 1000 nucleotides. U1 and U2 RNAs have been reported to be derived from transcription units that may be as long as 5 kb (85). Precursors to U4, U5, or U6 RNAs have not been reported. As yet none of the precursors has been sequenced.

Genes Coding for SnRNA

DNA clones prepared for U1 RNA include chicken U1 RNA (78), human U1 RNA (86–89), and rat U1 RNA (90). The clones are prepared in the usual cloning vectors and analyzed for sequence homology to the reported U1 RNA. No IVS were found in any of the U1, U2, or U3 genes studied; in addition, the genes are not clustered as reported earlier, but rather are dispersed throughout the genome (66, 78, 86–90). Genes for the U1 RNA and tRNA are found in proximity in the same Eco R1 fragment; the gene sequence is identical to the structure of human U1 RNA (89). For other "pseudogenes," the sequences do not match precisely with those of the U1 RNA (86–88). These pseudogenes are dispersed throughout the genome in an approximate ratio of 10:1 compared to the true genes (86); some of them are flanked by 16–19-nucleotide long direct repeats (87), which are: AGAAACAGGCTTTTGC for U1 pseudogene, TAAATAATCAG-GATGGAA for U2 pseudogene, and TAAAATGCTAATTATCCAA for U3 pseudogene. Genes for U2 RNA have been isolated from human (86) and mouse (R. C. Huang, personal communication). The mouse U2 RNA gene sequence is identical to the mouse U2 RNA sequence (8, 19) and does not contain any IVS.

The first snRNA gene to be isolated and sequenced was the U3 RNA gene from *Dictyostelium* (66). The gene coding for *Dictyostelium* U3 RNA is identical in sequence to *Dictyostelium* RNA. Recently, genes for U6 RNA have been cloned and sequenced along with their flanking regions (89a, 89b). One had a TATAAT 31 nucleotides upstream of the coding region. The flanking regions had heterogeneity immediately outside the genes. There is uncertainty about the presence of a Hogness or TATAAA box in small RNA genes. If there is, it is further upstream from the cap site than the 25 bp of most mRNA genes.

Other features of sequenced eukaryotic genes, such as the putative adenylation signal (AATAAA, 25 bp prior to the 3' end of the RNA) and the termination signal (TTT, at the 3' end of the RNA) are not found for the U1 gene (78). As expected, the adenylation signal is missing, since U1 RNA is not polyadenylated. It is interesting that the termination signal is missing, since the presence of this signal seems to be a general feature of genes transcribed by RNA polymerases II and III (78). The lack of a Hogness box and of the putative termination signal (78) is consistent with the possibility that U1 RNA and other snRNAs are transcribed as part of a larger primary transcript.

Are the genes for the isolated and sequenced snRNAs the same as those transcribed in vivo? This is, at present, difficult to answer for any individual member of a multigene family.

Early studies estimated the number of genes coding for U1, U2, and U3 RNAs by DNA-RNA hybridization (91, 92). In baby hamster cells, Engberg et al (91) have suggested that 2000 genes code for each snRNA. Marzluff et al (92) have analyzed mouse genome sequences for snRNAs and have suggested that there are 100–2000 copies per genome. However, these values are much higher than the approximately 10 determined by cloning studies, and probably include many pseudogenes. The genes for U-snRNAs are unusual in not having a defined TATAAA box, IVS, and the usual termination signals.

Recently, Nojima and Kornberg (unpublished) isolated genes and pseudogenes for mouse U1 and U2 snRNAs. In *Xenopus* oocytes, the U2 RNA gene was transcribed with apparent fidelity but the pseudogenes were inactive.

Subcellular Localization

Several studies have been made (24–29, 49, 51, 76, 82, 93–96) to find the subcellular localization of different small RNAs (Figure 2). The methods used to fractionate the nuclei from cytoplasm include aqueous (21–25, 49, 51) and nonaqueous methods (76, 93, 96), and the citric acid method (95). The results obtained by different investigators for the location of small RNAs vary slightly; however, the data point to the localization of small RNAs as shown in Figure 2. The U- RNPs (U1, U2, U4, U5, and U6) are nucleoplasmic, as shown by indirect immunofluorescence using specific antibodies directed against these RNPs (20). Similar studies show La 4.5 and 4.5I RNPs to be located in the nucleus and Y RNPs to be located in the cytoplasm (81).

Association of snRNAs with Precursor RNAs

Studies made to further locate the snRNAs show that all capped snRNAs are associated with precursor RNAs. U3 RNA is hydrogen bonded to nucleolar 28S and 32S RNAs, and this bonding is stable to treatment with phenol and SDS at 25° (97). U3 was the first snRNA to be shown associated

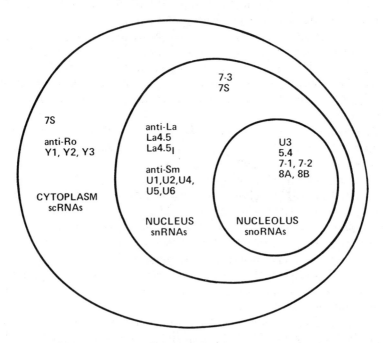

Figure 2 Diagrammatic representation of subcellular localization of small RNAs of rat. The three categories of small RNAs, based on subcellular localization, are small cytoplasmic RNAs (scRNAs), small nuclear RNAs (snRNAs), and small nucleolar RNAs (snoRNAs).

with precursor RNas. The nucleoplasmic U1 and U2 RNAs are found in hnRNP particles (98) as are other snRNAs (99–112a). U4, U5, and U6 RNAs are also bound to hnRNP particles (111). Other studies show U6 RNA in perichromatin granules (113, 114), 7S RNA bound to mRNA in polysomes (115), 7S and 7-3 RNA in polysomes (116), La 4.5 RNA hydrogen bonded to hnRNA (117), 7-3 RNA in the nuclear matrix (117a), and 7-1 and 8S RNA hydrogen bonded to nucleolar 28S RNA (95).

SEQUENCES OF UsnRNAs

U1 RNA

The initial sequence data of U1 RNA (see Table 1 for alternative nomenclature), which is in the highest concentration in the nucleus (Figure 3) revealed that its 5' cap contained the trimethyl G residue (118, 119). This was the first report of a cap structure on any RNA species, and was followed by studies on cap of mRNA and viral RNA (120, 121).

The U1 RNA sequence contains a large number of uridine residues many of which are clustered (118). Gel analysis of 4-8S RNA separated U1 RNA into two species (122), and the faster moving U1A RNA was sequenced (118). Using rapid sequencing methods (63, 78, 123a) a number of modifications were made of the original sequence of U1 RNA (Figure 3). The U1A RNA from the chicken, rat, and HeLa cells is highly conserved (63). Of the species examined chicken (63), rat, (8, 63) human, and insects (12) contain only one species of U1 RNA. Only mouse contains two U1 RNAs, which differ in the center of the molecule (19). The two forms of U1 RNA found (122) in Novikoff hepatoma appear to be isomers. In addition to establishing its unique structure, U1 RNA sequence analysis led to a concept of hydrogen bonding to specific pre-mRNA consensus IVS sequences and to analysis

$pm_3^{2,2,7}G$

p

	10	20	30	40	50
$p\underset{mm}{A}\underset{}{U}AC\Psi\Psi ACC$	UGGCAGGGGA	GAUACCAUGA	UCACGAAGGU	GGUUUUCCCA
		C	G C	

60	70	80	90	100
GGGCGAGGCU	UAUCCAUUGC	$\underset{m}{A}$CUCCGGAUG	UGCUGACCCC	UGCGAUUUCC
	C CC	G		

110	120	130	140	150
CCAAAUGCGG	GAAACUCGAC	UGCAUAAUUU	GUGGUAGUGG	GGGACUGCGU
U				

160	165			
UCGCGCUCUC	CCCUG$_{OH}$			
U				

U1 RNA

Figure 3 Nucleotide sequence of U1 RNA of Novikoff hepatoma (118, 123A). or rat brain (63). Human (63), and Chicken (63) nucleotide variants are shown below the main sequence.

of the secondary structures of U1 RNA, and later to analysis of homologies with other U-snRNAs. Multiple secondary structures are possible with similar stability numbers (123). With the techniques developed for specific immunoprecipitation of the U1 snRNP particles, it is possible to conduct direct studies on the conformational state of the U1 RNA in snRNP particles (124). When T1 RNase digestion is followed by immunoprecipitation of the U1 snRNP particles, only one major site specific cleavage is observed, at position 107. The additional fragments produced by RNase A cleavage permit a more detailed analysis of the RNA structure within the particles. The structure derived is in good agreement with one hypothetical structure determined by analysis of stability numbers (123). For these analyses, the computer program of Korn et al (125) was invaluable.

U2 RNA

The most remarkable feature of the U2 snRNA (7, 122, 126; Figure 4) is the large number of modifications in its 5' end (126). Detailed analysis of the 5' cap (127) produced both enzymatic and chemical data that unequivocally established the presence of the 5' trimethyl G and pyrophosphate linkage to the remainder of the molecule. The 5' structure of the U2 RNA (127) was the first for which detailed chemical evidence was provided. Among the unanswered questions raised in this study was whether the turnover rate of trimethyl G is the same or different from the remainder of the molecule and whether, as suggested by Fredrickson et al (74, 76), this snRNA molecule is involved in synthesis of hnRNA.

The presence of 2'-O-methylated nucleotides and the remarkable number of ψ residues on the 5' end of U2 RNA clearly indicates its chemical distinction from U1 RNA, and further that there are an extraordinary number of post-transcriptional modifications. No special functions of U2 RNA are known. Like U1 RNA, U2 RNA is in an Sm protein containing snRNP, which is in the extranucleolar chromatin.

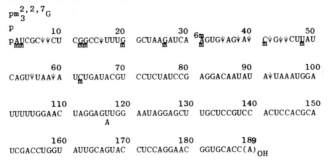

U2 RNA

Figure 4   Nucleotide sequence of U2 RNA of Novikoff hepatoma (8, 122, 126, 126a).

U3 RNA

U3 RNA (1,128) is of interest because of its specific localization in the nucleolus and its hydrogen bonding to nucleolar 35S and 28S RNA (97). This RNA was first purified by molecular sieve chromatography (26). U3 RNA is associated with protein, and is uniquely present in nucleoli (29). However, no significant concentration of U1 and U2 RNA is present in nucleoli. In addition, there is heterogeneity of the U3 RNA species (29).

The hydrogen bonding of U3 RNA to the 28S and 35S RNA is not stoichiometric and accordingly one suggestion is that U3 RNA is involved in the processing of 28S RNA (97).

Two of the three U3 RNA species found in Novikoff hepatoma cells have been sequenced (65, 129), and the minor differences in their sequences are shown in Figure 5. The third species of U3 RNA has not been completely sequenced, but it appears to be a minor variant of U3B RNA. The U3 RNA species are capped with trimethylguanosine (119, 130). Despite the two insertion/deletions in each RNA, U3A and U3B RNA have identical lengths of 214 nucleotides (129). When these two RNAs are aligned for maximum homology there are 17 base substitutions, including purine → purine, pyrimidine → pyrimidine, and purine → pyrimidine substitutions. The U3 RNA of HeLa cells consists mainly of one species that is similar but not identical to U3 RNA of Novikoff hepatoma cells (61).

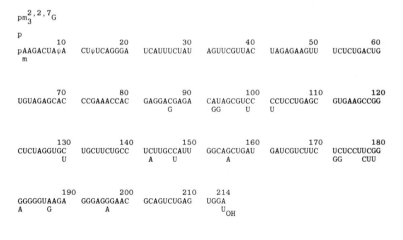

U3 RNA

Figure 5 Nucleotide sequence of U3 RNA of Novikoff hepatoma (65, 129). For Dictyostelium U3 RNA, see (66).

$$pm_3^{2,2,7}G$$

```
p        10          20          30          40          50
p m AGC C ΨUUGC G  CAGUGGCAGU  AUCGUAGCCA  AUGAGGUUUA  UCCGAGGCGC
  m m      m                               A

         60          70          80          90          100
GAUUAUUGCU  AAUUGAAAAC  UUΨUCCCAAΨ  ACCCCGCCAU  GACGACUUGA
                m                            G           C

  6
  m        110         120         130         140       145
AAUAUAGUCG  GCAUUGGCAA  UUUUUGACAG  UCUCUACGGA  GACUG(G)
                                                        OH
```

Sequence of U4 RNA

Figure 6 Nucleotide sequence of U4 RNA of Novikoff hepatoma (131) or rat (132). Mouse (133), human (132), and chicken (132) nucleotide variants are shown below the main sequence.

U4 and U5 RNA

The U4 RNA is another nucleoplasmic RNA that has important homologies to U1 RNA (131, 132). Its sequence (131–133) contains the same $m_3^{2,2,7}$ G cap of other U snRNAs and it contains a ψ near the 5' cap (Figure 6). In addition, position 63 is 2'-O-methylated and position 99 is m^6A. At present it appears that U4 RNA and possibly U5 RNA (Figure 7) are closely related to U1 RNA (132).

When U5 RNA was first purified, it was referred to as 5S RNA III (47). It has at least two (8) or more subspecies. Of the U1 to U6 RNAs, U5 RNA is most enriched in uridine (35%); it is capped with trimethylguanosine. Its complete sequence has been defined (7, 8, 134, 135; Figure 7).

U6 RNA

U6 RNA is associated with purified perichromatin granules (PCG) (113, 114). These dense granules are specifically juxtaposed to chromatin (114, 136). They appear in electron micrographs as dense cores surrounded by a white halo. Because of their association with newly synthesized mRNA

$$pm_3^{2,2,7}G$$

```
p        10          20          30          40          50
p AUACUCUGG  UUUCUCUUCA  GAUCGUAUAA  AUCUUUCGCC  UUUΨ ACΨAA A
 mm                                            m       m    m

         60          70          80          90          100
GAUΨUCCGUG  GAGAGGAACA  ACUCUGAGUC  UUAAACCAAU  UUUUUGAGGC
                            U                             U

         110         118
CUUGUCUUGG  CAAGGCU(A)
UC   CUCCAA         OH
```

Figure 7 Nucleotide sequence of U5 RNA of Novikoff hepatoma (8) or rat (134). Mouse (135), human (134), and chicken (134) nucleotide variants are shown below the main sequence.

```
                                                        6
       10          20          30          40        m    50
XpppGUGCUCGCU  UCGGCAGCAC  AUAUACUAAA  AΨUGGAACGA  ΨACAGAGₘAG


       60          70        2ₘ   80          90         100
AUUAGCAUGG   CCCCUGCGCA   AGGAUGACAC  GCAAAUΨCGU  GAAGCGUUCC
   mm        m mm          m          m
       108
    AUAUUUU(U)OH
```

Figure 8 Nucleotide sequence of U6 RNA of Novikoff hepatoma (137), and mouse (138).

(114, 136), they could be "carriers" or processing elements in or near the functional chromatin. Accordingly, U6 RNA and U6 RNP may be the first U-snRNPs to combine with newly synthesized hnRNP.

The structure of U6 RNA (137, 138; Figure 8) differs from that of the other U snRNAs in two major respects. First, the cap does not contain trimethyl G. It is not known what structure is linked to the nucleotide chain, but it is not a normal nucleotide.

Second, U6 RNA has several clusters of modified nucleotides in the center of the molecule (Figure 8); in other U snRNAs most modifications are in the 5' third of the structure. The U6 RNA contains several 2'-O-methylated nucleotides in addition to m^6A and m^2G. This molecule appears to be highly hydrogen bonded (137). Interestingly, the U1 to U5 RNAs have Am as their first nucleotide in the RNA sequence but U6 RNA has an unmodified Gp. When nucleotide sequences of the U snRNAs are analyzed, significant homologies are found (8, 138a), the most striking of which are between U1 and U4 RNA (Figure 9). Since U1, U2, and U4 to U6 RNPs contain one or more common proteins, these homologous regions may serve as binding sites for these protein(s). Similar homologies between U snRNAs

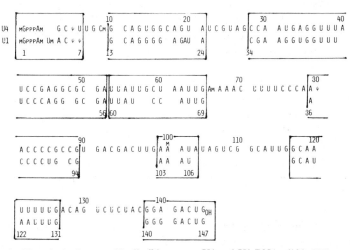

Figure 9 Homologies between Novikoff hepatoma U4 and U1 RNAs (131, 132).

have also been reported by Krol et al (132). Although earlier studies (6, 58) concluded that U snRNAs of different tissues are similar, their sequences show that there may be minor differences in snRNAs of different tissues (134). U5 RNA of chicken brain is one nucleotide shorter on the 3' end compared to chicken liver U5 RNA. Whether this difference is due to differential processing or due to expression of different genes is not known.

snRNA IN RIBONUCLEOPROTEINS

Isolation of snRNPs

Recognizing that snRNAs are probably functional as RNPs, efforts have been made to isolate U snRNPs (28, 40, 59); Raj et al (139) first isolated U1 and U2 RNPs and showed that they contain approximately 10 proteins. There are reported chromatographic methods for isolation of U1 and U2 RNPs (140). Partial purifications of small RNPs have also been reported (111, 113, 139–141).

The first attempts to isolate U-snRNPs made use of Sepharose gel filtration and sucrose density gradient centrifugation, using nuclear extracts (139). Other methods include molecular sieving, ammonium sulfate fractionation, DEAE-cellulose and phosphocellulose chromatography, CsCl density centrifugation, and affinity chromatography (142) with anti-snRNP antibodies.

Since some patients with autoimmune diseases produce antibodies directed against classes of small RNPs (17), several investigators have used antibody columns to isolate small RNPs (142, 143). Although the antibodies are highly specific, the problem appears to be recovery of the RNP particles in a functional form from the antibody columns. The immunological methods offer one approach to purification of the U snRNP particles, but it is difficult to recover undegraded particles.

Proteins Associated with snRNPs

In initial studies on the snRNP proteins, 11 proteins were identified on 2-D gels (139). These have molecular weights of 50,000–70,000 in the fractions that contained the U1 and U2 snRNP particles. However, it is not certain that all of these proteins are associated with the particles. Lerner et al (17) found 7 proteins in the immunoprecipitated particles, but these had low molecular weights of 10,000–30,000. These and larger proteins were found in snRNPs (142). This discrepancy is now being investigated to see whether the small proteins represent subunits or degradation products of the larger proteins or the larger proteins represent aggregates of more native species. Of the proteins associated with capped snRNPs, a 70,000-mol wt protein consisting of several 13,000-mol wt subunits, has been suggested to be the

Sm antigen (144, 145). The 8,000–14,000-mol wt proteins are tightly bound to the U1 RNA, since they do not dissociate in 0.5% sarkosyl (111). The antigenic protein associated with La-RNPs is a 68,000-mol wt protein (146).

Functions of U-snRNPs

The U snRNPs containing U1, U2, U4, U5, and U6 RNA are located in the nucleoplasm. Their function could be in processes relating to "splicing" of hnRNA in hnRNPs (11–13). The first clue to the role of the snRNPs came from reports that U1, U2, and other U snRNPs cosedimented with hnRNP particles (98). The second came from studies showing that some snRNAs are found hydrogen bonded to hnRNAs (99–110). Lerner et al (12) analyzed all the available IVS sequence near the splice consensus; sequences near the splice junctions have been derived by others (reviewed in 15–18).

Mechanisms of splicing must account for Breathnach and Chambon's "rule" that splice points have common sequences (GU . . . AG) (15). This "consensus structure" has been a crucial idea in the development of hypotheses relating to cleavage of hnRNA. Two questions emerged from studies with hnRNA: How does the cell make sense of the final product and why are mistakes relatively uncommon. When the sequences of known small RNAs are analyzed, the 5' end of U1 RNA has good complementarity to consensus sequences (Figure 1) (12).

The evidence for the involvement of U1 RNP in splicing of hnRNA is as follows: (a) the complementarity between conserved consensus sequences and U1 RNA; (b) U1 RNPs lacking the 5' terminus sequence are not associated with hnRNPs (12); (c) splicing of adenovirus hnRNA in HeLa cell nuclei is inhibited when they are preincubated with anti-Sm or anti-RNP antibodies (147). When the nuclei are incubated with anti-Ro or anti-La antibodies, splicing is not inhibited. Since other cellular functions, e.g. polyadenylation, are not inhibited, and since anti-RNA or anti-Sm antibodies react with and presumably inactivate the U snRNPs, it is likely that U snRNPs are required for splicing.

To determine whether IVS are removed in one or multiple steps and, if the latter, whether there is polarity, Avvedimento et al (148) analyzed the processing of collagen hnRNA. Several splicing steps are required for the removal of one IVS, and the polarity is 3' → 5'. This multistep removal of the IVS has been shown to be correlated to complementarity between U1 RNA and sites within the IVS.

In addition to these lines of evidence the data, e.g. the abundance of U1 RNP, which correlates positively with the prevalence of splicing, the presence of U1 RNP in hnRNP where splicing takes place, and the higher amounts of U1 RNP in rapidly dividing cells, are consistent with a role for

U1 RNP in splicing hnRNA. However, yeast and mitochondria, where splicing of hnRNA takes place, do not appear to have snRNAs (6). If so, U1 RNP-mediated splicing may not be the only mechanism for splicing of hnRNAs. Splicing of tRNA and rRNA precursors does not appear to involve a small RNA (15–19).

It seems likely that other snRNAs (U2, U4, U5, and U6) are also involved in processing of hnRNA, since they are associated with hnRNPs. Ohshima et al suggested that U2 RNA ensures specificity of splicing (148b). Daskal et al (113–144) have shown that of the snRNAs, only U6 RNA is associated with perichromatin granules. The U6 RNA may be involved in a nuclear function different from that of U2, U4, or U5 RNAs, possibly in formation of a precursor particle in the chromatin. Pederson & Bhorjee (149) have reported that snRNAs are covalently linked to DNA but this finding has not been confirmed.

Now that the sequences and secondary structures of the U- snRNA species have been defined, the primary tasks are the analyses of the U snRNP structures and their interactions with hnRNP structures. The molecular weight of the proteins of the snRNP particles vary depending upon the method of isolation of (19, 139). With one exception (139), there are no 2-D gel studies on the proteins of the snRNP particles. There are probably up to 20 as yet unidentified proteins associated with the U-snRNPs.

The major problem in the analysis of U snRNP proteins has been the development of suitable purification methods for the particles. When the particles are isolated on a sucrose density gradient, there are many contaminants. When chromatographic or immunochemical methods are used (140–142), the "core particles" have lost enzymes and structural elements. Thus, the problem of "nativeness" is a crucial one both from the point of view of retention of the active elements of the particles and prevention of contamination by nonparticulate structures of the nucleus.

It is important to obtain highly purified U-snRNPs that have specific nuclease functions, and possess ligase activity. This search could be implemented if (a) a proper assay system were available and (b) if the particles could be isolated in a native, uncontaminated form. Almost any defined spliceable hnRNA such as globin hnRNA (hnRNP) or ovomucoid hnRNA (hnRNP) would be satisfactory. Several studies are now in progress with cell free systems (158–161).

In addition to fidelity, there is a problem of rate of reaction. Electron microscopic studies (N. Domae, W. Schreier, R. Reddy, and H. Busch, unpublished) suggest that there is a dynamic equilibrium between the free, small particles in the nucleus and the snRNP-hnRNP complexed particles (143). In cells producing large amounts of hnRNP, such as growing and dividing cells, there are millions of snRNPs interacting with the hnRNP particles, many of which have multiple cleavage sites. The lack of tight binding of the U snRNP-hnRNP complexes has been noted by several groups (98–109). Many of the snRNPs are unattached, as are many hnRNP particles. At 37°, the hnRNP-snRNP complexes are unstable, which suggests that relatively weak binding forces exist between these two (102). When snRNP-hnRNP complexes are examined by electron microscopy, usually only one snRNP particle is found per hnRNP particle (143). From studies with ovomucoid RNA (162), the snRNP particles rapidly associate and dissociate from hnRNP particles randomly cleaving one IVS at a time.

Enzymatic Mechanisms of Precursor RNA Processing

Abelson (16) has pointed out that a variety of RNases are candidates for enzymes that specifically cleave hnRNA during its processing. Thus far only RNase III, P, Q (an exonuclease), T_2, and a few other enzymes exhibit specificity for tRNA precursor cleavage and for processing viral pre-mRNA.

The cleavage sites recognized by the RNases are dependent on RNA conformation (a) inherent in the nucleotide sequences (16, 227) and (b) resulting from bound protein in RNP. The RNA-directed signal is recognized by RNase III for 16S rRNA and 23S rRNA of the 30S pre-rRNA, and by RNase P and RNase "E" which recognize tRNA and 5S rRNA, respectively (16, 227). The RNP directed signal is recognized by RNase M5 for the maturation of p5S, and by RNase M16 for the maturation of p16S rRNA in E. coli (227). RNase III–like enzymes are present in eukaryotes (16, 280), but their functions are not known. The enzyme from chick embryos is associated with RNA, which is essential for 45S pre-rRNA processing (281).

In mitochondrial systems (278–290; see review 291) some special reactions have been found. Manipulation by mutation has been described for pre-mRNA of apocytochrome b from yeast mitochondria (287–290). Although mitochondrial snRNA has not been found, mitochondria carry out processing similar to the nuclear pre-mRNA. The processing pathways for pre-mRNA of apocytochrome b involve 34S pre-mRNA cleavage into 18S mRNA by 1. five splicing steps of which two independent steps are required to remove one of the four IVS, and 2. a preferred pathway composed of two sequences of events without an obligatory order (287–290). COB-mutants have two different processing pathways: 1. the preferred pathway in which a mutational block of splicing is defective but there is a bypass to subsequent steps, and 2. the obligatory pathway in multiple blocks, which can occur with polar effects in which the upstream mutations block not only one splicing site but also downstream splicing sites. There is also evidence for IVS coding for proteins, which may be involved in splicing (287–290).

As yet, no specific RNA cleavage enzyme or ligase has been found associated with U snRNP particles. The only small RNP shown to be involved in processing of precursor RNAs is RNase P (150–157). This endoribonuclease generates the 5' termini of mature tRNA molecules. The endonuclease consists of a 20,000-mol wt basic protein and a 300-nucleotide long RNA; neither is active alone, but they can be reconstituted to yield an active enzyme (155). The enzyme recognizes the structural conformation of its substrates, tRNA precursor molecules, rather than specific nucleotide sequences. The *E. coli* enzyme is best characterized, including fingerprinting of its RNA component (156). Similar activities have been reported in yeast, chick embryos (153), and human KB cells (157).

These RNases vary in their specificity for conformational states, specific attack sites, and interrelationships with specific proteins, but none has base sequence specificity as a requirement for cleavage. From the consensus sequences involved in the splicing of pre-mRNA, specific base sequences should be recognized by the splicing reactions of the snRNPs and their associated enzymes. Until "native" highly purified U snRNPs are available, it will not be possible to establish whether enzymes similar to RNase P are tightly bound to these particles or whether they are soluble factors in the nucleoplasm.

OTHER SMALL RNAs

La and Y RNPs

Anti-La antibodies precipitate distinct sets of snRNPs that are different from U snRNPs (81, 163). The mouse or rat RNAs are described below.

La 4.5 RNA was first shown (117) to be a group of RNAs 90–100 nucleotides long, hydrogen bonded to poly A–containing nuclear or cytoplasmic RNA from cultured Chinese hamster ovary cells. This RNA binds to nuclear precursor RNAs, presumably to their Alu family sequences, and these hybrids can be reconstituted (117). The RNA is also found in some RNA viruses (80) and is heterogenous at the 3' end in having varying numbers (80) of uridine residues. The sequence of La 4.5 RNA in mouse and hamster has been determined (164) as in Figure 10. A portion of this RNA, 14 nucleotides long, is present in the 300-nucleotide long highly repetitive Alu family DNA, and also in hnRNA (165). Many viruses have sequences common to this RNA (165).

Another RNP precipitated by anti-La antibodies contains 4.5S RNA I (81), the first snRNA to be sequenced (166; Figure 11). Rat 4.5I RNA has

```
        10          20          30          40          50
pppGCCGGUAGUG  GUGGCGCACG  CCGGUAGGAU  UUGCUGAAGG  AGGCAGAGGC
          U


        60          70          80          90
   AGAGGGAUCA  CGAGUUCGAG  GCCAGCCUGG  GCUACACAUU  UUUU
                                                      OH
```

Figure 10   Nucleotide sequence of La 4.5 RNA of mouse and hamster (164). Microheterogeneity was found at nucleotide 7.

a 50% sequence homology to La 4.5 RNA (R. Reddy, D. Henning, and H. Busch, unpublished results). La protein (antigen), which is bound to La 4.5 and 4.5I, is present in VA-RNP (163) and EBV-RNP (167). A different set of La RNAs is precipitated from human cells (81). The structure of human La RNAs is not known. Fingerprint analysis indicates La RNAs are less conserved than U snRNAs (81).

Y RNPs, first described to be antigenically nonoverlapping (163) are a subclass of La RNPs (81). Like La RNAs, Y RNAs are not conserved between mouse and human cell lines. Mouse cells contain two Y RNAs (Y1 and Y2) and human cells contain four to five Y RNAs (Y1 to Y5). All these Y RNAs are distinct RNAs with few structural similarities (81). The Novikoff hepatoma contains three Y RNAs (Y1–Y3) and Y2 and Y3 RNA have similar but not identical fingerprints (H. Busch, R. Reddy, D. Henning, and Y. C. Choi, unpublished results). Complementarity between one loop of 4.5S RNA$_I$ and the TATAT box was noted (167a), but its relevance to function is unknown.

7S RNA and Other Small RNAs

7S RNA, first found in RNA viruses (168) is a host coded, conserved RNA species (169). The sequence of 7S RNA from Novikoff hepatoma has been determined (170; Figure 12). 7S RNA is the only cellular small RNA that hybridizes to Alu family DNA sequences (171). 7S RNA is 85% homologous to the La 4.5 RNA and long stretches of Alu family DNA sequences. The homologous sequences between Alu family DNA sequences, mouse B1 sequences, 7S RNA, and La 4.5 RNA are shown in Figure. 13. The significance of these homologies is not clear; however, it is of interest that the DNA sequences at the origins of replication correspond to portions of the

```
        10          20          30          40        U 50
pppGGCUGGAGAG  AUGGC(UC)AGC  CGUUAAAGGC  UAGGCUCACA  ACCAAAAAUA


        60          70          80          90          98
   UAAGAGUUCG  GUUCCCAGCA  CCCACGGCUG  UCUCUCCAGC  CACCUUUU(U)
                                                             OH
```

Figure 11   Nucleotide sequence of La 4.5 I RNA of Novikoff hepatoma (166) modified. Microheterogeneity was found at nucleotide 47.

homologous sequences [(165); see also the review by W. R. Jelinek and C. W. Schmid in this volume]. These may be transposon related.

In addition to these small RNAs several other small RNAs include: RNA 7–1, 7–2 (95), 7–3 (51), and 8S (97), and RNase P RNA (156). There appear to be no additional abundant RNAs in higher eukaryotes other than the RNAs described in the size range of 90–400 nucleotides. However, RNAs with less than 10,000 copies per cell are difficult to detect. RNAs of the size 8–18S have been found and partially characterized (172, 173). Other studies have described RNA species specific to certain stages of tissue development (174, 175). Cloned Alu DNA fragments are transcribed in vitro by polymerase III to yield specific small RNAs. There is evidence that these RNAs exist in vivo (176–177a). 8S RNA contains 5.8S rRNA.

VIRAL RNAs

In 1966, Reich et al (37) compared the chemical and metabolic characteristics of a novel small RNA of adenovirus 2–infected KB cells with those of 5S RNA. They concluded that the former was synthesized during viral infection. Subsequently, other species of low-molecular-weight RNAs were identified in cells infected with SV-40 (178, 179), Epstein Barr (EB) virus (167), vesicular stomatitis virus (VSV) (180, 181), or VSV-defective interfering particles (182, 183). Virions of the retroviruses accumulate discrete classes of host small RNA (168, 169, 184); one class, tRNA, serves as a primer for viral genome replication (185). Small RNA molecules, both viroid and satellite, function as the causative agents of many plant diseases (186, 187), although the mechanisms of infection and propogation are not completely understood.

```
              10          20          30          40          50
      pppGCCGGGCGGU GCGCACGCCU GUAGUCCCAG CUACUCGGGA GGCUGAGACA

              60          70          80          90         100
      GGAGGAUCGC UUGAGUCCAG GAGUUCUGGG CUGUAGUGCG CUAUGCCGAU

             110         120         130         140         150
      CGGGUGUCCG CACUAAGUUC GGCAUCAAUA UGGUGACCUC CCGAGGGGGA

             160         170         180         190         200
      CCACCAGGUU GCCUAAGGAG GGGUGAACCG GCCCAGGUCG GAAACGGAGC

             210         220         230         240         250
      AGGUCAAAAC UCCCGUGCUG AUCAGUAGUG GGAUCGCGCC UGUGAAUAGC

             260         270         280         290    295
      CACUGCACUC CAGCCUGGGC AACAUAGCGA GACCCCGUCU CUUA(A)_{OH}
```

Figure 12   Nucleotide sequence of 7S RNA of Novikoff hepatoma (170).

```
7S RNA (rat)    pppGCCGGGC..  .GGUG.CGCA  CGCCUGUAGU  CCCAGCUACU  CGGGAGGCUG  AGACAGGCAG  AUCGCUUGAG  UCCAGGAGUU
                          10          20          30          40          50          60          70          80
Blur 8          AGGCUGGG.AG  URGUGGCUCA  CGCCUGUAAU  CCCAGAAUUU  UGGGAGGCCA  AGCAGGCAG   AUCACCUGAG  GUCAAGAGUU
(Human Alu)            GGCA.      UGAUGGCAAG  UGCCUGUAAU  CCCAGCUACU  UGGGAGGCUG  ACGAAGAGA   AUUGCUUAAA  CCUGGA
                                          150         160         170         180         190         200         210

B1 (mouse)      CCGGGCA.     UGGUGGUGCA  UGCCUUCAAU  CCCAGC.ACU  CGGGAGGCAG  AGGCAGGCGG  AUUUC.UGAG  UUCGAGCCCA
                                   10          20          30          40          50          60          70

La 4.5S (mouse) pppGCCGG.UAG  UGGUGGCCCA  CGCCGGUAGG  AUUUGCUCA.  .AGGAGGCAG  AGGCAGAGGG  AUCACACGAG  UUCGAGGCCA
                                   10          20          30          40          50          60          70
```

Figure 13 Homologies between human Alu DNA, mouse B1 DNA, rat 7S RNA, and mouse La 4.5 RNA. The sequences in rat 7S RNA, mouse B1 DNA, and mouse La 4.5S RNA are underlined where they are identical to the human Alu sequence.

Adenovirus Associated Small RNAs: VA-RNA

In addition to a 9S mRNA, the adenovirus 2 Ad-2 genome codes for two species of small RNA (188–190). The major viral-associated RNAs, VA RNA$_I$ and VA RNA$_{II}$, are primary transcripts of the viral genome (188–194). They are not products of the processing of larger precursors or degradation products (188, 195, 196). Their genes map close to position 30 on the r-strand of the adenovirus 2 genome, and are transcribed by RNA polymerase III (192, 193, 197). These genes have been transcribed accurately in vitro (191, 195).

Initial questions as to the number of species (189, 190) have been resolved. Mathews & Petterson (188) have investigated the small RNAs of the nuclei and cytoplasm of Ad 2–infected HeLa cells. By fingerprint analysis there are only two major species of small RNAs unique to those cells. Both of these hybridize to adenovirus 2 DNA. Analysis of the structures of the small RNA synthesized in vitro by nuclei of Ad 2–infected KB cells has led to the identical conclusion (197). The 156-nucleotide RNA synthesized in vitro is identical to VA RNA$_I$ sequenced by Ohe & Weissman (196), and the 140-nucleotide virus-coded RNA described by Mathews (198) and synthesized in vitro. Two other species of RNAs 140 nucleotides long also exist and one is probably a degradation product of VA RNA$_I$. The 200-nucleotide long VA RNA (V200) previously identified (193) is a run-through product of the VA RNA$_I$ gene toward a second termination site approximately 40 nucleotides downstream. Most of the fragments of the fingerprint of the V200 RNA can be predicted by the sequence of the VA$_I$ RNA gene (199, 200).

Heterogeneity of VA RNA$_I$ has also been reported by Vennstrom et al (201) who identified a 5' terminus of (pp)pApGpCp, the major 5' terminus of (pp)pGGGGC, and the 3' terminal heterogeneity of CUCCUU(U) in approximately equal molar amounts. The heterogeneities of the two major VA RNAs are better understood when correlated with the nucleotide sequence of the genes (202; Figure 14). The heterogeneity of the 5' sequence can be explained on the basis that RNA polymerase III can initiate at either of two positions in VA RNA$_I$ (Figure 14). The 3' heterogeneity results from termination at any of several contiguous thymidine residues at the 3' ends of both transcripts. From analysis of the sequences of the genes, it appears that the two genes have arisen from the duplication of an ancestral gene, with subsequent divergences that resulted in two RNAs with different secondary structure, and possibly more than one function (202). Recent studies on the variants of the VA RNAs appear to confirm this hypothesis (202a).

VA RNA$_I$ is found in preparations of hnRNA of Ad 2–infected cells (188), which is consistent with formation of an hnRNA-VA-RNA complex. In 1979, Murray & Holliday (11) proposed that VA RNA$_I$ could function as a splicer RNA. They predicted that 19 bases of VA RNA$_I$ could hybridize across a single splice point of the hexon pre-mRNA. In 1980, Mathews (203) reported that the VA RNAs could hybridize in vitro to unfractionated late virus mRNA and to cDNA of this mRNA, but not to cloned portions of the natural gene. The hybrids were constructed between filter bound DNA and RNA in solution. Thus the failure to observe hybrids to portions of the cloned natural genes may have been due to artificial constraints. These data were consistent with a role for VA RNA in splicing, or with the proposal that VA RNA could bind to spliced RNA and play a role in a subsequent metabolic step in mRNA metabolism. It was subsequently determined that VA RNA could be precipitated from extracts of Ad 2– infected HeLa cells by anti-La antibody, which suggested that VA-RNA is in an snRNP (12).

Two recent findings suggest that VA RNA is not involved in splicing viral hnRNA. As noted earlier, Yang et al (147) found that incubation of Ad 2–infected HeLa cell nuclei with anti-La antibody, does not prevent the processing of adenovirus mRNA. They found that anti-Sm and anti-RNP antibodies, which did not interact with the VA-RNA snRNP, inhibited the processing of VA-mRNA. If either small VA RNA is involved in splicing, it would have to associate with the hnRNA soon after polyadenylation. However, VA RNA is not found associated with the nascent adenovirus hnRNA (204, 205). The data argue against a role for Ad 2 VA RNA in processing adenovirus hnRNA, but do not preclude a role for these RNAs

VA RNA I

```
ACCGUGCAAA   AGGAGAGCCU   GUAAGCGGGC   ACUCUUCCGU   GGUCUGGUGG   AUAAAUUCGC   AAGGGUAUCA

UGGCGGACGA   CCGGGGUUCG   AACCCCGGAU   CCGGCCGUCC   GCCGUGAUCC   AUGCGGUUAC   CGCCCGCGUG

UCGAACCCAG   GUGUGCGACG   UCAGACAACG   GGGGAGCGCU   CCUUUUUGGCU  UCCUUCCAGG   CGCGGCGGCU

GCUGCGCUAG   CUUUUUUGGC   CACUGGCCGC   GCGCGGCGUA   AGCGGUUAGG   CUGGAAAGCG   AAAGCAUUAA
```

VA RNA II

```
GUGGCUCGCU   CCCUGUAGCC   GGAGGGUUAU   UUUCCAAGGG   UUGAGUCGCA   GGACCCCGG    UUCGAGUCUC

GGGCCGGCCG   GACUGCGGCG   AACGGGGGUU   UGCCUCCCCG   UCAUGCAAGA   CCCCGCUUGC   AAAUUCCUCC

GGAAACAGGG   ACGAGCCCCU   UUUUUGCUUU   UCCCAGAUGC   AUCCGGUGCU   GCGGCAGACG   CG
```

Figure 14   Nucleotide sequence of the VA RNA genes of Adenovirus-2 (202). The sequence is of the anticoding strand and is given as the RNA. The genes for the VA RNAs are underlined. VA$_{200}$ is underlined with a broken line extending to VA RNA$_1$. Alternative initiation and termination sites are shown with arrowheads.

in either transport of processed Ad 2 mRNA molecules from the nucleus or disruption of the metabolism of host cell hnRNA (206). Experiments with mutants of Ad 2, which have VA RNA genes mutated in vitro, suggest that these genes are essential for viability (206a). Recently, T. Shenk, (personal communication) has obtained evidence for a role of VA-RNA in translation as an inhibitor of host mRNA.

Other Viral RNAs

SIMIAN VIRUS-40 (SV-40) ASSOCIATED SMALL RNA A small RNA is synthesized late in the simian virus lytic infection (178, 207). Although it is synthesized from the SV-40 genome, it is not required for viral viability. This RNA is complementary to the 91 codon alternative reading frame of the 3' terminal portion of the early genome (179, 208). A third splice occurs in SV40 early mRNA late in the lytic cycle (209) coincident with the synthesis of the SV-40 associated small RNA, and suggests that the SV-40 snRNA plays an indirect role in this splicing event. An snRNA synthesized in SV-40 infected cells is capable of affecting transcription, but it has not been isolated (210).

EB VIRUS ASSOCIATED RNA Recently, Lerner et al (167) reported on two small RNAs encoded by the virus genome that are synthesized in large amounts in infected cells. They are precipitated with anti-La antibody, which indicates that they are in nuclear RNPs. These two RNAs, which have recently been sequenced (210a), are approximately 180 nucleotides long; they have a 5' pppA and lack poly(A).

RNAs ASSOCIATED WITH STANDARD AND DEFECTIVE VSV PARTI-CLES When cells are infected with standard and defective interfering (DI) VSV, or when the particles are incubated in vitro, a 47-nucleotide long RNA is synthesized (183, 211, 212). This RNA has been sequenced (212, 213); pppAp is the 5' end and it lacks poly(A) on the 3' end. This RNA could play a part in regulating the balance between RNA transcription and replication (214, 215).

Standard VSV synthesizes a plus-strand leader RNA that differs from the RNA synthesized by DI particles (180, 216). This RNA is synthesized when the virus associated, RNA-dependent RNA polymerase initiates transcription on the viral genome. The RNA is complementary to the 3' end of the genomic RNA and is apparently cleaved from the growing RNA chain before transcription of the viral mRNAs. Three other small RNAs, in addition to the 47-nucleotide leader sequence, are synthesized during in vitro transcription of VSV virions (181). These RNAs have chain lengths of 28, 42, and 70 nucleotides, contain (p) ppAA at their 5' terminal, and are not polyadenylated. The 42- and 28-nucleotide long RNAs contain 5' terminal sequences identical to those of the N and NS mRNAs (181). The authors hypothesize that these leader sequences are synthesized concurrently by multiple initiations of transcription, and that they are then elongated sequentially, which results in the synthesis of the mRNAs coding for the viral proteins N, NS, M, b, and L.

RETROVIRUS ASSOCIATED SMALL RNAs   Retrovirus virions contain several low-molecular-weight RNAs as well as the viral genomic RNA (217–219), none of which are viral encoded. However, they represent a discrete subclass of the complement of host small RNA (220). The association of tRNA with the viral genome and its function as the primer for initiation of the viral RNA-dependent DNA polymerase is well documented (220–224). The functions of the other viral associated low-molecular-weight RNAs are unknown. The 4.5S RNA species associated with Moloney leukemia virus and spleen focus–forming virus (80) differ from the host cell species in that they contain as many as thirty extra uridylate residues at their 3' termini. The 4.5S RNA associated with the genome of spleen focus forming virus is composed of more than thirty components (80) that vary in lengths of the poly(U). The viral genomic RNA is associated with the larger molecules (about 5S), and the cytoplasmic RNAs of the T3-K-1 cells are associated with the smaller (close to 4.5S) RNAs. When Moloney-murine leukemia virions, or spleen focus forming virions are isolated from vertebrate cells, other than cells derived from rodents, 4.5S RNA species are not found in the viral genomes (225). These data substantiate the finding that the 4.5S RNA molecules associated with the viral genome are provided by the host. Harada et al (225) have compared the 4.5S RNA species in the virions with those species that have been previously characterized. They concluded that the 4.5S RNA associated with the viral genome is not $4.5S_I$ RNA. The virus associated RNA is likely to be the same RNA as that found in snRNP recognized by anti-La antibody.

PRE-rRNA PROCESSING

Progress has been made in the study of pre-rRNA processing (15, 16, 226–230). Complete or nearly complete nucleotide sequences have been determined for many different rDNAs including those of human and mouse mitochondria, *Zea mays* chloroplast, *E. coli,* yeast, *Xenopus,* rat, mouse, and human. Some of these contain IVS that are cleaved out of the final rRNA products.

Table 3 shows examples of the evolutionary diversity of rDNA, which is characterized by various sizes, types, and arrangements and by the presence of spacers such as IVS, insertions, or gaps (230–247). Despite these wide evolutionary diversities there are common features:

(*a*) Conservation of gene polarity: 5'-16-18S RNA → 23-28S rRNA-3'.

(*b*) Prokaryotic genes contain 5S rDNA at the 3' end, interspersed spacer tDNAs, and smaller size high-molecular-weight rRNA.

(*c*) Eukaryotic genes have separated 5S rDNA and tDNA from rDNAs. They contain 5.8S rDNA and higher-molecular-weight rDNAs. The loss of 5S rRNA colinearity occurs in yeast (248) and *Dictyostelium* (249), which contain 5S rDNA in strands opposite to the coding rDNA strand.

(*d*) Chloroplast genes in higher plants contain 4.5S rDNA.

(*e*) Insect genes (Dipteran) have disjoined 5.8S rDNA (5.8 $rRNA_\alpha$ + 5.8S $rRNA_\beta$).

Initial Transcripts and Processing Intermediates

Pre-rRNA is generally large, and in some instances there are several precursors that are processed differently. Extensive studies have been made on 45S

Table 3 Diversity of rRNA genes (rDNA)

Species	Compartment	Order of gene	Pre-rRNA	Reference
Rat, mouse, man	Nucleolar	18S–5.8S–28S	45S	231
Xenopus	Nucleolar	18S–5.8S–28S	40S	230
Drosophila	Chromosomal	18S–5.8S–2S–28S$_\alpha$– Gap (and Insert)–28S$_\beta$	38S	230
Yeast	Chromosomal	17S–5.8S–26S	37S	231
Physarum	Extrachromosomal	19S–5.8S–28S$_a$–IVS$_a$– 28S$_b$–IVS$_b$–28S$_c$	40S	232
Tetrahymena	Extrachromosomal	17S–5.8S–26S$_a$–IVS$_a$– 28S$_b$	>35S	233
Leishmania	Extrachromosomal	16S–5.8S–26S$_\alpha$–Gap– 26S$_\beta$		234
Mouse, man	Mitochondrial	tRNA–12S–tRNA– 16S–tRNA		235, 236
Yeast	Mitochondrial	15S–tRNA–21S$_a$– IVS–21S$_b$		237
Neurospora	Mitochondrial	17S–tRNAs–25S$_a$– IVS–25S$_b$–tRNAs		238, 239
Zea mays, spinach	Chloroplastic	16S–tRNA–23S– 4.5S–5S		240–242
Euglena	Chloroplastic	16S–tRNAs–23S–5S		243, 244
Chlamydomonas	Chloroplastic	16S–tRNA–7S–3S– 23S$_a$–IVS–23S$_b$–5S		245, 246
Prokaryotes	Chromosomal	16S–tRNA–23S–5S– tRNA	30S	247

pre-rRNA of mammals, 40S of *Xenopus,* and 38S of *Drosophila.* The 35S pre-rRNA from *Tetrahymena* contains one IVS, and 40S pre-rRNA from *Physarum* contains two IVS (See Table 3). Mechanisms that process the pre-rRNA include specific cleavages (226), 5'-end trimming as found in 32S pre-rRNA of yeast (228) and 41S pre-rRNA from mammals (226), and splicing, as observed in Physarum (232).

LARGE CLEAVAGE INTERMEDIATES   At the initial cleavage stage, a set of specific sites at spacer junctions are cleaved. Initial cleavages of 30S pre-rRNA of *E. coli* produce tRNAs, p5S rRNA, p23S rRNA, and p16S rRNA (250). 32S pre-rRNA from yeast produces p17S rRNA (18S) and p26 rRNA (29S) (231, 248). In some organisms, the spacers between 18S rRNA and 28S rRNA have two or more defined cleavage sites that are not cleaved in a defined order. The alternate sites result in alternate pathways (226, 251); for example, in mammals 41S → 18S + 36S, and 41S → 20S + 32S (226).

Table 3 shows there are many variants of pre-28S (23S) rRNA among organisms, some of which undergo splicing reactions (Table 3).

Eukaryotic pre-rRNA ("32S") contains 5.8S rRNA at the 5' end. Similarly, pre-26S rRNA ("29S") of yeast has 5.8S rRNA at its 5' ends. The 29S pre-rRNA, is the precursor of 5.8S rRNA and 26S rRNA (252). Chloroplast pre-23S rRNA (1.65×10^6 daltons) from spinach is a precursor of both 4.5S rRNA and 28S rRNA (23S) (242).

Splicing occurs in mitochondrial and cholorplast pre-28S rRNA, both of which contain an IVS. Pre-21S rRNA ($5.1–5.4 \times 10^3$ bases) from yeast mitochondria also contains an IVS (1.2×10^3 bases) as does pre-25S rRNA (253) from *Neurospora* mitochondria (2.3×10^3 bases) (254). Prokaryotic pre-23S rRNA (p23S) from *E. coli* contains short spacers, 114 bases on the 5' end, and 71 bases on the 3' end (255).

PRE-18S (16S) rRNA   The pre-16S rRNA from *E. coli* (p16S) contains two flanking sequences, 146 bases on the 5' end and 43 bases on the 3' end (256). *E. coli* RNase III⁻ and P⁻ mutants produce a 20S pre-16S rRNA that contains ppp-leader sequences-16S-tRNA-tRNA (257). The precursor from yeast is 18S or p17S rRNA (252). In mammalian cells, pre-18S rRNA is 20S which has an identical 5'-end to 18S rRNA (226, 258).

MATURE rRNAs AND THEIR FORMATION (18S OR 16S rRNA)   Complete 18S or 16S rRNA sequences have been determined for human mitochondria (954 bases) (235), mouse mitochondria (236), *Zea mays* chloroplast (240), *E. coli* (247), yeast (259), and *Xenopus* (260). There is a 75% homology of mitochondrial and chloroplast 12–16S rRNA to *E. coli,* a substantial homology of yeast to *E. coli,* and greater than 70% homology of *Xenopus* 18S rRNA to yeast 17S rRNA. A universal homology has been observed in the 3' end portion containing -$m_2^6Am_2^6A$-. Eukaryotic 18S rRNA contains sequences with base modifications, including m^6A, m^7G and AIB -$m'\psi$ (259–261).

In addition to the primary structures, the secondary structures have been defined for prokaryotic 16S rRNA (262, 263). For more than 100 species, there are 37 helices, variable and conserved sequences, and a universal sequence (GCCGUAAACGAUG) located in positions 821–879.

28S (23S) rRNA   One form of 28S rRNA is a single continuous strand and the other is two discontinuous strands designated as 28S rRNA$_\alpha$ (the 5'-end portion) and 28S rRNA$_\beta$ (the 3' end portion).

The continuous 28S rRNA is derived from precursors with or without splicing. Splicing at one level of pre-rRNA has been demonstrated in *Physarum* (232) and *Tetrahymena* (233) and at another in pre-28S of yeast mitochondria (253). The 21S rRNA of yeast mitochondria is derived from pre-21S rRNA ($5.1–5.4 \times 10^3$ bases) by two alternative pathways: (*a*) 3'-end processing (800 bases) followed by splicing (1.2×10^3 bases) and (*b*) splicing followed by the 3'-end processing (253).

The discontinuous 28S rRNA$_\alpha$ and 28S rRNA$_\beta$ are derived from the nearly "mature" 28S rRNA (28S rRNA$_\alpha$-gap-28S rRNA$_\beta$) or the "newly synthesized 28S rRNA" by cleavages involved in the elimination of the "gap" (approximately 500 bases) (251, 264). This type of rRNA is observed in insects and thought to result from hidden breakage of newly synthesized 28S rRNA (264). Some prokaryotic ribosomal subunits contain a 14S rRNA and a 16S rRNA instead of the usual 23S rRNA (265).

Complete sequences of 28S or 23S rRNA are known for human mitochondria (1159 bases) (235), mouse mitochondria (1589 bases) (236), *Zea mays* chloroplast (241), and *E. coli* (2904 bases) (247). Partial sequences have been obtained for yeast (266) and *Xenopus* (267). There appears to be greater than 70% homology of mitochondria and chloroplast RNA to *E. coli* RNA.

Secondary structures have been constructed for 23S rRNAs from prokaryotes, chloroplast 23 rRNA from *Z. mays,* and mitochondrial 16S rRNA from human and mouse (268). 7 domains (I–VII) are observed; four are conserved (II, III, V, VI). It has been proposed that the domain V or VI is the site of the IVS in other organisms.

5.8S rRNA   One group theorizes that 5.8S rRNA resulted from the mutations of 5S rRNA from *E. coli* (269) while others propose that it evolved from the 5' end of 23S rRNA (270, 271). In yeast, 5.8S rRNA is cleaved from 7S RNA produced by 3' processing (252).

There are two forms of 5.8S rRNAs, the normal 5.8S rRNA (160 bases) and a cleaved form consisting of 5.8S rRNA$_\alpha$ (130 bases) and 5.8S rRNA$_\beta$ (30 bases), sometimes referred to as "normal" 5.8S rRNA (α) and 2S rRNA (β). The cleaved form has been observed in *Drosophila* (272) and *Sciara* (273), where pre-5.8S rRNA contains a spacer of 28 bases; the 28 base spacer is cleaved out during maturation. It has been suggested that the cleaved 5.8S rRNA is hydrogen bonded to a cleaved form of 28S rRNA$_{\alpha,\beta}$ (274). Recently, Erdmann compiled the 5.8S rRNA structures from 12 different species (275).

4.5S rRNA   4.5S rRNA (80–100 bases) is a small rRNA found in chloroplasts of higher plants. Its origin has been suggested to be by 3' fragmentation of progenitor to 23S rRNA from *E. coli* (271, 253). The primary sequences and structural homologies have been described for a number of higher plants (271, 276–279). In tobacco (279) and *Zea mays* (241) the linkage is 23S rRNA - spacer - 4.5S (103 bases)-spacer (256 bases)-5S rRNA (121 bases).

Role of snRNAs in rRNA Processing

U3 RNA and 8S RNA are two of the snRNAs located in the nucleolus. U3 and 8S RNAs are hydrogen bonded to precursor ribosomal RNAs. (51, 97) It has been suggested that U3 RNA or 8S RNAs have complementarity to regions where specific cleavage of 45S and other ribosomal RNA precursors are cleaved to yield mature 5.8, 18, and 28S RNAs.

Some evidence supports a role for nucleolus specific snRNAs in processing pre-rRNA (51, 97). Two types of small RNAs are associated with nucleolar 28S RNA. One, 5.8S rRNA, is associated with 28S RNA in an equimolar ratio, and the other snRNAs, U3 RNA and 8S snRNA, are in a molar ratio of 0.3 (97). An additional 7–1 snRNA is associated with nucleolar 28S RNA (95). Two species of nucleolar 28S RNAs may exist, 28S : 5.8S$\pm$ snRNAs (U3, 7S-1, and 8S). In addition, 32S pre-28S RNA is associated with U3 snRNA (97). Accordingly, U3 snRNA may exist as a stable complex with both nucleolar 28S RNA and 32S RNA, and may be an important component in processing.

Pre-rRNA Splicing

Direct evidence for splicing mechanisms has been obtained from in vitro systems using *Tetrahymena* nuclei or nucleoli. The initial ribosomal RNA transcript is cleaved at two positions 407 nucleotides apart and spliced to form a 35S pre-rRNA (282, 283). The excised linear IVS (407 bases) forms a circular product (282, 283). The double splice system of Physarum has neither a simultaneous attack nor a preferential site in splice order (232). To define the junction point, rRNA/IVS/rRNA, a number of structural determinations have been made. The systems studies include Tetrahymena (IVS = 407 bases, splice junctions TCTCT/AATTG——AATCG/TAAG) (284), Chlamydomonas = 870 bases, splice junctions CGT/AAA—— AGG/CGT (246), and yeast mitochondria (IVS = 1143 bases, splice junction GGATA/ATT——TGA/ACAGG) (285, 286). The splicing mechanism for the RNA polymerase I system differs from those of the RNA polymerase II and III systems (15). The functions of excised IVS are not known. The yeast mitochondrial IVS may be an mRNA because its structure contains the initiator codon AUG, codons for 235 amino acids, and the terminator codon UAA (285).

CONCLUSION AND PROSPECTS

Small RNAs discovered about 15 years ago are metabolically stable, conserved through evolution, and localized to specific subcellular compartments. The sequencing of small RNAs has resulted in the discovery of cap structures, and the nucleotide sequences of all the six capped snRNAs are defined, in some cases for several species. The discovery that patients with autoimmune diseases produce antibodies directed against small RNPs, and the hypothesis that U1 RNA may be involved in splicing hnRNAs, has

Table 4 Suggested functions of small RNAs

Function	References
Splicing of hnRNA	11–13
Processing tRNA precursors	150–157
Processing of ribosomal precursors	97, 281
Intranuclear or nucleocytoplasmic transport	6, 12
DNA replication	165, 295
Modulators of transcription	6, 97
Stimulator of transcription	210, 296
Genetic reprogramming	294
Chromosome organization	294
Part of nuclear or chromatin structure	51, 149
Structural role in hnRNP particle	102
Induces embryonic heart differentiation	174
Control of translation	297, 298
Acts as incompatability factor	300
Control of cell division	299
Involved in crossing-over during meosis	301

brought small RNAs to the attention of many investigators. All the capped snRNAs appear to be synthesized by polymerase II, and other small RNAs by polymerase III. Genes having identical sequences to U1, U2, and U3 RNA have been isolated and, interestingly, the human genome appears to contain more pseudogenes for small RNAs than real genes.

Table 4 lists the suggested functions for small RNAs. Apparently, the different small RNAs have different functions that are varied and diverse. There is conclusive evidence for the involvement of an RNA component as part of RNase P in processing of tRNA precursors.

We have to repeat that the role of snRNAs in the processing of hnRNA is only a hypothesis, with which, however, some experimental evidence accumulated to date is consistent. There are reports that suggest that the model is incomplete, and alternative splicing mechanisms may exist that do not utilize snRNA. The results of Spritz et al (292) indicate that regions other than consensus regions of introns can affect processing. The possibility exists that the processing site and its interaction with snRNA may be affected by the secondary structure(s) of the RNA(s) involved. Accordingly, all processing sites may not be identical. Naora (293) reported that a better match between U1 RNA and the splice junction of insulin pre-mRNA could be made with our reported model for the secondary structure of U1 RNA (124), sequences of the exon, and the intron portions of insulin pre-mRNA. In addition, mitochondria have not been found to contain small RNA molecules, yet mitochondrial messengers are spliced. Clearly, as in the biochemists epitaph "more work needs to be done," particularly to evaluate a transport function.

Literature Cited

1. Busch, H., Ro-Choi, T. S., Prestayko, A. W., Shibata, H., Crooke, S. T., El-Khatib, S. M., Choi, Y. C. Mauritzen, C. M. 1971. *Perspect. Biol. Med.* 15: 117–39
2. Busch, H., Choi, Y. C., Daskal, I., Inagaki, A., Olson, M. O. J., Reddy, R., Ro-Choi, T. S., Shibata, H., Yeoman, L. C. 1972. *Gene Transcription in Reproductive Tissue. Karolinska Symposia on Research Methods in Reproductive Endocrinology,* ed. E. Diczfalusy, Stockholm: Karolinska Inst. pp. 36–66
3. Ro-Choi, T. S., Busch, H. 1974. *Cell Nucleus* 3:151–208
4. Busch, H. 1976. *Perspect. Biol. Med.* 19:549–67
5. Naora, H. 1977. In *The Ribonucleic Acids,* ed. P. R. Stewart, D. S. Lethan, pp. 61–71. New York/Heidelberg/Berlin: Springer

6. Hellung-Larsen, P. 1977. *Low Molecular Weight RNA Components in Eukaryotic Cells.* Copenhagen: FADL's Forlag
7. Choi, Y. C., Ro-Choi, T. S. 1980. *Cell Biol.* 3:609–67
8. Reddy, R., Busch, H. 1981. *Cell Nucleus* 8:261–306
9. Weinberg, R. A. 1973. *Ann. Rev. Biochem.* 42:329–54
10. Zieve, G. 1981. *Cell* 25:296–97
11. Murray, V., Holliday, R. 1979. *FEBS Lett.* 106:6–7
12. Lerner, M. R., Boyle, J. A., Mount, S. M., Wolin, S. L., Steitz, J. A. 1980. *Nature* 283:220–24
13. Rogers, J., Wall, R. 1980. *Proc. Natl. Acad. Sci. USA* 77:1877–79
14. Roberts, R. 1980. *Nature* 283:132–33

15. Breathnach, R., Chambon, P. 1981. *Ann. Rev. Biochem.* 50:349–83
16. Abelson, J. 1979. *Ann. Rev. Biochem.* 48:1035–69
17. Crick, F. 1979. *Science* 204:264–71
18. Sharp, P. A. 1981. *Cell* 23:643–46
19. Lerner, M. R., Steitz, J. A. 1979. *Proc. Natl. Acad. Sci. USA* 76:5495–99
20. Lerner, E. A., Lerner, M. R., Janeway, C. A. Jr., Steitz, J. A. 1981. *Proc. Natl. Acad. Sci. USA* 78:2737–41
21. Lerner, M. R., Steitz, J. A. 1981. *Cell* 25:298–300
22. Douvas, A., Tan, E. M. 1981. *Cell Nucleus* 8:369–87
23. Tan, E. M. 1979. *Cell Nucleus* 7:457–77
24. Muramatsu, M., Hodnett, J. L., Busch, H. 1966. *J. Biol. Chem.* 241:1544–47
25. Hodnett, J. L., Busch, H. 1968. *J. Biol. Chem.* 243:6334–42
26. Nakamura, T., Prestayko, A. W., Busch, H. 1968. *J. Biol. Chem.* 243:1368–75
27. Moriyama, Y., Hodnett, J. L., Prestayko, A. W., Busch, H. 1969. *J. Mol. Biol.* 39:335–49
28. Prestayko, A. W., Busch, H. 1968. *Biochim. Biophys. Acta* 169:332–37
29. Prestayko, A. W., Tonato, M., Lewis, C., Busch, H. 1971. *J. Biol. Chem.* 246:182–87
30. Artman, M., Roth, J. S. 1971. *J. Mol. Biol.* 60:291–301
31. Heyden, H. W., Zachau, H. G. 1971. *Biochim. Biophys. Acta* 232:651–60
32. Chakravorthy, A., Busch, H. 1967. *Cancer Res.* 27:789–92
33. Higashi, K., Adams, H. R., Busch, H. 1966. *Cancer Res.* 2196–2201
34. Sporn, M. B., Dingman, C. W. 1963. *Biochim. Biophys. Acta* 68:389–400
35. Rosset, R., Monier, R. 1963. *Biochim. Biophys. Acta* 68:653–56
36. Hadjiolov, A. A., Venkov, P. V., Tsanev, R. 1966. *Anal. Biochem.* 17:263–67
37. Reich, P. R., Forget, B. G., Weissman, S. M., Rose, J. A. 1966. *J. Mol. Biol.* 17:428–39
38. Pene, J. J., Knight, E., Darnell, J. E. 1968. *J. Mol. Biol.* 33:609–23
39. Egyhazi, E., Daneholt, B., Edstrom, J. E., Labert, B., Ringborg, J. 1969. *J. Mol. Biol.* 44:517–32
40. Enger, M. D., Walters, R. A. 1970. *Biochemistry* 9:3551–62
41. Hellung-Larsen, P., Frederiksen, S. 1972. *Biochim. Biophys. Acta* 262:290–307
42. Hellung-Larsen, P., Frederiksen, S. 1977. *Comp. Biochem. Physiol. B* 58:273–81
43. Larsen, C. J., Galibert, F., Lelong, J. C., Boiron, M. 1967. *CR Acad. Sci.* D264:1523–26
44. Larsen, C. J., Galibert, F., Hampe, A., Boiron, M. M. 1968. *CR Acad. Sci.* 267:110–13
45. Larsen, C. J., Galibert, F., Hampe, A., Boiron, M. 1969. *Bull. Soc. Chim. Biol.* 51:649–68
46. Loening, U. E. 1967. *Biochem. J.* 102:251–57
47. Ro-Choi, T. S., Reddy, R., Henning, D., Busch, H. 1971. *Biochem. Biophys. Res. Commun.* 44:963–72
48. Ro-Choi, T. S., Moriyama, Y., Choi, Y. C., Busch, H. 1970. *J. Biol. Chem.* 245:1970–77
49. Weinberg, R. A., Penman, S. 1968. *J. Mol. Biol.* 38:289–304
50. Zapisek, W. F., Saponara, A. G., Enger, M. D. 1969. *Biochemistry* 8:1170–81
51. Zieve, G., Penman, S. 1976. *Cell* 8:19–31
52. Zieve, G., Benecke, B. J., Penman, S. 1977. *Biochemistry* 16:4520–25
53. Galibert, F., Lelong, J. C., Larsen, C. J., Boiron, M. 1967. *Biochim. Biophys. Acta* 142:89–98
54. Dingman, C. W., Peacock, A. 1968. *Biochemistry* 7:659–67
55. Moriyama, Y., Ip, P., Busch, H. 1970. *Biochim. Biophys. Acta* 209:161–70
56. Fredriksen, S., Tonnesen, T., Hellung-Larsen, P. 1971. *Arch. Biochem. Biophys.* 142:238–46
57. Fredriksen, S., Hellung-Larsen, P. 1972. *Exp. Cell Res.* 71:289–92
58. Rein, A., Penman, S. 1969. *Biochim. Biophys. Acta* 190:1–9
59. Rein, A. 1971. *Biochim. Biophys. Acta* 232:306–13
60. Yazdi, E., Gyorkey, F. 1971. *J. Natl. Cancer Inst.* 47:765–70
61. Nohga, K., Reddy, R., Busch, H. 1981. *Cancer Res.* 41:2215–20
62. Lerner, M. R., Boyle, J., Mount, S., Weliky, J., Wolin, S., Steitz, J. A. 1982. In *RNA Polymerase, tRNA and Ribosomes: Their Genetics and Evolution*, ed. S. Osawa, H. Ozeki, H. Uchida, T. Yura. Japan. In press
63. Branlant, C., Krol, A., Ebel, J. P., Lazar, E., Gallinaro, H., Jacob, M., Sri-Widada, J., Jeanteur, P. 1980. *Nucleic Acids Res.* 8:4143–54
63a. Mounts, S., Steitz, J. A. 1982. *Nucleic Acids Res.* 9:6351–68
64. Wise, J. A., Weiner, A. M. 1981. *J. Biol. Chem.* 256:956–63
64a. Takeishi, K., Kaneda, S. 1981. *J. Biochem.* 90:299–308

65. Reddy, R., Henning, D., Busch, H. 1979. *J. Biol. Chem.* 254:11097–11105
66. Wise, J. A., Weiner, A. M. 1980. Cell 22:109–18
66a. Diener, T. O. 1981. *Proc. Natl. Acad. Sci. USA* 78:5014–15
67. Marzluff, W. F., Murphy, E. C. Jr., Huang, R. C. C. 1974. *Biochemistry* 13:3689–96
68. McReynolds, L., Penman, S. 1974. *Cell* 1:139–45
69. Weinmann, R., Roeder, R. G. 1974. *Proc. Natl. Acad. Sci. USA* 71:1790–94
70. Roeder, R. G., Rutter, W. J. 1970. *Proc. Natl. Acad. Sci. USA* 65:675–82
71. Chesterton, C. J., Butterworth, P. H. W. 1971 *FEBS Lett.* 12:301–8
72. Reeder, R. H., Roeder, R. G. 1972. *J. Mol. Biol.* 70:433–41
73. Zybler, E. A., Penman, S. 1971. *Proc. Natl. Acad. Sci. USA* 68:2861–65
73a. Ro-Choi, T. S., Raj, N. B. K., Pike, L. M., Busch, H. 1976. *Biochemistry* 15:3823–28
74. Frederiksen, S., Hellung-Larsen, P., Gram-Jensen, E. 1978. *FEBS Lett.* 87:227–31
75. Gram-Jensen, E., Hellung-Larsen, P., Frederiksen, S. 1979. *Nucleic Acids Res.* 6:321–30
76. Eliceiri, G. L. 1980. *J. Cell Physiol.* 102:199–207
77. Tamm, I., Kikuchi, T., Darnell, J. E., Salditt-Georgieff, M. 1980. *Biochemistry* 19:2743–48
78. Roop, D. R., Kristo, P., Stumph, W. E., Tsai, M. J., O'Malley, B. W. 1981. *Cell* 23:671–80
79. Reichel, R., Benecke, B. 1980. *Nucleic Acids Res.* 8:225–33
80. Harada, F., Kato, N., Hoshino, H. O. 1979. *Nucleic Acids Res.* 7:909–18
81. Hendrick, J. P., Wolin, S. L., Rinke, J., Lerner, M. R., Steitz, J. A. 1981. *Mol. Cell. Biol.* 1:1138–49
82. Eliceiri, G. L. 1974. *Cell* 3:11–14
82a. Frederiksen, S., Hellung-Larsen, P. 1975. *FEBS Lett.* 58:374–78
83. Eliceiri, G. L., Sayavedra, M. S. 1976. *Biochem. Biophys. Res. Commun.* 72:507–12
84. Salditt-Georgieff, M., Harpold, M., Chen-Kiang, S., Darnell, J. E. Jr. 1980. *Cell* 19:69–78
85. Eliceiri, G. L. 1979. *Nature* 279:80–81
86. Denison, R., Van Arsdell, S., Bernstein, L., Weiner, A. 1981. *Proc. Natl. Acad. Sci. USA* 78:810–14
87. Van Arsdell, S. W., Denison, R. A., Bernstein, L. B., Weiner, A., Manser, T., Gesteland, R. F. 1981. *Cell* 26:11–17

88. Manser, T., Gesteland, R. F. 1981. *J. Mol. Appl. Genet.* 1:117–25
89. Lund, E., Dahlberg, J. E., Buckland, R., Cooke, H. 1982. *Proc. 10th ICN-UCLA Symp. Mol. Cell Biol.* 109:
89a. Hayashi, K. 1981. *Nucleic Acids Res.* 9:3379–88
89b. Ohshima, Y., Okada, N., Tani, T., Itoh, Y., Itoh, M. 1981. *Nucleic Acids Res.* 9:5145–58
90. Alonso, A., Krieg, L., Winter, H., Sekeris, C. E. 1980. *Biochem. Biophys. Res. Commun.* 95:148–55
91. Engberg, J., Hellung-Larsen, P., Frederiksen, S. 1974. *Eur. J. Biochem.* 41:321–28
92. Marzluff, W. F., White, E. L., Benjamin, R., Huang, R. C. C. 1975. *Biochemistry* 14:3715–24
93. Frederiksen, S., Flodgard, H., Hellung-Larsen, P. 1981. *Biochem. J.* 193:743–48
94. Gunning, P. W., Austin, L., Jeffrey, P. L. 1979. *J. Neurochem.* 32:1725–36
95. Reddy, R., Li, W.-Y., Henning, D., Choi, Y. C., Nohga, K., Busch, H. 1981. *J. Biol. Chem.* 256:8452–59
96. Gurney, T., Eliceiri, G. 1980. *J. Cell Biol.* 87:398–403
97. Prestayko, A. W., Tonato, M., Busch, H. 1971. *J. Mol. Biol.* 47:505–15
98. Sekeris, C. E., Niessing, J. 1975. *Biochem. Biophys. Res. Commun.* 62:642–50
99. Sekeris, C. E., Guialis, A. 1981. *Cell Nucleus* 8:247–59
100. Northemann, W., Klump, H., Heinrich, P. C. 1979. *Eur. J. Biochem.* 99:447–56
101. Northemann, W., Scheurlen, M., Gross, V., Heinrich, P. C. 1977. *Biochem. Biophys. Res. Commun.* 76:1130–37
102. Zieve, G., Penman, S. 1981. *J. Mol. Biol.* 145:501–23
103. Jacob, M., Devilliers, G., Fuchs, J-P., Gallinaro, H., Gattoni, R., Judes, C., Stevenin, J. 1981. *Cell Nucleus* 8:193–246
104. Deimel, B., Louis, C., Sekeris, C. E. 1977. *FEBS Lett.* 73:80–84
105. Gallinaro, H., Jacob, M. 1981. *Biochim. Biophys. Acta* 652:109–20
106. Gallinaro, H., Jacob, M. 1979. *FEBS Lett.* 104:176–82
107. Guimont-Ducamp, C., Sri-Widada, J., Jeanteur, P. 1977. *Biochimie* 59:755–58
108. Howard, E. F. 1978. *Biochemistry* 17:3228–36
109. Seifert, H., Scheurlen, M., Northe-/mann, W., Heinrich, P. C. 1979. *Biochim. Biophys. Acta* 564:55–66

534 BUSCH ET AL

110. Flytzanis, C., Alonso, A., Louis, C., Krieg, L., Sekeris, C. E. 1978. *FEBS Lett.* 96:201–6
111. Brunel, C., Sri-Widada, J., Lelay, M.-N., Jeanteur, P., Liautard, J.-P. 1981. *Nucleic Acids Res.* 9:815–30
112. Maxwell, E. S., Maundrell, K., Scherrer, K. 1980. *Biochem. Biophys. Res. Commun.* 97:875–82
112a. Calvet, J. P., Pederson, T. 1981. *Cell* 26:363–70
113. Daskal, Y., Komaromy, L., Busch, H. 1980. *Exp. Cell Res.* 126:39–46
114. Daskal, Y. 1981. *Cell Nucleus* 8:117–37
115. Walker, T. A., Pace, N. R., Erikson, R. L., Erikson, E., Behr, F. 1974. *Proc. Natl. Acad. Sci. USA* 71:3390–94
116. Gunning, P. W., Beguin, P., Shooter, E. M., Austin, L., Jeffrey, P. L. 1981. *J. Biol. Chem.* 256:6670–75
117. Jelinek, W., Leinwand, L. 1978. *Cell* 15:205–14
117a. Miller, T. E., Huang, C. Y., Pogo, A. O. 1978. *J. Cell Biol.* 76:692–704
118. Reddy, R., Ro-Choi, T. S., Henning, D., Busch, H. 1974. *J. Biol. Chem.* 249:6486–94
119. Ro-Choi, T. S., Reddy, R., Choi, Y. C., Raj, N. B., Henning, D. 1974. *Fed. Proc.* 33:1548
120. Rottman, F., Shatkin, A., Perry, R. 1974. *Cell* 3:1977–79
121. Shatkin, A. J. 1976. *Cell* 9:645
122. Shibata, H., Reddy, R., Henning, D., Ro-Choi, T. S., Busch, H. 1974. *Mol. Cell. Biochem.* 4:3–19
123. Krol, A., Branlant, C., Lazar, E., Gallinaro, H., Jacob, M. 1981. *Nucleic Acids Res.* 9:841–58
123a. Reddy, R., Henning, D., Busch, H. 1981. *Biochim. Biophys. Res. Commun.* 98:1076–83
124. Epstein, P., Reddy, R., Busch, H. 1981. *Proc. Natl. Acad. USA* 78:1562–66
125. Korn, L. J., Queen, C. L., Wegman, N. M. 1977. *Proc. Natl. Acad. Sci. USA* 74:4401–5
126. Shibata, H., Ro-Choi, T. S., Reddy, R., Choi, Y. C., Henning, D., Busch, H. 1975. *J. Biol. Chem.* 250:3909–20
126a. Reddy, R., Henning, D., Epstein, P., Busch, H. 1981. *Nucleic Acids Res.* 9:5645–57
127. Ro-Choi, T. S., Choi, Y. C., Henning, D., McCloskey, J., Busch, H. 1975. *J. Biol. Chem.* 250:3921–28
128. Busch, H., Smetana, K. 1970. In *The Nucleolus*, pp. 285–314. New York: Academic
129. Reddy, R., Henning, D., Busch, H. 1980. *J. Biol. Chem.* 255:7029–33

130. Reddy, R., Ro-Choi, T. S., Henning, D., Shibata, H., Choi, Y. C., Busch, H. 1972. *J. Biol. Chem.* 247:7245–50
131. Reddy, R., Henning, D., Busch, H. 1981. *J. Biol. Chem.* 256:3532–38
132. Krol, A., Branlant, C., Lazar, E., Gallinaro, H., Jacob, M. 1981. *Nucleic Acids Res.* 9:2699–2716
133. Kato, N., Harada, F. 1981. *Biochem. Biophys. Res. Commun.* 99:1477–85
134. Krol, A., Gallinaro, H., Lazar, E., Jacob, M., Branlant, C., 1981. *Nucleic Acids. Res.* 9:769–88
135. Kato, N., Harada, F. 1981. *Biochem. Biophys. Res. Commun.* 99:1468–76
136. Puvion, E., Moyne, G. 1981. *Cell Nucleus* 8:59–115
137. Epstein, P., Reddy, R., Henning, D., Busch, H. 1980. *J. Biol. Chem.* 255:8901–6
138. Harada, F., Kato, N., Nishimura, S. 1980. *Biochem. Biophys. Res. Commun.* 95:1332–40
138a. Busch, H., Reddy, R., Henning, D., Epstein, P. 1980. In *International Cell Biology*, ed. H. Schweiger, pp. 47–52. Berlin: Springer
139. Raj, N., Ro-Choi, T. S., Busch, H. 1975. *Biochemistry* 14:80–87
140. Fuchs, J. P., Jacob, M. 1979. *Biochemistry* 18:4202–8
141. Kinlaw, C. S., Berget, S. M. 1981. *Fed. Proc.* 40:1765
142. White, P. J., Gardner, W. D., Hoch, S. O. 1981. *Proc. Natl. Acad. Sci. USA* 78:626–30
143. Liew, T. S., Tan, E. M. 1979. *Arth. Rheum.* 22:635
144. Takano, M., Golden, S. S., Sharp, G. C., Agris, P. 1981. *Biochemistry* 20:5929–36
145. Takano, M., Agris, P. F., Sharp, G. C. 1981. *Clin. Res.* 28:559A
146. Teppo, A. M., Gripenberg, M., Kurk, P., Baklein, K., Helve, T., Wegelius, O. 1981. *Biochim. Biophys. Acta.* In press
147. Yang, V. W., Lerner, M., Steitz, J. A., Flint, S. J. 1981. *Proc. Natl. Acad. Sci. USA* 78:1371–75
148. Avvedimento, V. E., Vogeli, G., Yamada, Y., Maizel, J. E., Pastan, I., Crombrugghe, B. 1980. *Cell* 21:689–96
148a. Dickson, A., Ninomiya, Y., Bernard, P., Pesciotta, M., Parsons, J., Green, G., Eikenberry, E. F., de Crombrugghe, B., Vogeli, G., Pastan, I., Fietzek, P., Olsen, B. R. 1981. *J. Biol. Chem.* 16:8407–15
148b. Ohshima, Y., Itoh, M., Okada, N., Miyata, T. 1981. *Proc. Natl. Acad. Sci. USA* 78:4471–74

149. Pederson, T., Bhorjee, J. S. 1979. *J. Mol. Biol.* 128:451–80
150. Altman, S. 1978. *Int. Rev. Biochem.* 17:19–44
151. Altman, S., Bowman, E., Garber, R., Kole, R., Koski, R., Stark, B. C. 1980. *tRNA: Biological Aspects*, pp. 71–82. Cold Spring Harbor, NY: Spring Harbor Lab.
152. Bothwell, A. L. M., Garber, R. L., Altman, S. 1976. *J. Biol. Chem.* 251: 7709–16
153. Bowman, E. J., Altman, S. 1980. *Biochim. Biophys. Acta* 613:439–47
154. Garber, R. L., Siddiqui, M. A. Q., Altman, S. 1978. *Proc. Natl. Acad. Sci. USA* 75:635–39
155. Kole, R., Altman, S. 1979. *Proc. Natl. Acad. Sci. USA* 76:3795–99
156. Kole, R., Baer, M. F., Stark, B. C., Altman, S. 1980. *Cell* 19:881–87
157. Koski, R., Bothwell, A., Altman, S. 1976. *Cell* 9:101–16
158. Blanchard, J. M., Weber, J., Jelinek, W., Darnell, J. E. Jr. 1978. *Proc. Natl. Acad. Sci. USA* 75:5344–48
159. Harada, H., Igarashi, T., Muramatsu, M. 1980. *Nucleic Acids Res.* 8:587–99
160. Goldenberg, C. J., Raskas, H. J. 1980. *Biochemistry* 19:2719–23
161. Manley, J-L., Sharp, P., Gefter, M. 1979. *Proc. Natl. Acad. USA* 76:160–64
162. Tsai, M. J., Ting, A. C., Nordstrom, J. L., Zimmer, W., O'Malley, B. W. 1980. *Cell* 22:219–30
163. Lerner, M., Boyle, J., Hardin, J. A., Steitz, J. A. 1981. *Science* 211:400–2
164. Harada, F., Kato, N. 1980. *Nucleic Acids Res.* 1273–85
164a. Haynes, S. R., Toomey, T. P., Leinwand, L., Jelinek, W. R. 1981. *Mol. Cell. Biol.* 1:573–83
165. Jelinek, W. R., Toomey, T. P., Leinwand, L., Duncan, C. H., Biro, P. A., Choudary, P. V., Weissman, S. M., Rubin, C. M., Houck, C. M., Deininger, P. L., Schmid, C. W. 1980. *Proc. Natl. Acad. Sci. USA* 77:1398–402
166. Ro-Choi, T. S., Reddy, R., Henning, D., Takano, T., Taylor, C., Busch, H. 1972. *J. Biol. Chem.* 247:3205–22
167. Lerner, M. R., Andrews, N. C., Miller, G., Steitz, J. A. 1981. *Proc. Natl. Acad. Sci. USA* 78:805–9
167a. Gojobori, T., Nei, M. 1981. *J. Mol. Evol.* 17:245–50
168. Bishop, J. M., Levinson, W., Sullivan, D., Farrshier, L., Quintrell, N., Jackson, J. 1970. *Virology* 42:927–37
169. Erikson, E., Erikson, R. L., Henry, B., Pace, N. R. 1973. *Virology* 53:40–46

170. Li, W-Y., Reddy, R., Henning, D., Epstein, P., Busch, H. 1982. *J. Biol. Chem.* In press
171. Weiner, A. M. 1980. *Cell* 22:209–18
172. Savage, H., Grinchishin, V., Fang, W., Busch, H. 1974. *Physiol. Chem. Phys.* 6:113–26
173. Benecke, B., Penman, S. 1977. *Cell:* 12:939–46
174. Deshpande, A. K., Jakowlew, S. B., Arnold, H., Crawford, P. A., Siddiqui, M. A. Q. 1977. *J. Biol. Chem.* 252:6521–27
175. Gunning, P. W., Shooter, E. M., Austin, L., Jeffrey, P. L. 1981. *J. Biol. Chem.* 256:6663–69
176. Pan, J., Elder, J. T., Duncan, C. H., Weissman, S. M. 1981. *Nucleic Acids Res.* 9:1161–70
177. Duncan, C. H., Jagadeeswaran, P., Wang, R., Weissman, S. M. 1981. *Gene.* In press
177a. Haynes, S. R., Jelinek, W. R. 1981. *Proc. Natl. Acad. Sci. USA* 78:6130–34
178. Hutchinson, M. A., Hunter, T., Eckhart, W. 1979. *Cell* 15:65–77
179. Alwine, J. C., Khoury, G. 1980. *J. Virol.* 36:701–8
180. Colonno, R. J., Banerjee, A. K. 1976. *Cell* 8:197–204
181. Testa, D., Chands, P. K., Banerjee, A. K. 1980. *Cell* 21:267–75
182. Emerson, S. V., Dierks, P. M., Parson, J. T. 1977. *J. Virol.* 23:708–16
183. Rao, D. D., Huang, A. S. 1978. *Proc. Natl. Acad. Sci. USA* 76:3742–45
184. Peters, G. G., Harada, F., Dahlberg, J. E., Panet, A., Haseltine, W. A., Baltimore, D. 1977. *J. Virol.* 21:1031–42
185. Harada, F., Sawyer, R. C., Dahlberg, J. E. 1975. *J. Biol. Chem.* 250:3487–97
186. Diener, T. O. 1981. *Sci. Am.* 244:66–73
187. Koper, J. M., Tousignant, M. E., Lot, H. 1976. *Biochem. Biophys. Res. Commun.* 72:1237–43
188. Mathews, M. B., Pettersson, U. 1978. *J. Mol. Biol.* 119:293–328
189. Soderlund, H., Pettersson, U., Vennstrom, B., Philipson, L., Mathews, M. B. 1976. *Cell* 7:585–93
190. Raska, K., Schuster, L. M., Varrichio, F. 1976. *Biochem. Biophys. Res. Commun.* 69:79–84
191. Weinmann, R., Grendler, T. G., Raskas, H., Roeder, R. G. 1976. *Cell* 7:557–66
192. Weill, P. A., Segall, J., Harris, B., Ng, S. Y., Roeder, R. G. 1979. *J. Biol. Chem.* 254:6163–73
193. Weinmann, R., Raskas, H. J., Roeder, R. G. 1974. *Proc. Natl. Acad. Sci. USA* 71:3426–30

194. Price, R., Penman, S. 1972. *J. Virol.* 9:621–26
195. Ohe, K., Weissman, S. M., Cooke, N. 1969. *J. Biol. Chem.* 244:5320–32
196. Ohe, K., Weissman, S. M. 1979. *J. Biol. Chem.* 245:6991–7009
197. Harris, B., Roeder, R. G. 1978. *J. Biol. Chem.* 253:4120–27
198. Mathews, M. B. 1975. *Cell* 6:223–29
199. Celma, M. L., Pan, J., Weissman, S. M. 1977. *J. Biol. Chem.* 252:9032–42
200. Pan, J., Celma, M. L., Weissman, S. M. 1977. *J. Biol. Chem.* 252:9047–54
201. Vennstrom, B., Pettersson, U., Philipson, L. 1978. *Nucleic Acids Res.* 5:195–204
202. Akusjarvi, G., Mathews, M. B., Andersson, P., Vennstrom, B., Petterson, U. 1980. *Proc. Natl. Acad. Sci. USA* 77:2424–28
202a. Mathews, M. B., Grodzicker, T. 1981. *J. Virol.* 38:849–62
203. Mathews, M. B. 1980. *Nature* 285:575–77
204. Blanchard, J. 1980. *Biochem. Biophys. Res. Commun.* 97:524–29
205. Gallinaro, H., Gattoni, R., Stevenin, J., Jacob, M. 1980. *Biochem. Biophys. Res. Comm.* 95:20–26
206. Beltz, G., Flint, S. J. 1979. *J. Mol. Biol.* 131:353–73
206a. Shenk, T., Fowlkes, D. M., Thimmappaya, B., Weinberger, C. 1981. *Fed. Proc.* 40:1755
207. Alwine, J. C., Dhar, R., Khoury, G. 1980. *Proc. Natl. Acad. Sci. USA* 77:1379–83
208. Weissman, S. M. 1980. *Mol. Cell. Biochem.* 35:29–38
209. Mark, D. F., Berg, P. 1980. *Cold Spring Harbor Symp. Quant. Biol.* 44:55–62
210. Krause, M. O., Ringuette, M. J. 1977. *Biochem. Biophys, Res. Commun.* 76:786–803
210a. Rosa, M. D., Gottlieb, E., Lerner, R. M. and Steitz, J. A. 1981. *Mol. Cell Biol.* 1:785–96
211. Emerson, S. V., Dierks, P. M., Parsons, J. T. 1977. *J. Virol.* 23:708–16
212. Schubert, M., Keene, J. D., Lazzarini, R. A., Emerson, S. V. 1978. *Cell* 15:103–12
213. Semler, B. L., Perrault, J., Abelson, J., Holland, J. J. 1978. *Proc. Natl. Acad. Sci. USA* 75:4704–8
214. Rao, D. D., Huang, A. S. 1980. *J. Virol.* 36:756–65
215. Leppert, M., Rittenhouse, L., Perrault, J., Summers, D. F., Kolakofsky, D. 1979. *Cell* 18:735–48
216. Collonno, R. J., Banerjee, A. K. 1978. *Cell* 15:93–101

217. Bonar, R. A., Sverak, L., Bolognesi, D. P., Langlors, A. J., Beard, D., Beard, J. W. 1967. *Cancer Res.* 27:1138–57
218. Duesberg, P. H. 1968. *Proc. Natl. Acad. Sci. USA* 60:1511–18
219. Bishop, M. M., Levinson, W. E., Quintrell, N., Sullivan, D., Fanshier, L., Jackson, J. 1970. *Virology* 42:182–95
220. Sawyer, R. C., Harada, F., Dahlberg, J. E. 1974. *J. Virol.* 13:1302–11
221. Dahlberg, J. E., Sawyer, R. C., Taylor, J. M., Faras, A. J., Levinson, W. E., Goodman, H. M., Bishop, J. 1974. *J. Virol.* 13:1126–33
222. Harada, F., Taylor, J. M., Levinson, W. E., Bishop, J. M., Goodman, H. M. 1974. *J. Virol.* 13:1134–42
223. Harada, F., Peters, G. G., Dahlberg, J. E. 1979. *J. Biol. Chem.* 254:10979–85
224. Peters, G. G., Glover, C. 1980. *J. Virol.* 35:31–40
225. Harada, F., Kato, N., Hoshino, H. 1979. *Nucleic Acids Res.* 7:909–17
226. Perry, R. P. 1976. *Ann. Rev. Biochem.* 45:605–29
227. Schlessinger, D. 1980. *Ribosomes, Structure, Function and Genetics,* ed. G. Chambliss et al. pp. 767–80. Baltimore: Univ. Park Press
228. Planta, R. J., Meyerink, J. H. 1980. See Ref. 227, pp. 872–87
229. Boyington, J. E., Gillham, N. W., Lambowitz, A. M. 1980. See Ref. 227, pp. 903–49
230. Long, E. O., Dawid, I. B. 1980. *Ann. Rev. Biochem.* 49:727–64
231. Busch, H., Rothblum, L., eds. 1982. *The Cell Nucleus* Vols. 9, 10. In press
232. Gubler, U., Wyler, T., Seebeck, T., Braun, R. 1980. *Nucleic Acids Res.* 8:2647–64
233. Engberg, J., Din, N., Eckert, W. A., Kaffenberger, W., Pearlman, R. E. 1980. *J. Mol. Biol.* 142:289–313
234. Leon, W., Fouts, D. L., Manning, J. 1978. *Nucleic Acids Res.* 5:491–504
235. Anderson, S., Bankier, A. T., Barrell, B. G., de Bruijn, M. H. L., Coulson, A. R., Drouin, J., Eperon, I. C., Nierlich, D. P., Roe, B. A., Sanger, F., Schreier, P. H., Smith, A. J. H., Staden, R., Young, I. G. 1981. *Nature* 290:457–65
236. Van Etten, R. A., Walberg, M. W., Clayton, D. A. 1980. *Cell* 22:157–70
237. Locker, J., Morimoto, R., Synenki, R. M., Rabinowitz, M. 1980. *Curr. Genet.* 1:163–72
238. Hahn, U., Lazarus, C. M., Lunsdorf, H., Kuntzel, H. 1979. *Cell* 17:191–200
239. Heckman, J. E., Yin, S., Alzner-DeWeerd, B., RajBhandary, U. L. 1979. *J. Biol. Chem.* 254:12694–700

240. Schwarz, Z., Kossel, H. 1980. *Nature* 283:739–42
241. Edwards, K., Kossel, H. 1981. *Nucleic Acids. Res.* 9:2853–69
242. Hartley, M. R. 1979. *Eur. J. Biochem.* 96:311–20
243. Gray, P. W., Hallick, R. B. 1978. *Biochemistry* 17:284–89
244. Orozco, E. M. Jr., Rushlow, K. E., Dodd, J. R., Hallick, R. B. 1980. *J. Biol. Chem.* 255:10397–11003
245. Malnoe, P., Rochaix, J.-D. 1978. *Mol. Gen. Genet.* 166:269–75
246. Allet, B., Rochaix, J.-D. 1979. *Cell* 18:55–60
247. Brosius, J., Dull, T. J., Sleeter, D. D., Noller, H. F. 1981. *J. Mol. Biol.* 148:107–27
248. Planta, R. J., Retel, J., Klootwijk, J., Meyerink, J. H., DeJonge, P., Van Keulen, H., Brand, R. C. 1977. *Biochem. Soc. Trans.* 5:462–66
249. Maizels, N. 1976. *Cell* 9:431–38
250. Lund, E., Dahlberg, J. E. 1977. *Cell* 11:247–62
251. Long, E. O., Dawid, I. B. 1980. *J. Mol. Biol.* 138:873–78
252. Veldman, G. M., Brand, R. C., Klootwijk, J., Planta, R. J. 1980. *Nucleic Acids Res.* 8:2907–20
253. Merten, S., Synenki, R. M., Locker, J., Christianson, T., Rabinowitz, M. 1980. *Proc. Natl. Acad. Sci. USA* 77:1417–21
254. Mannella, C. A., Collins, R. A., Green, M. R., Lambowitz, A. M. 1979. *Proc. Natl. Acad. Sci. USA* 76:2635–39
255. Bram, R. J., Young, A. A., Steitz, J. A. 1980. *Cell* 19:393–402
256. Dahlberg, A. E., Dahlberg, J. E., Lund, E., Tokimatsu, H., Rabson, A. B., Calvert, P. C., Reynold, F., Zahalak, M. 1978. *Proc. Natl. Acad. Sci. USA* 75:3598–602
257. Gegenheimer, P., Apirion, D. 1980. *J. Mol. Biol.* 143:227–57
258. Egawa, K., Choi, Y. C., Busch, H. 1971. *J. Mol. Biol.* 56:565–77
259. Rubtsov, P. M., Musakhanov, M. M., Zakharyev, V. M., Krayev, A. S., Skryabin, K. G., Bayev, A. A. 1980. *Nucleic Acids Res.* 8:5779–94
260. Salim, M., Maden, B. E. H. 1981. *Nature* 291:205–8
261. Choi, Y. C., Busch, H. 1978. *Biochemistry* 17:2551–60
262. Woese, C. R., Magrum, L. J., Gupta, R., Siegel, R. B., Stahl, D. A., Kop, J., Crawford, N., Brosius, J., Gutell, R., Hogan, J. J. 1980. *Nucleic Acids. Res.* 8:2275–94
263. Goltz, C., Brimacombe, R. 1980. *Nucleic Acids Res.* 8:2377–96
264. Jordan, B. R. 1975. *J. Mol. Biol.* 98:277–80
265. MacKay, R. M., Zablen, L. B., Woese, C. R., Doolittle, W. F. 1979. *Arch. Microbiol.* 123:165–72
266. Bayev, A. A., Georgiev, O. J., Hadjiolov, A. A., Nikolaev, N., Skryabin, K. G., Zakharyev, V. M. 1981. *Nucleic Acids Res.* 9:789–99
267. Moss, T., Boseley, P. G., Birnstiel, M. L. 1980. *Nucleic Acids Res.* 8:467–85
268. Branlant, C., Krol, A., Machatt, M. A., Pouyet, J., Ebel, J.-P., Edwards, K., Kossel, H. 1981. *Nucleic Acids Res.* In press
269. Erdmann, V. A. 1976. *Prog. Nucleic Acid Res. Mol. Biol.* 18:45–90
270. Nazar, R. N. 1980. *FEBS Lett.* 119:212–14
271. Gerbi, S. A., Gourse, R. L., Clark, C. G. 1982. *Cell Nucleus* In press
272. Pavlakis, G. N., Jordan, B. R., Wurst, R. M., Vournakis, J. N. 1979. *Nucleic Acids Res.* 7:2213–38
273. Jordan, B. R., Latil-Dammotte, M., Jourdan, R. 1980. *Nucleic Acids Res.* 8:3565–773
274. Ishikawa, H. 1979. *Biochem. Biophys. Res. Commun.* 90:417–24
275. Erdmann, V. A. 1981. *Nucleic Acids Res.* 9:r25–r42
276. Bowman, C. M., Dyer, T. A. 1979. *Biochem. J.* 183:605–13
277. MacKay, R. M. 1981. *FEBS Lett.* 123:17–18
278. Wildeman, A. G., Nazar, R. N. 1980. *J. Biol. Chem.* 255:11896–11900
279. Takaiwa, F., Sugiura, M. 1980. *Mol. Gen. Genet.* 180:1–4
280. Ferrai, S., Yehle, C. O., Robertson, H. D. 1980. *Proc. Natl. Acad. Sci. USA* 77:2395–99
281. Denoya, C., Costa Giomi, P. C., Scodeller, E. A., Vasquez, C., La Torre, J. L. 1981. *Eur. J. Biochem.* 115:375–83
282. Carin, M., Jensen, B. F., Jentsch, K. D., Leer, J. C., Nielsen, O. F., Westergaard, O. 1980. *Nucleic Acids Res.* 8:5551–66
283. Brabowski, P. J., Zaug, A. J., Cech, T. R. 1981. *Cell* 23:467–76
284. Wild, M. A., Sommer, R. 1980. *Nature* 283:693–94
285. Dujon, B. 1980. *Cell* 20:185–97
286. Bos, J. L., Osinga, K. A., Vander Horst, G., Hecht, N. B., Tabak, H. F., Van Ommen, G. J. B., Borst, P. 1980. *Cell* 20:207–14
287. Haid, A., Grosch, G., Schmelzer, C., Schweyen, R. J., Kaudewitz, F. 1980. *Curr. Genet.* 1:155–61
288. Halbreich, A., Pajot, P., Foucher, M.,

Grandchamp, C., Slonimski, P. 1980. *Cell* 19:321–29
289. Van Ommen, G. J., Boer, P. H., Groot, G. S. P., DeHaan, M., Roosendaal, E., Grivell, L. A., Haid, A., Schweyen, R. J. 1980. *Cell* 20:173–83
290. Schmelzer, C., Haid, A., Grosch, G., Schweyen, R. J., Kaudewitz, F. 1981. *J. Biol. Chem.* 256:7610–19
291. Lewin, B. 1980. *Cell* 22:645–46
292. Spritz, R. A., Jagadeeswaran, P., Chowdary, P. V., Biro, P. A., Elder, J. T., DeRiel, J. K., Manley, J. L., Gefter, M. L., Forget, B. G., Weissman, S. M. 1981. *Proc. Natl. Acad. Sci. USA* 78:2455–59
293. Naora, H., Deacon, N. J. 1981. *Differentiation* 18:125–32

294. Goldstein, L., Wise, G. E., Ko, C. 1977. *J. Cell Biol.* 73:322–31
295. Pogo, A. O. 1981. *Cell Nucleus,* 8:337–67
296. Kanehisa, T., Oki, Y., Ikuta, K. 1974. *Biochim. Biophys. Acta* 277:584–89
297. Bester, A. J., Kennedy, D. S., Heywood, S. M. 1975. *Proc. Natl. Acad. Sci. USA* 72:1523–27
298. Rao, M. S., Blackstone, M., Busch, H. 1977. *Biochemistry* 16:2756–62
299. Howard, E., Stubblefield, E. 1971. *Exp. Cell Res.* 70:460–62
300. Tomizawa, J., Itoh, T., Selzer, G., Som, T. 1981. *Proc. Natl. Acad. Sci. USA* 78:1421–25
301. Hotta, Y., Stern, H. 1981. *Cell.* 27:309–19

Ann. Rev. Biochem. 1980. 49:1115–56
16

NUCLEOSOME STRUCTURE

James D. McGhee and Gary Felsenfeld

Laboratory of Molecular Biology, National Institute of Arthritis,
Metabolism and Digestive Diseases, National Institutes of Health,
Bethesda, Maryland 20205

CONTENTS

PROSPECTUS AND SUMMARY

Chromatin structure is determined largely by the basic proteins called
histones, responsible for the compaction of the DNA within the nucleus.
However, only the elementary subunit of chromatin structure, the nucleosome, has been identified and characterized in any detail, and it is the
nucleosome with which any review of chromatin structure must first concern itself.

The evidence for the existence of nucleosomes, and accounts of earlier investigations of their properties, are summarized in a number of reviews (1–3). Each nucleosome, excised from chromatin by mild nuclease digestion, contains about 200 base pairs of DNA complexed with histones. At the interior of the particle are the core histones: two molecules each of four species (H2A, H2B, H3, and H4). The DNA is wrapped around this core. A fifth species of histone, H1, is usually attached to the complex, but not as part of the core.

We review recent studies of nucleosome structure and chemistry which answer, albeit incompletely, the following questions: What is the path of the DNA on the surface of the nucleosome? What is the arrangement of histones with respect to each other and to DNA? What regions of the histones are most important in maintaining nucleosome structure? How easy is it for a nucleosome to change its shape or unfold? Do histone core complexes move along DNA by sliding or by exchange, and are there preferred points of interaction relative to DNA sequence? How do nucleosomes pack together to form the highly compacted structures present in the nucleus?

This detailed analysis is undertaken in the belief that much of the behavior of chromatin as a biologically active entity must be understood in terms of the properties of the nucleosome itself. It seems likely that during the next few years we will learn how the nucleosome is unpacked, modified, unfolded, or moved as it plays its role in the nucleus.

STRUCTURE OF THE CORE PARTICLE

Conformation of the DNA

The full nucleosome released by the initial attack of micrococcal nuclease contains 160–240 base pairs of DNA (1) but is not very stable. The nuclease begins at once to digest the terminal DNA originally derived from the spacer region between adjacent nucleosomes (4–9). In the course of this digestion, intermediates of somewhat greater stability are generated. Among these are particles with 160–170 base pairs of DNA which can be isolated with bound histone H1 (10, 11) (see below). Further digestion produces a more stable particle containing about 146 base pairs of DNA [recently revised upwards from the original estimate of 140 base pairs (12–15)] and an octamer consisting of two each of the four core histones, (H2A, H2B, H3, and H4), but no H1. This structure, the "nucleosome core particle." has been the object of most of the physical and chemical studies of subunit structure, since it is easily prepared and well-defined in content, and appears to contain that portion of the nucleosome DNA most tightly bound to the surface of the histone octamer.

The description of the core particle has changed very little during the past two years. The particle is roughly cylindrical with a diameter of 11 nm and a height of about 5.5 nm (16–20). The DNA is at the "outside," wrapped around the core histones. The topological changes induced in DNA by nucleosome binding suggest that the DNA is supercoiled. In the simplest model the DNA is wound around the cylindrical core to form between one and two left-handed superhelical turns. From the known elastic properties of DNA, one can estimate that about 25 kcal are required per mole of nucleosome core to bend 146 base pairs of DNA into a superhelix of the appropriate dimensions (21, 22).

While spectral studies agree that the conformation of nucleosome DNA is of the "B-genus" (23–26), there has been considerable discussion of the way in which this DNA may be deformed to permit superhelix formation. Although it has been suggested that abrupt bends or kinks in the DNA might occur at periodic intervals (27, 28), no special class of backbone conformations has been detected by either phosphorus or proton magnetic resonance (25, 26, 29–31). Moreover, one can build continuously deformed duplex structures that violate none of the known restrictions on bond angles or interatomic contacts in polynucleotides (22, 32). Thus, although the DNA coiling may not be completely uniform (14, 33), there is no positive evidence that supports the presence of kinks in nucleosomes.

A variety of endonucleases have been used to probe the structure of DNA in the nucleosome core: micrococcal nuclease (4–6, 34–37), pancreatic DNase (Dnase I) (37–45), spleen acid DNase (DNase II) (37, 39, 44, 46, 47), Aspergillus endonuclease (44), streptodornase (43), and the endonuclease of *Serratia marcescens* (48). Each of these enzymes can cut the nucleosome DNA at only a small fraction of the total 145 phosphodiester linkages in each chain. For each nuclease, a group of specific DNA fragments is produced which, on electrophoresis in denaturing gels, gives rise to a characteristic ladder pattern, with fragments spaced at roughly ten nucleotide intervals. The spacing is clearly related to the repeat of the DNA duplex, as first suggested by Noll (38), and supported by the observation that protein-free DNA, adsorbed to calcium phosphate, produces a pattern of similarly spaced fragments upon DNase I digestion. (49). The nuclease-cutting sites are limited in number since roughly speaking, the enzyme can only cut when the appropriate site on the DNA is facing away from the surface to which it is bound. However, the pattern of the nucleosome core digest is more complicated than this simple picture would suggest and thus contains information about the details of nucleosome structure.

An important feature of the nuclease digest gel patterns is the modulation in intensity of the bands. Of the 13 potential sites, some are cleaved less readily than others and have been identified by digesting nucleosomes labelled with ^{32}P at their 5' termini. In this way it has been found (40, 45) that the sites located at approximately 30, 60, 80, and 110 nucleotides from the 5' end are relatively protected from cleavage. The simplest way to explain this result is to assume that portions of the histones block these specific sites. Finch et al (17) have constructed such a model, based on Lutter's data (45) for the kinetics of DNase I digestion. They point out that there is a correlation between the cutting frequencies at pairs of sites located about 70–90 base pairs apart. Most notably, the sites located roughly 30 and 110 nucleotides from the 5' end are highly protected. They suggest that this result is consistent with a superhelix containing about 80 base pairs per turn, since sites on the DNA separated by this distance would be brought close together by supercoiling and might then share similar environments. The data are also consistent with the presence of an overall dyad axis: the absence of a cut at a particular site means that both strands of DNA are in similar environments at those points. Finally, the sense of the superhelix can be deduced if it is accepted that the twist of the DNA is the same in solution as on the nucleosome surface. In that case, a uniform toroidal supercoil must be left-handed if it is the structure which accounts for the topological effects of nucleosome binding (see section on higher order structure).

Recently, the spacing between DNase I cuts has been shown to be 10.4 nucleotides rather than 10.0 (12–15). This may reflect the true repeat of the DNA duplex on the nucleosome surface, or it may arise from some requirement that the nuclease attack from a preferred direction (12). Trifonov & Betteken (50) have assumed the former explanation is true, and have used the 10.4-nucleotide spacing to suggest an alternative explanation for the modulations in the DNase I cutting pattern. They point out that if the number of base pairs per turn of duplex is nonintegral, the orientation of susceptible bonds relative to the nucleosome surface will not be the same at each potential cutting site. If the enzyme can make only a limited angle of approach to the surface, then some potential cutting sites will not be oriented correctly and will not be cut—in other words, an interference pattern will be generated. For a uniform duplex with 10.4 base pairs per turn, and assuming that the enzyme must approach from a direction close to normal to the cylindrical surface, there is agreement between the predicted sites of greatest protection and those observed. Although such a mechanism must play some role in determining the susceptibility of various sites to cleavage, the situation is more complicated than suggested either by this model or by the other simple model involving site blocking by proteins. For example, digestion of both nucleosome cores (13) and artificial nucleosomes constructed with poly(dAdT)·poly(dAdT) (14) suggests that the spacing between cutting sites is smaller at the ends of the nucleosome than in the middle, so that the simplest models cannot be strictly valid.

Analysis using nucleases as probes is further complicated by the fact that different nucleases produce cutting patterns that differ in significant detail (37, 42, 44). Although all enzymes have potential cutting sites at approximately the same places, the exact positions of the cuts are different for each enzyme. Furthermore, cuts on opposite strands at a given site are offset or staggered by a distance that depends upon the enzyme (2, 37, 41, 42). For example, DNase I produces cuts with an average stagger of two nucleotides (5' and recessed), while DNase II produces an average stagger of four nucleotides. Though recent studies with DNase I show that several of the cuts may have a stagger of 1 or 3 nucleotides (13), it is clear that the pattern is not determined solely by the properties of the nucleosome. In the case of DNase II, the relative frequencies with which different sites are cut varies with ionic conditions (37), a result that is not easily explained by any of the simple models of nucleosome structure.

The DNase I cleavage site close to 60 nucleotides from the 5' end is responsive to a number of environmental variables. The cleavage rate at this site is enhanced by histone proteolysis (51), by histone acetylation (52), and by the binding of RNA polymerase (53); it is also much higher in nucleosomes containing poly(dGdC)· poly(dGdC) than in those containing poly(dAdT)· poly(dAdT) (15). Furthermore, this site is close to a region of heightened dimethylsulfate reactivity (54). The cause of this sensitivity is not known but in a 140-base pair core particle in which there are 80 base pairs of DNA per turn, this 60-base pair site must lie close to the DNA terminus.

The nuclease digestion patterns and the size of the nucleosome are best explained if there are about 80 base pairs per turn. However, results of electric dichroism studies have been interpreted as supporting models with about 100 base pairs per turn (55). This analysis, however, requires some assumptions about the uniformity of the DNA path, and implies that the average spacing per base pair must be less than 0.34 nm. Some types of nuclease digestion experiments also suggest that the core histones alone can protect up to 190 base pairs of DNA, considerably more than in the standard core particle (56).

A striking finding is that exonuclease III can also digest nucleosome DNA (57, 58). Starting at the 3'-end, up to 60 bases of the DNA can be removed from a mononucleosome and more than 100 bases from a dinucleosome (57). Although the opposite strand is left intact but unpaired, the particle apparently remains folded. Digestion is not uniform; discrete stopping points are observed at approximately 10-base intervals. Details of the digestion mechanism are not clear, in particular the degree to which the digestion is processive.

In view of their size, it is not too surprising that nucleases have restricted access to the core particle DNA. For example, micrococcal nuclease, by far the smallest nuclease used, is still somewhat larger than the average histone (59). Nonetheless, a substantial fraction of the DNA (3 or 4 out of every 10 base pairs) is accessible to cleavage by one or another of the nucleases. Smaller probes should show even greater accessibility. In an experiment analogous to that used to probe operator-repressor complexes (60, 61), 5'-labeled core particles were methylated with dimethylsulfate, the histones removed, and the DNA cleaved at the position of reacted guanines (54). The surprising result was the lack of protection of guanines in the large groove under conditions where protection of one or two bases in every ten would have been noticed. Raman spectroscopy, both of chromatin and of mononucleosomes, also indicates that histones are not bonded directly to guanine N7 (62). Further evidence that the large groove of core particle DNA is largely accessible comes from the observation that apparently normal core particles can be constructed using T4 DNA, in which roughly every third base pair has a glucose group protruding into the large groove (63). It has been suggested that the DNA large groove is a binding site for nonhistone proteins (62).

The dimethylsulfate reaction with core particles also indicates a lack of protection of adenine N3 in the minor groove, although it is difficult to establish precise protection limits. On the other hand, Goodwin et al (62) have reported that a Raman vibration that has been assigned to adenine N3 is perturbed in chromatin, as if some histone bonding may occur in the minor groove.

In summary, although nuclease digestion studies certainly show that the core particle DNA is held in a precise orientation with respect to the histone core, chemical probing and spectroscopy indicate that major portions of the DNA, especially around the large groove, still seem free and accessible to solvent. Moreover ^{31}P magnetic resonance studies (30, 31) are consistent with a considerable degree of local motion of the core particle DNA. Although regions of DNA that appear free within the isolated core particle may be occluded within the chromatin superstructure, the fact that the core particle digestion patterns can be used to predict nuclear digestion patterns quite accurately (13) makes this possibility less worrisome.

Conformation of the Histones

In this section, we consider observations which might elucidate the conformations and positions of histones within the nucleosome core particle. We include results from histone amino acid sequencing, from the physical complexes formed between purified histones, from chemical modifications of the core histones in vivo, from chemical and enzymic probing, and from chemical cross-linking studies.

The least ambiguous of these diverse findings are summarized in Figure 1. Each of the four core histones is represented by a straight line whose length is proportional to the number of amino acid residues in the molecule. Calf thymus histones are used as an example because all four core histones have been sequenced and because calf thymus chromatin is a favorite experimental substrate. Negatively charged residues (aspartic and glutamic acids) are designated by short lines extending toward the inside of the figure. Positively charged residues are represented by lines extending toward the outside, lysines by a straight line, and arginines by an arrow. Tyrosines (closed circles) are designated since they have been implicated in DNA binding in other systems (64, 65); histidines (open triangles) are shown since they are the obvious candidates for titration in the usual pH range; finally, phenylalanines are designated by open circles. The other symbols are explained in the text as they arise. This diagram is only a schematic summary of experimental results and implies nothing about nucleosome symmetry. In particular, since each histone core contains twice the number of histones displayed in Figure 1, it is possible that, for example, histone H4 could be cross-linked to two distinct, nonequivalent molecules of H2B.

Central to our discussion is the fact that several recent measurements agree that the core histones are present in equimolar amounts (66–68). The most careful study, by Albright et al (68), measured the histone content of electrophoretically separated nucleoprotein species and included the considerable effect of histone variants. The equal stoichiometry of the core histones is of course required by present nucleosome models and these recent results have removed the unsettling inequalities obtained when histone concentrations were measured by gel staining properties.

HISTONE SEQUENCES   Two general features of the core histones sequences have been recognized for some time (69–71): (a) they show a nonuniform distribution of amino acids such that the C-terminal two thirds of the molecules is similar in amino acid composition to a typical globular protein, but the N-terminal third is more basic (see Figure 1); and (b) all of them, and especially the arginine-rich histones H3 and H4, are highly conserved throughout evolution.

Two recent compilations of histone amino acid sequences have appeared (72, 73). For the arginine-rich histones, positions in the amino acid sequences where variations (both inter- and intra-species) have been detected are, for H3 positions, 41, 53, 89, 90, 96, and 127, and, for H4, 60, 73, and

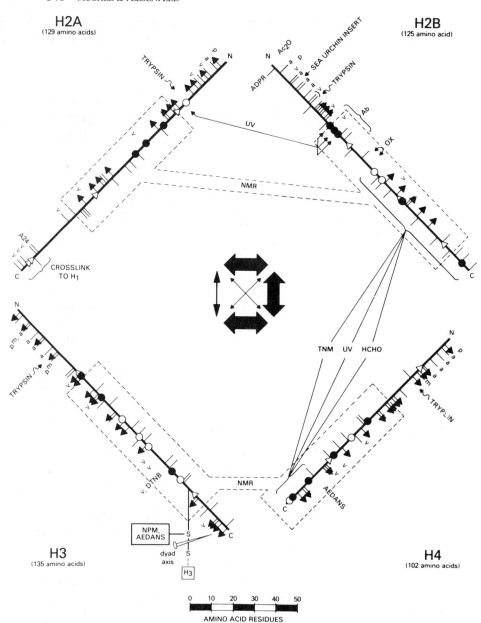

Figure 1   Schematic representation of the four histones of the nucleosome core particle.
Polypeptide backbones are represented by thick straight lines, drawn to the scale indicated at
the bottom of the figure. The slightly lysine rich histones, H2A and H2B, are at the top of
the figure; the arginine rich histones, H3 and H4, are at the bottom. For each histone, the N

77, (72–74). All correspond to conservative substitutions and are marked as v on Figure 1. Possible exceptions to this statement have been noted in yeast (75) and Tetrahymena (76). The slightly lysine-rich histones, H2A and H2B, are more variable than H3 and H4 and, for simplicity, Figure 1 shows only those amino acid alterations that occur within vertebrates. Thus generally conservative substitutions occur at positions 10, 16, 51, 87, 99, 121, 123, and 128 in vertebrate H2a and at 9–10, 21–26, 32, 38–39, 60, 75–77 in H2B (72–74). In histone H2B, where the greatest number of variants have been sequenced, most of the variability occurs in the basic N-terminal third of the molecule, whereas the C-terminal two-thirds is highly conserved (72). Histone H2B in sea urchins can be much longer than in vertebrates, and inserts of up to twenty amino acids can occur at a position corresponding to residues 14 and 15 of calf thymus H2B (72) (see Figure 1).

There are two divergent ways of interpreting these amino-acid sequence differences in terms of core particle structure. One point of view is that considering the evolutionary conservatism of histone sequences, positions at which histones can sustain alterations are those of the least structural consequence, perhaps on the nucleosome surface or at points of less-than-intimate contact with other histones. According to this view, histone sub-

---

terminus is toward the top of the figure, and the C terminus is toward the bottom. Calf thymus histones are used as the example. The thickness of the double-headed arrows in the centre of the figure represents the order of the interaction strengths determined for complex formation between histone pairs (73).

Several amino acid types are represented by the following symbols: straight lines toward the outside of the figure, lysines; arrows pointing out, arginines; straight lines toward the inside of the figure, aspartates or glutamates; closed circles, tyrosines; open circles, phenylalanines; open triangles, histidines. Other symbols are as follows: v, site of primary sequence variation, either intra- or interspecies, (for H3 and H4, all variants are included; for H2A and H2B, only variants within the vertebrates are shown); a, site of in vivo acetylation; p, site of in vivo phosphorylation; m, site of in vivo methylation; $A24$, site of attachment of ubiquitin to histone H2A (at lysine 119) to form the branched polypeptide A24 (112); *TRYPSIN*, approximate site of favored trypsin cleavage (137); Ac_2O, unhindered reaction between acetic anhydride and the N-terminal proline of histone H2B (130–132); *ADPR*, glutamate in position 2 of histone H2B is a strong acceptor of ADP-ribose (109); Ab, an antibody to peptide 26–50 of histone H2B still binds to nucleosomes (147); ox, oxidation of a methionine residue, probably corresponding to one of the positions marked, prevents complexing with histone H2A and weakens complexing of histone with histone H4 (80); *DTNB*, cysteine 96 may be exposed in chromatin to reaction with sulfhydryl reagents such as DTNB (125); *NPM, AEDANS,* fluorescent dyes that have been attached to cysteine 110 of histone H3 (122, 126, 127); AEDANS, a fluorescent dye has also been attached to methionine-84 of histone H4 (129); S--H$_3$, a disulfide cross-link can be formed between the two histone H3 molecules in a core particle (128); if the core particle does indeed have a dyad axis, it must pass through this bond; *NMR,* dashed lines enclose the peptides involved in pairwise interactions (85); *TNM, UV, HCHO,* cross-links demonstrated to form between the indicated regions of histones H2A, H2B, and H4 (148, 149).

types are simply examples of reiterated genes that are undergoing genetic drift. Quite the opposite view arises from consideration of the defined progression of these histone subtypes during development, at least in sea urchins (77–79). In this view, the different histone subtypes are intimately involved in pattern propagation during differentiation, perhaps by causing a different chromatin architecture in different cells. Thus variants may be expected to occur at positions of profound structural consequences. Our personal bias is toward the former view and we consider the positions of amino acid variations shown in Figure 1 as candidates for regions of the least structural importance.

HISTONE COMPLEXES   In this section, we consider experiments in which individually purified histone species are mixed in the absence of DNA and changes in various physical parameters are used as evidence of interaction. The presumption, largely justified by cross-linking studies, (described below) is that thermodynamically important interactions that occur between histone pairs in solution will also occur in chromatin.

The pioneering studies of Isenberg and co-workers (73) defined a hierarchy of pairwise interactions. Strong interactions occur between H2A and H2B and between H2B and H4 to form dimers; H3 and H4 interact to form a tetramer. The interaction between H2A and H3 is weak and interactions between H2A and H4 and H2B and H3 are even weaker. These interactions are indicated by the arrows in the centre of Figure 1, with the thickness of the arrows reflecting the order of the interaction strengths. This same interaction hierarchy has since been demonstrated with histones isolated from plants (80), fungi (75), and protists (76).

A remarkable finding is that mixed histone pairs, from peas and calf (81), yeast and calf (75), and tetrahymena and calf (76) exhibit interactions and interaction strengths almost identical to those within the same species. Furthermore, plant and animal histones, as well as their cross-complexes, can form the same specific cross-links (82). It has thus been argued (73, 81), that the interacting faces of the various histones have undergone at most one or two amino acid changes throughout evolution. An observation made by Spiker & Isenberg (80) indicates that a single residue can indeed be critical to complex formation: if pea histone H2B was treated so as to oxidize methionine residues, the interaction with H2A was abolished but the interaction with H4 was only weakened. Pea histone H2B has a single methionine residue (80) which, if it occurs in a position equivalent to one of the methionine residues of calf thymus H2B, would occur at either position 59 or 62, (indicated in Figure 1 as *ox*).

Bradbury and co-workers (83–85) have conducted a series of elegant experiments designed to determine which part of each histone is involved in complex formation. Chromatographically separated histone peptides are mixed, and physical parameters, in particular, characteristic perturbations in the aromatic regions of the proton magnetic resonance spectrum, are observed for evidence of complex formation. Their conclusions, indicated on Figure 1 by the notation *NMR*, are that histones H2A and H2B interact in the regions corresponding to residues 31–95 for H2A and 37–114 for H2B. Similarly, H3 and H4 interact through the region extending from residue 42–120 for H3 and 38–102 for H4. Thus the N-terminal regions, and in some cases, a short region near the C terminus, are not required for complex formation. Complex formation leads to a substantial increase in the content of alpha-helix but little increase in the content of beta-sheet (80, 83, 86).

IN VIVO MODIFICATIONS   The four core histones undergo five major types of covalent modification: acetylation, phosphorylation, methylation, ADP-ribosylation and, in the case of histone H2A, covalent linkage to the protein ubiquitin (for review see 73).

All four core histones undergo postsynthetic acetylation of one or more lysine residues in the N-terminal third of the molecule. These sites are marked as *a* in Figure 1; [for the exact site positions, see (87)]. Numerous correlations have been made between histone acetylation and gene activity (88, 89). The discovery (90) that hyperacetylated histones can be produced in vivo by butyrate inhibition of histone deacetylation (91, 92, 117) has allowed more direct experimental approaches to this problem. Histone hyperacetylation can alter the nuclease digestion properties of chromatin (52, 93–96) in ways reminiscent of the properties of active genes (97, 98). However, there are conflicting reports as to whether histone hyperacetylation substantially alters the patterns of gene expression (99–101).

From a physical point of view, acetylation of lysines (since it abolishes their positive charge) has long been suggested to weaken the binding between histones and DNA. Tetra-acetylation of H4 decreases its overall positive charge from $\sim$ 20 to $\sim$ 16; the charge of the N-terminal region of H4 would be decreased from $\sim$ 10 to $\sim$ 6. Mathis et al (95) have studied the effect of histone acetylation on core particle stability. Perhaps contrary to the conventional expectation, even extensive histone acetylation had little if any effect on the ability of core particles to exchange between different DNA molecules or to slide along the DNA.

All four core histones can be phosphorylated at some point during their lifespan; the modification consists of a phosphate ester attached to a serine residue, again in the N-terminal regions of the molecules. The sites of phosphorylation, as determined primarily in the developing trout testis (87), are marked as *p* on Figure 1.

Several biological activities have been associated with histone phosphorylation. Histones H2A and H4 undergo phosphorylation and dephosphorylation (along with more or less simultaneous acetylation and deacetylation)
at some point between histone synthesis and assembly into chromatin (87,
88, 102). Histone H2A phosphorylation has been correlated with cellular
content of heterochromatin (103) and increases after hormone stimulation
(104). Phosphorylation and dephosphorylation of H1 and H3 have been
associated with entry into and exit from the condensed stages of mitosis
(105).

Methylation sites on H3 and H4 are marked as *m* on Figure 1. The two
sites of H3 methylation are adjacent to the sites of H3 phosphorylation (87).

ADP-ribosylation of nuclear proteins has been implicated in numerous
biological processes (106). If nuclei from a variety of sources are incubated
in the presence of NAD, ADP-ribose groups are incorporated into all
histones (106–108), with histone H1 a major acceptor. Of the core histones,
H2B seems to be the most highly labeled (109, 110). Burzio et al (109) have
recently estimated that, in rat liver nuclei, approximately 15% of H2B
molecules can be modified with a single ADPR moiety. This is attached as
an ester to a glutamate residue at position 2 of the polypeptide chain,
(ADPR in Figure 1).

An unexpected observation in recent chromatin research is the finding by
Busch and co-workers (111, 112) that, in all cells so far examined, approximately 10% of histone H2A is conjugated to the 74-residue protein ubiquitin through an isopeptide linkage to lysine 119 of H2A (Fig 1). The resulting
branched "semihistone", termed A24, seems simply to replace H2A in the
nucleosome core as demonstrated by its presence in salt-stripped nucleosome cores (113) and by its ability to reconstitute into core particles and
to be specifically cross-linked to histone H2B in solution (114). Protein A24
is itself subject to in vivo modifications (115, 116).

With the possible exception of the formation of A24, the other in vivo
modifications can take place on preassembled chromatin (92, 105, 117–
119).

CHEMICAL PROBES OF HISTONE CONFORMATION   Only histone H3
contains cysteine, either one or two residues depending on the organism,
and this group has been a favorite target for chemical probing. Although
there are uncertainties in defining the intracellular oxidation state of the H3
sulfhydryls (120). Thomas et al (23) have shown, by Raman spectroscopy,
that in isolated chicken erythrocyte core particles the single cysteine residue
at position 110 exists as a free sulfhydryl rather than as a disulfide. Nonetheless, within the native core particle, cysteine 110 is unreactive to sulfhydryl
reagents (121–124). It becomes reactive at high concentrations of salt and/

or denaturants which suggests that cysteine 110 is buried in the intact nucleosome. Trypsin removal of the basic N-terminal regions of the histones does not cause increased sulfhydryl reactivity (123).

In native calf thymus chromatin, only one of the two H3 sulfhydryl groups is reactive to DTNB (dithio-bis-nitrobenzoic acid), whereas both become reactive at high salt (125). Since 110 is probably the unreactive residue, just as it is in chicken core particles, this suggests that cysteine 96 is on the nucleosome surface (*DTNB* in Figure 1). This region of histone H3 has sustained several sequence changes in evolution.

Fluorescent reporter groups have also been attached to cysteine 110 of chicken core particles (122, 126, 127). Again the finding is that native core particles ordinarily do not react but can be made to react if disrupted in urea and/or high salt. Most importantly, the particles can be induced to refold in apparently normal fashion, taking the dyes "inside." The fluorescent properties of the incorporated dye indicate a moderately hydrophobic environment (127). Moreover dyes on the two individual H3 molecules within a single core particle seem to be in close proximity, separated on the order of 1–2 nm (122, 127). Indeed, these two sulfhydryls can be linked together (128), (see below). The ability to incorporate bulky dye groups at cysteine 110 suggests the presence of a considerable cavity in the interior of the core particle. A fluorescent reporter group has also been attached to methionine-84 in histone H4 (129) (*AEDANS* in Figure 1). The spectral properties indicate an apolar environment of this group within reconstituted nucleosomes.

Lysines, long thought to be part of the primary DNA binding sites, are another group of histone residues that have been probed. Lysines have been acetylated by acetic anhydride (which removes the lysine positive charge) under conditions in which the chromatin structure should be minimally perturbed (130–132). The conclusion is that in calf thymus chromatin, the majority of the lysines are unreactive. The one exception is that the N-terminal proline residue of histone H2B seems to react normally (Ac_2O in Figure 1) and occurs in a region of sequence variability. Under harsher conditions, up to 70% of the lysines can be acetylated (133) but the structure of such highly modified chromatin may have been perturbed (134, 135).

A surprising finding is that up to 90% of the lysines in calf thymus chromatin can be reacted with ethyl acetimidate which, unlike acetylation, retains the lysine positive charge (136). With the exception of the apparent loosening of histone H1 binding, the chromatin structure, by a number of criteria, appeared to be unperturbed (136). This suggests that it is the charge of lysine that is important, rather than sterically exacting complementarities.

Another approach to determining the disposition of histones is by digestion with proteases, in particular with trypsin, with its specificity for lysine and arginine residues. Trypsin digestions have been conducted in nuclei, chromatin, and isolated nucleosomes (39, 137–140). All investigators agree that, of the core histones, histone H3 is the most susceptible to proteolysis, but the digestion rate of the other three histones seems to depend upon the experimental material. Weintraub & van Lente (137) demonstrated that a quite stable digestion product existed in which each of the inner histones had lost 20–30 amino acids from its N-terminal regions, i.e. the putative DNA binding sites which contain about one third of the histone positive charge. The approximate locations of these trypsin cleavages are indicated on Figure 1. The nucleosome core particles containing such histones show remarkably little change despite the loss of their tails (discussed below).

Although it is by no means clear whether histone proteolysis has any physiological significance (e.g. for chromatin remodeling during interphase), endogenous proteases have been found to cleave at least H2A and H3 in the same regions as trypsin (141). Also of uncertain physiological significance is the finding of a protease that cleaves histone H2A near the site of its potential joining to ubiquitin (142).

Several tyrosine residues are less accessible to iodination in chromatin than in isolated histones (143–145). The iodination of histone H4 has been the most thoroughly investigated but there is apparent disagreement about the order of reactivities of the individual tyrosines (144, 145). Raman spectroscopy suggests that all tyrosines in the core particle are acceptors of strong hydrogen bonds (23).

Immunoelectron microscopy reveals that at least some histone antigenic determinants are exposed in the core particle and provides a low resolution view of the relative positions of different histones (146). An antibody prepared against peptide 36–50 of histone H2B is still found to react with intact core particles (147). This region is designated as *Ab* on Figure 1 and includes the site of cross-linking to histone H2A (see the next section).

HISTONE-HISTONE CROSS-LINKS   Martinson and co-workers have developed several methods for making short, selective cross-links of high yield, both in nuclei and in isolated nucleosomes. A highly defined cross-link is induced by ultraviolet light between proline at position 26 of histone H2A and a tyrosine at either position 37, 40, or 42 of histone H2B (148). Cross-links can also be formed between histones H2B and H4 by tetranitromethane, ultraviolet light, and formaldehyde (149). These three cross-links are different and distinct but all occur between the C-terminal half of H2B and the C-terminal 18-amino acid residues of H4, as marked in Figure 1. A second formaldehyde cross-link may link the same C-terminal region of H2B to a region close to methionine 84 of H4 (149) i.e. the methionine labeled *AEDANS* in Figure 1. The expected H2A-H2B-H4 cross-linked trimers are also found. Thus histone H2B seems to have three distinct domains, a C-terminal region which interacts with H4, a middle region which interacts with H2A, and the highly basic N-terminal region (149).

Camerini-Otero & Felsenfeld (128) have shown that two molecules of histone H3, cross-linked by a disulfide between cysteines 110, can nonetheless still participate in the formation of an apparently native core particle (see Figure 1). There are two aspects of this cross-link that have theoretical interest: (*a*) it raises the possibility of such a cross-link in vivo [however, see (120)], and (*b*) if the core particle has a dyad axis, it must pass through this disulfide bond.

Finally, a "zero-length" cross-link can be formed between histones H3 and H4 (150) consistent with close apposition of these histones within the core particle. Numerous studies using long nonspecific cross-linking reagents have not been considered here, since the usual finding is that almost everything can be cross-linked to everything else.

RELATION OF HISTONES TO THE DNA   Cross-linking is the most direct method to establish which histone is interacting with a particular section of the core particle DNA. The most detailed report so far is by Mirzabekov et al (151) who have described an ingenious if complicated method to cross-link histone lysines to core particle DNA. An advantage of these reactions is that, ideally, the DNA chains break only at the position of the cross-link, enabling histones to be mapped to their positions on the DNA by means of two-dimensional gels. Interpretation of the gel autoradiographs suggests that cross-links can form in regions that are 6 base pairs long and spaced every ten base pairs, i.e. once every turn of the DNA helix. The linear order of the histones, proceeding from the 5'-end of the DNA, has been assigned as follows (151): H2B, H4, (H3+H2A), (H4+H3), H3, H2B, H2A, and H3. This order, it should be noted, supersedes the order described in the initial report (152). More recent studies (153) refine the histone order only slightly but introduce a further complication—in the central region of the core particle two different histone molecules can be cross-linked to the same 10-base pair region of DNA. This leads the authors to the conclusion that a particular histone-binding region may oscillate between the two DNA strands. The same cross-linking technique has been used to localize histone H3 to the 5'-end of the DNA (154), although this does not agree with the Mirzabekov scheme.

Photocross-linking is an attractive alternative for cross-linking histones to DNA, primarily because reactive intermediates are likely to be shortlived compared to those associated with the usual chemical procedures. A number of groups have demonstrated, both in chromatin and in isolated mononucleosomes, that histones can be photocross-linked to the DNA (155–157), although the cross-linking rates of individual histones are different in the several studies.

It is probably safe to conclude from the above studies that every core histone can be cross-linked to the core DNA; it would have been conceivable to have a histone interact only with other core histones and/or with extracore DNA. It is also likely that the histone-DNA map, like the histone-histone map, will remain tentative and incomplete until a consensus has been reached among a number of different cross-linking methods. From the earliest amino acid sequence determinations of histones, it was apparent that the molecules had a bipartite structure with basic N-terminal regions. These basic histone "tails," also referred to as "arms" or "fingers," were, quite reasonably, expected to be the primary binding sites to DNA (69), and indeed, the early physical studies of isolated histones bound to DNA substantiated this view (158). It has come as somewhat of a surprise that, at least within the core particle, these basic tails are probably not the major DNA binding sites. As summarized in Figure 1, these amino-terminal regions are the exclusive sites of in vivo modifications, and thus a considerable part of one's view of chromatin structure and function depends upon answering the question: where are the histone tails?

The earliest indication that the histone tails may not be the sine qua non of DNA attachment came from the study of Weintraub & van Lente (137), who showed that these histone regions could largely be removed by trypsin hydrolysis; yet the chromatin retained many of its original characteristics. Thus, depending on the exact cleavage site, 30–40% of the histone positive charge could be removed. These trypsin hydrolysis experiments have since been repeated on isolated core particles. Lilley & Tatchell (139) showed that, under trypsin digestion conditions such that essentially none, or at most one, of the inner histones was left intact, the digested core particle still sedimented at 9.7 S which is close to the rate expected if the DNA remained folded but each histone had lost 25% of its mass. Moreover, Whitlock & Stein (140) have demonstrated that similar particles can be reconstituted using trypsinized histones. Trypsinized dinucleosomes still yield a repeating pattern when digested by exonuclease III (57) and, as described above, trypsinized mononucleosomes still retain a buried sulfhydryl group (123).

Although trypsinized core particles seem to remain folded, there are some changes in their properties: (*a*) they show a lower, but still biphasic, thermal transition (139); (*b*) their circular dichroism spectrum becomes more DNA-like, though still markedly suppressed, (139, 140, 159) and (*c*) their digestion by DNaseI is more rapid, though still yielding the usual 10n pattern (51, 139, 140). However, the main conclusion remains that folded "core particles" can exist in the absence of histone tails. Indeed, both proton magnetic resonance (160) and polyanion binding studies (161) indicate that the histone tails may not be rigidly or completely bound within the core particle.

Even on long DNA, the basic histone tails appear to provide little of the free energy of interaction with DNA. Palter & Alberts (162) have measured the salt elution of core histones reconstituted onto high-molecular-weight DNA cellulose and find that both intact histones and trypsinized histones elute over the same salt range as found in native chromatin (163). Not surprisingly therefore, no difference in elution position could be detected between acetylated and nonacetylated histones. These results are complementary to those of Mathis et al (95) cited above.

Thus, the exact role that the histone tails play in chromatin architecture is somewhat puzzling. On the one hand, as suggested by the above experiments, the histone tails seem curiously expendable; on the other hand, especially for histones H3 and H4, the tails remain highly conserved in evolution. Perhaps they bind to their own core particle and/or spacer region only in conjunction with histone H1, or they bind to some other nucleosome, as for example on the next turn of a higher-order helix.

It is not yet clear what type of amino acid is involved in the DNA binding site. Trypsin cleavage of the histone tails removes a somewhat larger fraction of lysine residues (30–50%) than of arginine residues (20–30%). This may suggest that arginine residues are more likely to participate in DNA binding and receives some support from magnetic resonance studies (160, 164). In 2 M NaCl, the histones are dissociated from the DNA but remain complexed to each other; yet tyrosine iodination patterns (143) and Raman spectra (23) remain largely unchanged. Thus tyrosine residues are probably not involved in direct interaction with DNA.

An important insight into the overall histone conformation within the core particle has emerged from spectroscopic studies. Circular dichroism, laser raman (23, 159), and infrared spectroscopy (25) studies all support the fact that the histones within the core particle have a high content of alpha helix, estimated as 50%, with little or no beta-pleated sheet. Furthermore, trypsinization of core particles does not change the alpha-helix content (23, 159). Thus, the remaining globular regions of the histones must be 70–80% alpha-helical, comparable to myoglobin. We note with amusement that the total length of alpha-helix in the core particle is very close to 16×5.5 nms, a sufficient length to run from top to bottom of the core particle once for every turn of the DNA duplex. Perhaps the core particle is constructed like a barrel, with alpha helices (two per histone molecule) acting as the staves and the DNA acting as a helical hoop.

CONFORMATIONAL TRANSITIONS OF THE CORE PARTICLE   Since most if not all of the nuclear DNA is normally complexed with histones and since DNA is both replicated and transcribed, the nucleosomes cannot be entirely static structures. We consider two kinds of structural mobility: deformations of the core particle, and, in a later section, movement of the entire histone core with respect to the DNA. Both classes of reaction have been shown to occur.

Considerable insight into core particle structure was provided by the study of Weischet et al (165) who used hyperchromicity, circular dichroism, and calorimetry to monitor the thermal melting transitions of homogeneous, well-defined core particles. All three techniques showed transitions with two phases, both of which took place at higher temperatures than the melting of free DNA. The first phase of both the hyperchromic and the calorimetric transitions suggested that approximately 40 base pairs of DNA were completely melted. ^{31}P NMR studies verified that, after this first transition, there are indeed two classes of DNA backbone conformations (31, 33) and thus the absorbance changes do not correspond to an average change in all the DNA base pairs. By means of S1-nuclease digestion of core particles containing poly(dA-dT)·poly(dA-dT), Simpson has recently demonstrated that the DNA that denatures in the first transition corresponds to a 20-base pair region at each end of the core particle (166). The DNA in these terminal regions of the core particle may have a different twist than does DNA in the core particle interior (see above).

The first transition does not correspond to a change in the protein conformation as measured by circular dichroism. The second phase of the melting, on the other hand, seems to correspond to a coincident and complete destruction of both the DNA and the histone structure (165). Two further interesting observations were made. Changes in the DNA circular dichroism spectrum preceded the first hyperchromic melting; [see also the early change in the ^{31}P NMR spectrum (33) and CD spectrum (166) of core particles containing poly(dAdT)·poly(dAdT)]. Not only does this circular dichroism premelt have no calorimetric counterpart, but the enthalpy of the first hyperchromic and calorimetric transition corresponds almost completely to the enthalpy required to melt 40 base pairs of DNA, i.e. very little seems left over to break the histone-DNA bonding. Electron microscopy reveals possible configurations of these partially melted forms (167).

The nucleosome core can also undergo changes in shape as solvent conditions are changed. The most thoroughly studied of these is the change centered about 1–2 mM salt. Gordon et al (168) first reported that, as the salt concentration is lowered from 5 to 0.5 mM, the sedimentation coefficient of core particles decreases from 10.2 to 9.2S. Transitions in the same salt range have also been observed in the core particle diffusion coefficient (168), circular dichroism (169), (170), electric dichroism (171), and in the increased solvent accessibility of fluorescent probes (127, 129). Low ionic strength also prevents the formation of a cross-link between histones H2B and H4 but not between H2A and H2b (172). At least two or three ions are released in the unfolding transition (127, 171).

Thus the low ionic strength transition seems to involve a more extensive structural rearrangement than the freeing of the DNA termini found with thermal melting and it is difficult to arrive at a unique model. In particular, measurements of the hydrodynamic properties of charged macromolecules at low ionic strength present difficulties of interpretation, since the primary charge effect and large contributions to the second virial coefficient can alter sedimentation and diffusion at finite macromolecular concentrations (173). It may be, therefore, that the large frictional coefficients measured for the unfolded structure at low ionic strength owe something to these effects. In any case, models in which the core particle is rather fully unfolded at low salt (168) seem inconsistent with changes in the electric dichroism. Wu et al (171) have proposed, based on their transient electric dichroism studies, a more modest structural change in which the nucleosome core preserves its disc-like shape, but with an expanded radius of 8.9 nm, which contains the core complement of DNA in 0.9 superhelical turns. It must be kept in mind, however, that there have been some reports of fully unfolded nucleosomes in high-molecular-weight chromatin at very low ionic strength (see below).

There are indications from fluorescence measurements (127) that nucleosome cores undergo shape changes in 0.6M NaCl, near the point at which histones begin to dissociate. However, these shape changes are not seen as increases in reactivity of the sulfhydryl groups (123), i.e. the residues to which the fluorescent reporter groups are attached. A transition in the sedimentation coefficient at intermediate salt concentrations (168) has not been observed by other methods and has recently been identified as coming from nucleosome components containing DNA of larger than core size (169).

Induction of larger changes in the nucleosome shape requires more drastic conditions. Addition of urea in the range 6–8 M produces a particle with sedimentation coefficient ($s_{20,w}$) between 5 and 6 S, but in which the histones remain attached to the DNA. This is consistent with rather complete destruction of the histone-histone interactions which hold the nucleosome in a compact form (121). As discussed earlier, the sulfhydryl group of histone H3 becomes reactive over the same range of urea concentrations (121, 123). A potentially important generalization has been made from the denaturation of core particles by urea and by other denaturants, namely that the

DNA conformation changes coincidently with a change in the alpha-helix content of the core histones (174).

These experiments reveal the importance of both hydrophobic and electrostatic interactions in folding the nucleosome, but their relationship to the physiological properties of nucleosomes is unclear. Do nucleosomes in fact unfold when DNA must be replicated or transcribed? One paradigm of an unfolded structure has been proposed by Weintraub et al (175) and involves the opening of each nucleosome into two half-nucleosomes along the presumed dyad axis. Although none of the above physical studies of nucleosome cores shows that these particles can unfold into two half-nucleosomes, there is some evidence in support of such a reaction from electron microscopy of SV40 minichromosomes subjected to low ionic strength solvents (176). Under such conditions, it is possible to observe about twice the number of "beads" normally seen on minichromosomes. Each bead is smaller in diameter than the classical nucleosome core, but the overall contour length of the closed circular molecule is close to that of minichromosomes at higher ionic strength. These results are consistent with dissociation into half-nucleosomes, but asymmetric dissociation mechanisms cannot be ruled out. The same reservations apply to the experiments of Tatchell & van Holde (177) who have studied reconstituted complexes of the core histones with DNA that is 65-base pairs long. They find that at low ionic strength some of the particles sediment in the range 5–6S. As the authors point out, it will be necessary to determine the composition of the dissociated particles before drawing the conclusion that these slow-moving particles are indeed half-nucleosomes.

An observation that has often been cited as evidence for a dyad axis within the core particle is that, under the appropriate conditions, nuclear chromatin can be cleaved by DNase II twice per nucleosome repeat (47, 178). However, it remains to be demonstrated that the cleavage products have the histone composition expected for symmetric cleavage.

CORE PARTICLE ASSEMBLY   Isolated histones can be reconstituted onto high-molecular-weight DNA to yield a product with many of the properties of chromatin (34, 35, 39, 179). More recently, several methods have been used to reconstitute core particles from the separated components (21, 180–183). The reassembly can produce a high yield (80–90%) of particles that are fully native by a number of criteria.

Tatchell & Van Holde have recently investigated the accuracy of core particle reconstitution and its dependence on DNA length (177). By analysis of the DNase I digestion pattern of core particles reconstituted from histones and 5'-labeled DNA, it was concluded that core length DNA (144

± 5 base pairs) reassembled into a core particle with almost total fidelity; as the authors point out, a degeneracy in positioning of even five base pairs would have obliterated the digestion pattern. Reconstitution of end-labeled DNA fragments of average length 111, 125, or 161 base pairs yielded DNase I digestion patterns that were consistent with an asymmetric positioning of the DNA with respect to the histone core but with one end of the DNA fragment at the same position as one end of the DNA in a native core particle. No such unique DNA complex formed with 177-base pair long DNA. Tatchell & Van Holde (177) point out that their results can be explained if the strongest DNA binding sites are at the ends of the core particle. The DNA molecule then reassembles in such a way as to maximize binding interactions.

Such a mechanism of positioning of the core particle DNA is evidently independent of sequence, as the data suggest it must be; it is an "end effect" that will not be of importance in the positioning of nucleosomes on the very long DNA of the eukaryotic genome. However, some experiments with small DNA fragments of defined sequence suggest that in certain cases, nucleotide sequence may play a role in the placement of histones. Chao et al (184) have reconstituted histones onto a cloned DNA fragment 203-base pairs long, containing the *Escherichia coli lac* operator sequence, and find that there are two preferred positions for binding of histones. Similarly, the authors suggest that there are three preferred locations for histones on a *lac* operator-containing fragment 144-base pairs long. Although some contributions to the selectivity may also come from end effects, it is difficult to account for the data without also introducing some sequence selectivity. Whether there is any such sequence selectivity in vivo is a separate question and is discussed below.

The above studies concentrated solely on the final product of the reconstitution process, but other studies have investigated various aspects of the reassembly mechanism. Stein et al (181) demonstrated that core particles could be formed by adding free DNA to histone octamers that had been chemically cross-linked, thereby showing that many complicated assembly pathways, for example the ordered interaction of DNA with individual histones, were at least not necessary. Stein (182) has now examined in much more detail the reaction between core histones and core DNA, following a rapid decrease in salt concentration from 2.5 M to 0.6M. Although most of the nucleosomes are formed by direct combination of DNA molecules with free histone octamers, there is also a component containing one DNA molecule and two histone octamers; a second slower reaction pathway involves transfer of a histone octamer from this DNA-hexadecamer complex to a naked DNA molecule, forming two normal nucleosome core particles. It should be recalled that high-molecular-weight DNA with the normal abundance of nucleosomes still has the capacity to bind additional core histones (185).

The assembly and disassembly of the octameric histone core has also been investigated. An equimolar mixture of the core histones, dissolved in 2M NaCl, shows many of the properties of histones complexed with DNA (143). In particular, spectral properties indicative of histone conformation change little if at all upon core particle dissociation in high salt (23, 25, 186). Although it was initially thought that the complex consisted of two identical heterotypic tetramers, it is now clear that in 2 M NaCl, at sufficiently high histone concentrations, a complete histone octamer is the major species present (187–193). At lower histone concentrations, dissociation has been observed but the nature of the products remains the subject of controversy. Some investigators conclude that dissociation yields separate complexes of (H3 + H4) and (H2A + H2B), while others conclude that the product is a pair of heterotypic tetramers.

Detailed comparison of the results obtained in various laboratories is difficult because the equilibria depend upon histone concentration, pH, temperature, salt concentration, and even pressure (143, 190, 192, 193), and no two laboratories use exactly the same conditions. Furthermore, analysis of dissociating systems is dependent upon guessing the reaction scheme first. Nonetheless, at least some of the disagreement seems to be at the level of the raw data and we are therefore faced with the possibility that histones prepared by different methods, or from different organisms, or dissolved in different buffers may not behave identically.

Although all four core histones appear to be present in every nucleosome in vivo, H3 and H4 play a special role in stabilizing the structure (21, 35, 39, 85, 183, 190, 192, 194–196). Particles with the hydrodynamic and topological properties of a nucleosome core can be formed from H3 and H4 alone, complexed to DNA 145-base pairs long. Complexes have been isolated containing either a tetramer (two each) or an octamer (four each) of H3 and H4 for each DNA molecule of core size. Although there has been some disagreement about the DNA conformation in the tetramer complex, several kinds of evidence (197–199) suggest that the tetramer does not fold short DNA into a compact form, although it may distort it into a shallow supercoil (199). However, an octamer of H3 and H4 does compact the DNA into a structure with hydrodynamic properties very similar to those of a nucleosome (197, 198).

Chromatin reconstitution methods have traditionally involved lengthy dialysis from high salt with or without high urea concentrations, and it was not clear how nucleosome assembly might proceed in vivo. Laskey and co-workers (200, 201) have isolated a heat-stable acidic protein from

Xenopus eggs which facilitates nucleosome assembly in vitro. The assembly factor appears to complex stoichiometrically with an octamer of core histones and they suggest that it acts as a "molecular chaperone," by presenting the histone octamer in a properly receptive form to the DNA and perhaps at the same time discouraging nonproductive attachments. Germond et al (202) have recently reported that high concentrations of purified eukaryotic nicking-closing enzyme can also act as a nucleosome assembly factor.

Two recent developments may make it difficult to establish the importance of these assembly factors in vivo. First, conditions have been reported in which nucleosomes self-assemble simply by incubation at physiological ionic strengths (203, 204). Secondly, Stein et al (204) have discovered that acidic polypeptides, such as polyaspartate and polyglutamate, appear to act in the same manner as the in vivo assembly factors and to increase the rate of nucleosome formation.

CORE PARTICLE SLIDING   Chromatin can be depleted of histone H1 by washing in 0.6 M NaCl (205). However, limited micrococcal nuclease digestion of this material no longer yields the classic 200-base pair repeat associated with native chromatin. Rather the digestion pattern has a repeat length close to 140 base pairs, as if neighbouring core particles have slid together (178, 206, 207). This sliding occurs during the H1 stripping and does not occur as a consequence of the nuclease digestion since methods have now been developed to remove H1 without changing the nuclease digestion pattern (208–211).

A careful analysis of the digestion pattern reveals that, under certain incubation conditions, the repeat can actually be less than the core size, as if the cores can be "closer than close-packed" (206, 207). In particular, Tatchell & Van Holde (207) have isolated a series of "compact oligomers," or nucleoprotein species whose DNA sizes obey the relation "$120n + 20$," where n is an integer. The DNase I cleavage patterns of these particles are consistent with a sharing of 20 base pairs of DNA between neighboring histone octamers. The authors suggest that adjacent cores may be stacked on top of one another, with the dyad axes rotated 180° between neighbors. The ability of histone octamers to form these denser structures also suggests a potential for interactions between nucleosomes that may be important in the formation of higher order chromatin structures.

The studies that have just been described involve salt concentrations quite close to the point at which core histones begin to dissociate completely from DNA. There is also evidence, however, that histones can "slide" along DNA at lower ionic strength. Beard (212) has studied such migration by ligating a nucleosome-covered DNA molecule to a labeled, protein-free piece of DNA, and either examining the structure in the electron micro-scope, or measuring the appearance of label in nucleosomes generated by nuclease digestion. At 37°C and salt concentration of 0.15 M, nucleosomes are found on the naked DNA after several hours; from the position of the structures in the microscope, it seems likely that the histones reach the naked DNA by sliding, rather than by passing through an intermediate state in which octamers are free in solution. Furthermore, there is no detectable exchange of nucleosomes to added protein-free unligated DNA molecules under these conditions.

There are many possible mechanisms for movement of histone octamers along the DNA which do not involve a free octamer as an intermediate. It is possible that the sliding mechanism involves transient formation of some complex in which there is a local excess of histones, such as the ones described by Stein (182) which form during core particle reassembly. Such complexes could be formed by direct transfer of histones when nearby nucleosomes come into contact by flexion of the DNA in the spacer regions. Another conceivable mechanism may involve, for example, formation of a looped-out piece of DNA on the nucleosome surface, followed by propaga-tion of the loop (213). These are two of several possible reaction pathways which should have a low activation energy. The requirement in all such mechanisms is that intermediates make about as many strong interactions between histone and DNA as are made in the nucleosomes with which the reaction starts.

The idea that histone octamers transfer as a unit is consistent with the observation (214) that the octamer structure is conserved during DNA replication. Under such circumstances, the individual histone components do not dissociate in vivo, though the octamer may initially be assembled in a sequence of steps (215, 216).

THE NUCLEOSOME

Histone H1 and its Relation to the Core Particle

The next level of chromatin structure above the core particle is the mononu-cleosome, consisting of the core particle, the "spacer" or "linker" DNA, and the lysine-rich histone, H1. Most recent studies are agreed that the average nuclear content of H1 is close to one H1 per core particle, i.e. H1 is present at only half the stoichiometries of each of the core histones (67, 68, 217, 218). This immediately implies that the nucleosome, as the funda-mental repeating unit of chromatin, cannot have a dyad axis and indeed has the potential of assembling into polar structures. This of course says nothing about the dyad axis of the core particle itself. Although isolated mononu-cleosomes and dinucleosomes are consistently depleted in H1 (67, 68) this may well be due to losses that occur during nuclease digestion rather than to the existence of a class of nucleosomes which lack histone H1.

Histone H1 contains on the order of 210–220 amino acid residues and thus is considerably larger than a core histone (72, 73). Moreover, it is much less stringently conserved throughout evolution. Even within one tissue, there may be at least four H1 subtypes or primary sequence variants (see e.g. 219, 220); the relative concentration of these subtypes can vary considerably among different tissues within the same organism (72, 73, 221). The lysine-rich histone, H5, found in avian erythrocytes is smaller than H1 [196 amino acids in a goose; (222)] but patterns of sequence homology indicate that it can probably be regarded as an extreme variant of H1 rather than as a distinct histone type (222, 223).

Both sequence analysis and several physical studies have indicated that H1 has three distinct domains. The first forty amino acids from the N terminus and the last 100 or so amino acids at the C terminus are rich in lysine, alanine, and proline. The intervening section from amino acid position ~ 41–115 is more hydrophobic and typical of a globular protein (72, 73). Physical studies, both on H1, H5, and on their isolated peptides, demonstrate that, in the presence of salt, the entire central region exists as a highly defined structure, but the basic termini remain unstructured (220, 222, 224–228). Just as in the case of H2B discussed earlier, the majority of the evolutionary variants are in the basic terminal regions, with the central globular regions being quite highly conserved (72, 73).

Unlike the core histones, histone H1 shows no tendency to self-aggregate (220, 229); even the isolated globular region (of H5), with the basic tails removed, still does not aggregate (228). A most interesting finding is that different H1 subtypes can interact both strongly and selectively with the nonhistone chromosomal proteins HMG1 and HMG2 (230). Postsynthetic modifications of H1 are discussed in (73).

Early experiments suggested that histone H1 was somehow associated with the spacer or linker DNA between core particles (10, 231). Several more recent lines of evidence suggest that histone H1 is localized at the point where DNA enters and exits from the core particle. For example, H1-containing chromatin appears extended at low ionic strength, with DNA strands apparently entering and leaving the core particle on approximately the same tangent. If histone H1 is removed, the entering and leaving DNA strands now appear to be displaced from one another (232).

Many workers have found that a DNA fragment about 160-nucleotides long is a metastable intermediate on the digestion path from full-length mononucleosomes to core particles (4, 6, 180, 210, 233). Varshavsky et al (10) have described an electrophoretic species containing about 160 base pairs of DNA, all the core histones, and histone H1. Simpson (11) has recently isolated such a particle, which he names a "chromatosome" and which contains 168 base pairs of DNA, an octamer of core histones, and one molecule of H1 (or H5). The DNA length would thus correspond to two full turns around the histone core, as suggested earlier by Finch et al (17).

In most respects, the chromatosome behaves much like the standard core particle; however it is more resistant both to thermal denaturation (see also 180) and to DNase I digestion, especially at the DNA ends. The DNase I digestion pattern of end-labeled chromatosomes was interpreted as showing that an extra ten base pairs of DNA are added to each end of the core particle.

Thus it seems as though H1 binds to one side of the core particle and both protects and locks into position the entering and exiting DNA strands. It is not yet clear how H1 organizes the spacer DNA between adjacent core particles. Raman spectroscopy suggests that, like the core histones, H1 does not bind in the DNA major groove (62).

As mentioned above, H1 has little tendency to aggregate either with itself or with other histones (220, 229). On the other hand, H1 has been reported to form homopolymers if nuclei are reacted with nonspecific cross-linking reagents (150, 234, 235), which suggests that H1 molecules are adjacent within the higher order structure. A recent cross-linking study by Ring & Cole (236) has extended these previous observations. Nuclei were reacted with three different cross-linking reagents, ranging from a long nonspecific diimido ester to a carbodiimide, a "zero-length" condensing agent. The cross-linking spectrum seemed remarkably independent of the reagent. Analysis was focused on the dimer fraction where it was found that a minimum of 10% of the original H1 could be found as H1-H1 cross-links and a further 7% as cross-links with the core histones. Histone H3 was the most favored of the core histones to cross-link with H1 but cross-links to each of the other core histones were found. HMG proteins 14 and 17 (237, 238) were found cross-linked both to H1 and to the core histones.

Bonner & Stedman (239) have also treated nuclei with a carbodiimide and have described in some detail a "zero-length" cross-link formed between histone H1 and H2A. Peptide mapping has localized the cross-link between the globular region of H1 and the ten amino acids at the C terminus of H2A (W. Bonner, personal communication). This site on H2A must therefore be within several amino acid residues of the site of ubiquitin attachment in A24 (see above and Figure 1). Indeed it has also been found that H1 can be cross-linked to A24 (239). If this H1-A24 link can be formed in isolated nucleosomes, then the proposal that H1 is lacking in those nucleosomes containing A24 can probably be discarded.

A number of hydrodynamic and electron microscopic studies of H1-depleted chromatin have demonstrated that readdition of histone H1 leads to chromatin contraction (9, 241, 242). Furthermore, both H1-containing chromatin and isolated oligonucleosomes undergo a contraction in the salt range from 10–20 mM (232, 242–244). Stratling (244) has pointed out another salt-dependent property of H1, namely the ability to condense small oligomers into larger oligomers (e.g. to hold two dimers in a tetramer), a property that seems to be more specific than a simple aggregation.

Two other correlations should be kept in mind. The salt range in which chromatin contracts is close to that at which there is a transition from noncooperative to cooperative binding of H1 to DNA (245). Perhaps more importantly, H1 itself undergoes a folding transition in approximately the same range of salt concentrations (220).

Nucleosome Phasing

If a major role of the core histones is to render DNA more compact, it is to be expected a priori that the strength of interaction with DNA will not be particularly sensitive to nucleotide sequence. Every naturally occurring DNA molecule, regardless of base composition, appears able to form nucleosomes in a reconstitution experiment (14, 246), as do the alternating synthetic copolymers, poly(dAdT)·poly(dAdT) and poly(dGdC)·poly(dGdC) (14, 15). The only exceptions so far reported are the nonalternating polymers, polydA·polydT and polydG·polydC, but there is reason to suspect that their inability to form nucleosomes may be related to DNA-strand rearrangements occurring during reconstitution. To a first approximation at least, it appears that no naturally occurring DNA sequence binds the core histones with special strength or weakness. Although the bulk of nucleosomes are randomly placed (58), there is evidence suggesting that in a few cases there is a fixed relationship between the placement of the histones and the nucleotide sequence. This phenomenon is known as phasing, although this term has also been used to designate certain regularities in spacer length (247).

Phasing could be generated either by direct sequence specificity of nucleosome binding or by some unknown property of the chromatin assembly mechanism. An obvious place to look for such behavior in vivo is in the tandemly repeated sequences of the eukaryotic genome. If the sequence repeat is approximately 200-base pairs long, and if phasing occurs, the DNA length per nucleosome repeat should be the same as that of the sequence repeat.

Horz et al (248) have studied the nucleosome arrangement on mouse satellite DNA, which has a sequence repeat of 245 base pairs, compared to a bulk nucleosome repeat of 195 base pairs. They digested mouse nuclei with restriction endonucleases, and showed that the pattern of fragments was the one expected if only the DNA of the spacer was attacked, and if nucleosome core position had no fixed relationship to DNA sequence.

Recently, Gottesfeld & Melton (249) have analyzed the same system by isolating nucleosome oligomers by micrococcal nuclease digestion of mouse nuclei followed by filter-transfer hybridization using satellite DNA as probe. The nucleosome repeat of the satellite was shown to be identical to that of bulk DNA and quite different from the sequence repeat. It is clear that in this mouse satellite, no phasing is observed. Similar results have recently been obtained with the 5S genes in rat liver (250).

This result must be compared with the observation of Maio and his collaborators concerning the alpha DNA component of African green monkey nuclei (251, 252). These workers report that the nucleosome repeat on alpha DNA is 172 base pairs, which coincides with the sequence repeat (253) but is smaller than the repeat of the bulk material.

These results are consistent with phasing of the histone cores with respect to sequence, but they are not conclusive, since the alpha sequences in nucleosomes, even if distinct in size, may be positioned randomly (254). Furthermore, there are some digestion conditions in which the 172-base pair alpha nucleosome repeat disappears (252, 254). Fortunately, these contradictions have now been resolved by the incisive study of Fittler & Zachau (255), who have shown that micrococcal nuclease does indeed excise alpha-DNA nucleosomes 172-base pairs long, but that the same pattern is generated on naked alpha DNA. This arises from some DNA sequence specificity for cleavage by micrococcal nuclease, rather than from nucleosome phasing. In contrast, DNase II digestion of chromatin does not lead to discrete alpha DNA bands. The conclusion is that the nucleosomes of African green monkey alpha DNA are not uniquely phased.

Questions of sequence-specific nucleosome placement have also arisen in studies of the SV40 minichromosome. There is a region of the compact minichromosome near the origin of replication that is particularly accessible to restriction endonucleases (256–258). After formaldehyde fixation of the complex, only the DNA in this region is accessible to restriction endonucleases, and can be liberated from the minichromosome by sodium dodecyl sulfate (SDS), while the remainder of the DNA remains attached. It seems reasonable to conclude that the region is free of nucleosomes. This may result from the presence of some other protein which prevents the binding of histones during a crucial stage of assembly. It is known, for example, that the T antigen of SV40 binds to this region of the DNA.

The other possibility is that there is a section of DNA in SV40 that is reduced in binding affinity for nucleosomes. Recent electron microscopic studies of reconstituted SV40 minichromosomes show some tendency for the nucleosomes to avoid certain regions of the DNA (259). Exclusion is not absolute but the three regions involved include the one sensitive to restriction enzymes. The selectivity shows some correlation with DNA base composition; it does not require intact closed circular DNA and thus cannot depend upon the relief of supercoiling strain. Ponder & Crawford (260) have also reported partial phasing of nucleosomes in polyoma minichromosomes, both native and reconstituted.

Relation of Nucleosomes to their Neighbors

From the earliest studies it was apparent that the basic nucleosome repeat length was different in different tissues (reviewed in 1, 2). Assuming that two full turns of DNA (166 base pairs) wrapped around the histone core (see above) is the largest feature held in common by nucleosomes in all tissues, then the "spacer" length varies from zero base pairs in yeast and lower eukaryotes (261–264), up to ~ 80 base pairs in sea urchin sperm (265). Although nuclease digestion studies had suggested that there may be classes of different spacer lengths even within the same cell (261, 266), direct evidence for this intracellular variability has only recently been provided by Humphries et al (267). The 5S genes of *Xenopus* were shown to have a repeat length of 175 base pairs in erythrocytes, compared to the bulk repeat length of 189 base pairs. In contrast, the 5S repeat and the bulk repeat are the same both in *Xenopus* liver (267) and in rat liver (250).

Why the nucleosome repeat length varies among tissues is not yet understood. It has been proposed that shorter repeat lengths are correlated with higher levels of gene activity, and although this certainly appears to hold true in some instances (265, 268, 269), exceptions, at least in lower eukaryotes, have been noted (270, 271).

It was also proposed (268) that longer repeat lengths were associated with longer and/or more basic histone H1 molecules. Again, a clear exception has been noted; chicken erythrocytes have a longer repeat than other chicken tissues, yet the erythrocyte-specific lysine-rich histone H5 not only has almost exactly the same number of basic groups as the usual H1 molecule, but is also substantially shorter (222).

Chicken erythropoiesis affords clear evidence that the nucleosome repeat length can change during development. However it is not clear how exact the correspondence is between nucleosome repeat length and histone H5 deposition (272, 273).

It has been proposed that the lengths of the spacer regions are constrained to be integral multiples of the duplex repeat (247). This would account for the presence, in DNase I digests of nuclei, of single-strand DNA fragments greater than a full nucleosome-equivalent in length, which nonetheless fall on the regular ladder of DNase I fragments separated by 10.4 nucleotides. This extended ladder is seen most clearly in yeast chromatin which has little or no spacer (247) and can be generated if there is any considerable proportion of nucleosome cores in close contact, either with no spacer or in the rearranged form of the "compact oligomers" (206, 207). Careful quantitation of yields of the larger fragments in the ladder is therefore necessary, to show that the bulk of the spacer regions have quantized lengths. This is especially important, since experiments by Prunell & Kornberg (58) with isolated, terminally labeled nucleosome dimers have failed to reveal any such quantization. However, there is further evidence that adjacent nucleosomes may be separated by a fixed distance; this comes from the observation (57, 274) that extensive exonuclease-III digestion of isolated dinucleosomes is capable of yielding a series of discrete subdimer digestion products. However, the possibility that the constituent monomers slid together during digestion was not eliminated; moreover, the H1 content of these digested dimers was not specified.

HIGHER ORDER STRUCTURE

Packing of Nucleosomes

The next two levels of order beyond that of the nucleosome itself are the 10-nm fibril and the 25–30-nm chromatin fiber that have been observed in the electron microscope (275). The nucleosomes must pack in some more or less regular way into these higher order structures, but until recently there were few experimental measurements with sufficient resolution to permit distinguishing among proposed models. This state of affairs is now coming to an end; new data from light, X-ray, and neutron scattering, and from electron microscopy, now permit the elimination of many possible models, and suggest the general nature of the next two orders of structure beyond the nucleosome.

A large part of the difficulty in obtaining reproducible information about higher order structure stems from the variability in the methods of isolation of chromatin, and the solvent conditions used during measurement. The preparation of the intact 30-nm fiber requires the use of solvents of moderate ionic strength (e.g. at least 0.02 M NaCl), the presence of histone H1 in the complex, and the avoidance of large shearing forces during preparation.

The 30-nm fibers prepared in this way can be converted reversibly to 10-nm fibrils by lowering the ionic strength (243, 275, 276). The process of conversion has now been studied in some detail using both neutron scattering (277) and electron microsocopy (232, 242). Earlier studies using these techniques (276, 278) had suggested that the 10-nm fibril consisted of a linear array of nucleosomes, which was in turn supercoiled to form the 30-nm fiber or solenoid. This general picture is supported by the new results.

The recent neutron scattering experiments have been carried out by Suau et al (277) in a series of mixed D_2O-H_2O solvents which permit more or less independent determination of the radii of gyration of protein and nucleic acid components within the structure; the approach is similar to that used earlier in studying the mass distribution of individual nucleosomes (19, 279). Analysis of data for the 10-nm fibril, obtained at low ionic strength, reveals that the linear array of nucleosomes cannot be stacked face to face, since the radii of gyration predicted for such a model are considerably larger than those that are observed. The data are consistent with models in which the cylindrical edges of the core particles are in contact, with the faces oriented parallel or nearly parallel to the fibril axis. When the ionic strength is raised, a new family of structures begins to form. The neutron scattering data indicate that these have radii of gyration corresponding to a fiber about 30 nm in diameter, but with a variable mass per unit length. The structures become more compact as the ionic strength is raised; the most compact structures are consistent with a superhelical model with 6–7 nucleosomes per turn. The core particles are thought to be arranged with cylindrical diameters perpendicular to the axis of the superhelix. The authors suggest that the cylindrical axes of the nucleosome cores may be nearly parallel to the superhelix axis, so that each turn of the supercoil is relatively flat, with the faces of the nucleosomes, in successive turns, parallel to each other.

These results must be compared with recent electron microscopic studies (232) which have also examined the salt-dependent transition from 10-nm fibril to 30-nm fiber. At low ionic strength, the fibril appears in a zigzag configuration with a two-nucleosome repeat. As the salt concentration is raised, somewhat irregular solenoidal structures of larger diameter begin to form, and at the highest salt concentrations (0.06–0.1 M NaCl) most of the material is in a fiber ~ 25 nm in diameter. Cross-striations separated by about 11 nm can be seen in these fibers (280, 281), which suggests that the nucleosomes in successive turns may be packed edge-to-edge, rather than with their faces parallel as proposed by Suau et al (277). In this model, contact between adjacent nucleosome faces occurs within a single turn. Electron microscopic results at intermediate salt concentration are interpreted as showing structures of constant pitch, but variable diameter. This must again be compared with the neutron scattering results mentioned above, which suggest that the pitch varies and the diameter remains constant in the transitional region.

Though the details of their proposed structures differ, both groups of investigators agree that the nucleosomes of the 10-nm fibril are arranged edgewise, rather than with faces touching, and that the 30-nm fiber consists of a continuous superhelix of nucleosomes, with about six or seven nucleosomes in each turn. Such a model for the 30-nm fiber is also consistent (232) with the mass per unit length and other parameters that can be calculated from light scattering measurements (282).

Investigations of the salt-dependent transition between the 10-nm chromatin fibril and the 30-nm fiber reveal transitional structures that are interpreted as having either intermediate pitch or diameter. A variety of structures in this family can be seen in electron micrographs of mitotic chromosomes (281). As Dubochet & Noll (20) have shown in their electron microscopic studies of nucleosome aggregates, the core particle may be rather flexible. Perhaps, despite the preference of structural chemists for unique structures, the ability to form a variety of packed structures is an important property of the nucleosome.

The continuous superhelix model differs from the "superbead" model which envisions stable clusters of six or eight nucleosomes, strung together to form the 30-nm fiber. Such structures have been seen in the electron microscope (283–285). Renz et al have shown that, in a series of oligonucleosome fractions of increasing chain length, the affinity of the oligomers for H1 increases with size up to, but not beyond, the octanucleosome (243). The strongest evidence for an array of repeating superbeads is the series of peaks seen in sucrose gradients of mild chromatin digests (285, 286). However, the presence of ribonucleoprotein particles on these gradients has not yet been stringently eliminated and may explain some results (287).

Many of these observations are in any case reconcilable with the superhelix model. For example, the discontinuity in H1 binding may reflect the relative stability of a complete superhelical turn. To explain the release by nuclease of structures that are multiples of about eight nucleosomes, it is necessary to suppose that the nuclease, having made one double-stranded cut, tends to cut a second time on the same side of the supercoil, and thus at sites exactly one turn above or below the original cut (232). If the 30-nm fiber was packed in such a way that there is a preferred direction of approach by the nuclease, then higher multiples of the superbead unit could be generated by a mechanism analogous to that proposed for the intranucleosomal cuts made by DNase I.

Histone H1, which may lock individual nucleosomes in a "closed" conformation (see discussion of H1 above), may also stabilize the 30-nm fiber (232). It is not yet clear whether H1 is on the inside or on the outside of this "solenoid." On the one hand, H1 is the histone most susceptible to proteolysis; on the other hand, it has been observed that an antibody to H1 bound less to chromatin at high ionic strength than at low ionic strength (288), which suggests that H1 may be on the inside of the coil.

It is also not clear how variable lengths of spacer DNA are accommodated within the 30-nm fibril, although suggestions have been made (289). It is to be hoped that the above detailed structural studies will sooner or later be repeated with chromatin of differing nucleosome spacing, as well as with chromatin containing hypermodified histones (90).

Much attention has been given to the effects of core histone binding on the topology of DNA (290). If nucleosome cores are reconstituted on closed circular DNA, and the complex treated with nicking-closing enzyme, the DNA subsequently reisolated is found to have a linking number (relative to the fully relaxed circular form) that is proportional to the number of nucleosomes that were bound. The measured increment in linking number is about −1.25 per nucleosome core particle, and must in principle arise from some combination of changes in the twist and the writhe of the DNA. [The reader should refer to Crick's paper (290) for definitions of these terms; roughly speaking, the twist measures the number of base pairs per turn of duplex, while the writhe is related to the path of the space curve followed by the distorted duplex.]

The measured change in linking number of −1.25 is not easily reconciled with the simple models of the nucleosome core that have been proposed. If the DNA repeat on the surface of the nucleosome is 10.4 base pairs (12–15), and if this is identical to the repeat of free DNA in solution (291), then a nucleosome core structure with 1.75 regular, left-handed superhelical turns of DNA should produce a linking number change of −1.75. This discrepancy would disappear if the DNA repeat on the nucleosome surface were actually 10.0 base pairs per turn (12, 22). On the other hand, it would become more severe if the DNA repeat in solution were actually 10.0 base pairs per turn (292). Although there may be some other explanation of this discrepancy, such as experimental error, it is possible that the path of the spacer DNA in high-molecular-weight chromatin contributes to the linking number change.

The experimental results described above were obtained with small, closed, circular DNA molecules, using only the core histones. Therefore any topologically important aspects of structure requiring histone H1 cannot be included in the comparison. Muller et al (293) have reported that removing H1 from SV40 chromosomes does not lead to a change in linking number; however it was not certain that the H1-containing mini-chromosomes were initially accessible to relaxing enzyme.

It is possible that the core histones by themselves are sufficient to alter the topological properties of the spacer DNA. As Crick & Fuller have pointed out (290, 294), the path taken by the spacer DNA as it emerges from the core particle may have a large effect on the measured linking number change per nucleosome; the contributions of spacer and core DNA to the linking number are not even necessarily additive. However, it should not be assumed that the introduction, for example, of a zig-zag configuration such as that observed for the 10-nm fibril (232), will necessarily affect the topology.

Interactions between nucleosomes could also alter the path of the spacer DNA. However, at least under some conditions of reconstitution onto circular DNA, the core particles appear to be spaced fairly far apart (56) at low histone:DNA ratios, and there is no evidence from supercoiling experiments that the first few nucleosomes bound make anomalous contributions to linking number (179, 295, 296).

The tendency of core particles to slide together when histone H1 is removed (178, 206, 207) suggests that, under some conditions, neighboring core particles do interact with one another. Whether or not this affects supercoiling, it may play a role in stabilizing higher order structure. Although, as discussed earlier, the presence of histone H1 affects the ability of chromatin to form highly regular compact structures, many of the features of the 30-nm fiber can be seen, though in somewhat disordered form, in chromatin preparations stripped of H1 (232, 242). Furthermore, individual nucleosomes have been observed to form similar structures (276). If, as seems likely, interactions between nucleosome cores do help to stabilize higher order structure, it may be that some of the polypeptide chains of the histones (perhaps those that appear to contribute little to the stability of individual core particles) are important to the internucleosome contacts.

Chromosomes

To make a mitotic chromosome, the 30-nm fiber must be compacted another two orders of magnitude. The principle of supercoiling may be used again to generate these higher orders of compaction (297, 298) but the data do not reveal the details of this level of organization. Benyajati & Worcel (299) have shown that the long continuous DNA of the interphase chromosome is organized in looped domains containing on the order of 80 kilobase pairs, which appear to be held in place by RNA and protein. Nuclease digestion studies of chromatin also support models in which the nucleoprotein is organized in large-scale "domains." Igo-Kemenes et al (300), using restriction enzymes or micrococcal nuclease, have examined the kinetics of solubilization and the size of the DNA released from chromatin at very low levels of digestion. Their results are best explained by models in which cross-links between nucleoprotein fibers occur at intervals of about 30–70 kilobase pairs, and are certainly consistent with the results of Benyajati & Worcel (299).

Wu et al (301, 302) have used very limited digestion with nucleases (DNase I and micrococcal nuclease) to probe the higher order structure of *Drosophila* chromatin in the neighborhood of defined sequences of the genome. The analysis is carried out by filter transfer hybridization of the DNA digest, using as probes recombinant plasmids containing the DNA of the *Drosophila* heat shock loci. For each locus, a different pattern of discrete DNA bands is observed. DNase I produces fragments in the range 20–2 kilobase pairs; subsequent redigestion with restriction endonucleases generates some new, smaller, fragments not produced when the endonuclease acts alone on *Drosophila* DNA. Thus, at least some of the DNase I cleavages are at well-defined sites relative to the sequence being probed. Similar results are obtained using micrococcal nuclease, though the fragment sizes are smaller. In one case there is evidence for partial phasing of the nucleosomes relative to the DNA sequence. It is not yet clear how any of these regular structural features are related to the morphology of the chromosome described above. Most of the DNase I cleavage products are shorter than the domain or loop sizes deduced from microscopy or analysis of restriction endonuclease digests. The DNase I sensitivity of chromatin varies with the gene transcriptional activity (97, 303), but the structural basis of this sensitivity is not yet known. The DNase I sensitivity seen by Wu et al (301, 302) may have some other origin; it provides, in any case, strong evidence for long-range regularity of structure in the vicinity of a defined segment of the genome.

A detailed series of studies of metaphase chromosomes by Laemmli and his collaborators (304–306) shows that when the histones are removed, the DNA remains attached to a central scaffold (see also 307) of nonhistone proteins, from which the DNA projects as a series of loops between 30 and 100 kilobase pairs long. In certain aqueous solvents the histones remain attached to the DNA, but the swelling is sufficient to reveal a similar structure in which the loops are compacted into 10-nm fibrils but with the same 30–100 kilobase pair length of DNA in each loop (assuming the usual compaction ratio for 10-nm fibrils). Finally, chromosomes treated with hexylene glycol but not stripped of histones reveal a more contracted radially symmetric structure, in which the projecting fibers are about 50 nm in diameter, presumably related to the 30-nm solenoid described above. Marsden & Laemmli (306) suggest that these are the compacted form of the loops, still attached to the axial scaffold, and that this is the structure of an intact metaphase chromosome.

SUMMARY AND CONCLUSIONS

The involvement of histones in replication and the structure of transcriptionally active chromatin, require, and have received, recent comprehensive reviews (308, 309) and we do not discuss them here.

The chemistry of nucleosomes is relevant to both replication and transcription. In replication, the newly synthesized histone octamer is deposited shortly after synthesis. The order in which these histones are laid down and the way in which newly made histones segregate, both with respect to newly made DNA and parental histones, are all problems that depend for their solution on an understanding of nucleosome chemistry.

In the case of transcriptionally active chromatin, there is evidence that nucleosome-like structures are present, probably modified in composition and in structure compared to the usual nucleosome. The results of these chemical and microscopic investigations continue to provide encouragement that the study of nucleosome structure is a proper place to begin the study of active chromatin structure. Certainly it is clear that the simple presence or absence of the core histones is not likely to be the determining factor in the regulation of transcription (310–313). No doubt many modifications must occur in the composition of active chromatin, affecting the conformation or mobility of individual nucleosomes, as well as their ability to interact in the formation of higher order structures.

Literature Cited

1. Kornberg, R. D. 1977. *Ann. Rev. Biochem.* 46:931–54
2. Felsenfeld, G. 1978. *Nature* 271:115–22
3. Lilley, D. M. J., Pardon, J. F. 1979. *Ann. Rev. Genet.* 13:197–233
4. Sollner-Webb, B., Felsenfeld, G. 1975. *Biochemistry* 14:2915–20
5. Axel, R. 1975. *Biochemistry* 14:2921–25
6. Shaw, B. R., Herman, T. M., Kovacic, R. T., Beaudreau, G. S., Van Holde, K. E. 1976. *Proc. Natl. Acad. Sci. USA* 73:505–9
7. Simpson, R. T., Whitlock, J. P. Jr. 1976. *Nucleic Acids Res.* 3:117–27
8. Greil, W., Igo-Kemenes, T., Zachau, H. G. 1976. *Nucleic Acids Res.* 3:2633–44
9. Noll, M., Kornberg, R. D. 1977. *J. Mol. Biol.* 109:393–404
10. Varshavsky, A. J., Bakayev, V. V., Georgiev, G. P. 1976. *Nucleic Acids Res.* 3:477–92
11. Simpson, R. T. 1978. *Biochemistry* 17:5524–31
12. Prunell, A., Kornberg, R. D., Lutter, L. C., Klug, A., Levitt, M., Crick, F. H. C. 1979. *Science* 204:855–58
13. Lutter, L. C. 1979. *Nucleic Acids Res.* 6:41–56
14. Bryan, P. N., Wright, E. B., Olins, D. E. 1979. *Nucleic Acids Res.* 6:1449–65
15. Simpson, R. T., Kunzler, P. 1979. *Nucleic Acids Res.* 6:1387–1415
16. Langmore, J. P., Wooley, J. C. 1975. *Proc. Natl. Acad. Sci. USA* 72:2691–95
17. Finch, J. T., Lutter, L. C., Rhodes, D., Brown, R. S., Rushton, B., Levitt, M., Klug, A. 1977. *Nature* 269:29–36
18. Pardon, J. F., Worcester, D. L., Wooley, J. C., Cotter, R. I., Lilley, D. M. J., Richards, B. M. 1977. *Nucleic Acids Res.* 4:3199–3214
19. Suau, P., Kneale, G. G., Braddock, G. W., Baldwin, J. P., Bradbury, E. M. 1977. *Nucleic Acids Res.* 4:3769–86
20. Dubochet, J., Noll, M. 1978. *Science* 202:280–86
21. Camerini-Otero, R. D., Felsenfeld, G. 1977. *Nucleic Acids Res.* 4:1159–81
22. Levitt, M. 1978. *Proc. Natl. Acad. Sci. USA* 75:640–44
23. Thomas, G. J., Prescott, B., Olins, D. E. 1977. *Science* 197:385–88
24. Goodwin, D. C., Brahms, J. 1978. *Nucleic Acids Res.* 5:835–50
25. Cotter, R. I., Lilley, D. M. J. 1977. *FEBS Lett.* 82:63–68
26. Feigon, J., Kearns, D. R. 1979. *Nucleic Acids Res.* 6:2327–37
27. Crick, F. H. C., Klug, A. 1975. *Nature* 255:530–33
28. Sobell, H. M., Tsai, C. C., Gilbert, S. G., Jain, S. C., Sakore, T. D. 1976. *Proc. Natl. Acad. Sci. USA* 73:3068–72
29. Kallenbach, N. R., Appleby, D. W., Bradley, C. H. 1978. *Nature* 272:134–38
30. Klevan, L., Armitage, I. M., Crothers, D. M. 1979. *Nucleic Acids Res.* 6:1607–16
31. Shindo, H., McGhee, J. D., Cohen, J. E. 1980. *Biopolymers.* 19:523–39
32. Sussman, J. L., Trifonov, E. N. 1978. *Proc. Natl. Acad. Sci. USA* 75:103–7
33. Simpson, R. T., Shindo, H. 1979. *Nucleic Acids Res.* 7:481–92
34. Axel, R., Melchior, W. B. Jr., Sollner-Webb, B., Felsenfeld, G. 1974. *Proc. Natl. Acad. Sci. USA* 71:4101–5
35. Camerini-Otero, R. D., Sollner-Webb, B., Felsenfeld, G. 1976. *Cell* 8:333–47
36. Whitlock, J. P. Jr. 1977. *J. Biol. Chem.* 252:7635–39
37. Sollner-Webb, B., Melchior, W. B. Jr., Felsenfeld, G. 1978. *Cell* 14:611–27
38. Noll, M. 1974. *Nucleic Acids Res.* 1:1573–78
39. Sollner-Webb, B., Camerini-Otero, R. D., Felsenfeld, G. 1976. *Cell* 9:179–93
40. Simpson, R. T., Whitlock, J. P. Jr. 1976. *Cell* 9:347–53

41. Sollner-Webb, B., Felsenfeld, G. 1977. *Cell* 10:537-47
42. Lutter, L. C. 1977. *J. Mol. Biol.* 117:53-69
43. Noll, M. 1977. *J. Mol. Biol.* 116:49-71
44. Whitlock, J. P. Jr., Rushizky, G. W., Simpson, R. T. 1977. *J. Biol Chem.* 252:3003-6
45. Lutter, L. C. 1978. *J. Mol. Biol.* 124:391-420
46. Oosterhof, D. K., Hozier, J. C., Rill, R. L. 1975. *Proc. Natl. Acad. Sci. USA* 72:633-37
47. Altenburger, W., Horz, W., Zachau, H. G. 1976. *Nature* 264:517-22
48. Pospelov, V. A., Svetlikova, S. B., Vorobev, V. I. 1979. *Mol. Biol.* 12:796-806
49. Liu, L. F., Wang, J. C. 1978. *Cell* 15:979-84
50. Trifonov, E. N., Bettecken, T. 1979. *Biochemistry* 18:454-56
51. Whitlock, J. P. Jr., Simpson, R. T. 1977. *J. Biol. Chem.* 252:6516-20
52. Simpson, R. T. 1978. *Cell* 13:691-99
53. Lilley, D. M. J., Jacobs, M. R., Houghton, M. 1979. *Nucleic Acids Res.* 7:377-99
54. McGhee, J. D., Felsenfeld, G. 1979. *Proc. Natl. Acad. Sci. USA* 76:2133-37
55. Crothers, D. M., Dattagupta, N., Hogan, M., Klevan, L., Lee, K. S. 1978. *Biochemistry* 17:4525-33
56. Spadafora, C., Oudet, P., Chambon, P. 1978. *Nucleic Acids Res.* 5:3479-89
57. Riley, D., Weintraub, H. 1978. *Cell* 13:281-93
58. Prunell, A., Kornberg, R. D. 1977. *Cold Spring Harbor Symp. Quant. Biol.* 42:103-8
59. Cotton, F. A., Hazen, E. E. Jr. 1971. *The Enzymes* 4:153-75
60. Gilbert, W., Maxam, A., Mirzabekov, A. D. 1976. In *Control of Ribosome Synthesis,* The Alfred Benzon Symposium IX, ed. N. C. Kjeldgaard, O. Maaloe, pp. 139-48. Copenhagen: Munksgaard
61. Humayun, Z., Kleid, D., Ptashne, M. 1977. *Nucleic Acids Res.* 4:1595-1607
62. Goodwin, D. C., Vergne, J., Brahms, J., Defer, N., Kruh, J. 1979. *Biochemistry* 18:2057-64
63. Felsenfeld, G., McGhee, J. D., Sollner-Webb, B., Williamson, P. In *Eukaryotic Gene Regulation,* ed. C. F. Fox, pp. 541-49. New York: Academic
64. Anderson, R. A., Nakashima, Y., Coleman, J. E. 1975. *Biochemistry* 14:907-17
65. Pretorius, H. T., Klein, M., Day, L. A. 1975. *J. Biol. Chem.* 250:9262-69
66. Joffe, J., Keene, M., Weintraub, H. 1977. *Biochemistry* 16:1236-38
67. Rall, S. C., Okinaka, R. T., Strniste, G. F. 1977. *Biochemistry* 16:4940-44
68. Albright, S. C., Nelson, P. P., Garrard, W. T. 1979. *J. Biol. Chem.* 254:1065-73
69. DeLange, R. J., Smith, E. L. 1971. *Ann. Rev. Biochem.* 40:279-314
70. Huberman, J. A. 1973. *Ann. Rev. Biochem.* 42:355-78
71. Elgin, S. C. R., Weintraub, H. 1975. *Ann. Rev. Biochem.* 44:725-74
72. Von Holt, C., Strickland, W. N., Brandt, W. F., Strickland, M. S. 1979. *FEBS Lett.* 100:201-18
73. Isenberg, I. 1979. *Ann. Rev. Biochem.* 48:159-91
74. Urban, M. K., Franklin, S. G., Zweidler, A. 1979. *Biochemistry* 18:3952-60
75. Mardian, J. K. W., Isenberg, I. 1978. *Biochemistry* 17:3825-33
76. Glover, C. V. C., Gorovsky, M. A. 1978. *Biochemistry* 17:5705-13
77. Newrock, K. M., Alfageme, C. R., Nardi, R. V., Cohen, L. H. 1977. *Cold Spring Harbor Symp. Quant. Biol.* 42:421-31
78. Newrock, K. M., Cohen, L. H., Hendricks, M. B., Donnelly, R. J., Weinberg, E. S. 1978. *Cell* 14:327-36
79. Brandt, W. F., Strickland, W. N., Strickland, M., Carlisle, L., Woods, D., Von Holt, C. 1979. *Eur. J. Biochem.* 94:1-10
80. Spiker, S., Isenberg, I. 1977. *Biochemistry* 16:1819-26
81. Spiker, S., Isenberg, I. 1977. *Cold Spring Harbor Symp. Quant. Biol.* 42:157-64
82. Martinson, H. G., True, R. J. 1979. *Biochemistry* 18:1947-51
83. Moss, T., Cary, P. D., Abercrombie, B. D., Crane-Robinson, C., Bradbury, E. M. 1976. *Eur. J. Biochem.* 71:337-50
84. Bohm, L., Hayashi, H., Cary, P. D., Moss, T., Crane-Robinson, C., Bradbury, E. M. 1977. *Eur. J. Biochem.* 77:487-93
85. Bradbury, E. M., Moss, T., Hayashi, H., Hjelm, R. P., Suau, P., Stephens, R. M., Baldwin, J. P., Crane-Robinson, C. 1977. *Cold Spring Harbor Symp. Quant. Biol.* 42:277-86
86. Baker, C. C., Isenberg, I. 1976. *Biochemistry* 15:629-34
87. Dixon, G. H., Candido, E. P. M., Honda, B. M., Louie, A. J., Macleod, A. R., Sung, M. T. 1975. *Ciba Found. Symp.* 28:229-58
88. Ruiz-Carrillo, A., Wangh, L. J., Allfrey, V. G. 1975. *Science* 190:117-28
89. Allfrey, V. G. 1977. In *Chromatin and Chromatin Structure,* ed. H. J. Li, R. A.

Eckhardt, pp. 167–92. New York: Academic

90. Riggs, M. G., Whittaker, R. G., Neumann, J. R., Ingram, V. M. 1977. *Nature* 268:462–64

91. Hagopian, H. K., Riggs, M. G., Swartz, L. A., Ingram, V. M. 1977. *Cell* 12:855–60

92. Sealy, L., Chalkley, R. 1978. *Cell* 14:115–21

93. Sealy, L., Chalkley, R. 1978. *Nucleic Acids Res.* 5:1863–75

94. Vidali, G., Boffa, L. C., Bradbury, E. M., Allfrey, E. M. 1978. *Proc. Natl. Acad. Sci. USA* 75:2239–43

95. Mathis, D. J., Oudet, P., Wasylyk, B., Chambon, P. 1978. *Nucleic Acids Res.* 5:3523–47

96. Nelson, D., Perry, M. E., Chalkley, R. 1979. *Nucleic Acids Res.* 6:561–74

97. Weintraub, H., Groudine, M. 1976. *Science* 193:848–56

98. Bloom, K. S., Anderson, J. N. 1978. *Cell* 15:141–50

99. Reeves, R., Cserjesi, P. 1979. *J. Biol. Chem.* 254:4283–90

100. Rubenstein, P., Sealy, L., Marshall, S., Chalkley, R. 1979. *Nature* 280:692–93

101. Lilley, D. M. J., Berendt, A. R. 1979. *Biochem. Biophy. Res. Commun.* 90:917–24

102. Louie, A. J., Candido, E. P. M., Dixon, G. H. 1973. *Cold Spring Harbor Symp. Quant. Biol.* 38:803–19

103. Gurley, L. R., Walters, R. A., Barham, S. S., Deaven, L. L. 1978. *Exp. Cell. Res.* 111:373–83

104. Prentice, D. A., Taylor, S. E., Newmark, M. Z., Kitos, P. A. 1978. *Biochem. Biophy. Res. Commun.* 85:541–50

105. Gurley, L. R., D'Anna, J. A., Barham, S. S., Deaven, L. L., Tobey, R. A. 1978. *Eur. J. Biochem.* 84:1–15

106. Hayaishi, O., Ueda, K. 1977. *Ann. Rev. Biochem.* 46:95–116

107. Ord, M. G., Stocken, L. A. 1977. *Biochem. J.* 161:583–92

108. Riquelme, P. T., Burzio, L. O., Koide, S. S. 1979. *J. Biol. Chem.* 254:3018–28

109. Burzio, L. O., Riquelme, P. T., Koide, S. S., 1979. *J. Biol. Chem.* 254:3029–37

110. Okayama, H., Ueda, K., Hayaishi, O. 1978. *Proc. Natl. Acad. Sci. USA* 75:1111–15

111. Goldknopf, I. L., Busch, H. 1977. *Proc. Natl. Acad. Sci. USA* 74:864–68

112. Goldknopf, I. L., Busch, H. 1978. *The Cell Nucleus* 5:149–80

113. Goldknopf, I. L., Busch, H. 1977. *Proc. Natl. Acad. Sci. USA* 74:5492–95

114. Martinson, H. G., True, R., Burch, J. B. E., Kunkel, G. 1979. *Proc. Natl. Acad. Sci. USA* 76:1030–34

115. Okayama, H., Hayaishi, O. 1978. *Biochem. Biophy. Res. Commun.* 84:755–62

116. Goldknopf, I. L., Rosenbaum, F., Sterner, R., Vidali, G., Allfrey, V. G., Busch, H. 1979. *Biochem. Biophy. Res. Commun.* 90:269–77

117. Cousens, L. S., Gallwitz, D., Alberts, B. M. 1979. *J. Biol. Chem.* 254:1716–23

118. Duerre, J. A., Wallwork, J. C., Quick, D. P., Ford, K. M. 1977. *J. Biol. Chem.* 252:5981–85

119. Mullins, D. W. Jr., Giri, C. P., Smulson, M. 1977. *Biochem.* 16:506–13

120. Garrard, W. T., Nobis, P., Hancock, R. 1977. *J. Biol. Chem.* 252:4962–67

121. Olins, D. E., Bryan, P. N., Harrington, R. E., Hill, W. E., Olins, A. L. 1977. *Nucleic Acids Res.* 4:1911–31

122. Zama, M., Bryan, P. N., Harrington, P. E., Olins, A. L., Olins, D. E. 1977. *Cold Spring Harbor Symp. Quant. Biol.* 42:31–42

123. Wong, N. T. N., Candido, E. P. M. 1978. *J. Biol. Chem.* 253:8263–68

124. Maher, P., Candido, E. P. M. 1977. *Can. J. Biochem.* 55:404–7

125. Hyde, J. E., Walker, I. O. 1974. *Nucleic Acids Res.* 1:203–15

126. Dieterich, A. E., Axel, R., Cantor, C. R. 1977. *Cold Spring Harbor Symp. Quant. Biol.* 42:199–206

127. Dieterich, A. E., Axel R., Cantor, C. R. 1979. *J. Mol. Biol.* 129:587–602

128. Camerini-Otero, R. D., Felsenfeld, G. 1977. *Proc. Natl. Acad. Sci. USA* 74:5519–23

129. Lewis, P. N. 1979. *Eur. J. Biochem.* 99:315–22

130. Malchy, B., Kaplan, H. 1974. *J. Mol. Biol.* 82:537–45

131. Malchy, B. L., Kaplan, H. 1976. *Biochem. J.* 159:173–5

132. Malchy, B. L. 1977. *Biochem.* 16:3922–27

133. Tack, L. O., Simpson, R. T. 1979. *Biochem.* 18:3110–18

134. Simpson, R. T. 1971. *Biochemistry* 10:4466–70

135. Wong, T. K., Marushige, K. 1976. *Biochemistry* 15:2041–46

136. Tack, L. O., Simpson, R. T. 1977. *Biochemistry* 16:3746–53

137. Weintraub, H., Van Lente, F. 1974. *Proc. Natl. Acad. Sci. USA* 71:4249–53

138. Marks, D. B., Keller, B. J. 1976. *Arch. Biochem. Biophys.* 175:598–605

139. Lilley, D. M. J., Tatchell, K. 1977. *Nucleic Acids Res.* 4:2039–55

140. Whitlock, J. P. Jr., Stein, A. 1978. *J. Biol. Chem.* 253:3857–61
141. Brandt, W. F., Bohm, L., Von Holt, C. 1975. *FEBS Lett.* 51:88–93
142. Eickbush, T. H., Watson, D. K., Moudrianakis, E. N. 1976. *Cell* 9:785–92
143. Weintraub, H., Palter, K., Van Lente, F. 1975. *Cell* 6:85–110
144. Biroc, S. L., Reeder, R. H. 1976. *Biochemistry* 15:1440–48
145. Griffiths, G. R., Huang, P. C. 1979. *J. Biol. Chem.* 254:8057–66
146. Goldblatt, D., Bustin, M., Sperling, R. 1978. *Exp. Cell Res.* 112:1–14
147. Absolom, D., Van Regenmortel, M. H. V. 1978. *FEBS Lett.* 85:61–64
148. DeLange, R. J., Williams, L. C., Martinson, H. G. 1979. *Biochemistry* 18:1942–46
149. Martinson, H. G., True, R., Lau, C. K., Mehrabian, M. 1979. *Biochemistry* 18:1075–82
150. Bonner, W. M., Pollard, H. B. 1975. *Biochem. Biophy. Res. Commun.* 64:282–88
151. Mirzabekov, A. D., Shick, V. V., Belyavsky, A. V., Bavykin, S. G. 1978. *Proc. Natl. Acad. Sci. USA* 75:4184–88
152. Mirzabekov, A. D., Shick, V. V., Belyavsky, A. V., Karpov, V. L., Bavykin, S. G. 1977. *Cold Spring Harbor Symp. Quant. Biol.* 42:149–55
153. Shick, V. V., Belyavsky, A. V., Bavykin, S. G., Mirzabekov, A. D. 1980. *J. Mol. Biol.* In press
154. Simpson, R. T. 1978. *Nucleic Acids Res.* 5:1109–19
155. Sperling, J., Sperling, R. 1978. *Nucleic Acids Res.* 5:2755–73
156. Kunkel, G. R., Martinson, H. G. 1978. *Nucleic Acids Res.* 5:4263–72
157. Mandel, R., Kolomijtseva, G., Brahms, J. G. 1979. *Eur. J. Biochem.* 96:257–65
158. Boublik, M., Bradbury, E. M., Crane-Robinson, C., Rattle, H. W. E. 1971. *Nature New Biol.* 229:149–50
159. Olins, D. E. 1979. In *Chromatin Structure and Function*, ed. C. A. Nicolini, pp. 109–35. New York: Plenum
160. Cary, P. D., Moss, T., Bradbury, E. M. 1978. *Eur. J. Biochem.* 89:475–82
161. Villeponteau, B., Harary, I., Martinson, H. G. 1980. *Biochemistry.* In press
162. Palter, K., Alberts, B. M. 1979. *J. Biol. Chem.* 254:1160–69
163. Burton, D. R., Butler, M. J., Hyde, J. E., Phillips, D., Skidmore, C. J., Walker, I. O. 1978. *Nucleic Acids Res.* 5:3643–63
164. Nicola, N. A., Fulmer, A. W., Schwartz, A. M., Fasman, G. D. 1978. *Biochemistry* 17:1779–85

165. Weischet, W. O., Tatchell, K., Van Holde, K. E., Klump, H. 1978. *Nucleic Acids Res.* 5:139–60
166. Simpson, R. T. 1979. *J. Biol. Chem.* 254:10123–27
167. Seligy, V. L., Poon, N. H. 1978. *Nucleic Acids Res.* 5:2233–52
168. Gordon, V. C., Knobler, C. M., Olins, D. E., Schumaker, V. N. 1978. *Proc. Natl. Acad. Sci. USA* 75:660–63
169. Gordon, V. C., Schumaker, V. N., Olins, D. E., Knobler, C. M., Horwitz, J. 1979. *Nucleic Acids Res.* 6:3845–58
170. Fulmer, A. W., Fasman, G. D. 1979. *Biopolymers* 18:2875–91
171. Wu, H. M., Dattagupta, N., Hogan, M., Crothers, D. M. 1979. *Biochemistry* 18:3960–65
172. Martinson, H. G., True, R. J., Burch, J. B. E. 1979. *Biochemistry* 18:1082–89
173. Eisenberg, H. 1976. *Biophys. Chem.* 5:243–51
174. Zama, M., Olins, D. E., Prescott, B., Thomas, G. J. Jr. 1978. *Nucleic Acids Res.* 5:3881–97
175. Weintraub, H., Worcel, A., Alberts, B. 1976. *Cell* 9:409–17
176. Oudet, P., Spadafora, C., Chambon, P. 1977. *Cold Spring Harbor Symp. Quant. Biol.* 42:301–12
177. Tatchell, K., Van Holde, K. E. 1979. *Biochemistry* 18:2871–80
178. Steinmetz, M., Streeck, R. E., Zachau, H. G. 1978. *Eur. J. Biochem.* 83:615–28
179. Germond, J. E., Hirt, B., Oudet, P., Gross-Bellard, M., Chambon, P. 1975. *Proc. Natl. Acad. Sci. USA* 72:1843–47
180. Tatchell, K., Van Holde, K. E. 1977. *Biochemistry* 16:5295–303
181. Stein, A., Bina-Stein, M., Simpson, R. T. 1977. *Proc. Natl. Acad. Sci. USA* 74:2780–84
182. Stein, A. 1979. *J. Mol. Biol.* 130:103–34
183. Jorcano, J. L., Ruiz-Carrillo, A. 1979. *Biochemistry* 18:768–74
184. Chao, M. V., Gralla, J., Martinson, H. G. 1979. *Biochemistry* 18:1068–74
185. Voordouw, G., Eisenberg, H. 1978. *Nature* 273:446–48
186. Lilley, D. M. J., Pardon, J. F., Richards, B. M. 1977. *Biochemistry* 16:2853–60
187. Thomas, J. O., Kornberg, R. D. 1975. *Proc. Natl. Acad. Sci. USA* 72:2626–30
188. Thomas, J. O., Butler, P. J. G. 1977. *J. Mol. Biol.* 116:769–81
189. Chung, S. Y., Hill, W. E., Doty, P. 1978. *Proc. Natl. Acad. Sci. USA* 75:1680–84
190. Eickbush, T. H., Moudrianakis, E. N. 1978. *Biochemistry* 17:4955–64

191. Butler, A. P., Harrington, R., Olins, D. E. 1978. *Nucleic Acids Res.* 6:1509–20
192. Ruiz-Carrillo, A., Jorcano, J. L. 1979. *Biochemistry* 18:760–68
193. Philip, M., Jamaluddin, J., Sastry, R. V. R., Chandra, H. S. 1979. *Proc. Natl. Acad. Sci. USA* 76:5178–82
194. Oudet, P., Germond, J. E., Sures, M., Gallwitz, D., Bellard, M., Chambon, P. 1977. *Cold Spring Harbor Symp. Quant. Biol.* 42:287–300
195. Bina-Stein, M., Simpson, R. T. 1977. *Cell* 11:609–18
196. Bina-Stein, M. 1978. *J. Biol. Chem.* 253:5213–19
197. Simon, R. H., Camerini-Otero, R. D., Felsenfeld, G. 1978. *Nucleic Acids Res.* 5:4850–18
198. Stockley, P. G., Thomas, J. O. 1979. *FEBS Lett.* 99:129–35
199. Klevan, L., Dattagupta, N., Hogan, M., Crothers, D. M. 1978. *Biochemistry* 17:4533–40
200. Laskey, R. A., Mills, A. D., Morris, N. R. 1977. *Cell* 10:237–43
201. Laskey, R. A., Honda, B. M., Mills, A. D., Finch, J. T. 1978. *Nature* 275:416–20
202. Germond, J. E., Rouviere-Yaniv, J., Yaniv, M., Brutlag, D. 1979. *Proc. Natl. Acad. Sci. USA* 76:3779–83
203. Ruiz-Carrillo, A., Jorcano, J. L., Eder, G., Lurz, R. 1979. *Proc. Natl. Acad. Sci. USA* 76:3284–88
204. Stein, A., Whitlock, J. P. Jr., Bina-Stein, M. 1979. *Proc. Natl. Acad. Sci. USA* 76:5000–4
205. Ohlenbusch, H. H., Olivera, B. M., Tuan, D., Davidson, N. 1967. *J. Mol. Biol.* 25:299–315
206. Klevan, L., Crothers, D. M. 1977. *Nucleic Acids Res.* 4:4077–89
207. Tatchell, K., Van Holde, K. E. 1978. *Proc. Natl. Acad. Sci. USA* 75:3583–87
208. Cole, R. D., Lawson, G. M., Hsiang, M. W. 1977. *Cold Spring Harbor Symp. Quant. Biol.* 42:253–64
209. Lawson, G. M., Cole, R. D. 1979. *Biochemistry* 18:2160–66
210. Weischet, W. O., Allen, J. R., Riedel, G., Van Holde, K. E. 1979. *Nucleic Acids Res.* 6:1843–62
211. Allan, J., Staynov, D. Z., Gould, H. J. 1980. *Proc. Natl. Acad. Sci. USA.* 77:885–89
212. Beard, P. 1978. *Cell* 15:955–67
213. Cantor, C. R. 1976. *Proc. Natl. Acad. Sci. DNA* 73:3391–93
214. Leffak, I. M., Grainger, R., Weintraub, H. 1977. *Cell* 12:837–46
215. Worcel, A., Han, S., Wong, M. L. 1978. *Cell* 15:969–77
216. Senshu, T., Fukuda, M., Ohashi, M. 1978. *J. Biochem.* 84:985–88
217. Deleted in proof.
218. Hayashi, K., Hofstaetter, T., Yakuwa, N. 1978. *Biochemistry* 17:1880–83
219. Welch, S. L., Cole, R. D. 1979. *J. Biol. Chem.* 254:662–65
220. Smerdon, M. J., Isenberg, I. 1976. *Biochemistry* 15:4233–41
221. Seyedin, S. M., Kistler, W. S. 1979. *Biochemistry* 18:1371–75
222. Yaguchi, M., Ray, C., Seligy, V. 1980. *Biochem. Biophy. Res. Commun.* In press
223. Yaguchi, M., Roy, C., Dove, M., Seligy, V. 1977. *Biochem. Biophy. Res. Commun.* 76:100–6
224. Chapman, G. E., Hartman, P. G., Cary, P. D., Bradbury, E. M., Lee, D. R. 1978. *Eur. J. Biochem.* 86:35–44
225. Tancredi, T., Temussi, P. A. 1979. *Biopolymers* 18:1–7
226. Chapman, G. E., Aviles, F. J., Crane-Robinson, C., Bradbury, E. M. 1978. *Eur. J. Biochem.* 90:287–96
227. Crane-Robinson, C., Danby, S. E., Bradbury, E. M., Garel, A., Kovacs, A. M., Champagne, M., Daune, M. 1976. *Eur. J. Biochem.* 67:379–88
228. Aviles, F. J., Chapman, G. E., Kneale, G. G., Crane-Robinson, C., Bradbury, E. M. 1978. *Eur. J. Biochem.* 88:363–71
229. Sperling, R., Bustin, M. 1976. *Nucleic Acids Res.* 3:1263–75
230. Smerdon, M. J., Isenberg, I. 1976. *Biochemistry* 15:4242–47
231. Whitlock, J. P. Jr., Simpson, R. T. 1976. *Biochemistry* 15:3307–14
232. Thoma, F., Koller, T., Klug, A. 1979. *J. Cell Biol.* 83:403–27
233. Todd, R. D., Garrard, W. T. 1977. *J. Biol. Chem.* 252:4729–38
234. Olins, D. E., Wright, E. B. 1973. *J. Cell Biol.* 59:304–9
235. Chalkley, R. 1975. *Biochem. Biophy. Res. Commun.* 64:587–94
236. Ring, D., Cole, R. D. 1979. *J. Biol. Chem.* 254:11668–95
237. Goodwin, G. H., Johns, E. W. 1977. *Methods Cell Biol.* 16:257–67
238. Weisbrod, S., Weintraub, W. 1979. *Proc. Natl. Acad. Sci. USA* 76:630–34
239. Bonner, W. M., Stedman, J. D. 1979. *Proc. Natl. Acad. Sci. USA* 76:2190–94
240. Deleted in proof
241. Gaubatz, J., Hardison, R., Murphy, J., Eichner, M. E., Chalkley, R. 1977. *Cold Spring Harbor Symp. Quant. Biol.* 42:265–72
242. Thoma, F., Koller, T. 1977. *Cell* 12:101–7

243. Renz, M., Nehls, P., Hozier, J. 1977. *Proc. Natl. Acad. Sci. USA* 74:1879–83
244. Stratling, W. H. 1979. *Biochemistry* 18:596–603
245. Renz, M., Day, L. A. 1976. *Biochemistry* 15:3220–28
246. Bryan, P. N., Wright, E. B., Hsie, M. H., Olins, A. L., Olins, D. E. 1978. *Nucleic Acids Res.* 5:3603–17
247. Lohr, D., Tatchell, K., Van Holde, K. E. 1977. *Cell* 12:829–36
248. Horz, W., Igo-Kemenes, T., Pfeiffer, W., Zachau, H. G. 1976. *Nucleic Acids Res.* 3:3213–26
249. Gottesfeld, J. M., Melton, D. A. 1978. *Nature* 273:317–19
250. Baer, B. W., Kornberg, R. D. 1979. *J. Biol. Chem.* 254:9678–81
251. Maio, J. J., Brown, F. L., Musich, P. R. 1977. *J. Mol. Biol.* 117:637–55
252. Brown, F. L., Musich, P. R., Maio, J. J. 1979. *J. Mol. Biol.* 131:777–99
253. Rosenberg, H., Singer, M. F., Rosenberg, M. 1978. *Science* 200:394–402
254. Singer, D. S. 1979. *J. Biol. Chem.* 254:5506–14
255. Fittler, F., Zachau, H. G. 1979. *Nucleic Acids Res.* 7:1–13
256. Varshavsky, A. J., Sundin, O., Bohn, M. 1979. *Cell* 16:453–66
257. Varshavsky, A. J., Sundin, O., Bohn, M. 1978. *Nucleic Acids Res.* 5:3469–78
258. Scott, W. A., Wigmore, D. J. 1978. *Cell* 15:1511–18
259. Wasylyk, B., Chambon, P. 1980. *Nucleic Acids Res.* In press
260. Ponder, B. A. J., Crawford, L. V. 1977. *Cell* 11:35–49
261. Lohr, D., Kovacic, R. T., Van Holde, K. E. 1977. *Biochemistry* 16:463–71
262. Thomas, J. O., Furber, V. 1976. *FEBS Lett.* 66:274–79
263. Noll, M. 1976. *Cell* 8:349–55
264. Morris, N. R. 1976. *Cell* 8:357–63
265. Spadafora, C., Bellard, M., Compton, J. L., Chambon, P. 1976. *FEBS Lett.* 69:281–85
266. Todd, R. D., Garrard, W. T. 1977. *J. Biol. Chem.* 252:4729–38
267. Humphries, S. E., Young, D., Carroll, D. 1979. *Biochemistry* 18:3223–31
268. Morris, N. R. 1976. *Cell* 9:627–32
269. Thomas, J. O., Thompson, R. J. 1977. *Cell* 19:633–40
270. Lipps, H. J., Morris, N. R. 1977. *Biochem. Biophy. Res. Commun.* 74: 230–34
271. Lohr, D., Ide, G. 1979. *Nucleic Acids Res.* 6:1909–27
272. Wilhelm, M. L., Mazen, A., Wilhelm, F. X. 1977. *FEBS Lett.* 79:404–8
273. Weintraub, H. 1978. *Nucleic Acids Res.* 5:1179–88
274. Prunell, A. 1979. See Ref. 159, pp. 441–50
275. Ris, H., Kubai, D. F. 1974. *Ann. Rev. Genet.* 4:263–94
276. Finch, J. T., Klug, A. 1976. *Proc. Natl. Acad. Sci. USA* 73:1897–1901
277. Suau, P., Bradbury, E. M., Bradbury, J. P. 1979. *Eur. J. Biochem.* 97:593–602
278. Carpenter, B. G., Baldwin, J. P., Bradbury, E. M., Ibel, K. 1976. *Nucleic Acids Res.* 3:1739–46
279. Pardon, J. F., Worcester, D. L., Wooley, J. C., Tatchell, K., Van Holde, K. E., Richards, B. M. 1975. *Nucleic Acids Res.* 2:2163–76
280. Rattner, J. B., Hamkalo, B. A. 1979. *J. Cell Biol.* 81:453–57
281. Rattner, J. B., Hamkalo, B. A. 1980. *Chromosoma.* In press
282. Campbell, A. M., Cotter, R. I., Pardon, J. F. 1978. *Nucleic Acids Res.* 5: 1571–80
283. Renz, M., Nehls, P., Hozier, J. 1977. *Cold Spring Harbor Symp. Quant. Biol.* 42:245–52
284. Hozier, J., Renz, M., Nehls, P. 1977. *Chromosoma* 62:301–17
285. Stratling, W. H., Muller, U., Zentgraf, H. 1978. *Exp. Cell Res.* 117:301–11
286. Butt, T. R., Jump, D. B., Smulson, M. E. 1979. *Proc. Natl. Acad. Sci. USA* 76:1628–32
287. Walker, B. W., Lothstein, L., LeSturgeon, W. M. 1979. *J. Cell. Biol.* 83:a169
288. Takahashi, K., Tashiro, Y. 1979. *Eur. J. Biochem.* 97:353–60
289. Worcel, A., Benyajati, C. 1977. *Cell* 12:83–100
290. Crick, F. H. C. 1976. *Proc. Natl. Acad. Sci. USA* 73:2639–40
291. Wang, J. C. 1979. *Proc. Natl. Acad. Sci. USA* 76:200–3
292. Zimmerman, S. B., Pheiffer, B. H. 1979. *Proc. Natl. Acad. Sci. USA* 76:2703–7
293. Muller, U., Zentgraf, H., Eicken, I., Keller, W. 1978. *Science* 201:406–15
294. Fuller, F. B. 1978. *Proc. Natl. Acad. Sci. USA* 75:3557–61
295. Simpson, R. T., Whitlock, J. P. Jr., Bina-Stein, M., Stein, A. 1977. *Cold Spring Harbor Symp. Quant. Biol.* 42:127–36
296. Camerini-Otero, R. D., Sollner-Webb, B., Simon, R. H., Williamson, P., Zasloff, M., Felsenfeld, G. 1977. *Cold Spring Harbor Symp. Quant. Biol.* 42:57–75
297. Sedat, J., Manuelidis, L. 1978. *Cold Spring Harbor Symp. Quant. Biol.* 42:331–50

298. Bak, A. L., Zeuthen, J., Crick, F. H. C. 1977. *Proc. Natl. Acad. Sci. USA* 74:1595–99
299. Benyajati, C., Worcel, A. 1976. *Cell* 9:393–407
300. Igo-Kemenes, T., Greil, W., Zachau, H. G. 1977. *Nucleic Acids Res.* 4:3387–3400
301. Wu, C., Bingham, P. M., Livak, K. J., Holmgren, R., Elgin, S. C. R. 1979. *Cell* 16:797–806
302. Wu, C., Wong, Y. C., Elgin, S. C. R. 1979. *Cell* 16:807–14
303. Garel, A., Axel, R. 1976. *Proc. Natl. Acad. Sci. USA* 73:3966–70
304. Adolph, K. W., Cheng, S. M., Laemmli, U. K. 1977. *Cell* 12:805–16
305. Paulson, J. R., Laemmli, U. K. 1977. *Cell* 12:817–28
306. Marsden, M. P. F., Laemmli, U. K. 1979. *Cell* 17:849–58
307. Stubblefield, E., Wray, W. 1971. *Chromosoma* 32:262–94
308. Cremisi, C. 1979. *Microbiol. Rev.* 43:297–319
309. Mathis, D., Oudet, P., Chambon, P. 1980. *Prog. Nucleic Acid Res. Mol. Biol.* In press
310. Williamson, P., Felsenfeld, G. 1978. *Biochemistry* 17:5695–705
311. Wasylyk, B., Thevenin, G., Oudet, P., Chambon, P. 1979. *J. Mol. Biol.* 128:411–40
312. Brooks, T. L., Green, M. H. 1977. *Nucleic Acids Res.* 4:4261–77
313. Gariglio, P., Llopis, R., Oudet, P., Chambon, P. 1979. *J. Mol. Biol.* 131:75–105

SUBJECT INDEX

A

N-Acetoxyaminofluorene
Z-DNA and, 389
Acetylcholine receptor
crystallization of, 210
membrane-related helices
in, 230
ACTH
melanocyte-stimulating
hormone and, 101
Actin
MgATP hydrolyses by
myosins and, 238
myosin interaction site with,
244–246
Adenovirus 2 snRNA 522–524
nucleotide sequence of, 523
Adenylate kinase
secondary structure, 5
Adrenocorticotropic hormones,
100–102
Adriamycin
Z-DNA and, 389
Aflatoxin
Z-DNA and, 389
Alamethicin
ion-conductive channels in
membranes and, 208
structure of, 212–214
Alcohols
Z-DNA stability and, 369
Aldolase
secondary structure, 16, 17
Alteromonas nuclease
PM2 DNA and, 329
Amino acid sequences
detection of membrane-
spanning helices from,
228–229
membrane protein, 230–232
membrane-related
protein, 209
Amino acid sequence analysis
of membrane-related proteins,
218–232
Aminomethyltrioxsalen
rRNA structure analysis
and, 406
Antibodies
anti–Z-DNA, 745–748
Apomyoglobin
unfolding of, 151
Aspartate transcarbamylase
assembly pathways of, 160
Aspartokinase-homoserine
dehydrogenase I
domain folding in, 160

B

Bacteriophage lambda Cro
DNA-binding, 472–477
Bacteriophage lambda Cro-
operator complex, 473–477
Bacteriophage lambda repressor
DNA-binding, 477–484
Bacteriorhodopsin
amino acid sequence of,
230–231
crystallization of, 209–10
diffraction studies of, 211–12
hydrophobic moment plot
and, 224, 226–227
membrane-related helices
in, 230
membrane-spanning segments
in, 229
three-dimensional structure
of, 208
Bence Jones proteins
domain folding in, 160
equilibrium unfolding
transition of, 155
Blue copper proteins
NMR spectroscopy and, 115
Bovine pancreatic trypsin
inhibitor
NMR spectroscopy and, 115
BPTI
S-S bond formation in, 139
salt bridges in, 158
Butoxycarbonyl
insulin receptor affinities
and, 82

C

Caenorhabditis elegans
unc-54 myosin heavy chain
gene of, 250
Calcium-binding proteins
NMR spectroscopy and, 115
Carbonic anhydrase
denaturant-induced unfolding
and, 139
folding and refolding of,
150–51
Carboxypeptidase A
exterior sidechain and loop
motions in, 193
nonproline peptide bond
isomerization and, 148
Carcinogens
Z-DNA and, 388–89
Catch muscle
thick filaments of, 258–59

Cetus spp.
Z-DNA in, 382
Chloramphenicol
peptidyl transferase function
and, 433
resistance to
rRNA gene point mutations
and, 429
Chorionic somatomammotropin
amino acid sequences of, 88
Chromatin
histone cores of DNA wound
around, 287–88
Z-DNA and, 388
Chromosomes
Z-DNA in, 380–82
Chymotrypsin
salt bridges in, 158
slow- and fast-folding molecules of, 145
Chymotrypsinogen
slow- and fast-folding molecules of, 145
Circular dichroism
A form DNA and, 300
C form DNA and, 300
DNA, 316, 327
endorphins and, 102
enkephalins and, 101
glucagon and, 94, 142
glycoprotein hormones
and, 89
growth hormones and, 88–89
insulin conformation in solution and, 82–83
neurohypophyseal hormones
and, 97
polypeptide hormone con-
formation and, 77
relaxin and, 86
secretin and, 95–96
[13]C nuclear magnetic resonance
spectroscopy, 464–65
Collagen
NMR spectroscopy and, 115
Computer analysis
rRNA structure and, 404
Creatine kinase
NMR spectroscopy and, 116
Cytochromes
NMR spectroscopy and, 115
Cytochrome b_2
domains, 21
Cytochrome b_5
helical folding pattern, 22
Cytochrome b_{562}
helices, 10
Cytochrome c